OXFORD LOGIC GUIDES: 40

OXFORD LOGIC GUIDES

1. Jane Bridge: *Beginning model theory: the completeness theorem and some consequences*
2. Michael Dummett: *Elements of intuitionism*
3. A.S. Troelstra: *Choice sequences: a chapter of intuitionistic mathematics*
4. J.L. Bell: *Boolean-valued models and independence proofs in set theory* (*1st edition*)
5. Krister Seberberg: *Classical propositional operators: an exercise in the foundation of logic*
6. G.C. Smith: *The Boole–De Morgan correspondence 1842–1864*
7. Alec Fisher: *Formal number theory and computability: a work book*
8. Anand Pillay: *An introduction to stability theory*
9. H.E. Rose: *Subrecursion: functions and hierarchies*
10. Michael Hallett: *Cantorian set theory and limitation of size*
11. R. Mansfield and G. Weitkamp: *Recursive aspects of descriptive set theory*
12. J.L. Bell: *Boolean-valued models and independence proofs in set theory* (*2nd edition*)
13. Melvin Fitting: *Computability theory: semantics and logic programming*
14. J.L. Bell: *Toposes and local set theories: an introduction*
15. R.Kaye: *Models of Peano arithmetic*
16. J. Chapman and F. Rowbottom: *Relative category theory and geometric morphisms: a logical approach*
17. Stewart Shapiro: *Foundations without foundationalism*
18. John P. Cleave: *A study of logics*
19. R.M. Smullyan: *Gödel's incompleteness theorems*
20. T.E. Forster: *Set theory with a universal set: exploring an untyped universe*
21. C. McLarty: *Elementary categories, elementary toposes*
22. R.M. Smullyan: *Recursion theory for metamathematics*
23. Peter Clote and Jan Krajíček: *Arithmetic, proof theory, and computational complexity*
24. A. Tarski: *Introduction to logic and to the methodology of deductive sciences*
25. G. Malinowski: *Many valued logics*
26. Alexandre Borovik and Ali Nesin: *Groups of finite Morley rank*
27. R.M. Smullyan: *Diagonalization and self-reference*
28. Dov M. Gabbay, Ian Hodkinson, and Mark Reynolds: *Temporal logic: mathematical foundations and computational aspects: Volume 1*
29. Saharon Shelah: *Cardinal arithmetic*
30. Erik Sandewall: *Features and fluents: Volume I: a systematic approach to the representation of knowledge about dynamical systems*
31. T.E. Forster: *Set theory with a universal set: exploring an untyped universe* (*2nd edition*)
32. Anand Pillay: *Geometric stability theory*
33. Dov. M. Gabbay: *Labelled deductive systems*
34. Raymond M. Smullyan and Melvin Fitting: *Set theory and the continuum problem*
35. Alexander Chagrov and Michael Zakharyaschev: *Modal logic*
36. G. Sambin and J. Smith: *Twenty-five years of constructive type theory*
37. Maria Manzano: *Model theory*
38. Dov M. Gabbay: *Fibring logics*
39. Michael Dummett: *Elements of intuitionism* (*2nd edition*)
40. Dov M. Gabbay, Mark A. Reynolds, and Marcelo Finger: *Temporal Logic: mathematical foundations and occupational aspects: Volume 2*

Temporal Logic

Mathematical Foundations and Computational Aspects

Volume 2

DOV M. GABBAY
Department of Computer Science
King's College London

MARK A. REYNOLDS
School of Information Technology
Murdoch University, Australia

MARCELO FINGER
Institute of Mathematics and Statistics
University of Sao Paulo

CLARENDON PRESS • OXFORD

2000

OXFORD
UNIVERSITY PRESS

Great Clarendon Street, Oxford OX2 6DP

Oxford University Press is a department of the University of Oxford.
It furthers the University's objective of excellence in research, scholarship,
and education by publishing worldwide in

Oxford New York

Athens Auckland Bangkok Bogotá Buenos Aires Calcutta
Cape Town Chennai Dar es Salaam Delhi Florence Hong Kong Istanbul
Karachi Kuala Lumpur Madrid Melbourne Mexico City Mumbai
Nairobi Paris São Paulo Singapore Taipei Tokyo Toronto Warsaw

with associated companies in Berlin Ibadan

Oxford is a registered trade mark of Oxford University Press
in the UK and in certain other countries

Published in the United States
by Oxford University Press Inc., New York

First published 2000

A catalogue record for this book is available from the British Library

Library of Congress Cataloging in Publication Data
(Data available)

ISBN 0 19 853768 9

Typeset using LaTex
Printed in Great Britain
on acid-free paper by
T.J. International Ltd., Cornwall

PREFACE

We are happy to present the readers with Volume 2 of our temporal logic monograph. It took us four years to finish this volume. We have covered more or less the material promised in the tentative table of contents given in Volume 1. The algorithmic proof chapter is postponed to Volume 3.

We would welcome the readers to accompany us into the future.

Until later

D. Gabbay, M. Reynolds and M. Finger

Acknowledgements

We would like to thank Guillaume Malod for reading and commenting on the book and Mrs Jane Spurr for helping with production.

CONTENTS

TENTATIVE CONTENTS OF VOLUME 3

- Complexity of temporal logics
- Temporal logic and natural language
- Temporal representation in artificial intelligence
- Temporal logic and non-monotonic reasoning
- Time and conditional
- Proof theory for temporal logic
- Time and action
- Time and space
- Models of time travel; temporal causation
- Products of temporal logics
- Temporal logic and real time systems
- Translations into and from temporal logics
- Predicate Temporal Logic
- Skolem functions, epsilon symbols and Herbrand theorems in temporal logic.

1

INTRODUCTION AND OVERVIEW

1.1 Introduction

The sections below give an overview of the contents of this book. This volume, Volume 2 in the series, deals with the more applicable aspects of temporal logics. Volume 1 gave the basic concepts and methods and concentrated more on the expressive power of temporal languages, their axiomatization, their dimension and their decidability. Now we are ready to deal with some aspects of their applications. The first four chapters of the present volume continue with the more theoretical presentation, covering automata, branching time and labelled deduction. The rest of the book progresses in the direction of temporal databases, temporal execution and programming, actions and planning. For details, see the following sections of this chapter.

There are many more applications not covered in Volume 2—we leave these to Volume 3.

We now describe the contents of the book chapter by chapter.

1.2 Temporal logic and automata

Formulae in a temporal language classify structures: some structures are models of the formula, some structures are not. This is useful as a possibly infinite set of possibly infinite structures can be summarized to some extent by a small finite formula. When the temporal language is expressive enough then a formula, ϕ say, can be found which describes exactly the correct behaviour of a software system specification (for example) and a formula, ψ say, can be found which describes exactly the possible behaviour of a given program. If there are no models of $\psi \wedge \neg\phi$ then we know that the program correctly implements the software specification.

Automata equally summarize whole sets of temporal structures. Automata accept some structures and reject the rest. They are also often described in a succinct finite way. And, as we shall see, automata are as expressive (often more so) as temporal languages. We could thus use automata to answer the kinds of practical application questions that we use temporal logic for.

In Chapter 2 we shall demonstrate the close relationship between automata and temporal logic. However, we shall not do this in order to present automata as an alternative formal approach to practical applications. It is probably generally accepted that temporal languages are more convenient to deal with at a syntactical level. The use of the various results showing the equivalence of certain temporal logic properties to certain automata properties is that it is sometimes easier to answer a question about an automata property. One particular example is the emptiness question. We saw above that it is sometimes useful to decide whether a particular temporal formula has no models. This translates directly (after some serious work) into a question about whether certain (deterministic) automata are empty, i.e. accept no structures. There exists a straightforward way of answering the emptiness question for these automata.

In order to summarize the work of Chapter 2 it is worth looking ahead to the definition of automata. First, we consider the case of linear time. A Muller automaton can be defined as follows:

Definition 1.1. Let L be a finite set of atoms. A (finite state non-deterministic) L-automaton (i.e. one reading L) is a quadruple $M = (S, s_0, T, \varphi)$ where:

- S is a finite set (of states);
- $s_0 \in S$ is the initial state;
- $T \subseteq S \times \wp L \times S$ is a relation—the transition table;
- φ is a propositional formula written with propositional atoms s $(s \in S)$. φ is called the accepting or final condition.

Given a temporal structure in the form of a valuation for the atoms of L on the natural numbers, such an automaton may have many possible *runs* on the structure. These are just sequences of states that the automaton passes through as it successively reads the valuation of atoms at one time (i.e. natural number) and uses the transition table to pass into a new state at the next time. If the transition table is specific enough so that there is a unique new state that the automaton can move into in any given circumstance, then the automaton is said to be deterministic. The accepting condition ϕ determines whether such a run is an accepting run or not. We say that the automaton accepts the structure if there is such an accepting run.

Although much important earlier work was done by Muller and Büchi, the foundation for much of the work we need is the equivalence result of McNaughton [McNaughton, 1966]. We present McNaughton's proof that for any Muller automaton there is a deterministic Muller automaton which is equivalent (i.e. accepts exactly the same structures).

Our main interest in automata is as tools for solving problems to do with temporal logic. The relationship between automata and temporal logic can be described via the intermediary of the second-order monadic language S1S which we met in Volume 1.

We saw, in Section 8.7 of Volume 1 (see also Section 2.6 of this volume)', that S1S is really just a first-order logic (with equality and) with two binary relation symbols $\subseteq$ and $Succ$ as well as a set of constants which is just the set of atoms from a propositional temporal language. A temporal structure $(\mathbb{N}, <, h)$ serves as a structure for S1S with the power set of $\mathbb{N}$ as the domain, $\subseteq$ as usual, $Succ$ as the successor relation on singleton subsets of $\mathbb{N}$, and the interpretation of the atom p is just the set $h(p)$.

In Chapter 8 of Volume 1 we established that S1S was exactly as expressive as a certain, quite powerful, linear temporal logic which we called USF. In order to use the decision procedures (about automata) in Chapter 2 to provide decision procedures about temporal logics we need only know that the temporal logic of interest is less expressive than USF (or equivalently S1S)—as they mostly are—but it is worth briefly recalling here the definion of USF. We will meet USF in more detail in Chapter 10.

In fact we start with the very similar language UYF.

Notation: $\mathbb{N}$ will be the set $\{0, 1, 2, \ldots\}$ of natural numbers. 2^S will denote the set of all subsets of the set S. We often write $\bar{x}, \bar{a}, \ldots$, for *tuples*—finite sequences of variables, atoms, elements of a structure, etc. Other notations will be defined when required.

1.2.1 *Syntax of UYF*

We start by developing the syntax and semantics of the fixed point operator. This is not entirely a trivial task. We will fix an infinite set of propositional atoms, with which our formulae will be written; we write $p, q, r, s, \ldots$ for atoms.

Definition 1.2.

1. The set of formulae of UYF is the smallest class closed under the following:
   (a) Any atom q is a formula of UYF, as is $\top$.
   (b) If A is a formula so is $\sim A$.
   (c) If A is a formula so is YA. We read Y as 'yesterday'.
   (d) If A and B are formulae, so are $A \wedge B$ and $U(A, B)$. (The latter is read as 'Until'. $A \vee B$ and $A \to B$ are regarded as abbreviations.)
   (e) Suppose that A is a formula such that every occurrence of the atom q in A not within the scope of a φq is within the scope of a Y but not within the scope of a U. Then $\varphi q A$ is a formula. (The conditions ensure that $\varphi q A$ has fixed point semantics.)
2. The *depth of nesting* of φs in a formula A is defined by induction on its formation: formulae formed by clause (a) have depth 0; clause (e) adds 1 to the depth of nesting; clauses (b) and (c) leave it unchanged; and in clause (d), the depth of nesting of $U(A, B)$ and $A \wedge B$ is the maximum of the depths of nesting of A and B. So, for example, $\sim\varphi r(\sim Yr \wedge \varphi q Y(q \to r))$ has a nesting depth of 2.
3. A UYF-formula is said to be a YF-*formula* if it does not involve U.
4. Let A be a formula and q an atom. A *bound occurrence* of q in A is one in a subformula of A of the form $\varphi q B$. All other occurrences of q in A are said to be *free*. An occurrence of q in A is said to be *pure past* in A if it is in a subformula of A of the form YB but not in a subformula of the form $U(B, C)$. So $\varphi q A$ is well formed if and only if all free occurrences of q in A are pure past.

1.2.2 *Semantics of UYF*

An *assignment* is a map h providing a subset $h(q)$ of $\mathbb{N}$ for each atom q. If h, h' are assignments, and $\bar{q}$ a tuple of atoms, we write $h =_{\bar{q}} h'$ if $h(r) = h'(r)$ for all atoms r not occurring in $\bar{q}$. If $S \subseteq \mathbb{N}$ and q is an atom, we write $h_{q/S}$ for the unique assignment h' satisfying $h' =_q h$, $h'(q) = S$.

For each assignment h and formula A of UYF we will define a subset $h(A)$ of $\mathbb{N}$, the interpretation of A in $\mathbb{N}$. Intuitively, $h(A) = \{n \in \mathbb{N} \mid A$ is true at n under $h\} = \{n \in \mathbb{N} \mid \|A\|_n^h = 1\}$. We will ensure that, whenever $\varphi q A$ is well formed,

(∗) $h(\varphi q A)$ is the unique $S \subseteq \mathbb{N}$ such that $S = h_{q/S}(A)$.

Notation 1.3. If $S \subseteq \mathbb{N}$, we write $S + 1$ (or $1 + S$) for $\{s + 1 \mid s \in S\}$.

Definition 1.4. We define the semantics of UYF by induction. Let h be an assignment. If A is atomic then $h(A)$ is already defined. We set:

- $h(\top) = \mathbb{N}$;
- $h(\sim A) = \mathbb{N} \setminus h(A)$;
- $h(YA) = h(A) + 1$;
- $h(A \wedge B) = h(A) \cap h(B)$;
- $h(U(A, B)) = \{n \in \mathbb{N} \mid \exists m > n(m \in h(A) \wedge \forall m'(n < m' < m \to m' \in h(B)))\}$;
- finally, if $\varphi q A$ is well formed we define $h(\varphi q A)$ as follows. First define assignments h_n $(n \in \mathbb{N})$ by induction: $h_0 = h, h_{n+1} = (h_n)_{q/h_n(A)}$. Then $h(\varphi q A) \stackrel{\text{def}}{=} \{n \in \mathbb{N} \mid n \in h_n(A)\} = \{n \in \mathbb{N} \mid n \in h_{n+1}(q)\}$.

To establish (∗) we need a theorem.

Theorem 1.5 (Fixed point theorem).

1. *Suppose that A is any UYF-formula and $\varphi q A$ is well formed. Then if h is any assignment, there is a unique subset $S = h(\varphi q A)$ of $\mathbb{N}$ such that $S = h_{q/S}(A)$. Thus, regarding $S \mapsto h_{q/S}(A)$ as a map $\alpha : 2^{\mathbb{N}} \to 2^{\mathbb{N}}$ (depending on h, A), α has a unique fixed point $S \subseteq \mathbb{N}$, and we have $S = h(\varphi q A)$. For any h, $h(A) = h(q) \Leftrightarrow h(\varphi q A) = h(q)$.*
2. *If q has no free occurrence in a formula A and $g =_q h$, then $g(A) = h(A)$.*
3. *If $\varphi q A$ is well formed and r is an atom not occurring in A, then for all assignments h, $h(\varphi q A) = h(\varphi r A(q/r))$, where $A(q/r)$ denotes substitution by r for all free occurrences of q in A.*

We define USF using the first-order connectives Until and Since as well as the fixed point operator. The logic UYF is just as expressive as USF: Yq is definable in USF by the formula $S(q, \bot)$, while $S(p, q)$ is definable in UYF by $\varphi r Y(p \vee (q \wedge r))$. Using UYF allows easier proofs and stronger results.

Returning now to our automata results, we can easily use the fact (see Corollary 2.24) that it is very easy to tell if deterministic automata are empty (i.e. accept no structures) to decide whether a given temporal formula is satisfiable.

To do so we need to have a translation from formulae in S1S into a Muller automaton such that a structure is a model of the formula iff the corresponding automaton accepts the structure. Fortunately, this translation is straightforward: we can use McNaughton's result to help take care of the negation case and the rest is simple induction on the complexity of the S1S formula (see Theorem 2.23). To get our decidability result, we simply translate the temporal (USF) formula into S1S, then into a Muller automaton and into an equivalent deterministic automaton, and finally we check for emptiness.

The case of tree automata. In the second half of Chapter 2 we extend the above results to the case of branching time structures and the tree automata appropriate for accepting or rejecting them. Here we need to bring in a lot of new machinery to deal with the branching—which is a new form of non-determinacy. One very intuitive way to handle this is via games played between two players. Essentially one player represents the tree automaton choosing which new state to move into in accordance with its current state and the valuation of atoms at the current time point. The other player represents the environment choosing which branch to take in order to move from one time point to one of its successors.

There is no branching analogue of McNaughton's theorem. However, with some hard work a decidability result can be achieved for branching structures. This is the famous result on the decidability of S2S by M. Rabin [Rabin, 1969]. We met S2S, the second-order monadic logic of two successor functions, briefly in Volume 1. In Chapter 2 we define it and then present a sketch of Rabin's result. To prove Rabin's result we need to find a tree automaton equivalent to the complement of a given tree automaton.

1.3 Branching time

Continuing our investigation of branching time in Chapter 3, we step back from concentrating on the very particular discrete, height-limited and binary branching models for S2S and consider the question of temporal logic for general branching time. Philosophical considerations about the nature of time and the indeterminacy of the future motivate certain branching models of time. Analysis of natural language leads to similar proposals. The same problem with an indeterminate future also leads computer scientists and builders of artificial intelligences to try to model the progress of a computer program or the reasoning of an agent by such branching models. In Chapter 3 we survey the many specific approaches which result.

When it comes to supplying a formal modal language to cope with these temporal structures, one of the main methods of classifying the approaches is to consider whether the concept of a path or branch through the structure has any role to play in the semantics. In a traditional Kripke manner we can give semantics for modal languages by making the truth of a formula at a point of time just depend on the truth of other formulae at its various successors. Such an approach, in which we follow Prior in calling it Peircean, unfortunately has several drawbacks. The most important is that it fails to capture certain everyday notions of time in a branching setting—in particular the notion of an actual future—and that it fails to allow for nested temporal operators to coordinate their references to sequences of states (see Chapter 3 for details).

For these reasons both computer scientists and philosophers have turned to a more expressive approach to handling branching time. Such approaches, called Ockhamist, involve a new path component in the semantics. Formulae are true at points on particular paths in models. Our usual linear temporal operators operate along the paths while new path operators allow one to switch between one path and another (actually to quantify over paths).

After introducing these two general approaches we concentrate on computer science applications and in particular the specification reactive modules, i.e. programs (hardware etc.) which carry on interacting with their environment. We show that most behaviour that one would like to specify can be captured by supplying a linear temporal logic formula (with atomic expressions for both module and environment properties) and requiring that all possible behaviours of the module along with the environment satisfy the formula. This requirement, which we call implementability of the module specification, is quite different from satisfiability of the temporal formula. Nevertheless, it turns out that we can use results about both tree and linear automata to achieve a decision procedure for implementability.

1.4 Labelled deduction presentation of temporal logics

Chapter 4 introduces the discipline of *labelled deductive systems* (*LDSs*) and formulates the logic $Q\mathbf{K}_t$ (quantified $\mathbf{K}_t$) in this discipline. The basic idea of LDS is that the notion of a declarative unit is not the traditional formula A but a pair $t : A$ of a formula and a label, where in the case of temporal logic the label names a moment of time. A theory is a family of declarative units $\{t_i : A_i\}$ together with some earlier–later relations among the labels of the form $t < s$. A temporal model satisfies a theory if the labels can be mapped into the flow of time in such a way that their order relation is respected and for each $t : A$, A holds at the image moment of time assigned to t.

A proof theory for such labelled theories Δ are manipulation rules of formulae and labels defining the consequence notion $\Delta \vdash t : A$.

1.5 Temporal logic programming

The Horn clause fragment in classical logic has emerged as the computationally effective part of the language and the area of logic programming, based on this fragment, has had widespread spectacular applications throughout computer science, artificial intelligence and logic itself. A natural question to ask is what would be the corresponding 'Horn clause' computational fragment of temporal logic? The answer to this question is important if we want to do direct temporal logic programming. There are two ways of finding the 'Horn clause' fragment of temporal logic. The first method is to look at the traditional translation τ of temporal logic into classical logic (through the possible world semantics) and declare that A is a temporal Horn clause formula if its translation $\tau(A)$ is a Horn clause in classical logic. Thus, for example, the temporal formula $q \to Fp$ is a temporal Horn clause since its translation is

$$q(t) \to \exists s(t < s \wedge p(s))$$

which is equivalent to the database

$$\begin{array}{l} q(t) \to t < f(t) \\ q(t) \to p(f(t)) \end{array}$$

where $f(t)$ is a Skolem function.

The second approach is less immediate. We recall that the Horn clause fragment of classical logic can also be defined as the fragment of the language preserved under certain kinds of products of classical models. We can define similar kinds of products of temporal models and seek to identify the fragment of the temporal language preserved under such products.

We will not pursue the second approach. Thus Chapter 5 develops the first approach and studies the computational aspects of temporal Horn clauses.

1.6 Combining temporal logic systems

In Chapter 6 is a continuation of the work started in Volume 1, Chapter 14, on combining temporal logics. Here, four combination methods are described and studied with respect to the transfer of logical properties from the component one-dimensional temporal logics to the resulting combined two-dimensional temporal logic. Three basic logical properties are analysed, namely soundness, completeness and decidability.

Each combination method comprises three submethods that combine the languages, the inference systems and the semantics of two one-dimensional temporal logic systems, generating families of two-dimensional temporal languages with varying expressivity and varying degree of transfer of logical properties. The *temporalization method* and the *independent combination method* are shown to transfer all three basic logical properties. The method of *full join* of logic systems generates a considerably more expressive language but fails to transfer completeness and decidability in several cases. So a weaker method of *restricted join* is proposed and shown to transfer all three basic logical properties.

1.7 Extensional semantics

Claims about the usefulness of temporal logic for reasoning about program behaviour are frequently met. In Chapter 7, we give a reasonably detailed example of using temporal logic to prove that a program terminates with the correct results—the total correctness of the program.

There are two novel aspects to this example. One is that we go to some lengths to use an extensional temporal language, i.e. Lamport's TLA from [Lamport,

1994]. The other is that we prove the correctness of a program written in the multiset rewriting language Gamma of [Banâtre and Le Métayer, 1990].

TLA is a predicate temporal language. It is unusual in having atomic predicates based on consecutive pairs of 'externally visible states' instead of on machine states at particular instants. The atomic formulae are also syntactically restricted so that, semantically, they cannot differentiate between something happening and everything staying the same for a while and then that something happening. This results in the language being stuttering invariant. That is, the truth of formulae is preserved under the adding or removing of a finite number of repetitions of the same state of values of the observable variables. As a result of this syntactic restriction, TLA descriptions of the behaviour of programs are correct whether or not other programs are running in parallel and whether or not the original program is actually implemented by a lower level program.

The idea of stuttering invariance is that ticks of an internal clock in a program during which there are no changes in the state of externally visible variables should be ignored. They should not be used to distinguish one behaviour from another and should not be all that distinguishes a correct program from an incorrect one. In Chapter 7 we motivate the use of stuttering invariance and more generally the desire to base the semantics of programs on extensional definitions, i.e. on observable behaviour rather than abstract objects like internal states.

We also examine the use of the extensional language TLA. The formal language comes with a proof theory and a helpful general pattern for formalizing the behaviour of programs. This pattern uses special abbreviations, defined in the formal language, for common expressions such as fairness constraints. There are also particular proof rules appropriate for reasoning in the presence of these constraints. By considering certain unusual programming language expressions we find that we need to add to the set of these abbreviations and proof rules.

In the second half of Chapter 7 we examine a particular language and an example program. The language Gamma, which was proposed in [Banâtre and Le Métayer, 1990], is a minimal formalism for describing programs. It allows us to abstract away from much of the detail of the control structure and, in particular, to refrain from having to introduce any artificial sequentiality. Such a language is ideal for use with the following two-step method of parallel program construction. First, prove correct a high-level description of a class of algorithmic solutions to the problem. Then include more details of control to describe a refinement which runs efficiently on particular hardware. The benefits of using Gamma in systematic program construction are described more fully in [Banâtre and Le Métayer, 1990], [Creveuil, 1991], [Mussat, 1992].

In the chapter we give a formal account of Gamma but, essentially, Gamma achieves its minimality by using one data structure, the multiset, and one control structure, the rewrite. The multiset rewrites are often compared to chemical reactions in which certain collections of objects in the multiset react, being consumed and producing other objects as the reaction product. These reactions can occur quite locally in the multiset and herein lies the possibility of parallel implementations.

In our example we present a program consisting of two rules. One rule requires the replacement of a triple of numbers by a pair of the larger two. The other rule requires the replacement of two numbers by their product. We wish to show that if we proceed first of all by indefinitely repeating the use of the first rule until it can no longer proceed, and then use the second rule until it also can no longer proceed then we will end up with the product of the two largest numbers in the original multiset.

The temporal logic proof of correctness of this program proceeds on many levels as it is quite complicated. It involves temporal reasoning in TLA, reasoning about the semantic rules for the language Gamma and reasoning about multisets.

1.8 Intervals and planning

In Chapter 8 we will have a brief look at an area of artificial intelligence in which interval-based temporal logics are much used. This is the task of *planning*, which means deciding in what order to perform certain tasks in order to achieve a desired situation or *goal*.

Because these tasks or *actions* take time to accomplish in many of the usual planning applications it is best to use an interval-based temporal logic as a foundation for the reasoning. The planning problem then becomes one of ordering these temporally extensive actions taking full account of the possibility that actions can overlap in their execution. An appropriate temporal logic can be developed from *Allen's interval algebra* described in [Allen, 1981]. In this algebra we have 13 different relations which represent various ways in which two intervals of time can relate to each other in a linear flow of time.

In order to introduce these relations it is best to use a concrete representation. So suppose that $(T, <)$ is a linear order. Although it is not necessary to confine ourselves to open intervals let us say here that an interval in $(T, <)$ is a set of points $(a, b) = \{x \mid a < x < b\}$ from T lying strictly between some pair $a < b$. Then we have the following relations between intervals:

- equality
- precedence $(a, b) < (c, d)$ iff $b < c$
- (a, b) meets (c, d) iff $b = c$
- (a, b) overlaps (c, d) iff $a < c < b < d$
- (a, b) is during (c, d) iff $c < a < b < d$
- (a, b) starts (c, d) iff $c = a < b < d$
- (a, b) ends (c, d) iff $c < a < b = d$

as well as the inverses of the last six.

Armed with an interval basis we can now formally define a planning problem. We follow the description in [Hirsch, 1993] of a modified version of Allen and Koomen's planner [Allen and Koomen, 1983]. Each particular problem has four components: an interval network, a planning environment, a goal state and a stock of available actions. The interval network just contains the few intervals mentioned in the planning environment and the goal state and details the relationships between them. That is, for every pair of intervals, we list those of the 13 relations which may hold between the pair.

The planning environment describes the fixed information about the problem. It is a list of interval names with associated propositions indicating that the propositions hold on that interval. For example, it might detail what is true on some initial interval.

The goal state is also such a list of intervals with propositions. It indicates which propositions we want to end up being true at each interval.

An action is formalized as follows. There is a small set $\{X_0, X_1, \ldots, X_n\}$ of variables for intervals involved in the execution of the action. The action will have a name, a set of preconditions describing which propositions must be true on each of the intervals X_i, a set of postconditions describing what will end up being true on the intervals (if we 'do' the action at time X_0) and a set of temporal constraints describing the possible relationships between the intervals.

For example, an action might be called brewing tea, have preconditions tea bag in cup at X_1, postconditions black tea in cup at X_2 and constraints X_0 during X_1 and $X_0 < X_2$.

A solution to a planning problem will be a set of intervals from a linear order with names including those mentioned in the problem, a list of which actions are done when and a list of which propositions are true at which times such that:

- the interval network is consistent with the original one;
- the preconditions and postconditions of each of the actions apply correctly;
- each proposition is true at an interval because either it was declared so in the planning environment or it is the postcondition of one of the actions.

It is important to realize that two things are going on in achieving a solution. One is the introduction of actions to explain the truth of propositions at various times but equally important is the realization of the interval network. The latter often includes deciding that two interval names name the same interval.

An algorithm for finding solutions to such planning problems is presented in [Allen and Koomen, 1983]. The basic approach of the Allen and Koomen planner is to work simultaneously on the two tasks described above. If the goal state requires proposition p to be true at interval i, the planning environment states that p is true at interval j and the equality relation is possible between i and j, then we say that p at i has a *possible causal explanation*—we simply make $i = j$. The planner keeps attempting the following two activities until every goal assertion has a possible causal explanation:

- *introducing* actions to explain goal assertions which do not have potential causal explanations, and
- *collapsing* intervals which do have potential causal explanations into the same interval.

This is just the outline of a procedure which can often involve back-tracking to change decisions which have subsequently lead to dead-ends.

Some interesting new work on this planner concerns the problem of collapsing several pairs of intervals at once. An example in [Hirsch, 1993] shows that even if each pair separately is consistently collapsible, together they may not be. Suppose that the planning environment states that p is true at i_0 and q at i_1 while the goal requires p at i_1 and q at i_2. Suppose that we have i_0 before or equal to i_1 and i_1 before or equal to i_2 but i_0 before i_2. Then we have one potential causal explanation for p at i_1 by collapsing i_0 and i_1 and a potential causal explanation for q at i_2 by collapsing i_1 and i_2. However, we cannot consistently collapse both pairs. A modified planning algorithm which can deal with this problem is presented in [Hirsch, 1993].

This *collapsing problem* is related to the persistence problem which infects many other approaches to planning. To make the algorithms more efficient or even just to ensure that a particular solution is found, various algorithms make tentative assumptions about the persistence of properties over time or, equivalently, the identity of intervals in which they are true. These assumptions may need to be retracted during back-tracking. Thus we find *stretching* in TMM in [Dean and McDermott, 1987] and the *persistence axiom* in the event calculus in [Kowalski and Sergot, 1986]. In all these approaches there is plenty of scope for developing better algorithms to solve the planning problem.

1.9 Many dimensional systems and generalized quantifiers

Capter 9 udies many-imensional logics whose set of possible worlds has the form S^k, where $(S, <)$ is a Kripke model, S is a set of possible worlds and $<$ is the binary accessibility relation on S. We define, for $\bar{t}_i = (t^i_1, \ldots, t^i_k) \in S^k, i = 1, 2$, the notion of $\bar{t}_1 < \bar{t}_2$ to mean that $t^i_1 < t^i_2$ coordinate-wise for all i. We study satisfaction of various temporal connectives evaluated at points $\bar{t} \in S^k$ and give a proposed Hilbert system and completeness proof.

By way of application we interpret van Lambalgen's notion of a generalized quantifier in our many-dimensional temporal logic.

A generalized quantifier has the form $(Qx)A(x, y_1, \ldots, y_n)$ where A is a formula of predicate logic with the free variables $x, y_1, \ldots, y_n$. (Qx) is interpreted as ranging over all elements d in a set $V_{(y_1,\ldots,y_n)}$, dependent on the elements $y_1, \ldots, y_n$ (more precisely on what they are assigned in the model). Thus $\mathbf{m} \vDash (Qx)A(x, y_1, \ldots, y_n)$ iff for all $d \in V_{(y_1,\ldots,y_n)}, \mathbf{m} \vDash A(d, y_1, \ldots, y_n)$.

Any closed formula A of the quantifier language can be faithfully translated into many-dimensional temporal logic.

1.10 The declarative past and the imperative future

In Chapter 10, we describe a temporal logic with Since, Until and fixed point operators. The logic is based on the natural numbers as the flow of time and can be used for the specification and control of process behaviour in time. A specification formula of this logic can be automatically rewritten into an executable form. In this executable form it can be used as a program for controlling process behaviour. The executable form has the structure 'If A holds in the past then do B'. This structure shows that declarative and imperative programming can be integrated in a natural way.

There are at least two camps in computing, following two seemingly incompatible paradigms, the imperative and the declarative. Both paradigms are based on sound practical and intuitive grounds, and both have their strengths and weaknesses. This chapter suggests a natural meeting ground for these paradigms and shows that they are compatible and complementary in a natural way. The chapter proposes the theme of the declarative past and the imperative future, and describes an executable temporal logic in which both the imperative and the declarative themes play a part in the most natural way.

1.11 METATEM

Specification needs a formal language. Some early attempts at specifying systems included the idea in [Lamport, 1977] of listing *liveness* properties and *soundness* properties of the system. Liveness properties are characterized as stating that something should happen, while soundness properties state that some things should always hold. These distinctions are still important but, as we will see, a much more intricate description of the expected behaviour of the system is desired. For such purposes *temporal logic* is a very widely used and successful formalism (see, e.g. [Manna and Pnueli, 1992]). In temporal logic we use a surprisingly simple language capable of expressing and reasoning about processes changing over time.

More recently, temporal logic is also being used as a formalism in which programs are actually written. We see many examples of *executable temporal logic* from the initial developments in [Moszkowski, 1986] to the very complex METATEM framework we describe below and the temporal logic programming in [Abadi and Manna, 1989]. The idea is simply that the program can also be read as a sentence of a temporal language. This sentence describes how the program behaves when it runs. Thus executable temporal logics are *declarative* programming languages.

METATEM began with the simple idea of 'declarative past implies imperative future' in [Gabbay, 1989]. A complex expressive programming language took shape in [Barringer *et al.*, 1989] and an implementation of it was described in [Fisher and Owens, 1992]. Recently we have seen it applied to modelling rail networks in [Finger *et al.*, 1993], hospital intensive care wards in [Reynolds, 1997] (and Chapter 12) and developing into the concurrent programming language CMP in [Fisher, 1993].

As a higher level language it seems very appropriate for prototyping, while lower level, more efficient—and less expressive—temporal languages like TEMPURA ([Moszkowski, 1986]) might be appropriate for final implementations.

There are great advantages in using the unified approach of the same temporal formalism for specification, implementation and formal justification.

METATEM is really a paradigm for programming languages rather than one particular language. There are three bases:

- programs should be expressed in a temporal language;
- programs should be able to be read declaratively;
- the operation of the program should be interpretative with individual program clauses operating according to the idea that 'declarative past implies imperative future'.

Most versions of METATEM use the propositional temporal language PML or its first-order version FML with Until and Since.

The basic idea of declarative languages is that a program should be able to be read as a specification of a problem in some formal language and that running the program should solve that problem. Thus we will see that a METATEM program can easily be read as a temporal sentence and that running the program *should* produce a model of that sentence. This is an ideal situation and the correctness question is not usually so trivial: owing to such behaviour as 'looping', sometimes the program does not construct a model.

There are related completeness issues for such a declarative programming language:

- can one specify all desired problems in that subset of the formal language used for the programs; and
- does the program guarantee to come up with a solution if one exists?

These are important questions which we will consider but it is useful to remember that one can go too far in the quest for completeness. A very expressive language allows very difficult problems to be stated, and relying on the program interpreter to find a solution might lead to inefficiencies. Sometimes it is better to make the user rewrite the problem in a better way. However, although we restrict the language of programs, much work is going on to enable problems expressed in a more general way to be rewritten in the appropriate form.

The task of the METATEM program is to build a model satisfying the declared specification. This can sometimes be done by a machine following some arcane, highly complex procedure which eventually emerges with the description of the model. That would *not* be the METATEM approach. Because we are describing a programming language, transparency of control is crucial. It should be easy to follow and predict the program's behaviour and the contribution of the individual clauses must be straightforward.

Fortunately, these various disparate aims can be satisfied very nicely by the intuitively appealing idea of 'declarative past implies imperative future' [Gabbay, 1989]. The METATEM program *rule* is of the form $P \Rightarrow F$ where P is a strict past-time formula and F is a not necessarily strict future-time formula. The idea is that on the basis of the declarative truth of the past time P the program should go on to 'do' F.

A METATEM program is a list of such rules and, at least in the propositional case, it represents the universally temporally quantified conjunction of the corresponding formulae. The program is read declaratively as a specification: the execution mechanism should deliver a model of this formula. To do so it will indicate which propositions are true and which are false at time 0, then at time 1, then at time 2, etc.

To achieve this result one could follow the idea of [Pnueli and Rosner, 1989a] and compile the specification into a transducer which then builds a model. As we have said, METATEM, on the other hand, uses an interpretive approach. The general idea is to go through the whole list of rules $P_i \rightarrow F_i$ at each successive stage and make sure that F_i gets made true whenever P_i is. This would seem to be straightforward as the P_i only depend on the past—so we know when they are true—and the F_i only depend on propositions in the present and future so we can make them true. However, we will see below that there are subtleties.

There are many versions of METATEM in existence. The main distinctions are:

- propositional versus first-order languages;
- closed systems versus reactive modules, and
- single-threaded execution versus concurrency.

In Chapter 11, we examine some of the main variants. In Chapter 12, we give a detailed example of the METATEM approach in action in an intensive care ward of a hospital.

1.12 Non-monotonic coding of the declarative past

Chapter 13 gives a case study of how to code logically a declarative past database and use it within the framework of executable temporal logic. It will help the reader not only understand how our executable temporal logic works but also serves as a typical example of a coded logical database in the artificial intelligence literature.

1.13 A logical view of temporal databases

Traditional approaches to temporal databases are either not formal or deal only with static features, considering mainly a single state of the database. The temporal nature of data in temporal databases, however, requires a dynamic treatment of several features which are not dealt with in non-temporal systems.

In Chapters 14 and 15, a temporal logic framework is proposed to cope simultaneously with the static and the dynamic aspects of temporal data. The framework consists of a two-dimensional temporal model. One-dimensional temporal models deal with static aspects such as temporal querying and data representation, and are presented in this chapter. The second temporal dimension is introduced as a natural extension that captures the dynamics of temporal updates, and is described in the next chapter. The resulting two-dimensional temporal model is used to characterize formally the differences between valid-time and transaction-time, to characterize the notion of 'now' in a temporal database, and to give a formal model for bitemporal databases.

Dealing with temporal data is an activity that inherits the intrinsic complexity of modelling, querying and manipulating data in general. It also contains extra difficulties of dealing explicitly with time. This extra complexity is not trivial and deserves special attention.

It may seem trivial to add a temporal dimension to data simply by adding a time stamp to it. But immediately new issues appear, which do not manifest themselves in the case of non-temporal data and which have to be addressed. For example, it is not clear what that time stamp is intended to represent. Is it the time when the data was true in the universe of discourse, or is it the time when the system learned about that data? Traditionally, the former has been called the valid time of data and the latter the transaction time of data. But if it is one or the other, how does that affect the manipulation of the time-stamped data?

There are other problems which were once purely metaphysical but in the context of temporal data need to have their practical consequences analysed. Does the nature of time affect the modelling and manipulation of temporal data? If time is considered a discrete set of time points or a dense set of temporal intervals, can we 'say' the same things? Can we represent all the things that we want to express about time? At what cost? All these questions lead to the need for a formal, precise discussion of the concepts involved.

The area of temporal databases has been addressing these problems for more than a decade. In this chapter we propose to treat these exclusively temporal issues under a unique framework, namely a temporal logic perspective. Logic has always been present in relational databases as an underlying framework; the notion of a relation can be seen as an interpretation of a predicate in a logical (finite) model, and the relational calculus in a logical query language. When dealing with temporal databases, it is natural to extend the underlying framework to temporal logic. Unfortunately, apart from a few exceptions, temporal database presentations have tended to hide this underlying temporal logic basis. We hope to make it more explicit with this work.

What is particularly new about our approach is that we put the emphasis on the *dynamics of data*, differing from the traditional approach that focuses on the *statics* of a single state of the database. We intend to show that the stress put into the dynamics of data is fundamental in capturing the evolving nature of temporal data.

The static aspects are not ignored or relegated—quite the opposite. We show how a temporal logic approach can be developed to deal with both aspects of data. That is done by presenting the several *temporal dimensions* that a temporal model may have. The static aspects, such as querying and data representation, are dealt within a single temporal dimension. The latter is indeed not an issue in non-temporal databases, for it is obvious how to represent data; in temporal databases, however, there are uncountably many possible databases and only a countable set may ever be represented in a computer. So temporal data representation becomes an issue that requires special attention, to be dealt with by one-dimensional temporal logics.

The second temporal dimension naturally appears when we try to describe some dynamic aspects. Suppose there is a one-dimensional temporal database that stores either only the transaction-time of data or only the valid-time of data, but we do not know which. There is no query that can be posed to the database such that, according to its answer, it is possible to tell whether it is a valid-time or a transaction-time database. Since queries reveal only the static aspect of data, this indicates that valid time versus transaction-time is an implicitly dynamic issue. Indeed, the two-dimensional description of the database has several uses:

- It is used to present formally the evolution, i.e. the dynamic aspects, of a temporal database. We show that the two-dimensional aspects have to be taken into consideration even when a single (valid-time or transaction-time) temporal dimension is stored by the database.
- It is used to characterize the distinctions between valid-time and transaction-time in a formal, temporal logic framework. Such characterization is done via an axiomatization over the two-dimensional model.
- It is used to clarify the notion of 'now' in a temporal database. 'Now' is the place in time where the *database observer* is situated when accessing the database by querying or updating. It is shown how this notion can be captured as the diagonal of the two-dimensional model, and some properties of it are studied.
- Finally, it is shown how the two-dimensional model may be used as the formal basis for *bitemporal databases*, wherein both valid-time and transaction-time dimensions are stored.

The contents here can be perceived differently depending on the reader's background. For the logician, this chapter can be seen as a case of 'applied logic', where a temporal logic theory is applied to describe a temporal database. For the computer scientist, the database practitioner or anyone else interested in temporal information, this chapter provides a unified formal description of the principles underlying temporal databases.

1.14 Temporal conceptual-level databases

In Chapter 16 we describe the temporal database concepts already seen, viewed from a higher abstraction level. Such an abstraction level is called the *conceptual level* and it is normally used in the design of databases and database applications.

Our aim in this chapter is to show that all the database concepts can be raised to a higher abstraction level and that, indeed, we could manipulate the database exclusively from this level. Such a manipulation includes the generation of a conceptual-level schema, semantics, querying, updates and even active rules.

The starting point for the conceptual level is a structured schema which is claimed to be better suited to model the universe of discourse (UoD) of real applications. Such a structured schema is presented in the form of an *ERT diagram*. In particular, we focus on the several temporal features in the form of temporal marks that are included in ERT to cope with the nuances of the representation of temporal features.

The structural properties impose certain implicit constraints on the semantics of *instances* of the diagram. Those constraints are also reflected in what updates should be allowed and how updates are interpreted. Querying an instantiated diagram is done with the ERL temporal conceptual query language. ERL also has the capabilities of representing triggered active rules, allowing the modelling of complex data constraints as well as creating a conceptual level temporal active database.

Such a rich collection of features allows the development of real applications from the conceptual level. To enable such a development we need a translation mechanism that maps a conceptual-level database to an existing commercial relational system.

Such a translation mechanism is sketched in the relevant chapter in a special section. We can then proceed to describe in detail the elements used in the (valid-time) temporal modelling of a UoD using ERT diagrams; the emphasis will be on the modelling of temporal features and the semantics of ERT instances. It is then shown how such a diagram can be formally queried. The transformation of a conceptual-level database from a passive repository to an active one will be described, and the full ERL rule language, together with its active semantics, will be presented.

1.15 Temporal active databases: a background for changing the history and the detection of time paradoxes

In Chapter 17 we show how to create logical links between information associated to possibly distinct times in history. We can then discuss the effects that changes in history may have upon those temporal links.

For the purpose of establishing those links, we extend a valid-time database with *temporal rules*, and we provide those rules with an execution semantics; the resulting combined system is called an *active valid-time database*.

Temporal rules have the general form

$$\mathit{Condition} \to \mathit{Action}$$

which, intuitively speaking, implies that whenever *Condition* is verified in the database, then *Action* is forced to hold. On the formal side, we will see that rules are constraints posed on the two-dimensional evolution of the valid-time database.

For efficiency and practical reasons, it is usual to extend rules with *triggers*. Triggered rules are represented as

$$\mathit{Trigger}\text{: } \mathit{Condition} \to \mathit{Action}$$

with the intuition that both *Trigger* and *Condition* have to be satisfied for the *Action* to be executed. We will see that the basic differences between a *Condition* and a *Trigger* lie in their *persistency properties*, i.e. triggers are very short-lived entities which become false immediately after they are made true. We will see that triggers can help us control the serial execution of rules, therefore improving the efficiency of rule execution.

We have thus created the background in which to analyse the following problem:

What are the effects of changes in history?

The setting in which such a question can meaningfully be posed and answered is, of course, an active valid-time database. We have already seen that valid-time databases allow us to change information about any time in history. So we know the meaning of 'changing the history' and we know how to do it. Furthermore, with the extension of temporal databases with active rules, we know how to create logical links between temporal information associated to any points in history. Therefore we know how one time in history may 'affect' any other time. To answer the question above, it remains to define what are the 'effects of changing history'.

The goal of the final part of this chapter is to study precisely what the effects are of changing the history and how to detect them, whenever possible. For that we define a declarative *valid-time interpretation* of the temporal rules in an active database. We show that, under the execution semantics, the occurrence of updates at any time may cause an invalidation of the valid-time interpretation of rules at some time in the database state, generating a *time paradox*. In the same way that database updates were interpreted as *changes in history*, these time paradoxes are interpreted as *the effects of changing history* or *how changes in history affect other times*. We classify these time paradoxes and, in order to detect their occurrences, the notions of *temporal and syntactical dependences* of a rule are studied. These notions are then used to develop algorithms that perform the detection of time paradoxes.

1.16 Calendar logic

In Chapter 18 we link temporal logic to real-time calendar systems. Instead of the very general operators like 'sometimes in the future' or 'always in the future', we introduce relativized operators of the kind 'sometime next week' or 'always in office hours'. Calendar logic is developed in three stages. We start with a system for specifying everyday temporal notions like 'hours', 'weeks', 'weekend', 'office hour', 'holiday' within a given calendar system. In this system one can check whether a given point in time lies within an interval specified by a time term, or whether a subsumption relation holds between two time terms.

In the next stage these temporal notions are integrated into a propositional modal logic with two parameterized operators 'sometimes within τ' and 'always within τ', where τ may be one of the previously defined temporal notions. An example for a statement in this logic is

$$[2000, year] : \langle June(x_{year})\rangle election.$$

It expresses that some time in June in the year 2000 there is an election. $[2000, year]$ denotes the time region corresponding to the year 2000. $June(x_{year})$

is a time term. It denotes a function that takes a year coordinate as argument and returns a month coordinate. Temporal expressions like these are not built in, but they can be defined with some primitive constructors.

In the last stage, calendar logic will be combined with other (multi)modal logics, e.g. modal logics of knowledge and belief. A simple statement in this logic is

$$[1998, year] : [believe\_I]\langle x_{year} + 1\rangle\langle June(y_{year})\rangle election.$$

(In 1998 I believe that at some time in June the year after (i.e. 1999) there is an election.)

2

TEMPORAL LOGIC AND AUTOMATA

Ian Hodkinson

2.1 Introduction

This chapter[1] covers certain automata, their expressive power, non-deterministic versus deterministic automata, and their relation to games. We apply these results to prove the decidability of certain related monadic second-order logics ('S1S' and 'S2S'). These logics are equivalent to some temporal logics, and also very useful in proving decidability of a host of other temporal logics, as we saw in Chapter 15 of Volume 1 of the current work.

We consider two cases of linear and branching time, namely natural numbers time ($\mathbb{N}$), and time whose shape is the binary tree of height ω (or equivalently, the finite sequences of bits 0 and 1). The linear case was pioneered by J. R. Büchi. The appropriate automata here are the Büchi and Muller automata. We will cover a key result on equivalence of deterministic and non-deterministic Muller automata, from [McNaughton, 1966], and use it to prove decidability of S1S, the monadic second-order logic of one successor function. This is a result of Büchi [Büchi, 1962a]. Decidability in the branching time case was established by M. O. Rabin. In his seminal 1969 paper [Rabin, 1969], he proved decidability of the monadic second-order logic of two successor functions, known as S2S. We will present a version of the proof in [Gurevich and Harrington, 1982]. There are other proofs and related work by Büchi, A. Muchnik, A. and V. Yakhnis, and S. Zeitman [Büchi, 1983; Muchnik, 1984; Yakhnis and Yakhnis, 1990a; Zeitman, 1993].

For a survey of the area see [Gurevich, 1985]; see also [Büchi, 1983], which contains many interesting remarks on the history and methodology of this area.

There is a powerful technique not explicitly using automata that solves many of the problems addressed by automata theory. This 'composition' method can give new insights into the structure of temporal structures, analysing them by chopping them into blocks. It originated with Ramsey, Läuchli, Leonard, and Shelah [Ramsey, 1930; Läuchli and Leonard, 1966; Shelah, 1975]. We will not consider it here, but the Gurevich survey just mentioned covers it, and, in fact, its relationship to automata is very close.

[1]This was originally a set of lecture notes on automata theory and monadic second-order logic by Ian Hodkinson. During their development he was supported by SERC Advanced Fellowship B/ITF/266. Some editing and additions have been made in the final section by Mark Reynolds.

Uniformization is a more advanced topic, also not covered here. It has to do with implementation of (temporal) specifications. Two references are: for linear time, [Büchi and Landweber, 1969], and for branching time, [Gurevich and Shelah, 1983], using Cohen forcing.

2.2 Automata—Definitions

Why consider automata?

- They provide a natural bridge between logic and computer science. They are very expressive yet 'easy' to implement.
- A lot is known about them. Very powerful techniques are available.
- They permit analysis of the environment (cf. reactive systems, planning).
- They are in general use in theoretical computer science. Other kinds of automata have well-known applications in natural language theory and parsing for compilers.

What are they?

First we recall, from Section 8.5.1 of Volume 1, the notion of a *recursive system* $\rho = (\bar{q}, \bar{R})$, where:

- $\bar{q} = (q_i : i < k)$ is a sequence of propositional atoms;
- $\bar{R} = (R_i : i < k)$ is a sequence of Boolean combinations R_i of temporal formulae of the form $Y^m p$, where
    - p is an atom or $\top$,
    - 'Y' is read 'Yesterday',
    - $Y^m p = YY \dots Yp$ with m Ys,
    - if $p \in \{q_i : i < k\}$ then $m > 0$ (i.e. R_i is pure past in each q_j if it occurs).

For a set X, let $\wp X$ be the set of all subsets of X (the power set of X). Given a temporal structure $N = (\mathbb{N}, <, h)$, where $h : \{\text{atoms}\} \to \wp\mathbb{N}$, we define the semantics of these temporal formulae by:

- $h(\neg A) = \mathbb{N} \setminus h(A)$;
- $h(A \wedge B) = h(A) \cap h(B)$;
- $h(YA) = \{n + 1 : n \in h(A)\}$.

We then define a structure $N_\rho = (\mathbb{N}, <, h_\rho)$ recursively by:

- $m \in h_\rho(q_i)$ iff $m \in h_\rho(R_i)$, for all $i < k$, all $m \in \mathbb{N}$ (this is recursive in m);
- $h_\rho(p) = h(p)$ for all atoms $p \notin \{q_i : i < k\}$.

This is well defined, as R_i is pure past in each q_j if it occurs, and it delivers a unique h_ρ.

Recursive systems can be taken as the basis for the fixed point temporal logic USF (see Chapter 8 of Volume 1, and Chapter 10 here).

Example 2.1. If $\rho = (q, \neg Yq)$, then for all h, $h_\rho(q) = \{0, 2, 4, \dots\}$.

So the true–false values of the q_i at $m \in \mathbb{N}$ are defined recursively as a Boolean combination of their previous values (and the values of other atoms). Suppose that e is the maximum power of Y in any R_j. Then Fig. 2.1 shows the situation.

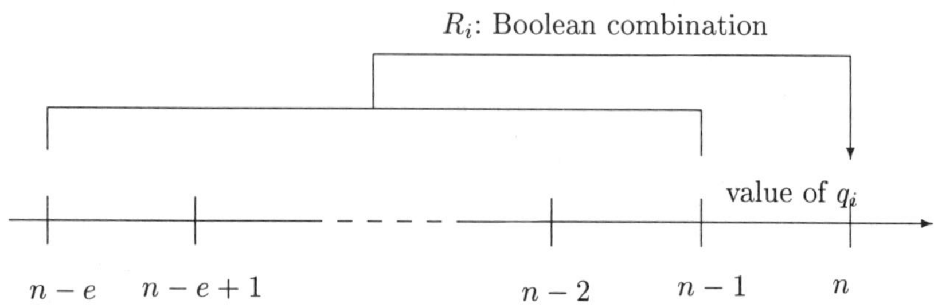

FIG. 2.1.

The figure does not show that R_i can use the values of the other atoms (other than $\{q_i : i < k\}$) at n itself. In any case, there are finitely many possible patterns of values of all the atoms at the points that R_i 'looks at'—the number is at most $x = 2^{m.(e+1)}$, where m is the total number of atoms involved in the R_i. Thus, there are at most x possible 'inputs' to the recursive system ρ. The 'output' of ρ—the values of the q_i at n—are determined by the input.

So we can view ρ as a *finite state machine*:

- its state at time n is the pattern of values of the atoms in the R_i at the points $n-1, n-2, \ldots, n-e$;
- it reads from the structure $(\mathbb{N}, <, h)$ the values at n of the atoms in the R_i but not in $\{q_i : i < k\}$;
- it then uses this information, and its state, to determine (using the R_i) the values of the atoms q_i at n;
- these values give its next state (at time $n+1$).

Now it repeats the above.

There are finitely many states. Note that ρ is deterministic in that its next state is determined by its current state and the 'input' of values at n.

Thus, finite state machines arise naturally in temporal logic.

We formalize the notion of a finite state machine more simply and generally as follows.

Definition 2.2. Let L be a finite set of atoms. A (finite state non-deterministic) L-automaton (i.e. one reading L) is a quadruple $M = (S, s_0, T, \varphi)$ where:

- S is a finite set (of states);
- $s_0 \in S$ is the initial state;
- $T \subseteq S \times \wp L \times S$ is a relation—the transition table;

- φ is a propositional formula written with propositional atoms s ($s \in S$). φ is called the accepting or final condition.

This definition is essentially due to Muller around 1963. The idea is as follows. Let $N = (\mathbb{N}, <, h)$ be an L-structure. We can equally view h as a map : $\mathbb{N} \to \wp L$, saying which atoms are true at time n, i.e. $h(n) = \{p \in L : p \text{ is true at time } n\}$. (We will view assignments this way whenever convenient.) M starts off in state s_0 at time 0. It will run along $\mathbb{N}$, from 0 to 1 to $2, \ldots$, reading information from N and changing state accordingly. Suppose that M is in state s_n at time n. M first looks to see which atoms are true at n (i.e. to see what $h(n)$ is). It then chooses a triple $(s_n, h(n), s') \in T$, and defines its next state s_{n+1} to be s'. So at time $n+1$, M is in state $s' = s_{n+1}$. It then repeats this for ever.

The result is a run $s_0, s_1, \ldots$ of M on N. Formally, a run of the automaton $M = (S, s_0, T, \varphi)$ on the temporal structure $N = (\mathbb{N}, <, h)$ is a sequence $s_i \in S$ ($i \in \mathbb{N}$) such that $(s_n, h(n), s_{n+1}) \in T$ for each $i \in \mathbb{N}$. (Note that the run starts with s_0.)

There are, in general, many such runs even on a single structure. This is because there could be more than one s' such that $(s_n, h(n), s') \in T$. What if there is no such s' at all? Then M grinds to a halt. This is inconvenient, so we usually assume that it cannot happen. We can (and will) assume that

$$\forall s \in S \, \forall L' \subseteq L \, \exists s' \in S((s, L', s') \in T),$$

i.e. there is at least one s' matching any s and L'. If there is exactly one s', then M is said to be *deterministic*. In this case, M does have a unique run on any N.

We do not say how M chooses the triple $(s, L', s') \in T$, if there is more than one. Many people think of M as 'guessing' a transition, and we will often use this term here, but just as many others dislike thinking of it this way. Mathematically, any possible choice is valid, and we just have to allow for all possible choices it could make.

Why do we need non-determinism at all? In a sense, we don't—see Section 2.4.

Example 2.3. Take $S = \{s_0, s_1\}$ and $T = \{(s', L', s'') : L' \subseteq L, s' \neq s''$ in $S\}$. Then in any N, there is only the run $\bar{s} = (s_0, s_1, s_0, s_1, \ldots)$. Compare $\rho = (q, \neg Y q)$ above.

Exercise 2.4. Show that any recursive system can be mimicked by a deterministic automaton.

We now discuss φ.

So far we have described a machine that does rather little: it simply runs along a temporal structure, responding to what it finds by changing state. We want to use automata to describe or analyse the structure. We do this by examining the properties of the possible runs on that structure.

Suppose that we have a yes/no question about temporal structures. A typical example might be: '$N \models A$?' for a formula A of some temporal logic. (Recall that $N \models A(n)$ means that A is true at n in N ($n \in \mathbb{N}$); $N \models A$ means $N \models A(n)$ for all $n \in \mathbb{N}$.) Can we design an automaton M whose runs on the 'yes-structures' are describably different from its runs on the 'no-structures'? If so, we can use M to check whether the answer to the question is yes or no.

So we consider yes/no questions about runs of automata.

Now there is no final (last) state in a run. So the question 'is the final state acceptable?', as used with Turing machines and automata for regular languages, is not meaningful here. The nearest analogue to a final state is *a state that occurs infinitely often in a run.* Our basic questions about runs will (therefore) take the form 'does such and such a state occur infinitely often in the run?' We will allow more complex combinations of such questions—in fact, we admit any Boolean combination.

Formally, let $M = (S, s_0, T, \varphi)$ be an automaton. A run $\bar{s} = (s_0, s_1, \ldots)$ of M gives rise to an assignment to the propositional atoms $s (s \in S)$, by:

$$\bar{s} \models s \text{ iff } \{n \in \mathbb{N} : s = s_n\} \text{ is infinite.}$$

We can thus define whether $\bar{s} \models \varphi$ for any propositional φ using these atoms, by induction on φ as usual in propositional logic.

Definition 2.5.

1. A run $\bar{s}$ of $M = (S, s_0, T, \varphi)$ is said to be accepting if $\bar{s} \models \varphi$. Otherwise, $\bar{s}$ is said to be rejecting.
2. M is said to accept an L-structure N iff there exists an accepting run of M on N. If there is no accepting run, M is said to reject N.
3. Two automata M_1, M_2 for L are said to be equivalent if for every L-structure N, M_1 accepts N iff M_2 does.

Examples 2.6.

1. Let $s \in S$. The accepting runs of (S, s_0, T, s) (for some $s \in S$) are precisely those runs in which s occurs infinitely often.
2. There are no accepting runs of $(S, s_0, T, \bigwedge_{s \in S} \neg s)$.
3. If φ and ψ are logically equivalent then (S, s_0, T, φ) and (S, s_0, T, ψ) are equivalent automata.

Exercise 2.7. Design an automaton that accepts a temporal structure $N = (\mathbb{N}, <, h)$ iff the temporal formula $p \to Fq$ is valid in N (true at every time $0, 1, \ldots$).

2.3 Variants of our automata

The widely used automaton of Büchi has the form $M = (S, s_0, T, X)$, where $T \subseteq S \times \Sigma \times S$ for some finite alphabet Σ, and $X \subseteq S$. Here, M is given a 'letter' from Σ at each stage, rather than a whole set of atoms that are true. There is no essential difference here—take $\Sigma = \wp L$ to convert from the automata above to Büchi automata, and to go the other way is also easy. There is, however, a more significant difference in the accepting conditions: a run of a Büchi automaton is accepting iff some state in X occurs infinitely often. This says essentially that 'φ' (as in the preceding section) must be of the restricted form $\bigvee_{s \in X} s$ (for arbitrary $X \subseteq S$). The restriction has the effect that not every Büchi automaton is equivalent to a deterministic Büchi automaton.

Exercise 2.8. Let p be a propositional atom. Design a Büchi automaton B that accepts a structure $N = (\mathbb{N}, <, h)$ iff p is true at only finitely many times in N. Show that there is no deterministic Büchi automaton equivalent to B. Find a deterministic automaton in the sense of Definition 2.2 that is equivalent to B.

A slightly different style of automaton was invented by Muller. Again, the traditional definintion has the automaton reading letters of an alphabet Σ, rather than tuples of atoms; we ignore this difference.

Definition 2.9. For a finite set L of propositional atoms, a Muller automaton is one of the form (S, s_0, T, F) where $T \subseteq S \times \wp L \times S$ as for our version, and $F \subseteq \wp S$. A run is accepting iff the set of states occurring infinitely often in the run is in F.

Both Büchi and Muller automata are equivalent to the version we give: for any M of one form there is an M' of the other that accepts exactly the same structures. We will need this result only for Muller automata.

Proposition 2.10.

1. *For any Muller automaton there is an equivalent one of ours (as in Definition 2.2).*
2. *For any one of our automata there is an equivalent Muller automaton.*

Further, in each case, the new automaton can be constructed effectively from the old one, and if the original automaton was deterministic then the new, equivalent one can be taken to be deterministic also.

Proof

1. Let a Muller automaton $M = (S, s_0, T, F)$ be given. If $X \subseteq S$, define φ_X to be $\bigwedge_{s \in X} s \wedge \bigwedge_{s \notin X} \neg s$. Then the automaton $(S, s_0, T, \bigvee_{X \in F} \varphi_X)$ is equivalent to M.
2. Take an automaton (S, s_0, T, φ). We may replace φ by any logically equivalent formula and obtain an equivalent automaton. So we can assume that φ is in disjunctive normal form $\bigvee_i \bigwedge_j \chi_{ij}$, where each χ_{ij} is atomic or negated atomic, and for each i and every atom $s (s \in S)$, either s or $\neg s$ occurs as a conjunct in $\bigwedge_j \chi_{ij}$. For each i, let

$$F_i = \{s \in S : \text{ for some } j, \chi_{ij} = s\}.$$

   Let F be the set of all these sets F_i. Then the Muller automaton (S, s_0, T, F) is equivalent to (S, s_0, T, φ).

In both cases, determinacy is preserved (since only the accepting condition changes), and the constructions are clearly effective. □

The difference between Muller automata and the automata defined in Definition 2.2 is so slight that we will pass from one to the other whenever convenient.

2.4 Elimination of non-determinism

In this section, we will prove the following theorem.

Theorem 2.11 (McNaughton, 1966). *For every automaton M (as defined in Definition 2.2), there is a deterministic automaton D that is equivalent to M and can be constructed effectively from M.*

The proof is adapted from that of McNaughton [McNaughton, 1966]. We will see several applications of the result later, e.g. to decidability questions.

We assume a fixed finite set L of atoms that all the automata use. We begin with some reductions. By Proposition 2.10, it is enough to show that every Muller automaton $M = (S, s_0, T, F)$ has a deterministic equivalent. Moreover, we can assume that $|F| = 1$. For suppose $F = \{X_1, \ldots, X_n\}$. Suppose that we had a deterministic automaton M_i equivalent to $(S, s_0, T, \{X_i\})$ for all $i \leq n$. We form a big deterministic automaton M_* equivalent to M, as follows. M_* just runs the M_i in parallel along a structure; it accepts the structure iff at least one of the M_i does. (It is not hard to formalize this: take the state set S_* (say) of M_* to be the product of the state sets S_i of the individual M_i, and its acceptance condition F_* to be the set of all subsets of S_* whose projection to some component i is in F_i, the accepting condition of M_i.) But one of the M_i accepts iff for some i, the original automaton M has a run such that the set of states occurring infinitely often is X_i. And that is iff M accepts the structure.

So it is enough to find a deterministic equivalent for $M = (S, s_0, T, X)$, where $X \subseteq S$ and a run of M is accepting iff the set of states occurring infinitely often in it is exactly X. If $X = \emptyset$ then no run of M is accepting, and so there is certainly an equivalent deterministic automaton, with, for example, one state. So we may assume $X \neq \emptyset$. Fix arbitrary $x \in X$.

The following trivial definition will be useful.

Definition 2.12.

1. Define M_x to be (S, x, T), i.e. M_x is the same as M above, but its initial state is x. We are not interested in the accepting condition, so we omit it.
2. Let G be any automaton. A *partial run* of G is a finite initial segment of a full run of G on some structure. That is, it is a finite sequence of the form $s_0 s_1 \ldots s_n$ for some run $s_0 s_1 \ldots$ of G. Let S be the state set of G. Then the set S^* of finite sequences of elements of S contains the set of partial runs. If $\xi, \eta \in S^*$ we write $\xi\eta$ for their concatenation: ξ followed by η.

Let σ be an accepting run of M. Then clearly, we can write σ as $\sigma_0\sigma_1\sigma_2\ldots$, where:

(a) σ_0 is a partial run of M ending in x;

(b) for each $i > 0$, the sequence $x\sigma_i$ is a partial run of M_x, σ_i consists only of elements of X, every element of X occurs at least once in σ_i, and σ_i ends in x.

Moreover, any such sequence $\sigma_0\sigma_1\sigma_2\ldots$ is an accepting run of M.

The reason why it is useful to write an accepting run σ as the concatenation of sequences of type (a) and type (b) is that we can find deterministic automata that can check whether an (a)- or a (b)-sequence is possible at any point. We prove this now.

Lemma 2.13. *We can design a deterministic automaton A such that if A runs along any structure N, we can tell from the state of A at n whether or not M has a partial run of type (a) (as above) from 0 to n.*

Proof Set $A = (\wp S, \{s_0\}, T')$, where T' is arranged so that the state of A at n is the set of possible states of M at n. Formally, if $Y \subseteq S$ and $L' \subseteq L$, write $Y * L'$ for $\{s' \in S : (s, L', s') \in T$ for some $s \in Y\}$. Then

$$T' = \{(Y, L', Y * L') : Y \subseteq S, L' \subseteq L\}.$$

A is clearly deterministic, and if, in its run on N, its state at $n \in \mathbb{N}$ is Y, then M has a partial run of type (a) from 0 to n iff $x \in Y$. □

Lemma 2.14. *We can design a deterministic automaton B such that if B runs along any structure N, we can tell from its state at n whether or not M_x has a partial run of type (b) from 0 to n (i.e. a run covering exactly X and ending in x at n).*

Proof The idea is as in the preceding lemma, but we have to include information about which states have arisen so far in the run. First, form a possibly non-deterministic automaton B' with state set $\wp S \times S$, initial state $(\emptyset, x)$ and transition table as follows:

$$\{((Y, s), L', (Y \cup \{s\}, s')) : s, s' \in S, Y \subseteq S, L' \subseteq L, (s, L', s') \in T\}.$$

Clearly, (Y, s) is a possible state of B' at n iff there is a possible run of M_x from 0 to n ending in state s and covering exactly the states in Y up to $n-1$. Hence, there is a type (b) run of M_x from 0 to n iff B' has a run from 0 to n ending in state (X, x).

But by the proof of the preceding lemma, there is a deterministic automaton B such that we can tell from its state at n whether or not B' has a run ending in state (X, x) at n. We call such a state of B a *yes-state*. B can be constructed effectively from B', and so from M_x, and B is in a yes-state at time n iff there is a run of M_x of type (b) from 0 to n, as required. □

We will build a deterministic automaton D that finds out if N sustains a run $\sigma_0\sigma_1\sigma_2\ldots$ of M as above. The rough idea is simple—it is the obvious thing to try. D starts off by firing off the automaton A at time 0. At each time $t \geq 0$, if

A says that there is a partial run of M up to t ending in x (type (a)), then D fires off a copy of B. These copies of B report back to D at each step, saying whether they are in a yes-state or not. Suppose a copy B_0 of B is launched at t_0 and reports 'yes' at time t_1: there is a run of M_x from t_0 to t_1 of type (b). Evidently there must also be a possible run of M from 0 to t_1 of type (a); A will be aware of this, so D will automatically launch a new copy B_1 of B at t_1. If, later, B_1 reports a possible (b)-run from t_1 to t_2 say, then again D will launch a B_2 at t_2. And so on. D accepts a structure N iff at the end of time, there has been an infinite sequence of launches end to end, each B_{i+1} launched when B_i was in a yes-state. Clearly, D and M are equivalent automata.

We have to implement this idea into a finite state deterministic automaton. The problem is that D seems to be firing off infinitely many copies of B. This will not do, as D would need infinitely many states to handle the situation. So we try the following: whenever two or more copies of B move into the same state (at the same time), then (as B is deterministic) they will all subsequently stay in the same state as each other. So we remove all of them except one, the one that was launched earliest. This one will represent the others from now on; and the other copies can then be used again. As B has finitely many states—say, b states—this will limit the necessary complement of copies of B to b.

The problem with this idea is that D loses track of which copies of B are launched when. So we have to find another way of checking the end-to-end condition.

We adopt the following 'program' for D. Convention: if $s_0 s_1 \ldots$ is a run of an automaton G, we say that G is in state s_i at time i (and not just before time i, as we said earlier). For simplicity, we actually use $b+1$ copies of B. Each copy of B has a 'bell' on it, and a 'space' to hold 'pointers' to other copies of B. These are notational conveniences and can all be easily formalized.

1. At time 0, D launches A (Lemma 2.13), which is then left running for ever.
   For each time $t \in \mathbb{N}$:
2. If A says that there is a run of M of type (a) from 0 to t, then D launches an idle copy B_t (say) of B. There will always be an idle copy available.
3. Then, to each currently running copy of B in a yes-state, if any, D attaches a pointer to B_t (i.e. a note saying 'B_t'). (Note that a copy of B may soon have several pointers attached.)

   For each currently active copy B^1 of B in turn (including any that has just been turned on):
4. If there is an 'older' copy B^0 of B in the same state as B^1 (i.e. B^0 was last launched before B^1 was last launched), then B^1 is turned off and any pointers it carries are discarded. From now on, B^0 will play the role of B^1, so any pointer to B^1 (on any remaining copy of B) is replaced by a pointer to B^0. Any resulting duplicate pointers are deleted.

   Finally, and critically, for each remaining active copy B^0 of B in turn:

5. If B^0 now carries a pointer to itself, then the pointer is discarded and the bell on B^0 is pinged.

Step 4 requires that D keep track of the order of launch of the currently active copies of B. Clearly, the maximum number of pointers required is b^2, so this can all be implemented in a finite state manner. We invite the reader to try. Step 5 occurs when B^0 was pointing to some younger copy of B which just moved into the same state as itself.

Consider a run of D on an arbitrary structure N. Let B_0 be a copy of B that was launched at $t_0 \in \mathbb{N}$. B_0 may or may not be turned off after t_0. But in any case, at all times $t \geq t_0$, after D has finished its work at t there will be a unique active copy of B in the same state that B_0 would have been in, had it remained active throughout. This is the whole point of the construction, and of course it uses the deterministic nature of B. We refer to this copy as the *descendant* of B_0 at time t.

Suppose that B_0 is a copy of B that was launched at t_0 and is in a yes-state at some later time $t_1 \geq t_0$. Then $t_1 > t_0$, and there is an (a)-run of M from 0 to t_0 and a (b)-run of M_x from t_0 to t_1. Hence (by joining them up) there is an (a)-run of M from 0 to t_1. So a copy B_1 of B is launched at t_1, and B_0 gets a pointer to B_1.

If B_1 is later replaced by a descendant B_2, B_0s pointer is also replaced by a pointer to B_2. If, later still, B_2 is removed in favour of B_3, B_0's pointer is changed again to indicate B_3. And so on. Hence B_0's pointer always points to the current descendant of B_1—unless that descendant is actually B_0 itself. In that case, step 5 applies, B_0 pings and the pointer is deleted. This is the only situation in which B_0 pings. Hence the following holds:

Lemma 2.15. *In the situation outlined, the following are equivalent:*

(a) B_0 pinged at time t;

(b) there is $t_0 < t' \leq t$ such that

(i) B_0 was in a yes-state at t',

(ii) a copy B' of B was launched at time t', and

(iii) B_0 became the descendant of B' at time t.

Definition 2.16. A run of D on a structure N is said to be accepting if there is a copy of B that is turned off finitely often and whose bell rings infinitely often.

Note that this can be formalized in the Muller framework.

Theorem 2.17. *D accepts a structure N iff M does.*

Proof $\Rightarrow$: If D accepts N, choose a copy B_0 of B that was turned off finitely often and pinged infinitely often. Let t_0 be the time that B_0 was last turned on. We define a sequence $t_0 < t_1 < \ldots$ in $\mathbb{N}$, and copies $B_0, B_1, \ldots$ of B, by induction, so that for all $i > 0$:

- B_0 was in a yes-state at t_i;
- (hence) a copy B_i of B was launched at t_i;

- B_0 is the descendant of B_{i-1} at t_i.

Suppose that we have defined $t_0, \ldots, t_n$ for $n \geq 0$. Now clearly, only finitely many copies of B are launched at times t with $t_0 \leq t \leq t_n$. Some of these copies—$B^1, \ldots, B^k$, say—may have B_0 as a descendant (their final descendant). We can wait until this has all happened: choose $t > t_n$ such that the descendant of each of $B^1, \ldots, B^k$ at $t-1$ is already B_0, and such that B_0 pings at t.

By Lemma 2.15, B_0 becomes the descendant of some B' at t, such that B' was launched at a time $t' > t_0$ when B_0 was in a yes-state. By choice of t we have $t' > t_n$. We define t_{n+1} to be t' and B_{n+1} to be B'.

Clearly, the three conditions are satisfied (the last one trivially, if $i = 1$). There is an (a)-run σ_0 of M from 0 to t_0, as a copy of B was launched at t_0. Now for all $i > 0$, the state of B_0 at t_i was a yes-state. But at t_i, B_0 was already the descendant of B_{i-1}, i.e. the state of (the descendant at t_i of) B_{i-1} at t_i was a yes-state. This means that there is a (b)-run $x\sigma_i$ of M_x from t_{i-1} to t_i inclusive, i.e. a run with σ_i covering all of X, none of $S \setminus X$, and ending in x. This holds for all $i > 0$. The partial runs $\sigma_0, \sigma_1, \ldots$ can now be pieced together to form an accepting run $\sigma_0\sigma_1\ldots$ of M on N. See Fig. 2.2.

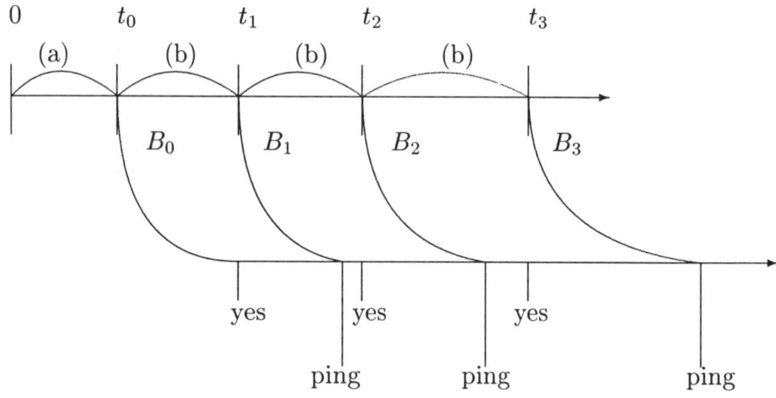

FIG. 2.2.

$\Leftarrow$: Assume that M accepts N. As we said, there is a run of M of the familiar form $\sigma_0\sigma_1\sigma_2\ldots$. Let t_i be the time of the end state of $\sigma_0\sigma_1\ldots\sigma_i$, so $t_0 < t_1 < \ldots$ in $\mathbb{N}$. Clearly for all i, $\sigma_0\ldots\sigma_i$ is a run of type (a). Hence D launched a copy B_i of B at t_i (all i). Note that B_i is in a yes-state at t_j for all $j > i$.

Claim 1: There is an infinite set $E \subseteq \mathbb{N}$ such that if $i, j \in E$ then B_i and B_j have a common descendant.

Proof of claim: Define $i \equiv j$ iff the descendants of B_i and B_j were eventually the same. $\equiv$ is reflexive and symmetric. But B is deterministic, so once the descendants of B_i and B_j become equal, they stay equal. It follows that $\equiv$ is transitive. So $\equiv$ is an equivalence relation on $\mathbb{N}$.

Recall that B has b states. Suppose there were a set of $b+1$ B's that (pairwise) never have the same descendants. Then at any time after they were all launched, their descendants must all be in different states. This is impossible, as B has only b states. Hence $\equiv$ has at most b classes. So we can take an infinite $\equiv$-class $E \subseteq \mathbb{N}$. This proves the claim.

By the claim, there is an infinite sequence $u_0 < u_1 < \ldots$ in $\mathbb{N}$ such that for all $i < j$ in $\mathbb{N}$:

- a copy B_i of B was launched at u_i;
- the descendant of B_i at u_j was in a yes-state at u_j;
- B_i and B_j already have the same descendant by u_{j+1}.

Take such a sequence with u_0 as small as possible.

Claim 2: B_0 is never turned off after u_0.

Proof of claim: If it were, then some copy B^* of B launched at $u' < u_0$ would move into the same state as B_0 at some time $u'' \geq u_0$. Choose n such that $u_n > u''$. Then the sequence $u' < u_n < u_{n+1} < \ldots$ satisfies the conditions above, contradicting the assumption that u_0 was least possible. This proves the claim.

Claim 3: B_0 pings infinitely often.

Proof of claim: Let $i > 0$. Clearly B_0 is in a yes-state at u_i. So it gets a pointer to B_i at time u_i. B_i and B_0 have the same descendant—namely, B_0—by time u_{i+1}. Let u^* be the time that B_0 becomes the descendant of B_i. By Lemma 2.15, B_0 pinged at time u^*. We have $u_i \leq u^* < u_{i+1}$. Hence B_0 pings between u_i and u_{i+1}, for all $i > 0$. So B_0 pings infinitely often. This proves the claim and, with it, the theorem. □

The construction of D from M is clearly effective. Hence, McNaughton's theorem is proved.

See [Safra, 1988] for a description of an algorithm which finds a deterministic (Muller) equivalent for any (Büchi) automaton and tries to be as efficient as possible in terms of the size of the resulting automaton.

2.5 Monadic second-order logic

We can now pick the fruit of the hard work. We will examine a powerful logic, S1S, used to describe temporal structures with flow of time $\mathbb{N}$. It is as powerful as USF (for details of USF see Chapter 8 of Volume 1, Chapter 10 of this volume, Chapter 2 in [Barringer *et al.*, 1996], or [Hodkinson, 1995]). It is also equivalent in expressive power to Muller automata. We will use this equivalence to prove its decidability.

The name S1S denotes the monadic second-order logic of one successor function—or equivalently, the monadic second-order logic of the structure $(\mathbb{N}, <)$, this being up to isomorphism the structure freely generated by a single function from a constant. We will actually present S1S as a first-order logic. Let L be a set of propositional atoms, as usual:

- the elements of L are regarded as constant symbols in S1S;
- the relation symbols are $=$, $\subseteq$ and $Succ$ (all binary);
- there are no function symbols;
- the syntax is standard first order.

Now we explain the semantics. Formulae of S1S are given meaning over temporal structures, of the form $N = (\mathbb{N}, <, h)$ as usual, where $h : L \to \wp\mathbb{N}$ is an assignment. However, we actually evaluate S1S-formulae over the fixed domain $D = \wp\mathbb{N}$. For each temporal structure N, we obtain a new structure, denoted D_N, with domain D. Each constant $q \in L$ is interpreted in D_N as $h(q) \in D$. The symbols '$=$' and '$\subseteq$' are interpreted as usual on D, and for $X, Y \in D$, $Succ(X, Y)$ holds iff for some $n \in \mathbb{N}$, $X = \{n\}$ and $Y = \{n+1\}$.

If $\varphi(x_1, \ldots, x_n)$ is a formula of S1S, and $X_1, \ldots, X_n \subseteq \mathbb{N}$, we write $N \models \varphi(X_1, \ldots, X_n)$ as an abbreviation for $D_N \models \varphi(X_1, \ldots, X_n)$. We extend this notation to sentences σ, so $N \models \sigma$ abbreviates $D_N \models \sigma$.

As is easily seen, S1S is essentially full monadic second-order logic over $\mathbb{N}$. This logic was introduced in the conventional way in Volume 1, section 8.7, using both first-order and second-order variables and quantifiers. The version given above is equivalent, but technically simpler, having only one kind of quantifier. We leave it as an exercise to translate one form into the other.

S1S is very expressive. We give some examples of its power.

Example 2.18. Here, N is an arbitrary temporal structure, and S is an arbitrary subset of $\mathbb{N}$.

1. Let $\sigma(x)$ be the formula $\exists y(Succ(x, y))$. Then $N \models \sigma(S)$ iff S is a singleton (has exactly one element in it).
2. Let $\zeta(x) = \sigma(x) \wedge \neg\exists y Succ(y, x)$. Then $N \models \zeta(S)$ iff $S = \{0\}$.
3. Let $\iota'(x, y)$ be the formula $x \subseteq y \wedge \forall zt(z \subseteq y \wedge Succ(t, z) \to t \subseteq y)$. Let $\iota(x, y)$ be the formula $\iota'(x, y) \wedge \forall z(\iota'(x, z) \to y \subseteq z)$. Then for all $n \in \mathbb{N}$, $N \models \iota(\{n\}, S)$ iff $S = \{0, 1, \ldots, n\}$.
4. Let $x < y$ be the formula
$$\sigma(x) \wedge \sigma(y) \wedge \forall zt(\iota(x, z) \wedge \iota(y, t) \to z \neq t \wedge z \subseteq t).$$
Then for all $n, n' \in \mathbb{N}$, $N \models \{n\} < \{n'\}$ iff $n < n'$ in the natural ordering on $\mathbb{N}$.
5. Let $\varepsilon(x)$ be $\exists y(\zeta(y) \wedge y \subseteq x) \wedge \forall yz(Succ(y, z) \to (y \subseteq x \leftrightarrow \neg(z \subseteq x)))$. Then $N \models \varepsilon(S)$ iff $S = \{0, 2, 4, \ldots\}$, the even numbers.
6. 'x is finite' can be expressed by $\exists y[\sigma(y) \wedge \forall z(y < z \to \neg(z \subseteq x))]$, where '$<$' is as in 4.

Just as temporal formulae using F and P, U and S, or any other first-order connectives, have a 'standard translation' into first-order logic (see Lemma 9.1.5

of Volume 1), so there is a standard translation of formulae of the fixed point temporal logic USF (and also of other fixed point temporal logics) into S1S (cf. Volume 1, Proposition 8.7.3).

Lemma 2.19. *Let A be an L-formula of* USF *(see Chapter 10). Then there is a formula $\alpha(x)$ of* S1S *such that for all temporal structures $N = (\mathbb{N}, <, h)$ and all $S \subseteq \mathbb{N}$, $N \models \alpha(S)$ iff $S = h(A)$. (Here, $h(A) = \{n \in \mathbb{N} : N \models A(n)\}$.) Moreover, α can be obtained effectively from A.*

Theorem 8.7.4 of Volume 1 shows that the converse is also true, so that USF and S1S are closely related.

Proof By induction on A. If A is $\top$ then $\alpha = \forall y(y \subseteq x)$ will do. If A is an atom $q \in L$, define α to be $x = q$. If $\alpha(z)$ works for A, then

$$\forall yz(\sigma(y) \wedge \alpha(z) \rightarrow (y \subseteq z \leftrightarrow \neg(y \subseteq x)))$$

works for $\neg A$. If $\beta(t)$ also works for B, then

$$\forall yzt(\alpha(z) \wedge \beta(t) \rightarrow (y \subseteq x \leftrightarrow y \subseteq z \wedge y \subseteq t))$$

works for $A \wedge B$. And $\forall abs(\alpha(a) \wedge \beta(b) \wedge \sigma(s) \rightarrow [s \subseteq x \leftrightarrow \exists t(t > s \wedge t \subseteq a \wedge \forall u(s < u < t \rightarrow u \subseteq b))])$ works for $U(A, B)$. The case of $S(A, B)$ is similar. Here, '$<$' is as in Example 2.18.

Finally, let q be a pure past atom in the formula A. So q occurs as a constant symbol in $\alpha(x)$. Write $\alpha(x, y)$ for the result of replacing q by a new variable y in α. Then $\exists y(x = y \wedge \alpha(x, y))$ works for $\varphi q A$. □

Example 2.20. Consider $\varphi q \neg Y q$, where $Yq = S(q, \bot)$: for $A = \neg Yq$, so that $\varphi q A$ defines the set of even numbers, we can take $\alpha(x) = \forall t[t \subseteq x \leftrightarrow \zeta(t) \vee \exists s(Succ(s,t) \wedge \neg(s \subseteq q))]$. Then the formula for $\varphi q A$ is just $\forall t[t \subseteq x \leftrightarrow \zeta(t) \vee \exists s(Succ(s,t) \wedge \neg(s \subseteq x))]$.

We now look at the connection of S1S with Muller automata. First, automata can be expressed by S1S.

Lemma 2.21. *Let M be a Muller automaton reading L (a finite set of atoms). Then there is an S1S sentence μ such that for all temporal L-structures N, M accepts N iff $N \models \mu$.*

Proof μ will say that there is an accepting run of M on N. Let $M = (S, s_0, T, F)$, where $S = \{s_0, s_1, \ldots, s_k\}$. μ uses variables $x_0, \ldots, x_k$; x_i represents the set of all times that M is in state s_i during a run. Then, informally, μ can be taken to be the sentence

$$\begin{aligned} \exists x_0, \ldots, x_k \ (&\text{the } x_i \text{ are pairwise disjoint} \\ &\wedge 0 \in x_0 \\ &\wedge \text{the } x_i \text{ represent a run of } M \text{ on } N \\ &\wedge \textstyle\bigvee_{X \in F}[\bigwedge_{s_i \in X}(x_i \text{ is infinite}) \wedge \bigwedge_{s_j \notin X}(x_j \text{ is finite})]). \end{aligned}$$

Here, 'x is infinite' can be expressed as in Example 2.18, and, given the other conditions, 'the x_i represent a run of M on N' can be written as

$$\forall zt \left(Succ(z,t) \to \bigvee_{\substack{i \leq k,\ L' \subseteq L \\ q \in L',\ r \in L \setminus L' \\ (s_i, L', s_j) \in T}} z \subseteq x_i \wedge z \subseteq q \wedge \neg(z \subseteq r) \wedge t \subseteq x_j \right).$$

□

Rather surprisingly, the converse also holds. This will yield decidability of S1S.

Theorem 2.22. *Let μ be an S1S sentence. Then there is a Muller automaton M such that for all temporal structures N, M accepts N iff $N \models \mu$. M can be obtained effectively from μ.*

Proof By induction on μ. The atomic cases are $p = q, p \subseteq q$ and $Succ(p,q)$ for $p, q \in L$. Figures 2.3 and 2.4 illustrate the automata for $p \subseteq q$ and $Succ(p,q)$; we leave the reader to verify that they work. Since $p = q$ is expressible as $p \subseteq q \wedge q \subseteq p$, we can omit this case.

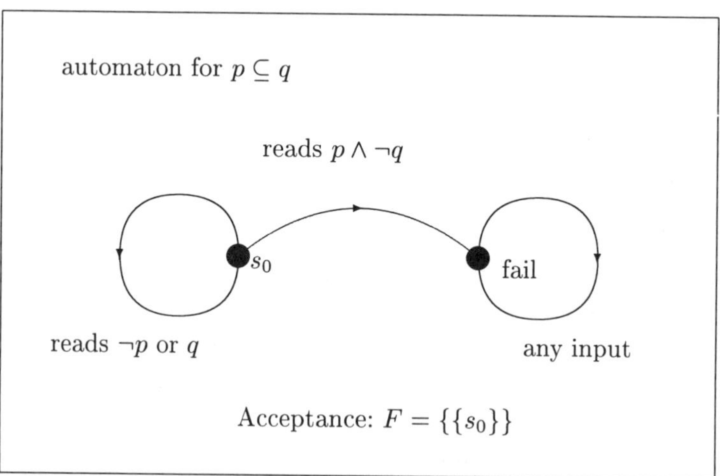

FIG. 2.3.

Now the induction steps. Consider $\mu \wedge \mu'$. If M works for μ and M' for μ', then we can run M and M' in parallel, as at the start of the proof of McNaughton's theorem (Theorem 2.11). The state set of the resulting automaton '$M \times M'$' is the product $S \times S'$ of the state sets S, S' of M and M', and a run is accepting iff

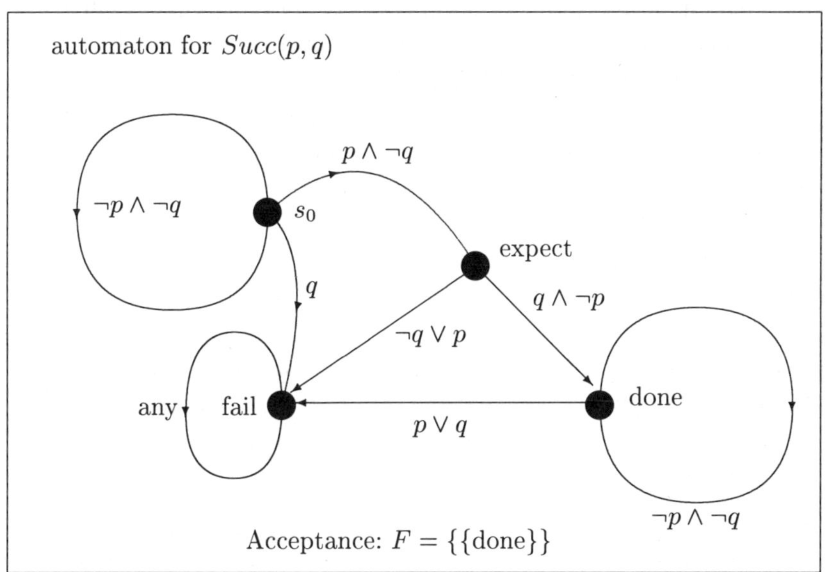

FIG. 2.4.

both the M-component and the M'-component are accepting. Clearly, $M \times M'$ has an accepting run iff both M and M' do, iff $N \models \mu \wedge \mu'$, as required.

Now we do $\neg\mu$, the key step. Inductively, let M be a Muller automaton appropriate for μ. Write M as (S, s_0, T, φ) for propositional φ (Proposition 2.10). The transformation of M into this form can be done effectively. Our first guess is that the M' for $\neg\mu$ would be $(S, s_0, T, \neg\varphi)$. But if M is non-deterministic, then on a given N it could have runs satisfying φ and other runs satisfying $\neg\varphi$! So M and M' would both accept M. Hence M' is not 'equivalent' to $\neg\mu$. But if M is deterministic, the M' above is 'equivalent' to $\neg\mu$. By McNaughton's theorem (Theorem 2.11), we can assume that M is deterministic; note that this theorem constructs the deterministic equivalent of an automaton effectively. Hence we can effectively find M' for $\neg\mu$.

Now consider $\mu = \exists x \nu(x)$. Choose a new atom $q \notin L$. By the inductive hypothesis, there is $M = (S, s_0, T, F)$ that reads the language $L \cup \{q\}$ and is equivalent to the sentence $\nu(q)$ (this last denoting the result of replacing x by q in the formula $\nu(x)$). Define $M' = (S, s_0, T', F)$, reading L, by

$$T' = \{(s, L', s') : L' \subseteq L, (s, L', s') \in T \text{ or } (s, L' \cup \{q\}, s') \in T\}.$$

M' is like M, but rather than reading the value of q from N, it guesses it! Clearly, any accepting run of M gives one for M'; and from an accepting run of M', we can read off a value of q at each time point such that M then has an accepting run on the resulting structure. It follows that M' accepts N iff $N \models \exists x \nu(x)$.

Notice that non-determinism is essential here, so McNaughton's result is really used—we could not manage with entirely deterministic automata. It also shows how lucky the guesses of a non-deterministic automaton can be! □

See [Vardi and Wolper, 1986] and [Emerson and Sistla, 1984] for recent results concerning the translation of temporal formulae into automata. Here the translation is made as efficient as possible in terms of its complexity and the size of the resulting automaton.

Corollary 2.23. *There is an algorithm that, given an arbitrary sentence μ of S1S, works out whether or not μ has a model. That is, S1S is decidable.*

Proof First use Theorem 2.22 to obtain a Muller automaton M equivalent to μ. Using the idea of the $\exists$-step of Theorem 2.22, let M' be an automaton like M but that guesses the values of each of the atoms of μ at each stage. M' has empty L—it doesn't read any values of atoms at all. So for any temporal structure N, M' accepts N iff there is some structure N' that M accepts, i.e. iff μ has a model.

Now take a deterministic M'' equivalent to M', using McNaughton's theorem again. M'' can be constructed effectively from μ. It has a unique run on any structure $N = (\mathbb{N}, <, h)$, and the run is independent of h. Suppose that $M'' = (S, s_0, T, F)$. Since S is finite, the run will become cyclic—it will 'loop'. The algorithm is now as follows: start M'' off, and watch the states as they emerge: $s_0, s_1, \ldots$. Stop at the first n such that $s_n = s_{n'}$ for some $n' < n$. (Clearly, $n \leq |S|$.) Write α for $s_0 s_1 \ldots s_{n'}$, and β for $s_{n'+1}, s_{n'+2}, \ldots, s_n$. Then the run of M'' is $\alpha\beta\beta\beta\ldots$. Hence, the states that come up infinitely often in the run are just those in β. Let X be the set of states in β. Output 'yes' if $X \in F$, and 'no', otherwise. Clearly, this algorithm always halts, and μ has a model iff the output is 'yes'. □

Corollary 2.24. *USF is decidable.*

Proof Use Lemma 2.19 and the corollary above. □

2.6 Games

In the rest of this chapter, we describe the relation between automata and certain games, examine automata in the presence of the environment and, using the argument given in [Gurevich and Harrington, 1982], outline a proof of Rabin's theorem [Rabin, 1969] that the logic S2S (the monadic second-order logic of the binary tree) is decidable. There is a great deal of work connected to this result: closely related papers include [Büchi, 1983; Muchnik, 1984; Yakhnis and Yakhnis, 1990a; Zeitman, 1993], as mentioned at the start of this chapter. The main lemma on the way is a determinacy result (Theorem 2.40 below); it was proved earlier than [Gurevich and Harrington, 1982] by Büchi.

The decidability of S2S is obviously useful for branching time models, but it has applications in many other cases, too, since S2S is extremely strong and many other systems can be encoded in it. See Volume 1, Chapter 15, for examples.

Automata will come in later; we begin with games.

2.6.1 *Definition of the game*

The game is as follows. MOVE is a fixed non-empty set (the alphabet; it is the set of 'pieces' used in the game). MOVE^* is the set of all finite sequences $a_0a_1\ldots a_{n-1}$ ($n < \omega$) where $a_i \in \mathsf{MOVE}$ for all $i < n$. (The length of such a sequence is n.) We write $\langle\rangle$ for the empty sequence (length 0). We regard any element of MOVE as a sequence of length 1, so that $\mathsf{MOVE} \subseteq \mathsf{MOVE}^*$. We write $a, b, \ldots$ for elements of MOVE.

If $p, q \in \mathsf{MOVE}^*$, we write pq for 'p followed by q' as sequences, i.e. the concatenation of p and q. We write $p \preceq q$, and say that p is an initial segment of q, if $q = pr$ for some $r \in \mathsf{MOVE}^*$. Then $(\mathsf{MOVE}^*, \preceq)$ is a tree.

Definition 2.25. A *game tree* is a subset A of MOVE^* satisfying:

1. if $p \preceq q \in A$ then $p \in A$ ('A is closed downwards');
2. $A \neq \emptyset$ (or equivalently, $\langle\rangle \in A$);
3. if $p \in A$ then for some $a \in \mathsf{MOVE}$, $pa \in A$.

We write p, q, etc., for elements of A; they are called *positions*. Each position is implicitly coloured with one of two colours: either red or green, say. We will not make this colouring explicit in the notation, since it plays rather little role in our arguments. It determines which player is to move in that position (see below).

Definition 2.26. A *game* is a tuple $G = (A, S_0, \ldots, S_k, \varphi, \pi)$, where A is a game tree, the S_i are subsets of A, φ is a formula of propositional logic written with the atoms $s_0, \ldots, s_k$, and $\pi \in \{\forall, \exists\}$ specifies who 'owns' φ.

Let $G = (A, S_0, \ldots, S_k, \varphi, \pi)$ be a game. Two players, $\forall$ ('automaton', a male) and $\exists$ ('environment', a female) play a game on A as follows. The play begins with the empty sequence $\langle\rangle$. If this is red, then $\forall$ moves, and if it is green, then $\exists$ moves. Whoever it is chooses $a_0 \in \mathsf{MOVE}$ such that $a_0 \in A$. Such a move is possible by condition 3 above. If a_0 is green, then $\exists$ moves next; otherwise, $\forall$ moves. Whoever it is chooses $a_1 \in \mathsf{MOVE}$ so that $a_0a_1 \in A$. The appropriate player (depending on the colour of a_0a_1) now chooses $a_2 \in \mathsf{MOVE}$ so that $a_0a_1a_2 \in A$. And so on.

A *branch* of A is a maximal set β of positions in A that is linearly ordered by $\preceq$. (Property 3 above is equivalent to saying that every branch of A is infinite.) The result of the play is a branch of A, namely the set

$$\beta = \{\langle\rangle, a_0, a_0a_1, a_0a_1a_2, \ldots\}$$

of all positions that arose during play.

Now we explain who wins the play. For each $i \leq k$, let

$$\beta \models s_i \text{ iff } \beta \cap S_i \text{ is infinite (each } i \leq k).$$

Hence, we can define whether or not $\beta \models \varphi$ in the usual way, by induction on φ. Player π wins the play with outcome β if $\beta \models \varphi$. The other player wins if $\beta \models \neg\varphi$.

Definition 2.27. Let A be a game tree. We say that two games $G = (A, S_0, \ldots, S_k, \varphi, \pi)$ and $G' = (A, S'_0, \ldots, S'_k, \varphi', \pi')$ are *equivalent* if for any branch β of A, β is a win for π in G iff it is a win for π' in G'.

For example, the games $(A, S_0, \ldots, S_k, \varphi, \forall)$ and $(A, S_0, \ldots, S_k, \neg\varphi, \exists)$ are equivalent.

2.6.2 *Strategies*

Roughly, a strategy for $\exists$ in the game $G = (A, S_0, \ldots, S_k, \varphi, \pi)$ is a set of rules telling her what move to make in any position in which it is her move. In such a (green) position, it always offers her at least one possible move. In a red position, it can only advise her to wait for $\forall$ to move. A strategy can be non-deterministic, if it offers her a choice of moves. A strategy is a very intuitive notion, and we will usually be quite informal about it.

Of course, if $\exists$ uses a strategy, the effect in general is that some positions in the game can never be reached. For example, viewing the game tree as laterally ordered in a left-to-right fashion, if $\exists$ uses a strategy that says 'go left' then many positions on the right of the tree will never come up during any play, however $\forall$ chooses to move. So more economically, a strategy σ for $\exists$ tells her how to move in any green position that could have been arrived at by using the strategy so far. We will write $A\lceil\sigma$ for the set of positions in A that can be reached if $\exists$ uses σ. The colour of a position $p \in A\lceil\sigma$ (red or green) is defined to be the colour of p in A.

Notice that $A\lceil\sigma$ is a game tree. It certainly is closed downwards and contains $\langle\rangle$. We check the third part of the definition. If $p \in A\lceil\sigma$ is red, then any $a \in \mathsf{MOVE}$ such that $pa \in A$ also satisfies $pa \in A\lceil\sigma$. If p is green, then σ will offer $\exists$ advice at p. If $a \in \mathsf{MOVE}$ is consistent with that advice, then $pa \in A\lceil\sigma$.

Hence, we obtain a new game $G\lceil\sigma$ by restricting G to these positions:

$$G\lceil\sigma = (A\lceil\sigma, S_0 \cap (A\lceil\sigma), \ldots, S_k \cap (A\lceil\sigma), \varphi, \pi).$$

A strategy for $\exists$ in G is said to be winning if $\exists$ wins any play in which she uses it. The notion of a (winning) strategy for $\forall$ is defined similarly.

Even if $\exists$ (say) has no winning strategy in G, it may be that she can win if play reaches a certain position. A position $p \in A$ is said to be winning for $\exists$ if she has a strategy that she can use if play has arrived at p, and which, if used from then on, wins for her regardless of the moves that $\forall$ makes. (So, for example, the starting position $\langle\rangle$ is winning for $\exists$ just when she has a winning strategy in G as a whole.) Note that p may not be a green position.

Certain tactical strategies are important here. Let $X \subseteq A$ be some set, and suppose that at some position p, $\exists$ has a 'strategy to hit X'. This is a strategy for $\exists$ that can be used once play reaches p, and ensures that however $\forall$ moves from then on, play will eventually reach a position $q \in X$ $(q \succeq p)$. It is easy to see that if at some position p $\exists$ has a strategy to reach some winning position, then p is already a winning position for her. For she can use the strategy to reach a winning position, and then pick up a strategy to win G. Hence, letting D denote the set of positions that are winning for $\exists$, if $p \notin D$ then $\exists$ has no strategy at p to hit D.

For such a p, $\forall$ is in a powerful position. For at p, he has the strategy:

- 'do not let $\exists$ get into a position where she does have a strategy to hit D'.

It is intuitively obvious that he can manage this: if he could not, $\exists$ would have a strategy at p to hit D. For example, if p is red, there must be some move that he can make so that $\exists$ has no strategy to hit D from the resulting position, for otherwise, $\exists$ would have the strategy at p: 'wait for $\forall$ to move, then pick up a strategy at the resulting position to hit D'. $\forall$'s strategy is to make such a move. If he follows such a strategy, then from p onwards play never reaches a position where $\exists$ has a strategy to hit D. In particular, it never reaches a position in D. On the other hand, by the definition of the strategy, if, even after following this strategy for a long while, $\forall$ makes a single move incompatible with it, $\exists$ will then have a strategy to hit D. This is an important point and we will use it several times below.

For brevity, we will call this strategy for $\forall$ 'avoid D'. In this context, there is nothing special about D: for any subset X of A, and $p \in A$, if $\exists$ has no strategy at p to hit X then $\forall$ has a well-defined strategy 'avoid X' at p, and if he uses it, play will never reach a position $q \succeq p$ with $q \in X$. Nor is there anything special about $\forall$: all these definitions can be made with the players swapped, *mutatis mutandis.*

2.7 Determinacy—an outline

We are going to prove that any game of the form described is *determined*—either $\forall$ or $\exists$ has a winning strategy in it. Further, the strategy will use only a bounded amount of information (about k^2 bits) about the history of the game so far. Since the strategies involved are rather devious, we will discuss them informally first, without considering how much information is required to implement them. In the next few sections, we examine the strategies in more detail.

Proposition 2.28. *Let $G = (A, S_0, \dots, S_k, \varphi, \pi)$ be a game. Then either $\forall$ or $\exists$ has a winning strategy in G.*

Proof The proof is by induction on the number n of atoms from $s_0, \dots, s_k$ that occur in φ.

First assume that $n = 0$. If φ is $\top$ then any strategy is winning for π, since every play β satisfies φ. If φ is $\bot$, then any strategy for the other player is winning.

Now suppose that $n > 0$, and inductively assume the theorem for all games with smaller n. By renaming the sets S_i if need be, and discarding any S_i where s_i does not occur in φ, we can suppose that the atoms involved in φ are $s_0, \dots, s_k$. Clearly, either $s_0 \wedge \cdots \wedge s_k \vdash \varphi$ or $s_0 \wedge \cdots \wedge s_k \vdash \neg\varphi$. Since $(A, S_0, \dots, S_k, \varphi, \pi)$ is equivalent to $(A, S_0, \dots, S_k, \neg\varphi, \pi')$, where π' is the other player from π, we can suppose the first possibility holds; and for clarity of notation, let us be specific and assume that the player π is $\exists$ (if not, then just rename the players and swap the red–green colours of positions in A). So we are considering the game $G = (A, S_0, \ldots, S_k, \varphi, \exists)$, and $s_0 \wedge \cdots \wedge s_k \vdash \varphi$. We want to prove that either $\forall$ or $\exists$ has a winning strategy in this game.

Let $D \subseteq A$ be the set of all positions that are winning for $\forall$ in G. If $\langle\rangle \in D$, we are finished. So assume not. We will describe a winning strategy for $\exists$ in G.

Certainly, $\forall$ has no strategy at $\langle\rangle$ to hit D. Hence, $\exists$ has the strategy 'avoid D' in G. She uses this strategy throughout. She is resolute in this, regardless of later complexities. So the original game G effectively reduces to

$$G' = G\lceil\text{'avoid } D\text{'} = (A', S'_0, \ldots, S'_k, \varphi, \exists),$$

where $A' = A\lceil$'avoid D' is the set of all positions of G that can arise in plays in which $\exists$ uses 'avoid D', and $S'_i = S_i \cap A'$ for each i.

We claim that in this game G', no position p is winning for $\forall$. It is obvious that p is not winning for $\forall$ in G; but in G', there is a nagging worry that he may be able to turn the fact that $\exists$ is using a certain strategy to his advantage. However, if $\forall$ did have a winning strategy τ in G' at p, then *in* G he can use τ from p onwards unless $\exists$ makes a move incompatible with her strategy 'avoid D'. If she never does so, he will win G', and as its winning condition is the same as for G, he will win G, too. If she does, then as we saw, $\forall$ has at that point a strategy in G to hit D, so that he has a strategy to win G from this position. It follows that the original position p was winning for him in G. This contradiction proves the claim.

The winning condition for $\exists$ in G' is still φ, so to show that she has a winning strategy in G, it suffices to find her a winning strategy in G'. This we now do.

1. In G', she tries to hit each set S'_i $(0 \leq i \leq k)$ in cyclic order infinitely often. That is, she begins by choosing, if she can, a strategy at $\langle\rangle$ to hit S'_0 in the future (i.e. to hit $S'_0 \setminus \{\langle\rangle\}$). She follows this strategy until play hits S'_0. Then she chooses a strategy at the current position, p, to hit S'_1 in the future—that is, to hit $S'_1 \setminus \{p\}$—and uses it until S'_1 is hit. She then picks up a strategy to hit S'_2 in the future, and so on, until S'_k is hit. Then she chooses a strategy to hit S'_0 in the future again, and so on, for ever.
   If she can continue this throughout play, she will win, because the resulting branch β will satisfy $\beta \models s_0 \wedge \cdots \wedge s_k$, so that $\beta \models \varphi$. So let us consider the other case.
2. Suppose that at some stage, in position p, say, she attempts to choose a strategy to hit S'_i, but no such strategy exists. Then $\forall$ has the strategy 'avoid S'_i' at p in G'.
   He may or may not decide to use this strategy from now on. For example, he may use it for a while, but then drop it (i.e. make a move that it does not permit). As we saw, the definition of the strategy 'avoid S'_i' implies that at the instant he drops it, $\exists$ has a strategy to hit S'_i. So she can return to trying to hit the S'_j cyclically, as described in 1 above. Because what matters in G' is whether the resulting play intersects the S'_j infinitely often, the delay caused by $\forall$'s tactic is of no consequence.

We see that $\forall$ is in a hopeless position unless he does use the strategy 'avoid S_i'' from p onwards.
So assume that he does. So after p, $\exists$ never has a strategy to hit S_i'. Certainly, play never hits S_i' after p. $\exists$ can take advantage of this, and use the inductive hypothesis.
Let $A'_{p,i}$ be the game tree consisting of all positions of G' that may arise in plays which reach p and in which $\forall$ uses the strategy 'avoid S_i'' from then on. For such plays, G' reduces to the game

$$G_i = (A'_{p,i}, S'_0 \cap A'_{p,i}, \dots, S'_k \cap A'_{p,i}, \varphi, \exists).$$

As play never hits S_i' after p, for any play β of G_i we have $\beta \models \neg s_i$. Hence, G_i is equivalent to

$$(A'_{p,i}, S'_0 \cap A'_{p,i}, \dots, S'_k \cap A'_{p,i}, \varphi(s_i/\bot), \exists).$$

Here, $\varphi(s_i/\bot)$ denotes the result of replacing the atom s_i in φ by $\bot$.
By the inductive hypothesis, one of the players has a winning strategy in this game. Who could it be? If $\forall$ had such a strategy—τ, say—then because 'avoid S_i'' is also his strategy (in G'), he could use the following double strategy in G' from p onwards:

- avoid S_i' (so reducing G' to G_i), and also use τ.

That would win G' for him, so that p would be a winning position for $\forall$ in G'. But this contradicts our assumption that no position in G' is winning for him.
Hence it must be $\exists$ who has a winning strategy in G_i. Let σ_i be such a strategy. Because we are assuming that $\forall$ uses the strategy 'avoid S_i'' in G' from p onwards, $\exists$ will be able to apply σ_i in G'. If she does so, she will win G_i, and hence she will win G'.
So $\exists$'s strategy in case 2 is as follows:

(a) While there is no strategy to hit S_i', use σ_i.
(b) If at some point there is such a strategy, return to case 1 above.

Note that the distinguishing condition is recognizable by $\exists$, so that she can tell which of these cases applies without needing psychological insight into which strategy $\forall$ is using.

As we have argued, this strategy is winning for $\exists$. We have shown that if $\forall$ has no winning strategy in G, then $\exists$ does have one. So G is determined, as required. □

2.8 Strategies using limited memory

For the arguments using automata, still to come, we need to show not only that in any game $(A, S_0, \dots, S_k, \varphi, \pi)$, one of the players has a winning strategy, but that there is such a strategy that uses only a bounded amount of 'memory'. This will be proved in Theorem 2.40 in the next section, and is the most technical and ingenious part of the proof of [Gurevich and Harrington, 1982]. Here, we set up the necessary concepts and clear away some obstacles to the proof.

Recall that a strategy advises its owner ($\forall$ or $\exists$) what move to make at a position p in which they are to move. We must be more precise now about the positions at which a strategy is defined.

A strategy need not be defined at every position. A strategy that is defined at position p provides a non-empty subset F of MOVE, if its owner is to move at p, and the advice 'wait until it is your move', if not. In the first case, if the owner takes the strategy's advice, play moves to a position pa for some $a \in F$. In the second case, the other player can of course make any move, so the new position could be any of the form pa for $a \in \mathsf{MOVE}$. In either case, the strategy should continue to be defined at the new position, so that it is usable in the game for as long as its advice is followed. However, there are exceptions: the strategy 'hit D' (for some set D of positions), if defined at p and followed from there, is only guaranteed to be defined until play reaches D.

This can be formalized by defining a strategy to be a function $\sigma : A \to \wp\mathsf{MOVE}$. The set of moves that σ advises at p is just $\sigma(p)$; we can say that σ is defined at p if $\sigma(p) \neq \emptyset$; and if p is the opponent's position and σ is defined at p, then $\sigma(p) = \{a \in \mathsf{MOVE} : pa \in A\}$. That way, a strategy σ is followed for a while if at each successive position p, the move made (by either player) is in $\sigma(p)$.

But we will not usually be so formal in the presentation here.

The advice given by the strategies considered so far can depend completely on the current position p; and every position is different! We seek strategies that do not have such strong dependence, at least on the past. Such a 'limited memory' strategy will give the same advice in two positions that have similar histories, so long as their *future* is the same. If not, the advice can be quite different. Thus, the advice still depends totally on the future. It will turn out that this is adequate for our purposes, essentially because although an automaton cannot remember too much about the past, it can 'guess' the future.

In $\exists$'s strategy in Proposition 2.28, the main use of memory is in working out which sets S_i to hit next. Memory may also be needed by the strategies that are obtained inductively. We need some definitions to capture the precise amount of memory needed for all this. It helps here if we alter our games very slightly.

Definition 2.29. In this section, games will have the form $G = (A, \Phi, S_s\ (s \in \Phi), \varphi, \pi, \delta)$. Here

1. A is a game tree;
2. Φ is a finite set of propositional atoms;
3. $S_s \subseteq A$ for each $s \in \Phi$, the S_s being pairwise disjoint;
4. φ is a propositional Φ-formula;
5. $\pi \in \{\forall, \exists\}$;

6. δ is an enumeration of Φ. For example, if $\Phi = \{1, 2, 3, 4, 5\}$, δ could be 13524. We call such an enumeration a *Φ-display.*

The game is played as before. If β is a play, we let $\beta \models s$ iff $\beta \cap S_s$ is infinite $(s \in S)$. π wins β iff $\beta \models \varphi$. The display δ plays no role in the game itself. It will be used in strategies, for 'memory initialization'.

For the rest of this section, fix an arbitrary game $G = (A, \Phi, S_s(s \in \Phi), \varphi, \pi, \delta)$.

Definition 2.30.

1. If ε is a Φ-display and $s \in \Phi$, we define $\varepsilon * s$ ('ε updated with s') to be the display obtained by moving s to the right-hand end of ε, keeping the order of the rest of ε intact, e.g. if $\Phi = \{1, 2, 3, 4, 5\}$, we have $13524 * 3 = 15243$.
2. If $p \in A$, we define the display $LAR_G(p)$ (standing for 'latest appearance record') by induction on the length of p. We omit the suffix 'G' where it is evident from the context.
   - $LAR(\langle\rangle) = \delta$, where δ is as given in G.
   - For $p \in A$ and $a \in \mathsf{MOVE}$ such that $pa \in A$, we let
     $$LAR(pa) = \begin{cases} LAR(p) * s, & \text{if } s \in \Phi \text{ and } pa \in S_s \\ LAR(p), & \text{if } pa \notin S_s \text{ for all } s \in \Phi \end{cases}.$$
     Note that the 's' in the first clause is unique, as the S_s are pairwise disjoint. So the LAR is well defined.

   Thus, $LAR(p)$ records the most recent order of appearance of the sets S_s at positions leading up to p. The reader might like to check that a strategy to hit every S_s infinitely often could work by always trying to hit S_s, where s is the leftmost entry of the LAR of the current position. LARs originate in [McNaughton, 1966]; see the discussion in [Büchi, 1983].
3. Positions p, q in G are said to have the *same future* if for every $r \in \mathsf{MOVE}^*$, $pr \in A$ iff $qr \in A$, and in that case, pr, qr have the same colour, and $pr \in S_s$ iff $qr \in S_s$ for each $s \in \Phi$.
4. We define an equivalence relation $\sim_G$ on A (depending on G) by $p \sim_G q$ iff p, q have the same future and $LAR_G(p) = LAR_G(q)$. We omit the suffix 'G' where possible.
5. Let $X \subseteq A$. We say that X is $\sim$-closed if for all positions $p, q \in A$, if $p \sim q$ then $p \in X$ iff $q \in X$.

A useful illustration of the properties of $\sim$ is given by the following simple lemmas.

Lemma 2.31. *Let $p, q \in A$, $s \in \mathsf{MOVE}^*$, and suppose that $ps \in A$ and $p \sim q$. Then $qs \in A$ and $ps \sim qs$.*

Proof Exercise. $\square$

Lemma 2.32. *Suppose that $A' \subseteq A$ is another game tree, where the colour (red or green) of any $p \in A'$ is the same as its colour in A. Assume that A' is $\sim$-closed. Let $S'_s = A' \cap S_s$ for each $s \in \Phi$, and let G' be the game*

$$G' = (A', \Phi, S'_s\ (s \in \Phi), \varphi, \pi, \delta).$$

Then for all positions $p, q \in A'$, if $p \sim_G q$ then $p \sim_{G'} q$.

Proof Take $p, q \in A'$ with $p \sim_G q$. It is easy to show by induction on the length of p that $LAR_{G'}(p) = LAR_G(p)$, and similarly for q. Hence, $LAR_{G'}(p) = LAR_{G'}(q)$.

Let $s \in \mathsf{MOVE}^*$, and suppose that $ps \in A'$. Then $ps \in A$, and as $p \sim_G q$, Lemma 2.31 gives $qs \in A$ and $ps \sim_G qs$. As A' is $\sim$-closed, $qs \in A'$. Clearly, ps, qs have the same colour, and for each $s \in \Phi$, $ps \in S'_s$ iff $qs \in S'_s$. Thus, p and q have the same future in G'. So $p \sim_{G'} q$. □

Definition 2.33. A strategy σ on A (for $\forall$ or $\exists$) in the game G is said to be *(G)-finitely based* if whenever p, q are positions in G with $p \sim_G q$, then σ is defined at p iff it is defined at q, and in that case it gives the same advice to its owner at p as at q. Formally, $\sigma(p) = \sigma(q)$ whenever $p \sim q$ in G.

In the next section, we will extend Proposition 2.28 to yield a finitely based strategy for one of the players. The argument will be much the same as before, but we have to be far more careful over the details.

The strategies of Proposition 2.28 often needed to hit or avoid a set. We now need to be precise about how such strategies are defined.

Definition 2.34. Let $X \subseteq A$.

1. We define the rank $rk^\exists_X(p)$ (an ordinal or ∞) of each position $p \in A$, as follows. Formally, we are defining unary relations '$rk^\exists_X(p) \geq \alpha$' on A by induction on $\alpha \geq 1$.
   - $rk^\exists_X(p) = 0$ iff $p \in X$
   - $rk^\exists_X(p) \geq 1$ iff $p \notin X$
   - for limit ordinals γ, $rk^\exists_X(p) \geq \gamma$ iff $rk^\exists_X(p) \geq \alpha$ for all $\alpha < \gamma$
   - $rk^\exists_X(p) \geq \alpha + 1$ iff $rk^\exists_X(p) \geq \alpha$ and
     - * if p is red, then for every $a \in \mathsf{MOVE}$ such that $pa \in A$, we have $rk^\exists_X(pa) \geq \alpha$;
     - * if p is green, then there is $a \in \mathsf{MOVE}$ such that $pa \in A$ and $rk^\exists_X(pa) \geq \alpha$.

   We then let $rk^\exists_X(p) = \alpha$ if $rk^\exists_X(p) \geq \alpha$ and $rk^\exists_X(p) \not\geq \alpha + 1$. That is, $rk^\exists_X(p)$ is the least ordinal α such that $rk^\exists_X(p) \not\geq \alpha + 1$.
   We set $rk^\exists_X(p) = \infty$ if $rk^\exists_X(p) \geq \alpha$ for all ordinals α.
2. We define $rk^\forall_X(p)$ similarly, swapping 'red' and 'green' above.
3. The strategy 'avoid X' for $\exists$ is defined by:
   - 'avoid X'$(p) = \{a \in \mathsf{MOVE} : pa \in A \text{ and } rk^\exists_X(pa) = \infty\}$.

It is defined at any position p with $rk^{\exists}_X(p) = \infty$.

4. The strategy 'hit X' for $\forall$ is defined by:
   - 'hit X'$(p) = \{a \in \mathsf{MOVE} : pa \in A$ and $rk^{\exists}_X(pa) < rk^{\exists}_X(p)\}$.

   It is defined at any position p with $0 < rk^{\exists}_X(p) < \infty$.
5. All these definitions are made for the other players similarly.

Exercise 2.35. Show that if X is empty then $rk^{\exists}_X(p) = \infty$ for all positions p.

Show that if MOVE is finite then $rk^{\exists}_X(p)$ is a natural number or ∞, for every p. In this case, the definition of rank only needs natural numbers and the 'limit ordinal' clause can be deleted.

Lemma 2.36. *Let p be a position in G, and let $X \subseteq A$.*

1. *$\forall$ has a strategy at p to hit X iff $rk^{\exists}_X(p) < \infty$. In that case, the strategy 'hit X' for $\forall$ is defined. If $\forall$ uses it from p, play will eventually hit X, the strategy remaining defined until this occurs.*
2. *$\exists$ has a strategy to avoid X iff $rk^{\exists}_X(p) = \infty$. In that case, the strategy 'avoid X' for $\exists$ is defined, and if she uses it from p it will remain defined, and play will never hit X.*

Proof

1. Suppose that $rk^{\exists}_X(p) = \alpha < \infty$. If p is red, there is $a \in \mathsf{MOVE}$ such that $pa \in A$ and $rk^{\exists}_X(pa) < \alpha$. Hence the strategy 'hit X' for $\forall$ is defined at p, and if he uses it at p, the rank of the resulting position has dropped (is $< \alpha$). If p is green, then every $a \in \mathsf{MOVE}$ with $pa \in A$ is such that $rk^{\exists}_X(pa) < \alpha$. Thus, whatever move $\exists$ makes, the rank of the new position has dropped. In either case, by the same argument, the strategy remains defined at the new position. Since there is no infinite decreasing sequence of ordinals, if $\forall$ continues to use the strategy, a position of rank 0 will eventually be reached—that is, play hits X.
2. Suppose now that $rk^{\exists}_X(p) = \infty$. Then for every ordinal α, $rk^{\exists}_X(p) > \alpha$. If p is red, then for every $a \in \mathsf{MOVE}$ with $pa \in A$, $rk^{\exists}_X(pa) \geq \alpha$. This holds for all α, so that $rk^{\exists}_X(pa) = \infty$. If p is green, then for every α there is some $a \in \mathsf{MOVE}$ with $pa \in A$ and $rk^{\exists}_X(pa) \geq \alpha$. Since MOVE is a set, the set-theoretic axiom of replacement implies that for some $a \in \mathsf{MOVE}$, $pa \in A$ and $rk^{\exists}_X(pa) \geq \alpha$ for all α—that is, $rk^{\exists}_X(pa) = \infty$.

   So in either case, the strategy is defined at p, if $\forall$ is to move at p then the rank of the new position will still be ∞, and if $\exists$ is to move, she can move to keep the rank at ∞. So the strategy remains defined at the new position, and $\exists$ can continue to use it. If she does so, every subsequent position has rank ∞ so is certainly not in X. Hence, the strategy avoids X.

We can now see that if $\forall$ has any strategy at p to hit X then $rk^{\exists}_X(p) < \infty$, or else $\exists$ would have a strategy to avoid X—impossible. So the strategy 'hit X', as defined above, will always serve to hit X if $\forall$ has any strategy at all to do this. A similar argument gives the other case. □

We now show that if X is $\sim$-closed, these two strategies are finitely based. This is intuitively clear—at position p, whether a player can hit or avoid a set depends only on the future of p. The LAR is not used here.

Lemma 2.37. *Let $X \subseteq A$ be a $\sim$-closed set. Then:*

1. *$rk^{\exists}_X(p) = rk^{\exists}_X(q)$, for all positions $p \sim q$ in G.*
2. *The strategies 'avoid X' for $\exists$, and 'hit X' for $\forall$, are finitely based.*

Similar results hold when swapping the players.

Proof

1. We show by induction on ordinals $\alpha \geq 1$ that whenever $p \sim q$, $rk^{\exists}_X(p) \geq \alpha$ iff $rk^{\exists}_X(q) \geq \alpha$. If $\alpha = 1$ this is clear because X is $\sim$-closed. For limit ordinals, the result follows easily by the inductive hypothesis. Assume it for α, and let $p \sim q$ in G. Suppose that $rk^{\exists}_X(p) \geq \alpha + 1$. We show the same for q.
   By definition of the rank, $rk^{\exists}_X(p) \geq \alpha$, and inductively, $rk^{\exists}_X(q) \geq \alpha$.
   Assume that p is red. Then so is q. Let $a \in \mathsf{MOVE}$ and suppose that $qa \in A$. Then as $p \sim q$, we have $pa \in A$; by definition of the rank, $rk^{\exists}_X(pa) \geq \alpha$. By Lemma 2.31, $pa \sim qa$. So by the inductive hypothesis, $rk^{\exists}_X(qa) \geq \alpha$. Hence, $rk^{\exists}_X(q) \geq \alpha + 1$, also.
   The case where p is green is similar.
2. Using the first part, if $p \sim q$ in G, then $rk^{\exists}_X(p) = rk^{\exists}_X(q)$, and for every $a \in \mathsf{MOVE}$ we have $pa \in A$ iff $qa \in A$, and in that case, $pa \sim qa$ and so $rk^{\exists}_X(pa) = rk^{\exists}_X(qa)$. Since the strategies 'avoid X' and 'hit X' are defined at any position in terms of the $rk^{\exists}_X$-rank of the position and of its immediate successors, it follows that they are finitely based.

$\square$

Now we describe how to 'sew' several finitely based strategies into a single finitely based strategy. This will be needed twice in the proof of Theorem 2.40.

The first step is to observe that a finitely based strategy that is winning when used from some position will also win when used from any $\sim$-related position. (It may not be winning from other positions.) This is the content of the following lemma.

Lemma 2.38. *Let σ be a finitely based strategy for $\forall$ (say) in G. Suppose that σ is defined at some position $p \in A$, and $\forall$ wins any play of G that reaches p and in which he uses σ from then on.*

Then there is a finitely based strategy σ' for $\forall$ in G, defined at p, such that if a play of G reaches a position at which σ' is defined, then $\forall$ can use σ' from then on, and if he does, he will win.

Proof Let X be the set of all positions $q \in A$ such that if $\forall$ starts at p and plays according to σ then $\exists$ can play so as to arrive eventually at some position $r \in A$ with $r \sim q$. Let σ' be the 'restriction' of σ to X (formally, $\sigma'(q) = \sigma(q)$ if

$q \in X$, and $\sigma'(q) = \emptyset$, otherwise). It is clear that X is $\sim$-closed; and as σ was finitely based, it follows that σ' is also finitely based.

Clearly, σ' is defined at p. Suppose that $\forall$, $\exists$ play G, and play arrives at a position q at which σ' is defined. Certainly, $q \in X$; let r be as above, for q. Suppose that $a \in \sigma'(q)$. Then $a \in \sigma(r)$. Since σ is winning when used from p, and r is a position in such a play, σ remains defined at ra. By Lemma 2.31, $qa \sim ra$, so σ is also defined at qa. By the definition of σ', σ' is also defined at qa, and indeed, 'ra' is a suitable choice of 'r' for the position qa.

A simple induction over time now shows that $\forall$ can use σ' from q onwards, and that if he does, subsequent successive positions of play take the form $qq_0, qq_1, qq_2, \ldots$, where $q_n \in \mathsf{MOVE}^*$ for each $n < \omega$, and:

- $rq_0, rq_1, rq_2, \ldots$ are successive positions of a 'parallel' play β of G that reached p and in which σ was used from then on,
- $rq_n \sim qq_n$ (all n).

As σ wins any such play, β is a win for him in G. Let β' be the original play through q ending $qq_0, qq_1, qq_2, \ldots$. Then for each $s \in L$, $\beta \cap S_s$ is infinite iff $\beta' \cap S_s$ is. Hence β' also is a win for $\forall$.

So σ' is a winning strategy for $\forall$ in G once play reaches any position at which it is defined, as required. □

The second step is to join up many finitely based strategies into a single one.

Lemma 2.39. *Let $X \subseteq A$ be a $\sim_G$-closed set. Suppose that for each $x \in X$, $\forall$ has a finitely based strategy σ_x that, if he uses it from x onwards, wins G for him however $\exists$ chooses to play. Then he has a finitely based strategy that is defined (at least) at every position in X and wins any play in which it is used from some point onwards.*

Proof By the preceding lemma, we can suppose that each σ_x is winning for $\forall$ from any position at which it is defined.

Let $<$ be a well ordering of X—for example, a lexicographical order based on a well order of MOVE. We propose a new strategy σ in G, namely:

- at $p \in A$, use σ_x for the $<$-least $x \in X$ such that σ_x is defined at p;
- if there is no such x, leave σ undefined.

Since σ_x is defined at $x \in X$, σ is certainly defined at every position in X. Note that it is finitely based (in G). For, suppose that $p, q \in A$, and $p \sim q$. Then any σ_x defined at p is also defined at q. So σ gives the same advice at p and q.

Clearly, $\forall$ can continue to use σ for ever, once it is defined. For if it is defined and used at $p \in A$ and involves using σ_x there, then σ_x, being a winning strategy whenever it is used, remains defined at the next position after q. So σ remains defined at the new position and can be used again.

In a play β of G, suppose that $\forall$ uses σ from some position onwards. Then at each step, he uses a strategy σ_x for some $x \in X$. As we saw, σ_x remains defined at the new position, so if σ employs $\sigma_{x'}$ there, $x' \leq x$. So the successive x form a non-increasing sequence in X with respect to $<$. Since $<$ is a well order, the 'x' eventually becomes constant. After that point, $\forall$ will use a single strategy σ_x for ever. But σ_x wins any play in which it is used permanently after some stage. Hence β is a win for $\forall$ in G.

So σ is a winning strategy for $\forall$ in G once play reaches any position at which it is defined, as required. □

2.9 Determinacy with limited memory

We now have a relatively clear run to proving the main theorem.

Theorem 2.40. *Let $G = (A, \Phi, S_s(s \in \Phi), \varphi, \pi, \delta)$ be a game. One of the players ($\forall$ or $\exists$) has a finitely based winning strategy in G.*

Proof The proof is similar to that of Proposition 2.28, and is by induction on $|\Phi|$, the set of atoms that φ can use. If this is zero, the trivial strategy 'do what you like' is winning for one of the players, and this strategy is certainly finitely based.

Inductively, assume the result for all games with winning formulae with fewer atoms than φ, and suppose as before that $\pi = \exists$ and $\bigwedge \Phi \vdash \varphi$.

Let D be the set of positions p in G such that $\forall$ has some finitely based strategy that wins any play of G that reaches p and in which it is used from then onwards.

Lemma 2.41. *D is $\sim_G$-closed.*

Proof Use Lemma 2.38. □

Lemma 2.42. *Let $p \in A \setminus D$. Then $\forall$ has no strategy at p to hit D.*

Note that the lemma is not restricted to finitely based strategies.

Proof As in Proposition 2.28, the idea is that if $\forall$ had such a strategy, he could use it until play hit D and then pick up a finitely based winning strategy from that position. Hence, $p \in D$, a contradiction. However, there is a catch: it is only a contradiction if this overall strategy is finitely based. We have to prove this.

As D is $\sim$-closed, by Lemma 2.39 there is a single finitely based strategy σ that $\forall$ can use once the position enters D, and will always win G for him. Consider now the following strategy Σ for $\forall$ at a position $q \in A$:

1. if $q \notin D$, use 'hit D' (reduce $rk_D^{\exists}$), if it is defined;
2. if $q \in D$, use σ.

Then Σ is finitely based. For let $q \sim r$ in G. As D is $\sim$-closed (Lemma 2.41), $q \in D$ iff $r \in D$. So the same clause (1 or 2) applies to q and r. Because both strategies involved in the two clauses are finitely based (for clause 1, as D is $\sim$-closed, Lemma 2.37 applies), it follows that Σ is finitely based, too.

Assume for contradiction that $\forall$ has a strategy at p to hit D. By Lemma 2.36, the strategy 'hit D' for $\forall$ is defined at p. Hence, Σ is defined at p, and it is clearly winning for $\forall$ if he uses it from there. So $p \in D$—a contradiction, as required. □

Let $A' = A \setminus D$. If $\langle\rangle \in D$, there is nothing to prove, so we can assume that $\langle\rangle \in A'$. Let the colour (red or green) of a position in A' be the same as its colour in A.

Corollary 2.43. *A' is a game tree.*

Proof This is because A' is the restriction of A by the strategy 'avoid D' for $\exists$. □

Let

$$S'_s = S_s \cap A' \ (s \in \Phi), \qquad G' = (A', \Phi, S'_s \ (s \in \Phi), \varphi, \exists, \delta).$$

By the corollary, G' is a game. Note that A' is $\sim_G$-closed (because D is).

Lemma 2.44. *$\forall$ has no G'-finitely based winning strategy at any position in G'.*

Proof If he had such a strategy at position $r \in A'$, say τ, then let σ be a G'-finitely based strategy to win G from any position in D (cf. Lemma 2.39), and consider the following strategy Σ in G. At position $p \in A$:

1. If $p \in A'$, use τ, if defined.
2. If $p \in D$, use σ.

This strategy is finitely based. For if $p \sim_G q$ in G, then:

1. If $p \in A'$, then also $q \in A'$. As A' is $\sim_G$-closed in A, Lemma 2.32 applies, yielding $p \sim_{G'} q$. So τ gives the same advice (if any) at p and q. Hence, Σ agrees on p, q in this case.
2. If $p \in D$ then $q \in D$. As σ is G-finitely based, its advice at p, q is the same, so the same holds for Σ.

We claim that Σ is winning for $\forall$ in G once play reaches r. Consider a play β of G in which this happens, and in which Σ is used from r onwards. If β remains within A' after r, then by case 1, τ is used from r onwards. Because τ is winning if used in this way, β is a win for $\forall$. If β leaves A' at some point after r, then case 2 is used. By definition of D and σ, β remains in D from then on. So σ is used from then on in the play, and this wins for $\forall$, too. This proves the claim.

Hence, $r \in D$, a contradiction to $D \cap A' = \emptyset$. □

Lemma 2.45. *Suppose that $\exists$ has a G'-finitely based winning strategy in G'. Then she has a G-finitely based winning strategy in G.*

Proof Let σ be a G'-finitely based winning strategy for $\exists$ in G'. Then σ is also a winning strategy for her in G. For, it is defined at $\langle\rangle$. Suppose (inductively) that $\exists$ has used it in G as far as the position p, and that $p \in A'$. If p is green, then σ will certainly advise her to move to a position $pa \in A'$. If it is red, then by definition of A', $\forall$ has no move to arrive at a position outside A'. Hence the new position will still be in A'. Thus, $\exists$ can use σ in a play of G, and if she does, the play will be a play of G'. She will win this play; and as the winning condition of G' is the same as for G, she will win G, too.

It remains to check that σ is G-finitely based. Let $p \sim_G q$ in G, and suppose that σ is defined at p. σ is only defined at positions in A', so $p \in A'$. As A' is $\sim$-closed, $q \in A'$. By Lemma 2.32, $p \sim_{G'} q$. Hence, σ gives the same advice at q as at p. □

So to prove the theorem, it suffices to prove that $\exists$ has a G'-finitely based winning strategy in G'. We work within G' from now on. To avoid carrying the '$'$' with us,

(†) *we now replace the original game G by this game G'.*

This amounts to assuming that D is empty.

$\exists$ will attempt to hit each set S_s infinitely often, as in Proposition 2.28. She will use the leftmost entry in the LAR of the current position to decide which S_s to aim for. We consider first what to do if no strategy to hit the next S_s is available.

Fix $s \in \Phi$, and let A_s be the set of all positions $p \in A$ such that the leftmost entry of $LAR_G(p)$ is s and $\exists$ has no strategy at p in the game G to hit S_s. Note that A_s is $\sim_G$-closed in A. Suppose that A_s is non-empty. Let $r \in A_s$, and define

$$\begin{aligned} A^r_s &= \{p \in \mathsf{MOVE}^* : rq \in A_s \text{ for all } q \preceq p \text{ in } \mathsf{MOVE}^*\}, \\ S^r_{s,t} &= \{p \in A^r_s : rp \in S_t\}, \text{ for } t \in \Phi \setminus \{s\}. \end{aligned}$$

Let $p \in A^r_s$ inherit the colour (red or green) of rp in A. Then A^r_s is a game tree. To see this, note that it is the set of all positions above r that can be reached by a play of G in which $\forall$ uses the strategy 'avoid S_s' from r onwards, and that the leftmost entry of the LAR in G of each such position is still s (because play never again enters S_s).

Let

$$G^r_s = (A^r_s, \Phi \setminus \{s\}, S^r_{s,t}\ (t \in \Phi \setminus \{s\}), \varphi(s/\bot), \exists, \delta^r_s),$$

where δ^r_s is the result of deleting the atom s from $LAR_G(r)$.

Lemma 2.46. *For any position $p \in A^r_s$, $LAR_G(rp)$ is s concatenated with $LAR_{G^r_s}(p)$.*

Proof An easy induction on the length of p. (The games are given an initial LAR, δ, to allow this to be shown.) □

Lemma 2.47. *Let τ be a finitely based winning strategy for one of the players in G^r_s. Then there is a G-finitely based strategy τ^* for the same player in G that is defined at r and wins G whenever its owner uses it in a play from r onwards.*

Proof τ^* is obtained by 'extending' τ to G, as follows:

- at position $p \in A$, pick $q \in A^r_s$ such that $rq \sim_G p$, and do whatever τ advises in G^r_s at q, if it is defined there;
- if there is no such q, leave τ^* undefined at p.

We claim that the first clause is well defined. Since τ is G^r_s-finitely based, it suffices to prove that if $p, q \in A^r_s$ and $rp \sim_G rq$ then $p \sim_{G^r_s} q$.

The proof extends that of Lemma 2.32 to this more complicated situation. Take such p, q. By Lemma 2.46, we have $LAR_{G^r_s}(p) = LAR_{G^r_s}(q)$. To check that p, q have the same future in G^r_s, we show by induction on the length of $s \in \mathsf{MOVE}^*$ that $ps \in A^r_s$ iff $qs \in A^r_s$, and in that case, ps, qs have the same colour and are in the same sets $S^r_{s,t}$. This is clear if $s = \langle\rangle$. Assume the result for shorter s, and suppose that $ps \in A^r_s$. Then for all $s' \preceq s$, $ps' \in A^r_s$, so inductively, $qs' \in A^r_s$, too. Certainly, $rps \in A$; by Lemma 2.31, $rqs \in A$ and $rps \sim_G rqs$. Since A_s is $\sim$-closed, $rqs \in A_s$. So by definition of A^r_s, $qs \in A^r_s$. Because $rps \sim_G rqs$, they have the same colour in A^r_s (red or green), and for each $t \in \Phi \setminus \{s\}$, $ps \in S^r_{s,t}$ iff $qs \in S^r_{s,t}$. Thus, p and q indeed have the same future in G^r_s. So $p \sim_{G^r_s} q$, proving the claim.

Clearly, τ^* is finitely based. Now the owner of τ (say $\forall$) can use τ^* in G if play reaches r. If he does so, play corresponds to a play of G^r_s, which he will win. That is, if β is the resulting play of G^r_s, $\beta \models \varphi(s/\bot)$. Let β^* be the full play of G: $\beta^* = \{rp : p \in \beta\}$. Then for all $p \in \beta$ and $t \in \Phi \setminus \{s\}$, $p \in S^r_{s,t}$ iff $rp \in S_t$. Because β^* only has finitely many positions $\preceq r$, we have $\beta^* \models t$ iff $\beta \models t$ for each $t \in \Phi \setminus \{s\}$. So $\beta^* \models \varphi(s/\bot)$, also. But $\beta^* \cap S_s$ is finite, because $A^r_s \cap S^r_s = \emptyset$. Hence, $\beta^* \models s \leftrightarrow \bot$, so $\beta^* \models \varphi$. So $\forall$ won G. □

Lemma 2.48. *$\exists$ has a finitely based winning strategy σ^r_s in G^r_s.*

Proof By the inductive hypothesis, $\forall$ or $\exists$ has such a strategy. Suppose for contradiction that it is $\forall$; let τ be one. By the lemma, τ^* is a finitely based winning strategy for him in G, defined at r. This contradicts Lemma 2.42 (cf. (†) above). □

By Lemma 2.47, σ^r_s can be extended to a G-finitely based strategy $(\sigma^r_s)^*$ that wins G for $\exists$ whenever play reaches r. This holds for all $r \in A_s$.

As A_s is $\sim_G$-closed, by Lemma 2.39, there is a single G-finitely based strategy σ_s for $\exists$ that can be used by her in any play of G that hits A_s, and will win for her.

This holds for each $s \in \Phi$ such that $A_s \neq \emptyset$.

Now, the following final strategy for $\exists$ in G is G-finitely based, and wins G for her. Let $p \in A$ be a position, and let s be the leftmost (or 'oldest') entry in $LAR_G(p)$. The strategy's advice at p is as follows.

1. Use 'hit S_s' if this is defined at p.
2. If 'hit S_s' is not defined at p, then use σ_s.

This strategy is finitely based in G. For let $p \sim q$ in G. As $LAR_G(p) = LAR_G(q)$, the 's' above will be the same for each, so $\exists$ will aim for the same S_s at p as at q. By Lemma 2.37, the strategy 'hit S_s' is defined at p iff it is at q. So the strategy above applies the same case to p as to q.

The first case of the strategy is known to be G-finitely based, by Lemma 2.37 again. The second case is G-finitely based by its definition. Hence, indeed, the strategy is finitely based in G.

Consider a play β of G in which $\exists$ uses the strategy. It is easily checked, using the definition of 'LAR', that if case 1 is used infinitely often in β, then play will indeed hit each S_s infinitely often, so $\exists$ wins. For, during any interval of play in which case 2 is used, the leftmost entry of the LAR in G of the position remains constant, because the corresponding set is not hit. Hence, when case 1 is resumed, the same set is aimed at as before the interruption (or perhaps it is even hit immediately). So β is a win for $\exists$ in G. If case 2 is eventually used for ever, we saw that this wins G for $\exists$. Thus, we have found a finitely based winning strategy for $\exists$ in G, and the proof of Theorem 2.40 is complete. □

2.10 Tree automata

We now define the appropriate automata for S2S, the monadic second-order logic of the binary tree. In the next section, we use the limited memory determinacy result just proved to obtain the complementation lemma for tree automata. This will lead to every S2S sentence having an equivalent tree automaton, and will yield decidability of S2S.

Let L be a finite set of atoms.

Definition 2.49 (Tree automaton). An L-tree automaton is a 4-tuple $M = (Q, P_0, P, \varphi)$, where Q is a finite set of states, $P_0 \subseteq Q$ is the set of initial states, $P \subseteq Q \times \wp L \times \{0,1\} \times Q$ is the transition table, and φ is a propositional formula written with atoms $\{s : s \in Q\}$.

Let T be the binary tree $\{0,1\}^*$, and let $h : T \to \wp L$ be an assignment. (So $p \in L$ is true at $t \in T$ iff $t \in h(p)$; in this case we write '$(T, h) \models p(t)$'.) Let M be an L-tree automaton. The players $\forall$ and $\exists$ play a game $\Gamma(M, h)$ on (T, h) as follows.

$\forall$ chooses:	$\exists$ chooses:
$s_0 \in P_0$	
	$e_0 \in \{0,1\}$
$s_1 \in Q$ with $(s_0, h(\langle\rangle), e_0, s_1) \in P$	
	$e_1 \in \{0,1\}$
$s_2 \in Q$ with $(s_1, h(e_0), e_1, s_2) \in P$	
	$e_2 \in \{0,1\}$
$s_3 \in Q$ with $(s_2, h(e_0e_1), e_2, s_3) \in P$	...

and so on. If $\forall$ is unable to choose s_n at some stage, he loses the game. Assume not. The result of play is then a branch $\{\langle\rangle, e_0, e_0e_1, \ldots\}$ of T and a sequence $\bar{s} = s_0s_1\ldots$ of states. $\forall$ wins the play iff $\bar{s} \models \varphi$ (as usual, if $\bar{s} = s_0s_1s_2\ldots$ is an infinite sequence from Q, and $s \in Q$, we let $\bar{s} \models s$ iff $\{n \in \mathbb{N} : s_n = s\}$ is infinite; the notion of $\bar{s} \models \varphi$ is then defined by induction on φ).

We think of $\exists$ pushing the automaton M up a branch of the tree, step by step. The branch is not fixed in advance, for at each stage, she can choose which way to send M by picking 0 or 1 as her next move. The automaton then has to choose a new state to enter, consistent with its current state, the atoms true at the current position, and her choice of direction.

So what M sees, during a 'run', is almost just a linear temporal structure, as for the Muller automata in our study of S1S; and its run and acceptance condition are also very similar to the Muller case. But at any stage, M is never sure *which* temporal structure it will be running on! This limits the usefulness of guessing. But it is given $\exists$'s choice (note the dependence on e in the transition table), so at least it knows which way it has been sent so far and which way it is about to be sent.

Definition 2.50. The automaton M is said to *accept* (T, h) iff $\forall$ has a winning strategy for this game. Otherwise, M is said to reject (T, h).

As with linear time, we disallow automata that may 'grind to a halt': that is, automata in which $\forall$ may not be able to move at some stage. In order to prevent such occurrences, we suppose that $P_0 \neq \emptyset$ and that for all $s \in Q$, $e \in \{0, 1\}$, and $L' \subseteq \wp L$, there exists $s' \in Q$ such that $(s, L', e, s') \in P$.

In order to prove that every automaton has a non-halting equivalent, we can simply introduce a 'failure' state which traps the progress of the automaton and does not contribute to its winning.

2.10.1 *Variations*

The automata we use are also seen in [Pnueli and Rosner, 1989a], but this is only one approach and there are several different ways of presenting tree automata to be seen in the literature. For the most part the apparent differences are really only differences in presentation and we can find equivalent automata of each kind. Just to be clear about this we say that two automata are *equivalent* if they accept precisely the same labelled trees. To say that two types of automaton are equivalent is to say that each automaton in one form has an equivalent in the other form and vice versa.

The main variations are in the way of defining acceptance criteria for specifying a winning play in the game $\Gamma(M, h)$. The differences here are sometimes significant. There are three main variations:

Büchi as in [Büchi, 1962a]. Büchi automata just have a set F of *final states* to define acceptance. Büchi acceptance means that at least one of these states comes up infinitely often. Thus Büchi acceptance can be specified by requiring φ to be just a disjunction of atoms.
In general this makes Büchi automata weaker. For example, there is no Büchi automaton which can recognize the set of $\{a, b\}$-trees which contain no branches with an infinite number of bs.

Muller as in [Gurevich and Harrington, 1982]. The acceptance criterion of a Muller automaton is specified by a set F of subsets of Q. For acceptance the set of states which come up infinitely often has to be a member of F. As in Proposition 2.10, this can be shown to be equivalent to our condition.

Rabin acceptance is specified by a set Ω of pairs of subsets of Q. We require there to be a pair $(U, V) \in \Omega$ such that some state in U appears infinitely often but no state in V appears infinitely often.

It is clear that each Rabin automaton has an equivalent one of ours—just translate the acceptance set into a formula in the obvious way. With a little more work (exercise) we can show that each of our automata has a Rabin equivalent.

Some of the results we mention concerning binary tree automata generalize to n-ary tree automata: $\exists$ chooses from $\{0, 1, \ldots, n-1\}$ as the automaton progresses along a branch of $\{0, 1, \ldots, n-1\}^*$. (Certainly, the corresponding monadic second-order logic 'SnS' is decidable, and it can be interpreted in S2S. Decidability even holds for 'SωS'.)

2.10.1.1 *Determinism* The automaton M is said to be deterministic iff:

- $|P_0| = 1$
- and for all $s \in Q$, for all $e \in \{0, 1\}$, for all $L' \subseteq L$, there is exactly one $s' \in Q$ such that $(s, L', e, s') \in P$.

In general not every automaton has a deterministic equivalent but we will see that it is useful for us mostly to deal only with deterministic automata.

Exercise 2.51. Find a tree automaton (as in Definition 2.49) for the language $\{q\}$ that accepts (T, h) iff $(T, h) \models q(t)$ for some $t \in T$. Show that there is no deterministic automaton with this property.

2.11 Complementation for tree automata

We will sketch a proof of the following theorem. T continues to denote the binary tree $\{0, 1\}^*$.

Theorem 2.52. *For any tree automaton M, there is a new tree automaton M' such that for all h, M rejects (T, h) iff M' accepts (T, h). M' can be constructed effectively from M.*

That is, tree automata are closed under complementation. For automata on $\mathbb{N}$, we did this using McNaughton's theorem. However, this will not work here as not every tree automaton is equivalent to a deterministic automaton.

The idea of the proof is very roughly this. If M rejects (T, h), this is in some sense an existential statement—*there exists* a branch of T that M will reject. To obtain M', we have to turn this into an equivalent universal statement (*for all* branches β, M' accepts β). Theorem 2.40 will permit us to do this, for it allows us to attach a 'signpost' to each node of T, saying which direction (0 or 1) $\exists$ should choose at t in order to carry on along a rejecting branch. The required universal statement will now take the form 'for all branches β, if β follows the signposts all the way then it is a rejecting branch'.

We now give a somewhat more formal proof.

Proof Let $M = (Q, P_0, P, \varphi)$ be an L-tree automaton, and let $h : T \to \wp L$ be given. We build a game tree A from h and the automaton M. We set $\mathsf{MOVE} = \{0,1\} \cup Q$ (we assume $\{0,1\} \cap Q = \emptyset$). Then A is defined as the smallest tree satisfying:

- A is closed downwards ($p \preceq q \in A \Rightarrow p \in A$);
- if $p \in A$ has odd length then p ends in some $s \in Q$;
- if p has even length > 0 then p ends in 0 or 1;
- if p has odd length then $p0, p1 \in A$;
- $P_0 \subseteq A$; and
- suppose $p \in A$ has even length > 0, so $p = rse$ where $r \in A$, $s \in Q$, and e is 0 or 1. Let $\tau(r)$ be the sequence of 0,1-entries from r, in order. Then $ps' \in A$ iff $(s, h(\tau(r)), e, s') \in P$.

A position in A is defined as green if it has odd length (ends in a state), and red otherwise (ends in 0 or 1, or is $\langle\rangle$). We define $S_s = \{p \in A : p \text{ ends in } s\} \subseteq A$, for each $s \in Q$. Clearly, these are pairwise disjoint.

Because we agreed our automata cannot grind to a halt, A is a game tree. There is a one–one correspondence between plays of the automaton game $\Gamma(M, h)$ and plays of the Section 2.6 game $G = (A, S_s \ (s \in Q), \varphi, \forall)$ on A:

$\forall$ chooses:	$\exists$ chooses:
$s_0 \in P_0$	
	e_0 with $s_0e_0 \in A$
s_1e_0 with $s_0e_0s_1 \in A$	
	e_1 with $s_0e_0s_1e_1 \in A$
s_2e_0 with $s_0e_0s_1e_1s_2 \in A$	...

$\forall$ wins a play of $\Gamma(M, h)$ iff $\forall$ wins the corresponding play of G. And $\forall$ has a winning strategy for both games or neither.

Suppose now that M rejects (T, h). So $\forall$ has no winning strategy for G. Choose an arbitrary Q-display δ. Then $\exists$ has a finitely based winning strategy σ for the game

$$G(M, h) = (A, Q, S_s \ (s \in Q), \varphi, \forall, \delta).$$

For otherwise, by Theorem 2.40, $\forall$ would have a (finitely based) winning strategy in this game—contradiction.

As above, define the projection $\tau : A \to T$, by: $\tau(p)$ is the sequence of 0,1-entries from p, in order. It is not hard to see that if $p, q \in A$ are green, have the same LAR, and $\tau(p) = \tau(q)$, then $p \sim_G q$. Hence, σ, being finitely based and not being needed on red positions, can be regarded as a map $: T \times \{Q\text{-displays}\} \to \wp\{0,1\}$. This can be represented by attaching to each node $t \in T$ a function $\sigma_t : \{Q\text{-displays}\} \to \wp\{0,1\}$. And this can be expressed in logic as follows. Introduce a new atom q_ρ for each possible function $\rho : \{Q\text{-displays}\} \to \wp\{0,1\}$. Set $L' = L \cup \{\text{all } q_\rho\}$. We extend h to a map $h' : T \to \wp L'$ by $h'(t) = h(t) \cup \{q_{\sigma_t}\}$. So h' encodes $\exists$'s winning strategy into T.

Now *any* possible extension h' of h to L' such that the truth sets of the q_ρ are pairwise disjoint defines a strategy for $\exists$ in $\Gamma(M,h)$:

- If you are at $t \in T$, and the current LAR is λ (this depends on the recent states of M that $\forall$ chooses, which you know all about!), then let q_ρ be the unique extra atom true at t, and let $e \in \rho(\lambda)$. So $e = 0$ or 1. Then choose e as your next move.

Not all such strategies (encodable by the new q_ρ atoms) are winning for $\exists$. But we showed that if M does not accept (T,h) then there is some winning strategy for $\exists$ of this form.

How can we tell whether an extension h' gives a winning strategy? Let β be a branch of T. So β contains a single word t_n of $\{0,1\}$ of length n, for all $n \in \mathbb{N}$, and the t_n are linearly ordered by 'initial segment' ($\preceq$). Then $(\beta,\preceq) \cong (\mathbb{N},\leq)$. So β gives in a natural way a linear L'-structure N_β, based on $\mathbb{N}$. Formally, if $\theta : \mathbb{N} \to \beta$ is an order-preserving bijection, then $N_\beta = (\mathbb{N}, h' \circ \theta)$. However, the directional information of β is missing from N_β. So we introduce one more atom, d, and interpret it in N_β by: $N_\beta \models d(n)$ iff t_{n+1} ends in 1 (all $n \in \mathbb{N}$).

Then N_β is capable of recording the progress of all plays of $\Gamma(M,h)$ in which $\exists$ chose branch β. If information about how $\forall$ played is added, we can see whether $\exists$ used the strategy given by h', and whether she won. So we can use the N_β to determine whether h' is a winning strategy for her. We obtain:

Proposition 2.53. *h' encodes a winning strategy for $\exists$ in $\Gamma(M,h)$ iff $N_\beta \models \psi$ for every branch β of T, where ψ is the following sentence:*

'For all subsets $S_s \subseteq \mathbb{N}$ ($s \in Q$), and all subsets $\lambda_\varepsilon \subseteq \mathbb{N}$, where ε ranges over all Q-displays, suppose that the following hold:

1. *the S_s are pairwise disjoint;*
2. *the S_s represent possible states of M as chosen by $\forall$ in a play of $\Gamma(M,h)$ in which $\exists$ chose branch β.*
   Formally, for $n \in \mathbb{N}$, let s_n be the unique state $s \in S$ such that $n \in S_s$. Then $s_0 \in P_0$ and $(s_n, X_n, e_n, s_{n+1}) \in P$ for each n, where $X_n = \{p \in L : N_\beta \models p(n)\}$, and e_n is 1 if $N_\beta \models d(n)$ and 0 otherwise.
3. *the λ_ε represent the corresponding LARs at each point, obtained from the S_s using initial value δ.*
   That is, the λ_ε are pairwise disjoint, $0 \in \lambda_\delta$, and for all $n \in \mathbb{N}$, if $n \in \lambda_\varepsilon$ then $n+1 \in \lambda_{\varepsilon * s_{n+1}}$, where the s_n are as above.
4. *The directional information (the values of d) corresponds to the values of q_ρ, i.e. in the corresponding play of $\Gamma(M,h)$, $\exists$ used the strategy given by the q_ρ.*
   Formally, for $n \in \mathbb{N}$, suppose that $n \in \lambda_\varepsilon$ and $N_\beta \models q_\rho(n)$. Then if $N_\beta \models d(n)$ then $1 \in \rho(\varepsilon)$, and if $N_\beta \models \neg d(n)$ then $0 \in \rho(\varepsilon)$.

Then, making the propositional atom s true iff S_s is infinite (all $s \in Q$), $\neg\varphi$ holds.'

Proof If h' gives a winning strategy for $\exists$, assume the antecedent of ψ holds in N_β for some β. Then N_β can be 'pulled back' to (T, h), to give a play of $\Gamma(M, h)$ in which $\exists$ used the strategy given by h'. $\forall$ lost this play, so the consequent of ψ holds in N_β.

Conversely, let $\forall$ and $\exists$ play $\Gamma(M, h)$, $\exists$ using the strategy given by h'. Let the branch eventually chosen by $\exists$ be β. The play can be pushed forward into N_β, and because $N_\beta \models \psi$, it follows that $\exists$ won. Hence the strategy given by h' is winning for her. □

But ψ can be written as a sentence of S1S! So by Theorems 2.11 and 2.22, there is a deterministic Muller automaton D that accepts an $L' \cup \{d\}$-structure N iff ψ holds. D can be obtained effectively from ψ, and so from M. We can convert D into a (deterministic) L'-tree automaton D' by simply letting it read d from the tree (i.e. according to which way $\exists$ sends it), rather than from the structure N. So D' is an L'-tree automaton.

Proposition 2.54. *D' accepts (T, h') iff for every branch β of T, if N_β is the corresponding $L' \cup \{d\}$-structure then D accepts N_β.*

Proof Because D is deterministic, $\forall$ never has a choice in the game $\Gamma(D', h')$. So if in a play of this game, $\exists$ ends up choosing the branch β, the successive states that D' passes through during its run correspond exactly to the successive states that D enters during a run on N_β.[2] Because the accepting conditions of D and D' are the same, D' accepts (T, h') iff D accepts every N_β, as required. □

Hence Propositions 2.53 and 2.54 D' accepts (T, h') iff h' encodes a winning strategy for $\exists$. Now let M' be a version of D' that guesses the values of the atoms q_ρ. So clearly, M' accepts (T, h) iff for some h' extending h, D' accepts (T, h'), iff h' describes a winning strategy for $\exists$ on (T, h). And this last is iff M rejects (T, h) (we explained '$\Leftarrow$' above, and '$\Rightarrow$' is obvious).

Since M' can clearly be constructed effectively from M, this completes the proof of Theorem 2.52. □

2.12 S2S

We briefly describe this second-order logic and sketch its decidability—Rabin's famous result of [Rabin, 1969].

Let $T = \{0, 1\}^*$ be as above, and let $D = \wp T$. Let L be a finite set of propositional atoms. As for S1S, we use a first-order signature $L' = L \cup \{=, \subseteq, \mathit{left}, \mathit{right}\}$, the last four being binary relation symbols. Any (T, h) where $h : T \to \wp L$ is an assignment makes D into an L'-structure denoted (D, h), via:

[2]This would fail if D were non-deterministic, because it could choose different states in a run on different structures N_β and N_γ, while D' does not have this luxury. For example, perhaps D accepts iff it guesses correctly at each time n whether there is a sea-battle at $n + 1$. It can accept any N_β by always guessing the correct prediction. But on the tree, not only would D' have to guess at each point whether there is a sea-battle tomorrow, but also which tomorrow $\exists$ will choose; and she can postpone her decision on this until she has seen its prediction!

1. $=$ and $\subseteq$ are interpreted as usual;
2. if $X, Y \in D$, $D \models left(X, Y)$ iff $X = \{t\}$ and $Y = \{t0\}$ for some $t \in T$;
3. if $X, Y \in D$, $D \models right(X, Y)$ iff $X = \{t\}$ and $Y = \{t1\}$ for some $t \in T$;
4. if $q \in L$, then q is a constant symbol, interpreted as $\{t \in T : q \in h(t)\}$.

The logic S2S is built on the first-order language L' as described. Quantification is over D, i.e. over subsets of T. Hence the logic is really second order, as for S1S. We write $(T, h) \models \sigma$ (for a sentence σ of S2S) to mean $(D, h) \models \sigma$.

S2S is very expressive indeed. We give one exercise to illustrate this; see Volume 1, Chapter 15 for more examples.

Exercise 2.55. This exercise is to show that S3S, S4S, ..., SωS can be encoded in S2S, so that nothing essentially new happens for ternary trees etc.

Let $3 \leq n \leq \omega$. Define the monadic second-order logic of n successors, SnS, by replacing the relation symbols *left, right* by n binary relation symbols $succ_i$ ($i < n$). Define the semantics of SnS using the tree $T_n = \{0, 1, \ldots, n-1\}^*$ in the obvious way.

Define $\theta_n : T_n \to T$ by induction:

$$\theta_n(\langle\rangle) = \langle\rangle, \quad \text{and} \quad \theta_n(ti) = \theta_n(t)\overbrace{11\ldots1}^{i \text{ times}}0 \quad \text{for } t \in T_n \text{ and } i < n.$$

For any assignment $h : T_n \to \wp L$, define an assignment $h' : T \to \wp L$ by

$$h'(t) = \begin{cases} h(u), & \text{if } u \in T_n \text{ and } \theta_n(u) = t, \\ \emptyset, & \text{if } t \notin rng(\theta_n). \end{cases}$$

1. Show that θ_n is an injection, so that h' is well defined.
2. Find a formula $\rho_n(x)$ of S2S such that for any $S \subseteq T$, $T \models \rho_n(S)$ iff $S \subseteq rng(\theta_n)$.

For any formula φ of SnS, define the S2S-formula φ' by induction:

- $(x = y)'$ is $x = y$, and $(x \subseteq y)'$ is $x \subseteq y$;
- $succ_i(x, y)'$ is $\exists z_0 \ldots z_i(z_0 = x \wedge \bigwedge_{j<i} succ_1(z_j, z_{j+1}) \wedge succ_0(z_i, y))$;
- $(\neg\varphi)'$ is $\neg\varphi'$ and $(\varphi \wedge \psi)'$ is $\varphi' \wedge \psi'$;
- $(\exists x\varphi)'$ is $\exists x(\rho_n(x) \wedge \varphi')$.

3. Show that for any L-sentence σ of SnS and any assignment $h : T_n \to \wp L$, $(T_n, h) \models \sigma$ iff $(T, h') \models \sigma'$.
4. Deduce that σ (as above) has a model in the SnS semantics iff σ' has a model in the S2S semantics.

As for S1S, if M is a tree automaton then there is a sentence σ of S2S such that for all h, $(T, h) \models \sigma$ iff M accepts (T, h). The proof is much as for S1S, and we leave it as an exercise, since we will not use it to obtain decidability of S2S. What we will use is the converse:

Proposition 2.56. *For every sentence σ of* S2S *in signature L, there is an L-tree automaton M such that for any assignment $h : T \to \wp L$, M accepts (T, h) iff $(T, h) \models \sigma$. M can be obtained effectively from σ.*

Proof By induction on σ. If σ is atomic, there are four cases: $p = q$, $p \subseteq q$, $left(p, q)$ and $right(p, q)$ for $p, q \in L$. We ignore the first of these, because $p = q$ is equivalent to $p \subseteq q \wedge q \subseteq p$ and so can be handled by the induction.

A deterministic automaton M for $p \subseteq q$ is easy to construct. It has two states, 'OK so far' and 'fail'. The set P_0 of initial states is just {OK so far}. If, at $t \in T$ and in state 'OK so far', it finds that p is true at t and q is false, it passes to state 'fail'. Otherwise, it stays in state 'OK so far'. Once in state 'fail', it stays there. The acceptance formula φ is just 'OK so far' (cf. Fig. 2.3 of Theorem 2.22.)

Clearly, if $(T, h) \models p \subseteq q$, then however $\exists$ plays in $\Gamma(M, h)$, $\forall$ will (must) always choose state 'OK so far' as the next state. Thus, he wins any play, so M accepts (T, h). But if not, so that there is $t \in T$ with $p \in h(t)$ and $q \notin h(t)$, then $\exists$ can move so as to drive the position towards such a t. Once there, $\forall$ is forced to choose 'fail' as his next state and as all subsequent states. Thus, he loses, showing that $\exists$ has a winning strategy $\Gamma(M, h)$ and that M rejects (T, h).

It is harder to find a tree automaton for $left(p, q)$. The following will suffice:

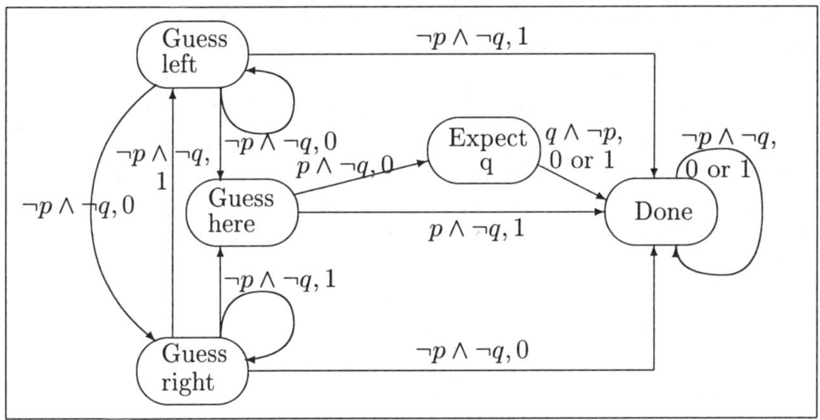

FIG. 2.5.

We use the same style of diagram as in Theorem 2.22. The set P_0 of initial states of this automaton is {Guess left, Guess right, Guess here}. The acceptance formula φ is 'Done'. Note that the automaton has another state, 'Fail', which is not shown in the diagram for lack of space. In any state shown, if the atoms true at the current time do not fit any transition shown, then the automaton passes to the 'Fail' state, where it stays from then on. This is exactly as for automata that get stuck, as mentioned in Definition 2.49.

We sketch the idea behind the construction of this automaton, which we call M.

Suppose first that $(T, h) \models left(p, q)$. Then there is a unique $t \in T$ with $(T, h) \models p(t)$, and $(T, h) \models q(t0)$, this being the only place where q holds. We describe a winning strategy for $\forall$ in $\Gamma(M, h)$.

In a play of $\Gamma(M, h)$, $\forall$ chooses 'Guess left', 'Guess right' or 'Guess here' at position $u \in T$, depending on whether $t \succeq u0$, $t \succeq u1$ or $t = u$, respectively. Initially, $u = \langle\rangle$, so exactly one of these possibilities holds. Suppose at some position u he selects 'Guess left', signifying that $t \succeq u0$. $\exists$ can see his choice, of course. If she chooses 0 as her move, $\forall$ must make another choice, according as whether $t \succeq u00$, $t \succeq u01$ or $t = u0$. In this way, he always states where he thinks t is, relative to the current position, so long as $\exists$ moves so that play eventually arrives at t.

If this happens, he will have chosen 'Guess here' as his last move. As $(T, h) \models p \wedge \neg q(t)$, if $\exists$ chooses 0 he must select 'Expect q'. At $t0$, his last move was therefore 'Expect q'. Since $(T, h) \models q \wedge \neg p(t0)$ and $(T, h) \models \neg p \wedge \neg q(v)$ for all $v \succ t0$, he may choose 'Done', from then on. Thus, he wins.

If, at any stage, $\exists$ does not choose as just outlined, $\forall$ can pass to 'Done' immediately. In essence, if she does not force him to justify all his predictions, he can resign with honour. Notice that as soon as this happens, no further positions in the tree will satisfy p or q. So $\forall$ can continue to choose 'Done' for ever. Thus, $\forall$ has a winning strategy in this case, and M accepts (T, h).

Now suppose $(T, h) \models \neg(p \subseteq q)$. This can happen in various ways; we sketch a winning strategy for $\exists$ in $\Gamma(M, h)$ for each of them. The idea is that $\forall$'s successive choices of 'Guess left', 'Guess right', 'Guess here' are treated as *predictions* of where the unique site t (as in the other case, above) should be. $\exists$'s strategy is to challenge these predictions and force $\forall$ to admit that they were wrong.

If no position $t \in T$ satisfies p (under h), then $\exists$ pushes play in the direction that $\forall$ predicts, at each stage. That is, if at some position, he selects state 'Guess left', then she will choose 0 next round. If he selects 'Guess right', she will choose 1. (If he selects 'Guess here' he will lose immediately.) Since no place where p holds is ever found, $\forall$ is forced to continue choosing 'Guess left' or 'Guess right' for ever, and so φ will be false in the outcome. A similar strategy applies if no position satisfies q.

If more than one position satisfies p, and there are two such positions $t, u \in T$ with $t \prec u$, then $\exists$ takes play up towards u. When play passed through t, $\forall$ must have chosen state 'Expect q', 'Done' or 'Fail'. One of these states must be the current state on arrival at u. In any case, the appearance of another site (u) satisfying p will force him to choose 'Fail' from then on. So he loses.

If, on the other hand, more than one position satisfies p but no such t, u exist, then $\exists$ can select two distinct $\preceq$-minimal elements $t, u \in T$ satisfying p. Let v be their greatest lower bound. $\exists$ takes play towards v. On arrival at v, perhaps $\forall$ has already lost, e.g. because a q was seen, or because he chose 'Guess here' already; in that case, he will be choosing 'Fail' from now on. If not, he may have chosen 'Done', because $\exists$ already disregarded one of his predictions. In that case, $\exists$ can take play towards either t or u; on arrival, $\forall$ will not be able to choose 'Done' and will have to choose 'Fail'. But perhaps he is still predicting Left or Right. In that case, *she inspects ∀'s prediction and proceeds to ignore it,* by choosing 0 if he last chose 'Guess right', and 1, if he chose 'Guess left'. Then, ∀ will select 'Done', reducing to the case just before. ∃ then takes play up towards whichever of t, u is still accessible. On arrival, ∀ will lose.

Notice that ∃'s *reacting* to ∀ is essential to ensure rejection of trees such as the one where p is true at 0 and 1 only, and q is true at 00 and 10 only. Similar considerations apply if more than one point of T satisfies q.

Finally, suppose that both p, q define singletons of T; let q be true at $t \in T$ only. Clearly, if ∃ leads play towards t, she will win.

Now we cover the inductive cases. The cases of $\wedge$ and $\exists$ are as for S1S. Finally, complementation. Let σ be a sentence, and suppose inductively that the tree automaton M accepts (T, h) iff $(T, h) \models \sigma$. By Theorem 2.52, there is a tree automaton M' that accepts (T, h) iff M rejects (T, h). M' can be constructed effectively from M. Clearly, M' works for $\neg\sigma$. Hence every S2S sentence σ has an equivalent automaton, which can be found effectively from σ. □

Corollary 2.57. S2S *is decidable.*

Proof We want an algorithm that, given any L-sentence σ of S2S, says whether or not there is $h : T \to \wp L$ such that $(T, h) \models \sigma$.

We may effectively construct a tree automaton M equivalent to σ. Let M' be the version of M that guesses all atoms of L. We want to know if M' accepts the (unique) tree $(T, \emptyset)$, where there are no atoms: $L = \emptyset$, and h is the empty map. It does so iff σ has a model.

Now, either ∀ or ∃ has a winning strategy in the game $G(M', \emptyset)$ of Theorem 2.52 (with arbitrary initial LAR). Moreover, the strategy can be taken to be finitely based. So its value at a node $t \in T$ depends only on the LAR and on the rest of the tree above t. But here, there are no atoms, so the tree above any two nodes t, u is the same. So the strategy depends only on the LAR.

There are finitely many possible strategies for M' and for ∃ that depend only on the LAR. Play them off against each other, and see which player has one that wins against all the strategies of the other player. Note that the strategies can be refined to deterministic ones, which are still finitely based. As there are no atoms to consider, any play between two such strategies eventually loops, so that we can find out in finite time whether one LAR-based strategy beats another. Hence this is a genuine algorithm—it terminates. If the player having a winning strategy is ∀ then σ has a model; otherwise not. □

Exercise 2.58. Deduce that SnS ($3 \leq n < \omega$) and SωS, as in Exercise 2.55, are decidable.

2.13 Transducers

Later in the book we will refer to a certain type of automaton which only recognizes one tree. It is useful in being able to describe this tree in a constructive manner.

A tree L-automaton $M = (Q, P_0, P, \varphi)$ is a *deterministic transducer* iff:

- for all $q \in Q$, there is $b_q \in \wp L$, such that for all $a \in \wp L$, for all $e \in \{0,1\}$, for all $q' \in Q$, if $(q, a, e, q') \in P$ then $a = b_q$;
- for all $q \in Q$, for all $e \in \{0,1\}$, there is $q(a,e) \in Q$ such that for all $q' \in Q$, if $(q, b_a, e, q') \in P$ then $q' = q(a,e)$;
- there is some $q_0 \in Q$ such that $P_0 = \{q_0\}$;
- $\varphi = \top$.

A deterministic transducer M gives us a function $f_M : (\{0,1\}^*) \to \wp L$ defined as follows. Put $f_M(\langle\rangle) = b_{q_0}$. Now for $\sigma = e_0{}^\wedge \ldots {}^\wedge e_i$, define a sequence $q_0, q_1, \ldots, q_i \in Q$ by $q_{j+1} = q(q_j, e_j)$. Let $f_M(\sigma) = b_{q_{i+1}}$.

Notice that this function defines a labelled tree $(\{0,1\}^*, f_M)$ and that this tree is the only L-tree accepted by M.

Furthermore, notice that, given M, we have a simple algorithm for effectively calculating the label on a node of this tree: just follow the sequence $q_0, q_1, \ldots, q_i$ of states that M goes through when given directions $e_0, \ldots, e_i$ to reach the node $e_0{}^\wedge \ldots {}^\wedge e_i$. It is easy to find $b_{q_{i+1}}$ from the table for M.

As we have seen, the basic decidability result underlying Rabin's result is the decidability of the emptiness question of tree automata. We want a procedure which tells us whether or not the set of labelled trees accepted by any given automaton is empty or not. As well as the method summarized above, it is worth mentioning a result from [Pnueli and Rosner, 1989a]:

Theorem 2.59. *The emptiness problem on infinite trees is decidable in deterministic time $O((nm)^{cm})$ for some constant c, where n is the number of states and m is the number of pairs.*

Furthermore, the same deciding algorithm can, in the case that the automaton A is non-empty, yield a deterministic transducer D which recognizes some tree t also recognized by A and $|D| = O((nm)^{3m})$.

The transducer part of the result is useful for effectively giving us a tree which is recognized by the given non-empty automaton. This can be used for the synthesis of programs which build models of a temporal formula in the case of reactive modules. See [Pnueli and Rosner, 1989a] for details.

3

BRANCHING TIME

3.1 Introduction

In this chapter we shall consider logics and languages that are appropriate for describing events in branching time. Much of the motivation for considering time to be branching comes from the idea of using different branches to represent different possible histories of some part of the world. Thus, we are really looking at logics which combine temporal aspects with features of a modal logic of possibility and necessity.

Although, the term 'branching' is sometimes used more generally, we shall follow much common usage and restrict our attention to flows of time in which branching out only happens towards the future: the futures of incomparable times never join up. Thus, any particular point of time has one linear past but, perhaps, many futures. If branching is used to capture the idea of alternative possibilities then we have restricted ourselves to so-called logics of historical necessity. As well as being used in this way in philosophy, branching time temporal logics are also used in artificial intelligence and in computer science. In AI, reasoning agents may need to consider different possible futures in formulating their plans for action. In software engineering, a reactive module may react differently in different possible environments or it may just be non-deterministic in its own behaviour. We can use branching time temporal logics to describe this behaviour.

Even restricting ourselves to branching towards the future, we still find many different shapes of branching structures. In particular applications of such temporal logics, we might find it useful to restrict our attention further to particular subclasses. We will see that the branching may be discrete, the amount of branching may be bounded, the height of the tree structure may be bounded, etc. Each of these conditions and all combinations of such conditions give rise to different logics.

We also produce different logics by varying the language of the logic. As in the case with linear time temporal logics, we may choose to have more or less expressive languages by building the syntax from different sets of temporal operators. Of course, for particular applications, it is not always best to choose the most expressive language: if a language can express many properties then it may be quite complicated to reason with. In general it is better to try and use the simplest language which is adequately expressive for a particular domain of interest.

3.2 Several approaches

It is widely held, and it has been widely held since at least the time of Aristotle, that the past is fixed but the future indeterminate. Thus a tree, as we now define, seems to capture the shape of time. A *tree* is a set of (time) points T ordered by a binary relation $<$ which satisfies the following requirements:

- $(T, <)$ is irreflexive;
- $(T, <)$ is transitive;
- the past of any point is linear $\forall t, u, v \in T$, if $u < t$ and $v < t$ then $u < v$, $u = v$ or $v < u$;
- $(T, <)$ is connected in fact, $\forall x, y \in T$, $\exists z \in T$ such that $z < x$ and $z < y$.

In this chapter we will only consider propositional languages. So, to get a branching structure, we need only add a valuation of atomic propositions as sets of time points.

We say that $z \in T$ is the *root* of a tree $(T, <)$ iff for all $x \in T$, $z \leq x$.

Sometimes we will limit our attention to particular classes of trees. We have *discrete* trees which satisfy the conditions that:

- for every $x \in T$ except the root if it exists, there is $y < x$ such that for all $z < x$, $z \leq y$; and
- for each $x \in T$, there is a set S of immediate successors of x which are pairwise incomparable (i.e. for all $y, z \in S$, neither $y < z$ nor $z < y$) such that for all $z > x$, there is $y \in S$ such that $y \leq z$.

A *branch* in a tree $(T, <)$ is a linearly ordered subset b of T that has no external upper bound and no missing internal points (i.e. if $x < y < z$ and $x, z \in b$ then $y \in b$). A *history* in a tree $(T, <)$ is a maximal linearly ordered subset of T.

Discrete rooted trees have height ω iff all the histories have the order type of the natural numbers.

Sometimes it will be important to look at trees in which the amount of branching is limited. We say that a discrete tree has branching factor at most b if and only if each point has at most b successors.

The question of the appropriate modal–temporal language to use in order to express such branching structures is closely related to the long-running problem about the nature of the relationship between time and determinism. The ancient Greeks were seriously discussing this problem as Aristotle's famous sea-battle example shows. In medieval times, the same sorts of problems were raised by many theologians concerned about the seeming contradiction between human freedom and God's omniscience. In more modern times, C. S. Peirce realized that it is important to take time into account when thinking logically, and that there is an important distinction between the future and the past in this respect. The first systematic modern account of branching time is that of Prior in [1968] in which two broad approaches to branching time are described: these are known as the Ockhamist and Peircean approaches.

We will introduce these two approaches below, but it is worth mentioning that there is another general approach to the indeterminate future—Łukasiewicz's three-valued logic with an 'undetermined' truth value—which we will not consider. A very useful survey of the history of temporal logic, in which all these approaches can be found, is in [Øhrstrøm and Hasle, 1996].

3.2.1 *The Peircean approach*

As usual we can place a valuation for a set of propositional atoms on a tree frame $(T, <)$. In the manner which Prior called Peircean in [Prior, 1967], we can then build up temporal languages using operators with tables to specify their semantics. For example, we could use a tomorrow operator X if the tree $(T, <)$ is discrete as well as an 'always-going-to-be' operator G. If $V : \mathcal{L} \longrightarrow 2^T$ is a valuation then we would have the semantics of the language with X defined by such clauses as:

- $(T, <, V), t \models p$ iff $t \in V(p)$;
- $(T, <, V), t \models XA$ iff for all $s > t$ such that there are no u such that $t < u < s$, we have $(T, <, V), s \models A$;
- $(T, <, V), t \models GA$ iff for all $s > t$ we have $(T, <, V), s \models A$.

In Peircean branching time the question of a sea-battle occurring tomorrow, for example, can be analysed using a unary modality X. If p is a proposition standing for 'there being a sea-battle today', then Xp can be read as 'tomorrow there will be a sea-battle'. If it is a contingency that there will be a sea-battle tomorrow then we will have $\neg Xp \wedge \neg X \neg p$ true today.

Although Prior was apparently satisfied with this system, there are two major criticisms of it as well as a minor one.

One criterion is that it fails to represent many everyday instances of reasoning involving time adequately. There is no provision made for a concept of plain future in between necessary future and possible future. I might want to say that there will be a sea-battle tomorrow but it is possible that there will not, i.e. I am making a claim about the contingent future. In a branching temporal structure we can easily model such a situation by having one branch starting from today and moving next to a time when p is true and another branch from today containing a next point when p is not true. However, there is no facility for indicating an 'actual' future.

The other major problem with this approach is only apparent when we need to make complicated nestings of operators. This happens, for example, when complicated reasoning about program behaviour needs to be done in computer science applications. Consider an example in which we want to be able to say formally that 'it is possible that our machine eventually halts after proceeding in such a way that a red light is always immediately followed by a green light'. Now such a requirement could easily be stated using a purpose-built operator but it is certainly not straightforward to express this using a nesting of X operators within an Until operator. This is because the X operators along the way must not be used to say that a green light is necessary after a red light, just that it must occur on the path which is heading towards the halting state.

A minor problem with the Peircean approach concerns the question of 'futurity within the present' recently raised by Ming Xu [Zanardo, 1996]. In the Peircean approach we must suppose that the truth of atomic propositions is entirely point based. Thus we cannot have a proposition representing, for example, 'Stefano chooses a red tent'. This proposition, call it p, contains an element of futurity. To see this, consider the proposition q representing 'Stefano choosing a blue tent'—under the assumption that p and q cannot both happen. In natural language we might want to say 'Stefano now chooses a red tent but it is possible that he is choosing a blue one'. We cannot find a consistent Peircean model of this, however, as only the 'today' in which Stefano chooses a red tent is allowed to be in the past of the 'tomorrow' in which Stefano has just chosen a red tent.

Thus, the Peircean approach must be seen as an inadequate account of branching time for many purposes and for these three reasons we turn next to a more sophisticated approach. However, it should be recalled that it is often the case that a simple approach is best for a simple application. An axiomatization for this logic appears in [Zanardo, 1990].

3.2.2 *The Ockhamist approach*

Concerns about the actual future, about relativizing the semantics to a particular branch or about non-trivial possibilities for the present state of the world lead us to want to consider languages with semantics which pay attention to particular branches in the tree. Thus we turn to the other system which Prior studied in detail, his so-called Ockhamist approach. This was inspired to some extent by the fourteenth-century theologian William of Ockham, who believed (amongst other things!) that future contingents could still be true even if humans did not know that they were.

Truth in the Ockhamist system is defined relative to points on histories. There are several equivalent ways of formalizing this and several variations on these. One way is to introduce the set $\mathcal{H}$ of all histories on a tree $(T, <)$ and a valuation V which maps atoms to sets of pairs $(t, \pi) \in T \times \mathcal{H}$ such that $t \in \pi$. Then we define truth of propositions via:

- $(T, <, V), t, \sigma \models p$ iff $(t, \sigma) \in V(p)$;

recursively we build up a propositional logic:

- $(T, <, V), t, \sigma \models \neg A$ iff we do not have $(T, <, V), t, \sigma \models A$;
- $(T, <, V), t, \sigma \models A \wedge B$ iff both $(T, <, V), t, \sigma \models A$ and $(T, <, V), t, \sigma \models B$;

and we add some temporal operators such as:

- $(T, <, V), t, \sigma \models XA$ iff there is $s \in \sigma$ such that $t < s$, there is no u with $t < u < s$ and $(T, <, V), s, \sigma \models A$;
- $(T, <, V), t, \sigma \models FA$ iff there is $s \in \sigma$ such that $t < s$, $(T, <, V), s, \sigma \models A$;
- $(T, <, V), t, \sigma \models \Box A$ iff for all histories τ through t, $(T, <, V), t, \tau \models A$.

In this approach we can render 'there will be a sea-battle tomorrow but it is not necessary' as $Xp \wedge \Diamond X \neg p$ once we have defined 'tomorrow' X in the obvious way and defined the abbreviation $\Diamond A$ as $\neg\Box\neg A$. In [Belnap and Green, 1994], there is an interesting discussion of philosophical problems with the concept of actual future.

Futurity in the present (*à la* Ming Xu) can be easily handled with a slight variation by having valuations which, like more complicated formulae, depend on histories as well as points for their truth. See [Di Maio and Zanardo, 1994] and [Prior, 1967, pp. 123–124] for discussions on this. Also note that if we do not allow such valuations then the usual substitution rule becomes invalid for Ockhamist logic: for example, $p \to \Box p$ is a validity whose validity is not preserved under substitutions. An interesting new approach to futurity within the present can be found in [Zanardo, 1996].

Note that in the literature such as in [Thomason, 1984] and [Di Maio and Zanardo, 1994] there are various other ways of presenting the semantics for variations of Ockhamist logis. For example, we can base the semantics on truth at branches with an initial point. There is no known effective axiomatization for Prior's Ockhamist logic.

3.2.3 *Bundled Ockhamist logics*

For various reasons, including the current lack of an axiomatization, more general Ockhamist branching time logics have been suggested in [Burgess, 1979], [Zanardo, 1985] and [Øhrstrøm and Hasle, 1996]. These involve restricting our attention to some, but not necessarily all, of the histories in a tree structure. The histories of interest are called *bundles* and so we have *bundled Ockhamist logics.* In [Øhrstrøm and Hasle, 1996] such a logic is called *Leibnizian* as it is argued that Leibniz had a conception of possible worlds which equates to our notion of possible alternative histories.

There are also philosophical reasons to allow the set of possible histories to be smaller than the set of all histories in a tree. Consider the following example based on one in [Øhrstrøm and Hasle, 1996] in the style of those in [Nishimura, 1979]. It may be claimed that it is consistent to believe both the following:

- life on earth will necessarily end; and
- it is necessarily always the case that if there is life on earth then it is possible for there to be life on earth in the next second.

However, it is readily seen (under certain very plausible assumptions about the height and discreteness of the tree of possible states of the universe), that these statements are inconsistent in the Ockhamist model. This is because the second statement implies that there must be a history in the tree in which life goes on for ever.

In computer science applications bundled logics can sometimes be useful when only histories—that is, computation runs—which obey certain fairness criteria are actually considered in the semantics.

When dealing with bundled logics, the original Ockhamist logics as described in the last section are sometimes called *full* Ockhamist logics as the bundle of histories which define the semantics is in fact the set of all histories in the frame. All of the validities of bundled Ockhamist logics are validities of the full logic. However, there are validities of the full logic which are not validities in the bundled logic. The following formula seems quite an important example of such a formula:

$$\begin{aligned} &\Diamond Fp \wedge (\Box G(p \rightarrow \Diamond Fp)) \\ &\rightarrow (\Diamond (Fp \wedge G(p \rightarrow Fp))). \end{aligned}$$

For details, including semantics and axiomatizations, of bundled logics see Section 7.7 of Volume 1 or see the comprehensive account in [Zanardo, 1996].

3.3 Computer science and AI applications

As reasoning about non-determinism and time is fundamental to both artificial intelligence and computer science, it is not surprising that branching time is widely used in these areas. We will look briefly at AI uses before spending the rest of the chapter showing how branching time is used in formal approaches to software engineering.

In describing or modelling the reasoning of an agent or agents faced with an unpredictable environment, it is clear that branching time temporal logic might have a role to play. In many applications, in particular those involving planning, it is useful for time to involve even a branching past (see e.g. [McDermott, 1982]) but the reasoning of an agent which observes THE past and acts to help determine one future among many possibilities can best be represented using (future) branching time.

Examples of such applications can be found in [Ladner and Reif, 1986] where the individual agents may or may not be able to reason about the branching nature of the universe of possibilities which their system can exhibit, and in [Rao and Georgeff, 1993] in which the branches represent choices of actions by the agent and alternative environmental possibilities are rendered by having separate whole trees.

Of course, in all these applications it is necessary to have other modal operators in the language to allow reasoning about the agent's knowledge or intentions. The interaction of the various modalities including the temporal operators can often be very complex. A very useful guide to the many logics which can arise in such ways can be found in [Halpern and Vardi, 1989]. Here, the authors classify 96 different logics of knowledge and time and investigate their decidability and computational complexity. The logics are classified according to both language employed and assumptions about the underlying distributed system. So we have logics which are based on linear as well as branching time, logics which assume that agents do not forget, logics which assume that there is a fixed initial state, etc.

For more details on such applications see [Xu, 1994], [Belnap, 1996] and [Chellas, 1992].

As with linear time temporal logics, one of the main applications for branching time temporal logics within computer science is as a formalism for specifying the behaviour of systems. When a desired behaviour is formally specified, we can proceed to check whether a given system actually implements the specification (i.e. is correct with respect to it) or we might try to synthesize a correct implementation from the specification. It is hoped that, to some extent, both of these activities will be able to be assisted automatically.

In the next section we will meet some of the main branching time temporal logics which are used in specification and verification. These include CTL which was first described in [Clarke and Emerson, 1981] and CTL* which was proposed in [Emerson and Halpern, 1986]. However, branching time logics for computer science were being proposed as early as those in [Lamport, 1980], [Emerson and Clarke, 1980] and [Abrahamson, 1980]. These proposals led to a productive debate on the merits of using branching time logics compared with linear time logics for the particular purposes of specification and verification.

Without going into the history or details of the debate, we can give a brief summary of its nature. For more details see [Lamport, 1980], [Pnueli, 1985] and [Emerson and Halpern, 1986]. The main questions debated concern efficiency of reasoning and expressiveness of the rival approaches.

One of the main arguments for using branching time is that simple branching logics like CTL are adequate to express a certain set of useful properties but, in contrast to the exponential complexities of model checking with a linear temporal logic, model checking with CTL is of linear complexity. We will examine this task in the next section.

However, reasoning with and model checking with the much more expressive CTL* is as hard or harder than with a linear temporal logic. In fact, CTL* includes the usual (future-only) linear temporal logic. It is probably now generally accepted that the linear fragment is sufficient for verifying the correctness of given programs—even concurrent ones. Correctness usually means that all possible runs of the program satisfy some linear temporal formula. Below we will show that this is true even for systems which are usually specified by a branching logic.

On the positive side of branching time logics, as pointed out in [Emerson and Halpern, 1986], there are a few disadvantages to only using linear time temporal logics. One is that such a logic cannot express the property of there being a run of a system which satisfies a certain linear time formula. This property is just the negation of a correctness property but not being able to state or prove the incorrectness of a program or system may be a serious problem. Another disadvantage of linear time temporal logics is that they are not able to express properties concerning the existence of alternative computation pathways. This can be a problem if the specifier wants positively to require certain forms of non-determinism.

For these reasons, there has been much research into the use of branching time temporal logics in computer science as we will see in the next section.

Let us now look at an example in which branching time turns out not to be necessary despite first appearances.

Consider two boxes α and β each sporting an array of three buttons a, b and c. With box α we find that we can press button a and then either button b or button c. No matter which choice of b or c we make, we find that whenever we use box α it allows us to press a followed by that choice of b or c. Box β, on the other hand, always allows us to press a followed by one of b or c but it sometimes only allows us to press b after a and at other times only allows us to press c after a.

Following Milner in [Milner, 1989], we might like to distinguish α and β using a branching semantics to capture the difference in possible behaviours. See Fig. 3.1.

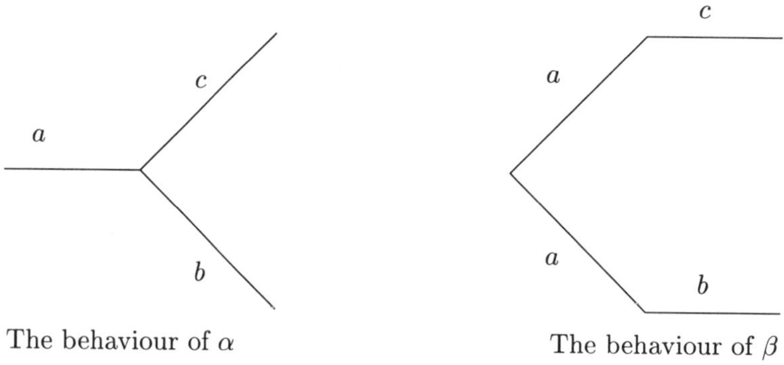

FIG. 3.1. The behaviours of α and β.

It is clear that we can give such a branching time semantics to describe the behaviour of buttoned boxes like α and β, and that we may wish to specify a class of modules which includes α but does not include β. Since the set of branch labellings of our two boxes are identical, $\{ab, ac\}$ in fact, it seems that a linear time temporal language will not be adequate for the task and we will have to use a branching time language.

However, as we will now see, it is possible to specify α in a linear way. The answer is that basically our formalization above was a little confused over the extent of observable behaviour. Let us be more careful. Let A be the event of the environment (the button presser) trying to press button a. Similarly for B and C. Let a be the event of button a actually being depressed and similarly for b and c. Then it is clear that box β can exhibit a run in which $A \wedge a$ is followed by $B \wedge \neg b$. Box α has no such run. If we want to include α and disallow β then we must just rule out such a run. In fact, the Until and Since linear time formula

$$U(A \wedge U(B, \neg(A \vee B \vee C)), \neg(A \vee B \vee C))$$
$$\rightarrow U(A \wedge U(B \wedge b, \neg(A \vee B \vee C), \neg(A \vee B \vee C))$$

will do.

3.4 CTL*

The first branching time logics proposed for use in computer science were 'Peircean' in the sense that the truth of formulae was evaluated at points on a branching structure. These early languages included those in [Lamport, 1980], [Queille and Sifakis, 1982] and [Abrahamson, 1980] as well as CTL ([Clarke and Emerson, 1981]) and UB ([Ben-Ari *et al.*, 1981]) which we will meet below.

In response to criticisms that CTL was not sufficiently expressive, the language CTL* was introduced in [Emerson and Halpern, 1986] to extend and include both CTL and the usual linear time temporal logic. CTL*, which we will define formally below, has since become the most widely used branching time temporal logic in computer science. It is an Ockhamist-style logic, with truth being evaluated at points on paths.

The logics mentioned above are all restricted to having only future-time temporal connectives. Recently, a logic PCTL* was introduced in [Laroussinie and Schnoebelen, 1994] which also incorporates past-time connectives. Although this does not necessarily increase the expressive power, it does allow natural and convenient expression of practical properties.

As PCTL* extends CTL*, we will introduce PCTL* first and then describe how to restrict it to other languages.

We build formulae out of atomic propositions (including $\top$) recursively using the classical $\neg$ and $\wedge$, the unary operator E and the binary operators U and S.

As well as the usual classical abbreviations we have $X\phi = U(\phi, \bot)$, $Y\phi = S(\phi, \bot), F\phi, P\phi, G\phi, H\phi, \overset{\infty}{F}\phi = GF\phi$, and $A\phi = \neg E(\neg\phi)$.

Formulae of PCTL* are evaluated at nodes on branches of labelled discrete rooted trees of height ω. We write $\mathcal{T}, x, \sigma \models \phi$ whenever ϕ holds at the node x on the branch σ of structure $\mathcal{T} = (T, <, h)$. More formally this is defined recursively by:

$\mathcal{T}, x, \sigma \models p$	iff	$t \in h(p)$;
$\mathcal{T}, x, \sigma \models \top$;		
$\mathcal{T}, x, \sigma \models \neg\phi$	iff	$\mathcal{T}, x, \sigma \not\models \phi$;
$\mathcal{T}, x, \sigma \models \phi \wedge \chi$	iff	$\mathcal{T}, x, \sigma \models \phi$ and $\mathcal{T}, x, \sigma \models \chi$;
$\mathcal{T}, x, \sigma \models U(\phi, \chi)$	iff	there is a node y in the branch σ such that $x < y$ and $\mathcal{T}, y, \sigma \models \phi$ and for all nodes z, if $x < z < y$ then $\mathcal{T}, z, \sigma \models \chi$;
$\mathcal{T}, x, \sigma \models S(\phi, \chi)$	iff	there is a node y in the branch σ such that $y < x$ and $\mathcal{T}, y, \sigma \models \phi$ and for all nodes z, if $y < z < x$ then $\mathcal{T}, z, \sigma \models \chi$;
$\mathcal{T}, x, \sigma \models E\phi$	iff	there is a branch τ containing x such that $\mathcal{T}, x, \tau \models \phi$.

In many accounts of CTL*, we see reference to the path/state dichotomy. The truth of some formulae of PCTL* does not depend on the path on which they are evaluated. Some of these can be identified syntactically. We define the set of *state* formulae recursively as follows:

- atomic propositions are state formulae;
- if ϕ and ψ are state formulae then so are $\neg\phi$ and $\phi \wedge \psi$;
- if ϕ is any formula then $E\phi$ is a state formula.

It is easy to check that these formulae are path independent. In order to emphasize this property of state formulae general PCTL* formulae are often called *path* formulae.

The language CTL*, which we have seen is the most common branching time temporal logic for computer science applications, contains just the formulae of PCTL* which do not contain the past operator S. It is clear that the usual (future-only) language of linear time is a sublanguage.

To be entirely accurate, it should be noted that CTL* is usually defined to contain both the *non-strict* Until operator (which we will call) U^*

$$U^*(\alpha, \beta) \equiv \alpha \vee (\beta \wedge U(\alpha, \beta))$$

and the tomorrow operator X. However, it is clear that the sets of operators are easily interdefinable.

The early 'Peircean' logic UB consists of just those formulae of CTL* built from atoms recursively using only the classical connectives and the unary constructs $EX\phi$, $EF\phi$, $AX\phi$ and $AF\phi$.

CTL formulae are obtained from atoms by recursively using the classical connectives and $EX\phi$, $AX\phi$, $EU(\phi, \psi)$ and $AU(\phi, \psi)$.

There is another similar language ECTL whose formulae are obtained from atoms by recursively using the classical connectives and $EX\phi$, $AX\phi$, $EU(\phi, \psi)$, $AU(\phi, \psi)$, $E \overset{\infty}{F} (\phi)$ and $A \overset{\infty}{F} (\phi)$.

Notice that all formulae of ECTL (and so of CTL and UB) are state formulae. We could just evaluate them at nodes of structures without specifying a branch.

We can attempt to make these languages more expressive—while still keeping all their formulae as state formulae—by allowing classical connectives in between the temporal connectives and the path connective above. This gives us languages UB$^+$, CTL$^+$ and ECTL$^+$. UB$^+$, for example, contains the formula $A(Fq \to Xp)$.

We have the following comparisons in expressive power:

$$UB < UB^+ < CTL \equiv CTL^+ < ECTL < ECTL^+ < CTL^*.$$

See [Emerson, 1996] for details. Obviously if we add past time operators then we are not decreasing expressive power. However, it is shown in [Hafer and Thomas, 1988] and [Laroussinie and Schnoebelen, 1994] that adding past operators does not increase expressiveness: it just makes it easier and more convenient to express useful properties.

Note that there are future-only branching time logics even more expressive than CTL*. As we did to linear time temporal logics in Chapter 8 of Volume 1, we find the addition of fixed point operators to branching temporal logics a natural extension. One very useful logic based on the use of fixed points is the μ-calculus of [Kozens, 1983] and [Emerson and Clarke, 1980]. See [Emerson, 1996] and [Stirling, 1992] for excellent surveys.

In many of the most useful applications of branching time temporal logics to computer science it is convenient to formalize the application in a slightly different way than the tree structures above. Although the trees are still definable it is useful to regard a set of states with a transition relation as the more basic object. If we have a set S of states and a binary relation $R \subseteq S \times S$ then we can get a tree by looking at all the finite sequences $s_0, s_1, s_2, \ldots, s_n$ which begin with a particular state s_0 and go on satisfying $(s_i, s_{i+1}) \in R$. The set T of such finite sequences can be ordered by prefixing $s_0, s_1, \ldots, s_n < s_0, s_1, \ldots, s_n, s_{n+1}, \ldots, s_m$. A *fullpath* of (S, R) is just an infinite sequence $s_0, s_1, s_2, \ldots$ with each $(s_i, s_{i+1}) \in R$. Given a valuation V for the propositional atoms on the states, it is possible to define truth directly for PCTL* formulae at points s_i on fullpaths $s_0, s_1, s_2, \ldots$.

Although this is just a different, but equivalent, way of presenting discrete ω-height branching logics, this transition system approach can be easily generalized to take account of applications in which there are several different named transitions between states. This allows convenient representation of fairness conditions on transitions. See the discussions on transition systems in [Stirling, 1992], [Pratt, 1976], [Pnueli, 1977], [Abrahamson, 1979], [Lamport, 1980], and [Emerson and Clarke, 1980].

3.4.1 *Reasoning with branching time logics*

The main computer science applications of branching time logics involve the following tasks: deciding the validity (or satisfiability) of a formula, checking that a given finite transition system is a model of a given formula and, given a satisfiable formula, constructing, or *synthesizing*, a finite transition system model for it.

In Chapter 15 of Volume 1, we have already encountered techniques for deciding the validity of branching time formulae. These include Rabin's result ([Rabin, 1969]) on S2S as well as the tree decidability result in [Gurevich and Shelah, 1985]. However, for the specific cases of discrete trees of height ω, and the languages CTL or CTL*, several less complex procedures have been developed. We have a tableau-based deterministic exponential time complete procedure for CTL satisfiability in [Emerson and Clarke, 1982]. In [Emerson and Sistla, 1984] we find a deterministic double-exponential time complete procedure for CTL* satisfiability. The basic idea here is to find a tree automaton which accepts the models of the given CTL* formula exactly, to find a deterministic equivalent and to check whether that is empty (i.e. accepts no tree structures).

We saw that one of the main arguments for using branching time logics in computer science is that the model-checking procedure is very efficient. The task here is to represent a given (finite state) system as a Kripke structure and check that it is a model of a given specification, i.e. formula. See, for example, [Clarke *et al.*, 1986], where, for CTL formulae, the complexity of the procedure (essentially computing a fixed point set of states) is linear in the size of the specification and the size of system. CTL* is much more complex to check as it needs a recursion involving checking of all paths from a particular state at each round. See [Emerson, 1996] for a summary.

In the rest of this chapter we turn to consider the question of synthesis.

3.5 Reactive modules

There is a lot of theory and practice surrounding the justification of the traditional input-to-output transformational computer program. The rest of this chapter is not about such programs: it is about complex systems which carry on in an ongoing interaction with their environment. This includes such systems as automated, distributed share dealing and ambulance dispatching ones as well as many other examples ranging from drink vending machines to Gosplan.

As usual, it is assumed that a formal description or specification of the requirements for the system is made. So we need a formal language. There is a very wide range of such formalisms available from early ideas of listing *liveness* and *soundness* properties of the system to process algebras and Petri nets. Each approach has its advocates. However, it appears that temporal loigc is one formalism which is attracting much attention and which, unlike other alternatives, makes a clear distinction between specification and internal implementation details. In temporal logic we use a surprisingly simple and natural language capable of expressing and reasoning about the intricate details of the expected behaviour of processes over time. It is natural because it derives from the temporal components of natural language and that it is expressive is more than adequately shown in the fairly recent introduction [Manna and Pnueli, 1992] which catalogues a long list of temporal specifications for common properties and often used constructs.

Modularity and division of labour are important aspects here. They are even more important if, as in [Barringer, 1987], we expect the building of the system to be carried out in a modular manner. The difficulty of combining an account of modularity was once recognized as a disadvantage of the temporal logic approach to specifications, but the clear and neat system outlined in [Manna and Pnueli, 1992] shows that this problem has now been comfortably solved. Within this paradigm, the following distinction plays a pivotal role: there are concerns about the behaviour of the individual components—or *reactive modules*—and there are concerns about how they are put together. Provided that two modules exhibit *interface compatibility* we can simply put them together to form a system and deduce the overall behaviour as just the conjunction of the individual behaviours (we summarize this result below). Interface compatibility concerns the question of whether a module or its environment (possibly including another module) has control of a particular communication channel.

In this section we consider the questions of *implementability* of a reactive module specification and *synthesis* of a module from a specification. These are, respectively, the question of whether it could be at all possible to build a module which implements a given specification and how we can automatically describe the building of a module to implement a given specification.

One factor which is relevant to these questions is how, precisely, the module communicates with its environment. Some possible variations are the synchronicity of output from various modules (and/or the environment) and whether the output has to be noticed or not. In a synchronous communication protocol the module sends out its output simultaneously with the arrival of input from the environment. In an asynchronous protocol the module sends its output out whenever it wants to and the environment sends the module messages whenever it wants to. In a shared variable system, the communication channels are variables whose values can be changed by their owners and read by other modules. In a message-passing protocol, messages are sent which queue up to get read.

We will see below that a particular specification may or may not be implementable, depending on the communication protocol chosen.

In [Pnueli and Rosner, 1989a] and [Pnueli and Rosner, 1989b] we find these questions of implementability and synthesis answered in the case of a shared variable communication protocol in the synchronous and asynchronous case respectively. In both these papers a decision procedure is given for the question of implementability of a given specification and a method is described for automatically synthesizing an implementation in case the specification is implementable.

3.5.1 *Formal approach to reactive modules*

For us, a reactive module will interact with its environment via some kind of interface. In the simple situation which we consider here, the interface will consist of a finite number of communication channels each of which can convey a binary message at a particular time. Some of the channels are under the control of the module while the rest are under the control of their environment.

Once we have the ability to define a reactive module, it is not a great step further to generalize the idea of a distributed system with several modules acting concurrently. This distributed system may be closed or it may itself be immersed in an environment. The modules within the system are just the ordinary reactive modules we have described above but the environment of a particular module is now the extra-system environment in combination with all the other modules. But the whole system is itself just an ordinary module as well. In fact the possibility of a hierarchy of modules opens up here. At its depth are atomic modules implemented, perhaps, in some programming language, but at all other heights a module at one level turns out to be a distributed system of modules when we look into it. In this chapter we will be concentrating on the problems of implementing an individual module but it is important to consider the possible nature of its environment.

Although the exact details of our problem of implementabilty have to wait until we specify the details of the communication interface, we can see that the behaviour of a particular module (and its environment) can be described in our temporal logic. We simply choose a propositional atom for each of the interface channels and use its truth at a particular time to represent its channel being 'on' or being used at that time. Thus, a formula of the temporal logic built from these atoms can be a formal specification of the desired behaviour of a module.

As well as the actual temporal formula, the question of control of the communication channels is most important for determining whether a temporal specification is implementable or not and we will require the details of control to be part of the overall specification. For example, if the channel q is under control of the module then we can build a module guaranteeing that the formula Fq is satisfied. However, if the channel q is under the control of the environment then Fq cannot be guaranteed to be satisfied.

In [Manna and Pnueli, 1992] we find the formalization of these ideas of control of communication channels: a *module specification* is defined as a pair consisting of:

- an *interface specification* being details of who controls the various channels; and
- a temporal logic formula representing the desired behaviour of the module in its environment.

The focus on interfaces is crucial for considering composing modules and doing modular proofs of correctness of systems. When putting modules together in one system it is important to realize that sometimes this is not possible because of their respective interface specifications: for example, when both A and B have control of p. We say that a set of module specifications are *interface compatible* if and only if the details of channel control contained in the various interface specifications are not contradictory.

In case we have several modules $M_1, \ldots, M_n$ acting together in a system (perhaps with its own extra-system environment) we write $M_1 || \ldots || M_n$ to represent the system. We do not want to try to formalize this here as it is not our main concern, but the interested reader is referred to [Manna and Pnueli, 1992] for details.

As in [Manna and Pnueli, 1992] we say that a temporal sentence ϕ is *modularly valid* for a module M if and only if for any module E which is interface compatible with M, any behaviour of $M||E$ is a model of ϕ.

In [Manna and Pnueli, 1992] we have the very useful claim that:

> If formulas ϕ_1 and ϕ_2 are modularly valid over modules M_1 and M_2, respectively, then $\phi_1 \wedge \phi_2$ is modularly valid over $M_1||M_2$.

In proofs of correctness of a complex system, one can make great use of this rule.

We must now leave the investigation of complex systems and concentrate on single modules. In particular we want to know when, for a given formula ϕ, there is any module M such that M is modularly valid over ϕ. This is the question of the implementability of ϕ (along with an interface specification).

Notice that the question of implementability is not the same as the question of satisfiability. If q is under the control of the environment then there are models of Fq (i.e. Fq is satisfiable) but Fq is not *implementable* as we cannot build a module to guarantee a model.

We define implementability more carefully in the next section but here we will see that we must be even more careful about the nature of the communication channels before the issue is quite clear.

Consider the formula $G(q \rightarrow Fp) \wedge G(p \rightarrow Pq)$ with q under the control of the environment and p under the control of the module. Is this specification implementable? Well, the answer is that it depends. It depends on whether we can guarantee that the module will notice whether q has been made true or not.

In [1989b], Pnueli and Rosner consider the situation in which the communication channels are 'shared variables'. This means that both the module and the environment can look at the contents of the channel (its truth or falsity in the Boolean case) but only one of them can change the contents as specified by the interface specification. The specification above is not implementable in this set-up. M might not notice if the environment makes q true quickly and then false again shortly after. No physically realizable module could check often enough to guarantee not to miss such an event in the face of any mischievous environment.

On the other hand, suppose now that the communication channels are such that messages can be sent down them and are guaranteed to be delivered. In this case, if q true means that a q message is sent, then we can eventually guarantee to send a p message. This is the message passer protocol. In this protocol the above specification *is* implementable.

It is worth looking briefly at message passing in its full generality, before we return to a very specific case of it. We only require the guaranteed eventual delivery of a message—changing the value of a shared variable does *not* count as sending a message because the value could be changed back again before anyone notices. Within this paradigm are many examples already mentioned in the literature. They include channel systems such as that in [Hoare, 1985] and the actor model of [Agha, 1986] in which messages pass from a specified sender to a specified receiver. There are also broadcast message-passing arrangements such as that CMP in [Fisher, 1993] in which sent messages get delivered to a whole set of modules which have previously indicated general interest in messages of that type.

In the next section we will be considering the question of implementing message passers. There, this will be done on an abstract a level as possible making only the most essential concessions to physical limitations. However, it is interesting to note that many ways of actually building such module systems exist. As well as the broadcast message system in [Fisher, 1993] we have actors of [Agha, 1986] in a direct implementation in [Lieberman, 1986] and actors used as the method of introducing concurrency into such languages as ABCL [Yonezawa, 1990] and concurrent Prolog in [Prasad, 1991] and [Kahn *et al.*, 1987].

Another type of object-based message-passing system is the Linda parallel programming paradigm of [Gelerter *et al.*, 1985]. With its central tuple space this provides another mechanism for communication between modules—here called processors. The processors themselves can be programmed in a variety of languages but only interact with the tuple space through special Linda operators.

Although we will be looking at questions concerning a single message-passing module, one final aspect of the communication protocol is relevant. This is the question of whether all the modules send out their messages at once or whether they send out messages whenever they want to. This affects the individual module as in the latter, *asynchronous*, case messages will be arriving (from other modules and the extra-system environment) at all sorts of times while the module is composing its next message to send.

In the former *synchronous* case the module receives a whole bag of messages and sends out a whole bag of messages as the clock ticks. It is sensible to use a temporal logic of discrete time here—with such operators as 'tomorrow' (defined by Xp iff $U(p, \perp)$) and its dual 'yesterday'. With only some minor modifications (to take into account the fact that a module's message sent at a particular moment cannot depend on the messages received at that moment) the results in [Pnueli and Rosner, 1989a] can be used here.

In the asynchronous case, there are still discrete intervals in which no module sends any messages but to use a discrete frame of time to underlie the reasoning is clearly a mistake when we are considering putting the modules in parallel with other systems. For example, the module M which sends out a p message every second will be represented as exhibiting quite different behaviours depending on the behaviour of the environment. If the model of time is the natural numbers and i the time of the ith move then we may have M sending the message p every even time (if the environment sends a message every half second), we may have M sending p on all but those times i divisible by 10 (if the environment sends out messages at real time 9.5 seconds and every 10 seconds after that) or we may have even more seemingly bizarre behaviours—all from the same well-behaved module.

For these reasons, in this case we should specify module behaviour in a language appropriate for real numbers time and some of the work would concern converting real-time behaviour of the basically discrete type we expect from modules into the properly discrete representation needed for reasoning on certain types of discrete trees. Using discrete trees such as those used in [Pnueli and Rosner, 1989a] will allow us to represent possibilities by branching.

3.5.2 *Definition of implementation in synchronous case*

In the synchronous case, the distinction between message passing and shared variables is not important. We consider a finite set of propositions: some under the control of the environment, the rest under the control of the module. Say that the finite set P contains the environment's propositions and the finite set Q contains the modules propositions.

As we are considering the synchronous case, we deal with natural numbers time. A run of a module will thus be a structure $(\mathbb{N}, <, \nu)$ where ν is a labelling of $\mathbb{N}$ by sets of atoms from $P \cup Q$. Note that we could just as well use a valuation $g : (P \cup Q) \to \mathbb{N}$ of the atoms instead of the labelling ν but the labelling approach will fit in better with our labelling of trees below. Without loss of generality we may identify a module with its set of possible runs. Of course, we expect the module to exhibit some sort of behaviour no matter what the environment does, so we require that for each labelling $\mu : \mathbb{N} \to 2^P$, there is a labelling $\nu : \mathbb{N} \to 2^Q$ such that $(\mathbb{N}, <, \mu \cup \nu)$ is a possible run of the module. Here we define $\mu \cup \nu$ via $(\mu \cup \nu)(n) = \mu(n) \cup \nu(n)$.

Definition 3.1. Suppose that P and Q are disjoint finite sets of atoms and that ϕ is a linear temporal formula using only the atoms from $P \cup Q$.

We say that $\phi(P, Q)$ is implementable (by a synchronous reactive module) iff there exists a set M of labellings $\pi : \mathbb{N} \to 2^{P \cup Q}$ such that:

- for each labelling $\mu : \mathbb{N} \to 2^P$, there is a labelling $\nu : \mathbb{N} \to 2^Q$ such that $(\mu \cup \nu) \in M$; and
- for each $\pi \in M$, $(\mathbb{N}, <, \pi) \models \phi$.

For example, if $P = \{p\}$ and $Q = \{q\}$ then the following formulae are implementable:

$$G(p \leftrightarrow Xq)$$
$$G(S(p,p) \leftrightarrow q);$$

but these are not:

$$G(p \leftrightarrow q)$$
$$G(q \leftrightarrow Xp).$$

3.5.3 *A reminder of tree automata*

For a particular $k > 0$, we introduce the k-ary infinite tree. The tree $\mathcal{T}_k$ is just the set of all sequences from the alphabet $\Delta_k = \{\beta_0, \ldots, \beta_{k-1}\}$ including the empty sequence ε. We write $\sigma^\wedge \rho$ for the concatenation of sequence σ followed by sequence ρ.

For historical reasons we will switch now to a language Σ of letters rather than keep using a language of propositional atoms. The nodes of trees will be labelled by a single letter from Σ. In order to apply the results in this section we will later have to take the alphabet Σ to be 2^P where P is the set of atomic messages or propositions.

If Σ is a finite alphabet then a k-ary Σ-tree is a pair $(\mathcal{T}_k, \nu)$ where ν is a map from $\mathcal{T}_k$ into Σ. Call ν a Σ-labelling of $\mathcal{T}_k$. If Σ is actually the direct product $\Delta \times \Xi$ of other alphabets, and the label $\mu \times \pi$ of $\mathcal{T}_k$ is given by $(\mu \times \pi)(x) = (\mu(x), \pi(x))$, then we introduce $(\mathcal{T}_k, (\mu \times \pi))$ as $(\mathcal{T}_k, \mu, \pi)$.

For the purposes of this chapter it is most convenient to use the Pnueli–Rosner approach to tree automata which can easily be shown to be of equal expressive power as the Rabin tree automata we met in the last chapter.

A k-ary Σ-*tree automaton* is a 4-tuple $M = (S, T, S_0, \phi)$ where:

- S is a finite non-empty set called the set of *states*;

- $T \subseteq S \times \Sigma \times \Delta_k \times S$ is the *transition table*;
- $S_0 \subseteq S$ is the *initial state set*; and
- ϕ is a propositional formula built from atoms in S called the *acceptance condition.*

Tree automata get to work on k-ary Σ-trees. Below we use a game to define whether or not the tree automaton M accepts the tree $L = (\mathcal{T}_k, \nu)$.

The game $\Gamma(M, L)$ is played between the automaton M and a player called *Pathfinder* on the tree L. The game goes on for ω moves (starting at move 1). The ith move consists of M choosing a state q_i from S followed by Pathfinder choosing a direction $\delta_i \in \Delta_k$. M must choose q_i so that:

- $q_1 \in S_0$; and
- for each $i \geq 1$, $(q_i, \nu(\delta_1 {}^\wedge \ldots {}^\wedge \delta_{i-1}), \delta_i, q_{i+1}) \in T$.

We can view a play of the game as being directed along the branch $\delta_1 {}^\wedge \delta_2 {}^\wedge \ldots$ of $\mathcal{T}_k$. We will say that M is in state q_i at node $\delta_1 {}^\wedge \ldots {}^\wedge \delta_{i-1}$ of this branch.

Provided M can always find a state q_i, into which to move on the ith move, a play of the game gives rise to a whole sequence $q_1, q_2, q_3, \ldots$ of states along the branch $\delta_1 {}^\wedge \delta_2 {}^\wedge \ldots$. The criterion for deciding the winner of a play is determined by this sequence as follows. We say that M has won the play $q_1 \delta_1 q_2 \delta_2 \ldots$ if and only if $q_1 q_2 \ldots \models \phi$ where $\models$ is defined by induction on the construction of ϕ from the atomic conditions $q_1 q_2 \ldots \models s$ iff $s = q_i$ for infinitely many i. Otherwise Pathfinder has won. If M cannot move at any stage we also deem that Pathfinder has won.

We say that M *accepts* L if and only if there is a winning strategy for the player M in the game $\Gamma(M, L)$. This means there must be some function f which tells M which state to move into at each node $x \in \mathcal{T}_k$ in such a way that playing $f(\varepsilon), f(\delta_1), f(\delta_1 {}^\wedge \delta_2), \ldots$ wins the play for M along the branch $\delta_1 {}^\wedge \delta_2 {}^\wedge \ldots$.

Note that in the case of 1-ary tree automata, when $\Delta_1 = \{\beta_0\}$ and $\mathcal{T}_1$ is effectively the natural numbers, then we are just dealing with linear automata. However, for ease of conversion, we leave these automata as 1-ary tree automata.

3.5.4 *Unique character entry as tree*

Now we present some techical results about making a tree automaton from a linear automaton.

A deterministic 1-ary $(\Delta_k \times \Sigma)$-tree automaton satisfies the unique character entry condition iff

UCE for each state $q \in S$, there is a letter $a_q \in \Sigma$ such that for all $p \in S$, for all $\delta \in \Delta_k$, for all $b \in \Sigma$, if $(p, (\delta, b), \beta_0, q) \in T$ then $b = a_q$.

Lemma 3.2. *There is an algorithm which given any deterministic linear automaton constructs an equivalent deterministic linear automaton which satisfies UCE.*

Proof Given a deterministic 1-ary Σ-tree automaton M, we build a 1-ary Σ-tree automaton M' simply by giving it a state (a, s) for each state s of M and each $a \in \Sigma$. For each transition (s, a, β_0, s') of M, and each letter $b \in \Sigma$, let M' have a transition $((b, s), a, \beta_0, (a, s'))$. Thus M' acts like M but the state also notes which letter has just been read.

If we give M' initial state (a_0, s_0) where s_0 is the (unique) initial state of M and a_0 is any letter from Σ, and replace each atom s in the acceptance condition of M by $\bigvee_{a \in \Sigma}(a, s)$ to get the acceptance condition of M', then it is clear that M' is deterministic, equivalent to M and satisfies UCE. □

Lemma 3.3. *Let M be any deterministic 1-ary $(\Delta_k \times \Sigma)$-tree automaton which satisfies UCE and δ_0 be any element of Δ_k.*

Then we can construct a k-ary Σ-tree automaton N such that for all Σ-trees $(\mathcal{T}_k, \nu)$, N accepts $(\mathcal{T}_k, \nu)$ iff M accepts each of the branches $(b, <, \mu, \nu)$ of the k-ary $(\Delta_k \times \Sigma)$-tree $(\mathcal{T}_k, \mu, \nu)$ with Δ_k-labelling μ given by

$$\mu(\varepsilon) = \delta_0$$
$$\textit{and} \;\; \mu(\delta_1 \,{}^\wedge \ldots {}^\wedge \delta_i) = \delta_i.$$

Proof Suppose that $M = (S, T, S_0, \phi)$ where $S_0 = \{s_0\}$. Define the k-ary Σ-tree automaton $N = (S, U, S'_0, \phi)$ as follows. The transition table U contains (p, a_p, δ, q) iff $(p, (\delta, a_q), \beta_0, q) \in T$. The initial states of N are some of the second ones in M: $S'_0 = \{s \in S \mid (s_0, (\delta_0, a_s), \beta_0, s) \in T\}$.

($\Rightarrow$) Take a branch b in a tree $(\mathcal{T}_k, \nu)$ accepted by N. Suppose that the elements of b are, in order, $\varepsilon, \delta_1, \delta_1 {}^\wedge \delta_2, \ldots, \delta_1 {}^\wedge \ldots {}^\wedge \delta_i, \ldots$ say.

The winning strategy for N describes a sequence $s_1, s_2, \ldots$ of states to be used along this branch. Thus we know $s_1 \in S'_0$ and for all $i \geq 1$, $(s_i, \nu(\delta_1 {}^\wedge \ldots {}^\wedge \delta_{i-1}), \delta_i, s_{i+1}) \in U$. By definition of U this implies that for all $i \geq 1$, $\nu(\delta_1 {}^\wedge \ldots {}^\wedge \delta_{i-1}) = a_{s_i}$ and $(s_i, (\delta_i, a_{s_{i+1}}), \beta_0, s_{i+1}) \in T$. It follows that, for all $i \geq 0$,

$$\nu(\delta_1 {}^\wedge \ldots {}^\wedge \delta_i) = a_{s_{i+1}} \tag{1}$$

and so for all $i \geq 1$, $(s_i, (\delta_i, \nu(\delta_1 {}^\wedge \ldots {}^\wedge \delta_i)), \beta_0, s_{i+1}) \in T$, i.e.

$$(s_i, (\mu \times \nu)(\delta_1 {}^\wedge \ldots {}^\wedge \delta_i), \beta_0, s_{i+1}) \in T. \tag{2}$$

Now since $s_1 \in S'_0$, we have $(s_0, (\delta_0, a_{s_1}, \beta_0, s_1)) \in T$ but by (1) we thus have $(s_0, (\delta_0, \nu(\varepsilon)), \beta_0, s_1) \in T$ which means $(s_0, (\mu \times \nu)(\varepsilon), \beta_0, s_1) \in T$. Along with (2), this implies that $s_0, s_1, s_2, \ldots$ is a legitimate sequence of states for M to choose in the game along the branch b of $(\mathcal{T}_k, \mu, \nu)$. Since its infinitary behaviour is the same as that of $s_1, s_2, \ldots$ as used in the winning strategy of N, we have that M accepts the branch as required.

($\Leftarrow$) Suppose that M accepts each of the branches of $(\mathcal{T}_k, \mu, \nu)$. We describe a winning strategy for N on $(\mathcal{T}_k, \nu)$. The winning strategy σ will, for each $i \geq 0$, when given a sequence $\delta_1, \delta_2, \ldots, \delta_i$, choose $\sigma(\delta_1, \delta_2, \ldots, \delta_i) \in S$.

We define σ recursively, so that for all $i \geq 0$ the following condition $C(i)$ holds. For any sequence $\delta_1, \ldots, \delta_i$ of elements from Δ_k, if we define $s_1, \ldots, s_{i+1}$ by $s_j = \sigma(\delta_1 {}^\wedge \ldots {}^\wedge \delta_{j-1})$ then $s_1, s_2, \ldots, s_{i+1}$ is the unique choice of states satisfying the condition that for all j such that $0 \leq j \leq i$, $(s_j, (\mu \times \nu)(\delta_1 \wedge \ldots \wedge \delta_j), \beta_0, s_{j+1}) \in T$.

First, $\sigma(\varepsilon)$. Do this by using a winning strategy for M along any branch of $(\mathcal{T}_k, \mu, \nu)$, say $\varepsilon, \beta_0, \beta_0 \wedge \beta_0, \ldots, \beta_0^i, \ldots$. This will give us a sequence $q_0, q_1, \ldots$ of states such that $q_0 = s_0$ and for all $i \geq 0$, $(q_i, (\mu \times \nu)(\beta_0^i), \beta_0, q_{i+1}) \in T$. In particular, if we put $\sigma(\varepsilon) = q_1$ note that we have $C(0)$. The uniqueness follows from the determinism of M.

To define $\sigma(\delta_1, \delta_2, \ldots, \delta_i, \delta_{i+1})$ assume that we have already defined $s_j = \sigma(\delta_1, \delta_2, \ldots, \delta_j)$ (for each $j \leq i$) such that $C(i)$ holds. Now use the winning strategy for M along any branch of $(\mathcal{T}_k, \mu, \nu)$ which begins with $x_0 = \varepsilon, x_1 = \delta_1, x_2 = \delta_1 \wedge \delta_2, \ldots, x_i = \delta_1 \wedge \ldots \wedge \delta_i$. Let us say we use the branch which goes on $x_{i+1} = \delta_1 \wedge \ldots \wedge \delta_i \wedge \beta_0, x_{i+2} = \delta_1 \wedge \ldots \wedge \delta_i \wedge \beta_0^2, \ldots$. This will tell us a sequence $q_0, q_1, q_2, \ldots$ of states to use. But then $q_0 = s_0$ and for all $j \geq 0$, $(q_j, (\mu \times \nu)(x_j), \beta_0, q_{j+1})$. By our induction hypothesis this means that for each $j \leq i$, $q_j = s_j$. We simply put $\sigma(\delta_1 \wedge \ldots \wedge \delta_i \wedge \delta_{i+1}) = q_{i+1}$ and we have condition $C(i+1)$ as required. Again uniqueness follows by the determinism of M.

Now let us check that this strategy is a winning strategy. Suppose that we use it along the branch $x_0 = \varepsilon, x_1 = \delta_1, x_2 = \delta_1 \wedge \delta_2, \ldots$. For all i let $s_i = \sigma(x_i)$. By UCE and $C(0)$, $\nu(\varepsilon) = a_{s_1}$ and so $s_1 \in S_0'$. Also, for each $i > 0$, $(s_i, (\mu \times \nu)(x_i), \beta_0, s_{i+1}) \in T$ so that by definition of μ, $(s_i, (\delta_i, \nu(x_i)), \beta_0, s_{i+1}) \in T$. Thus UCE tells us that for each $i > 0$, $\nu(x_i) = a_{s_{i+1}}$ and by definition of U, $(s_i, a_{s_i}, \delta_i, s_{i+1}) \in U$. Thus for all $i \geq 1$, $(s_i, \nu(x_{i-1}), \delta_i, s_{i+1}) \in U$ and we can see that σ has given us a legitimate sequence of states. That this sequence of states satisfies ϕ follows from the fact that it is also the winning play for M along the branch $x_0, x_1, \ldots$. □

3.5.5 *Full trees*

We will need a 2^Q-tree which exhibits all possible labelling on the successors of each node. If $Q = \{q_0, \ldots, q_{m-1}\}$ is a set with m elements then we will clearly need a k-ary tree where $k = 2^m$. One slight complication is that there are k different trees of this sort depending on the label of the root. Thus define the *full* k-ary 2^Q-tree $(\mathcal{T}_k, f_Q^X)$ with root labelled by $X \subseteq Q$ as follows. Let $\gamma_Q : \Delta_k \longrightarrow 2^Q$ be any bijection: call it the full 2^Q-tree bijection. Then the labelling f_Q^X is given by

- $f_Q^X(\varepsilon) = X$
- $f_Q^X(\delta_1 \wedge \ldots \wedge \delta_i) = \gamma_Q(\delta_i)$.

3.5.6 *An important lemma*

The following result is a slight generalization of a result due to Pnueli and Rosner which does the bulk of work in their paper [Pnueli and Rosner, 1989a].

If P and Q are disjoint sets of atoms, and ψ is a linear temporal formula using only atoms from P and Q, define

$$X_Q = \bigwedge_{\sigma \subseteq Q} \left(\bigwedge_{q \in \sigma} \Diamond X q \wedge \bigwedge_{q \notin \sigma} \Diamond X \neg q \right)$$

and then put

$$\text{impl}(Q, P, \psi) = A\psi \wedge \Box(X_Q \wedge G X_Q).$$

Lemma 3.4. *Suppose that P and Q are disjoint sets of propositional atoms, that the size of Q is m and that $k = 2^m$. Then there is an algorithm A which does the following.*

Suppose that ψ is a linear temporal logic formula using only atoms from P and Q.

Given ψ, A will decide whether or not impl(Q, P, ψ) is satisfiable CTL.*

Furthermore, if it is satisfiable then A will construct an algorithm B which, given a 2^Q-labelling λ_Q on $\mathcal{T}_k$, calculates (in the manner described below) a 2^P-labelling λ_P such that for all branches b,

$$(\mathcal{T}_k, V)\varepsilon, b \models \psi$$

where V is the variable assignment given by

$$\forall q \in Q, V(q) = \{\rho \in \mathcal{T}_k \mid r \in \lambda_Q(\rho)\}$$
$$\forall p \in P, V(p) = \{\rho \in \mathcal{T}_k \mid r \in \lambda_P(\rho)\}.$$

B will calculate λ_P from λ_Q in the sense that for any $x = \delta_1 \hat{} \ldots \hat{}\delta_n \in \mathcal{T}_k$, if we input the sequence $\lambda_Q(\varepsilon)$, $\lambda_Q(\delta_1), \lambda_Q(\delta_1 \hat{}\delta_2), \ldots, \lambda_Q(\delta_1 \hat{} \ldots \hat{}\delta_n)$ then B will output $\lambda_P(x)$.

The complexity of deciding the satisfiability of impl(Q, P, ψ) is at most double exponential in the size of ψ and the size of the algorithm B is also of order double exponential.

Proof

1. Here we describe operation of algorithm A.
   Using the method of Section 2.6, construct a Büchi automaton A_1 for ψ. It accepts a 1-ary $2^{P \cup Q}$-tree $(\mathcal{T}_1, \nu)$ iff $(\mathcal{T}_1, <, \nu), \varepsilon \models \psi$. A_1 is a 1-ary $2^{P \cup Q}$-tree automaton.
   Use the algorithm of Safra [Safra, 1988] to find a deterministic 1-ary Rabin $2^{P \cup Q}$-tree automaton A_2 which is equivalent to A_1. It is clear that we can regard A_2 as just a 1-ary $(2^Q \times 2^P)$-tree automaton.
   Using the full 2^Q-tree bijection $\gamma_Q : \Delta_k \longrightarrow 2^Q$, construct a 1-ary $(\Delta_k \times 2^P)$-tree automaton A_3 much the same as A_2 except that the transition table of A_3 has $(q, (\delta, S), \beta_0, q')$ wherever the transition table of A_2 has $(q, (\gamma_Q(\delta), S), \beta_0, q')$. This means that if μ is a Δ_k-labelling of $\mathcal{T}_1$, then A_3 accepts $(\mathcal{T}_1, \mu, \nu)$ iff A_2 accepts $(\mathcal{T}_1, \gamma_Q \circ \mu, \nu)$.
   Use Lemma 3.2 to construct a deterministic 1-ary $(\Delta_k \times 2^P)$-tree automaton A_4 which satisfies UCE and is equivalent to A_3.

By Lemma 3.3, for each $\delta_0 \in \Delta_k$ we have a determinsitic k-ary 2^P-tree automaton $A_5(\delta_0)$ which accepts a 2^P-tree $(\mathcal{T}_k, \nu)$ iff A_4 accepts all the branches $(b, <, \mu_{\delta_0}, \nu)$ of the k-ary $(\Delta_k \times 2^P)$-tree $(\mathcal{T}_k, \mu_{\delta_0}, \nu)$ with Δ_k-labelling μ_{δ_0} given by $\mu_{\delta_0}(\varepsilon) = \delta_0$ and $\mu_{\delta_0}(\delta_1 \wedge \ldots \wedge \delta_i) = \delta_i$.

Now A uses the algorithm of Theorem 2.59 to check for the emptiness of each $A_5(\delta_0)$. If all of the $A_5(\delta_0)$ are empty then A reports that impl(Q,P,ψ) is *not* satisfiable. Otherwise, A reports that it is. Note that in this latter case the algorithm of Theorem 2.59 has given us, for some $\delta_0 \in \Delta_k$, a deterministic transducer $D(\delta_0)$ which accepts some tree also accepted by $A_5(\delta_0)$.

2. Now we check that A gives us the correct answer about the satisfiability of the formula. First some useful results.

Note X If b is a branch of $\mathcal{T}_k$, $\delta_0 \in \Delta_k$, $X = \gamma_Q(\delta_0)$ and α is a P-labelling then A_4 accepts $(b, \mu_{\delta_0}, \alpha)$ iff A_1 accepts (b, f_Q^X, α). This follows from the way we constructed A_2, A_3 and A_4 and from the fact that $\gamma_Q \circ \mu_{\delta_0} = f_Q^X$.

Note Y If b is a branch of $\mathcal{T}_k$, η is a Q-labelling and θ is a P-labelling then A_1 accepts (b, η, θ) iff $\mathcal{T}_k, \varepsilon, b, V \models \psi$ where V is the variable assignment

$$\begin{aligned} \forall q \in Q, V(q) &= \{\rho \in \mathcal{T}_k \mid r \in \eta(\rho)\} \\ \forall p \in P, V(p) &= \{\rho \in \mathcal{T}_k \mid r \in \theta(\rho)\}. \end{aligned}$$

This follows immediately from the definition of A_1.

Definition 3.5. For a Q-labelling η of $\mathcal{T}_k$, define the map $I_\eta : \mathcal{T}_k \longrightarrow \mathcal{T}_k$ by $I_\eta(\varepsilon) = \varepsilon$ and $I_\eta(\delta_1 \wedge \ldots \wedge \delta_i) = \tau_1 \wedge \ldots \wedge \tau_i$ where each $\tau_j = \gamma_Q^{-1}(\eta(\delta_1 \wedge \ldots \wedge \delta_j))$. It is clear that I_η maps branches to branches.

Note Z Suppose that $\delta_0 \in \Delta_k$, $X = \gamma_Q(\delta_0)$, η is a Q-labelling such that $\eta(\varepsilon) = X$, α is a P-labelling such that $(\mathcal{T}_k, \alpha)$ is accepted by $A_5(\delta_0)$ and $\theta = \alpha \circ I_\eta$.

Then for all branches b of $\mathcal{T}_k$,

$$(\mathcal{T}_k, V), \varepsilon, b \models \psi$$

where V is the variable assignment

$$\begin{aligned} \forall q \in Q, V(q) &= \{\rho \in \mathcal{T}_k \mid r \in \eta(\rho)\} \\ \forall p \in P, V(p) &= \{\rho \in \mathcal{T}_k \mid r \in \theta(\rho)\}. \end{aligned}$$

To prove Note Z, suppose that $A_5(\delta_0)$ accepts $(\mathcal{T}_k, \alpha)$ so that A_4 accepts each branch of $(\mathcal{T}_k, \mu_{\delta_0}, \alpha)$. In particular, A_4 accepts $(I_\eta(b), \mu_{\delta_0}, \alpha)$. By Note X, A_1 accepts $(I_\eta(b), f_Q^X, \alpha)$ but the sequence of labels along this branch is just the same as on (b, η, θ) as assuming $b = \varepsilon, \delta_1, \delta_1 \wedge \delta_2, \ldots$, we have $\eta(\varepsilon) = X = f_Q^X(\varepsilon) = f_Q^X(I_\eta(\varepsilon))$ and

$$\begin{aligned} \eta(\delta_1 \wedge \ldots \wedge \delta_i) &= \gamma_Q(\gamma_Q^{-1}(\eta(\delta_1 \wedge \ldots \wedge \delta_i))) \\ &= \gamma_Q(\tau_i) \\ &= f_Q^X(\tau_1 \wedge \ldots \wedge \tau_i) \\ &= f_Q^X(I_\eta(\delta_1 \wedge \ldots \wedge \delta_i)) \end{aligned}$$

where $I_\eta(\delta_1 \wedge \ldots \wedge \delta_i) = \tau_1 \wedge \ldots \wedge \tau_i$.

Thus A_1 accepts (b, η, θ) and Note Y tells us that

$$(\mathcal{T}_k, \eta, \theta), \varepsilon, b \models \psi$$

where V is the variable assignment

$$\forall q \in Q, V(q) = \{\rho \in \mathcal{T}_k \mid r \in \eta(\rho)\}$$
$$\forall p \in P, V(p) = \{\rho \in \mathcal{T}_k \mid r \in \theta(\rho)\}.$$

This is as required.

2a. First consider if all $A_5(\delta_0)$ are empty. Suppose for contradiction that

$$(\mathcal{T}_k, \eta, \theta), \varepsilon, b \models \text{impl}(Q, P, \psi).$$

Let $\delta_0 = \gamma_Q^{-1}(\eta(\varepsilon))$. Define λ_P in terms of θ, μ_{δ_0} and η.
There are no trees $(\mathcal{T}_k, \lambda_P)$ accepted by $A_5(\delta_0)$. By construction of $A_5(\delta_0)$, A_4 cannot accept each of the branches of $(\mathcal{T}_k, \mu_{\delta_0}, \lambda_P)$.
Suppose that b is a branch not accepted. Then show there is a branch of $\mathcal{T}_k, \eta, \theta$ also not accepted by A1. Thus

$$(\mathcal{T}_k, V), \varepsilon, b \not\models \psi$$

where V is the variable assignment given by

$$\forall q \in Q, V(q) = \{\rho \in \mathcal{T}_k \mid r \in f_Q(\rho)\}$$
$$\forall p \in P, V(p) = \{\rho \in \mathcal{T}_k \mid r \in \lambda_P(\rho)\}.$$

Thus

$$\text{impl}(Q, P, \psi)$$

is not satisfiable.

2b. Now consider the case when some of the $A_5(\delta_0)$ are empty.
Thus, for some $\delta_0 \in \Delta_k$, there is a 2^P-labelling λ_P on $\mathcal{T}_k$ which is accepted by $A_5(\delta_0)$.
Now, given any 2^Q-labelling j_Q on $\mathcal{T}_k$, we define a 2^P-labelling $j_P = \lambda_P \circ I_{j_Q}$.
Note Z tells us that for all branches b,

$$\mathcal{T}_k, \varepsilon, b, V \models \psi$$

where V is the variable assignment given by

$$\forall q \in Q, V(q) = \{\rho \in \mathcal{T}_k \mid r \in j_Q(\rho)\}$$
$$\forall p \in P, V(p) = \{\rho \in \mathcal{T}_k \mid r \in j_P(\rho)\}.$$

Thus

$$\text{impl}(Q, P, \psi)$$

is satisfiable.

3. We now show how to construct B from the $D(\delta_0)$. Given $x = \delta_1 \hat{} \ldots \hat{} \delta_n \in \mathcal{T}_k$, we feed $\lambda_Q(\varepsilon), \lambda_Q(\delta_1), \lambda_Q(\delta_1 \hat{} \delta_2), \ldots, \lambda_Q(x)$ into B. Let $X = \lambda_Q(\varepsilon)$ and suppose $\delta_0 = \gamma_Q^{-1}(X)$.
   B calculates $\tau = I_{\lambda_Q}(x) = \gamma_Q^{-1}(\lambda_Q(\delta_1 \hat{} \ldots \hat{} \delta_i))$ which has the same Q-history in $(\mathcal{T}_k, f_Q^X)$ as x has in $(\mathcal{T}_k, \lambda_Q)$.
   The sequence τ is fed into the algorithm of $D(\delta_0)$ and the set $f_P(\tau) \in 2^P$ is output. This set is also what B outputs as $\lambda_P(x)$.
   By the construction of $D(\delta_0)$, $A_5(\delta_0)$ accepts $(\mathcal{T}_k, f_P)$. But $\lambda_P = f_P \circ I_{\lambda_Q}$ so Note Z tells us that for all branches b of $\mathcal{T}_k$,

$$\mathcal{T}_k, \varepsilon, b, V \models \psi$$

   where V is the variable assignment given by

$$\begin{aligned} \forall q \in Q, V(q) &= \{\rho \in \mathcal{T}_k \mid r \in \lambda_Q(\rho)\} \\ \forall p \in P, V(p) &= \{\rho \in \mathcal{T}_k \mid r \in \lambda_P(\rho)\}. \end{aligned}$$

   This is as required.
4. It is now clear that the complexity of deciding the satisfiability of $\text{impl}(Q, P, \psi)$ is at most double exponential in the size of ψ and the size of the algorithm B is also of order double exponential although its complexity is linear in the length of its input sequence.

□

3.5.7 *Summary*

Theorem 3.6. *Suppose that P and Q are disjoint finite sets. Then we can describe an algorithm X which, when given a linear temporal logic formula ϕ using just atoms from P and Q:*

- *decides whether or not $\phi(P, Q)$ is implementable by a synchronous reactive module;*
- *and if so, describes how to construct such a machine.*

Proof X decides whether or not

$$\text{impl}(P, Q, \psi)$$

is satisfiable.

By the result of Lemma 3.4 there is an algorithm to decide the satisfiability of this formula and this is equivalent to deciding whether or not $\phi(P, Q)$ is implementable by a sequential register machine.

As described in the lemma, there is an algorithm for constructing the machine M in case the formula is a validity. □

4

LABELLED DEDUCTION PRESENTATION OF TEMPORAL LOGICS

4.1 Introducing LDS

This chapter develops proof theory for temporal logic within the framework of labelled deductive systems [Gabbay, 1996].

To motivate our approach consider a temporal formula $\alpha = FA \wedge PB \wedge C$. This formula says that we want A to hold in the future (of now), B to hold in the past and C to hold now. It represents the following temporal configuration:

- $t < d < s, t \vDash B, d \vDash C$ and $s \vDash A$

where d is now and t, s are temporal points.

Suppose we want to be very explicit about the temporal configuration and say that we want another instance of B to hold in the past of s but not in the past or future of d, i.e. we want an additional point r such that:

- $r < s, \sim (r = d \vee r < d \vee d < r)$ and $r \vDash B$.

The above cannot be expressed by a formula. The obvious formula $\beta = F(A \wedge PB) \wedge PB \wedge C$ will not do. We need extra expressive power. We can, for example, use an additional atom q and write

$$\gamma = q \wedge Hq \wedge Gq \wedge F(A \wedge P(B \wedge \sim q)) \wedge PB \wedge C.$$

This will do the job.

However, by far the simplest approach is to allow names of points in the language and write $s : A$ to mean that we want $s \vDash A$ to hold. Then we can write a theory Δ as

$$\Delta = \{t < d < s, t : A, d : C, s : B, r : A, \sim (r < d \vee r = d \vee d < r)\}.$$

Δ is satisfied if we find a model (S, R, a, h) in which d can be identified with $g(d) = a$ and t, s, r can be identified with some points $g(t), g(s), g(r)$ such that the above ordering relations hold and the respective formulae are satisfied in the appropriate points.

The above language has turned temporal logic into a labelled deductive system (LDS). It has brought some of the semantics into the syntax.

But how about proof theory?

Consider $t : FFPB$. This formula does hold at t (because B holds at r). Thus we must have rules that allow us to show that

$$\Delta \vdash FFPB.$$

It is convenient to write Δ as:

Assumptions	Configuration
$t : B$	$t < d < s$
$d : C$	$\sim (r < d \vee d = r \vee d < r)$
$s : A$	$r < s$
$r : B$	

and give rules to manipulate the configuration until we get $t : FFPB$.

Thus formally a temporal database is a set of labelled formulae $\{t_i : A_i\}$ together with a configuration on $\{t_i\}$, given in the form of an earlier–later relation $<$. A query is a labelled formula $t : Q$. The proof rules have the form

$$\frac{t_1 : A_1; \ldots ; t_n : A_n, \text{ configuration}}{s : B \text{ configuration}'}$$

where the configuration is a set of conditions on the ordering of $\{t_i\}$.

The use of labels is best illustrated via examples:

Example 4.1. This example shows the LDS in the case of modal logic. Modal logic has to do with possible worlds. Thus we think of our basic database (or assumptions) as a finite set of information about possible worlds. This consists of two parts. The configuration part, the finite configuration of possible worlds for the database, and the assumptions part which tells us what formulae hold in each world. The following is an example of a database:

	Assumptions	**Configuration**
(1)	$t : \Box\Box B$	$t < s$
(2)	$s : \Diamond(B \to C)$	

The conclusion to show (or query) is

$$t : \Diamond\Diamond C.$$

The derivation is as follows:

(3) From (2) create a new point r with $s < r$ and get $r : B \to C$.

We thus have

Assumptions	**Configuration**
(1), (2), (3)	$t < s < r$

(4) From (1), since $t < s$ get $s : \Box B$.

(5) From (4) since $s < r$ get $r : B$.

(6) From (5) and (3) we get $r : C$.
(7) From (6) since $s < r$ get $s : \Diamond C$.
(8) From (7) using $t < s$ we get $t : \Diamond\Diamond C$.

Discussion The object rules involved are:

$\Box E$ rule:

$$\frac{t < s; t : \Box A}{s : A}$$

$\Diamond I$ rule:

$$\frac{t < s, s : B}{t : \Diamond B}$$

$\Diamond E$ rule:

$$\frac{t : \Diamond A}{\text{create a new point } s \text{ with } t < s \text{ and deduce } s : A}.$$

Note that the above rules are not complete. We do not have rules for deriving, for example, $\Box A$. Also, the rules are all for intuitionistic modal logic.

The metalevel considerations which determine which logic we are working in, may be properties of $<$, e.g. $t < s \wedge s < r \to t < r$, or linearity, e.g. $t < s \vee t = s \vee s < t$ etc.

There are two serious problems in modal and temporal theorem proving. One is that Skolem functions for $\exists x \Diamond A(x)$ and $\Diamond \exists x A(x)$ are not logically the same. If we 'Skolemize' we get $\Diamond A(c)$. Unfortunately it is not clear where c exists, in the current world $(\exists x = c \Diamond A(x))$ or the possible world $(\Diamond \exists x = cA(x))$.

If we use labelled assumptions then $t : \exists x \Diamond A(x)$ becomes $t : \Diamond A(c)$ and it is clear that c is introduced at t.

On the other hand, the assumption $t : \Diamond \exists x A(x)$ will be used by the $\Diamond E$ rule to introduce a new point $s, t < s$ and conclude $s : \exists x A(x)$. We can further 'Skolemize' at s and get $s : A(c)$, with c introduced at s. We thus need the mechanism of remembering or labelling constants as well, to indicate where they were first introduced.

Example 4.2. Another example has to do with the Barcan formula

	Assumption	Configuration
(1)	$t : \forall x \Box A(x)$	$t < s$

We show

(2) $s : \forall x A(x)$.

We proceed intuitively

(3) $t : \Box A(x)$ (stripping $\forall x$, remembering x is arbitrary).
(4) Since the configuration contains $s, t < s$ we get

$$s : A(x).$$

(5) Since x is arbitrary we get

$$s : \forall x A(x).$$

The above intuitive proof can be restricted.

The rule

$$\frac{t : \Box A(x), t < s}{s : A(x)}$$

is allowed only if x is instantiated.

To allow the above rule for arbitrary x is equivalent to adopting the Barcan formula axiom

$$\forall x \Box A(x) \to \Box \forall x A(x).$$

Example 4.3. To show $\forall x \Box A(x) \to \Box \forall x A(x)$ in the modal logic where it is indeed true.

(1) Assume $t : \forall x \Box A(x)$.
We show $\Box \forall x A(x)$ by the use of the metabox:

	create α,	$t < \alpha$
(2)	$t : \Box A(x)$	from (1)
(3)	$\alpha : A(x)$	from (2) using a rule which allows this with x a variable.
(4)	$\alpha : \forall x A(x)$	universal generalization.

(5) Exit: $t : \Box \forall x A(x)$.

This rule has the form

Create α,	$t < \alpha$
Argue to get	$\alpha : B$
Exit with	$t : \Box B$

4.2 LDS semantics

We can now formally define a simplified version of LDS, sufficient for our temporal logics. The reader is referred to [Gabbay, 1996] for full details.

An algebraic LDS is built up from two components: an algebra $\mathcal{A}$ and a logic **L**. To make things specific, let us assume that we are dealing with a particular algebraic model $\mathcal{A} = (S, <, f_1, \ldots, f_k)$, where S is the domain of the algebra, $<$ is a strict order, i.e. irreflexive and transitive, relation on S and $f_1, \ldots, f_k$ are function symbols on S of arities $r_1, \ldots, r_k$ respectively. The sequence $\Sigma = (<, f_1, \ldots, f_k)$ is called the signature of $\mathcal{A}$. It is really the language of $\mathcal{A}$ in logical terms but we use Σ to separate it from **L**. We assume the functions are *isotonic*, i.e. they are either monotonic up or monotonic down in each variable, namely for each coordinate x in f we have that either

$$\forall x, y (x < y \to f(\ldots, x, \ldots) < f(\ldots, y, \ldots))$$

or

$$\forall x, y(x < y \to f(\dots, y, \dots) < f(\dots, x, \dots))$$

holds.

A typical algebra is a binary tree algebra where each point x in the tree has two immediate successor points $r_1(x)$ and $r_2(x)$ and one predecessor point $p(x)$. $<$ is the (branching) earlier–later relation and we have $p(x) < x < r_i(x), i = 1, 2$.

The general theory of LDS (see [Gabbay, 1996, Section 3.2]) requires a source of labels and a source of formulae. These together are used to form the declarative units of the form $t : A$, where t is a label and A is a formula.

The labels can be syntactical terms in some algebraic theory. The algebraic theory itself can be characterized either syntactically by giving axioms which the terms must satisfy or semantically by giving a class of models (algebras) for the language.

The formulae part of an LDS declarative unit is defined in the traditional way in some language $\mathbf{L}$.

An LDS database (theory) Δ is a set of terms and their formulae (i.e. a set of declarative units) with some relationships between the terms. In a general LDS, the terms themselves are syntactical and one always has to worry whether the required relations between the terms of Δ are possible (i.e. are consistent).

If, however, we have a semantic theory for the labels characterized by one single model (algebra), then we can take the labels to be elements of this model and consistency and relationships among the labels (elements) of Δ will always be clear—they are as dictated by the model. This represents a temporal logic with a concrete specific flow of time (e.g. integers, rationals, reals, etc.).

We therefore present for the purpose of this chapter, a concrete definition of LDS based on a single model as an algebra of labels.

Definition 4.4 (Concrete algebraic LDS).

1. A concrete algebraic LDS has the form $(\mathcal{A}, \mathbf{L})$ where:
   (a) $\mathcal{A} = (S, <, f_1, \dots, f_k)$ is a concrete algebraic model. The elements of S are called labels.
   (b) $\mathbf{L}$ is a predicate language with connectives $\sharp_1, \dots, \sharp_m$ with arities $r_1, \dots, r_m$, and quantifiers $(Q_1 x), \dots, (Q_{m'} x)$. The connectives can be some well-known modalities, binary conditional, etc., and the quantifiers can be some known generalized or traditional quantifiers. We assume that the traditional syntactical notions for $\mathbf{L}$ are defined. We also assume that $\mathbf{L}$ has only constants, no function symbols and the constants of $\mathbf{L}$ are indexed by elements of the algebra $\mathcal{A}$, i.e. have the form $c_i^t, t \in S, i = 1, 2, 3, \dots$.
2. A declarative unit has the form $t : A$, where A is a wff and $t \in S$, or it has the form $t : c_i^s$. The unit $t : A$ intuitively means 'A holds at label t' and the unit $t : c_i^s$ means 'the element c_i which was created at label s does exist in the domain of label t'.
3. A database has the form $\Delta = (D, \mathbf{f}, d, U)$ where $D \subseteq S$ is non-empty and $\mathbf{f}$ is a function associating with each $t \in D$ a set of wffs $\mathbf{f}(t) = \Delta_t$.

U is a function associating with each $t \in D$ a set of terms U_t. $d \in D$ is a distinguished point in D. The theory Δ can be displayed by writing $\{t : A_1, t : A_2, s : B, r : c_3^s, \ldots\}$, where $t : A$ indicates $A \in \mathbf{f}(t)$ and $r : c^s$ indicates $c^s \in U_r$.

Definition 4.5 (Semantics for LDS). Let $(\mathcal{A}, \mathbf{L})$ be a concrete algebraic LDS, with algebra $\mathcal{A}$ and language $\mathbf{L}$ with connectives $\{\sharp_1, \ldots, \sharp_m\}$ and quantifiers $\{Q_1, \ldots, Q_{m'}\}$ where $\sharp_i$ is r_i place.

1. A semantical interpretation for the LDS has the form $\mathcal{I} = (\Psi_0(x, X), \Psi_1, \ldots, \Psi_i(x, X_1, \ldots, X_{r_i}), \ldots, \Psi_m, \Psi'_1(x, Z), \ldots, \Psi'_{m'}(x, Z))$ where Ψ_i is a formula of the language of $\mathcal{A}$, possibly second order, with the single free element variable x and the free set variables X as indicated, and Ψ'_i have a single free element variable x and free binary relation variable Z. We need to assume that Ψ_i and Ψ'_j have the property that if we substitute in them for the set variables closed under $\leq$ then the element variable coordinate is monotonic up under $\leq$. In symbols:
   - $\bigwedge_j \forall x, y(x \in X_j \wedge x \leq y \to y \in X_j) \to [x \leq y \to \Psi(x, X_j) \to \Psi(y, X_j)]$.
2. A model for the LDS has the form $\mathbf{m} = (V, h, g, d)$ where $d \in S$ is the distinguished world, V is a function associating a domain V_t with each $t \in S$, h is a function associating with each n-place atomic predicate P a subset $h(t, P) \subseteq V_t^n$.
   g is an assignment giving each variable x an element $g(x) \in V_d$ and for each constant c_i^s an element $g(c_i^s) \in V_s$.
3. Satisfaction is defined by structural induction as follows:
   - $t \vDash P(b_1, \ldots, b_n)$ iff $(b_1, \ldots, b_n) \in h(t, P)$;
   - $t \vDash \exists x A(x)$ iff for some $b \in V_t, t \vDash A(b)$;
   - $t \vDash A \wedge B$ iff $t \vDash A$ and $t \vDash B$;
   - $t \vDash \sim A$ iff $t \nvDash A$;
   - $t \vDash \sharp_i(A_1, \ldots, A_{r_i})$ iff $\mathcal{A} \vDash \Psi_i(t, \hat{A}_1, \ldots, \hat{A}_{r_i})$, where $\hat{A} = \{s \in S \mid s \vDash A\}$;
   - $t \vDash (Q_i y)A(y)$ iff $\mathcal{A} \vDash \Psi'_i(t, \widehat{\lambda y A(y)})$, where $\widehat{\lambda y A(y)} = \{(t, y) \mid t \vDash A(y)\}$.
4. We say A holds in $\mathbf{m}$ iff $\mathcal{A} \vDash \Psi_0(d, \hat{A})$.
5. The interpretation $\mathcal{I}$ induces a translation $*$ of $\mathbf{L}$ into a two-sorted language $\mathbf{L}^*$ based on the two domains S (of the algebra $\mathcal{A}$) and $U = \bigcup_t V_t$ (of the predicates of $\mathbf{L}$) as follows:
   - each atomic predicate $P(x_1, \ldots, x_n)$ (interpreted over the domain U) is translated into

     $$[P(x_1, \ldots, x_n)]_t^* = P^*(t, x_1, \ldots, x_n),$$

     where P^* is a two-sorted predicate with one more variable t ranging over S and $x_1, \ldots, x_n$ ranging over U;

- $[A \wedge B]_t^* = [A]_t^* \wedge [B]_t^*$;
- $[\sim A]_t^* = \sim [A]_t^*$;
- $[\sharp_i(A_1, \ldots, A_{r_i}]_t^* = \Psi_i(t, \lambda s[A_1]_s^*, \ldots, \lambda s[A_{r_i}]_s^*)$;
- $[(Q_i y)A(y)]_t^* = \Psi_i'(t, \lambda y \lambda s[A(y)]_s^*)$;
- Let $||A||^* = \Psi_0(d, \lambda t[A]_t^*)$.

6. It is easy to show by induction that:
   - $t \vDash A$ iff $[A]_t^*$ holds in the naturally defined two-sorted model.

The reader should compare this definition with [Gabbay, 1996, Definition 3.2.6] and with Chapter 5 of Volume 1.

Gabbay's book on LDS contains plenty of examples of such systems.

In the particular case of temporal logic, the algebra has the form $(D, <, d)$, where D is the flow of time and $<$ is the earlier–later relation. $d \in D$ is the present moment (actual world).

4.3 Sample temporal completeness proof

The previous section presented the LDS semantics. This section will choose a sample temporal logical system and present *LDS* proof rules and a completeness theorem for it. We choose the modal logic $\mathbf{K}_t$ of Section 3.2 of Volume 1. This is the propositional logic with H and G complete for all Kripke frames $(S, R, a), a \in S$, such that R is transitive and irreflexive. A wff A is a theorem of $\mathbf{K}_t$ iff for all models (S, R, a, h) with assignment h, we have $a \vDash A$.

We want to turn $\mathbf{K}_t$ into a quantified logic $Q\mathbf{K}_t$. We take as semantics the class of all models of the form (S, R, a, V, h) such that V_t for $t \in S$ is the domain of world t. The following is assumed to hold:

- $tRs \wedge sRs'$ and $a \in V_{s'}$ and $a \in V_t$ imply $a \in V_s$ (i.e. elements are born, exist for a while and then possibly die).

Definition 4.6 (Traditional semantics for $Q\mathbf{K}_t$).

1. A $Q\mathbf{K}_t$ Kripke structure has the form (S, R, a, V, h), where S is a non-empty set of possible worlds, $R \subseteq S^2$ is the irreflexive and transitive accessibility relation and $a \in S$ is the actual world. V is a function giving for each $t \in S$ a non-empty domain V_t.
   - Let $V_S = \bigcup_{t \in S} V_t$.

   h is the assignment function assigning for each t and each n-place atomic predicate P its extension $h(t, P) \subseteq V_t^n$, and for each constant c of the language its extension $h(c) \in V_S$.

   Each n-place function symbol of the language of the form $f(x_1, \ldots, x_n)$ and t is assigned a function $h(f) : V_S^n \mapsto V_S$.

   Note that function symbols are rigid, i.e. the assignment to a constant c is a fixed rigid element which may or may not exist at a world t.[3]

[3]Had we wanted non-rigid semantics we would have stipulated that the extension of a function symbol is $h(t, f) : V_t^n \mapsto V_t$. There is no technical reason for this restriction and our

2. Satisfaction $\vDash$ is defined in the traditional manner.
   (a) h can be extended to arbitrary terms by the inductive clause $h(f(x_1, \ldots, x_n)) = h(f)(h(x_1), \ldots, h(x_n))$.
   (b) We define for atomic P and terms $x_1, \ldots, x_n$

$$t \vDash_h P(x_1, \ldots, x_n) \text{ iff } (h(x_1), \ldots, h(x_n)) \in h(t, P).$$

   (c) The cases of the classical connectives are the traditional ones.
   (d) $t \vDash_h \exists x A(x)$ iff for some $a \in V_t, t \vDash_h A(a)$.
   (e) $t \vDash_h \forall x A(x)$ iff for all $a \in V_t, t \vDash_h A(a)$.
3. $t \vDash_h GA(a_1, \ldots, a_n)$ (resp. $t \vDash HA$) iff for all s, such that tRs (resp. sRt) and $a_1, \ldots, a_n \in V_s, s \vDash A$.
4. $t \vDash U(A(a_1, \ldots, a_n), B(b_1, \ldots, b_k))$ iff for some s, tRs and $a_i \in V_s, i = 1, \ldots, n$, we have $s \vDash A$ and for all s', tRs' and $s'Rs$ and $b_j \in V_{s'}, j = 1, \ldots, k$, imply $s' \vDash B$. (The mirror image holds for $S(A, B)$.)
5. Satisfaction in the model is defined as satisfaction in the actual world.

Note that the logic $Q\mathbf{K}_t$ is not easy to axiomatize traditionally.
We now define an LDS corresponding to the system $Q\mathbf{K}_t$.

Definition 4.7 (The algebra of labels).

1. Consider the first-order theory of one binary relation $<$ and a single constant d. Consider the axiom $\partial = \forall x \sim (x < x) \wedge \forall xyz(x < y \wedge y < z \rightarrow x < z)$. Any classical model of this theory has the form $\mathbf{m} = (S, R, a, g)$, where S is the domain, R is a binary relation on S giving the extension of the syntactical '$<$' and g gives the extension of the variables and of d. $g(d)$ equals a and is the interpretation of the constant 'd'. Since $(S, R, a, g) \vDash \partial$, we have that R is irreflexive and transitive.
2. Let $U = \{t_1, t_2, \ldots\}$ be a set of additional constants in the predicate language of $<$ and d. Let $\mathcal{A}$ be the set of all terms of the language. By a diagram $\Delta = (D, <, d)$, with $D = (D_1, D_2)$, we mean a set $D_1 \subseteq \mathcal{A}, d \in D_1$ of constants and variables and a set D_2 of formulae $\varphi(t, s)$ of the form $t < s, \sim (t < s), t = s, t \neq s$, with constants and variables from D_1.
3. A structure $\mathbf{m} = (S, R, a, g)$ is a model of Δ iff the following hold:
   (a) $g : D_1 \mapsto S$, with $g(d) = a$;
   (b) R is irreflexive and transitive (i.e. it is a model of ∂);
   (c) whenever $\varphi(t, s) \in D_2$ then $\varphi(g(t), g(s))$ holds in the model.

methods still apply. We are just choosing a simpler case to show how LDS works. Note that by taking $h(t, P) \subseteq V_t^n$ as opposed to $h(t, P) \subseteq V_S^n$, we are introducing peculiarities in the semantic evaluation. $t \vDash P(a_1, \ldots, a_n)$ becomes false if not all a_i are in V_t. We can insist that we give values to $t \vDash A(a_1, \ldots, a_n)$ only if all elements are in V_t, but then what value do we give to $t \vDash GA$? One option is to let $t \vDash GA(a_1, \ldots, a_n)$ iff for all s such that tRs and such that all $a_i \in V_s$ we have $s \vDash A$.

Anyway, there are many options here and a suitable system can probably be chosen for any application area.

4. Note that g assigns elements of S also to the variables of D_1. Let x be a constant or a variable. Denote by $g =_x g'$ iff for all variables and constants $y \neq x$ we have $g(y) = g'(y)$.

Definition 4.8 (LDS language for $Q\mathbf{K}_t$).

1. Let $\mathbf{L}$ be the predicate modal language with the following:
   (a) Its connectives and quantifiers are $\wedge, \vee, \rightarrow, \bot, \top, \forall, \exists, G, F, H, P$.
   (b) Its variables are $\{x_1, x_2, \ldots\}$.
   (c) It has atomic predicates of different arities.
   (d) It has function symbols $e_1, e_2, \ldots,$ of different arities.
   (e) Let $\mathcal{A}$ be the language of the algebra of labels of Definition 4.7. We assume that $\mathcal{A}$ may share variables with $\mathbf{L}$ but its constants are distinct from the constants of $\mathbf{L}$. For each constant $t \in U$ of $\mathcal{A}$ and each natural number n, we assume we have in $\mathbf{L}$ a sequence of n-place function symbols

   $$c^t_{n,1}(x_1, \ldots, x_n), c^t_{n,2}(x_1, \ldots, x_n) \ldots$$

   parameterized by t.
   Thus in essence we want an infinite number of Skolem functions of any arity parameterized by any $t \in U$. The elements of $\mathcal{A}$ are our labels.

   The LDS language is presented as $(\mathcal{A}, \mathbf{L})$.
2. A declarative unit is a pair $t : A$, where t is a constant from $\mathcal{A}$ and A is a wff of $\mathbf{L}$.
   Note that because some of the function symbols of $\mathbf{L}$ are parameterized from U, we can get labels in A as well. For example,

   $$t : P(x, c^t_{1,1}(x))$$

   is a declarative unit.
3. A configuration has the form $(D, <, \mathbf{f}, d, U)$, where $(D, <, d)$ is a diagram as defined in Definition 4.7 and $\mathbf{f}$ and U are functions associating with each $t \in D_1$ a set $\mathbf{f}(t)$ of wffs of $\mathbf{L}$ and a set U_t of terms of $\mathbf{L}$.
   We also write:
   $t : A$ to indicate that $A \in \mathbf{f}(t)$;
   $t : c$ to indicate that $c \in U_t$.
   Configurations can also be presented graphically. See for example Fig. 4.1.

Definition 4.9 (LDS semantics for $Q\mathbf{K}_t$).

1. A model for an LDS language $(\mathcal{A}, \mathbf{L})$ has the form $\mathbf{n} = (S, R, a, V, h, g)$ where (S, R, a, V, h) is a traditional $Q\mathbf{K}_t$ model for the language $\mathbf{L}$ and g is an assignment from the labelling language $\mathcal{A}$ into S, giving values in S to each label and each variable. The following must hold:

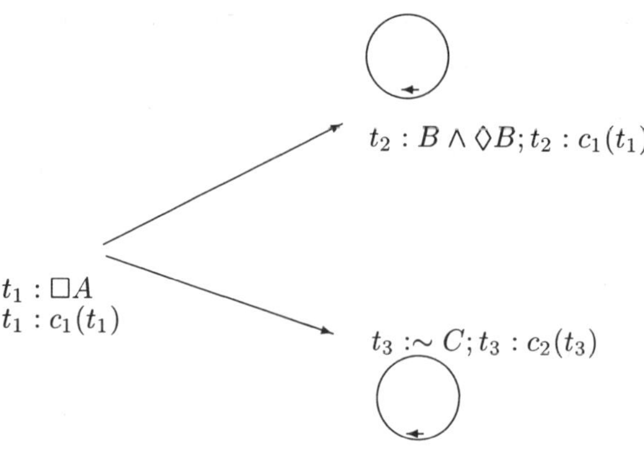

FIG. 4.1.

(a) for a variable or a constant x common to both $\mathcal{A}$ and $\mathbf{L}$, $g(x) = h(x)$;
(b) for any $c = c^t_{n,k}(x_1, \ldots, x_n)$ we have $h(x) \in V_{g(t)}$;
(c) let $t_1(x), t_2(y)$ be two terms of $\mathbf{L}$ containing the subterms x and y. Assume that ∂ (when augmented with the function symbols of $\mathbf{L}$ and equality axioms) satisfies

$$\partial \vdash_{\mathcal{A}} (x = y) \rightarrow (t_1 = t_2);$$

then if $g(x) = g(y)$ then $h(t_1) = h(t_2)$.
(d) $g(d) = a$.

2. Let $\Delta = (D, <, \mathbf{f}, d, U)$ be a modal configuration in a language $\mathbf{L}$. We define the notion of $\mathbf{n} \vDash \Delta$ to mean that the following hold:
   (a) for every $t \in D$ and $A \in \mathbf{f}(t)$ we have $g(t) \vDash_h A$ in (S, R, a, V, h), according to Definition 4.6;
   (b) $(S, R, a, g) \vDash (D, <, d)$ according to Definition 4.7;
   (c) if $x \in D_1$ is a variable then for all $g' =_x g$ we have $\mathbf{n}' = (S, R, a, V, h, g') \vDash \Delta$, provided $\mathbf{n}'$ is an acceptable model (satisfying the restrictions in 1). This means that the free variables occuring in D are interpreted universally.
3. Let $\mathbf{n} = (S, R, a, V, h, g)$ be a model in a language $(\mathcal{A}, \mathbf{L})$. Let $(\mathcal{A}', \mathbf{L}')$ be an extension of the language. Then $\mathbf{n}'$ is said to be an extension of $\mathbf{n}$ to $(\mathcal{A}', \mathbf{L}')$ if the restriction of V', h', g' to $(\mathcal{A}, \mathbf{L})$ equals V, h, g respectively.
4. Let Δ be a temporal configuration in $(\mathcal{A}, \mathbf{L})$ and Δ' in $(\mathcal{A}', \mathbf{L}')$. We write $\Delta \vDash \Delta'$ iff any model $\mathbf{n}$ of Δ can be extended to a model $\mathbf{n}'$ of Δ'.
5. We write $\varnothing \vDash \Delta$ or equivalently $\vDash \Delta$ iff for any model $\mathbf{n}, \mathbf{n} \vDash \Delta$.

Definition 4.10 (LDS proof rules for $Q\mathbf{K}_t$). Let $\Delta = (D, <, \mathbf{f}, d, U)$ and $\Delta' = (D', <', \mathbf{f}', d, U')$ be two temporal configurations. We say that Δ' is obtained from Δ by the application of a single forward proof rule if one of the following cases hold (let $\Diamond \in \{P, F\}$, and $\Box \in \{G, H\}$).

1. **$\Diamond$ introduction case**
   For some $t, s \in D_1, t < s \in D_2$, $A \in \mathbf{f}(s)$ and Δ' is the same as Δ except that $\mathbf{f}'(t) = \mathbf{f}(t) \cup \{\Diamond A\}$.
   Symbolically we can write $\Delta' = \Delta_{[t<s;s:A]}$ for F and $\Delta_{[s<t,s:A]}$ for P.
2. **$\Diamond$ elimination case**
   For some $t \in D_1$ and some A, $\Diamond A \in \mathbf{f}(t)$ and for some new atomic $s \in U$, that does not appear in Δ, we have that Δ' is the same as Δ except that $D_1' = D_1 \cup \{s\}$. $D_2' = D_2 \cup \{t < s\}$ for $\Diamond = fF$ (resp. $s < t$ for $\Diamond = P$). $\mathbf{f}'(s) = \{A\}$.
   Symbolically we can write $\Delta' = \Delta_{[t:\Diamond A;s]}$.
3. **$\Box$ elimination case**
   For some $t, s \in D_1$ such that $t < s \in D_2$ for $\Box = G$ (resp. $s < t$ for $\Box = H$) we have $\Box A \in \mathbf{f}(t)$ and Δ' is like Δ except that $\mathbf{f}'(s) = \mathbf{f}(s) \cup \{A\}$.
   Symbolically we can write $\Delta' = \Delta_{[t<s;t:GA]}$, and $\Delta' = \Delta_{[s<t;t:HA]}$.
4. **Local classical case**
   Δ' is like Δ except that for $t \in D_1$ we have $\mathbf{f}'(t) = \mathbf{f}(t) \cup \{A\}$, where A follows from $\mathbf{f}(t)$ using classical logic inference only.[4]
   Symbolically we can write $\Delta' = \Delta_{[t \vdash A]}$.
5. **Local $\forall$ elimination**
   For some $t \in D_1$ we have $\forall x A(x) \in \mathbf{f}(t)$ and $c \in U_t$, Δ' is like Δ except $\mathbf{f}'(t) = \mathbf{f}(t) \cup \{A(c)\}$.
   Symbolically we can write $\Delta' = \Delta_{[t:\forall x A(x);x/c]}$.
6. **Local $\exists$ elimination**
   For some $t \in D_1$ and $\forall x_1, \ldots, x_n \exists y A(x_1, \ldots, x_n, y) \in \mathbf{f}(t)$ and some *new* function symbol $c^t(x_1, \ldots, x_n)$ we have $\mathbf{f}'(t) = \mathbf{f}(t) \cup \{\forall x_1, \ldots, x_n A(x_1, \ldots, x_n, c^t(x_1, \ldots, x_n))\}$. Otherwise Δ' is like Δ.
   Symbolically we can write $\Delta' = \Delta_{[t:\forall x_1,\ldots,x_2 \exists y A;y/c^t]}$.
7. **Visa rules**
   (a) For $t, s \in D_1, t < s < t' \in D_2$ and $c \in U_t$ and $c \in U_{t'}$, Δ' is like Δ except that $U_s' = U_s \cup \{c\}$.
   (b) $U_t' = U_t \cup \{c^t\}$. In other words any c^t can be put in the domain of world t.

   Symbolically we write $\Delta' = \Delta_{[t:c,t':c \text{ to } s:c]}$ and $\Delta' = \Delta_{[t:c^t]}$ respectively.
8. **Inconsistency rules**
   (a) If $t, s \in D_1$ and $\bot \in \mathbf{f}(t)$ then let Δ' be like Δ except that $\mathbf{f}'(s) = \mathbf{f}(s) \cup \{\bot\}$ and $\mathbf{f}'(x) = \mathbf{f}(x)$, for $x \neq s$.

[4]Every formula of $Q\mathbf{K}_t$ can be presented in the form $B(Q_1/\Box_1 A_1, \ldots, Q_n/\Box_n A_n)$ where $\Box_i \in \{G, H\}$ and $B(Q_1, \ldots, Q_n)$ is a modal free classical formula and A_i are general $Q\mathbf{K}_t$ formulae. P/A means the substitution of A for P. A set of formulae of the form $B_i(Q_j^i/\Box_j A_j^i)$ proves using classical rules only a formula $B(Q_j/\Box_j A_j)$ iff $\{B_i(Q_j^i)\}$ proves classically $B(Q_j)$.

Symbolically we write $\Delta' = \Delta_{[t:\bot \text{ to } s:\bot]}$.

(b) If $(D, <, d)$ is classicaly inconsistent and $s \in D_1$, we let Δ' be as in (a) above.
Symbolically we write $\Delta' = \Delta_{[\bot \text{ to } s:\bot]}$.

9. **$\Box$ introduction rule**
We say Δ' is obtained from Δ by a single-level $n+1$ introduction rule if the following holds. For some $t \in D_1$ we have $\mathbf{f}'(t) = \mathbf{f}(t) \cup \{\Box A\}$, $\Box \in \{G, H\}$ and Δ_2 follows from Δ_1 using a sequence of applications of single, forward rules or of single-level $m \leq n$ introduction rules, and Δ_1 is like Δ except that

$$D_1^1 = D_1 \cup \{t_1\}, D_2^1 = D_2 \cup \{t < t_1\} \text{ for } \Box = G$$
$$(\text{and respectively } D_2^1 = D_2 \cup \{t_1 < t\} \text{ for } \Box = H)$$

where t_1 is a completely new constant label and $\mathbf{f}^1(t_1) = \{\top\}$.
Δ_2 is like Δ_1 except that $\mathbf{f}^2(t_1) = \mathbf{f}^1(t_1) \cup \{A\}$.
Symbolically we write $\Delta' = \Delta_{[t:\Box A]}$.

10. **Diagram rule**
Δ' is an extension of Δ as a diagram, i.e. $D \subseteq D', \mathbf{f}'; (t) = \mathbf{f}(t), t \in D_1$ and $\mathbf{f}'(t) = \{\top\}$, for $t \in D'_1 - D_1$, and we have $(D, <, d) \vdash_{\mathcal{A}} (D', <, d)$.
Symbolically we write $\Delta' = \Delta_{[D \vdash D']}$.
Note that for our logic, $Q\mathbf{K}_t$, this rule just closes $<$ under transitivity.

11. We write $\Delta \vdash \Delta'$ iff there exists a sequence of steps of any level leading from Δ to Δ'. We write $\vdash \Delta'$ if $\Delta_0 \vdash \Delta'$, for Δ_0 the theory with $\{d\}$ only with $\mathbf{f}(d) = \{\top\}$.
Let t be a label of Δ. We write $\Delta \vdash t : A$ iff for any Δ' such that $\Delta \vdash \Delta'$ there exists a Δ'' such that $\Delta' \vdash \Delta''$ and $A \in \mathbf{f}''(t)$. In other words, Δ can be proof-theoretically manipulated until A is proved at node t.

12. Notice that if $\Delta \vdash \Delta'$ then the language $\mathbf{L}'$ of Δ' is slightly richer in labels and Skolem functions than the language $\mathbf{L}$.
Note also that if $\Delta \vdash \Delta'$ then there is a sequence of symbols $\pi_1, \ldots, \pi_n$ such that $\Delta' = \Delta_{\pi_1, \ldots, \pi_n}$.
In fact Δ' is uniquely determined up to symbolic isomorphism by the sequence $\pi_1, \ldots, \pi_n$.

13. **Local cut rule**
$\vdash$ of item 11 above is without the following *local cut rule*. Let $\Delta_{t:B}$ denote the database obtained from Δ by adding B at label t. Then $\Delta_{t:B} \vdash \Delta'$ and $\Delta_{t:\sim B} \vdash \Delta'$ imply $\Delta \vdash \Delta'$.

14. The consequence $\Delta \vdash \Delta'$ can be implicitly formulated in a 'sequent'-like form as follows:
   - $\Delta \vdash \Delta_\pi$ (axiom)
     for π as in any of items 1–10 above.
   - $\dfrac{\Delta \vdash \Delta'; \Delta' \vdash \Delta''}{\Delta \vdash \Delta''}$.

- Local cut rule.

Definition 4.11 (Inconsistency).

1. A theory $\Delta = (D, <, \mathbf{f}, d, U)$ is immediately inconsistent iff either $(D, <, d)$ is inconsistent as a classical diagram or for some $t \in D_1, \bot \in \mathbf{f}(t)$.
2. Δ is inconsistent iff $\Delta \vdash \Delta'$ and Δ' is immediately inconsistent.
3. Note that if we do not provide the following inconsistency rules, namely

$$\frac{t : \bot}{s : \bot}$$

and

$$\frac{(D, <, d) \text{ is inconsistent}}{s : \bot},$$

then two inconsistent modal configurations cannot necessarily prove each other.

Theorem 4.12 (Soundness). *Let Δ be in $\mathbf{L}$ and Δ' in $\mathbf{L}'$. Then $\Delta \vdash \Delta'$ implies $\Delta \vDash \Delta'$. In words, if $\mathbf{n}$ is a model in $\mathbf{L}$ such that $\mathbf{n} \vDash \Delta$, then $\mathbf{n}$ can be extended to $\mathbf{n}' \vDash \Delta'$, where $\mathbf{n}'$ is like $\mathbf{n}$ except that the assignments are extended to $\mathbf{L}'$. In particular $\vdash \Delta'$ implies $\vDash \Delta'$.*

Proof By induction on the number of single steps proving Δ' from Δ. We can assume Δ is consistent. It is easy enough to show that if Δ' is obtained from Δ by a single step then there is an $\mathbf{n}' \vDash \Delta'$.

Let $\mathbf{n} = (S, R, a, V, h, g)$ be a model of $\Delta = (D, <, \mathbf{f}, d, U)$. Let Δ' be obtained from Δ by a single proof step and assume $\mathbf{n} \vDash \Delta$. We show how to modify $\mathbf{n}$ to an $\mathbf{n}'$ such that $\mathbf{n}' \vDash \Delta'$.

We follow the proof steps case by case:

1. **$\Diamond$ introduction case**
   Here take $\mathbf{n}' = \mathbf{n}$.
2. **$\Diamond$ elimination case**
   Here Δ' contains a completely new constant $s \in D'_1$. We can assume g is not defined for this constant. Since $\mathbf{n} \vDash \Delta$, we have $g(t) \vDash \Diamond A$ and so for some $b \in S, g(t)Rb$ and $b \vDash A$. Let g' be like g except $g'(s) = b$. Then $\mathbf{n}' \vDash \Delta'$.
3. **$\Box$ elimination case**
   Here let $\mathbf{n}' = \mathbf{n}$.
4. **Local classical case**
   Let $\mathbf{n}' = \mathbf{n}$.
5. **Local $\forall$ elimination**
   Let $\mathbf{n}' = \mathbf{n}$.
6. **Local $\exists$ elimination**
   We can assume the new function $c^t(x_1, \ldots, x_n)$ is new to the language of Δ and that h is not defined for c^t. Since in $\mathbf{n}$, $t \vDash \forall x_1, \ldots, x_n \exists y A(x_1, \ldots, x_n, y)$,

for every $(x_1, \ldots, x_n) \in V_t^n$ a $y \in V_t$ exists such that $t \vDash A(x_i, y)$. Thus an assignment $h'(c^t)$ can be given to c^t by defining it to be this y for $(x_1, \ldots, x_n) \in V_t^n$ and to be a fixed element of V_t for tuples $(x_1, \ldots, x_n)$ not in V_t^n. Let $\mathbf{n}'$ be like $\mathbf{n}$ except we take h' instead of h.

7. **Visa rules**
   (a) Take $\mathbf{n}' = \mathbf{n}$.
   (b) Take $\mathbf{n}' = \mathbf{n}$.
   (c) Take $\mathbf{n}' = \mathbf{n}$.
8. Inconsistency rules are not applicable as we assume Δ is consistent.
9. Let $\mathbf{n}_1$ be a model for the language of Δ_1 and assume $\mathbf{n}_1 \vDash \Delta_1$. Then by the induction hypothesis there exists $\mathbf{n}_2 \vDash \Delta_2$, where $\mathbf{n}_2$ extends $\mathbf{n}_1$.
   Assume now that $\mathbf{n} \nvDash \Delta'$. Then $t \nvDash \Box A$, and hence for some b such that $tRb, b \nvDash A$.
   Let $\mathbf{n}_1$ be defined by extending g to $g_1(t_1) = b$. Clearly $\mathbf{n}_1 \vDash \Delta_1$. Since $\Delta_1 \vdash \Delta_2$, g_1 can be extended to g_2 so that $\mathbf{n}_2 \vDash \Delta_2$, i.e. $b \vDash A$, but this contradicts our previous assumption. Hence $\mathbf{n} \vDash \Delta'$.
10. Let $\mathbf{n}' = \mathbf{n}$.

The above completes the proof of soundness. □

Theorem 4.13 (Completeness theorem for LDS proof rules). *Let Δ be a consistent configuration in* $\mathbf{L}$*; then Δ has a model $\mathbf{n} \vDash \Delta$.*

Proof We can assume that there are an infinite number of constants, labels and variables in $\mathbf{L}$ which are not mentioned in Δ. We can further assume that there are an infinite number of Skolem functions $c^t_{n,i}(x_1, \ldots, x_n)$ in $\mathbf{L}$, which are not in Δ, for each t and n.

Let $\delta(n)$ be a function such that

$$\delta(n) = (t_n, B_n, t'_n, \alpha_n, k_n),$$

where t_n, t'_n are labels or variables, α_n a term of $\mathbf{L}$ and B_n a formula of $\mathbf{L}$, and k_n a number between 0 and 7.

Assume that for each pair of labels t, t' of $\mathbf{L}$ and each term α and each formula B of $\mathbf{L}$ and each $0 \leq k \leq 7$ there exist an infinite number of numerals m such that $\delta(m) = (t, B, t', \alpha, k)$.

Let $\Delta_0 = \Delta$. Assume we have defined $\Delta_n = (D^n, <, \mathbf{f}^n, d, U^n)$ and it is consistent.

We now define Δ_{n+1}. Our assumption implies that $(D^n, <, d)$ is classically consistent as well as each $\mathbf{f}^n(t)$. Consider $(t_n, B_n, t'_n, \alpha_n, k_n)$. It is possible that the formulae and labels of $\delta(n)$ are not even in the language of Δ_n, but if they are we can carry out a construction by case analysis.

Case $k_n = 0$

This case deals with the attempt to add either the formula B_n or its negation $\sim B_n$ to the label t_n in D^n. We need to do that in order to build eventually a Henkin model. We first try to add B_n and see if the resulting database $(\Delta_n)_{t_n : B_n}$

is LDS-consistent. If not then we try to add $\sim B_n$. If both are LDS-inconsistent then Δ_n itself is LDS-inconsistent, by the local cut rule (see item 13 of Definition 4.10).

The above is perfectly acceptable if we are prepared to adopt the local cut rule. However, if we want a system without cut then we must try to add B_n or $\sim B_n$ using possibly other considerations, maybe a different notion of local consistency, which may be heavily dependent on our particular logic. The kind of notion we use will most likely be correlated to the kind of traditional cut elimination available to us in that logic.

Let us motivate the particular notion we use for our logic, while at the same time paying attention to the principles involved and what possible variations are needed for slightly different logics.

Let us consider an LDS-consistent theory $\Delta = (D, <, \mathbf{f}, d, U)$ and an arbitrary B. We want a process which will allow us to add either B or $\sim B$ at node t (i.e. form $\Delta_{t:B}$ or $\Delta_{t:\sim B}$) and make sure that the resulting theory is LDS-consistent. We do not have the cut rule and so we must use a notion of local consistency particularly devised for our particular logic.

Let us try the simplest notion, that of classical logic consistency. We check whether $\mathbf{f}(t) \cup \{B\}$ is classically consistent. If yes, take $\Delta_{t:B}$, otherwise take $\Delta_{t:\sim B}$. However, the fact that we have classical consistency at t does not necessarily imply that the database is LDS-consistent. Consider $\{t : FA \to \sim B, s : A; t < s\}$. Here $\mathbf{f}(t) = FA \to \sim B$. $\mathbf{f}(s) = A$. If we add B to t then we still have local consistency but as soon as we apply the F introduction rule at t we get inconsistency.

Suppose we try to be more sophisticated. Suppose we let $\mathbf{f}_*(t) = \mathbf{f}(t) \cup \{F^k X \mid X \in \mathbf{f}(s)$, for some $k, t <^k s\} \cup \{Y \mid G^k Y \in \mathbf{f}(s)$, for some $k, s <^k t\}$ and try to add either B or $\sim B$ to $\mathbf{f}_*(t)$ and check for classical consistency. If neither is consistent then for some X_i, Y_k, X'_i, Y'_j we have

$$\begin{aligned} &\mathbf{f}(t), X_i, Y_j \vdash \sim B \\ &\mathbf{f}(t), X'_i, Y'_j \vdash B. \end{aligned}$$

Hence $\mathbf{f}_*(t)$ is inconsistent. However, we do have LDS rules that can bring the X and Y into $\mathbf{f}(t)$ which will make Δ *LDS*-inconsistent.

So at least one of B and $\sim B$ can be added. Suppose $\Delta_{t:B}$ is locally consistent. Does that make it LDS-consistent?

Well, not necessarily. Consider

$$\{t < s; t : F^k A \to \sim B; s : A\}$$

and assume we have a condition ∂_1 on the diagrams

$$\partial_1 = \forall xy(x < y \to \exists z(x < z < y)).$$

We would have to apply the diagram rule k times at the appropriate labels to get inconsistency.

It is obvious from the above discussion that to add B or $\sim B$ at node t we have to make it consistent with all possible wffs that can be proved using LDS rules to have label t (i.e. to be at t).

Let us define then, for t in Δ,

$$\Delta_{(t)} = \{A \mid \Delta \vdash t : A\}.$$

We say B is locally consistent with Δ at t, or that $\mathbf{f}(t) \cup \{B\}$ is locally consistent iff it is classically consistent with $\Delta_{(t)}$. Note that if $\Delta_{(t)} \vdash A$ classically then $\Delta \vdash t : A$.

We can now proceed with the construction:

1. if $t_n \notin D_1^n$ then $\Delta_{n+1} = \Delta_n$;
2. if $t_n \in D_1^n$ and $\mathbf{f}^n(t_n) \cup \{B_n\}$ is locally consistent, then let $\mathbf{f}^{n+1}(t_n) = \mathbf{f}^n(t_n) \cup \{B_n\}$.
   Otherwise $\mathbf{f}^n(t_n)$ is locally consistent with $\sim B_n$ and we let $\mathbf{f}^{n+1}(t_n) = \mathbf{f}^n(t_n) \cup \{\sim B_n\}$. For other $x \in D_1^n$, let $\mathbf{f}^{n+1}(x) = \mathbf{f}^n(x)$. Let $\Delta_{n+1} = (D^n, <, d, \mathbf{f}^{n+1}, U^n)$.

We must show the following:

- if Δ is LDS-consistent and $\Delta_{t_i:B_i}$ is obtained from Δ by simultaneously adding the wffs B_i to $\mathbf{f}(t_i)$ of Δ and if for all i $\mathbf{f}(t_i) \cup \{B_i\}$ is *locally*-consistent, then $\Delta_{t_i:B_i}$ is LDS-consistent.

To show this, assume that $\Delta_{t_i:B_i}$ is LDS-inconsistent. We show by induction on the complexity of the inconsistency proof that Δ is also inconsistent.

Case one step

In this case $\Delta_{t_i:B_i}$ is immediately inconsistent. It is clear that Δ is also inconsistent, because the immediate inconsistency cannot be at any label t_i, and the diagram is consistent.

Case $(l+1)$ steps

Consider the proof of inconsistency of $\Delta_{t_i:B_i}$. Let π be the first proof step leading to this inconsistency. Let $\Delta' = (\Delta_{t_i:B_i})_\pi$. π can be one of several cases as listed in Definition 4.10. If π does not touch the labels t_i, then it can commute with the insertion of B_i, i.e. $\Delta' = (\Delta_\pi)_{t_i:B_i}$ and by the induction hypothesis Δ_π is LDS-inconsistent and hence so is Δ.

If π does affect some label t_i, we have to make a case analysis:

- $\pi = [t_i < s; s : A]$, i.e. $\Diamond A$ is put in t. In this case consider $(\Delta_\pi)_{t_i:B_i}$. If adding B_i is still locally consistent then by the induction hypothesis $(\Delta_\pi)_{t_i:B_i}$ is consistent. But this is Δ' and so Δ' is consistent. If adding B_i is locally inconsistent, this means that $\Delta_{(t_i)} \cup \{B_i\}$ classically proves $\bot$, contrary to assumption.
- $\pi = [s < t_i; s : GA]$ (resp. $\pi = [t_i < s; s : HA]$) i.e. A is put in t_i. The reasoning is similar to the previous case.
- $\pi = [t_i : \Box A]$, $\Box \in \{G, H\}$, the reasoning is similar to previous cases.
- $\pi = [t_i \vdash A]$, similar to the previous ones.

- π = classical quantifier rules. This case is also similar to previous ones;
- π = visa rule. This means some new constants are involved in the new inconsistency from $\Delta_{(t_i)} \cup \{B_i\}$. These will turn into universal quantifiers and contradict the assumptions.
- The inconsistency rules are not a problem.
- The diagram rule does not affect t_i.
- π applies to B_i itself. Assume that $B_i = \Box C_i$ where $\Box$ is G (resp. H) and that C_i is put in some $s, t_i < s$ (resp. $s < t_i$).
  Let us first check whether C_i is locally consistent at s. This will not be the case if $\Delta_{(s)} \vdash \sim C_i$. This would imply $\Delta_{(t)} \vdash \Diamond \sim C_i$ contradicting the fact that B_i is consistent with $\Delta_{(t)}$. Thus consider $\Delta_{t_i:B_i;s:C_i}$. This is the same as Δ'. It is *LDS*-inconsistent, by a shorter proof; hence by the induction hypothesis Δ is *LDS*-consistent;
- B_i is $\Diamond C_i$ and π eliminates $\Diamond$, i.e. a new point s is introduced with $t_i < s$ for $\Diamond = F$ (resp. $s < t_i$ for $\Diamond = P$) and $s : C_i$ is added to the database.
  We claim that C_i is locally consistent in s, in Δ^+, where Δ^+ is the result of adding s to Δ but not adding $s : C_i$. Otherwise $\Delta^+_{(s)} \vdash \sim C_i$, and since s is a completely new constant and $\Delta^+ \vdash s : \ \sim C_i$ this means that $\Delta \vdash t_i : \Box \sim C_i$, a contradiction. Hence C_i is locally consistent in $\Delta^+_{(s)}$. Hence by the induction hypothesis if $\Delta^+_{s:C_i;t_i:B_i}$ is LDS-inconsistent so is Δ^+. If Δ^+ is LDS-inconsistent then $\Delta^+ \vdash s :\sim C_i$ and hence $\Delta \vdash t : \Box \sim C_i$ contradicting the local consistency of B_i at t.
- π applies to B_i and π adds $\Diamond B_i$ to $s < t_i$ for $\Diamond = F$ (resp. $t_i < s$ for $\Diamond = P$).
  Again we claim $\Diamond B_i$ is locally consistent at s. Otherwise $\Delta_{(s)} \vdash \Box \sim B_i$ and so B_i would not be locally consistent at t_i. We now consider $\Delta_{t_i:B_i;s:\Diamond B_i}$ and get a contradiction as before.
- π is a use of a local classical rule, i.e. $\mathbf{f}(t) \cup \{B_i\} \vdash C_i$, and π adds C_i at t_i. We claim we can add $B_i \wedge C_i$ at t, because it is locally consistent. Otherwise $\Delta_{(t)} \vdash B_i \to \ \sim C_i$ contradicting consistency of $\Delta_{(t)} \cup \{B_i\}$ since $\mathbf{f}(t) \vdash B_i \to C_i$.
- π is a Skolemization on B_i or an instantiation from B_i. All these classical operations are treated as in the previous case.

Case $k_n = 1$

1. If $t_n, t'_n \in D^n_1$ and $t_n < t'_n \in D^n_2$ and $B_n \in \mathbf{f}^n(t'_n)$ then let $\mathbf{f}^{n+1}(t_n) = \mathbf{f}^n(t_n) \cup \{\Diamond B_n\}$ and $\mathbf{f}^{n+1}(x) = \mathbf{f}^n(x)$, for $x \neq t_n$. Let $\Delta_{n+1} = (D^n, <, \mathbf{f}^{n+1}, d, U^n)$.
2. Otherwise let $\Delta_{n+1} = \Delta_n$.

Case $k_n = 2$

1. If $t_n \in D^n_1$ and $B_n = \Diamond C \in \mathbf{f}^n(t_n)$, then let s be a completely new constant and let $D^{n+1}_1 = D^n_1 \cup \{s\}, D^{n+1}_2 = D^n_2 \cup \{t_n < s\}$ for $\Diamond = F$ (resp. $s < t_n$ for $\Diamond = P$). Let $\mathbf{f}^{n+1}$ be like $\mathbf{f}^n$ or D_1 and let $\mathbf{f}^{n+1}(s) = \{C\}$. Let

$\Delta_{n+1} = (D^{n+1}, <, \mathbf{f}^{n+1}, d, U^n)$ and let the new domain at s, U_s^{n+1}, contain all free variables of C.

2. Otherwise let $\Delta_{n+1} = \Delta_n$.

Case $k_n = 3$

1. If $t_n, t'_n \in D_1^n$ and $t_n < t'_n \in D_2^n$ and $B_n = \Box C$ and $\Box = G$ (resp. $t'_n < t_n$ and $\Box = H$) and $B_n \in \mathbf{f}^n(t_n)$ and all free variables of C are in the domain $U_{t'_n}^n$ then let $\mathbf{f}^{n+1} = \mathbf{f}^n(x)$, for $x \neq t'_n$ and $\mathbf{f}^{n+1}(t'_n) = \mathbf{f}^n(x) \cup \{C\}$.

   Let $\Delta_{n+1} = (D^n, <, \mathbf{f}^{n+1}, d, U^n)$.

2. Otherwise let $\Delta_{n+1} = \Delta_n$.

Case $k_n = 4$

1. We have $t_n \in D_1^n$ and $\mathbf{f}^n(t_n) \vdash B_n$ classically. Let $\mathbf{f}^{n+1}(x) = \mathbf{f}^n(x)$ for $x \neq t_n$ and let $\mathbf{f}^{n+1}(t_n) = \mathbf{f}^n(t_n) \cup \{B_n\}$.

   Let $\Delta_{n+1} = (D^n, <, \mathbf{f}^{n+1}, d, U^n)$.

2. Otherwise let $\Delta_{n+1} = \Delta_n$.

Case $k_n = 5$

1. $t_n \in D_1^n$ and $B_n = \forall x C(x) \in \mathbf{f}^n(t_n)$ and $\alpha_n \in U^n$. Then let $\mathbf{f}^{n+1}(t_n) = \mathbf{f}^n(t_n) \cup \{C(\alpha_n)\}$ and $\mathbf{f}^{n+1}(x) = \mathbf{f}^n(x)$ for $x \neq t_n$ and $\Delta_{n+1} = (D^n, <, \mathbf{f}^{n+1}, d, U^n)$.

2. Otherwise let $\Delta_{n+1} = \Delta_n$.

Case $k_n = 6$

1. $t_n \in D_1^n$ and $B_n \in \mathbf{f}^n(t_n)$ and $B_n = \exists u C(u, y_1, \ldots, y_k)$.

   Let $c^{t_n}(y_1, \ldots, y_k)$ be a completely new Skolem function of this arity not appearing in Δ_n and let $\mathbf{f}^{n+1}(t_n) = \mathbf{f}^n(t_n) \cup \{C(c^t(y_1, \ldots, y_k), y_1, \ldots, y_k)\}$. This is consistent by classical logic. Let $U_{t_n}^{n+1} = U_{t_n}^n \cup \{c^t(y_1, \ldots, y_k)\}$. Let U^{n+1} and $\mathbf{f}^{n+1}$ be the same as U^n and $\mathbf{f}^n$ respectively, for $x \neq t_n$. Take $\Delta_{n+1} = (D^n, <, \mathbf{f}^{n+1}, d, U^{n+1})$.

2. Otherwise let $\Delta_{n+1} = \Delta_n$.

Case $k_n = 7$

1. If $t_n, t'_n < s_n \in D_1^n$ and $t_n < t'_n < s_n \in D_2^n$ and $\alpha_n \in U_{t_n}^n \cap U_{s_n}^n$ then let $U_{t'_n}^{n+1} = U_{t'_n}^n \cup \{\alpha_n\}$ and $U_x^{n+1} = U_x^n$ for $x \neq t'_n$. Let $\Delta_{n+1} = (D^n, <, \mathbf{f}^n, d, U^{n+1})$.

2. Otherwise let $\Delta_{n+1} = \Delta_n$.

Let Δ_∞ be defined by $D_i^\infty = \bigcup_n D_i^n$, $\mathbf{f}^\infty = \bigcup_n \mathbf{f}^n$, $U^\infty = \bigcup_n U^n$.

Δ_∞ is our Henkin model. Let $\mathbf{n} = (S, R, a, V, h, g)$, where

$$\begin{aligned}
&S = D_1^\infty \\
&R = \{(x,y) \mid x < y \in D_2^\infty\} \\
&a = d \\
&V_t = U_t \\
&V = U^\infty \\
&g = \text{ identity;}
\end{aligned}$$

then $h(t,P) = \{(x_1,\ldots,x_n) \mid P(x_1,\ldots,x_n) \in \mathbf{f}^\infty(t)\}$ for $t \in S$ and P n-place atomic predicate, and $x_1,\ldots,x_n \in V_t$. □

Lemma 4.14.

1. $\mathbf{n}$ *is an acceptable structure of the semantics.*
2. *For any* t *and* B, $t \vDash B$ *iff* $B \in \mathbf{f}^\infty(t)$.

Proof

1. We need to show that R is irreflexive and transitive. This follows from the construction and the diagram rule.
2. We prove this by induction. Assume $\Diamond A \in \mathbf{f}^\infty(t)$. Then for some n we have $t \in D_1^n$ and $\Diamond A \in \mathbf{f}^n(t)$. Thus at some $n' \geq n$ we put $s \in D_1^{n'}, t < s \in D_2^{n'}$ for $\Diamond = F$ (resp. $s < t$ for $\Diamond = P$) and $A \in \mathbf{f}^{n'}(s)$.
   Assume $\Diamond A \notin \mathbf{f}^\infty(t)$. At some $n, B_n = \Diamond A$ and had $\mathbf{f}^n(t) \cup \{\Diamond A\}$ been consistent, B_n would have been put in $\mathbf{f}^{n+1}$.
   Hence $\Box \sim A \in \mathbf{f}^{n+1}$.
   Assume $t < s$ (resp. $s < t$), $s \in D_1^\infty$. Hence for some $n'' \geq n'$ we have $\delta(n'') = (t, \Box \sim A, s, -, 3)$. At this stage $\sim A$ would be in $\mathbf{f}^{n''+1}(s)$. Thus for all $s \in S$ such that tRs (resp. sRt) we have $s \vDash \sim A$.

The classical cases follow the usual Henkin proof.

This completes the proof of the lemma and the proof of Theorem 4.13. □

4.4 Label-dependent connectives

We saw in Volume 1 that *since* and *until* cannot be defined from $\{G,H,F,P\}$ but if we allow names for worlds we can write

$$\sim q \wedge Gq \wedge Hq \to [U(A,B) \leftrightarrow F(A \wedge H(P \sim q \to B))].$$

We can introduce the label-dependent connectives $G^x A, H^x A$ meaning

$$\begin{aligned}
&t \vDash G^x A \text{ iff for all } y(t < y < x \to y \vDash A); \\
&t \vDash H^x A \text{ iff for all } y(x < y < t \to y \vDash A).
\end{aligned}$$

We can then define

$$t : U(A,B) \text{ as } t : F(A \wedge H^t B).$$

Let $F^x A$ be $\sim G^x \sim A$ and $P^x A$ be $\sim H^x \sim A$. Then

$$\begin{aligned}
&t \vDash F^x A \text{ iff for some } y, t < y < x \text{ and } y \vDash A \text{ hold.} \\
&t \vDash P^x A \text{ iff for some } y, x < y < t \text{ and } y \vDash A \text{ hold.}
\end{aligned}$$

Consider $t \vDash G^x\bot$ and $t \vDash F^x\top$. The first holds iff $\sim\exists y(t < y < x)$ which holds if either $\sim(t < x)$ or x is an immediate successor of t. The second holds if $\exists y(t < y < x)$.

Label-dependent connectives are very intuitive semantically since they just restrict the temporal range of the connectives. There are many applications where such connectives are used. In the context of LDS such connectives are also syntactically natural and offer no additional complexity costs.

We have the option of defining two logical systems. One is an ordinary predicate temporal logic (which is not an LDS) where the connectives G, H are labelled. We call it $LQ\mathbf{K}_t$ (next definition). This is the logic analogous to $Q\mathbf{K}_t$. The other system is an LDS formulation of $LQ\mathbf{K}_t$. This system will have (if we insist on being pedantic) two lots of labels: labels of the LDS and labels for the connectives. Thus we can write $t : F^xA$, where t is an LDS label from the labelling algebra $\mathcal{A}$ and x is a label from L; when we give semantics, both t and x will get assigned possible worlds. So to simplify the LDS version we can assume $L = \mathcal{A}$.

Definition 4.15 (The logic $LQ\mathbf{K}_t$).

1. Let L be a set of labels, and for each $x \in L$, let G^x and H^x be temporal connectives. The language of $LQ\mathbf{K}_t$ has the classical connectives, the traditional connectives G, H and the labelled connectives G^x, H^x, for each $x \in L$.
2. An $LQ\mathbf{K}_t$ model has the form(S, R, a, V, h, g), where (S, R, A, V, h) is a $Q\mathbf{K}_t$ model (see Definition 4.6) and $g : L \mapsto S$, assigning a world to each label. Satisfaction for G^x (resp. H^x) is defined by

   (3x) $t \vDash_{h,g} G^x(a_1, \ldots, a_n)$ iff for all s such that $tRs \wedge sRg(x)$ such that $a_1, \ldots, a_n \in B_s$, we have $s \vDash_{h,g} A$.

   The mirror image condition is required for H^x.

Definition 4.16 (LDS version of $LQ\mathbf{K}_t$). Our definition is in parallel to Definitions 4.7–4.9. We have the added feature that the language $\mathbf{L}$ of the LDS allows for the additional connectives F^t, P^t, G^t, H^t, where t is from the labelling algebra. For this reason we must modify the LDS notion of an $LQ\mathbf{K}_t$ theory and require that all the labels appearing in the connectives of the formulae of the theory are also members of D_1, the diagram of labels of the theory.

Definition 4.17 (LDS proof rules for $LQ\mathbf{K}_t$). We modify Definition 4.10 as follows:

1. F^x introduction case
   For some $t, s \in D_1, t < s < x \in D_2$ and $A \in \mathbf{f}(s)$, Δ' is the same as Δ except that $\mathbf{f}'(t) = \mathbf{f}(t) \cup \{F^xA\}$.
   Symbolically we write
   $$\Delta' = \Delta^x_{[t<s<x;s:A]}.$$

2. F^x elimination case
   For some $t \in D_1$ and some A, $F^x A \in \mathbf{f}(t)$, and for some new atomic $s \in U$ that does not appear in Δ, we have that Δ' is the same as Δ except that

$$\begin{aligned} D_1' &= \Delta_1 \cup \{s\}. \\ D_2' &= D_2 \cup \{t < s < x\}. \\ \mathbf{f}'(s) &= \{A\}. \end{aligned}$$

   Note that it may be that $\Delta_2 \cup \{t < s < x\}$ is inconsistent in which case Δ is inconsistent.
3. G^x elimination case
   For some $t \in D_1$ such that $t < s < x \in D_2$ we have $G^x A \in \mathbf{f}(t)$ and Δ' is like Δ except that $\mathbf{f}'(s) = f(s) \cup \{A\}$.

The P^x, H^x rules are the mirror images of all the above and all the other rules remain the same.

4. G^x, H^x introduction case
   This case is the same as in Definition 4.10 except that in the tex twe replace

$$D_2^1 = D_2 \cup \{t < t_1\}$$

   by $D_2^1 = D_2 \cup \{t < t_1 < x\}$ for the case of G^x and the mirror image for the case of H^x.
   Similarly we write $\Delta' = \Delta_{[t:\Box^x A]}$.

Theorem 4.18 (Soundness and completeness). *The LDS version of $LQ\mathbf{K}_t$ is sound and complete for the proposed semantics.*

Proof The soundness and completeness are proved along similar lines to the $Q\mathbf{K}_t$ case see Theorems 4.12 and 4.13. □

5

TEMPORAL LOGIC PROGRAMMING

5.1 Introduction

We can distinguish two views of logic, the declarative and the imperative. The declarative view is the traditional one, and it manifests itself both syntactically and semantically. Syntactically a logical system is taken as being characterized by its set of theorems. It is not important how these theorems are generated. Two different algorithmic systems generating the same set of theorems are considered as producing the same logic. Semantically a logic is considered as a set of formulae valid in all models. The model $\mathcal{M}$ is a static semantic object. We evaluate a formula φ in a model and, if the result of the evaluation is positive (notation $\mathcal{M} \models \varphi$), the formula is valid. Thus the logic obtained is the set of all valid formulae in some class $\mathcal{K}$ of models.

In contrast to the above, the imperative view regards a logic syntactically as a dynamically generated set of theorems. Different generating systems may be considered as different logics. The way the theorems are generated is an integral part of the logic. From the semantic viewpoint, a logical formula is not *evaluated* in a model but performs *actions* on a model to get a *new* model. Formulae are accepted as valid according to what they do to models. For example, we may take φ to be valid in $\mathcal{M}$ if $\varphi(\mathcal{M}) = \mathcal{M}$. (i.e. $\mathcal{M}$ is a fixed point of φ).

Applications of logic in computer science have mainly concentrated on the exploitation of its declarative features. Logic is taken as a language for describing properties of models. The formula φ is evaluated in a model $\mathcal{M}$. If φ holds in $\mathcal{M}$ (evaluation successful) then $\mathcal{M}$ has property φ. This view of logic is, for example, most suitably and most successfully exploited in the areas of databases and in program specification and verification. One can present the database as a deductive logical theory and query it using logical formulae. The logical evaluation process corresponds to the computational querying process. In program verification, for example, one can describe in logic the properties of the programs to be studied. The description plays the role of a model $\mathcal{M}$. One can now describe one's specification as a logical formula φ, and the query whether φ holds in $\mathcal{M}$ (denoted $\mathcal{M} \vdash \varphi$) amounts to verifying that the program satisfies the specification. These methodologies rely solely on the declarative nature of logic.

Logic programming as a discipline is also declarative. In fact it advertises itself as such. It is most successful in areas where the declarative component is dominant, e.g. in deductive databases. Its procedural features are not imperative (in our sense) but computational. In the course of evaluating whether $\mathcal{M} \vdash \varphi$, a procedural reading of $\mathcal{M}$ and φ is used. φ does not imperatively act on $\mathcal{M}$, the declarative logical features are used to guide a procedure—that of taking steps for finding whether φ is true. What does not happen is that $\mathcal{M}$ and φ are read imperatively, resulting in some action. In logic programming such actions (e.g. assert) are obtained by side-effects and special non-logical imperative predicates and are considered undesirable. There is certainly no conceptual framework within logic programming for allowing only those actions which have logical meaning.

Some researchers have come close to touching upon the imperative reading of logic. Belnap and Green [1994] and the later so-called data semantics school regard a formula φ as generating an action on a model $\mathcal{M}$, and changing it. See [van Benthem, 1996]. In logic programming and deductive databases the handling of integrity constraints borders on the use of logic imperatively. Integrity constraints have to be maintained. Thus one can either reject an update or do some corrections. Maintaining integrity constraints is a form of executing logic, but it is logically *ad hoc* and has to do with the local problem at hand. Truth maintenance is another form. In fact, under a suitable interpretation, one may view any resolution mechanism as model building which is a form of execution. In temporal logic, model construction can be interpreted as execution. Generating the model, i.e. finding the truth values of the atomic predicates in the various moments of time, can be taken as a sequence of execution.

As the need for the imperative executable features of logic is widespread in computer science, it is not surprising that various researchers have touched upon it in the course of their activity. However, there has been no conceptual methodological recognition of the imperative paradigm in the community, nor has there been a systematic attempt to develop and bring this paradigm forward as a new and powerful logical approach in computing.

The area where the need for the imperative approach is most obvious and pressing is temporal logic. In general terms, a temporal model can be viewed as a progression of ordinary models. The ordinary models are what is true at each moment of time. The imperative view of logic on the other hand also involves step-by-step progression in virtual 'time', involving both the syntactic generation of theorems and the semantic actions of a temporal formula on the temporal model. Can the two intuitive progressions, the semantic time and the action (transaction) time, be taken as the same? In the case of temporal logic the answer is 'yes'. We can act upon the models in the same time order as their chronological time. This means acting on earlier models first. In fact intuitively a future logical statement can be read (as we shall see) both declaratively and imperatively. Declaratively it describes what should be true, and imperatively it describes the actions to be taken to ensure that it becomes true. Since the chronology of the action sequence and the model sequence are the same, we can virtually *create* the future model by our actions. The logic USF, presented in Chapter 10, was the first attempt at promoting the imperative view as a methodology, with a proposal for its use as a language for controlling processes.

The purpose of this chapter is twofold:

1. to present a practical, sensible, logic programming machine for handling time and modality;
2. to present a general framework for extending logic programming to non-classical logics.

Point 1 is the main task of this chapter. It is done within the framework of 2.

Horn clause logic programming has been generalized in essentially two major ways:

1. using the metalevel features of ordinary Horn clause logic to handle time while keeping the syntactical language essentially the same;
2. enriching the syntax of the language with new symbols and introducing additional computation rules for the new symbols.

The first method is basically a simulation. We use the expressive power of ordinary Horn clause logic to talk about the new features. The *Demo* predicate, the *Hold* predicate and other metapredicates play a significant role.

The second method is more direct. The additional computational rules of the second method can be broadly divided into two:

2.1 Rewrites

2.2 Subcomputations

The rewrites have to do with simplifying the new syntax according to some rules (basically eliminating the new symbols and reducing goals and data to the old Horn clause language) and the subcomputations are the new computations which arise from the reductions.

Volume 1 examined the possibilities of handling time in classical logic. It examined the first approach and compared it with the second approach. We summarize the conclusions of Volume 1 in the following:

Given a temporal set of data, this set has the intuitive form:

'$A(x)$ is true at time t'.

This can be represented in essentially two ways (in parallel to the two methods discussed):

1. adding a new argument for time to the predicate A, writing $A^*(t,x)$ and working within an ordinary Horn clause computational environment;
2. leaving time as an external indicator and writing '$t : A(x)$' to represent the above temporal statement.

To compare the two approaches, imagine that we want to say the following:

'If $A(x)$ is true at t, then it will continue to be true'.

The first approach will write it as

$$\forall s(A^*(t,x) \wedge t < s \rightarrow A^*(s,x)).$$

The second approach has to talk about t. It would use a special temporal connective 'G' for 'always in the future'. Thus the data item becomes

$$t : A(x) \to GA(x).$$

It is equivalent to the following in the first approach:

$$A^*(t,x) \to \forall s[t < s \to A^*(s,x)].$$

The statement 'GA is true at t' is translated as

$$\forall s(t < s \to A^*(s)).$$

The second part of this chapter introduces temporal connectives and wants to discover what kind of temporal clauses for the new temporal language arise in ordinary Horn clause logic when we allow time variables in the atoms (e.g. $A^*(t,x), B^*(s,y)$) and allow time relations like $t < s$, $t = s$ for time points. This would give us a clue as to what kind of temporal Horn clauses to allow in the second approach. The computational tractability of the new language is assured, as it arises from Horn clause computation. Skolem functions have to be added to the Horn clause language, to eliminate the F and P connectives which are existential. All we have to do is change the computational rules to rely on the more intuitive syntactical structure of the second approach.

5.2 Temporal Horn clauses

Our problem for this section is to start with Horn clause logic with the ability to talk about time through time coordinates, and see what expressive power in term of connectives (P, F, G, H, etc.) is needed to do the same job. We then extend Prolog with the ability to compute directly with these connectives. The final step is to show that the new computation defined for P, F, G, H is really ordinary Prolog computation modified to avoid Skolemization.

Consider now a Horn clause written in predicate logic. Its general form is of course $\bigwedge$ atoms $\to$ atom. If our atomic sentences have the form $A(x)$ or $R(x,y)$ or $Q(x,y)$ then these are the atoms one can use in constructing the Horn clause. Let us extend our language to talk about time by following the first approach; that is, we can add time points, and allow special variables t, s to range over a flow of time $(T, <)$ (T can be the set of integers, for example) and write $Q^*(t,x,y)$ instead of $Q(x,y)$, where $Q^*(t,x,y)$ (also written as $Q(t,x,y)$, abusing notation) can be read as

$$'Q(x,y) \text{ is true at time } t.$$

We allow the use of $t < s$ to mean 't is earlier than s'. Recall that we do not allow mixed atomic sentences like $x < t$ or $x < y$ or $A(t,s,x)$ because these would read as

'John loves 1980 at 1979' or
'John $<$ 1980' or
'John $<$ Mary'.

Assume that we have organized our Horn clauses in such a manner: what kind of time expressive power do we have? Notice that our expressive power is potentially increased. We are committed, when we write a formula of the form $A(t,x)$, to t ranging over time and x over our domain of elements. Thus our model theory for classical logic (or Horn clause logic) does not accept any model for $A(t,x)$, but only models in which $A(t,x)$ is interpreted in this very special way. Meanwhile let us examine the syntactical expressive power we get when we allow for this two-sorted system and see how it compares with ordinary temporal and modal logics, with the connectives P, F, G, H.

When we introduce time variables t, s and the earlier–later relation into the Horn clause language we are allowing ourselves to write more atoms. These can be of the form

$$A(t,x,y)$$
$$t < s$$

(as we mentioned earlier, $A(t,s,y), x < y, t < y, y < t$ are excluded).

When we put these new atomic new sentences into a Horn clause we get the following possible structures for Horn clauses. $A(t,x), B(s,y)$ may also be *truth*.

($a0$) $A(t,x) \wedge B(s,y) \to R(u,z)$.
Here $<$ is not used.

($a1$) $A(t,x) \wedge B(s,y) \wedge t < s \to R(u,z)$.
Here $t < s$ is used in the body but the time variable u is not the same as t, s in the body.

($a2$) $A(t,x) \wedge B(s,y) \wedge t < s \to R(t,z)$.
Same as ($a1$) except the time variable u appears in the body as $u = t$.

($a3$) $A(t,x) \wedge B(s,y) \wedge t < s \to R(s,z)$.
Same as ($a1$) with $u = s$.

($a4$) $A(t,x) \wedge B(s,y) \to R(t,z)$.
Same as ($a0$) with $u = t$, i.e. the variable in the head appears in the body.

The other two forms (b) and (c) are obtained when the head is different: (b) for time independence and (c) for a pure $<$ relation.

(b) $A(t,x) \wedge B(s,y) \to R1(z)$

(b$'$) $A(t,x) \wedge B(s,y) \wedge t < s \to R1(z)$.

(c) $A(t,x) \wedge B(s,y) \to t < s$.

(d) $A((1970, x)$, where 1970 is a *constant* date.

Let us see how ordinary temporal logic with additional connectives can express directly, using the temporal connectives, the logical meaning of the above sentences. Note that if time is linear we can assume that one of $t < s$ or $t = s$ or $s < t$ always occur in the body of clauses because for linear time

$$\vdash \forall t \forall s[t < s \vee t = s \vee s < t],$$

and hence $A(t, x) \wedge B(s, y)$ is equivalent to

$$(A(t, x) \wedge B(t, y)) \vee (A(t, x) \wedge B(s, y) \wedge t < s) \vee (A(t, x) \wedge B(s, y) \wedge s < t).$$

Ordinary temporal logic over linear time allows the following connectives:

Fq, read: 'q will be true'
Pq, read: 'q was true'
$\Diamond q = q \vee Fq \vee Pq$
$\Box q = \sim \Diamond \sim q$
$\Box q$ is read: 'q is always true'.

If $[A](t)$ denotes, in symbols, the statement that A is true at time t, then we have

$$[Fq](t) \equiv \exists s > t([q](s))$$
$$[Pq](t) \equiv \exists s < t([q](s))$$
$$[\Diamond q](t) \equiv \exists s([q](s))$$
$$[\Box q](t) \equiv \forall s([q](s)].$$

Let us see now how to translate into temporal logic the Horn clause sentences mentioned above.

Case (a0) Statement (a0) reads

$$\forall t \forall s \forall u[A(t, x) \wedge B(s, y) \rightarrow R(u, z)].$$

If we push the quantifiers inside we get

$$\exists t A(t, x) \wedge \exists s B(x, y) \rightarrow \forall u R(u, z),$$

which can be written in the temporal logic as

$$\Diamond A(x) \wedge \Diamond B(y) \rightarrow \Box R(z).$$

If we do not push the $\forall u$ quantifier inside we get $\Box(\Diamond A(x) \wedge \Diamond B(y) \rightarrow R(z))$.

Case (a1) The statement (a1) can be similarly seen to read (we do not push $\forall t$ inside)

$$\forall t\{A(t, x) \wedge \exists s[B(s, y) \wedge t < s] \rightarrow \forall u R(u, z)\}$$

which can be translated as: $\Box\{A(x) \wedge FB(y) \rightarrow \Box R(z)\}$. Had we pushed $\forall t$ to the antecedent we would have got

$$\exists t[A(t, x) \wedge \exists s(B(s, y) \wedge t < s)] \rightarrow \forall u R(u, z),$$

which translates into

$$\Diamond[A(x) \wedge FB(y)] \rightarrow \Box R(z).$$

Case (a2) The statement ($a2$) can be rewritten as

$$\forall t[A(t,x) \wedge \exists s(B(s,y) \wedge t < s) \rightarrow R(t,z)],$$

and hence it translates to $\Box(A(x) \wedge FB(y) \rightarrow R(z))$.

Case (a3) Statement ($a3$) is similar to ($a2$). In this case we push the external $\forall t$ quantifier in and get

$$\forall s[\exists t[A(t,x) \wedge t < s] \wedge B(s,y) \rightarrow R(s,z)],$$

which translates to

$$\Box[PA(x) \wedge B(y) \rightarrow R(z)].$$

Case (a4) Statement ($a4$) is equivalent to

$$\forall t[A(t,x) \wedge \exists s B(s,y) \rightarrow R(t,z)],$$

and it translates to

$$\Box(A(x) \wedge \Diamond B(y) \rightarrow R(z)).$$

Case (b) The statement (b) is translated as

$$\Diamond(A(x) \wedge \Diamond B(y)) \rightarrow R1(z).$$

Case (b′) The statement (b′) translates into

$$\Diamond(A(x) \wedge FB(y)) \rightarrow R1(z).$$

Case (c) Statement (c) is a problem. It reads $\forall t \forall s[A(t,x) \wedge B(s,y) \rightarrow t < s]$; we do not have direct connectives (without negation) to express it. It says for any two moments of time t and s if $A(x)$ is true at t and $B(y)$ true at s then $t < s$. If time is linear then $t < s \vee t = s \vee s < t$ is true and we can write the conjunction

$$\sim \Diamond(A(x) \wedge PB(y)) \wedge \Diamond(A(x) \wedge B(y)).$$

Without the linearity of time how do we express the fact that t *should be* '$<$-related to s'?

We certainly have to go beyond the connectives $P, F, \Diamond, \Box$ that we have allowed here. We will not do that; see [Gabbay, 1998b].

Case (d) $A(1970, x)$ involves a constant, naming the date 1970. The temporal logic will also need a propositional constant *1970*, which is true *exactly* when the time is 1970, i.e.

$$\mathbf{M}t \vDash 1970 \text{ iff } t = 1970.$$

Thus (d) will be translated as $\Box(\mathit{1970} \rightarrow A(x))$. *1970* can be read as the proposition 'The time now is 1970'.

The above examples show what temporal expressions we can get by using Horn clauses with time variables as an *object* language. We are not discussing here the possibility of 'simulating' temporal logic in Horn clauses by using the Horn clause as a *metalanguage*. Horn clause logic can do that to any logic as can be seen from Hodges [Hodges, 1985].

Definition 5.1. The language contains $\wedge, \rightarrow, F$ (it will be the case) P (it was the case) and $\Box$ (it is always the case).

We define the notions of:
Ordinary clauses;
Always clauses;
Heads;
Bodies;
Goals.

1. A *clause* is either an always clause or an ordinary clause.
2. An *always clause* is $\Box A$ where A is an ordinary clause.
3. An *ordinary clause* is a head or an $A \rightarrow H$ where A is a body and H is a head.
4. A *head* is either an atomic formula or FA or PA, where A is a conjunction of ordinary clauses.
5. A *body* is an atomic formula, a conjunction of bodies, an FA or a PA, where A is a body.
6. A goal is any body.

Example 5.2.

$$a \rightarrow F((b \rightarrow Pq) \wedge F(a \rightarrow Fb))$$

is an acceptable clause.

$$a \rightarrow \Box b$$

is not an acceptable clause.

The reason for not allowing $\Box$ in the head is computational and not conceptual. The difference between a (temporal) logic programming machine and a (temporal) automated theorem prover is *tractability*. Allowing disjunctions in heads or $\Box$ in heads crosses the boundary of tractability. We can give computational rules for richer languages and we will in fact do so in later sections, but we will lose tractability; what we will have then is a theorem prover for full temporal logic.

Examples 5.3.

$$P[F(FA(x) \wedge PB(y) \wedge A(y)) \wedge A(y) \wedge B(x)] \rightarrow P[F(A(x) \rightarrow FP(Q(z) \rightarrow A(y)))]$$

is an ordinary clause. So is $a \rightarrow F(b \rightarrow Pq) \wedge F(a \rightarrow Fb)$, but not $A \rightarrow \Box b$.

First let us check the expressive power of this *temporal Prolog.* Consider

$$a \to F(b \to Pq).$$

This is an acceptable clause. Its predicate logic reading is

$$\forall t[a(t) \to \exists s > t[b(s) \to \exists u < sq(u)]].$$

Clearly it is more directly expressive than the Horn clause *Prolog* with time variables. Ordinary Prolog can rewrite the above as

$$\forall t(a(t) \to \exists s(t < s \wedge (b(s) \to \exists u(u < s \wedge q(u))))),$$

which is equivalent to

$$\forall t(\exists s \exists u(a(t) \to t < s \wedge (b(s) \to u < s \wedge q(u)))).$$

If we Skolemize with $\mathbf{s}_0(t)$ and $\mathbf{u}_0(t)$we get the clauses

$$\forall t[(a(t) \to t < \mathbf{s}_0(t)) \wedge (a(t) \wedge b(\mathbf{s}_0(t)) \to \mathbf{u}_0(t) < \mathbf{s}_0(t)) \wedge \\ (a(t) \wedge b(\mathbf{s}_0(t)) \to q(\mathbf{u}_0(t)))].$$

The following are representations of some of the problematic examples mentioned in the previous section.

($a1$) $\Box(A(x) \wedge FB(y) \to \Box R(z))$.
This is not an acceptable always clause but it can be equivalently written as

$$\Box(\Diamond(A(x) \wedge FB(y)) \to R(z)).$$

($a2$) $\Box(A(x) \wedge FB(y) \to R(z))$.
(b′) $\Diamond(A(x) \wedge FB(y)) \to R1(z)$.

(b) can be written as the conjunction below using the equation

$$\Diamond q = Fq \vee Pq \vee q : \\ (A(x) \wedge FB(y) \to R1(z)) \wedge \\ (F(A(x) \wedge FB(y)) \to R1(z)).$$

($a2$) Can be similarly written.
(c) can be written as

$$\forall t, s(A(x)(t) \wedge B(y)(s) \to t < s).$$

This is more difficult to translate. We need negation as failure here and write

$$\Box(A(x) \wedge PB(y) \to \bot) \\ \Box(A(x) \wedge B(y) \to \bot)$$

From now on we continue to develop the temporal logic programming machine.

5.3 LDS—Labelled Deductive System

This section will use the labelled deductive methodology of the previous chapter as a framework for developing the temporal Prolog machine. We begin by asking ourselves what is a temporal database? Intuitively, looking at existing real temporal problems, we can say that we have information about things happening at different times and some connections between them. Figure 5.1 is such an example.

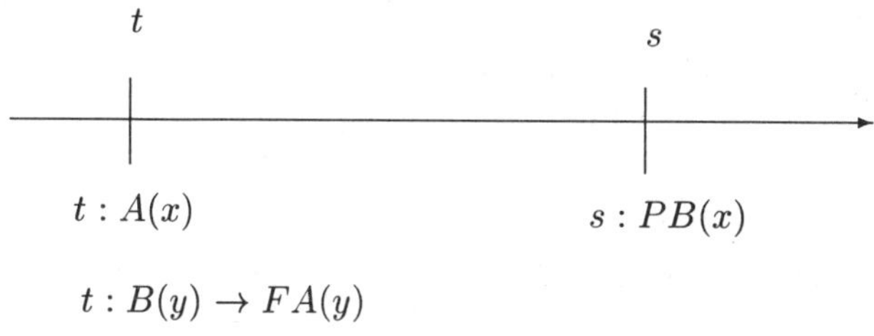

FIG. 5.1. A temporal configuration.

The diagram shows a finite set of points of time and some labelled formulae which are supposed to hold at the times indicated by the labels. Notice that we have labelled not only assertions but also Horn clauses showing dependences across times. Thus at time t it may be true that B will be true. We represent that as $t : FB$. The language we are using has F and P as connectives. It is possible to have more connectives and still remain within the Horn clause framework. Most useful among them are '$t : F^s A$' and '$t : P^s A$', reading '$t < s$ and $s : A$' and '$s < t$ and $s : A$'. In words: 'A will be true at time $s > t$'.

The temporal configuration comprises two components.

1. A (finite) graph $(\rho, <)$ of time points and the temporal relationships between them.
2. With each point of the graph we associate a (finite) set of clauses and assertions, representing what is true at that point.

In Horn clause computational logic, there is an agreement that if a formula of the form $A(x) \to B(x)$ appears in the database with x free then it is understood that x is universally quantified. Thus we assume $\forall x(A(x) \to B(x))$ is in the database. The variable x is then called universal (or type 1). In the case of modal and temporal logics, we need another type of variable, called type 2 or a Skolem variable. To explain the reason, consider the item of data

$$`t : FB(x)'.$$

This reads, according to our agreement,

'$\forall x F B(x)$ true at t.'

For example, it might be the sentence: t: 'Everyone will leave'.

The time in the future in which $B(x)$ is true depends on x. In our example, the time of leaving depends on the person x. Thus, for a given unknown (uninstantiated) u, i.e. for a given person u which we are not yet specifying, we know there must be a point t_1 of time (t_1 is dependent on u) with $t_1 : B(u)$. This is the time in which u leaves.

This u is by agreement not a type 1 variable. It is a u to be chosen later. Really u is a Skolem constant and we do not want to and cannot read it as $t_1 : \forall u B(u)$. Thus we need two types of variables. The other alternative is to make the dependency of t_1 on u explicit and to write

$$t_1(x) : B(x)$$

with x a universal type 1 variable, but then the object language variable x appears in the world indices as well. The world indices, i.e. the t, are external to the formal clausal temporal language, and it is simpler not to mix the t and the x. We chose the two types of variable approach. Notice that when we ask for a goal $?G(u)$, u is a variable to be instantiated, i.e. a type 2 variable. So we have these variables anyway, and we prefer to develop a systematic way of dealing with them.

To explain the role of the two types of variables, consider the following classical Horn clause database and query:

$$\begin{array}{l} A(x,y) \to B(x,y) \; ?B(u,u) \\ A(a,a). \end{array}$$

This means 'Find an instantiation u_0 of u such that $\forall x, y[A(x,y) \to B(x,y)] \wedge A(a,a) \vdash B(u_0,u_0)$'. There is no reason why we cannot allow for the following

$$\begin{array}{l} A(u,y) \to B(x,u) \; ?B(u,u) \\ A(a,a). \end{array}$$

In this case we want to find a u_0 such that

$$\forall x, y[A(u_0,y) \to B(x,u_0)] \wedge A(a,a) \vdash B(u_0,u_0)$$

or to show

$$\vdash \exists u\{[\forall x, y[A(u,y) \to B(x,u)] \wedge A(a,a) \to B(u,u)]\}$$

u is called a type 2 (Skolem) variable and x, y are universal type 1 variables. Given a database and a query of the form $\Delta(x,y,u)?Q(u)$, success means $\vdash \exists u[\forall x, y\Delta(x,y,u) \to Q(u)]$.

The next sequence of definitions will develop the syntax of the temporal Prolog machine. A lot depends on the flow of time. We will give a general definition (Definition 5.6 below), which includes the following connectives:

$\Box$ Always.
F It will be the case.
P It was the case.
G It will always be the case (not including now).
H It has always been the case (up to now and not including now);.
$\bigcirc$ Next moment of time (in particular it implies that such a moment of time exists).
$\bullet$ Previous moment of time (in particular it implies that such a moment of time exists).

Later on we will also deal with S (Since) and U (Until).

The flows of time involved are mainly three:

- general partial orders $(T, <)$;
- linear orders;
- the integers or the natural numbers.

The logic and theorem provers involved, even for the same connectives, are different for different partial orders. Thus the reader should be careful to note in which flow of time we are operating. Usually the connectives $\bigcirc$ and $\bullet$ assume we are working in the flow of time of integers.

Having fixed a flow of time $(T, <)$, the temporal machine will generate finite configurations of points of time according to the information available to it. These are denoted by $(\rho, <)$. We are supposed to have $\rho \subseteq T$ (more precisely, ρ will be homomorphic into T), and the ordering on ρ will be the same as the ordering on T. The situation gets slightly complicated if we have a new point s and we do not know where it is supposed to be in relation to known points. We will need to consider all possibilities. Which possibilities do arise depend on $(T, <)$, the background flow of time we are working with. Again we should watch for variations in the sequel.

Definition 5.4. Let $(\rho, <)$ be a finite partial order. Let $t \in \rho$ and let s be a new point. Let $\rho' = \rho \cup \{s\}$, and let $<'$ be a partial order on ρ'. Then $(\rho', <, t)$ is said to be a (one new point) *future* (resp. *past*) *configuration of* $(\rho, <, t)$ iff $t <' s$ (resp. $s <' t$) and $\forall xy \in \rho(x < y \leftrightarrow x <' y)$.

Example 5.5. Consider a general partial flow $(T, <)$ and consider the subflow $(\rho, <)$.

The possible future configurations (relative to $T, <$) of one additional point s are displayed in Fig. 5.2.

For a finite $(\rho, <)$ there is a finite number of future and past non-isomorphic configurations. This finite number is exponential in the size of ρ. So in the general case without simplifying assumptions we will have an intractable exponential computation. A configuration gives all possibilities of putting a point in the future or past.

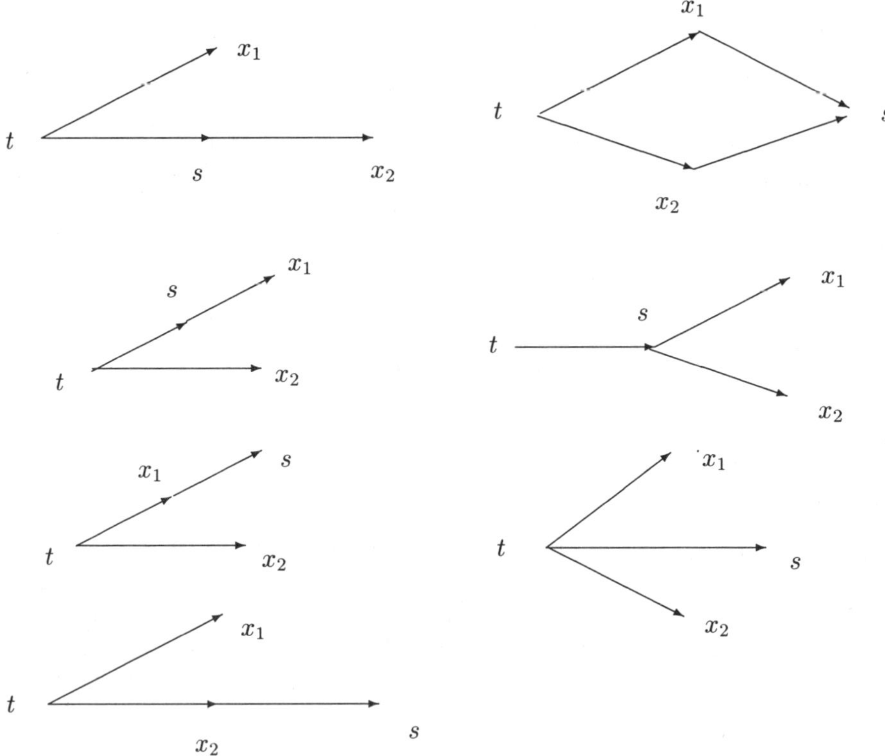

FIG. 5.2.

In the case of an ordering in which a next element or a previous element exists (like $t+1$ and $t-1$ in the integers) the possibilities for configurations are different. In this case we must assume that we know the *exact* distance between the elements of $(\rho, <)$.

For example, in the configuration $\{t < x_1, t < x_2\}$ of Fig. 5.3 we may have the further following information as part of the configuration:

$$t = \bullet^{18} x_1$$

$$t = \bullet^{6} x_2$$

so that we have only a finite number of possibilities for putting s in.

Note that although $\bullet$ operates on propositions, it can also be used to operate on points of time, denoting the predecessor function.

Definition 5.6. Consider a temporal Prolog language with the following connectives and predicates:

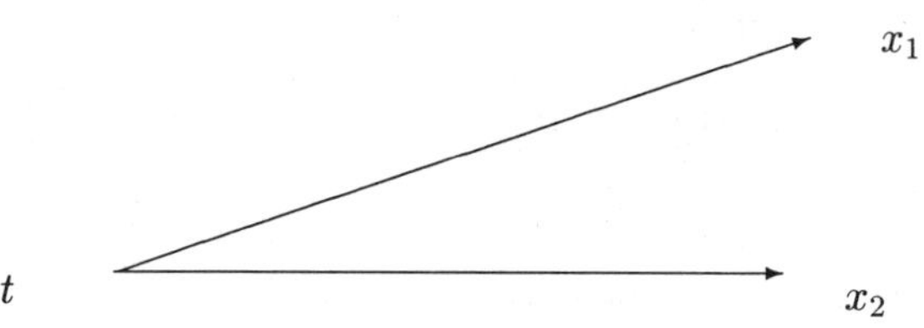

FIG. 5.3.

1. atomic predicates;
2. function symbols and constants;
3. two types of variables:
   universal variables (type 1) $V = \{x_1, y_1, z_1, x_2, y_2, z_2, \ldots\}$
   and Skolem variables (type 2) $U = \{u_1, v_1, u_2, v_2, \ldots\}$;
4. the connectives $\wedge, \rightarrow, \vee, F, P, \bigcirc, \bullet, \Box$ and $\neg$.

FA reads: it will be the case that A.
PA reads: it was the case that A.
$\bigcirc A$ reads: A is true tomorrow (if a tomorrow exists; if tomorrow does not exist then it is false).
$\bullet A$ reads: A was true yesterday (if yesterday does not exist then it is false).
$\neg$: represents negation by failure.

We define now the notions of an *ordinary clause*, an *always clause*, a *body*, a *head* and a *goal*.

1. A *clause* is either an always clause or an ordinary clause.
2. An *always clause* has the form $\Box A$, where A is an ordinary clause.
3. An *ordinary clause* is a head or an $A \rightarrow H$, where A is a body and H is a head.
4. A *head* is an atomic formula or an FA or a PA or an $\bigcirc A$ or an $\bullet A$, where A is a finite conjunction of ordinary clauses.
5. A *body* is an atomic formula or an FA or a PA or an $\bigcirc A$ or an $\bullet A$ or $\neg A$ or a conjunction of bodies where A is a body.
6. A *goal* is a body whose variables are *all* Skolem variables.
7. A disjunction of goals is also a goal.

Remark 5.7. Definition 5.6 included all possible temporal connectives. In practice different systems may contain only some of these connectives. For example, a modal system may contain only $\Diamond$ (corresponding to F) and $\Box$. A future discrete system may contain only $\bigcirc$ and F etc.

Depending on the system and the flow of time, the dependences between the connectives change. For example, we have the equivalence

$$\Box(a \to \bigcirc b) \text{ and } \Box(\bullet a \to b)$$

whenever both $\bullet a$ and $\bigcirc b$ are meaningful.

Definition 5.8. Let $(T, <)$ be a flow of time. Let $(\rho, <)$ be a finite partial order. A *labelled temporal database* is a set of labelled ordinary clauses of the form $(t_i : A_i), t \in \rho$, and always clauses of the form $\Box A_i, A_i$ a clause. A labelled goal has the form $t : G$, where G is a goal.

Δ is said to be a labelled temporal database over $(T, <)$ if $(\rho, <)$ is homomorphic into $(T, <)$.

Definition 5.9. We now define the computation procedure for the temporal Prolog for the language of Definitions 5.6 and 5.8. We assume a flow of time $(T, <)$. $\rho \subseteq T$ is the finite set of points of time involved so far in the computation. The exact computation steps depend on the flow of time. It is different for branching, discrete linear, etc. We will give the definition for linear time, though not necessarily discrete. Thus the meaning of $\bigcirc A$ in this logic is that there exists a next moment and A is true at this next moment. Similarly for $\bullet A$. $\bullet A$ reads: there exists a previous moment and A was true at that previous moment.

We define the success predicate $\mathbf{S}(\rho, <, \Delta, G, t, G_0, t_0, \Theta)$ where $t \in \rho, (\rho, <)$ is a finite partial order and Δ is a set of labelled clauses $(t : A), t \in \rho$.

$\mathbf{S}(\rho, <, \Delta, G, t, G_0, t_0, \Theta)$ reads: the labelled goal $t : G$ succeeds from Δ under the substitution Θ to all the type 2 variables of G and Δ in the computation with starting labelled goal $t_0 : G_0$.

When Θ is known, we write $\mathbf{S}(\rho, <, \Delta, G, t, G_0, t_0)$ only.

We define the simultaneous success and failure of a set $\mathbf{\Pi}$ of metapredicates of the form $\mathbf{S}(\rho, <, \Delta, G, t, G_0, t_0)$ under a substitution Θ to type 2 variables. To explain the intuitive meaning of success or failure, assume first that Θ is a substitution which grounds all the Skolem type 2 variables. In this case $(\mathbf{\Pi}, \Theta)$ succeeds if by definition all $\mathbf{S}(\rho, <, \Delta, G, t, G_0, t_0, \Theta) \in \mathbf{\Pi}$ succeed and $(\mathbf{\Pi}, \Theta)$ fails if at least one of $\mathbf{S} \in \mathbf{\Pi}$ fails. The success or failure of $\mathbf{S}$ for a Θ as above has to be defined recursively. For a general Θ, $(\mathbf{\Pi}, \Theta)$ succeeds, if for some Θ' such that $\Theta\Theta'$ grounds all type 2 variables $(\mathbf{\Pi}, \Theta\Theta')$ succeeds. $(\mathbf{\Pi}, \Theta)$ fails if for all Θ' such that $\Theta\Theta'$ grounds all type 2 variables we have that $(\mathbf{\Pi}, \Theta\Theta')$ fails. We need to give recursive procedures for the computation of the success and failure of $(\mathbf{\Pi}, \Theta)$. In the case of the recursion, a given $(\mathbf{\Pi}, \Theta)$ will be changed to a $(\mathbf{\Pi}', \Theta')$ by taking $\mathbf{S}(\rho, <, \Delta, G, t, G_0, t_0) \in \mathbf{\Pi}$ and replacing it by $\mathbf{S}(\rho', <', \Delta', G', t', G_0, t_0)$. We will have several such changes and thus get several $\mathbf{\Pi}'$ by replacing several $\mathbf{S}$ in $\mathbf{\Pi}$. We write the several possibilities as $(\mathbf{\Pi}'_i, \Theta'_i)$. If we write $(\mathbf{\Pi}, \Theta)$ to mean $(\mathbf{\Pi}, \Theta)$ succeeds and $\sim (\mathbf{\Pi}, \Theta)$ to read $(\mathbf{\Pi}, \Theta)$ fails, then our recursive computation rules have the form: $(\mathbf{\Pi}, \Theta)$ succeeds (or fails) if some Boolean combination of $(\mathbf{\Pi}'_i, \Theta'_i)$ succeeds (or fails). The rules allow us to pick an element in $\mathbf{\Pi}$, e.g. $\mathbf{S}(\rho, <, \Delta, G, t, G_0, t_0)$, and replace it with one or more elements to obtain the different $(\mathbf{\Pi}'{}_i, \Theta'_i)$, where Θ'_i is obtained from Θ. In case of failure we require that Θ grounds all type 2 variables. We do not define failure for a non-grounding Θ.

To summarize the general structure of the rules is:
$(\mathbf{\Pi}, \Theta)$ succeeds (or fails) if some Boolean combination of the successes and failures of some $(\mathbf{\Pi}'_i, \Theta'_i)$ holds and $(\mathbf{\Pi}, \Theta)$ and $(\mathbf{\Pi}'_i, \Theta'_i)$ are related according to one of the following cases:

Case I If $\mathbf{\Pi} = \emptyset$ then $(\mathbf{\Pi}, \Theta)$ succeeds (i.e. the Boolean combination of $(\mathbf{\Pi}_i, \Theta_i)$ is *truth*).

Case II $(\mathbf{\Pi}, \Theta)$ fails if for some $\mathbf{S}(\rho, <, \Delta, G, t, G_0, t_0)$ in $\mathbf{\Pi}$ we have G is atomic and for all $\Box(A \to H) \in \Delta$ and for all $(t : A \to H) \in \Delta, H\Theta$ does *not* unify with $G\Theta$. Further, for all Ω and s such that $t = \Omega s$ and for all $s : A \to \Omega H$ and all $\Box(A \to \Omega H)$ we have $H\Theta$ does not unify with $G\Theta$, where Ω is a sequence of $\bigcirc$ and $\bullet$.

Remark 5.10. We must qualify the conditions of the notion of failure. If we have a goal $t : G$, with G atomic, we know for sure that $t : G$ finitely fails under a substitution Θ, if $G\Theta$ cannot unify with any head of a clause. This is what the condition above says. What are the candidates for unification? These are either clauses of the form $t : A \to H$, with H atomic, or $\Box(A \to H)$, with H atomic.

Do we have to consider the case where H is not atomic? The answer depends on the flow of time and on the configuration $(\rho, <)$ we are dealing with. If we have, say, $t : A \to FG$ then if $A \to FG$ is true at t, G would be true (if at all) in some s, $t < s$. This s is irrelevant to our query $?t : G$. Even if we have $t' < t$ and $t' : A \to FG$ and A true at t', we can still ignore this clause because we are not assured that any s such that $t' < s$ and G true at s would be the desired t (i.e. $t = s$).

The only case we have to worry about is when the flow of time and the configuration are such that we have, for example, $t' : A \to \bigcirc^5 G$ and $t = \bigcirc^5 t'$.

In this case we must add the following clause to the notion of failure: for every s such that $t = \bigcirc^n s$ and every $s : A \to \bigcirc^n H$, $G\Theta$ and $H\Theta$ do not unify.

We also have to check what happens in the case of always clauses.

Consider an integer flow of time and the clause $\Box(A \to \bigcirc^5 \bullet^{27} H)$. This is true at the point $s = \bullet^5 \bigcirc^{27} t$ and hence for failure we need that $G\Theta$ does not unify with $H\Theta$.

The above explains the additional condition on failure.

The following conditions 1–10, 12–13 relate to the success of $(\mathbf{\Pi}, \Theta)$ if $(\mathbf{\Pi}'_i, \Theta'_i)$ succeed. Condition (11) uses the notion of failure to give the success of negation by failure. Conditions 1–10, 12–13 give certain alternatives for success. They give failure if each one of these alternatives ends up in failure.

1. **Success rule for atomic query:**
   $\mathbf{S}(\rho, <, \Delta, G, t, G_0, t_0) \in \mathbf{\Pi}$ and G is atomic and for some head H, $(t : H) \in \Delta$ and for some substitutions Θ_1 to the universal variables of H and Θ_2 to the existential variables of H and G we have $H\Theta_1\Theta\Theta_2 = G\Theta\Theta_2$ and $\mathbf{\Pi}' = \mathbf{\Pi} - \{\mathbf{S}(\rho, <, \Delta, G, t, G_0, t_0)\}$ and $\Theta' = \Theta\Theta_2$.

2. **Computation rule for atomic query:**
$\mathbf{S}(\rho, <, \Delta, G, t, G_0, t_o) \in \mathbf{\Pi}$ and G is atomic and for some $(t : A \to H) \in \Delta$ or for some $\Box(A \to H) \in \Delta$ and for some Θ_1, Θ_2, we have $H\Theta_1\Theta\Theta_2 = G\Theta\Theta_2$ and $\mathbf{\Pi}' = (\mathbf{\Pi} - \{\mathbf{S}(\rho, <, \Delta, G, t, G_0, t_0)\}) \cup \{\mathbf{S}(\rho, <, \Delta, A\Theta_1, t, G_0, t_0)\}$ and $\Theta' = \Theta\Theta_2$.

The above rules deal with the atomic case. Rules 3, 4 and 4* deal with the case the goal is FG. The meaning of 3, 4 and 4* is the following. We ask FG at t. How can we be sure that FG is true at t? There are two possibilities, (a) and (b):

(a) We have $t < s$ and at $s : G$ succeeds. This is rule 3;

(b) Assume that we have the fact that $A \to FB$ is true at t. We ask for A and succeed and hence FB is true at t. Thus there should exist a point s' in the future of t where B is true. Where can s' be? We don't know where s' is in the future of t. So we consider all future configurations for s'. This gives us all future possibilities where s' can be. We assume for each of these possibilities that B is true at s' and check whether either G follows at s' or FG follows at s'. If we find that for all future constellations of where s' can be $G \vee FG$ succeeds in s' from B, then FG holds at t. Here we use the transitivity of $<$. Rule 4a gives the possibilities where s' is an old point s in the future of t; Rule 4b gives the possibilities where s' is a new point forming a new configuration. Success is needed from *all* possibilities.

3. **Immediate rule for** F**:**
$\mathbf{S}(\rho, <, \Delta, FG, t, G_0, t_0) \in \mathbf{\Pi}$ and for some $s \in \rho$ such that $t < s$ we have $\mathbf{\Pi}' = (\mathbf{\Pi} - \{\mathbf{S}(\rho, <, \Delta, FG, t, G_0, t_0)\}) \cup \{\mathbf{S}(\rho, <, \Delta, G, s, G_0, t_0)\}$ and $\Theta' = \Theta$.
4. **First configuration rule for** F**:**
$\mathbf{S}(\rho, <, \Delta, FG, t, G_0, t_0) \in \mathbf{\Pi}$ and for some $(t : A \to F \wedge_j B_j) \in \Delta$ and some Θ_1, Θ_2 we have both (a) and (b) below are true. A may not appear in which case we pretend $A =$ truth.
   (a) For *all* $s \in \rho$ such that $t < s$ we have that $\mathbf{\Pi}'_s = (\mathbf{\Pi} - \{\mathbf{S}(\rho, <, \Delta, FG, t, G_0, t_0)\}) \cup \{\mathbf{S}(\rho, <, \Delta, E\Theta_1, t, G_0, t_0)\} \cup \{\mathbf{S}(\rho, <, \Delta \cup \{(s : B_j\Theta_1) \mid j = 1, 2, \ldots\}, D, s, G_0, t_0)\}$ succeeds with $\Theta'_s = \Theta\Theta_2$ and $D = G \vee FG$ and $E = A$.
   (b) For all future configurations of $(\rho, <, t)$ with a new letter s, denoted by the form $(\rho_s, <_s)$, we have that

$$\begin{aligned}\mathbf{\Pi}'_s = \quad & (\mathbf{\Pi} - \{\mathbf{S}(\rho, <, \Delta, FG, t, G_0, t_0\}) \\ & \cup \{\mathbf{S}(\rho, <, \Delta, E\Theta_1, t, G_0, t_0)\} \\ & \cup \{\mathbf{S}(\rho_s, <_s, \Delta \cup \{(s : B_j) \mid j = 1, 2, \ldots\}, D, s, G_0, t_0)\}\end{aligned}$$

   succeeds with $\Theta'_s = \Theta\Theta_2$ and $D = G \vee FG$ and $E = A$.

The reader should note that conditions 3, 4a and 4b are needed only when the flow of time has some special properties. To explain by example, assume we have the configuration of Fig. 5.4 and $\Delta = \{t : A \to FB, t' : C\}$ as data, and our query is $?t : FG$.

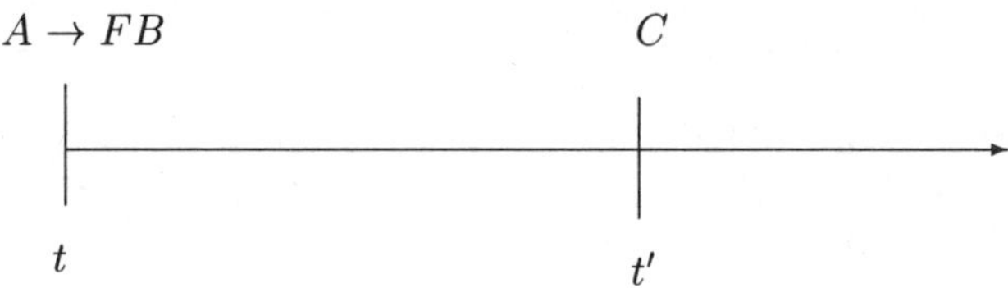

FIG. 5.4.

Then according to rules 3, 4 we have to check and succeed in all the following cases:

1. from rule 3 we check $\{t' : C, t : A \to FB\}?t' : G$;
2. from rule 4a we check $\{t' : C, t : A \to FB, t' : B\}?t' : G$;
3. from rule 4b we check $\{t' : C, t : A \to FB, s : B\}?s : G$;

for the three configurations of Fig. 5.5.

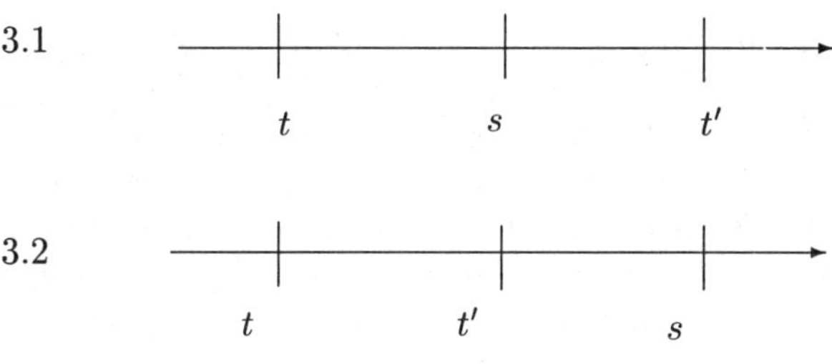

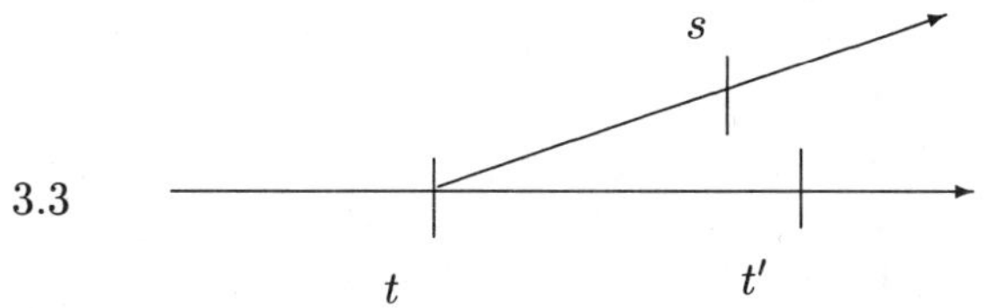

FIG. 5.5.

If time is linear, configuration 3.3, shown in Fig. 5.5, does not arise and we are essentially checking 3.1, 3.2 of Fig. 5.5 and the case 4a corresponding to $t' = s$.

If we do not have any special properties of time, success in case 3.2 is required. Since we must succeed in all cases and 3.2 is the case with *least* assumptions, it is enough to check 3.2 alone.

Thus for the case of no special properties of the flow of time, case 4 can be replaced by case 4 general below:

4 **general** $\mathbf{S}(\rho, <, \Delta, FG, t, G_0, t_0) \in \mathbf{\Pi}$ and for the future configuration $(\rho_1, <_1)$ defined as $\rho_1 = \rho \cup \{s\}$ and $<_1 = < \cup \{t < s\}$, s a new letter, we have that: $\mathbf{\Pi}'s = (\mathbf{\Pi} - \{\mathbf{S}(\rho, <, \Delta, FG, t, G_0, t_0)\} \cup \{\mathbf{S}(\rho, <, \Delta, E\Theta_1, t, G_0, t_0)\} \cup \{\mathbf{S}(\rho_1, <_1, \Delta \cup \{(s : B_j) \mid j = 1, 2, \ldots\}, D, s, G_0, t_0)\}$ succeeds with $\Theta'_s = \Theta\Theta_1$ and $D = G \vee FG$ and $E = A$.

4*. **Second configuration rule for F:** For some $\mathbf{S}(\rho, <, \Delta, FG, t, G_0, t_0)$ and some $\Box(A \to F \wedge_j B_j) \in \Delta$ and some $\Theta_1\Theta_2$ we have both cases 4a and 4b above true with $E = A \vee FA$ and $D = G \vee FG$.

4* **general** Similar to (4 **general**) for the case of general flow.

5. This is the the mirror image of 3 with 'PG' replacing 'FG' and '$s < t$' replacing '$t < s$'.

6; 6* This is the mirror image of 4 and 4* with 'PG' replacing 'FG', '$s < t$' replacing '$t < s$' and 'past configuration' replacing 'future configuration'.

6 **general** This is the image of 4 **general**.

We now give the computation rules 7-10 for $\bigcirc$ and $\bullet$ for orderings in which a next point and/or previous points exist. If $t \in T$ has a next point we denote this point by $s = \bigcirc t$. If it has a previous point we denote it by $s = \bullet t$. For example, if $(T, <)$ is the integers then $\bigcirc t = t + 1$ and $\bullet t = t - 1$. If $(T, <)$ is a tree then $\bullet t$ always exists, except at the root, but $\bigcirc t$ may or may not exist. For the sake of simplicity we must assume that if we have $\bigcirc$ or $\bullet$ in the language then $\bigcirc t$ or $\bullet t$ always exist. Otherwise we can sneak negation in by putting $(t : \bigcirc A) \in \Delta$ when $\bigcirc t$ does not exist!

7. **Immediate rule for $\bigcirc$:** $\mathbf{S}(\rho, <, \Delta, \bigcirc G, t, G_0, t_0) \in \mathbf{\Pi}$ and $\bigcirc t$ exists and $\bigcirc t \in \rho$ and $\Theta' = \Theta$ and $\mathbf{\Pi}' = (\mathbf{\Pi} - \{\mathbf{S}(\rho, <, \Delta, \bigcirc G, t, G_0, t_0)\}) \cup \{\mathbf{S}(\rho, <, \Delta, G, \bigcirc t, G_0, t_0)\}$.
8. **Configuration rule for $\bigcirc$:** $\mathbf{S}(\rho, <, \Delta, \bigcirc G, t, G_0, t_0) \in \mathbf{\Pi}$ and for some Θ_1, Θ_2 some $(t : A \to \bigcirc \wedge_j B_j) \in \Delta$ and $\mathbf{\Pi}' = (\mathbf{\Pi} - \{\mathbf{S}(\rho, <, \Delta, \bigcirc G, t, G_0, t_0)\} \cup \{\mathbf{S}(\rho, <, \Delta, A\Theta_1, t, G_0, t_0)\} \cup \{\mathbf{S}(\rho \cup \{\bigcirc t\}, <', \Delta \cup \{(\bigcirc t : B_j)\}, G, \bigcirc t, G_0, t_0)\}$ succeeds with $\Theta' = \Theta\Theta_2$, and $<'$ is the appropriate ordering closure of $< \cup \{(t, \bigcirc t)\}$.
   Notice that case 8 is parallel to case 4. We do not need 8a and 8b because of $\bigcirc t \in \rho$; then what would be case 8b becomes 7.
9. The mirror image of 7 with '$\bullet$' replacing '$\bigcirc$'.
10. The mirror image of 8 with '$\bullet$' replacing '$\bigcirc$'.

11. **Negation as failure rule:**
$\mathbf{S}(\rho, <, \Delta, \neg G, t, G_0, t_0) \in \mathbf{\Pi}$ and Θ grounds every type 2 variable and the computation for success of $\mathbf{S}(\rho, <, \Delta, G, t, \Theta)$ ends up in failure.
12. **Disjunction rule:**
$\mathbf{S}(\rho, <, \Delta, G_1 \vee G_2, t, G_0, t_0) \in \mathbf{\Pi}$ and $\mathbf{\Pi}' = (\mathbf{\Pi} - \{\mathbf{S}(\rho, <, \Delta, G_1 \vee G_2, t, G_0, t_0)\}) \cup \{\mathbf{S}(\rho, <, \Delta, G_i, t, G_0, t_0)\}$ and $\Theta' = \Theta$ and $i \in \{1, 2\}$.
13. **Conjunction rule:**
$\mathbf{S}(\rho, <, \Delta, G_1 \wedge G_2, t, G_0, t_0) \in \mathbf{\Pi}$ and $\mathbf{\Pi}' = (\mathbf{\Pi} - \{\mathbf{S}(\rho, <, \Delta, G_1 \wedge G_2, t, G_0, t_0)\}) \cup \{\mathbf{S}(\rho, <, \Delta, G_i, t, G_0, t_0) \mid i \in \{1, 2\}\}$.
14. **Restart rule:**
$\mathbf{S}(\rho, <, \Delta, G, t, G_0, t_0) \in \mathbf{\Pi}$ and $\mathbf{\Pi}' = (\mathbf{\Pi} - \{\mathbf{S}(\rho, <, \Delta, G, t, G_0, t_0)\}) \cup \{\mathbf{S}(\rho, <, \Delta, G_1, t_0, G_0, t_0)\}$ where G_1 is obtained from G_0 by substituting completely new type 2 variables u_i' for the type 2 variables u_i of G_0, and where Θ' extends Θ by giving $\Theta'(u_i') = u_i'$ for the new variables u_i'.
15. **To start the computation:**
Given Δ and $t_0 : G_0$ and a flow $(T, <)$, we start the computation with $\mathbf{\Pi} = \{\mathbf{S}(\rho, <, \Delta, G_0, t_0, G_0, t_0)\}$, where $(\rho, <)$ is the configuration associated with Δ, over $(T, <)$ (Definition 5.8).

Let us check some examples.

Example 5.11.
Data:

1. $t : a \to Fb$
2. $\Box(b \to Fc)$
3. $t : a$.

Query: $?t : Fc$
Configuration: $\{t\}$

Using rule 4* we create a future s with $t < s$ and ask the two queries (the notation $A?B$ means that we add A to the data 1, 2, 3 and ask $?B$).

4. $?t : C \vee Fb$
and
5. $s : c?s : c \vee Fc$
5 succeeds and 4 splits into two queries by rule 4.
6. $?t : a$
and
7. $s' : b?s' : b$.

Example 5.12.
Data:

1 $t : FA$
2 $t : FB$

Query: $t : F\varphi$ where $\varphi = (A \wedge B) \vee (A \wedge FB) \vee (B \wedge FA)$.

The query will fail in any flow of time in which the future is not linear. The purpose of this example is to examine what happens when time is linear. Using 1 we introduce a point s, with $s : A$, and query from s the following:

$$?s : \varphi \vee F\varphi$$

If we do not use the restart rule, the query will fail. Now that we are at a point s there is no way to go back to t. We therefore cannot reason that we also have a point $s' : B$ and $t < s$ and $t < s'$ and that because of linearity $s = s'$ or $s < s'$ or $s' < s$. However, if we are allowed to restart, we can continue and ask $t : F\varphi$ and now use the clause $t : FB$ to introduce s'. We now reason using linearity in rule 4 that the configurations are:

$t < s < s'$
or $t < s' < s$
or $t < s = s'$

and φ succeeds at t for each configuration.

The reader should note the reason for the need to use the restart rule. When time is just a partial order, the two assumptions $t : FA$ and $t : FB$ do not interact. Thus when asking $t : FC$, we know that there are two points $s_1 : A$ and $s_2 : B$;, see Fig. 5.6.

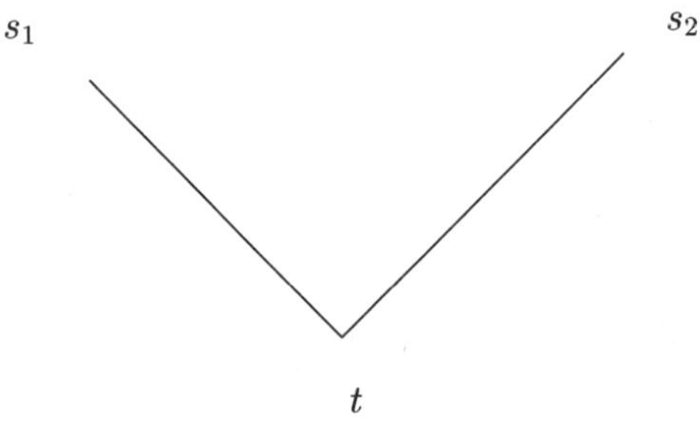

FIG. 5.6.

C can be true in either one of them. $s_1 : A$ has no influence on $s_2 : B$. When conditions on time (such as linearity) are introduced, s_1 does influence s_2 and hence we must introduce both at the same time. When one does forward deduction one can introduce both s_1 and s_2 going forward. The backward rules do not allow for that. That is why we need the restart rule. When we restart, we keep all that has been done (with, for example, s_1) and have the opportunity to restart with s_2. The restart rule can be used to solve the linearity problem for classical logic only. Its side-effect is that it turns intuitionistic logic into classical logic; see Gabbay's paper on N-Prolog [Gabbay, 1985] and [Gabbay and Olivetti, 1999].

In theorem proving based on intuitionistic logic where disjunctions are allowed, forward reasoning cannot be avoided. See the next example.

It is instructive to translate the above into Prolog and see what happens there.

Example 5.13.

1. $t : FA$ translates into $(\exists s_1 > t)A^*(s_1)$.
2. $t : FB$ translates into $(\exists s_2 > t)A^*(s_2)$.
   The query translates into the formula $\psi(t)$:
   $\psi = \exists s > t[A^*(s) \wedge B^*(s)] \vee \exists s_1 > t[A^*(s_1) \wedge \exists s_2 > s_1 B^*(s_2)] \vee \exists s_2 > t[B^*(s_2) \wedge \exists s_1 > s_2 A^*(s_2)]$
   which is equivalent to the disjunction of:
   (a) $[t < s \wedge A^*(s) \wedge B^*(s)]$;
   (b) $t < s_1 \wedge s_1 < s_2 \wedge A^*(s_1) \wedge B^*(s_2)$;
   (c) $t < s_1 \wedge s_2 < s_1 \wedge A^*(s_1) \wedge B^*(s_2)$.

All of (a), (b), (c) fail from the data, unless we add to the data the disjunction

$$\forall xy(x < y \vee x = y \vee y < x).$$

Since this is not a Horn clause, we are unable to express it in the database.

The logic programmer might add this as an integrity constraint. This is wrong as well. As an integrity constraint it would require the database to indicate which of the three possibilities it adopts, namely:

$x < y$ is in the data;
or $x = y$ is in the data;
or $y < x$ is in the data.

This is stronger than allowing the disjunction in the data.

The handling of the integrity constraints corresponds to our metahandling of what configurations $(\rho, <)$ are allowed depending on the ordering of time. By labelling data items we are allowing for the metalevel considerations to be done separately on the labels.

This means that we can handle properties of time which are not necessarily expressible by an object language formula of the logic. In some cases (finiteness of time) this is because they are not first order; in other cases (irreflexivity) it is because there is no corresponding formula (axiom) and in still other cases because of syntactical restrictions (linearity).

We can now make clear our classical versus intuitionistic distinction. If the underlying logic is classical then we are checking whether $\Delta \vdash \psi$ in classical logic. If our underlying logic is intuitionistic, then we are checking whether $\Delta \vdash \psi$ in intuitionistic logic where Δ and ψ are defined below.

Δ is the translation of the data together with the axioms for linear ordering, i.e. the conjunction of:

1. $\exists s_1 > tA^*(s_1)$;
2. $\exists s_2 > tB^*(s_2)$;

3. $\forall xy(x < y \vee x = y \vee y < x)$;
4. $\forall x \exists y(x < y)$;
5. $\forall x \exists y(y < x)$;
6. $\forall xyz(x < y \wedge y < z \rightarrow x < z)$;
7. $\forall x \neg(x < x)$.

ψ is the translation of the query as given above.

The computation of Example 5.12, using restart, answers the question $\Delta \vdash ? \psi$ in classical logic. To answer the question $\Delta \vdash ? \psi$ in intuitionistic logic we cannot use restart, but must use forward rules as well.

Example 5.14. See Example 5.12 for the case that the underlying logic is intuitionistic. *Data* and *Query* as in Example 5.12.

Going forward, we get:

3 $s : A$ from 1;
4 $s' : B$ from 2.

By linearity,

$$t < s < s'$$
$$\text{or } t < s' < s$$
$$\text{or } t < s = s'.$$

ψ will succeed for each case.

Our language does not allow us to ask queries of the form $\Box G(x)$, where x are all universal variables (i.e. $\forall x \Box G(x)$). However, such queries can be computed from a database Δ. The *only* way to get *always* information out of Δ for a general flow of time is via the always clauses in the database. Always clauses are true everywhere, so if we want to know what else is true everywhere, we ask it from the always clauses. Thus to ask

$$?\Box G(x), x \text{ a universal variable}$$

we first Skolemize and then ask

$$\{X, \Box X \mid \Box X \in \Delta\} ? G(c)$$

where c is a Skolem constant.

We can add a new rule to Definition 5.9:

16. **Always rule**:
$\mathbf{S}(\rho, <, \Delta, \Box G, t, G_0, t_0) \in \mathbf{\Pi}$ and $\mathbf{\Pi}' = (\mathbf{\Pi} - \{\mathbf{S}(\rho, <, \Delta, \Box G, t, G_0, t_0)\}) \cup \{\mathbf{S}(\{s\},$
$\emptyset, \Delta', G', s, g', s)\}$ where s is a completely new point and G' is obtained from G by substituting new Skolem constants for all the universal variables of G and
$$\Delta' = \{B, \Box B \mid \Box B \in \Delta\}.$$

We can use 16 to add another clause to the computation of Definition 5.9, namely:

17. $\mathbf{S}(\rho, <, \Delta, F(A \wedge B), t, G_0, t_0) \in \mathbf{\Pi}$ and $\mathbf{\Pi}' = (\mathbf{\Pi} - \{\mathbf{S}(\rho, <, \Delta, F(A \wedge B), t, G_0, t_0)\}) \cup \{\mathbf{S}(\rho, <, \Delta, FA, t, G_0, t_0), \mathbf{S}(\rho, <, \Delta, \Box B, t, G_0, t_0)\}$.

Example 5.15.

Data	Query	Configuration
$\Box a$	$t : F(a \wedge b)$	$\{t\}$
$t : Fb$		

First computation
Create $s, t < s$ and get

Data	Query	Configuration
$\Box a$	$s : a \wedge b$	$t < s$
$t : Fb$		
$s : b$		

$s : b$ succeeds from the data. $s : a$ succeeds by rule 2, Definition 5.9.

Second computation
Use rule 17. Since $?\Box a$ succeeds ask for Fb and proceed as in the first computation.

5.4 Different flows of time

We now check the effect of different flows of time on our logical deduction (computation). We consider a typical example.

Example 5.16.

Data	Query	Configuration
$t : FFA$	$?t : FA$	$\{t\}$

The possible world flow is a general binary relation.

We create by rule 4b of Definition 5.9 a future configuration $t < s$ and add to the database $s : FA$. We get

Data	Query	Configuration
$t : FFA$	$?t : FA$	$t < s$
$s : FA$		

Again we apply rule 4a of Definition 5.9 and get the new configuration with $s < s'$ and the new item of data $s' : A$. We get

Data	Query	Configuration
$t : FFA$	$?t : FA$	$t < s$
$s : FA$		$s < s'$
$s' : A$		

Whether or not we can proceed from here depends on the flow of time. If $<$ is transitive, then $t < s'$ holds and we can get $t : FA$ in the data by rule 3.

Actually by rule 4* we could have proceeded along the following sequence of deduction. Rule 4* is especially geared for transitivity.

Data	Query	Configuration
$t : FFA$	$t : FA$	t

Using rule 4* we get

Data	Query	Configuration
$t : FFA$	$s : FA \vee FFA$	$t < s$
$s : FA$		

The first disjunct of the query succeeds.

If $<$ is not transitive, rule 3 does not apply, since $t < s'$ does not hold.

Suppose our query were $?t : FFFA$.

If $<$ is reflexive then we can succeed with $?t : FFFA$ because $t < t$.

If $<$ is dense (i.e. $\forall xy(x < y \rightarrow \exists z(x < z \wedge z < y))$) we should also succeed because we can create a point z with $t < z < s$.

$z : FFA$ will succeed and hence $t : FFFA$ will also succeed.

Here we encounter a new rule (density rule), whereby points can always be 'landed' between existing points in a configuration.

We now address the flow of time of the type natural numbers, $\{1, 2, 3, 4, \ldots\}$. This has the special property that it is generated by a function symbol **s**:

$$\{1, \mathbf{s}(1), \mathbf{ss}(1), \ldots\}.$$

Example 5.17.

Data	Query	Configuration
$\Box(q \rightarrow \bigcirc q)$	$1 : F(p \wedge q)$	$\{1\}$
$1 : \bigcirc q$		
$1 : Fp$		

If time is the natural numbers, the query should succeed from the data. If time is not the natural numbers but, for example, $\{1, 2, 3, \ldots, w, w+1, w+2, \ldots\}$ then the query should fail.

How do we represent the fact that time is the natural numbers in our computation rule? What is needed is the ability to do some induction. We can use rule 4b and introduce a point t with $1 < t$ into the configuration and even say that $t = n$, for some n. We thus get

Data	Query	Configuration
$\Box(q \rightarrow \bigcirc q)$	$1 : F(p \wedge q)$	$1 < n$
$1 : \bigcirc q$		
$1 : Fp$		
$n : p$		

Somehow we want to derive $n : q$ from the first two assumptions. The key reason for the success of $F(p \wedge q)$ is the success of $\Box q$ from the first two assumptions. We need an induction axiom on the flow of time.

To get a clue as to what to do, let us see what Prolog would do with the translations of the data and goal.

Translated data

$$\forall t[1 \leq t \wedge Q^*(t) \to Q^*(t+1)]$$
$$Q^*(1)$$
$$\exists t P^*(t).$$

Translated query

$$\exists t(P^*(t) \wedge Q^*(t)).$$

After we Skolemize, the database becomes:

1. $1 \leq t \wedge Q^*(t) \to Q^*(t+1)$
2. $Q^*(1)$
3. $P^*(c)$

and the query is

$$P^*(s) \wedge Q^*(s).$$

We proceed by letting $s = c$. We ask $Q^*(c)$ and have to ask after a slightly generalized form of unification $?1 \leq c \wedge Q^*(c-1)$.

Obviously this will lead nowhere without an induction axiom. The induction axiom should be that for *any* predicate $PRED$

$$PRED(1) \wedge \forall x[1 \leq x \wedge PRED(x) \to PRED(x+1)] \to \forall x PRED(x).$$

Written in Horn clause form this becomes

$$\exists x \forall y[PRED(1) \wedge [1 \leq x \wedge PRED(x) \to PRED(x+1)] \to PRED(y)].$$

Skolemizing gives us

4. $PRED(1) \wedge (1 \leq d \wedge PRED(d) \to PRED(d+1)) \to PRED(y)$

where d is a Skolem constant.

Let us now ask the query $P^*(s) \wedge Q^*(s)$ from the database with 1–4. We unify with clause 3 and ask $Q^*(c)$. We unify with clause 4 and ask $Q^*(1)$ which succeeds and ask for the implication

$$?1 \leq d \wedge Q^*(d) \to Q^*(d+1).$$

This should succeed since it is a special case of clause 1 for $t = d$.

The above shows that we need to add an induction axiom of the form

$$\bigcirc x \wedge \Box(x \to \bigcirc x) \to \Box x.$$

Imagine that we are at time t, and assume $t' < t$. If A is true at t' and $\Box(A \to \bigcirc A)$ is true, then A is true at t.

We thus need the following rule:

18. **Induction rule:**
$t : F(A \wedge B)$ succeeds from Δ at a certain configuration if the following conditions all hold.

1. $t : FB$ suceed.
2. For some $s < t, s : A$ succeeds.
3. $m : \bigcirc A$ succeeds from the database Δ', where $\Delta' = \{X, \Box X \mid \Box X \in \Delta\} \cup \{A\}$ and m is a completely new time point and the new configuration is $\{m\}$.

The above shows how to compute when time is the natural number. This is not the best way of doing it. In fact, the characteristic feature involved here is that the ordering of the flow of time is a Herbrand universe generated by a finite set of function symbols. FA is read as 'A is true at a point generated by the function symbols'. This property requires a special study. See Chapter 11.

5.5 A theorem prover for modal and temporal logics

This section will briefly indicate how our temporal Horn clause computation can be extended to an automated deduction system for full modal and temporal logic. We present computation rules for propositional temporal logic with $F, P, \bigcirc, \bullet, \wedge$ $\rightarrow$ and $\bot$. We intend to approach predicate logic in Volume 3 as it is relatively complex. The presentation will be intuitive.

Definition 5.18. We define the notions of a *full clause*, a *body* and a *head*.

(a) A *full clause* is an atom q or $\bot$ or $B \rightarrow H$, or H where B is a body and H is a head.

(b) A *body* is a conjunction of full clauses.

(c) A *head* is an atom q or $\bot$ or FH or PH or $\bigcirc H$ or $\bullet H$, where H is a body.

Notice that negation by failure is not allowed. We used the connectives $\wedge, \rightarrow, \bot$. The other connectives, $\vee$ and $\sim$, are definable in the usual way: $\sim A = A \rightarrow \bot$ and $A \vee B = (A \rightarrow \bot) \rightarrow B$. The reader can show that every formula of the language with the connectives $\{\sim, \wedge, \vee, F, G, P, H\}$ is equivalent to a conjunction of full clauses. We use the following equivalences:

$$A \rightarrow (B \wedge C) = (A \rightarrow B) \wedge (A \rightarrow C);$$

$$A \rightarrow (B \rightarrow C) = A \wedge B \rightarrow C;$$

$$GA = F(A \rightarrow \bot) \rightarrow \bot;$$

$$HA = P(A \rightarrow \bot) \rightarrow \bot.$$

Definition 5.19. A database is a set of labelled full clauses of the form $(\Delta, \rho, <)$, where $\rho = \{t \mid t : A \in \Delta,$ for some $A\}$. A query is a labelled full clause.

Definition 5.20. The following is a definition of the predicate $\mathbf{S}(\rho, <, \Delta, G, t, G_0, t_0)$, which reads: the labelled goal $t : G$ succeeds from $(\Delta, \rho, <)$ with parameter (initial goal) $t_0 : G_0$.

1(a) $\mathbf{S}(\rho, <, \Delta, q, t, G_0, t_0)$ for q atomic or $\bot$ if for some $t : A \rightarrow q$, $\mathbf{S}(\rho, <, \Delta, A, t, G_0, t_0)$.

(b) If $t : q \in \Delta$ or $s : \bot \in \Delta$ then $\mathbf{S}(\rho, <, \Delta, q, t, G_0, t_0)$.
(c) $\mathbf{S}(\rho, <, \Delta, \bot, t, G_0, t_0)$ if $\mathbf{S}(\rho, <, \Delta, \bot, s, G_0, t_0)$.
This rule says that if we can get a contradiction from any label, it is considered a contradiction of the whole system.

2. $\mathbf{S}(\rho, <, \Delta, G, t, G_0, t_0)$ if for some $s : A \to \bot$, $\mathbf{S}(\rho, <, \Delta, A, s, G_0, t_0)$.
3. $\mathbf{S}(\rho, <, \Delta, t, FG, G_0, t_0)$ if for some $s \in \rho, t < s$ and $\mathbf{S}(\rho, <, \Delta, G, s, G_0, t_0)$.
4. $\mathbf{S}(\rho, <, \Delta, FG, t, G_0, t_0)$ if for some $t : A \to FB \in \Delta$ we have that both (a) and (b) below hold true:
   (a) For all $s \in \rho$ such that $t < s$ we have $\mathbf{S}(\rho, <, \Delta^*, s, D, G_0, t_0)$ and $\mathbf{S}(\rho, <, \Delta, E, t, G_0, t_0)$ hold, where $\Delta^* = \Delta \cup \{s : B\}$ and $D \in \{G, FG\}$ and $E \in \{A, FA\}$.
   Note: The choice of D and E is made here for the case of transitive time. In modal logic, where $<$ is not necessarily transitive, we take $D = G, E = A$. Other conditions on $<$ correspond to different choices of D and E.
   (b) For all future configurations of $(\rho, <, t)$ with a new letter s, denoted by $(\rho_s, <_s)$, we have $\mathbf{S}(\rho_s, <_s, \Delta^*, s, D, G_0, t_0)$ and $\mathbf{S}(\rho_s, <_s, \Delta, E, t, G_0, t_0)$ hold, where Δ^*, E, D are as in (a).
5. This is the mirror image of 3.
6. This is the mirror image of 4.

7(a) $\mathbf{S}(\rho, <, \Delta, A_1 \wedge A_2, t, G_0, t_0)$ if both $\mathbf{S}(\rho, <, \Delta, A_i, t, G_0, t_0)$ hold for $i = 1, 2$.
(b) $\mathbf{S}(\rho, <, \Delta, A \to B, t, G_0, t_0)$ if $\mathbf{S}(\rho, <, \Delta \cup \{t : A\}, B, t, G_0, t_0)$.

8. **Restart rule**:
   $\mathbf{S}(\rho, <, \Delta, G, t, G_0, t_0)$ if $\mathbf{S}(\rho, <, \Delta, G_0, t_0, G_0, t_0)$.

If the language contains $\bigcirc$ and $\bullet$ then the following are the relevant rules.

9. $\mathbf{S}(\rho, <, \Delta, \bigcirc G, t, G_0, t_0)$ if $\bigcirc t$ exists and $\bigcirc t \in \rho$ and $\mathbf{S}(\rho, <, \Delta, G, \bigcirc t, G_0, t_0)$.
10. $\mathbf{S}(\rho, <, \Delta, \bigcirc G, t, G_0, t_0)$ if for some $t : A \to \bigcirc B \in \Delta$ both $\mathbf{S}(\rho, <, \Delta, A, t, G_0, t_0)$ and $\mathbf{S}(\rho \cup \{\bigcirc t\}, <', \Delta \cup \{\bigcirc t : B\}, G, \bigcirc t, G_0, t_0)$ hold where $<'$ is the appropriate ordering closure of $< \cup \{t < \bigcirc t\}$.
11. This is the mirror image of 9 for $\bullet$.
12. This is the mirror image of 10 for $\bullet$.

Example 5.21. (Here $\Box$ can be either G or H.)

	Data	Query	Configuration
1.	$t : \Box a$	$?t : \Box b$	$\{t\}$
2.	$t : \Box(a \to b)$		t is a constant

Translation:

	Data	Query	Configuration
1.	$t : F(a \to \bot) \to \bot$	$t : F(b \to \bot) \to \bot$	$\{t\}$
2.	$t : F((a \to b) \to \bot) \to \bot$		

Computation

The problem becomes

	Additional data	Current query	Configuration
3.	$t : F(b \to \bot)$	$?t : \bot$	$\{t\}$
from 2		$?t_0 : F((a \to b) \to \bot)$	

From 3 using ** create a new point s:

	Additional data	Current query	Configuration
4.	$s : b \to \bot$	$?s : (a \to b) \to \bot$	$t < s$

Add $s : a \to b$ to the database and ask

5.	$s : (a \to b)$	$?s : \bot$	

From 4 and 5 we ask:

$$?s : a.$$

From computation rule 2 and clause 1 of the data we ask

$$?t : F(a \to \bot).$$

From computation rule 2 we ask

$$?s : a \to \bot$$

We add $s : a$ to the data and ask

	Additional data	Current query	Configuration
6.	$s : a$	$?s : \bot$	$t < s$

The query succeeds.

5.6 Modal and temporal Herbrand universes

This section deals with the soundness of our computation rules. In conjunction with soundness it is useful to clarify the notion of modal and temporal Herbrand models. For simplicty we deal with temporal logic with P, F only and transitive irreflexive time or with modal logic with one modality $\Diamond$ and a general binary accessibility relation $<$. We get our clues from some examples:

Example 5.22. Consider the database

1. $t : a \to \Diamond b$
2. $\Box(b \to c)$
3. $t : a$.

The constellation is $\{t\}$.

If we translate the clauses into predicate logic we get:

1. $a^*(t) \to \exists s > tb^*(s)$
2. $\forall x[b^*(x) \to c^*(x)]$
3. $a^*(t)$.

Translated into Horn clauses we get after skolemising:

1.1 $a^*(t) \to b^*(s)$
1.2 $a^*(t) \to t < s$
2 $b^*(x) \to c^*(x)$
3 $a^*(t)$.

t, s are Skolem constants.

From this program, the queries

$$a^*(t), \neg b^*(t), \neg c^*(t), \neg a(s), b(s), c^*(s)$$

all succeed. $\neg$ is negation by failure.

It is easy to recognize that $\neg a^*(s)$ succeeds because there is no head which unifies with $a^*(s)$. The meaning of the query $\neg a^*(s)$ in terms of modalities is the query $\Diamond\neg a$.

The question is: how do we recognize syntactically what fails in the modal language? The heads of clauses can be whole databases and there is no immediate way of syntactically recognizing which atoms are not heads of clauses.

Example 5.23. We consider a more complex example:

1. $t : a \to \Diamond b$
2. $\Box(b \to c)$
3. $t : a$
4. $t : a \to \Diamond d$.

We have added clause 4 to the database in the previous example. The translation of the first three clauses will proceed as before. We will get

1.1 $a^*(t) \to c^*(s)$
1.2 $a^*(t) \to t < s$
2 $b^*(x) \to c^*(x)$
3 $a^*(t)$.

We are now ready to translate clause 4. This should be translated like clause 1 into

4.1 $a^*(t) \to d^*(r)$
4.2 $a^*(t) \to t < r$.

The above translation is correct if the set of possible worlds is just an ordering. Suppose we know further that in our modal logic the set of possible worlds is linearly ordered. Since $t < s \wedge t < r \to s = r \vee s < r \vee r < s$, this fact must be reflected in the Horn clause database. The only way to do it is to add it as an integrity constraint.

Thus our temporal program translates into a Horn clause program with integrity constraints.

This will be true in the general case. Whether we need integrity constraints or not will depend on the flow of time.

Let us begin by translating from the modal and temporal language into Horn clauses. The labelled wff $t : A$ will be translated into a set of formulae of predicate logic denoted by $Horn(t, A)$. $Horn(t, A)$ is supposed to be logically equivalent to A. The basic translation of a labelled atomic predicate formula $t : A(x_1 \ldots x_n)$ is $A^*(t, x_1 \ldots x_n)$. A^* is a formula of a two-sorted predicate logic where the first sort ranges over labels and the second sort over domain elements (of the world t).

Definition 5.24. Consider a temporal predicate language with connectives P and F, and $\neg$ for negation by failure.

Consider the notion of labelled temporal clauses, as defined in Definition 5.6.

Let $Horn(t, A)$ be a translation function associating with each labelled clause or goal a set of Horn clauses in the two-sorted language described above. The letters t, s which appear in the translation are Skolem constants. They are assumed to be *all different*.

We assume that we are dealing with a general transitive flow of time. This is to simplify the translation. If time has extra conditions, i.e. linearity, additional integrity constraints may need to be added. If time is characterized by non-first-order conditions (e.g. finiteness) then an adequate translation into Horn clause logic may not be possible.

The following are the translation clauses:

1. $Horn(t, A(x_1 \ldots x_n)) = A^*(t, x_1 \ldots x_n)$, for A atomic;
2. $Horn(t, FA) = \{t < s\} \cup Horn(s, A)$
   $Horn\ (t, PA) = \{s < t\} \cup Horn(s, A)$;
3. $Horn(t, A \wedge B) = Horn(t, A) \cup Horn(t, B)$;
4. $Horn(t, \neg A) = \neg \bigwedge Horn(t, A)$;
5. $Horn(t, A \to F \wedge B_j) = \{\bigwedge Horn(t, A) \to t < s\} \cup \bigcup_{B_j} \{\bigwedge Horn(s, A) \wedge C \to D \mid (C \to D) \in Horn(s, B_j)\}$;
6. $Horn(t, A \to P \wedge B_j) = \{\bigwedge Horn(t, A) \to s < t\} \cup \bigcup_{B_j} \{\bigwedge Horn(s, A) \wedge C \to D \mid (C \to D) \in Horn(s, B_j)\}$;
7. $Horn(t, \Box A) = Horn(x, A)$ where x is a universal variable.

Example 5.25. To explain the translation of $t : A \to F(B_1 \wedge (B_2 \to B_3))$, let us write it in predicate logic. $A \to F(B_1 \wedge (B_2 \to B_3))$ is true at t if A true at t implies $F(B_1 \wedge (B_2 \to B_3))$ is true at t. $F(B_1 \wedge (B_2 \to B_3))$ is true at t if for some s, $t < s$ and $B_1 \wedge (B_2 \to B_3)$ are true at s.

Thus we have the translation

$$A^*(t) \to \exists s(t < s \wedge B_1^*(s) \wedge (B_2^*(s) \to B_3^*(s))).$$

Skolemizing on s and writing it in Horn clauses we get the conjunction

$$A^*(t) \to t < s$$
$$A^*(t) \to B_1^*(s)$$
$$A^*(t) \wedge B_2^*(s) \to B_3^*(s).$$

Let us see what the translation *Horn* does: *Horn* $(t, A \to F(B_1 \wedge (B_2 \to B_3))) = \{\bigwedge Horn(t,A) \to t < s\} \cup \{\bigwedge Horn(t,A) \to Horn(s,B_2)\} \cup \{\bigwedge Horn(t,A) \wedge \bigwedge Horn(s,B_2) \to \bigwedge Horn(s,B_3)\} = \{A^*(t) \to t < s, A^*(t) \to B_2^*(s), A^*(t) \wedge B_2^*(s) \to B_3^*(s)\}$.

We prove soundness of the computation of Definition 5.9, relative to the Horn clause computation for the Horn database in classical logic. In other words, if the translation *Horn*(t, A) is accepted as sound, as is intuitively clear, then the computation of $\mathbf{S}(\rho, <, \Delta, G, t, G_0, t_0, \Theta)$ can be translated isomorphically into a classical Horn clause computation of the form *Horn*(t, Δ)?*Horn*(t, G), and the soundness of the classical Horn clause computation would imply the soundness of our computation.

This method of translation will also relate our temporal computation to that of an ordinary Horn clause computation.

The basic unit of our temporal computation is $\mathbf{S}(\rho, <, \Delta, G, t, G_0, t_0, \Theta)$. The current labelled goal is $t : G$ and $t_0 : G_0$ is the original goal. The database is $(\rho, <, \Delta)$ and Θ is the current substitution. $t_0 : G_0$ is used in the restart rule. For a temporal flow of time which is ordinary transitive $<$, we do not need the restart rule. Thus we have to translate $(\rho, <, \Delta)$ to classical logic and translates $t : G$ and Θ to classical logic and see what each computation step of $\mathbf{S}$ of the source translates into the classical logic target.

Definition 5.26. Let $(\rho, <)$ be a constellation and let Δ be a labelled database such that

$$\rho = \{t \mid \text{ for some } A, t : A \in \Delta\}.$$

Let *Horn*$((\rho, <), \Delta) = \{t < s \mid t, s \in \rho \text{ and } t < s\} \cup \bigcup_{t:A\in\Delta} Horn(t, A)$.

Theorem 5.27 (Soundness). $\mathbf{S}(\rho, <, \Delta, G, t, \Theta)$ *succeeds in temporal logic if and only if in the sorted classical logic* *Horn*$((\rho, <), \Delta)$?*Horn*(t, G) *succeeds with* Θ.

Proof The proof is by induction on the complexity of the computation tree of $\mathbf{S}(\rho, <, \Delta, G, t, \Theta)$.

We follow the inductive steps of Definition 5.8. The translation of $(\mathbf{\Pi}, \Theta)$ is a conjunction of Horn clause queries, all required to succeed under the same subsitution Θ.

Case I The empty goal succeeds in both cases.

Case II $(\mathbf{\Pi}, \Theta)$ fails if for some $\mathbf{S}(\rho, <, \Delta, G, t)$, we have G is atomic and for all $\Box(A \to H) \in \Delta$ and all $t : A \to H \in \Delta$, $G\Theta$ and $H\Theta$ do not unify. The reason they do not unify is because of what Θ substitutes to the variables u_i.

The corresponding Horn clause predicate programs are

$$\bigwedge Horn(x, A) \to H^*(x)$$

and

$$\bigwedge Horn(t, A) \to H^*(t)$$

and the goal is $?G^*(t)$.

Clearly, since x is a general universal variable, the success of the two-sorted unification depends on the other variables and Θ. Thus unification does *not* succeed in the classical predicate case iff it does not succeed in the temporal case.

Rules 1 and 2 deal with the atomic case: the query is $G^*(t)$ and in the database among the data are

$$\bigwedge Horn(t, A) \to H^*(t) \text{ and } \bigwedge Horn(x, A) \to H^*(x)$$

for the cases of $t : A \to H$ and $\Box(A \to H)$ respectively.

For the Horn clause program to succeed $G^*(t)$ must unify with $H^*(t)$. This will hold if and only if the substitution for the domain variables allows unification, which is exactly the condition of Definition 5.8.

Rules 3, 4(general) and 4*(general) deal with the case of a goal of the form $?t : FG$. The translation of the goal is $t < u \wedge \bigwedge Horn(u, G)$ where u is an existential variable.

Rule 3 gives success when for some $s, t < s \in \Delta$ and $?s : G$ succeeds. In this case let $u = s$; then $t < u$ succeeds and $\bigwedge Horn(s, G)$ succeeds by the induction hypothesis.

We now turn to the general rules 4(general) and 4*(general). These rules yield success when for some clause of the form

$$t : A \to F \wedge B_j$$

or

$$\Box(A \to F \wedge B_j).$$

$\Delta ?t : A$ succeeds and $\Delta \cup \{(s : B_j)\} ?s : G \vee FG$ both succeed. s is a new point.

The translation $\bigwedge Horn(t, A)$ succeeds by the induction hypothesis.

The translation of

$$t : A \to F \wedge B_j$$

or

$$\Box(A \to F \wedge B_j)$$

contains the following database:

1. $\bigwedge Horn(t, A) \to t < s$.
2. For every B_j and every $C \to D$ in $Horn(s, B_j)$ the clause $\bigwedge Horn(s, A) \wedge C \to D$.

Since $\wedge Horn(t, A)$ succeeds we can assume we have in our database:

1* $t < s$;

2* $C \to D$, for $C \to D \in Horn(s, B_j)$ for some j.

These were obtained by substituting *truth* in 1 and 2 for $\bigwedge Horn(t, A)$.

The goal is to show $t < u \wedge \bigwedge Horn(u, G)$.

Again for $u = s, t < u$ succeeds from (1*) and by the induction hypothesis, since $\Delta \cup \{s : B_j\}?s : G \vee FG$ is successful, we get $\bigcup_j Horn(s, B_j)? \bigwedge Horn(s, G) \vee (s < u' \wedge \bigwedge Horn(u', G))$ should succeed, with u' an existential variable.

However, 2* is exactly $\bigcup_j Horn(s, \bigwedge B_j)$. Therefore we have shown that rules 4(general) and 4*(general) are sound.

Rules 6(general) and 6*(general) are sound because they are the mirror images of 4(general) and 4*(general).

The next relevant rules for our soundness cases are 11–13. These follow immediately since the rules for $\wedge, \vee, \neg$ are the same in both computations.

Rule 14, the restart rule, is definitely sound. If we try to show in general that $\Delta \vdash A$ then since in classical logic $\sim A \to A$ is the same as A ($\sim$ is classical negation) it is equivalent to show $\Delta, \sim A \vdash A$.

If $\sim A$ is now in the data, we can *at any time* try to show A instead of the current goal G. This will give us A (shown) and $\sim A$ (in Data) which is a contradiction, and this yields *any goal* including the current goal G.

We have thus completed the soundness proof. □

5.7 Tractability and persistence

We defined a temporal database Δ essentially as a finite piece of information telling us which temporal formulae are true at what times. In the most general case, for a general flow of time $(T, <)$, all a database can do is to provide a set of the form $\{t_i : A_i\}$, meaning that A_i is true at time t_i and a configuration $(\{t_i\}, <)$, giving the temporal relationships among $\{t_i\}$. A query would be of the form $?t : Q$, where t is one of the t_i. The computation of the query from the data is in the general case exponential, as we found in Section 5.2, from the case analysis of clause 4 of Definition 5.9 and from Example 5.5. We must therefore analyse the reasons for the complexity and see whether there are simplifying natural assumptions, which will make the computational problem more tractable.

There are three main components which contribute to complexity:

1. The complexity of the temporal formulae allowed in the data and in the query. We allow $t : A$ into the database, with A having temporal operators. So, for example, $t : FA$ is allowed and also $t : \bigcirc A$. $t : FA$ makes life more difficult because it has a hidden Skolem function in it. It really means $\exists s[t < s \text{ and } (s : A)]$. This gives rise to case analysis, as we do not know in general where s is. See Example 5.13 and Examples 5.22 and 5.23. In this respect $t : \bigcirc A$ is a relatively simple item. It says $(t + 1) : A$. In fact any temporal operator which specifies the time is relatively less complex. In practice, we do need to allow data of the form $t : FA$. Sometimes we know an event will take place in the future but we do not know when.

The mere fact that A is going to be true can affect our present actions. A concrete example where such a case may arise is when someone accepts a new appointment beginning next year, but has not yet resigned from their old position. We know they are going to resign but we do not know when;

2. The flow of time itself gives rise to complexity. The flow of time may be non-Horn clause (e.g. linear time which is defined by a disjunctive axiom

$$\forall xy[x < y \vee y < x \vee x = y].$$

This complicates the case analysis of 1 above.

3. Complexity arises from the behaviour. If atomic predicates get truth values at random moments of time, the database can be complex to describe. A very natural simplifying assumption in the case of temporal logic is *persistence.* If atomic statements and their negations remain true for a while then they give rise to less complexity. Such examples are abundant. For example, people usually stay at their residences and jobs for a while. So for example, any payroll or local tax system can benefit from persistence as a simplifying assumption. Thus in databases where there is a great deal of persistence, we can use this fact to simplify our representation and querying. In fact, we shall see that a completely different approach to temporal representation can be adopted when one can make use of persistence.

   Another simplifying assumption is *recurrence.* Saturdays, for example, recur every week, so are paydays. This simplifies the representation and querying. Again, a payroll system would benefit from that.

We said at the beginning that a database Δ is a finitely generated piece of temporal information stating what is true and when. If we do not have any simplifying assumptions, we have to represent Δ in the form $\Delta = \{t_i : A_i\}$ and end up needing the computation rules of Section 5.2 to answer queries.

Suppose now that we adopt all three simplifying assumptions for our database. We assume that the A_i are only atoms and their negations, we further assume that each A_i is either persistent or recurrent, and let us assume, to be realistic, that the flow of time is linear. Linearity does not make the computation more complicated in this particular case, because we are not allowing data of the form $t : FA$, and so complicated case analysis does not arise. In fact, together with persistence and recurrence, linearity becomes an additional simplifying assumption!

Our aim is to check what form our temporal logic programming machine should take in view of our chosen simplifying assumptions.

First note that the most natural units of data are no longer of the form:

$$t : A$$

reading A is true at t, but either of the form

$$[t, s] : A, [t < s]$$

reading A is true in the closed interval $[t, s]$, or the form

$$t||d : A$$

reading A is true at t and recurrently at $t+d$, $t+2d, \ldots$, that is, every d moments of time.

A is assumed to be a literal (atom or a negation of an atom) and $[t, s]$ is supposed to be a maximal interval where A is true. In $t||d$, d is supposed to be the minimal cycle for A to recur. The reasons for adopting the notation $[t, s] : A$ and $t||d : A$ are not mathematical but simply intuitive and practical. This is the way we think about temporal atomic data when persistence or recurrence is present. In the literature there has been a great debate on whether to evaluate temporal statements at points or intervals. Some researchers were so committed to intervals that they tended, unfortunately, to disregard any system which uses points. Our position here is clear and intuitive. First perform all the computations using intervals. Evaluation at points is possible and trivial. To evaluate $t : A$, i.e. to ask $?t : A$ as a query from a database, compute the (maximal) intervals at which A is true and see whather t is there. To evaluate $[t, s] : A$ do the same, and check whether $[t, s]$ is a subset.

The query language is left in its full generality. i.e. we can ask queries of the form $t : A$ where A is unrestricted (e.g. $A = FB$ etc.). It makes sense also to allow queries of the form $[t, s] : A$, although exactly how we are going to find the answer remains to be seen. The reader should be aware that the data representation language and the query language are no longer the same. This is an important factor. There has been a lot of confusion, especially among the AI community, in connection with these matters. We shall see later that as far as computational tractability is concerned, the restriction to persistent data allows one to strengthen the query language to full predicate quantification over time points.

At this stage we might consider allowing recurrence within an interval, i.e. we allow something like

'A is true every d days in the interval $[t, s]$.'

We can denote this by

$$[t||d, s] : A$$

meaning A is true at $t, t + d, t + 2d$, as long as $t + nd \leq s, n = 1, 2, 3, \ldots$.

We may as well equally have recurrent intervals. An example of that would be taking a two-week holiday every year. This we denote by

$$[t, s]||d : A, \quad t < s, \quad (s - t) < d,$$

reading A is true at the intervals $[t, s], [t + d, s + d], [t + 2d, s + 2d]$, etc.

The reader should note that adopting this notation takes us outside the realm of first-order logic. Consider the integer flow of time. We can easily say that q is true at all even numbers by writing $[0, 0]||1$ as a truth set for q and $[1, 1||1$ as a truth set for $\sim q$ (i.e. q is true at 0 and recurs every 1 unit and $\sim q$ is true at 1 and recurs every 1 unit).

The exact expressive power of this language is yet to be examined. It is connected with the language USF of Chapter 10.

The above seem to be the most natural options to consider. We can already see that it no longer makes sense to check how the computation rules of Definition 5.9 simplify for our case. Our case is so specialized that we may as well devise computation rules especially for it. This should not surprise us. It happens in mathematics all the time. The theory of Abelian groups, for example, is completely different from the theory of semigroups, although Abelian groups are a special case of semigroups. The case of Abelian groups is so special that it does not relate to the general case any more.

Let us go back to the question of how to answer a query from our newly defined simplified databases. We start with an even more simple case, assuming only persistence and assuming that the flow of time is the integers. This simple assumption will allow us to present our point of view of how to evaluate a formula at a point or at an interval. It will also ensure we are still within what is expressible in first-order logic. Compare this with Chapter 13.

Assume that the atom q is true at the maximal intervals $[Xx_n, y_n], x_n \leq y_n < x_{n+1}$. Then $\sim q$ is true at the intervals $[y_n + 1, x_{n+1} - 1]$, a sequence of the same form, i.e. $y_n + 1 \leq x_{n+1} - 1$ and $x_{n+1} - 1 < y_{n+1} + 1$.

It is easy to compute the intervals corresponding to the truth values of conjunctions: we take the intersection:

If $I_j = \bigcup_n [x_n^j, y_n^j]$ then $I_1 \cap I_2 = \bigcup_n [x_n, y_n]$ and the points x_n, y_n can be effectively linearly computed. Also, if I_j is the interval set for A_j, the interval set for $U(A_1, A_2)$ can be effectively computed.

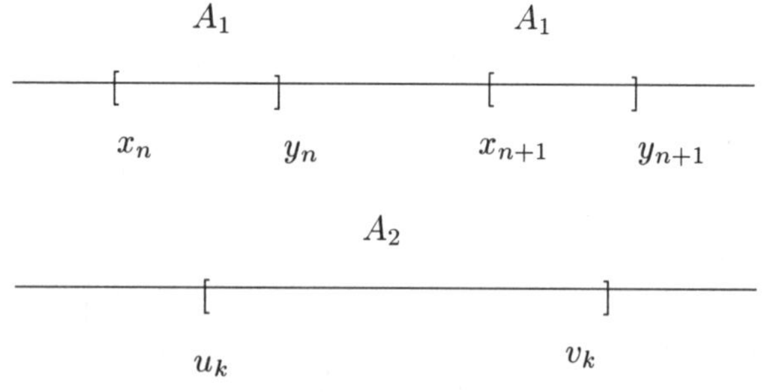

FIG. 5.7.

In Fig. 5.7, $U(A_1, A_2)$ is true at $[u_k, y_n - 1], [u_k, y_{n+1} - 1]$ which simplifies to the maximal $[u_k, y_{n+1} - 1]$.

The importance of the above is that we can regard a query formula of the full language with Until and Since as an operator on the model (database) to give a new database. If the database Δ gives for each atom or its negation the set of intervals where it is true, then a formula A operates on Δ to give the new set of intervals Δ_A; thus to answer $\Delta?t : A$ the question we ask is $t \in \Delta_A$. The new notion is that the query operates on the model.

This approach was adopted by I. Torsun and K. Manning when implementing the query language USF. The complexity of computation is polynomial (n^2). Note that although we have restricted the database formulae to atoms, we discovered that for no additional cost we can increase the query language to include the connectives Since and Until. As we have seen in Volume 1, in the case of integers the expressive power of Since and Until is equivalent to quantification over time points.

To give the reader another glimpse of what is to come, note that intuitively we have a couple of options:

1. We can assume persistence of atoms and negation of atoms. In this case we can express temporal models in first-order logic. The query language can be full Since and Until logic. This option does not allow for recurrence. In practical terms this means that we cannot generate or easily control recurrent events. Note that the database does not need to contain Horn clauses as data. Clauses of the form $\Box$(present $\text{wff}_1 \rightarrow$ present wff_2) are redundant and can be eliminated (this has to be properly proved!). Clauses of the form $\Box$(past $\text{wff}_1 \rightarrow$ present wff_2) are not allowed as they correspond to recurrence;
2. This option wants to have recurrence, and is not interested in first-order expressibility. How do we generate recurrence?
   The language USF (which was introduced for completely different reasons) allows one to generate the database using rules of the form $\Box$(past formula $\rightarrow$ present or future formula).

The above rules, together with some initial items of data of the form $t : A$, A a literal, can generate persistent and recurrent models.

6

COMBINING TEMPORAL LOGIC SYSTEMS

This chapter is a continuation of the work started in Volume 1, Chapter 14 (which also appeared as [Finger and Gabbay, 1992]) on combining temporal logics. Here, four combination methods are described and studied with respect to the transfer of logical properties from the component one-dimensional temporal logics to the resulting combined two-dimensional temporal logic. Three basic logical properties are analysed, namely soundness, completeness and decidability. Meanwhile Gabbay's [Gabbay, 1998] book *Fibring Logics* has appeared.

Each combination method comprises three submethods that combine the languages, the inference systems and the semantics of two one-dimensional temporal logic systems, generating families of two-dimensional temporal languages with varying expressivity and varying degrees of transfer of logical properties. The *temporalization method* and the *independent combination method* are shown to transfer all three basic logical properties. The method of *full join* of logic systems generates a considerably more expressive language but fails to transfer completeness and decidability in several cases. So a weaker method of *restricted join* is proposed and shown to transfer all three basic logical properties.

6.1 Introduction

We are interested in describing systems in which two distinct temporal 'points of view' coexist. Descriptions of temporal systems under a single point of view, i.e. one-dimensional temporal systems, abound in the literature. These one-dimensional temporal logics differ from each other in several ways. They differ on the approach, whether proof theoretic, model theoretic or algebraic. They differ on the ontology of time adopted, whether time is represented as a set of points, intervals or events. They can also differ on the properties assigned to flows of time, whether linear or branching time, discrete or dense, continuous or allowing gaps. In this chapter we contemplate both proof- and model-theoretic presentations of temporal logics on a point-based ontology. Most of the results presented assume that the flow of time is linear.

The motivation for combining logics came from the study of applications of two-dimensional temporal logics [Finger, 1992]. We were aware of Venema's negative results concerning the unaxiomatizability of two-dimensional temporal logics over the upper semiplane of $\mathbb{N} \times \mathbb{N}$, $\mathbb{Z} \times \mathbb{Z}$ and $\mathbb{R} \times \mathbb{R}$ [Venema, 1990] (see also Proposition 6.17 below). However, for our purposes, the full expressivity of Venema's two-dimensional language was not required, and a weaker language provided the appropriated expressivity.

It then became clear that this weaker two-dimensional language could be generalized and a family of languages resulting from adding a (second) temporal dimension *externally* to a temporal logic system was thus obtained. This process was formalized in [Finger and Gabbay, 1992], where it was called *temporalization*, and several results were obtained concerning the transfer of logical properties from the component logical system to the combined one. As a result, a family of temporalized logic systems was obtained, the properties of which can be derived from the properties of the component logic systems via the transfer results.

The next step comes from the observation that there may be several distinct ways in which two temporal logic systems can be combined, thus generating several families of combined two-dimensional temporal systems. Different combination methods may be presented by the distinct interactions between related parts of the two logic systems involved, leading to two-dimensional systems based on distinct languages with distinct semantical structure, expressive power and other properties (that may be transferred or not from the component systems).

Several cases in which two distinct temporal dimensions (or temporal 'points of view') can coexist are described next, motivating several different methods for combining two temporal logics. We will also attempt to relate these methods to recent, mostly unpublished work on combining two generic (not necessarily temporal) logic systems, e.g. [Gabbay, 1994; Gabbay, 1992].

6.1.1 *First case: external time*

One temporal point of view can be *external* to the other. The external point of view is seen as describing the temporal evolution of a system $\mathcal{S}$, when system $\mathcal{S}$ is itself a temporal description. Suppose $\mathcal{S}$ is described using a temporal logic T and suppose that the external point of view is given in a possibly distinct logic $\overline{\mathsf{T}}$. For example, consider an agent A, whose temporal beliefs are expressed in logic $\overline{\mathsf{T}}$, that we want to allow to reason about the temporal beliefs of an agent B, which are expressed in a possibly distinct logic T. This is illustrated in Fig. 6.1.

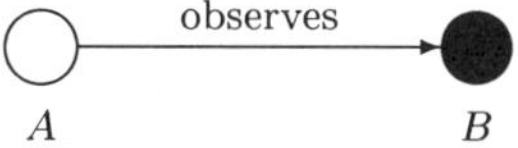

FIG. 6.1. One agent externally observing the other.

Agent A's beliefs are external to agents B's beliefs, so that $\overline{\mathsf{T}}$ is externally describing the evolution of T. The external temporal point of view $\overline{\mathsf{T}}$ is then applied to the internal system T, in a process called *temporalization* or *adding a temporal dimension to a logic system*, defined in [Finger and Gabbay, 1992]. The resulting combined logic system $\overline{\mathsf{T}}(\mathsf{T})$ is illustrated in Fig. 6.2.

The temporalization associates every time point in $\overline{\mathsf{T}}$ with a temporal description in T, where those T-descriptions need not all be identical. Given the logical properties of T and $\overline{\mathsf{T}}$, what can be said about the logical properties of $\overline{\mathsf{T}}(\mathsf{T})$?

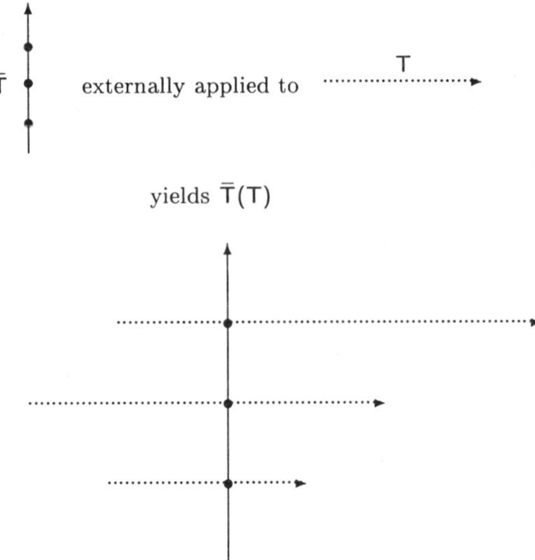

FIG. 6.2. The combined flow of time resulting from temporalization.

In terms of a generic combination of logics, the temporalization method can be matched with a process called *fuzzling* or *layering*, which is characterized by the fact that the formulae of system T can be substituted for the atoms of system $\bar{\mathsf{T}}$. In (Kripke-)semantical terms, this means that every possible world of $\bar{\mathsf{T}}$ is associated to a whole model of T; see [Gabbay, 1992].

6.1.2 *Second case: independent agents*

Suppose now that agent A has the ability of referring to agent B's temporal beliefs and vice versa. The agents are therefore observing each other, as illustrated in Fig. 6.3.

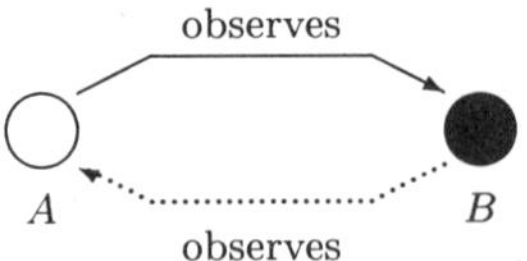

FIG. 6.3. Independent interaction of agents.

The agents' beliefs are then capable of interacting with each other through several levels of cross-reference, as in the sentence 'A believes that B believes that A believes that ...'. A new combination method for T and $\bar{\mathsf{T}}$ is needed in order to represent such a sentence as a formula, which is called the *independent combination*, $\bar{\mathsf{T}} \oplus \mathsf{T}$. Since a formula of $\bar{\mathsf{T}} \oplus \mathsf{T}$ has a finite nature, it can be unravelled in a finite number of alternating temporalizations, as illustrated in

Fig. 6.4.

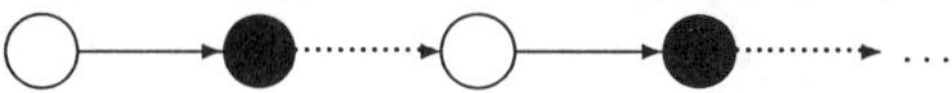

FIG. 6.4. Unravelling the independent combination.

Figure 6.4 suggests a way of analysing the properties of the independent combination method using the temporalization method as an intermediary step. It will turn out that the independent combination method is the (infinite) union of all finite alternated temporalizations. An illustration of a possible independently combined flow of time is presented in Fig. 6.5.

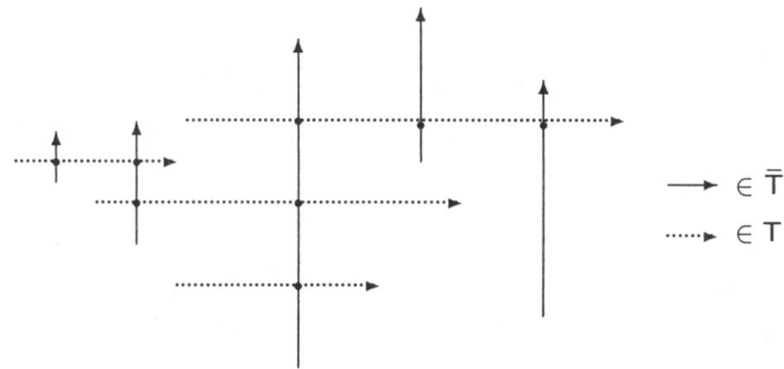

FIG. 6.5. Independently combined flow of time.

In terms of a generic combination of two logics, this process can be matched to the *dovetailing* process of [Gabbay, 1992], whereby atoms of T can be substituted by formulae of $\bar{\mathsf{T}}$ and vice versa. The semantical counterpart is obtained by providing each possible world with two distinct accessibility relations, $<$ and $\overline{<}$, so that from every possible world it is possible to reach another possible world via either $<$ or $\overline{<}$.

6.1.3 *Third case: two-dimensional plane*

Yet another distinct situation can be found where we have the coexistence of two distinct temporal 'points of view'. This time a single agent with temporal reasoning capabilities is considered, and we want to be able to describe the evolution of the agent's own beliefs. This is, perhaps, better illustrated by considering the agent as a temporal database where each piece of information is associated to a valid time (or interval). For example, consider the traditional database relation $employee(Name, Salary, Manager)$. Suppose the following is in the database at March 94:

Name	*Salary*	*Dept*	*Start*	*End*
Peter	1000	R&D	Apr 93	Mar 94

where the attributes start and end represent the end points of the valid interval associated with the information. We assume that Peter's salary has not changed since April 93. Suppose that in April 94 Peter receives a retroactive promotion dating back to the beginning of the year, increasing his salary to 2000. The whole database evolution is illustrated in Fig. 6.6, where only the value of Peter's salary is indicated at each point.

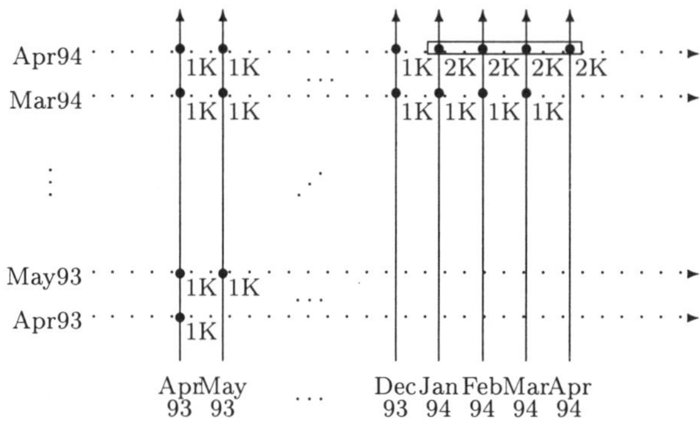

FIG. 6.6. Two-dimensional temporal database evolution.

If T represents valid time and $\overline{\mathsf{T}}$ represents transaction time, we have guaranteed a two-dimensional plane $\overline{\mathsf{T}} \times \mathsf{T}$ in order to represent the database evolution.

Another application of the two-dimensional plane (or its NW-semiplane) is in the representation of intervals on a line, as presented in [Venema, 1990].

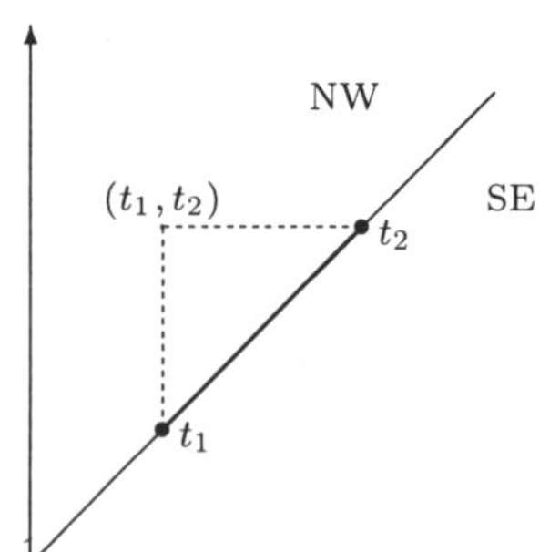

FIG. 6.7. Two-dimensional representation of intervals.

In Fig. 6.7 a line is considered as the diagonal of a two-dimensional plane and an interval $[t_1, t_2]$ on that line is represented by the point (t_1, t_2) on the NW-semiplane.

The combination between two temporal systems that generates a combined flow of time that is isomorphic to a two-dimensional plane is called the *join* of two logical systems. We have adopted the term *join of logics* here (instead of the previously used *interlacing of logics* in [Finger, 1994]) to be in accordance with the concept as defined in the larger context of generic combination of logics [Gabbay, 1992]. Although the language generated in this process is the same as that of the independent combination (for the case of two temporal logics), the semantical interaction between T and $\overline{\mathsf{T}}$ is a lot stronger; this is due to the fact that the temporal operators of the two logics are commutative in the join. As will be seen in Section 6.7, it is necessary to restrict this interaction to obtain the transfer of logical properties. The restriction will be applied to the type of operators allowed in one of the logics involved in the restricted join.

6.1.4 *Aims of this chapter*

In this chapter we study those three situations of coexistence of 'two temporal points of view' as the result of a combination of two linear, one-dimensional temporal logics.

In this sense this is a continuation of the work started in Volume 1 (Chapter 14, also [Finger and Gabbay, 1992]) on the combination of temporal logics. There, a process for adding a temporal dimension to a logic system was described, in which a temporal logic T is *externally* applied to a generic logic system L, generating a combined logic $\mathsf{T}(\mathsf{L})$.

We explore several methods for systematically combining two temporal logics, T and $\overline{\mathsf{T}}$, thus generating for each method a new family of *two-dimensional temporal logics*.

A great number of (one-dimensional) temporal logics exist in the literature to deal with the great variety of properties one may wish to express about flows of time. When building two-dimensional temporal logics, the combination of two classes of flows of time generates an even greater number of possible systems to be studied. Furthermore, as we will see, there are several distinct classes of temporal logics that may be considered as two dimensional, each generated by a distinct combination method. It is, therefore, desirable to study if it is possible to *transfer the properties* of the long-known and studied (one-dimensional) temporal logic system to the two-dimensional case.

So the main goal is to study, for each combination method, the transfer of logical properties from component one-dimensional temporal systems to a combined two-dimensional one.

We concentrate on the transfer of three basic properties of logic systems, namely soundness, completeness and decidability. This by no means implies that those are the only properties whose transfer deserves to be studied, but, as has already been noted in [Finger and Gabbay, 1992] for the temporal case, and in [Kracht and Wolter, 1991; Fine and Schurz, 1991] for the monomodal case, the transfer of completeness serves as a basis for the transfer of several other properties of logical systems.

As for the methods for combining two temporal logics, we consider the following:

(a) the temporalization method, i.e. the external application of a temporal logic to another temporal system, also known as adding a temporal dimension to a logic system;

(b) the independent combination of two temporal systems;

(c) the full join of two temporal systems, where flows of time are considered over a two-dimensional plane;

(d) the restricted join of two temporal systems, a combination method that restricts the previous one but generates nice transfer results.

We proceed as follows. Section 6.2 presents the basic notions of one- and two-dimensional temporal logics. Section 6.3 discusses combinations of logics in general terms, so that in the rest of the chapter we can present special cases of combination methods. Section 6.4 briefly examines the transfer results obtained for the temporalization method in [Finger and Gabbay, 1992]. Section 6.5 studies the independent combination method. Section 6.6 deals with the full join method and Section 6.7 with its restricted version. Section 6.8 analyses the properties of a two-dimensional diagonal on the model generated by the full and restricted join methods. In Section 6.9 we discuss the results of this work.

6.2 Preliminaries

For the purposes of this chapter, a logic system is composed of three elements:

(a) A language, normally given by a set of formation rules generating well-formed formulae over a signature and a set of logical connectives.

(b) An inference system, i.e. a relation, $\vdash$, between sets of formulae, normally represented by upper case Greek letters $\Delta, \Gamma, \Sigma, \Psi, \Phi$ and a single formula, normally represented by upper case letters $A, B, C, ...$; the fact that A is inferred from a set Δ is indicated by $\Delta \vdash A$. When Δ is a singleton, $\Delta = \{B\}$, the notation is abused and we write $B \vdash A$.

(c) The semantics of formulae over a class $\mathcal{K}$ of model structures. The fact that a formula A is true of or holds at a model $\mathcal{M} \in \mathcal{K}$ is indicated by $\mathcal{M} \models A$.

In providing a method for combining two logics into a third one, it will be necessary to provide three submethods that combine the languages, inference systems and semantics of the component logic systems. The component systems considered here will be one-dimensional linear US-temporal logics. Their language is built from a countable signature of propositional letters $\mathcal{P} = \{p_1, p_2, \ldots\}$, the Boolean connectives $\wedge$ (conjunction) and $\neg$ (negation), the two-place temporal operation U (Until) and S (Since), possibly renamed, and the following formation rules:

- every propositional letter is a formula;
- if A and B are formulae, so are $\neg A$ and $A \wedge B$;
- if A and B are formulae, so are $U(A, B)$ (reads 'until A is true in the future, B will be true') and $S(A, B)$ (reads 'since A was true in the past, B has been true');

- nothing else is a formula.

The *mirror image* of a formula is another temporal formula obtained by swapping all occurrences of U and S, e.g. the mirror image of $U(A, S(B, C))$ is $S(A, U(B, C))$.

The other Boolean connectives $\vee$ (disjunction), $\rightarrow$ (material implication), $\leftrightarrow$ (material bi-implication) and the constants $\bot$ (false) and $\top$ (true) can be derived in the standard way. Similarly, the one-place temporal operator F ('sometime in the future'), P ('sometime in the past'), G ('always in the future') and H ('always in the past') can be defined in terms of U and S.

To provide the semantics of temporal formulae we have to consider a (one-dimensional) *flow of time*, $\mathcal{F} = (T, <)$, where T is a set of time points and $<$ is an order over T. A temporal valuation $h : T \rightarrow 2^{\mathcal{P}}$ associates every time point with a set of propositional letters, i.e. $h(t)$ is the set of propositions that are true at time t. (Equivalently, and perhaps more usually, a valuation could be defined as a function $h : \mathcal{P} \rightarrow 2^T$, associating every propositional letter to a set of time points in which it holds true; see [Burgess, 1984; Gabbay *et al.*, 1994].) A model structure $\mathcal{M} = (T, <, h)$ consists of a flow of time $(T, <)$ and a temporal assignment h and, for the purposes of combination of logics, we consider a 'current world' $t \in T$ as part of the model. $\mathcal{M}, t \models A$ reads 'A is true at t over model $\mathcal{M}$'. Classes of models are normally defined by restrictions over the order relation $<$ of the flow of time.

The semantics of temporal formulae is given by:

$\mathcal{M}, t \models p$	iff $p \in \mathcal{P}$ such that $p \in h(t)$.
$\mathcal{M}, t \models \neg A$	iff it is not the case that $\mathcal{M}, t \models A$.
$\mathcal{M}, t \models A \wedge B$	iff $\mathcal{M}, t \models A$ and $\mathcal{M}, t \models B$.
$\mathcal{M}, t \models S(A, B)$	iff there exists an $s \in T$ with $s < t$ and $\mathcal{M}, s \models A$ and for every $u \in T$, if $s < u < t$ then $\mathcal{M}, u \models B$.
$\mathcal{M}, t \models U(A, B)$	iff there exists an $s \in T$ with $t < s$ and $\mathcal{M}, s \models A$ and for every $u \in T$, if $t < u < s$ then $\mathcal{M}, u \models B$.

The following restriction will be applied throughout this presentation. Flows of time will always be considered to have the properties:

(a) irreflexivity: $\forall t \neg(t < t)$;

(b) transitivity: $\forall t, s, u(t < s \wedge s < u \rightarrow t < u)$;

(c) totality: $\forall t, s(t = s \vee t < s \vee s < t)$.

The class of all flows respecting the restrictions above is the class $\mathcal{K}_{lin}$ of linear flows of time. We also represent the class of all models based on linear flows as $\mathcal{K}_{lin}$. Further restrictions can be applied to the nature of flows of time so that several other linear subclasses can be formed, e.g. the classes of dense ($\mathcal{K}_{dense}$), discrete ($\mathcal{K}_{dis}$), $\mathbb{Z}$-like, $\mathbb{Q}$-like and $\mathbb{R}$-like flows of time. The linearity property allows for the definition of the 'at all times' temporal connective $\square$:

$$\Box A = A \wedge GA \wedge HA.$$

In the case of discrete flows of time, the operators 'next time', $\bigcirc$, and 'previous time', $\bullet$, are also defined:

$$\bigcirc A = U(A, \bot)$$
$$\bullet A = S(A, \bot)$$

The inference systems will be considered to be finite axiomatization, i.e. a pair $(\Sigma, \mathcal{I})$ where Σ is a finite set of formulae called *axioms* and $\mathcal{I}$ is a set of inference rules. Consider the Burgess–Xu axiomatization for $\mathcal{K}_{lin}$ [Burgess, 1982; Xu, 1988] consisting of the following axioms:

A0 all classical tautologies
A1a $G(p \to q) \to (U(p,r) \to U(q,r))$
A2a $G(p \to q) \to (U(r,p) \to U(r,q))$
A3a $(p \wedge U(q,r)) \to U(q \wedge S(p,r), r)$
A4a $U(p,q) \to U(p, q \wedge U(p,q))$
A5a $U(q \wedge U(p,q), q) \to U(p,q)$
A6a $(U(p,q) \wedge U(r,s)) \to$
$(U(p \wedge r, q \wedge s) \vee U(p \wedge s, q \wedge s) \vee U(q \wedge r, q \wedge s))$

plus their mirror images (**b** axioms). The inference rules are:

Subst Uniform substitution, i.e. let $A(q)$ be an axiom containing the propositional letter q and let B be any formula; then from $\vdash A(q)$ infer $\vdash A(q \backslash B)$ by substituting all appearances of q in A by B.
MP *Modus ponens*: from $\vdash A$ and $\vdash A \to B$ infer $\vdash B$.
TG Temporal generalization: from $\vdash A$ infer $\vdash HA$ and $\vdash GA$.

A formula A is deducible from the set of formulae Δ, $\Delta \vdash A$, if there exists a finite sequence of formulae $B_1, \ldots, B_n = A$ such that every B_i is

(a) a formula in Δ; or
(b) an axiom; or
(c) obtained from previous formulae in the sequence through the use of an inference rule.

We write $\vdash A$ for $\emptyset \vdash A$, i.e. only items (b) and (c) above are used in the deduction of A, in which case A is said to be a *theorem*. A set of formulae Δ is *inconsistent* if $\Delta \vdash \bot$, otherwise it is *consistent*. A formula A is consistent if $\{A\}$ is consistent.

On the semantical side, a set of formulae Δ is *satisfiable* over a class of models $\mathcal{K}$ if there exists a model $\mathcal{M} \in \mathcal{K}$ with a $t \in T$ such that, for every $B \in \Delta$, $\mathcal{M}, t \models B$. A formula A is *valid* over $\mathcal{K}$, $\mathcal{K} \models A$, if for every model $\mathcal{M} = (T, <, h) \in \mathcal{K}$ and every $t \in T$, $\mathcal{M}, t \models A$. The expression $\Delta \models A$ represents that every model satisfying Δ also satisfies A.

An inference system is *sound* with respect to a class of models $\mathcal{K}$ iff every theorem is a valid formula, i.e. $\vdash A$ implies $\mathcal{K} \models A$. An inference system is (*weakly*) complete over $\mathcal{K}$ if every theorem $\vdash A$ is valid, $\mathcal{K} \models A$, or, equivalently, if every consistent formula is satisfied over $\mathcal{K}$. Strong completeness states that whenever $\Delta \models A$ then $\Delta \vdash A$, for a possibly infinite Δ. Let $\mathsf{L} = \langle \mathcal{L}, \vdash, \models \rangle$ be a logic system with language L, inference system $\vdash$ and semantics $\models$. We say that L is *decidable* if there exists an algorithm (decision procedure) that determines, for every $A \in \mathcal{L}$, whether A is a theorem or not. The *validity problem* for L is to determine whether some $A \in \mathcal{L}$ is a valid formula or not.

We have the following results:

Theorem 6.1 (Burgess–Xu [Burgess, 1982; Xu, 1988]). *The Burgess–Xu axiomatization presented above is sound and complete over the class $\mathcal{K}_{lin}$.*

Theorem 6.2 ([Rabin, 1969]). *The logic $US = \langle \mathcal{L}_{\mathsf{US}}, \vdash_{US}, \models_{US} \rangle$ is decidable over K_{lin}.*

6.3 Combining logics

As we have mentioned earlier, the combination of two one-dimensional temporal logics will generate a two-dimensional temporal logic. Throughout this presentation, we refer to one of the temporal dimensions as the *horizontal dimension* and the other one as the *vertical dimension*; the symbols related to the vertical dimension are normally obtained by putting a bar on top of the corresponding horizontal ones, e.g. T and $\overline{\mathsf{T}}$, $F$and $\overline{F}$, $<$ and $\overline{<}$.

There are two distinct criteria for defining a modal/temporal logic system as two-dimensional:

(i) if the alphabet of the language contains two non-empty, disjoint sets of corresponding modal or temporal operators, Φ and $\overline{\Phi}$, each set associated to a distinct flow of time, $(T, <)$ and $(\overline{T}, \overline{<})$, then the system is two dimensional;

(ii) if the truth value of a formula is evaluated with respect to two time points, then the system is two dimensional. In this case, we even have the distinction between strong and weak interpretation of formulae that, as a consequence, generates different notions of valid formulae (a formula is valid if it holds in all models for all pairs of time points). Under the *strong interpretation*, the truth value of atoms depends on both dimensions, giving rise to the notion of *strongly valid formulae* when the evaluation of formulae is inductively extended to all connectives. In the *weak interpretation*, the truth value of atoms depends only on the one dimension, e.g. the horizontal dimension, giving rise to the notion of *weakly valid formulae.* Usually for this notion of two-dimensionality, both time points refer to the same flow of time, so we may also have the notion of (weak/strong) *diagonally valid* formulae by restricting validity to the case where both dimensions refer to the same point, i.e. A is diagonally valid iff $\mathcal{M}, t, t \models A$ for all $\mathcal{M}$ and t; see [Gabbay *et al.*, 1994] for more details.

Criterion (i) will be called the *syntactic criterion* for two-dimensionality, although it is not completely syntactic, i.e. it depends on the semantic notion of flows of time; criterion (ii) will be called the *semantic criterion* for two-dimensionality.

Note that both cases can yield, as an extreme case, one-dimensional temporal logic. In (i), by making $T = \overline{T}$ and $\overline{<} = (<)^{-1} = (>)$, i.e. by taking two flows with the same set of time points such that one order is the inverse of the other, the future operators $\Phi = \{F, G, U\}$ are associated with $(T, <)$ and the past operators $\overline{\Phi} = \{P, H, S\}$ are associated with $(T, >)$. In (ii), this is achieved by fixing one dimension to a single time point so that it becomes redundant.

These two distinct approaches to the two-dimensionality of a system are independent. In fact, we will see in Section 6.5 a system that contains two distinct sets of operators over two classes of flows of time, but its formulae are evaluated at a single point. On the other hand, there are several temporal logics in the literature satisfying (ii) but not (i), containing a single set of temporal operators in which formulae are evaluated according to two or more time points in the same flow, e.g. [Aqvist, 1979; Kamp, 1971; Gabbay *et al.*, 1994].

A logic system that respects both the syntactic and the semantic criteria for two-dimensionality is called ***broadly two dimensional***, and this will be the kind of system we will be aiming to achieve through combination methods. We consider in this work only strong evaluation and validity; the weak interpretation generates systems with the expressivity of only monadic first-order language [Gabbay *et al.*, 1994], but for broadly two-dimensional systems we are interested in the expressivity of dyadic first-order language, although it is known that no set of temporal operators can be expressively complete over dyadic first-order language [Venema, 1990]. (A modal/temporal language is *expressively complete* over a class of first-order formulae if, for any first-order formula A in that class, there exists a modal/temporal formula B such that A is first-order equivalent to B^*, where B^* is the standard first-order translation of B; see [Gabbay *et al.*, 1994].) Venema's [1990] two-dimensional temporal logic, Segerberg's [Segerberg, 1973] two-dimensional modal logic and the temporalization of a temporal logic are all broadly two dimensional; so are the combined logics in Sections 6.6 and 6.7.

In the study of one-dimensional temporal logics (1DTLs) several classes of flows of time are taken into account. When we move to 2DTLs, the number of such classes increases considerably, and every pair of one-dimensional classes can be seen as generating a different two-dimensional class. The study of 2DTLs would benefit much if the properties known to hold for 1DTLs could be systematically transferred to 2DTLs, avoiding the repetition of much of the work that has been published in the literature. This is a strong motivation for considering methods of combination of 1DTLs with 2DTLs and studying the transfer of logical properties through each method. Also in favour of such an approach is the fact that the results concerning 2DTLs are then presented in a general, compact and elegant form.

In providing a method to combine two 1DTLs $\overline{\mathsf{T}}$ and T we have to pay attention to the following points:

(a) A method for combining logics $\overline{\mathsf{T}}$ and T is composed of three submethods, namely a method for combining the languages of $\overline{\mathsf{T}}$ and T, a method for combining their inference systems and a method for combining their semantics.

(b) We study the combined logic system with respect to the way certain logical properties of $\overline{\mathsf{T}}$ and T are transferred to the two-dimensional combination. We focus here on the properties of soundness, completeness and decidability of the combined system given those of the component ones.

(c) The combined language should be able to express some properties of the interaction between the two dimensions; otherwise the combination is just a partial one, and the two systems are not fully combined. For example, it is desirable to express formulae like $F\overline{F}A \leftrightarrow \overline{F}FA$ and $P\overline{F}A \leftrightarrow \overline{F}PA$ that are not in the temporalized language of $\overline{\mathsf{T}}(\mathsf{T})$.

(d) If we want to strengthen the interaction between the two systems, some properties of the interaction between the two dimensions are expected to be theorems of the combining system, e.g. the commutativity of horizontal and vertical future operators such as $F\overline{F}A \leftrightarrow \overline{F}FA$ and $P\overline{F}A \leftrightarrow \overline{F}PA$. These are called the *interaction axioms* in [Gabbay, 1992].

(e) We want the combination method to be as independent as possible from the underlying flows of time.

All methods of combination must comply with item (a). The method for combining the languages of $\overline{\mathsf{T}}$ and T includes the choice of which sublanguage of $\overline{\mathsf{T}}$ and T is going to be part of the combined two-dimensional language, as well as the way in which this combination is done; in this presentation we will work, in the most general case, with the standard languages of S and $U, \bar{S}$ and $\bar{U}$, but we also consider some sublanguages, e.g. the sublanguage generated by a set of derived operators, as the vertical 'previous' ($\overline{\bullet}$) and 'next' ($\overline{\bigcirc}$) in Section 6.7. In combining the inference systems of $\overline{\mathsf{T}}$ and T, we will assume that they are both an extension of classical logic and that they are presented in the form of a regular, normal axiomatic system $(\Sigma, \mathcal{I})$, where Σ is a set of axioms and $\mathcal{I}$ is a set of inference rules. One important requirement is that the combined system be a *conservative extension* of the two components. The conservativeness property states that if A is a formula in the language of L and L^* is a logic system extending L (i.e. the language of L is a sublanguage of the language of L^*) then A will be a theorem of L^* only if it is a theorem of L already; conservativeness guarantees that no new information about the original system L is present in the extended one L^*.

The combined semantics has to deal with the structure of the combined model, the evaluation of two-dimensional formulae over those structures and also with the combinations of classes of flows of time.

Items (b), (c), (d) and (e) may conflict with each other. In fact, the rest of this chapter shows that this is the case, as we try to compromise between expressivity, independence of the underlying flow of time and the transfer of logical properties.

6.4 Temporalizing a logic

The first of the combination methods, known as 'adding a temporal dimension to a logic system' or simply 'temporalizing a logic system', has been extensively discussed in [Finger and Gabbay, 1992].

Temporalization is a methodology whereby an arbitrary logic system L can be enriched with temporal features to create a new system $\mathsf{T}(\mathsf{L})$. The new system is constructed by combining L with a pure propositional temporal logic T (such as linear temporal logic with 'Since' and 'Until') in a special way.

Although we are only interested here in temporalizing an already temporal system, so as to generate a 2DTL, the original method is more general and is applicable to a generic logic L; L is constrained to be an extension of classical logic, i.e. all propositional tautologies must be valid in it, but such a constraint does not affect us, for we are assuming that both temporal systems T and L are extensions of $US/\mathcal{K}_{lin}$. The language of a temporalized system is based on the US language and on a subset of the language of L, $\mathcal{L}_\mathsf{L}$. The set $\mathcal{L}_\mathsf{L}$ is partitioned in two sets, BC_L and ML_L. A formula $A \in \mathcal{L}_\mathsf{L}$ belongs to the set of *Boolean combinations*, BC_L, iff it is built up from other formulae by the use of one of the Boolean connectives $\neg$ or $\wedge$ or any other connective defined only in terms of those; it belongs to the set of *monolithic formulae* ML_L otherwise.

The result of temporalizing over $\mathcal{K}$ the logic system L is the logic system $\mathsf{T}(\mathsf{L})/\mathcal{K}$. The alphabet of the temporalized language uses the alphabet of L plus the two-place operators S and U, if they are not part of the alphabet of L; otherwise, we use $\overline{S}$ and $\overline{U}$ or any other proper renaming.

Definition 6.3 (Temporalized formulae). The set $\mathcal{L}_{\mathsf{T}(\mathsf{L})}$ of formulae of the logic system L is the smallest set such that:

1. if $A \in ML_\mathsf{L}$, then $A \in \mathcal{L}_{\mathsf{T}(\mathsf{L})}$;
2. if $A, B \in \mathcal{L}_{\mathsf{T}(\mathsf{L})}$ then $\neg A \in \mathcal{L}_{\mathsf{T}(\mathsf{L})}$ and $(A \wedge B) \in \mathcal{L}_{\mathsf{T}(\mathsf{L})}$;
3. if $A, B \in \mathcal{L}_{\mathsf{T}(\mathsf{L})}$ then $S(A, B) \in \mathcal{L}_{\mathsf{T}(\mathsf{L})}$ and $U(A, B) \in \mathcal{L}_{\mathsf{T}(\mathsf{L})}$.

Note that, for instance, if $\blacksquare$ is an operator of the alphabet of L and A and B are two formulae in $\mathcal{L}_\mathsf{L}$, the formula $\blacksquare U(A, B)$ is *not* in $\mathcal{L}_{\mathsf{T}(\mathsf{L})}$. The language of $\mathsf{T}(\mathsf{L})$ is independent of the underlying flow of time, but not its semantics and inference system, so we must fix a class $\mathcal{K}$ of flows of time over which the temporalization is defined; if $\mathcal{M}_\mathsf{L}$ is a model in the class of models of L, $\mathcal{K}_\mathsf{L}$, for every formula $A \in \mathcal{L}_\mathsf{L}$ we must have either $\mathcal{M}_\mathsf{L} \models A$ or $\mathcal{M}_\mathsf{L} \models \neg A$. In the case where L is a temporal logic we must consider a 'current time' o as part of its model to achieve that condition.

Definition 6.4 (Semantics of the temporalized logic). Let $(T, <) \in \mathcal{K}$ be a flow of time and let $g : T \to \mathcal{K}_{\mathsf{L}}$ be a function mapping every time point in T to a model in the class of models of L. A model of $\mathsf{T(L)}$ is a triple $\mathcal{M}_{\mathsf{T(L)}} = (T, <, g)$ and the fact that A is true in $\mathcal{M}_{\mathsf{T(L)}}$ at time t is written as $\mathcal{M}_{\mathsf{T(L)}}, t \models A$ and defined as

$\mathcal{M}_{\mathsf{T(L)}}, t \models A,\ A \in ML_{\mathsf{L}}$	iff $g(t) = \mathcal{M}_{\mathsf{L}}$ and $\mathcal{M}_{\mathsf{L}} \models A$.
$\mathcal{M}_{\mathsf{T(L)}}, t \models \neg A$	iff it is not the case that $\mathcal{M}_{\mathsf{T(L)}}, t \models A$.
$\mathcal{M}_{\mathsf{T(L)}}, t \models (A \wedge B)$	iff $\mathcal{M}_{\mathsf{T(L)}}, t \models A$ and $\mathcal{M}_{\mathsf{T(L)}}, t \models B$.
$\mathcal{M}_{\mathsf{T(L)}}, t \models S(A, B)$	iff there exists $s \in T$ such that $s < t$ and $\mathcal{M}_{\mathsf{T(L)}}, s \models A$ and for every $u \in T$, if $s < u < t$ then $\mathcal{M}_{\mathsf{T(L)}}, u \models B$.
$\mathcal{M}_{\mathsf{T(L)}}, t \models U(A, B)$	iff there exists $s \in T$ such that $t < s$ and $\mathcal{M}_{\mathsf{T(L)}}, s \models A$ and for every $u \in T$, if $t < u < s$ then $\mathcal{M}_{\mathsf{T(L)}}, u \models B$.

Figure 6.2 illustrates a temporalized model. The inference system of $\mathsf{T(L)}/\mathcal{K}$ is given by the following:

Definition 6.5 (Axiomatization for $\mathsf{T(L)}$). An axiomatization for the temporalized logic $\mathsf{T(L)}$ is composed of:

- the axioms of $\mathsf{T}/\mathcal{K}$;
- the inference rules of $\mathsf{T}/\mathcal{K}$;
- for every formula A in $\mathcal{L}_{\mathsf{L}}$, if $\vdash_{\mathsf{L}} A$ then $\vdash_{\mathsf{T(L)}} A$, i.e. all theorems of L are theorems of $\mathsf{T(L)}$. This inference rule is called **Persist**.

Example 6.6 (Temporalizing propositional logic). Consider classic propositional logic $\mathsf{PL} = \langle \mathcal{L}_{\mathsf{PL}}, \vdash_{\mathsf{PL}} \rangle$. Its temporalization generates the logic system $\mathsf{T(PL)} = \langle \mathcal{L}_{\mathsf{T(PL)}}, \vdash_{\mathsf{T(PL)}} \rangle$.

It is not difficult to see that $\mathcal{L}_{\mathsf{T(PL)}} = \mathcal{L}_{\mathsf{US}}$ and $\vdash_{\mathsf{T(PL)}} = \vdash_{\mathsf{US}}$, i.e. the temporalized version of PL over any $\mathcal{K}$ is actually the temporal logic $\mathsf{T} = \mathsf{US}/\mathcal{K}$. With respect to $\mathcal{M}_{\mathsf{T(L)}}$, the function g actually assigns, for every time point, a PL model.

Example 6.7 (Temporalizing US-temporal logic). If we temporalize over $\mathcal{K}$ the one-dimensional logic system $\mathsf{US}/\mathcal{K}$ we obtain the two-dimensional logic system $\mathsf{T(US)} = \langle \mathcal{L}_{\mathsf{T(US)}}, \vdash_{\mathsf{T(US)}} \rangle = \mathsf{T}^2(\mathsf{PL})/\mathcal{K}$. In this case we have to rename the two-place operators S and U of the temporalized alphabet to, say, $\overline{S}$ and $\overline{U}$.

In order to obtain a model for $\mathsf{T(US)}$, we must fix a 'current time', o, in $\mathcal{M}_{\mathsf{US}} = (T_1, <_1, g_1)$, so that we can construct the model $\mathcal{M}_{\mathsf{T(US)}} = (T_2, <_2, g_2)$ as previously described. Note that, in this case, the flows of time $(T_1, <_1)$ and $(T_2, <_2)$ need not to be the same. $(T_2, <_2)$ is the flow of time of the upper level temporal system whereas $(T_1, <_1)$ is the flow of time of the underlying logic which, in this case, happens to be a temporal logic.

The logic system we obtain by temporalizing US-temporal logic is the 2DTL described in [Finger, 1992].

Example 6.8 (n-dimensional temporal logic). If we repeat the process started in the last two examples, we can construct an n-dimensional temporal logic $\mathsf{T}^n(\mathsf{PL})/\mathcal{K}$ (its alphabet including S_n and U_n) by temporalizing an $(n-1)$-dimensional temporal logic.

Every time we add a temporal dimension, we are able to describe changes in the underlying system. Temporalizing the system L once, we are creating a way of describing the history of L; temporalizing for the second time, we are describing how the history of L is viewed in different moments of time. We can go on indefinitely, although it is not clear what is the purpose of doing so.

To present the transfer results we restrict the logic systems to $\mathsf{L} = US/\mathcal{K}$ and $\mathsf{T} = \bar{U}\bar{S}/\overline{\mathcal{K}}$, where $\mathcal{K}, \overline{\mathcal{K}} \subseteq \mathcal{K}_{lin}$. We write $\bar{U}\bar{S}(US)$ instead of $\mathsf{T}(\mathsf{L})$ and the generated class of models is referred to as $\overline{\mathcal{K}}(\mathcal{K})$. For this system, we enumerate a series of results that are proved in [Finger and Gabbay, 1992]. Those results will be useful for the discussion of the independent combination method.

Theorem 6.9 (Transfer via temporalization). *Let $\bar{U}\bar{S}/\overline{\mathcal{K}}$ and $US/\mathcal{K}$ be two logic systems such that $\overline{\mathcal{K}}, \mathcal{K} \subseteq \mathcal{K}_{lin}$.*

- *(a) If $\bar{U}\bar{S}$ is sound with respect to $\overline{\mathcal{K}}$ and US is sound with respect to $\mathcal{K}$, then $\bar{U}\bar{S}(US)$ is sound w.r.t. $\overline{\mathcal{K}}(\mathcal{K})$.*
- *(b) If $\bar{U}\bar{S}$ is complete w.r.t. $\overline{\mathcal{K}}$ and US is complete w.r.t. $\mathcal{K}$ then $\bar{U}\bar{S}(US)$ is complete w.r.t. $\overline{\mathcal{K}}(\mathcal{K})$.*
- *(c) If $\bar{U}\bar{S}$ is complete w.r.t. $\mathcal{K}$, then $\bar{U}\bar{S}(US)$ is a conservative extension of both $\bar{U}\bar{S}$ and US.*
- *(d) If $\bar{U}\bar{S}$ is complete and is decidable over $\overline{\mathcal{K}}$ and US is complete and decidable over $\mathcal{K}$ then $\bar{U}\bar{S}(US)$ is decidable over $\overline{\mathcal{K}}(\mathcal{K})$.*

6.5 Independent combination

We have seen in the previous section how to add a temporal dimension to a logic system. In particular, if a temporal logic is itself temporalized we obtain a 2DTL. Such a logic system is, however, very weakly expressive; if US is the internal (horizontal) temporal logic in the temporalization process (F is derived in US), and $\bar{U}\bar{S}$ is the external (vertical) one ($\overline{F}$ is defined in $\bar{U}\bar{S}$), we cannot express that vertical and horizontal future operators commute,

$$F\overline{F}A \leftrightarrow \overline{F}FA.$$

In fact, the subformula $F\overline{F}A$ is not even in the temporalized language of $\bar{U}\bar{S}(\mathsf{US})$, nor is the whole formula. In other words, the interplay between the two dimensions is not expressible in the language of the temporalized $\bar{U}\bar{S}(\mathsf{US})$.

The idea then is to define a new method for combining logic systems that puts together all the expressivity of the two component logic systems in an independent way; for that we assume that the language of a system is given by a set of formation rules.

Definition 6.10. Let $Op(\mathsf{L})$ be the set of non-Boolean operators of a generic logic L. Let $\bar{\mathsf{T}}$ and T be logic systems such that $Op(\mathsf{T}) \cap Op(\bar{\mathsf{T}}) = \emptyset$. The *fully combined language* of logic systems $\bar{\mathsf{T}}$ and T over the set of atomic propositions $\mathcal{P}$ is obtained by the union of the respective set of connectives and the union of the formation rules of the languages of both logic systems.

Let the operators U and S be in the language of US and $\overline{U}$ and $\overline{S}$ be in that of $\bar{U}\bar{S}$. Note that the renaming of the temporal operator is done prior to the combination, so that the combined systems contains the set of Boolean operators $\{\neg, \wedge\}$ coming from both components, plus the set of temporal operators $\{U, S, \overline{U}, \overline{S}\}$. Their fully combined language over a set of atomic propositions $\mathcal{P}$ is given by:

- every atomic proposition is in it;
- if A, B are in it, so are $\neg A$ and $A \wedge B$;
- if A, B are in it, so are $U(A, B)$ and $S(A, B)$;
- if A, B are in it, so are $\overline{U}(A, B)$ and $\overline{S}(A, B)$.

In general, we do not want any non-Boolean operator to be shared between the two languages, for this may cause problems when combining their axiomatizations. For example (this example is due to Ian Hodkinson), if a generic operator $\Box$ belongs to both temporal logic systems such that T contains axiom $q \leftrightarrow \Box q$ and system $\bar{\mathsf{T}}$ contains axiom $\neg q \leftrightarrow \Box q$, the union of their axiomatizations will result in an inconsistent system even though each system might have been itself consistent. To avoid such a behaviour the restriction $Op(\mathsf{T}) \cap Op(\bar{\mathsf{T}}) = \emptyset$ was imposed on the fully combined language of $\bar{\mathsf{T}}$ and T.

This new combination method is called *independent* because it takes the independent union of the axiomatization of its two component systems, and it is based on their fully combined language.

Definition 6.11. Let US and $\bar{U}\bar{S}$ be two US-temporal logic systems defined over the same set $\mathcal{P}$ of propositional atoms such that their languages are independent. The *independent combination* $US \oplus \bar{U}\bar{S}$ is given by the following:

- The fully combined language of US and $\bar{U}\bar{S}$.
- If $(\Sigma, \mathcal{I})$ is an axiomatization for US and $(\overline{\Sigma}, \overline{\mathcal{I}})$ is an axiomatization for $\bar{U}\bar{S}$, then $(\Sigma \cup \overline{\Sigma}, \mathcal{I} \cup \overline{\mathcal{I}})$ is an axiomatization for $US \oplus \bar{U}\bar{S}$. Note that, apart from the classical tautologies, the set of axioms Σ and $\overline{\Sigma}$ are supposed to be disjoint, but not the inference rules.
- The class of independently combined flows of time is $\mathcal{K} \oplus \overline{\mathcal{K}}$ composed of biordered flows of the form $(\tilde{T}, <, \overline{<})$ where the connected components of $(\tilde{T}, <)$ are in $\mathcal{K}$ and the connected components of $(\tilde{T}, \overline{<})$ are in $\overline{\mathcal{K}}$, and $\tilde{T}$ is the (not necessarily disjoint) union of the sets of time points T and $\overline{T}$ that constitute each connected component; such a biordered flow of time has been discussed in [Kracht and Wolter, 1991] for the case of the independent combination of two monomodal systems.

A model structure for $US \oplus \bar{U}\bar{S}$ over $\mathcal{K} \oplus \overline{\mathcal{K}}$ is a 4-tuple $(\tilde{T}, <, \mathrel{\overline{<}}, g)$, where $(\tilde{T}, <, \mathrel{\overline{<}}) \in \mathcal{K} \oplus \overline{\mathcal{K}}$ and g is an assignment function $g : \tilde{T} \to 2^{\mathcal{P}}$. An independently combined model is illustrated in Fig. 6.5.

The semantics of a formula A in a model $\mathcal{M} = (\tilde{T}, <, \mathrel{\overline{<}}, g)$ is defined as the union of the rules defining the semantics of $US/\mathcal{K}$ and $\bar{U}\bar{S}/\overline{\mathcal{K}}$. The expression $\mathcal{M}, t \models A$ reads that the formula A is true in the (combined) model $\mathcal{M}$ at the point $t \in \tilde{T}$. The semantics of formulae is given by induction in the standard way:

$\mathcal{M}, t \models p$ iff $p \in g(t)$ and $p \in \mathcal{P}$;
$\mathcal{M}, t \models \neg A$ iff it is not the case that $\mathcal{M}, t \models A$;
$\mathcal{M}, t \models A \wedge B$ iff $\mathcal{M}, t \models A$ and $\mathcal{M}, t \models B$;
$\mathcal{M}, t \models S(A, B)$ iff there exists an $s \in \tilde{T}$ with $s < t$ and $\mathcal{M}, s \models A$ and for every $u \in \tilde{T}$, if $s < u < t$ then $\mathcal{M}, u \models B$;
$\mathcal{M}, t \models U(A, B)$ iff there exists an $s \in \tilde{T}$ with $t < s$ and $\mathcal{M}, s \models A$ and for every $u \in \tilde{T}$, if $t < u < s$ then $\mathcal{M}, u \models B$;
$\mathcal{M}, t \models \bar{S}(A, B)$ iff there exists an $s \in \tilde{T}$ with $s \mathrel{\overline{<}} t$ and $\mathcal{M}, s \models A$ and for every $u \in \tilde{T}$, if $s \mathrel{\overline{<}} u \mathrel{\overline{<}} t$ then $\mathcal{M}, u \models B$;
$\mathcal{M}, t \models \bar{U}(A, B)$ iff there exists an $s \in \tilde{T}$ with $t \mathrel{\overline{<}} s$ and $\mathcal{M}, s \models A$ and for every $u \in \tilde{T}$, if $t \mathrel{\overline{<}} u \mathrel{\overline{<}} s$ then $\mathcal{M}, u \models B$.

Note that, despite the combination of two flows of time, formulae are evaluated according to a single point. The independent combination generates a system that is two dimensional according to the first criterion but fails the second one, so it is not broadly two dimensional.

The following result is due to [Thomason, 1980] and is more general than the independent combination of two US-logics.

Proposition 6.12. *With respect to the validity of formulae, the independent combination of two modal logics is a conservative extension of the original ones.*

Note that we have previously defined conservative extension in proof-theoretic terms; completeness for the independently combined case will lead to the conservativeness with respect to derivable theorems.

As usual, we will assume that $\mathcal{K}, \overline{\mathcal{K}} \subseteq \mathcal{K}_{lin}$, so $<$ and $\mathrel{\overline{<}}$ are transitive, irreflexive and total orders; similarly, we assume that the axiomatizations are extensions of $US/\mathcal{K}_{lin}$.

The temporalization process will be used as an inductive step to prove the transfer of soundness, completeness and decidability for $US \oplus \bar{U}\bar{S}$ over $\mathcal{K} \oplus \overline{\mathcal{K}}$. Let us first consider the *degree of alternation* of a $(US \oplus \bar{U}\bar{S})$-formula A for US, $dg(A)$, and $\bar{U}\bar{S}$, $\overline{dg}(A)$.

$$
\begin{array}{l|l}
dg(p) = 0 & \overline{dg}(p) = 0 \\
dg(\neg A) = dg(A) & \overline{dg}(\neg A) = \overline{dg}(A) \\
dg(A \wedge B) = max\{dg(A), dg(B)\} & \overline{dg}(A \wedge B) = max\{\overline{dg}(A), \overline{dg}(B)\} \\
dg(S(A,B)) = max\{dg(A), dg(B)\} & \overline{dg}(\overline{S}(A,B)) = max\{\overline{dg}(A), \overline{dg}(B)\} \\
dg(U(A,B)) = max\{dg(A), dg(B)\} & \overline{dg}(\overline{U}(A,B)) = max\{\overline{dg}(A), \overline{dg}(B)\} \\
dg(\overline{S}(A,B)) = 1 + max\{\overline{dg}(A), \overline{dg}(B)\} & \overline{dg}(S(A,B)) = 1 + max\{dg(A), dg(B)\} \\
dg(\overline{U}(A,B)) = 1 + max\{\overline{dg}(A), \overline{dg}(B)\} & \overline{dg}(U(A,B)) = 1 + max\{dg(A), dg(B)\}.
\end{array}
$$

Any formula A of $US \oplus \bar{U}\bar{S}$ can be seen as a formula of some finite number of alternating temporalizations of the form $\mathsf{US}(\bar{U}\bar{S}(\mathsf{US}(\ldots)))$; more precisely, A can be seen as a formula of $\mathsf{US}(\mathsf{L}_n)$, where $dg(A) = n$, $\mathsf{US}(\mathsf{L}_0) = \mathsf{US}$, $\bar{U}\bar{S}(\mathsf{L}_0) = \bar{U}\bar{S}$, and $\mathsf{L}_{n-2i} = \bar{U}\bar{S}(\mathsf{L}_{n-2i-1})$, $\mathsf{L}_{n-2i-1} = US(\mathsf{L}_{n-2i-2})$, for $i = 0, 1, \ldots, \lceil \frac{n}{2} \rceil - 1$. This fact is illustrated in Fig. 6.4. The following lemma actually allows us to see the independent combination as the (infinite) union of a finite number of alternating temporalizations of US and $\bar{U}\bar{S}$; it will also be used in the proof of the transfer of completeness and decidability (given completeness) for $US \oplus \bar{U}\bar{S}$.

Lemma 6.13. *Let US and $\bar{U}\bar{S}$ be two complete logic systems. Then, A is a theorem of $US \oplus \bar{U}\bar{S}$ iff it is a theorem of $\mathsf{US}(\mathsf{L}_n)$, where $dg(A) = n$.*

Proof If A is a theorem of $\mathsf{US}(\mathsf{L}_n)$, all the inferences in its deduction can be repeated in $US \oplus \bar{U}\bar{S}$, so it is a theorem of $US \oplus \bar{U}\bar{S}$.

Suppose A is a theorem of $US \oplus \bar{U}\bar{S}$; let $B_1, \ldots, B_m = A$ be a deduction of A in $US \oplus \bar{U}\bar{S}$ and let $n' = max\{dg(B_i)\}$, $n' \geq n$. We claim that each B_i is a theorem of $\mathsf{US}(\mathsf{L}_{n'})$. In fact, by induction on m, if B_i is obtained in the deduction by substituting into an axiom, the same substitution can be done in $\mathsf{US}(\mathsf{L}_{n'})$; if B_i is obtained by temporal generalization from B_j, $j < i$, then by the induction hypothesis, B_j is a theorem of $\mathsf{US}(\mathsf{L}_{n'})$ and so is B_i; if B_i is obtained by *modus ponens* from B_j and B_k, $j, k < i$, then by the induction hypothesis, B_j and B_k are theorems of $\mathsf{US}(\mathsf{L}_{n'})$ and so is B_i.

So A is a theorem of $\mathsf{US}(\mathsf{L}_{n'})$ and, since US and $\bar{U}\bar{S}$ are two complete logic systems, by Theorem 6.9, each of the alternating temporalizations in $\mathsf{US}(\mathsf{L}_{n'})$ is a conservative extension of the underlying logic; it follows that A is a theorem of $\mathsf{US}(\mathsf{L}_n)$, as desired. □

The transfer of soundness, completeness and decidability follows directly from this result.

Theorem 6.14 (Independent combination). *Let US and $\bar{U}\bar{S}$ be two sound and complete logic systems over the classes $\mathcal{K}$ and $\overline{\mathcal{K}}$, respectively. Then their independent combination $US \oplus \bar{U}\bar{S}$ is sound and complete over the class $\mathcal{K} \oplus \overline{\mathcal{K}}$. If US and $\bar{U}\bar{S}$ are complete and decidable, so is $US \oplus \bar{U}\bar{S}$.*

Proof Soundness follows immediately from the validity of axioms and inference rules. For completeness, suppose that A is a consistent formula in $US \oplus \bar{U}\bar{S}$; by Lemma 6.13, A is consistent in $\mathsf{US}(\mathsf{L}_n)$, so we construct a temporalized model for it, and we obtain a model $(\tilde{T}_1, <_1, g_1, o_1)$ over $\mathcal{K}(\overline{\mathcal{K}}(\mathcal{K}(\ldots)))$, where o_1 is the 'current time' necessary for the successive temporalizations. We show now how it can be transformed into a model over $\mathcal{K} \oplus \overline{\mathcal{K}}$.

Without loss of generality, suppose that US is the outermost logic system in the multilayered temporalized system $\mathsf{US}(\bar{U}\bar{S}(\mathsf{US}(\ \ldots\ )))$, and let n be the number of alternations. The construction is recursive, starting with the outermost logic. Let $i \leq n$ denote the step of the construction; if i is odd, it is a US-temporalization, otherwise it is a $\bar{U}\bar{S}$-temporalization. At every step i we construct the sets $\tilde{T}_{i+1}$, $<_{i+1}$ and $\overline{<}_{i+1}$ and the function g_{i+1}.

We start the construction of the model at step $i = 0$ with the temporalized model $(\tilde{T}_1, <_1, g_1, o_1)$ such that $(\tilde{T}_1, <_1) \in \mathcal{K}$, and we take $\overline{<}_1 = \emptyset$. At step $i < n$, consider the current set of time points $\tilde{T}_i$; according to the construction, each $t \in \tilde{T}_i$ is associated to:

- a temporalized model $g_i(t) = (\tilde{T}^t_{i+1}, <^t_{i+1}, g^t_{i+1}, o^t_{i+1}) \in \mathcal{K}$ and take $\overline{<}^t_{i+1} = \emptyset$, if i is even; or
- a temporalized model $g_i(t) = (\tilde{T}^t_{i+1}, \overline{<}^t_{i+1}, g_{i+1}, o^t_{i+1}) \in \overline{\mathcal{K}}$ and take $<^t_{i+1} = \emptyset$, if i is odd.

The point t is made identical to $o^t_{i+1} \in \tilde{T}^t_{i+1}$, so as to add the new model to the current structure. Note that this preserves the satisfiability of all formulae at t. Let $\tilde{T}_{i+1}$ be the (possibly infinite) union of all $\tilde{T}^t_{i+1}$ for $t \in \tilde{T}_i$; similarly, $<_{i+1}$ and $\overline{<}_{i+1}$ are generated. And finally, for every $t \in \tilde{T}_{i+1}$, the function g_{i+1} is constructed as the union of all g^t_{i+1} for $t \in \tilde{T}_i$.

Repeating this construction n times, we obtain a combined model over $\mathcal{K} \oplus \overline{\mathcal{K}}$, $\mathcal{M} = (\tilde{T}_n, <_n, \overline{<}_n, g_n)$, such that for all $t \in \tilde{T}_n$, $g_n(t) \subseteq \mathcal{P}$. Since satisfiability of formulae is preserved at each step, it follows that $\mathcal{M}$ is a model for A, and completeness is proved.

For decidability, suppose we want to decide whether a formula $A \in US \oplus \bar{U}\bar{S}$ is a theorem. By Lemma 6.13, this is equivalent to deciding whether $A \in \mathsf{US}(\mathsf{L}_n)$ is a theorem, where $n = dg(A)$. Since $US/\mathcal{K}$ and $\bar{U}\bar{S}/\overline{\mathcal{K}}$ are both complete and decidable, by successive applications of Theorem 6.9(b) and (d), it follows that the following logics are decidable: $\mathsf{US}(\bar{U}\bar{S})$, $\bar{U}\bar{S}(\mathsf{US}(\bar{U}\bar{S})) = \bar{U}\bar{S}(\mathsf{L}_2)$, ..., $\bar{U}\bar{S}(\mathsf{L}_{n-1}) = \mathsf{L}_n$; so a last application of Theorem 6.9(b) and (d) yields that $\mathsf{US}(\mathsf{L}_n)$ is decidable. □

6.6 Full join

With respect to the generation of two-dimensional systems, the method of independent combination has two main drawbacks. First, it generates logic systems whose formulae are evaluated at one single time point, not generating a broadly two-dimensional logic. Second, since the method independently combines the two component logic systems, no interaction between the dimensions is provided. As a consequence, although a formula like $F\overline{F}A \leftrightarrow \overline{F}FA$ is expressible in its language, it will not be valid, as can easily be verified, for it expresses an interplay between the dimensions. We therefore introduce the notion of a *two-dimensional plane model*.

Definition 6.15. Let $\mathcal{K}$ and $\overline{\mathcal{K}}$ be two classes of flow of time. A *two-dimensional plane model* over the *fully combined class* $\mathcal{K} \times \overline{\mathcal{K}}$ is a 5-tuple $\mathcal{M} = (T, <, \overline{T}, \overline{<}, g)$, where $(T, <) \in \mathcal{K}$, $(\overline{T}, \overline{<}) \in \overline{\mathcal{K}}$ and $g : T \times \overline{T} \to 2^{\mathcal{P}}$ is a two-dimensional assignment. The semantics of the horizontal and vertical operators are independent of each other.

$$\begin{array}{lll} \mathcal{M}, t, x \models S(A, B) & \text{iff} & \text{there exists } s < t \text{ such that } \mathcal{M}, s, x \models A \text{ and} \\ & & \text{for all } u,\ s < u < t,\ \mathcal{M}, u, x \models B. \\ \mathcal{M}, t, x \models \overline{S}(A, B) & \text{iff} & \text{there exists } y \overline{<} x \text{ such that } \mathcal{M}, t, y \models A \text{ and} \\ & & \text{for all } z,\ y \overline{<} z \overline{<} x,\ \mathcal{M}, t, z \models B. \end{array}$$

Similarly for U and $\overline{U}$, the semantics of atoms and boolean connectives remain the standard one. A formula A is (strongly) valid over $\mathcal{K} \times \overline{\mathcal{K}}$ if for all models $\mathcal{M} = (T, <, \overline{T}, \overline{<}, g)$, for all $t \in T$ and $x \in \overline{T}$ we have $\mathcal{M}, t, x \models A$.

With respect to the expressivity of fully combined two-dimensional languages, Venema [1990] has shown that no finite set of two-dimensional temporal operators is expressively complete over the class of linear flows with respect to dyadic first-order logic—despite the fact that US-temporal logic is expressively complete with respect to monadic first-order logic over $\mathbb{N}$ and over $\mathbb{R}$, and that, with additional operators (the Stavi operators), we can get expressive completeness over $\mathbb{Q}$ and $\mathcal{K}_{lin}$ [Gabbay, 1981b]. So expressive completeness is not transferred by neither the full join nor any other combination method.

It is easy to verify that the following formulae expressing the commutativity of future and past operators between the two dimensions are valid formulae in two-dimensional plane models.

I1 $F\overline{F}A \leftrightarrow \overline{F}FA$

I2 $F\overline{P}A \leftrightarrow \overline{P}FA$

I3 $P\overline{F}A \leftrightarrow \overline{F}PA$

I4 $P\overline{P}A \leftrightarrow \overline{P}PA$.

Therefore, if we want to satisfy both the syntactic and the semantic criteria for two-dimensionality, we may define the method of *full join* containing the fully combined language of US and $\bar{U}\bar{S}$ and their fully combined class of models. The question is whether there is a method for combining their axiomatizations so as to generate a *fully joined axiomatization* that transfers the properties of soundness, completeness and decidability. The answer, however, is 'not in general'. In some cases we can obtain the transfer of completeness, in some other cases it fails. To illustrate that, we consider completeness results over classes of the form $\mathcal{K} \times \mathcal{K}$.

We start by defining some useful abbreviations. Let p be a propositional atom; define

$$\begin{aligned} hor(p) &= \Box(p \wedge \overline{H}\neg p \wedge \overline{G}\neg p) \\ ver(p) &= \overline{\Box}(p \wedge H\neg p \wedge G\neg p). \end{aligned}$$

It is clear that $hor(p)$ makes p true along the horizontal line and false elsewhere; similarly for $ver(p)$ with respect to the vertical.

The axiomatization of $US \times \bar{U}\bar{S}$ over $\mathcal{K}_{lin} \times \mathcal{K}_{lin}$ extends that of $US \oplus \bar{U}\bar{S}$ over $\mathcal{K}_{lin} \oplus \mathcal{K}_{lin}$ by including the join axioms **I1**–**I4** and the following inference rules:

IR1 if $\vdash hor(p) \to A$ and p does not occur in A, then $\vdash A$;
IR2 if $\vdash ver(p) \to A$ and p does not occur in A, then $\vdash A$.

IR1 and **IR2** are two-dimensional extensions of the irreflexivity inference rule (IRR) defined in [Gabbay, 1981a] for the one-dimensional case: if $\vdash p \wedge H\neg p \to A$ and p does not occur in A, then $\vdash A$.

Theorem 6.16 (2D-completeness). *There is a sound and complete axiomatization over the class of full two-dimensional temporal models over $\mathcal{K}_{lin} \times \mathcal{K}_{lin}$.*

The proof consists of a Henkin-style construction of a two-dimensional grid, where each point is a maximally consistent set. The basic step of the construction is the elimination of 'defects' from the grid, i.e. adding new points to the grid for a semantic condition that fails for the grid. The final model is obtained as the (infinite) union of all steps, and the grid thus constructed is shown to be a $\mathcal{K}_{lin} \times \mathcal{K}_{lin}$ model for an original consistent formula. The full details of the proof can be found in [Finger, 1994], but owing to space limitations (the full proof takes up to 10 pages) we omit it here. If $\mathcal{K}_{dis}$ is the class of all linear and discrete flows, [Finger, 1994] also shows completeness results for the classes $\mathcal{K}_{dis} \times \mathcal{K}_{dis}$, $\mathbb{Q} \times \mathbb{Q}$, $\mathcal{K}_{lin} \times \mathcal{K}_{dis}$, $\mathcal{K}_{lin} \times \mathbb{Q}$ and $\mathbb{Q} \times \mathcal{K}_{dis}$.

The negative result is the following.

Proposition 6.17 (2D-unaxiomatizability). *There are no finite axiomatizations for the (strongly) valid two-dimensional formulae over the classes $\mathbb{Z} \times \mathbb{Z}$, $\mathbb{N} \times \mathbb{N}$ and $\mathbb{R} \times \mathbb{R}$.*

This proposition follows directly from Venema's proof that the valid formulae over the upper half two-dimensional semiplane are not enumerable for $\mathbb{Z} \times \mathbb{Z}$, $\mathbb{N} \times \mathbb{N}$ and $\mathbb{R} \times \mathbb{R}$, which in its turn was based on [Halpern and Shoham, 1986]. Since there are sound, complete and decidable US-temporal logics over $\mathbb{Z}$, $\mathbb{N}$ and $\mathbb{R}$ [Reynolds, 1992; Buchi, 1962b; Rabin, 1969; Burgess and Gurevich, 1985], the general conclusion on full join is the following.

Theorem 6.18 (Full join). *In the general case completeness and decidability do not transfer through full join.*

It has to be noted that 2DTLs seem to behave like modal logics in the following sense. We can see the result of the independent combination of US and $\bar{U}\bar{S}$ as generating a 'minimal' combination of the logics, i.e. one without any interference between the dimensions. The addition of extra axioms, inference rules or an extra condition on its models has to be studied on its own, just as adding a new axiom to a modal logic or imposing a new property on its accessibility relation has to be analysed on its own.

The full join method illustrates the conflict between the generality of a method and its ability to achieve the transfer of logical properties. Next, we restrict the join method so as to recover the transfer of logical properties.

6.7 Restricted join

The fact that the transfer of logical properties fails for the join of two US-temporal logics does not mean that the join of any two temporal logic systems fails to achieve this transfer. We restrict the vertical logic system to a temporal logic $\bar{\mathsf{N}}\bar{\mathsf{P}}$ with operators $\overline{\bigcirc}$ for Next time and $\overline{\bullet}$ for Previous time; the formation rules for the formulae of $\bar{\mathsf{N}}\bar{\mathsf{P}}$ are the standard ones. This is a restriction of the $\bar{U}\bar{S}$-language for $\overline{\bigcirc}$ and $\overline{\bullet}$ can be defined in terms of $\overline{U}$ and $\overline{S}$, namely by

$$\overline{\bigcirc}A =_{def} \overline{U}(A, \bot);$$
$$\overline{\bullet}A =_{def} \overline{S}(A, \bot).$$

Not only is the expressivity of the language reduced this way, but also the underlying flow of time is now restricted to a discrete one; in fact, we concentrate our attention on integer-like flows of time.

Let $h : \mathbb{Z} \to \mathcal{P}$ be a temporal assignment over the integers so that the semantics of $\bar{\mathsf{N}}\bar{\mathsf{P}}$ over the integers is the usual one for atoms and Boolean operators and

$$(\mathbb{Z}, <, h), t \models \overline{\bigcirc}A \quad \text{iff} \quad (\mathbb{Z}, <, h), t+1 \models A;$$
$$(\mathbb{Z}, <, h), t \models \overline{\bullet}A \quad \text{iff} \quad (\mathbb{Z}, <, h), t-1 \models A.$$

An axiomatization for $\mathsf{NP}/\mathbb{Z}$ is given by the classical tautologies plus:

NP1 $\overline{\bigcirc}\,\overline{\bullet}p \to p$;
NP2 $\overline{\bigcirc}\neg p \leftrightarrow \neg\overline{\bigcirc}p$;
NP3 $\overline{\bigcirc}(p \wedge q) \to \overline{\bigcirc}p \wedge \overline{\bigcirc}q$;
NP4 the mirror image of **NP1**–**3** obtained by swapping $\overline{\bigcirc}$ and $\overline{\bullet}$.

The rules of inference are the usual substitution, *modus ponens* and temporal generalization (from A infer $\overline{\bigcirc}A$ and $\overline{\bullet}A$).

The converse of each axiom can be straightforwardly derived, so the formulae on both sides of the $\to$-connective are actually equivalent. It follows that every $\bar{\mathsf{N}}\bar{\mathsf{P}}$-formula can be transformed into an equivalent one by 'pushing in' the temporal operators, e.g. by following the arrows in the axioms, and by 'cancelling' the occurrences of $\overline{\bigcirc}$ and $\overline{\bullet}$ in a string of temporal operators, e.g. $\overline{\bigcirc}\,\overline{\bullet}\,\overline{\bullet}\,\overline{\bigcirc}\,\overline{\bullet}p$ is equivalent to $\overline{\bullet}p$. The resulting $\bar{\mathsf{N}}\bar{\mathsf{P}}$-*normal form* formula is a Boolean combination of formulae of the form $\overline{\bigcirc}^k p$ and $\overline{\bullet}^l q$, where p and q are atoms, $k, l \in \mathbb{N}$ and $\overline{\bigcirc}^k$ is a sequence of $\overline{\bigcirc}$-symbols of size k, similarly for $\overline{\bullet}^l$; it is useful sometimes to consider k negative or 0, so we define $\overline{\bigcirc}^{-k}A = \overline{\bullet}^k A$ and $\overline{\bigcirc}^0 A = A$. As an example, the formula $\overline{\bigcirc}\,\overline{\bigcirc}(\overline{\bullet}\,\overline{\bullet}\,\overline{\bullet}(p \wedge q) \vee p)$ has normal form $(\overline{\bullet}p \wedge \overline{\bullet}q) \vee \overline{\bigcirc}\,\overline{\bigcirc}p$. The existence of such a normal form gives us the very simple proofs for completeness and decidability of $\bar{\mathsf{N}}\bar{\mathsf{P}}/\mathbb{Z}$ that we outline next.

For completeness, let Σ be a possibly infinite consistent set of $\bar{\mathsf{N}}\bar{\mathsf{P}}$-formulae and assume all formulae in the set are in normal form. Σ can be seen as a consistent set of propositional formulae where each maximal subformulae of the form $\overline{\bigcirc}^k p$ is understood as a new propositional atom, so let h_0 be a propositional valuation assigning every extended atom into $\{true, false\}$. For $n \in \mathbb{Z}$, let $h(n) = \{p \in \mathcal{P} \mid h_0(\bar{\bigcirc}^n p) = true\}$. Clearly $(\mathbb{Z}, <, h)$ is a model for the original set.

For decidability, let A be a formula of $\bar{\mathsf{N}}\bar{\mathsf{P}}$ and let A^* be its normal form; clearly there exists an algorithm to transform A into A^*. By considering subformulae of the form $\bar{\bigcirc}^k p$ as new atoms, k possibly negative, we apply any decision procedure for propositional logic to A^*. A is an $\bar{\mathsf{N}}\bar{\mathsf{P}}$-valid formula iff A^* is a propositional tautology.

Definition 6.19. The *restricted join* of temporal logic systems $US/\mathcal{K}$ and $\bar{\mathsf{N}}\bar{\mathsf{P}}/\mathbb{Z}$ is the 2DTL system $US \times \bar{\mathsf{N}}\bar{\mathsf{P}}$ given by:

- the fully combined language of US and $\bar{\mathsf{N}}\bar{\mathsf{P}}$;
- the two-dimensional plane model over $\mathcal{K} \times \mathbb{Z}$, equipped with the broadly two-dimensional semantics;
- the union of the axioms of $US/\mathcal{K}$ and $\bar{\mathsf{N}}\bar{\mathsf{P}}/\mathbb{Z}$ plus the join axioms

$$\bar{\bigcirc} U(p,q) \rightarrow U(\bar{\bigcirc} p, \bar{\bigcirc} q);$$
$$\bar{\bigcirc} S(p,q) \rightarrow S(\bar{\bigcirc} p, \bar{\bigcirc} q);$$

  plus their duals obtained by swapping $\bar{\bigcirc}$ with $\bar{\bullet}$; the inference rules are just the union of the inference rules of both component systems.

What has therefore been restricted in the interlacing process is the expressivity of the language over the vertical dimension, which also restricted the underlying flow of time to a discrete one. The following gives us a normal form for $US \times \bar{\mathsf{N}}\bar{\mathsf{P}}$.

Lemma 6.20. *Let A be a formula of $US \times \bar{\mathsf{N}}\bar{\mathsf{P}}$. There exists a normal form formula A^* equivalent to A, such that all the occurrences of $\bar{\bigcirc}$ and $\bar{\bullet}$ in it are in the form $\bar{\bigcirc}^k p$ and $\bar{\bullet}^l q$, where p and q are atoms.*

Proof First we show that the converses of the join axioms are theorems too. For that, note that U and S respect the *congruence property*, i.e. if $A \leftrightarrow C$ and $B \leftrightarrow D$ then $U(A,B) \leftrightarrow U(C,D)$ and $S(A,B) \leftrightarrow S(C,D)$. Also note that

(equiv) $\vdash (p \leftrightarrow \bar{\bigcirc}\bar{\bullet} p)$ and $\vdash (p \leftrightarrow \bar{\bullet}\bar{\bigcirc} p)$.

The transitivity of the $\rightarrow$-operator connects the steps in the proof of the formula $U(\bar{\bigcirc} p, \bar{\bigcirc} q) \rightarrow \bar{\bigcirc} U(p,q)$ below:

$$\begin{array}{rll} U(\bar{\bigcirc} p, \bar{\bigcirc} q) & \rightarrow \bar{\bigcirc}\bar{\bullet} U(\bar{\bigcirc} p, \bar{\bigcirc} q) & \text{by } \mathbf{equiv}; \\ & \rightarrow \bar{\bigcirc} U(\bar{\bullet}\bar{\bigcirc} p, \bar{\bullet}\bar{\bigcirc} q) & \text{by join axiom}; \\ & \rightarrow \bar{\bigcirc} U(p,q) & \text{by } \mathbf{equiv} \text{ and congruence.} \end{array}$$

It follows that $U(\bar{\bigcirc} p, \bar{\bigcirc} q) \leftrightarrow \bar{\bigcirc} U(p,q)$. It is completely analogous to show the converse of other join axioms, so we omit the details.

Given A in the language of $US \times \bar{\mathsf{N}}\bar{\mathsf{P}}$, the equivalence between both sides of the join axioms allows for 'pushing in' the vertical operators $\bar{\bigcirc}$ and $\bar{\bullet}$, so a simple induction on the number of nested temporal operators in A shows an algorithmic way to generate an equivalent formula A^* in the desired normal form. □

Theorem 6.21. *Let US be a logic system complete over the class $\mathcal{K} \subseteq \mathcal{K}_{lin}$. Then the two-dimensional system $US \times \bar{\mathsf{N}}\bar{\mathsf{P}}$ is complete over $\mathcal{K} \times \mathbb{Z}$.*

Proof Consider a $US \times \bar{\mathsf{N}}\bar{\mathsf{P}}$-consistent formula A and assume it is in the normal form. So we can see A as a US-formula over the extended set of atoms $\bar{\bigcirc}^k$, k possibly negative or 0. From the completeness of $US/\mathcal{K}$ there exists a one-dimensional model $(T, <, h_{US})$ for A at a point $o \in T$, where $(T, <) \in \mathcal{K}$. Define the two-dimensional assignment

$$h(k,t) = \{p \in \mathcal{P} \mid \bar{\bigcirc}^k p \in h_{US}(t)\}.$$

Clearly, $(T, <, \mathbb{Z}, <_{\mathbb{Z}}, h)$ is a two-dimensional plane $US \times \bar{\mathsf{N}}\bar{\mathsf{P}}$-model for A at $(o, 0)$. □

Corollary 6.22. *If $US/\mathcal{K}$ is strongly complete, so is $US \times \bar{\mathsf{N}}\bar{\mathsf{P}}/\ \mathcal{K} \times \mathbb{Z}$.*

Theorem 6.23 (Decidability via restricted join). *If the logic system US is decidable over $\mathcal{K}$, so is $US \times \bar{\mathsf{N}}\bar{\mathsf{P}}$ over $\mathcal{K} \times \mathbb{Z}$.*

Proof The argument of the proof is the same as that of the decidability of NP. All we have to do is note that there exists an algorithmic way to convert a combined two-dimensional formula into its normal form, so it can be seen as a US-formula and we can apply the US-decision procedure to it. □

So by restricting the expressivity and the underlying class of flows of time, we can obtain the transfer of the basic logical properties via restricted join. It should not be difficult to extend these results to $\mathbb{N}$ instead of $\mathbb{Z}$, although we do not explore this possibility here.

It is also worth noting that the restricted join method answers a conjecture posed by Venema [1990] on the existence of some expressively limited 2DTL over $\mathbb{Z} \times \mathbb{Z}$ that was 'well behaved' in the sense of having the completeness and decidability properties.

6.8 The two-dimensional diagonal

We now study some properties of the diagonal in two-dimensional plane models. The diagonal is a privileged line in the two-dimensional model intended to represent the sequence of time points we call 'now', i.e. the time points on which an historical observer is expected to traverse. The observer is, therefore, on the diagonal when he or she poses a query (i.e. evaluates the truth value of a formula) on a two-dimensional model. The diagonal is illustrated in Fig. 6.8.

So let δ be a special atom and consider the formulae:

D1 $\Diamond\delta \wedge \bar{\Diamond}\delta$;

D2 $\delta \rightarrow (G\neg\delta \wedge H\neg\delta \wedge \overline{G}\neg\delta \wedge \overline{H}\neg\delta)$;

D3 $\delta \rightarrow (\overline{H}G\neg\delta \wedge \overline{G}H\neg\delta)$.

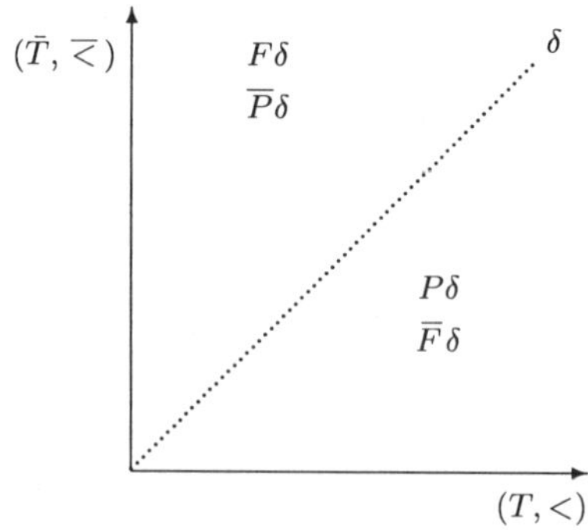

FIG. 6.8. The two-dimensional diagonal.

Let $Diag = \Box\overline{\Box}(\mathbf{D1} \wedge \mathbf{D2} \wedge \mathbf{D3})$. The intuition behind $Diag$ is the following. **D1** implies that the two-dimensional diagonal can always be reached in both vertical and horizontal directions; **D2** implies that there are no two diagonal points on the same horizontal line and on the same vertical line; and **D3** implies that the diagonal goes in the direction SW–NE. We say that $Diag$ characterizes a two-dimensional diagonal in the following sense.

Proposition 6.24. *Let $\mathcal{M} = (T, <, \overline{T}, \overline{<}, g)$ be a full two-dimensional model over $\mathcal{K} \times \overline{\mathcal{K}}$, $\mathcal{K}, \overline{\mathcal{K}} \subseteq \mathcal{K}_{lin}$, and let δ be a propositional letter. Then the following are equivalent.*

(a) $\mathcal{M}, t, x \models Diag$, for some $t \in T$ and $x \in \overline{T}$.

(b) $\mathcal{M}, t, x \models Diag$, for all $t \in T$ and $x \in \overline{T}$.

(c) There exists an isomorphism $i : T \to \overline{T}$ such that $\mathcal{M}, t, x \models \delta$ iff $x = i(t)$.

Proof It is straightforward to show that (a) $\iff$ (b) and (c) $\Longrightarrow$ (a); we show only (b) $\Longrightarrow$ (c). So assume that $\mathcal{M}, t, x \models Diag$, for all $t \in T$ and $x \in \overline{T}$. Define

$$i = \{(t, x) \in T \times \overline{T} \mid \mathcal{M}, t, x \models \delta\}.$$

All we have to show is that i is an isomorphism.

- i, i^{-1} are functions such that $dom(i) = T$ and $dom(i^{-1}) = \overline{T}$. Suppose that both (t, x_1) and (t, x_2) are in i; then $\mathcal{M}, t, x_1 \models \delta$ and $\mathcal{M}, t, x_2 \models \delta$. By linearity of $\overline{T}$, $x_1 = x_2$, $x_1 \overline{<} x_2$ or $x_2 \overline{<} x_1$, but **D2** eliminates the latter two; **D1** gives us that $dom(i) = T$. Similarly, the linearity of T and **D2** gives us that i^{-1} is a function and **D1** gives us that $dom(i^{-1}) = \overline{T}$.
- $i(t) = x$ iff $i^{-1}(x) = t$ follows directly from the definition. So i is a bijection.
- i preserves ordering. Suppose $t_1 < t_2$; by the linearity of $\overline{T}$ we have three possibilities:
  - $i(t_1) = i(t_2)$ contradicts i is a bijection.
  - $i(t_2) \overline{<} i(t_1)$ contradicts **D3**.

* $i(t_1)\overline{<}i(t_2)$ is the only possible option.

Therefore i is an isomorphism, which proves the result. □

This result shows that by adding **D1**–**D3** to the axiomatization over the two-dimensional plane $\mathcal{K}_{lin} \times \mathcal{K}_{lin}$ of Section 6.6 gives us completeness over the class of models of the form $(T, <, T, <, g)$, where $(T, <) \in \mathcal{K}_{lin}$. It follows from [Halpern and Shoham, 1986], however, that such a logic system is undecidable.

The diagonal is interpreted as the sequence of time points we call 'now'. The diagonal divides the two-dimensional plane into two semiplanes. The semiplane that is to the (horizontal) left of the diagonal is 'the past', and the formula $F\delta$ holds over all points of this sem-plane. Similarly, the semiplane that is to the (horizontal) right of the diagonal is 'the future', and the formula $P\delta$ holds over all points of this semiplane. Fig. 6.8 illustrates this fact. If we assume that *Diag* holds over $\mathcal{M}$ such that i is the isomorphism defined in Lemma 6.24, $t < s$ iff $i(t)\overline{<}i(s)$, then:

$\mathcal{M}, t, x \models F\delta$ iff there exists $s > t$ such that $\mathcal{M}, s, x \models \delta$ and $i(s) = x$;
iff there exists $y = i(t)\overline{<}x$ such that $\mathcal{M}, t, y \models \delta$;
iff $\mathcal{M}, t, x \models \overline{P}\delta$.

Similarly, it can be shown that:

$\mathcal{M}, t, x \models P\delta$ iff $\mathcal{M}, t, x \models \overline{F}\delta$.

It follows that the following formula is valid for $US \times \overline{US}$ over $\mathcal{K}_{lin} \times \mathcal{K}_{lin}$:

$$Diag \rightarrow ((F\delta \leftrightarrow \overline{P}\delta) \wedge (P\delta \leftrightarrow \overline{F}\delta)).$$

As a consequence, $\overline{P}\delta$ holds over all points of the 'past' semiplane and $\overline{F}\delta$ holds over all points of the 'future' semiplane, as is indicated in Fig. 6.8.

The formula *Diag* is in the language of $US \times \overline{US}$ but not in the language of $US \times \bar{\mathsf{N}}\bar{\mathsf{P}}$, for *Diag* contains the vertical temporal operators $\overline{G}$, $\overline{H}$, $\overline{\Box}$ and $\overline{\Diamond}$. To characterize a two-dimensional diagonal in $US \times \bar{\mathsf{N}}\bar{\mathsf{P}}$ we do the following. We say that a formula A *holds over* or *is valid over* a two-dimensional model $\mathcal{M}$ if for every $t \in T$ and every $x \in \overline{T}$, it is the case that $\mathcal{M}, t, x \models A$. Consider the formulae

d1 $\Diamond\delta$
d2 $\delta \rightarrow (G\neg\delta \wedge H\neg\delta)$
d3 $\delta \leftrightarrow \bar{\bigcirc}\bigcirc\delta$

where δ is a proposition. Those formulae are all in the language of $US \times \bar{\mathsf{N}}\bar{\mathsf{P}}$, for *Diag* (so also in the language of $US \times \overline{US}$) and they can characterize the two-dimensional diagonal due to the following property.

Proposition 6.25. *Let $\mathcal{M}$ be a two-dimensional plane model over $\mathbb{Z} \times \mathbb{Z}$. Then the formula* **D1** $\wedge$ **D2** $\wedge$ **D3** *holds over $\mathcal{M}$ iff* **d1** $\wedge$ **d2** $\wedge$ **d3** *holds over $\mathcal{M}$.*

Proof From Proposition 6.24 we know that **D1** $\wedge$ **D2** $\wedge$ **D3** holds over $\mathcal{M}$ iff the relation i defined as

$$i = \{(t, x) \in \mathbb{Z} \times \mathbb{Z} \mid \mathcal{M}, t, x \models \delta\};$$

is an isomorphism in $\mathbb{Z}$. So all we have to do is to prove that i as defined above is an isomorphism iff **d1** $\wedge$ **d2** $\wedge$ **d3** holds over $\mathcal{M}$. The *only if* part is a straightforward verification that for all x and t in $\mathbb{Z}$, $\mathcal{M}, t, x \models \mathbf{d1} \wedge \mathbf{d2} \wedge \mathbf{d3}$.

For the *if* part, assume **d1** $\wedge$ **d2** $\wedge$ **d3** holds over $\mathcal{M}$. Then:

(a) **d1** gives us that for every x there exists a t such that $\mathcal{M}, t, x \models \delta$;

(b) **d2** gives us that for every $x, t, t', t \neq t'$, $\mathcal{M}, t, x \models \delta$ implies $\mathcal{M}, t', x \not\models \delta$;

(c) **d3** gives us that for every x, t, $\mathcal{M}, t, x \models \delta$ iff $\mathcal{M}, t+1, x+1 \models \delta$ iff for every $n \in \mathbb{Z}$, $\mathcal{M}, t+n, x+n \models \delta$.

The first two items give us that $i^{-1} : \mathbb{Z} \to \mathbb{Z}$ is a function. To show that i is also a function, suppose that $(t, x_1), (t, x_2) \in i$. By linearity of $\mathbb{Z}$, it follows that either $x_1 < x_2$ or $x_2 < x_1$ or $x_1 = x_2$. Let $x_1 - x_2 = m$; then, by the third item above, $(t + m, x_2 + m = x_1) \in i$, so $t = (t + m)$ and $m = 0$. It follows that $x_1 = x_2$, so $i : \mathbb{Z} \to \mathbb{Z}$ is a function. Directly by the definition of i, it follows that i is a bijection.

By the third item above, if $i(t_1) = x_1$ and $i(t_2) = x_2$, then $t_1 - t_2 = x_1 - x_2$. It follows that i is order preserving and hence an isomorphism, which completes the proof. □

It would be desirable to generalize the idea of a diagonal as the sequence of 'now' moments to any pair of flows of time that are not necessarily isomorphic. For that, we would have to create an order between the points of the two flows, i.e. we would have to merge the flows.

So let $(T, <)$ and $(\overline{T}, \overline{<})$ be two flows of time such that T and $\overline{T}$ are disjoint. Then there always exist a flow $(T', <')$ and a mapping $f : T \cup \overline{T} \to T'$ such that f is one-to-one and order preserving. The *f-merge of* $(T, <)$ *and* $(\overline{T}, \overline{<})$ is the flow of time consisting of the image of f ordered by the restriction of $<'$ to the image of f. An example of an f-merge is shown in Fig. 6.9, where $f(y)$ is made equal, via merge, to $f(\bar{x})$, and on the merged flow the order is preserved, i.e. originally $x < y$ and $\bar{x} \overline{<} \bar{y}$ and on the f-merged flow $f(x) <' f(y) = f(\bar{x}) <' f(\bar{y})$.

We can then construct a two-dimensional model with two copies of the f-merged flow, in which we can define a diagonal over $(T', <') \times (T', <')$ as shown in Fig. 6.10. Another particularly interesting situation arises when the f-merged flow $(T', <')$ is identical to one of the component flows, e.g. $(\overline{T}, \overline{<})$, so that f is an *embedding* of $(T, <)$ into $(\overline{T}, \overline{<})$. In this case, the flow $(T, <)$ could be viewed as a more 'abstract' representation of $(\overline{T}, \overline{<})$ wherein several details, i.e. pieces of information, points in time, are ignored.

The f-merge construction serves as motivation to another method for combining two 1DTLs, this time generating another one-dimensional logic. This could be achieved over the class of all f-merges of its two-component flows of time

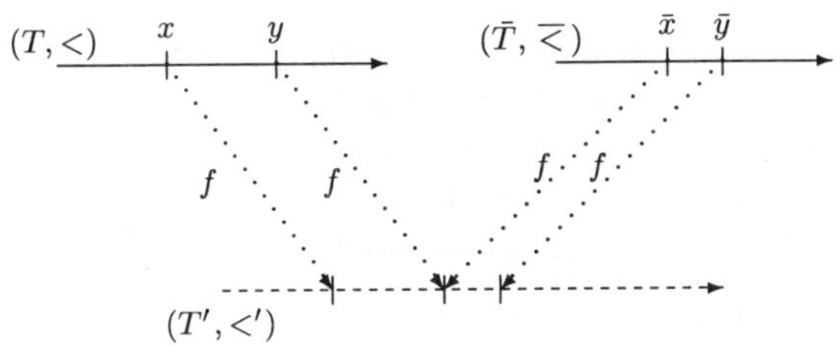

FIG. 6.9. The f-merge.

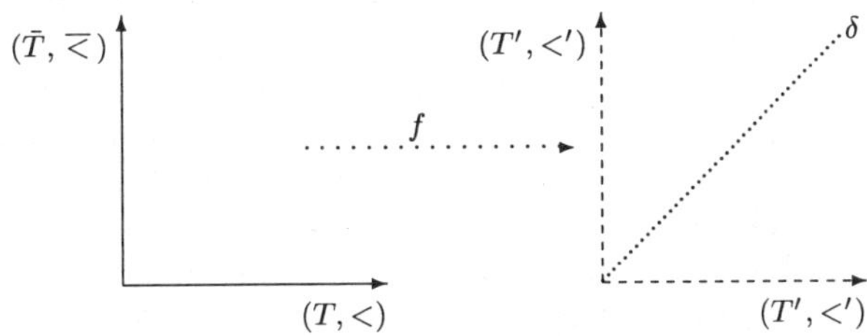

FIG. 6.10. The diagonal of two distinct flows.

or subclasses of it. We could then study the transfer of logical properties in the same way as we have done in this and the previous section, but those matters remain beyond the scope of this work.

6.9 Conclusions

This study dealt with the combination of two logic systems in order to obtain a new logic system. The issues were:

- Several methods of combination of two logic systems were presented. Each combination involved at least one temporal logic system. Each method had a particular discipline for combining the language, the semantics and the inference system of two logic systems. Each combination generated a single logic system.
- The study of the transfer of logical properties from the component systems into their combined form has been the major point in the analysis of combination methods. The basic logical properties whose transfer was analysed were soundness, completeness and decidability; for some combination methods, the transfer of other properties was also investigated such as conservativeness and the compactness property (in the form of strong completeness).
- The investigation of four basic methods has been accomplished. The temporalization method and the independent combination method were shown to transfer all basic properties, although they do not generate an expressive enough system to be called fully two dimensional. The full join method

does generate a fully two-dimensional temporal system, but in many cases it failed to transfer even the completeness property. As a compromise, it was shown that a restricted join method, although generating 2DTLs that were not as expressive and generic as the fully interlaced one, accomplishes the transfer of all basic logical properties.

Another contribution of our analysis was to answer a question raised by Venema [Venema, 1990] on the existence of a fragment of the two-dimensional plane temporal logic that, in his own words, was 'better behaved' than the two-dimensional plane system with respect to completeness and decidability properties. We have shown that the 2DTL systems obtained by restricted join are an example of such fragments.

Another question raised by Venema in that work remains open: namely, whether there exists a complete axiomatization over the two-dimensional model using only canonical inference rules, i.e. without using the special inference rules **IR1** and **IR2**. This problem seems to be a very hard one. Nevertheless we succeeded in extending Venema's completeness result that originally hold for only two-dimensional flows built from two identical one-dimensional flows, to any two-dimensional flow built from any flow in the classes $\mathcal{K}_{lin}$, $\mathcal{K}_{dis}$, $\mathcal{K}_{dense}$ and $\mathbb{Q}$.

6.9.1 *Comparisons and possible extensions*

With respect to the combination of logics, the works found in the literature that most closely approximate ours in spirit and aim are those of Kracht and Wolter [1991] and Fine and Schurz [1991]. Both works concentrated on monomodal logics, and investigated the transfer of logical properties for the special case which we called independent combination. However, their work does seem to suggest further directions for our studies. First, they analysed the transfer of many other properties from two logic systems to their combined form, e.g. finite model property and interpolation. Second, both works did not concentrate only on linear systems and they were able to extend their results to any class of underlying Kripke frames. Third, Fine and Schurz's work generalized the independent combination method to more than two monomodal logics.

The two papers cited above suggest several extensions to our work. Note, however, that the temporalization method was easily shown to be extendible to many temporal logic systems. The focus on linear flows of time was due to database applications of 2DTLs as in [Finger and Gabbay, 1992; Finger, 1994], but we believe that this restriction may be lifted without losing the transfer results of the temporalization and independent combination methods. These have to be further investigated and the transfer of any other logical property has to be analysed on its own.

The generalization of combination methods other than the independent combination method to modal logics is another area for further work. As noted in [Finger and Gabbay, 1992], the temporalization process is directly extendible to monomodal logics. It may even be the case that, for monomodal logics, the full join method achieves transfer of completeness over several classes of fully two-dimensional Kripke frames using only canonical inference rules, as is suggested by the results in [Segerberg, 1973].

The complexity class of the decision problem for the combined logic is another interesting subject for study. For the independent combination of monomodal logics, such a study was done by Spaan [Spaan, 1993] and the conclusion was that the satisfiability problem of an independently combined logic is reducible to that of one of the component logics, or it is PSPACE-hard or it is in NP. We believe a similar result can be obtained for the temporalization and the independent combination of temporal logics, although the details have not yet been worked out. The complexity of the full and restricted join methods still has to be studied.

All the systems dealt with in this chapter were extensions of classical logic. It is possible that the temporalization process preserves its transfer properties even in the case when the underlying system is not an extension of classical logic. What if the external temporal logic is non-classical itself? The same question applies to other combination methods. Do they transfer logical properties when one or both of the combined temporal of modal logics is not classical? This question is better posed in the generic framework involving labelled deductive systems (LDS); in such a framework, one finds that to obtain the transfer of completeness one does not need the full power of classical logic but only some weaker form of monotonicity. LDS supports a general method of combination called *fibring* that depends on the choice of a fibring function. A fibring function maps the truth value of atoms in one logic's semantics with the semantics of formulae in another logic's semantics. Gabbay's *dovetailing* process, obtained with a certain class of fibring functions, is similar to the independent combination method extended to logics respecting those weaker conditions of monotonicity. More work on this area is needed to clarify exactly how fibring is related to existing combination methods. See Gabbay's book on fibring, [Gabbay, 1998].

There are also other possible types of combinations of 1DTLs that may be explored. As pointed out in Section 6.8, two linear flows of time can be merged into another one; the question then is how to combine two 1DTLs into another 1DTL over the merged flow.

7

EXTENSIONAL SEMANTICS

In Volume 1, and in some of the preceding chapters, we have seen how temporal logic can be used to reason about program behaviour. In describing a new programming language (or a new way of constructing reactive systems) we can often use temporal logic to specify the desired behaviour of any given program. Thus we are giving temporal semantics to the language. There are two main ways of approaching this task: intensionally or extensionally. Real programs will exhibit *extensional* actions, i.e. observable changes in values of program variables. However, specifiers may want to describe the behaviour in terms of states and transitions between them. These intensional states are abstract objects in their own right and are not necessarily determined by the values they associate with program variables. An example of this approach is the *fair transitions system* of [Manna and Pnueli, 1981] and [Pnueli, 1986] which generalizes such approaches as CSP and Petri nets. If such intensional states determine values for the observable variables, which they generally do, then it is straightforward to associate an extensional action—a change in variable values—with each transition between intensional states. However, the converse relationship is more problematic and there is not necessarily exactly one or indeed any one intensional transition associated with each extensional action.

In this chapter we present and investigate the use of an extensional temporal language, Lamport's temporal logic of actions (TLA) [Lamport, 1994] to describe the semantics of certain programming languages such as the Gamma programming language [Hankin *et al.*, 1992]. TLA is a logic designed for specifying and reasoning about concurrent systems. Gamma is a minimal language based on conditional multiset rewriting which is intended to allow the systematic design and construction of programs for highly parallel machines. It should be straightforward to describe a Gamma program in TLA.

We show that, in fact, this is not quite straightforward. The problems are to do with the fact that TLA is an extensionally based approach to program semantics. This means that the semantics is not based on the changes in internal states or on some abstract intensional model of the changes in internal states of the machine. Instead, the semantics is based on the externally visible values of a set of program variables.

This is achieved by forcing the formulae of TLA to be stuttering invariant. That is, their truth is preserved under adding or removing a finite number of repetitions of the same state of values of the observable variables. As a result of this syntactic restriction, TLA descriptions of the behaviour of programs are correct whether or not other programs are running in parallel and whether or not the original program is actually implemented by a lower level program.

Stuttering invariance also means that stuttering itself cannot be described. Unfortunately, rewrite languages such as Gamma generally allow trivial rewrites which do not change anything and performing such a rewrite is just one form of stuttering. A general account of the semantics of such a language must sometimes require the fairness or liveness of trivial rewrites. We show that existing ways of using TLA have trouble expressing this.

TLA is not just a formal language but is a whole approach to reasoning about program behaviour. The formal language comes with its semantics, of course, but also with a proof theory and, importantly, a helpful general pattern for formalizing the behaviour of programs. This template involves abbreviations in the formal language which are supposed to capture the idea of fairness of program actions. There is an abbreviation $\mathrm{WF}(\mathcal{A})$ for weak fairness of an action $\mathcal{A}$ and a similar one $\mathrm{SF}(\mathcal{A})$ for strong fairness.

We show that although any particular intensionally defined program action, a say, (e.g. a may be a rewrite) may correspond closely, and intuitively, to an extensionally defined TLA action $\mathcal{A}$, say, the correspondence does not generally mean that the weak fairness of a is formalized by $\mathrm{WF}(\mathcal{A})$. Thus, if we know that our language requires weak fairness of the action (corresponding to) $\mathcal{A}$, we cannot just plug $\mathrm{WF}(\mathcal{A})$ into Lamport's general template for program behaviour.

Fortunately, we are able to propose a new formal abbreviation to capture weak fairness and prove its correctness.

The case of strong fairness is similar, but worse. We show that in order to require strong fairness of an action $\mathcal{A}$, we cannot generally use $\mathrm{SF}(\mathcal{A})$. Nor, in contrast to the case with weak fairness, can we even construct some complex combination of the abbreviations SF and WF to express the property.

However, once again, here we are able to supply a new correct abbreviation to capture strong fairness.

TLA is designed to be used to reason about program behaviour and program correctness. Thus, in Section 7.7.3, we provide useful proof rules to support reasoning with the two new abbreviations.

In Section 7.5, we introduce the rewrite language Gamma and in Section 7.6 we apply the new abbreviations to provide a TLA description of its semantics. A TLA semantics for Gamma is quite useful. Because Gamma is a minimal specification language, it is intended that the construction of an actual parallel program proceeds from the Gamma program towards a more detailed *scheduling* of the individual program behaviours (see [Chaudron and de Jong, 1996]). The temporal semantics will provide a vital underlying framework in which to ensure that this implementation is done correctly. We also briefly consider languages which vary from Gamma in requiring slightly different fairness properties.

In summary, although we show that some of the abbreviations, formalizations and templates associated with TLA are inadequate for the special requirements of rewrite languages, we are able to provide generally correct and useful formalizations within the general framework of TLA and we conclude that it is a very convenient paradigm in which to be precise about the various subtly different languages and to reason about such languages.

7.1 TLA

TLA is a logic designed for specifying and reasoning about concurrent systems. Thus, TLA is a formal language with a semantics and a proof theory. In [Lamport, 1994] we also see a technique for using TLA to describe concurrent algorithms. Below we give a brief introduction to these two aspects of TLA. The interested reader is referred to [Lamport, 1994] for more details.

One of the main motivations behind several of the interesting technical properties of TLA is the requirement that it describes the behaviour of a program independently of what other separate processes are concurrently doing and independently of the level of refinement of description of basic program steps. As observed in [Lamport, 1983], these goals can be achieved if the description of the program is invariant under stuttering, i.e. under adding or removing steps which do not change the state of the program. Despite being based on a quite expressive temporal language (called RTLA), TLA itself is syntactically restricted so that it cannot itself describe stuttering and therefore it is stuttering invariant.

Like many of the temporal logics used in computer science (see, e.g. [Gabbay *et al.*, 1980]), TLA uses the 'always' operator $\Box$ and its dual 'sometime' $\Diamond$. Because it allows one to describe stuttering, the other usual operator 'next' $\bigcirc$ and the extra expressivity gained by using it are shunned by TLA. However, this causes a problem as a program is most conveniently described by detailing how it moves from one state to the *next*. TLA gets around this problem by allowing a very limited and hidden use of the 'next' operator within its most basic construct—the action.

More general uses of 'next' operators are not allowed and, as we will see, this requires a fairly strict format for describing the behaviour of a program in terms of soundness and fairness properties.

7.1.1 *The formal language*

As with most temporal logics, TLA describes the changes over time in the values of a certain set of variables. A particular assignment of values to the variables constitutes a *state*. A sequence $s_1, s_2, \ldots$ is a *behaviour*. Most temporal logics are built from basic units describing states to allow formulae to describe sequences of states. However, for the reasons mentioned above, the basic unit for semantics in TLA is the *action*. This is a relation between successive states in the behaviour. If we describe a state in terms of the values of such variables as X, Y and Z, then we describe an action using the variables X, Y and Z along with a new disjoint set X', Y' and Z'. These new primed variables represent the values of their unprimed versions at the next step. For example, the action $Y' = X + Y$ holds whenever we move from one state into the next and have the value of Y increased by the value of X. Note that this example shows that actions can be non-deterministic in general.

An action is any Boolean-valued expression containing constants, variables and primed variables. We evaluate an action as the holding of a pair of states whenever it is true when viewed as a relation with variables interpreted as their values in the first state and primed variables interpreted as the values of their unprimed versions in the second state.

More formally we first define a function (s, t) on terms for each pair of states (s, t):

- $(s, t)c = c$ for constant c;
- $(s, t)x = s(x)$ for variable x;
- $(s, t)x' = t(x)$ for variable x;
- $(s, t)f(u) = f((s, t)u)$ for any function f.

We then define whether action $\mathcal{A}$ holds between the pair of states s and t—written $s[\![\mathcal{A}]\!]t$—by induction on the construction of $\mathcal{A}$:

- $s[\![u = v]\!]t$ holds iff $(s, t)u = (s, t)v$;
- $s[\![\mathcal{A} \wedge \mathcal{B}]\!]t$ iff $s[\![\mathcal{A}]\!]t$ and $s[\![\mathcal{B}]\!]t$;
- $s[\![\neg\mathcal{A}]\!]t$ iff we do not have $s[\![\mathcal{A}]\!]t$.

A particularly simple kind of action is one that refers only to variables and constants and not to primed variables. We call such an action a *predicate*. To evaluate whether it holds in a state we need only look at the values of the variables in that state. For any action $\mathcal{A}$, we define a special predicate Enabled $\mathcal{A}$ which holds in a state s if and only if there is some state t such that having s followed immediately by t makes the action $\mathcal{A}$ true—the truth of this predicate does not depend on which state actually does follow s in the behaviour.

As with most temporal logics we then build up formulae to describe behaviours from the basic formulae—in this case actions and predicates. Formulae of TLA hold (or do not hold) of behaviours, a basic formula being evaluated at the start of the behaviour. As we have mentioned above, we use the 'always' operator $\Box$—we say $\Box\phi$ holds of a behaviour if and only if ϕ holds of all suffixes of that behaviour.

Unfortunately, being able to write $\Box(X' = X + 1)$ would allow us to require that the value of X kept increasing during a behaviour without stuttering. This formula is *not* stuttering invariant and Lamport does not want it to be syntactically well formed. Since we want TLA to be closed under applying $\Box$, we have to disallow actions themselves as TLA formulae. Instead, we do allow $\Box[\mathcal{A}]_f$ where

$$[\mathcal{A}]_f = \mathcal{A} \vee (f = f').$$

$\mathcal{A}$ is any action and f any expression containing constants and variables. f is called a *state function* and f' is obtained from f by replacing all the variables in f by their primed versions. Since any stuttering step will leave the value of f unchanged, the truth of the TLA formula $\Box[\mathcal{A}]_f$ is clearly invariant under adding or removing stuttering steps during a behaviour.

Syntax of TLA. Formulae of TLA are built from predicates and basic formulae of the form $\Box[\mathcal{A}]_f$ recursively using the operators $\neg$, $\wedge$ and $\Box$.

Semantics of TLA. We say that formula ϕ holds of behaviour $\sigma = s_0, s_1, s_2, \ldots$ if and only if $\sigma[\![\phi]\!]$ as defined recursively below:

- for predicate p, $\sigma[\![p]\!]$ iff p is true of state s_0;
- for action $\mathcal{A}$ and state function f, $\sigma[\![\Box[\mathcal{A}]_f]\!]$ iff for all $n \geq 0$, either $s_n[\![\mathcal{A}]\!]s_{n+1}$ or $s_n[\![f = f']\!]s_{n+1}$;
- $\sigma[\![\neg\phi]\!]$ iff we do not have $\sigma[\![\phi]\!]$;
- $\sigma[\![\phi \wedge \psi]\!]$ iff we have both $\sigma[\![\phi]\!]$ and $\sigma[\![\psi]\!]$;
- $\sigma[\![\Box\phi]\!]$ iff for all $n \geq 0$, we have $(s_n, s_{n+1}, s_{n+2}, \ldots)[\![\phi]\!]$.

See [Lamport, 1994] for more details.

Also see [Lamport, 1994] on how to extend this simple TLA to allow for the use of quantification of variable symbols. Variable symbols come in two types: rigid variables (often called constants because they do not change their value over time) and non-rigid or program variable symbols which we have met above. In this chapter, we make some use of rigid variable symbols, mostly as free variables but also occasionally implicitly under quantification. It is very straightforward to define truth for formulae involving quantification of rigid variables: truth of formulae about a behaviour is defined with respect to a rigid interpretation on the values of the free rigid variables and quantification just involves a possible switch to a slightly different interpretation. Quantification of non-rigid variable symbols does not appear here. Therefore we will not delve into the slightly complicated semantics for quantification of non-rigid variables which is needed to preserve stuttering invariance. See [Pnueli and Rosner, 1989b] or [Lamport, 1994] for discussions of these matters.

TLA has some abbreviations:

$$\Diamond\phi = \neg\Box\neg\phi$$
$$\text{and } \langle\mathcal{A}\rangle_f = \mathcal{A} \wedge (f' \neq f).$$

Note that because

$$\begin{aligned}\Diamond\langle\mathcal{A}\rangle_f &= \Diamond(\mathcal{A} \wedge (f' \neq f)) \\ &= \Diamond\neg(\neg\mathcal{A} \vee (f' = f)) \\ &= \neg\Box[\neg\mathcal{A}]_f\end{aligned}$$

we do allow $\Diamond\langle\mathcal{A}\rangle_f$ as a TLA formula. It is useful for showing that interesting things do happen.

7.1.2 *Applying TLA to programs*

As with other temporal logics, TLA is able to describe properties of behaviours ranging from very detailed descriptions of the intricate behaviour of a program to simple, natural temporal properties which we wish programs to exhibit. Here we will concentrate on giving a complete description of the behaviour of a given program, i.e. providing a TLA formula which is guaranteed to be satisfied by all possible behaviours of the program but which is sufficiently detailed so that from it we can deduce all interesting temporal properties of those behaviours.

Because of the restricted syntax of TLA, [Lamport, 1994] gives a strict template for expressing this TLA description of a program. Essentially it consists of the conjunction of a specification of the initial state, a list of the actions which can possibly ever get the program from one state into the next and a list of fairness properties which guarantee that some of those actions occur every so often.

The state of the running of the program is defined by the values of some finite set of TLA variables. As well as variables explicitly mentioned in the program this set may need to contain other variables representing the location of control etc. The initial condition is simply a TLA predicate consisting of a conjunction of predicates of the form $X = x_0$ saying that the variable X has the value x_0.

Suppose that each possible step in any behaviour of the program can be described by one of the actions $\mathcal{A}_1, \mathcal{A}_2, \ldots, \mathcal{A}_n$ and all of these actions are possible. We call the disjunction

$$\mathcal{N} = \mathcal{A}_1 \vee \mathcal{A}_2 \vee \cdots \vee \mathcal{A}_n$$

the program's next-state relation. If f is the n-tuple of all program variables then the TLA formula $\Box[\mathcal{N}]_f$ ensures that apart from stuttering steps, all the steps in a behaviour consist of possible program actions. Provided we have taken care to specify program control movements in these actions, then this formula will ensure that the actions happen in the right order.

Of course, the above soundness formula does not guarantee that anything much happens at all in a behaviour: remember that it is possible that $f = f'$ always holds. In order to ensure liveness and other, more complicated fairness conditions, [Lamport, 1994] suggests we add a conjunction of specific fairness formulae. It is argued that these can all be written in the form of either weak or strong fairness of particular subsets of the program actions $\mathcal{A}_1, \mathcal{A}_2, \ldots, \mathcal{A}_n$.

In order to express such conditions conveniently, TLA has some more abbreviations:

$$\begin{aligned} \mathrm{WF}_f(\mathcal{A}) &= \Box\Diamond\langle\mathcal{A}\rangle_f \vee \Box\Diamond\neg\mathrm{Enabled}\langle\mathcal{A}\rangle_f \\ \mathrm{SF}_f(\mathcal{A}) &= \Box\Diamond\langle\mathcal{A}\rangle_f \vee \Diamond\Box\neg\mathrm{Enabled}\langle\mathcal{A}\rangle_f. \end{aligned}$$

So, for example, if we require that at least one of actions $\mathcal{A}_1$ or $\mathcal{A}_2$ happen if their disjunction is eventually continuously enabled then we write

$$\mathrm{WF}_f(\mathcal{A}_1 \vee \mathcal{A}_2).$$

Similarly, if we require that $\mathcal{A}_n$ happens eventually if it is infinitely often enabled then we write

$$\mathrm{SF}_f(\mathcal{A}_n).$$

Note that $\Diamond\langle\mathcal{A}\rangle_f$ means that an actual $\mathcal{A}$ action with a change in program variables will eventually happen: any stuttering will not go on for ever.

So we have ([Lamport, 1994], p. 885) a very useful *template*

T1: TLA formulae that represent programs can always be written ... as a conjunction $\mathit{Init} \wedge \Box[\mathcal{N}]_f \wedge F$, where

Init is a predicate specifying the initial values of variables.

$\mathcal{N}$ is the program's *next-state relation*, the action whose steps represent executions of individual atomic operations.

f is the n-tuple of all program variables.

F is the conjunction of formulae of the form $\mathrm{WF}_f(\mathcal{A})$ and/or $\mathrm{SF}_f(\mathcal{A})$, where $\mathcal{A}$ is an action representing some subset of the program's atomic operations.

In the next section we will look more carefully at the idea of a formula such as this *representing* a program.

7.2 Stuttering invariance

In this section we say why it is useful to have a stuttering invariant language, say exactly what this means, show that TLA is such a language and consider what it means for a formula to describe the behaviour of a program when we are dealing with a stuttering invariant language.

That a formula Φ in a temporal logic *represents* the program p is usually taken to mean that the possible behaviours of p are exactly the models of Φ. The use of such a concept is in showing that the program is guaranteed to exhibit certain behaviour—such as termination, for example. To show this, we first express the desired behaviour as a formula ψ. Then we need to show that ψ is a logical consequence of Φ in the logic. This can be argued for directly (i.e. semantically) or we can use the proof rules such as those in Volume 1 or those for TLA in [Lamport, 1994].

The concept of a possible behaviour of a program is usually (i.e. in traditional computer science temporal logics) taken to mean a possible sequence of states of values of the program variables as the individual steps of the program are run through. This approach has several related and well-known problems or deficiencies. The three we consider concern compositionality, refinement and extensionality.

Suppose that we model the behaviours of a program p by using temporal structures in the usual way. We need to find a formula which describes exactly those structures. Suppose further that we use a discrete flow of time in our structures with the instants of time corresponding exactly to the steps in our model of how the program proceeds. We have a class $\mathcal{S}$ of structures capturing the possible behaviours of p and, if we have a sufficiently expressive temporal language, we will have a formula ϕ whose models are exactly the structures in $\mathcal{S}$. In some sense we have captured the behaviour of p in ϕ.

However, consider the following context. Suppose that we wish to use the program as part of some modular system in which other programs are working in parallel. For the sake of simplicity, suppose that p is a closed program whose behaviour is unaffected by the behaviour of the other programs. The temporal models of the overall system will involve steps at much more frequent intervals than the steps of p. Although the values of the variables under the control of p are not changing during these intermediate steps, there is no guarantee that the behaviour of the overall system will satisfy ϕ.

Next, consider the situation of refinement of p in the sense of [Turski and Maibaum, 1987], i.e. we have another 'more detailed' program q which supposedly implements p. We will want to show that all of q's possible behaviours produce possible behaviours of p. In general, q may use new program variables as well as those mentioned in p. During a behaviour of q there may be many little steps in between the steps that correspond to the steps of p. During these steps, the values of new variables are changing while the values of the variables mentioned in p are unchanged. We might hope that the temporal structures which model q's behaviour also satisfy ϕ but there is no guarantee that this will happen.

Both of these problems with temporal semantics involve conflict over the meaning of the 'next' instant.

In fact, there are strong philosophical and practical reasons why we should not make program semantics so sensitive to the notion of 'next' instant anyway. The concept of program steps is an intentional one and the actual observable effect of a program should not refer to it. So another important motivation for using a stuttering invariant semantics for the behaviour of programs is the desire for extensionality. Suppose that an observer of the behaviour of a system or program can only see changes in the values of certain parameters—say the program variables in W. When the values of the variables in W are constant for a while there is no way for the observer to know whether the program is stepping or not. The observer cannot see actual 'steps' of the program happening unless they coincide with some kind of change in observable parameters. Any other kind of 'step' is likely to be an abstract notion dreamt up by the designer of the program—or some other theoretician—and cannot contribute to the effect of the program.

There are several solutions to these problems of concurrent composition, refinement (or granularity of temporal structure) and extensionality. In [Barringer *et al.*, 1984] it is suggested that we use a dense model of time in which there are no next instants to argue about. In [Fiadeiro and Maibaum, 1994] the suggestion is to be happy about using different formulae at different levels of abstraction. In TLA and also in the logic MTL in [Mokkedem and Méry, 1994] the solution is stuttering invariance. In this chapter, we are considering this latter approach. In order to see how it works we need some definitions.

Consider behaviours defined by the changes in values of a set $\mathcal{V}$ of (program) variables. The basic concept is stuttering relative to a set $W \subseteq \mathcal{V}$. We say that behaviour $\tau = (t_0, t_1, \ldots)$ is a *W-stuttering reduct* of behaviour $\sigma = (s_0, s_1, \ldots)$, written $\tau \preceq_W \sigma$, iff there exists a map $g : \mathbb{N} \longrightarrow \mathbb{N}$ such that:

- if $i < j$ then $g(i) < g(j)$;
- $g(0) = 0$; and
- for all i, j, if $g(i) \leq j < g(i+1)$ then for all $v \in W$, $s_j(v) = t_i(v)$.

We say that behaviours σ and τ *differ only up to W-stuttering* iff there exists behaviour ρ such that $\rho \preceq_W \sigma$ and $\rho \preceq_W \tau$.

Finally, we say that a TLA formula is invariant under adding or removing W-stuttering or just *W-stuttering invariant* if its truth is not affected by differing up to W-stuttering, i.e. ϕ is W-stuttering invariant iff for all behaviours σ, τ, such that σ and τ differ only up to W-stuttering, we have

$$\sigma[\![\phi]\!] \text{ iff } \tau[\![\phi]\!].$$

It is a very straightforward induction on the construction of TLA formulae to prove that:

Lemma 7.1. *For all TLA formulae ϕ with free program variables amongst W, ϕ is W-stuttering invariant.*

So the syntactic restriction imposed on TLA is quite sufficient to do the required work of eliminating sensitivity to stuttering. As we have said above, this is because there is no use of the $\bigcirc$ operator apart from its very restricted and implicit use within the concept of an action: the TLA action $X' = X + 1$ corresponds to the formula $\exists u((X = u) \wedge \bigcirc(X = u + 1))$ in a traditional-style temporal logic using a rigid variable symbol u.

See [Lamport, 1994] for details on adding quantification of program variable symbols while remaining stuttering invariant.

Let us now consider how having a stuttering invariant language helps with the three problems mentioned above. The solution actually involves two moves:

E1: the class of temporal structures which are counted as possible behaviours of a program is assumed to be closed under extension of the language and adding or removing finite amounts of stuttering;

E2: the formulae which we use to describe programs, properties or specifications are stuttering invariant.

It is worth emphasizing that under assumption E1, if we have a program p and we can only observe changes in its variables W, and we observe a run of p, then a particular W-behaviour (i.e. sequence of W states) σ will be a possible formal description of the run if and only if there is any τ such that σ and τ differ only up to W-stuttering. Runs of p are not associated with individual behaviour now but with equivalence classes of behaviours. Programs are now associated with unions of such equivalence classes.

After prescribing these two conditions, reasoning about program correctness can go on just as it does under the traditional approach, i.e. program properties are deduced from program descriptions. However, under these conditions, if we have a formula describing a property of a program then it will still apply to the program no matter what other programs we have running in parallel, no matter how we have correctly implemented the program, and no matter what sort of intentional idea an observer has about program steps when the program variables are not changing.

7.3 The problem with fairness

In this section we show that there is a problem with a naive use of the template T1. We will examine the reason for this problem.

7.3.1 *A problem example*

As we will see in Section 7.9, problems with expressing fairness properties in TLA show up commonly with rewrite languages. Here we will consider an example from a (slightly) more ordinary programming language.

Consider the program q

```
loop ( ( X:= 0 or 1 ) or ( X:=0 ) )
```

in a language which allows infinite looping, non-deterministic choice and non-deterministic assignment statements with the obvious meaning.

For this example, let us suppose that there is a weak fairness property of programs in this language when such a combination of looping and non-deterministic choice appears. On the other hand we require no special fairness within non-deterministic assignment. Thus, during the infinite looping when this program is executed, there are an infinite number of assignments of the first pattern and an infinite number of assignments of the second pattern.

When started with `X` having the value 0 this program has the possible behaviour of repeatedly looping through the successive states $0, 0, 0, \ldots$. Call this behaviour σ.

In order to describe this program formally in TLA we might use the variable X and consider two actions related to the two assignment statements:

- let $\mathcal{A} = (X' = 0) \vee (X' = 1)$
- and let $\mathcal{B} = (X' = 0)$.

A naive use of the template T1 produces

$$(X = 0) \wedge \Box[\mathcal{A} \vee \mathcal{B}]_X \wedge \mathrm{WF}_X(\mathcal{A}) \wedge \mathrm{WF}_X(\mathcal{B})$$

as a TLA formalization of the behaviour of q when it is begun in the state $\mathtt{X} = 0$. However, we have a problem.

Notice that during the behaviour σ, the action $\langle \mathcal{A} \rangle_X = (\mathcal{A} \wedge (X \neq X'))$ is continuously enabled. That is, $\Box\Diamond\neg\mathrm{Enabled}\langle \mathcal{A} \rangle_X$ does not hold. However, the action $\langle \mathcal{A} \rangle_X$ never occurs; only the action $\mathcal{A}$ does. This possible behaviour of the program q violates the condition $\mathrm{WF}_X(\mathcal{A})$.

Thus, our naive application of the TLA template T1 to the simple program q does *not* work.

7.3.2 *Why the naive application is flawed*

It is clear from the example above that, in general, for an action $\mathcal{A}$ using program variables f, the formalization $\mathrm{WF}_f(\mathcal{A})$ does not capture the weak fairness of action $\mathcal{A}$.

Let us look more closely at why not.

Actions in general may exhibit 'changing' or 'unchanging' forms. Suppose that we are interested in action $\mathcal{A}$ within the context of (tuple of all) program variables f and next-state relation $\mathcal{N}$. All the variables in $\mathcal{A}$ appear in f—and we take this to mean that for any program variable X, if either X or X' appears free in $\mathcal{A}$ then X is listed in f. The 'changing' form of $\mathcal{A}$ is the action $\mathcal{A} \wedge (f \neq f')$. The 'unchanging' form is $\mathcal{A} \wedge (f = f')$. This latter form of an action is discounted by Lamport as not very important but it is quite common in rewrite languages.

It is not hard to show that under the circumstances when $\mathcal{A}$ necessitates changes in f, i.e. $\vdash \mathcal{A} \rightarrow (f \neq f')$, then $\mathrm{WF}_f(\mathcal{A}) = \Box\Diamond\langle\mathcal{A}\rangle_f \vee \Box\,\Diamond\neg\mathrm{Enabled}\langle\mathcal{A}\rangle_f$ is an adequate formalization of the weak fairness of $\mathcal{A}$ itself. A similar result holds for strong fairness. However, when $\mathcal{A}$ may allow unchanging steps, the formalization $\mathrm{WF}_f(\mathcal{A})$ of the weak fairness of $\mathcal{A}$ does not accept the legitimate behaviour of the program doing 'unchanging' $\mathcal{A}$ steps infinitely often.

In trying to overcome this problem, we might first note that it is a manifestation of an important third cause of stuttering distinct from either concurrency or refinement. This is stuttering due to intensionally defined program steps which do not change the program variables. In the particular example above we have infinitely much of this type of stuttering.

Maybe this is not a serious problem. It might be objected that the program q above is rather pathological. Does it matter if we cannot describe *its* behaviour? A full treatment of the semantics of any language should allow us to describe the behaviour of any syntactically correct program, and from that description deduce that a particular program is 'pathological' or having any other property. A much more complex program than q might look like a plausible solution to a particular problem but, under certain conditions after being initiated with certain data values, the program might exhibit such behaviour. This is why we want a general account of semantics.

There are several possible ways of proceeding. One way is to assume that any program has a counter which increases on each step of the computation. This will ensure that we (and TLA) will notice the steps—they will not be stuttering steps. It may also be possible to institute some system of flagging steps to make them noticeable. Similarly, we could suppose that program data values are tagged with labels recording the history of their generation. Exactly this machinery is supposed by the labelled deductive system (LDS) (see [Gabbay, 1996]) approach to reasoning about program languages such as Gamma which is currently being developed by Dov Gabbay.

The above approaches all involve changes to the language, though. If we have a perfectly well-defined programming language, then it seems reasonable that TLA should be able to describe *its* behaviour.

In the rest of this chapter we will see how to do just that but, for now, let us summarize the findings of this section.

In general we want to be able to express the property of fairness with respect to a certain action. This action may have a direct TLA expression as the TLA action $\mathcal{A}$, but we have seen that the weak fairness of that action is not in general captured by $\mathrm{WF}_f(\mathcal{A})$. This raises the questions of whether fairness can be captured by another simple formulation and whether fairness can be captured in TLA at all.

7.4 New fairness abbreviations

In this section we will introduce completely general TLA formalisms for rendering the fairness of an action $\mathcal{A}$. This will work whether or not $\mathcal{A}$ has an 'unchanging' form.

7.4.1 *Intensional versus extensional*

Let us first be clear about weak fairness, the property of real programs which we are trying to capture with a formalism. The most important distinction here concerns the fact that real programs will exhibit *extensional* actions while specifiers may want to specify the fairness of *intensionally* defined actions. This is an inappropriate thing for specifiers to want to do if they want to use a formalism based on the observable changes in values of program variables.

The 'intensional' semantics we mention for programming languages often can be seen in terms of states and transitions between them. These intensional states are abstract objects in their own right and are not necessarily determined by the values they associate with program variables. An example of this approach is the fair transitions system of [Manna and Pnueli, 1981] and [Pnueli, 1986] which generalizes such approaches as CSP and Petri nets. If such intensional states determine values for the observable variables, which they generally do, then it is straightforward to associate an extensional action—a change in variable values—with each transition between intensional states. However, the converse relationship is more problematic and there is not necessarily only one or even one intensional transition associated with each extensional action. Thus, even if extensional action $\mathcal{A}$ is the observed effect of intensional transition a, the TLA Enabled($\mathcal{A}$) has little to do with the enablement of the transition a.

This is an interesting problem for those who wish to translate abstract intensional semantics into extensional semantics but it is not the concern of this chapter. We will guarantee that enough is observable of the computations we consider for this problem not to arise. The enablement of an action associated with such an intensional object as a rewrite will be determined exactly by the values of the observable variables. In other situations it may be necessary to introduce new program variables referring to program control states in order to force the simple relationship between intensional and extensional enablement. In such situations it is possible to hide the control variables under a TLA quantifier as they will not be externally observable. We will not pursue this topic here.

Under the assumption that intensional and extensional enablement have been made to correspond directly, we proceed to consider the fairness of extensional actions.

7.4.2 *Weak fairness in extensional languages*

Weak Fairness

Suppose that we are looking at an action $\mathcal{A}$ as part of a program with observable variables W. If we associate a run of the program with a sequence σ of states of values of the variables from W then we must—in the context of E1—also allow that the run is associated with any sequence τ such that σ and τ differ only up to W-stuttering.

If we are not careful with our definition of weak fairness then two such behaviours might differ as to whether they are fair for a particular action. For example, suppose that we have a program whose only variable is X taking integer values. Say that $\mathcal{A}$ is the action of doubling X. Consider σ to be the behaviour given by the sequence $1, 0, 1, 0, 1, 0, 1, \ldots$ of X-values. This does not seem to be fair for $\mathcal{A}$. Yet the behaviour τ given by $1, 0, 0, 1, 0, 0, 1, 0, 0, 1, \ldots$ does seem to be fair for $\mathcal{A}$. Now σ and τ differ only up to $\{X\}$-stuttering and could well both be formal models of the observation of the same run of the program. If the property of fairness is a property of program runs then this run must either be fair or be unfair with respect to the action $\mathcal{A}$.

Such considerations motivate the following definition:

Definition 7.2. A run of a program (with observable variables W) formalized by W-behaviour σ is weakly fair for the action $\mathcal{A}$ iff there is a behaviour τ such that σ and τ differ only up to W-stuttering and either:

- for infinitely many $i = 0, 1, 2, \ldots$ we have $\tau_i[\![\mathcal{A}]\!]\tau_{i+1}$; or
- for infinitely many i there is no state s such that $\tau_i[\![\mathcal{A}]\!]s$ holds.

It might be objected that we could have equally well made the definition of fairness of runs consistent by taking the existence of an unfair formalization to mean the run is unfair. However, this would have the result that no run would be fair for a trivial rewrite unless it eventually became continuously disabled. It still might be objected that no one would really ever want to require fairness of a trivial rewrite. The answer to this objection is, of course, that general languages (such as Gamma) require general fairness criteria of which trivial rewrites are a possible instance.

7.4.3 *VWF*

For an action $\mathcal{A}$ and tuple of all program variables f, define

$$\begin{aligned}\mathrm{VWF}_f(\mathcal{A}) = \ & \Box\Diamond\langle\mathcal{A}\rangle_f \\ & \vee \Box\Diamond\neg\mathrm{Enabled}(\mathcal{A}) \\ & \vee \Box\Diamond\mathrm{Enabled}(\mathcal{A} \wedge (f = f')).\end{aligned}$$

We might call this disjunction the *very weak fairness* of $\mathcal{A}$. It is really just an attempt to formalize the usual weak fairness requirement on action $\mathcal{A}$ but it is weaker than $\mathrm{WF}_f(\mathcal{A})$.

Notice that the third disjunct of $\mathrm{VWF}_f(\mathcal{A})$ says that 'unchanging' $\mathcal{A}$ steps are enabled infinitely often. It does not say that they happen infinitely often.

We cannot say that in TLA because $\Box\Diamond(\mathcal{A} \wedge (f = f'))$ is not a formula of TLA. However, $\mathrm{VWF}_f(\mathcal{A})$ is a formula of TLA: the predicate $\mathrm{Enabled}(\mathcal{B})$ is a formula for any action $\mathcal{B}$.

But is the enablement of unchanging $\mathcal{A}$ steps the same as them actually occurring? To see that $\mathrm{VWF}_f(\mathcal{A})$ does represent the weak fairness of $\mathcal{A}$ we can use Definition 7.2 above of the latter concept in the context of stuttering invariant semantics and its requirement E1:

Lemma 7.3. *Suppose that we formalize the runs of a program using just the variables listed in the tuple f and that $\mathcal{A}$ is an action mentioning only those variables.*

Then a run formalized by f-behaviour σ is weakly fair for $\mathcal{A}$ iff $\sigma[\![\mathrm{VWF}_f(\mathcal{A})]\!]$.

Proof $(\Rightarrow)$

Suppose that a run formalized by f-behaviour σ is weakly fair for $\mathcal{A}$.

Thus there is a behaviour τ such that σ and τ differ only up to f-stuttering and either:

- for infinitely many $i = 0, 1, 2, \ldots$ we have $\tau_i[\![\mathcal{A}]\!]\tau_{i+1}$; or
- for infinitely many i there is no state s such that $\tau_i[\![\mathcal{A}]\!]s$ holds.

In the second case it is clear that $\tau[\![\Box\Diamond\neg\mathrm{Enabled}\langle\mathcal{A}\rangle_f]\!]$.

In the first case we have either changing $\mathcal{A}$ steps happening infinitely often or unchanging $\mathcal{A}$ steps happening infinitely often. Thus either $\tau[\![\Box\Diamond\langle\mathcal{A}\rangle_f]\!]$ or $\tau[\![\Box\Diamond\mathrm{Enabled}(\mathcal{A} \wedge (f = f'))]\!]$.

In all cases $\tau[\![\mathrm{VWF}_f(\mathcal{A})]\!]$. But σ and τ agree on all TLA formulae with free variables from f so $\sigma[\![\mathrm{VWF}_f(\mathcal{A})]\!]$ as required.

$(\Leftarrow)$

Suppose that σ formalizes a run of the program and $\sigma[\![\mathrm{VWF}_f(\mathcal{A})]\!]$. Thus

- $\langle\mathcal{A}\rangle_f$ happens infinitely often in σ; or
- $\neg\mathrm{Enabled}\langle\mathcal{A}\rangle_f$ happens infinitely often in σ; or
- $\mathrm{Enabled}(\mathcal{A} \wedge (f = f'))$ happens infinitely often in σ.

In the first two cases it is clear that the run formalized by σ is weakly fair to $\mathcal{A}$.

In the third case we can construct a behaviour τ from σ by adding stuttering at the steps where $\mathrm{Enabled}(\mathcal{A} \wedge (f = f'))$ holds. Because $\mathrm{Enabled}(\mathcal{A} \wedge (f = f'))$ and $f = f'$ both hold between these new stuttering steps it is clear that $\mathcal{A}$ also then holds. It is also clear that τ formalizes the run. Thus we have the result. □

For example, look at the program q of Section 7.4.1. The behaviour σ in which X always has the value 1 clearly is a model of $\mathrm{VWF}_X(\mathcal{A})$ as $\mathrm{Enabled}(\mathcal{A} \wedge (X = X'))$ always holds.

As another concrete example, look at the program k:

```
loop ( ( X:=X) or (X:=X+1))
```

which loops continuously, non-deterministically choosing to assign an integer value to the counter `X` according to either of the rules p or q given by

$$p : \texttt{X:=X}$$
$$q : \texttt{X:=X+1}.$$

In TLA we can decide that the only program variable is X and the next step actions of this program are $\mathcal{P} = (X' = X)$ and $\mathcal{Q} = (X' = X + 1)$. There are several fairness requirements we could impose: suppose that we require weak fairness of $\mathcal{P}$ actions. Now k does have a behaviour σ in which both $\mathcal{P}$ and $\mathcal{Q}$ actions occur infinitely often. By definition the $\mathcal{P}$ steps do not change X. If we remove all stuttering from this behaviour to obtain the behaviour τ then τ seems unfair. However, we have defined weak fairness on actual runs of the program so that it does not make this distinction between formal representations. The actual run is fair because it can be represented by the behaviour σ. Now both σ and τ are models of $\mathrm{VWF}_X(\mathcal{P})$ as they are both models of $\Box\Diamond\mathrm{Enabled}(\mathcal{P} \wedge (X = X'))$.

7.4.4 *VWF in terms of WF*

So if we want to specify the weak fairness of an action $\mathcal{A}$ and it is at all possible that the action has an unchanging form then it is necessary to formalize this as $\mathrm{VWF}_f(\mathcal{A})$. Does this mean that the template T1 is wrong? No: the template does not imply a claim that the weak fairness of $\mathcal{A}$ is captured by $\mathrm{WF}_f(\mathcal{A})$. It just claims that any fairness requirement can be written as a conjunction of formulae of the form $\mathrm{WF}_f(\beta)$ or $\mathrm{SF}_f(\delta)$ for actions β or δ which are each subsets of the program's atomic operations. As observed by Lamport (private communication) the TLA formula $\mathrm{VWF}_f(\mathcal{A})$ is logically equivalent to $\mathrm{WF}_f(\mathcal{A} \wedge \neg\mathrm{Enabled}(\mathcal{A} \wedge (f = f')))$ for any $\mathcal{A}$ and it is easy to see that $\mathcal{A}^+ = \mathcal{A} \wedge \neg\mathrm{Enabled}(\mathcal{A} \wedge (f = f'))$ is a subaction of the program's next-state relation $\mathcal{N}$ (as $\mathcal{A}$ is).

Thus we could have expressed the required fairness property in the example in Section 4.1 as $\mathrm{WF}_X(\mathcal{A} \wedge \neg\mathrm{Enabled}(\mathcal{A} \wedge (X = X')))$ and then the TLA account of the behaviour would have been correct. The advantages we have in using VWF instead are that:

- it is $\mathrm{VWF}_f(\mathcal{A})$ which captures the weak fairness of $\mathcal{A}$; and
- we know this is correct immediately from the general result of Lemma 7.3.

In this section we have met a generally correct way of capturing the weak fairness of any action.

7.4.5 *Strong fairness*

Now we carry over the concepts and approach to the case of strong fairness. Again we have a definition of strong fairness of an extensional action, a mismatch between this concept and the abbreviation SF_f and a new correct abbreviation ASF_f.

Importantly, we also have a difference between the strong and weak case: the correct formalism for strong fairness cannot be expressed according to the template T1. Here we really do need the new abbreviation.

7.4.6 *Definition of strong fairness*

As in the case of weak fairness we have the following:

Definition 7.4. A run of a program (with observable variables W) formalized by W-behaviour σ is strongly fair for the action $\mathcal{A}$ iff there is a behaviour τ such that σ and τ differ only up to W-stuttering and either:

- for infinitely many $i = 0, 1, 2, \ldots$ we have $\tau_i[\![\mathcal{A}]\!]\tau_{i+1}$; or
- there is some i such that for all $j > i$, there is no state s such that $\tau_j[\![\mathcal{A}]\!]s$ holds.

Note that again we have defined strong fairness with respect to an underlying run of a program, not just a particular formal account of it.

7.4.7 *ASF*

It is not hard to see that, in the context of possibly trivial rewrite rules, the abbreviation SF suffers from the same problems as WF: requiring $\mathrm{SF}_f(\mathcal{A})$ will force the strong fairness of $\langle\mathcal{A}\rangle_f$. As before, the problems are easily solved by adding an extra disjunct. So let us introduce the abbreviation

$$\begin{aligned}\mathrm{ASF}_f(\mathcal{A}) = \quad & \Box\Diamond\langle\mathcal{A}\rangle_f \\ & \vee \Diamond\Box\neg\mathrm{Enabled}(\mathcal{A}) \\ & \vee \Box\Diamond\mathrm{Enabled}(\mathcal{A} \wedge (f' = f))\end{aligned}$$

which we might call *almost strong fairness* of the action $\mathcal{A}$. A similar argument to that seen in Section 7.5 proves the following:

Lemma 7.5. *A run of a program associated with f-behaviour σ is strongly fair for action $\mathcal{A}$ iff $\sigma[\![\mathrm{ASF}_f(\mathcal{A})]\!]$.*

7.4.8 *ASF not expressible in terms of SF*

The question arises whether $\mathrm{ASF}_f(\mathcal{A})$ is, like $\mathrm{VWF}_f(\mathcal{A})$, just a useful piece of syntax for a property which, with a little more complication, could be written in terms of the usual TLA fairness abbreviations. It turns out that in this case the answer is 'no'.

Lemma 7.6. *In general,* ASF *cannot be rewritten in the template form T1.*

Proof Consider the example of a program with one program variable X which can take on values from $\{1, 2, 3, 4\}$. Suppose that $\mathcal{A}$ is the action

$$((X = 1) \wedge (X' = 1)) \vee ((X = 2) \wedge (X' = 3)).$$

Suppose also that there are other program actions but that all we require is the strong fairness of action $\mathcal{A}$. For definiteness suppose that the next-state relation allows any values for X and X'.

Suppose for contradiction that the strong fairness of $\mathcal{A}$ is able to be represented in the form T1. Thus there is some number $n \geq 0$, some number $m \geq 0$, some collection of actions $\beta_1, \ldots, \beta_n$ and another collection of actions $\delta_1, \ldots, \delta_m$ such that the formula

$$\alpha = \bigwedge_{i=1}^{n} \mathrm{WF}_X(\beta_i) \wedge \bigwedge_{i=1}^{m} \mathrm{SF}_X(\delta_i)$$

exactly captures the requirement. Note that in case either $n = 0$ or $m = 0$ then the corresponding conjunction is assumed to be $\top$.

In what follows we present behaviours just as sequences of the values of X. Actions can be thought of as just binary relations on $\{1, 2, 3, 4\}$. Also, for any action γ, define $\pi_1(\gamma) = \{x \mid \text{ there is some } y \text{ such that } (x, y) \in \gamma\}$.

There may be no δ_is (i.e. $m = 0$) but if there are then we assume that none are the empty action. Otherwise, we can just throw these out.

Now consider any δ_i assuming there are some. From the fact that $\sigma = 3, 3, 3, \ldots$ is strongly fair for $\mathcal{A}$, we gather that $\pi_1(\langle\delta_i\rangle_X) \subseteq \{1, 2, 4\}$, because $\sigma[\![\mathrm{SF}_X(\delta_i)]\!]$, but $\langle\delta_i\rangle_X$ never holds during σ. It must eventually be continuously disabled.

Similarly $4, 4, 4, \ldots$ is strongly fair for $\mathcal{A}$ so $\pi_1(\langle\delta_i\rangle_X) \subseteq \{1, 2, 3\}$.

And again, $1, 1, 1, \ldots$ is strongly fair for $\mathcal{A}$ too so $\pi_1(\langle\delta_i\rangle_X) \subseteq \{2, 3, 4\}$.

Putting these results together tells us that

$$\text{Condition R:} \qquad \pi_1(\langle\delta_i\rangle_X) \subseteq \{2\}.$$

Now consider the behaviour $\sigma = 1, 2, 4, 1, 2, 4, 1, 2, 4, \ldots$. This is also strongly fair for $\mathcal{A}$. This is because our use of strong fairness allows the addition of finite stuttering steps. From R (and the assumption that it is not empty) we know that δ_i is enabled infinitely often. For $\mathrm{SF}_X(\delta_i)$ to hold we thus need $\langle\delta_i\rangle_X$ to occur infinitely often. By condition R we know that this must be when $(2, 4)$ happens i.e. $\{(2, 4)\} \subseteq \langle\delta_i\rangle_X$.

A similar argument about the fair $1, 2, 1, 2, 1, 2, \ldots$ tells us that $\{(2, 1)\} \subseteq \langle\delta_i\rangle_X$.

And if we look at the fair $2, 3, 2, 3, 2, 3, \ldots$ we deduce that $\{(2, 3)\} \subseteq \langle\delta_i\rangle_X$.

Putting all these results together we deduce that if there are any δ_is then they are either $D = \{(2, 1), (2, 3), (2, 4)\}$ or $D' = \{(2, 1), (2, 2), (2, 3), (2, 4)\}$.

Next consider the possibility that $n > 0$. Once again, discard the empty β_is.

From the fact that $\sigma = 3, 3, 3, \ldots$ is strongly fair for $\mathcal{A}$, we gather that $\pi_1(\langle\beta_i\rangle_X) \subseteq \{1, 2, 4\}$. This is because $\sigma[\![\mathrm{WF}_X(\beta_i)]\!]$, but $\langle\beta_i\rangle_X$ never holds during σ. It must eventually be continuously disabled.

As with the δ_i we can thus put together several such observations to deduce that $\pi_1(\langle\beta_i\rangle_X) \subseteq \{2\}$.

Now let us switch to consider a non-model of the strong fairness of $\mathcal{A}$. Consider $\sigma = 2, 4, 2, 4, 2, 4, \ldots$. This is not fair as $\mathcal{A}$ is infinitely often enabled but never occurs, even allowing for extra stuttering. So we know $\sigma[\![\neg\alpha]\!]$.

Now $\sigma[\![\mathrm{SF}_X(D)]\!]$ and $\sigma[\![\mathrm{SF}_X(D')]\!]$ so there must, after all, be a β_i conjunct which produces the falsity. That is, we have some $i = 1, \ldots, n$ such that $\sigma[\![\neg\mathrm{WF}_X(\beta_i)]\!]$. But this cannot be as no β_i is eventually continuously enabled. Since $\pi_1(\langle\beta_i\rangle_X) \subseteq \{2\}$, β_i is disabled when $X = 4$.

This is our contradiction. □

So the template T1 is clearly wrong in this case: in general, strong fairness requirements cannot be formalized in the format T1. Fortunately, we know that we can simply use the new abbreviations in the format T1 and propose a new correct general format which also has the advantage of being much easier to apply. We present this format in the conclusion.

7.4.9 *Machine closure*

We have introduced a new formalization for the strong fairness of an action. Although we have shown that the formalization exactly captures the intuitive idea of strong fairness, it is worth checking that the formalization conforms to established technical versions of fairness. One useful idea is that of machine closure from [Abadi and Lamport, 1992]. In what follows we identify finite behaviours $s_0, s_1, s_2, \ldots, s_n$ with the eventually continuously stuttering infinite behaviour $s_0, s_1, s_2, \ldots, s_n, s_n, s_n, \ldots$. If Π and L are properties of behaviours then (Π, L) being *machine closed* means that:

- Π is a safety property, i.e. Π holds of a behaviour iff it holds of all its finite prefixes; and
- L is any property such that any finite behaviour satisfying Π can be extended to an infinite behaviour satisfying $\Pi \wedge L$.

The machine closure of (Π, L) is supposed to capture the idea of L being a fairness property for a program which always satisfies Π.

It is quite straightforward to show the following:

Lemma 7.7. *If $\mathcal{N}$ is an action with free variables from tuple f, if I is a finite or countably infinite index set, if for all $i \in I$, $\mathcal{A}_i$ is a subaction of $\mathcal{N}$, and if L is the intersection of properties of form either* $\mathrm{VWF}_f(\mathcal{A}_i)$ *or* $\mathrm{ASF}_f(\mathcal{A}_i)$*, then* $(\square[\mathcal{N}]_f, L)$ *is machine closed.*

To show this just take a finite behaviour satisfying $\square[\mathcal{N}]_f$ and extend it by clever scheduling of the actions $\mathcal{A}_i$ that have been enabled but most neglected.

7.5 Gamma

In this section we will introduce a particular rewrite language so that we can then apply our new formalisms to it in the next section. Thus we will demonstrate the new constructs in a context where they are really necessary.

The language Gamma, which was proposed in [Banâtre and Le Métayer, 1990], is a minimal formalism for describing programs. It allows us to abstract away from much of the detail of the control structure and, in particular, to refrain from having to introduce any artificial sequentiality. Such a language is ideal for use with the following two-step method of parallel program construction. First, prove correct a high-level description of a class of algorithmic solutions to the problem. Then include more details of control to describe a refinement which runs efficiently on particular hardware. The benefits of using Gamma in systematic program construction are described more fully in [Banâtre and Le Métayer, 1990], [Creveuil, 1991], and [Mussat, 1992].

Below we will give a brief formal account of Gamma. For more details see [Hankin *et al.*, 1992]. Essentially, Gamma achieves its minimality by using one data structure, the multiset, and one control structure, the rewrite. The multiset rewrites are often compared to chemical reactions in which certain collections of objects in the multiset react, being consumed and producing other objects as the reaction product. These reactions can occur quite locally in the multiset and herein lies the possibility of parallel implementations. Descriptions of parallel implementations of Gamma appear in [Banâtre *et al.*, 1988], and [Creveuil, 1992].

Despite these implementations it should be emphasized that Gamma is not really itself a programming language. A particular Gamma 'program' can be implemented in many ways. The program describes a whole gamut of possible behaviours and although it is straightforward to implement them there is a lot of non-determinism in general. Our task in coming up with a temporal semantics for Gamma is to find a formula which describes for a particular program exactly this large class of its possible behaviours.

Although the chemical reaction metaphor suggests a very specific class of algorithms which can be described by Gamma, it has been shown that many algorithms in many different application areas are amenable to being described in this way. There is a long series of examples illustrating the Gamma style of programming in [Banâtre and Le Métayer, 1993] and [Mussat, 1992]. Graph problems, string processing problems, the knapsack problem and the shortest path problem are all described in these references.

7.5.1 *More formal definitions*

A particular Gamma program will rewrite multisets of elements from some domain. In what follows we will assume an unspecified but fixed domain D. Let $\mathcal{M}(D)$ be the set of multisets of elements from D. This can be defined rigorously by taking the set of all tuples of elements of D of all finite lengths and factoring out by an equivalence based on permutations of ordering. We assume that operations of multiset addition $+$ and multiset subtraction $-$ and the relation of inclusion $\subseteq$ are defined. Often, for a tuple $\overline{d}$ of elements from D we will write $\overline{d}$ to mean also the multiset equivalence class of $\overline{d}$. For example, if m is a multiset and $\overline{d}$ a tuple then we might write $\overline{d} \subseteq m$ if each element of D appears as many times in m as in $\overline{d}$ and we might write $\overline{d} + m$ for the multiset in which each element of D appears as many times as it does in $\overline{d}$ and in m added together.

We assume that some formal language(s) allows us to describe a set $\mathbf{A}$ of tuple-valued functions (of given arities) on D and a set $\mathcal{R}$ of relations (of given arities) on D. In order to simplify notation we do not distinguish between the formal syntax for the function or relation and its interpretation. We also suppose that for each m-ary function A in $\mathbf{A}$, there is some number $r > 0$ such that for all m-tuples $\overline{d}$ from D, $A(\overline{d})$ is an r-tuple from D.

A *basic program* is a pair $(R, A) \in \mathcal{R} \times \mathbf{A}$ in which R and A have the same arity. If this arity is n then we often write the basic program (R, A) as

$$x_1, \ldots, x_n \rightarrow A(x_1, \ldots, x_n) \Leftarrow R(x_1, \ldots, x_n).$$

The *programs* of Gamma are built recursively from basic programs using the parallel and sequential composition operators:

- if (R, A) is a basic program then (R, A) is a program;
- if p and q are programs then so is $p \circ q$ and $p + q$.

The set of programs thus depends on $\mathcal{R}$ and $\mathbf{A}$ and so indirectly on the choice of D. Let $\mathcal{P}(\mathcal{R}, \mathbf{A})$ be this set.

We will not give a completely formal description of the semantics of a Gamma program here. The program operates by rewriting a multiset of elements from D beginning from some given initial multiset. It proceeds in a series of such rewriting steps until (and if) the program finishes as described below, leaving the resulting multiset as the final product of the computation.

One single rewrite step of the basic program

$$x_1, \ldots, x_n \to A(x_1, \ldots, x_n) \Leftarrow R(x_1, \ldots, x_n)$$

consists of the computer simply choosing any n-elements $d_1, \ldots, d_n$ of the multiset which satisfy $R(d_1, \ldots, d_n)$, removing them from the multiset and replacing them with $A(d_1, \ldots, d_n)$. If there are no such elements then the program finishes.

The program $p+q$ keeps repeating rewriting steps of either p or q until both p and q are unable to rewrite. There is no fairness requirement that Gamma must allow a mixture of p rewrites and q rewrites even in a non-terminating repetition of $p + q$ steps.

The program $p \circ q$ rewrites as q until (and if) q finishes and then it rewrites as p. It finishes if p does then.

We can be a little more precise by introducing the relation Γ on $\mathcal{P}(\mathcal{R}, \mathbf{A}) \times \mathcal{M}(D) \times \mathcal{P}(\mathcal{R}, \mathbf{A}) \times \mathcal{M}(D)$ (by recursion on the construction of Gamma programs) as follows:

- for n-ary basic program (R, A), $\Gamma((R, A), m, (R, A), k)$ iff there is some n-tuple $\overline{d} \subseteq m$ such that $R(\overline{d})$ holds and $k = (m - \overline{d}) + A(\overline{d})$;
- $\Gamma(p + q, m, r, k)$ iff either
    - $\Gamma(p, m, s, k)$ and $r = s + q$
    - or $\Gamma(q, m, s, k)$ and $r = p + s$;
- $\Gamma(p \circ q, m, r, k)$ iff either
    - $\Gamma(q, m, s, k)$ and $r = p \circ s$
    - or $r = p$ and there are no s, l such that $\Gamma(q, m, s, l)$ holds.

Also define the relation Ω on $\mathcal{P}(\mathcal{R}, \mathbf{A}) \times \mathcal{M}(D)$ as follows. For any p, m we write $\Omega(p, m)$ iff there are no q, k such that $\Gamma(p, m, q, k)$.

When the Gamma program p_0 is started on a multiset m_0 then the operation of the program is defined by a sequence of steps which transform both the multiset and the active program. Thus a particular state in a computation is specified by a configuration which is a pair (p, m) consisting of the active program p and the current multiset m. Starting with (p_0, m_0) the program proceeds through $(p_1, m_1), (p_2, m_2), \ldots, (p_n, m_n)$ in such a way that $\Gamma(p_i, m_i, p_{i+1}, m_{i+1})$ holds for each $i = 0, 1, \ldots, (n-1)$. Note that this may not be a deterministic behaviour as for some (p, m) there may be several (q, k) such that $\Gamma(p, m, q, k)$ holds. The sequence may actually go on for ever or the program may halt at step (p_n, m_n) iff $\Omega(p_n, m_n)$ holds. In the latter case, m_n is the result of the computation.

The semantics of Gamma is given more formally in the style of Plotkin's structural operational semantics in [Hankin *et al.*, 1992]. There, the relation $\Gamma(p, m, p', m')$ is written as $(p, m) \to (p', m')$ and $\Omega(p, m)$ as $(p, m) \to m$. In this chapter we use the Γ notation so as to avoid a notational clash when it is incorporated in Lamport's logic. We also present the definition of Γ in a recursive manner instead of the usual proof rule format so that it is more clear how to fit the definition into a fixed point formulation (more about this later).

It is worth noting that there is no non-determinism about a Gamma program finishing. The program will finish exactly at the step when the active program p and the multiset m are such that $\Omega(p, m)$ holds, i.e. when there are no Γ steps which apply to (p, m).

A nice example of a Gamma program is given by $p_3 \circ (p_1 + p_2)$, where

$$\begin{aligned} p_3 &= (x, y \to x + y \Leftarrow \text{True}) \\ p_1 &= (0 \to 1 \Leftarrow \text{True}) \\ p_2 &= (x \to x - 1, x - 2 \Leftarrow (x > 1)). \end{aligned}$$

This calculates the nth Fibonacci number when started with the multiset containing just the natural number n.

7.6 Gamma in TLA

In this section we will describe Gamma using TLA. We will see that we need to use the new formal expressions for fairness.

It is straightforward to base the semantics on the idea of a configuration (p, m) as introduced in Section 7.5. Thus we use a program variable M whose value at a particular time represents the current value of the multiset and a program variable P whose value represents the part of the original program which has not been finished with. P will start off with value the original Gamma program but its value does change as sequential operators are proceeded through. So P is really representing the control state of the computation. At the end of the computation—if there is one—P will take on and keep the value 'finished'. For example, recall the Fibonacci program $p_3 \circ (p_1 + p_2)$ from the previous section. As Gamma runs on this program, the variable P starts as $p_3 \circ (p_1 + p_2)$ and keeps this value for a while. Then it changes to having the value p_3 for a while before changing to 'finished' forever after.

The value of M starts off as the initial multiset, changes to reflect changes in the multiset and, if the program does finish, the value of M ends up as the final multiset and keeps that value unchanged for ever.

To describe how one configuration (p, m) changes into the next (q, n) we just need to use the relations Γ and Ω as defined in the last section.

Thus, Lamport's universal template for representing programs tells us the first part of the TLA representation for Gamma. The behaviour of P and M when Gamma program p_0 operates on multiset m_0 is described by the TLA formula Φ_1:

$$\begin{aligned}
\text{Init} &= (P = p_0) \wedge (M = m_0) \\
\mathcal{S}_1 &= \Gamma(P, M, P', M') \\
\mathcal{S}_2 &= \Omega(P, M) \wedge (M = M') \wedge (P' = \textit{finished}) \\
\mathcal{S} &= \mathcal{S}_1 \vee \mathcal{S}_2 \\
\Phi_1 &= \text{Init} \wedge \Box[\mathcal{S}]_{(P,M)}.
\end{aligned}$$

Missing from this definition is a TLA rendering of the 4-ary relation Γ and the 2-ary relation Ω. With a simple use of quantification of rigid variable symbols, it is clear that we can define Ω in terms of Γ. To define Γ itself, though, presents a more difficult problem. Our definition in Section 7.5 is a recursive one. Now TLA, as defined in [Lamport, 1994], is not prescriptive about how we formalize actions: any Boolean-valued expression involving variables and primed variables is acceptable. Perhaps the best way to provide such an expression for Γ is to use some kind of least fixed point operator. We do not need any temporal aspect to this operator: a simple functional operator such as those considered in [Gurevich and Shelah, 1986] would suffice. It could be used to construct recursively defined actions out of variables and primed variables. We will not pursue this task here; instead see [Reynolds, 1996] for details.

Also missing from the above is any discussion of the data structures of programs and multisets. We need to choose function symbols and constants in order to construct multisets and Gamma programs as data objects. We need to make sure that our rendering of Γ, Ω and even $=$ is appropriate for these data structures. Again we do not pursue this task here.

And the fairness part of the representation is missing too. We consider that next.

We have introduced two important actions $\mathcal{S}_1$ and $\mathcal{S}_2$ which a Gamma program does. In order to complete the description of the semantics, we must decide whether Gamma requires weak or strong fairness of each of these actions or no fairness at all.

A short argument (see [Reynolds, 1996] for details) shows that, because of the straightforward way that the enablement of these two actions changes as the program runs, for each of these two actions, either a weak or strong fairness requirement will suffice.

Thus we can immediately use our new abbreviations to express these requirements.

However, let us back-track a moment and see why the old abbreviation WF will not do in this case.

Consider the Gamma program q:

$$\begin{aligned}
&(1 \to 1 \Leftarrow \top) \\
+\, &(1 \to 0 \Leftarrow \top).
\end{aligned}$$

When started with a multiset containing just the single element 1, this program has the possible behaviour of repeatedly removing 1 from the multiset, rewriting it as 1 and replacing 1 in the multiset. Thus we have a sequence of states $\sigma = (p_0, m_0), (p_1, m_1), \ldots$ in which, for all i, p_i is q and m_i is $\{1\}$.

During the behaviour σ, the action $\langle \mathcal{S}_1 \rangle_{(P,M)} = (\mathcal{S}_1 \wedge ((P \neq P') \vee (M \neq M')))$ is continuously enabled. That is, $\text{Enabled}\langle \mathcal{S}_1 \rangle_{(P,M)}$ is always true. However, the action $\langle \mathcal{S}_1 \rangle_{(P,M)}$ never occurs; only the action $\mathcal{S}_1$ does. This possible behaviour of the program q violates the condition $\text{WF}_{(P,M)}(\mathcal{S}_1)$.

The problem with using WF in a description of the behaviour of Gamma is thus that the condition $\text{WF}_{(P,M)}(\mathcal{S}_1)$ is so strong that it disallows a very particular, but perfectly legitimate, class of behaviours. These are behaviours which show no change ever after a time even though 'changing' steps are enabled continuously. Such behaviour is legitimate when allowed 'unchanging' steps keep happening for ever.

But this is just the type of behaviour that we have managed to allow with the new formalization VWF. So $\text{VWF}_{(P,M)}(\mathcal{S}_1)$ will render the weak fairness of $\mathcal{S}_1$ steps.

Note that we could actually have used WF to capture the fairness requirement on $\mathcal{S}_2$ steps as they always do change the state. However, it is equally correct to use VWF.

So the final version of the semantics of Gamma is:

$$\begin{aligned}
\text{Init} &= (P = p_0) \wedge (M = m_0) \\
\mathcal{S}_1 &= \Gamma(P, M, P', M') \\
\mathcal{S}_2 &= \Omega(P, M) \wedge (M = M') \wedge (P' = \textit{finished}) \\
\mathcal{S} &= \mathcal{S}_1 \vee \mathcal{S}_2 \\
\Phi(p_0, m_0) &= \text{Init} \wedge \Box[\mathcal{S}]_{(P,M)} \wedge \text{VWF}_{(P,M)}(\mathcal{S}_1) \wedge \text{VWF}_{(P,M)}(\mathcal{S}_2).
\end{aligned}$$

$\Phi(p_0, m_0)$ represents the Gamma program p_0 operating on the multiset m_0.

7.7 Using the semantics for formal reasoning

Another very important use for this formal semantics for Gamma is to enable programmers to reason about the correctness of a Gamma program with respect to some formally specified properties. It is generally recognized that temporal languages are capable of expressing most of these properties which are of interest to programmers. See, for example, the book by Manna and Pnueli [1992] or Volume 1 for lists of such properties. For example, termination can be expressed in TLA by $\Diamond(P = \text{'finished'})$. Establishing correctness thus becomes a matter of proving that one temporal formula—expressing the property—is a logical consequence of the formula Φ which describes the behaviour of the program.

[Lamport, 1994] provides us with a relatively complete proof system in which to conduct this reasoning. The system is necessarily incomplete because the underlying data domain in a general TLA application may be able to express arithmetic. However, the rules and axioms given are enough to derive all temporal properties assuming all properties of the data domain can be established.

For reasoning about a Gamma program, the TLA rules themselves need the support of many levels of other proof rules as the data domain from the point of view of the TLA program variables is not the set of multiset elements. It is, in fact, a set of data objects, multisets of those objects (these are the values of the program variable M) and Gamma program syntactic terms (these are the values of the program variable P). So a formal proof that a particular program guarantees certain behaviour will usually involve reasoning on more than just the temporal level. In general, there will be reasoning about the object domain, reasoning about multisets of these objects, reasoning about the functions and relations which define the rewrite rules, reasoning about program syntax, reasoning about the one'-tep Γ relation as well as reasoning about temporal behaviour, and all this reasoning will involve first-order logic, recursive definitions and mathematical induction.

We cannot give a complete system of reasoning because the rules needed will depend on the actual object domain in a particular application and in any case, as the language should be able to express arithmetic, there is no complete system possible. In the following sections we will give examples of the sorts of rules needed and illustrate their use by presenting a rough sketch of and selected excerpts from a formal proof of the total correctness of a Gamma program.

As mentioned, we suppose that rules

$$\vdash \phi \quad \text{where } \phi \text{ is a tautology in first-order logic,}$$

the rules of *modus ponens* and universal generalization and a rule of mathematical induction are available.

7.7.1 *Formal reasoning about objects, multisets and Gamma syntax*

There are two problems with supplying a formal system for reasoning about multisets and Gamma syntax:

- any such system must be incomplete because the language will be strong enough to express arithmetic; and
- properties of the multisets relevant to a particular Gamma program will depend on properties of the underlying data object domain.

Thus, a particular application of the Gamma semantics will need its own selection of definitions, rules and axioms for the object domain and multisets of objects. Often it will only be a few simple properties of these sets which are needed and so the particular rules will be easily invented. In this section we just describe minimum requirements of the system and give examples of the sorts of rules which might be usefully added for certain purposes.

We will need to quantify over both programs and multisets. So both these types of objects are supposed to be in the domain along with the data objects.

Any formalization of the data domain will be application specific. We often need a 1-ary relation **is_an_object** to pick out the data objects from amongst the other objects. If the object domain is just the natural numbers then arithmetic and ordering relations are often useful.

For reasoning about multisets we will need a 1-ary relation **is_a_multiset** to identify the multisets. It is also useful to have a 2-ary function **cons** to build a new multiset by adding an object to an old one. We can identify the empty multiset as **empty_ms** and use relation symbols **subms**, **add** and **remove** to represent the submultiset relation, and relations of adding and subtracting multisets.

Many applications need natural numbers to be in the domain, even if the actual object domain is something completely different. These are quantified over and related to various properties of multisets. We will often need arithmetic. For example, we may need to suppose that there is a 2-ary relation **size** relating multisets and numbers. It will be supported by axioms such as

$$\mathbf{size}(\mathbf{empty\_ms}, 0)$$

and

$$\forall x, d, n.(\mathbf{is\_an\_object}(d) \wedge \mathbf{size}(x, n)) \Rightarrow \mathbf{size}(\mathbf{cons}(d, x), n + 1).$$

For reasoning about program syntax we are going to use the 1-ary relations **is_a_program** and **rule** and 2-ary functions $+$ and $\circ$ with the obvious intended interpretations. For particular applications we may need to know such things as $\forall p, q, r, s.(p + q \neq r \circ s)$ and $\forall p, q, r.\ (\mathbf{rule}(r) \Rightarrow r \neq p + q)$, etc.

7.7.2 *Formal reasoning about one step*

In order to reason about an individual step of the Gamma program we need to have proof rules about rewrite rules and about the relations Γ and Ω.

Rewrites There are several alternative approaches to expressing formally the condition and action of a particular rewrite rule. Here we just suppose that there is a 3-ary relation **rewrite** which holds of triple (p, m, m') if and only if p represents a rewrite rule which can change the multiset m into the multiset m' in one step. Then we can just use a few constant symbols to stand for particular rewrite rules.

For example, if constant p is used to represent the rule

$$(x, y, z \rightarrow x, y \Leftarrow ((x \geq z) \wedge (y \geq z)))$$

then we may define

$$\begin{array}{ll} \forall m, m', n, x, y, z. & \\ [\quad \mathbf{rewrite}(p, m, m') & \\ \quad \Leftrightarrow \mathbf{subms}(\mathbf{cons}(x, \mathbf{cons}(y, \mathbf{cons}(z, \mathbf{empty\_ms}))), m) & \\ \quad\quad \wedge (x \geq z) \wedge (y \geq z) & \\ \quad\quad \wedge \mathbf{remove}(\mathbf{cons}(x, \mathbf{cons}(y, \mathbf{cons}(z, \mathbf{empty\_ms}))), m, n) & \\ \quad\quad \wedge \mathbf{add}(\mathbf{cons}(x, \mathbf{cons}(y, \mathbf{empty\_ms})), n, m') &]. \end{array}$$

Rules for Γ *and* Ω Our definition of Γ in Section 7.5 is a recursive one so it is not clear how it can be given a first-order definition or, indeed, if that is possible. On the other hand, it is straightforward to define Ω in terms of Γ:

$$\text{OMEGA:} \qquad \Omega(P, M) = \neg\exists q.\exists n.\Gamma(P, M, q, n).$$

To define Γ itself we first introduce a formula in the first-order language of programs and multisets along with a new 4-ary relation symbol G.

Define

$$\begin{aligned}
& X_G(r, m, r', m') \\
\equiv\ & [\text{rule}(r) \wedge \text{rewrite}(r, m, m') \wedge (r' = r)] \\
& \vee [\exists p, q, p'.(r = p + r) \wedge G(p, m, p', m') \wedge (r' = p' + q)] \\
& \vee [\exists p, q, q'.(r = p + q) \wedge G(q, m, q', m') \wedge (r' = p + q')] \\
& \vee [\exists p, q, q'.(r = p \circ q) \wedge G(q, m, q', m') \wedge (r' = p \circ q')] \\
& \vee [\exists p, q.(r = p \circ q) \wedge (m = m') \wedge (r' = p) \wedge \neg\exists q', k.G(q, m, q', k)].
\end{aligned}$$

Then we know that

$$\text{GAM:} \qquad \forall r, m, r', m'.[\Gamma(r, m, r', m') \leftrightarrow X_\Gamma(r, m, r', m')].$$

For the purposes of formal proof theory this turns out to be an adequate rule for reasoning about the relation Γ. Whether it is an adequate formal definition is another question.

A fixed point definition of Gamma The rule GAM is an adequate rule for reasoning about Γ but it does also actually act as a formal definition of Γ. It is a very straightforward induction on the construction of Gamma programs to show that there is only one relation between programs and multisets which satisfies the formula GAM: namely, the relation Γ defined in Section 7.5. This is true whatever fixed set of rewrite rules is assumed and whatever data object domain is chosen.

Of course, an equivalence which, like GAM, mentions the new relation symbol on both sides of the $\Leftrightarrow$ is not generally adequate as a complete definition of a new predicate. One need only consider the equivalence

$$\forall p, m, p', m',\ [G(p, m, p', m') \Leftrightarrow G(p, m, p', m')]$$

to see this.

In fact Beth's definability theorem says that if an equivalence does happen to define a new relation then there is a first-order formula in the original language which also defines that new relation. Beth's theorem is not applicable to the case of Γ, however, and we cannot give a first-order formula to define Γ because the equivalence GAM is not a general definition but only works in the standard model of multisets (i.e. in which all multisets are of finite size).

For our purposes in this chapter it is not necessary to give a formal definition of Γ in terms of the formal language of multisets and programs which we have introduced in the last section. We have a rigorous mathematical definition of Γ (in Section 7.5) to specify exactly the behaviour of Gamma programs and we have a proof rule GAM which is adequate for formal reasoning about the formal rendering of that relation.

However, it is not too difficult to give a formal definition for the sake of completeness. We have seen above that a formal version of the statement 'Γ is the solution to the equivalence GAM' would be an adequate definition.

Instead of devising a formal language just for this one definition, we can use a more general technique. We can turn to such fixed point languages as those proposed to extend first-order logic in the paper by Gurevich and Shelah [1986]. Because such equivalences as GAM may, as we have seen, have many solutions (or none), Gurevich and Shelah suggest limiting the equivalences to those of the form

$$\forall \overline{x}.\ [G(\overline{x}) \Leftrightarrow Y(G, \overline{x})]$$

where the relation symbol G only appears positively (i.e. nested under an even number of negation symbols) in the formula Y. Under these conditions (and supposing that $\overline{x}$ is an n-tuple) the operator

$$\begin{aligned} &F : D^n \to D^n \\ &\overline{d} \in (F\sigma) \text{ iff } D \models Y(\sigma, \overline{d}) \end{aligned}$$

will be monotone for any first-order structure D in the language of Y. Thus F will have a least fixed point in D^n. In this way we can define the term $LFP_G Y \overline{x}$ to pick out an n-ary relation on any such D. In this paper [Gurevich and Shelah, 1986] we also see that nesting one fixed point operator inside another is legitimate.

In using this language to define Γ we are faced with the problem that the natural defining formula X_G contains a negative appearance of G. The easiest way around this problem is to use one fixed point operator to define Ω and then to define Γ in terms of Ω. The result is

$$\begin{aligned} \Omega(p, m) = \ LFP_Z[\quad & (\mathbf{rule}(p) \wedge \neg \exists m'.\ \mathbf{rewrite}(p, m, m')) \\ & \vee (\exists r, s.\ ((p = r + s) \wedge Z(r, m) \wedge Z(s, m)))] \end{aligned}$$

and

$$\begin{aligned} &\Gamma(p, m, p', m') = \\ &LFP_G[\quad [\text{rule}(r) \wedge \text{rewrite}(r, m, m') \wedge (r' = r)] \\ &\qquad \vee [\exists p, q, p'.(r = p + r) \wedge G(p, m, p', m') \wedge (r' = p' + q)] \\ &\qquad \vee [\exists p, q, q'.(r = p + q) \wedge G(q, m, q', m') \wedge (r' = p + q')] \\ &\qquad \vee [\exists p, q, q'.(r = p \circ q) \wedge G(q, m, q', m') \wedge (r' = p \circ q')] \\ &\qquad \vee [\exists p, q.(r = p \circ q) \wedge (m = m') \wedge (r' = p) \wedge \Omega(p, m)]]. \end{aligned}$$

It is easy to see that these definitions give us the desired relations on the standard model of multisets and programs: if Γ is the only fixed point of the corresponding function then it is the least fixed point.

Table 7.1 *Some proof rules for propositional temporal logic.*

STLref: $\vdash \Box F \Rightarrow F$	STLgen: $\dfrac{\vdash \phi}{\vdash \Box\phi}$
STLtrans: $\vdash \Box\phi \Rightarrow \Box\Box\phi$	STLdistr: $\vdash \Box(\phi \Rightarrow \psi) \Rightarrow (\Box\phi \Rightarrow \Box\psi)$

7.7.3 *Formal reasoning about behaviours*

The various formal systems we have introduced in the last few sections are really just to play a supporting role in the formal proof of a program property. In general we will be trying to show that $\vdash \Phi \Rightarrow \Psi$ where Φ is the TLA description of the behaviour of Gamma along with some initial conditions and Ψ is a TLA formula expressing the desired correctness property. Thus, we are to show that a particular formula is a tautology in the TLA logic. In this section we will introduce rules for establishing temporal tautologies and for reducing the question of whether a TLA formula is a tautology onto a question framed in one of the supporting formal systems.

TLA abbreviations One preliminary task is to decode the various TLA abbreviations. From Section 7.1, we know what $[\mathcal{A}]_f$ and $\langle\mathcal{A}\rangle_f$ are. We should also be clear that for any predicate X, the abbreviation X' stands for X with each appearance of each program variable V replaced by V'.

We have given a semantic definition of the predicate Enabled$\mathcal{A}$. To reason with it we need to know that if $\overline{V}$ are the program variables which appear in $\mathcal{A}$ in either unprimed or primed form, then Enabled$\mathcal{A}$ is just $\exists\overline{w}.\ \mathcal{B}$ where $\mathcal{B}$ is got from $\mathcal{A}$ by replacing each appearance of V_i' by w_i. Here, the variables $\overline{w}$ are just rigid variables.

Simple temporal logic Much of the temporal reasoning needed is just simple, usually propositional, temporal logic. There are many published versions of complete proof systems for such logics. See, for example, Volume 1.

In Table 7.1 we give a selection of such rules for future reference.

TLA rules As well as these rules, Lamport [Lamport, 1994] gives some special rules for dealing with TLA specific constructs such as actions and primed variables.

For example,

$$\text{TLA1:} \qquad \vdash \Box P \equiv P \wedge \Box[P \Rightarrow P']_P.$$

The collection of such rules provides a relatively complete proof system but the rules are sometimes a little low level to support comfortable reasoning. Thus Lamport supplies some additional rules tailored for reasoning about program properties when the program is described via his template.

New TLA rule Since we have strayed slightly from the template in introducing the new abbreviation VWF in order to describe Gamma, it is worth supplying a new proof rule for use with this abbreviation.

The new rule appropriate for very weak fairness is:

VWF1:

$$\begin{array}{l} P \wedge [\mathcal{N}]_f \Rightarrow (P' \vee Q') \\ P \wedge \langle \mathcal{N} \wedge \mathcal{A} \rangle_f \Rightarrow Q' \\ P \wedge \text{Enabled}(\mathcal{N} \wedge \mathcal{A} \wedge (f = f')) \Rightarrow Q \\ P \Rightarrow \text{Enabled}\mathcal{A} \\ \hline \Box[\mathcal{N}]_f \wedge \text{VWF}_f(\mathcal{A}) \Rightarrow \Box(P \rightarrow \Diamond Q). \end{array}$$

It is quite easy to show that this rule is sound. The rule simply expresses the following fact. Suppose that:

- we are considering a behaviour which only evolves through action $\mathcal{N}$ but is weakly fair towards action $\mathcal{A}$;
- if property P holds before action $\mathcal{N}$ then either P will continue to hold or Q will hold;
- a non-stuttering version of $\mathcal{A} \wedge \mathcal{N}$ will convert a state satisfying property P into one satisfying property Q;
- if a stuttering $\mathcal{A} \wedge \mathcal{N}$ is possible then Q holds anyway;
- $\mathcal{A}$ is enabled when P is true.

Then, it is clear that property Q must eventually hold if ever P holds.

Many program properties can be established via this line of reasoning and we demonstrate the usefulness of this rule in the example in the next section.

Similarly, a new rule appropriate for almost strong fairness is:

ASF1:

$$\begin{array}{l} P \wedge [\mathcal{N}]_f \Rightarrow (P' \vee Q') \\ P \wedge \langle \mathcal{N} \wedge \mathcal{A} \rangle_f \Rightarrow Q' \\ P \wedge \text{Enabled}(\mathcal{N} \wedge \mathcal{A} \wedge (f = f')) \Rightarrow Q \\ \Box P \wedge \Box[\mathcal{N}]_f \wedge \Box F \Rightarrow \text{Enabled}\mathcal{A} \\ \hline \Box[\mathcal{N}]_f \wedge \text{ASF}_f(\mathcal{A}) \wedge \Box F \Rightarrow \Box(P \rightarrow \Diamond Q). \end{array}$$

It is also quite easy to show that this rule is sound.

7.8 An example

Let us look briefly at using such rules in reasoning about a Gamma program. Suppose that

$$p : (x, y, z \to x, y \Leftarrow ((x \geq z) \wedge (y \geq z))$$
$$q : (x, y \to xy \Leftarrow \text{True}))$$

and that we wish to use the program $q \circ p$ for finding the product of the largest two numbers in a multiset of numbers. We wish to use the TLA description of Gamma to prove the total correctness of this program.

We introduce the predicates:

- $\mathit{largest\_two}(a, b, m)$ = a and b are the largest elements in the multiset m;
- $\mathbf{size}(m, n)$ = the multiset m contains exactly n elements; and
- $B(a, b, m)$ = a and b are the only elements in the multiset m.

These are assumed to be defined in terms of basic predicates for multisets.

Using these the formalization of the total correctness property is

$$\begin{aligned} \forall a, b, m_0, p_0.\ [\ & \Phi \wedge \mathbf{size}(m_0, n_0) \wedge (2 \leq n_0) \\ & \wedge \mathit{largest\_two}(a, b, m_0) \wedge (p_0 = q \circ p) \\ \Rightarrow\ & \Diamond((P = \text{'finished'}) \wedge (M = \mathbf{cons}(a * b, \mathbf{empty\_ms})))]. \end{aligned}$$

We also define the following predicates:

- $X_n = \mathit{largest\_two}(a, b, M) \wedge \mathbf{size}(M, n) \wedge (P = q \circ p)$; and
- $Y = B(a, b, M) \wedge (P = q)$.

Our proof will proceed by showing that:

- X_{n_0} being true always will eventually lead to X_2 holding:

$$\vdash \Phi \Rightarrow \Box(X_{n_0} \Rightarrow \Diamond X_2);$$

- and that X_2 holding will eventually lead to Y holding:

$$\vdash \Phi \Rightarrow \Box(X_2 \Rightarrow \Diamond Y);$$

- and this will eventually lead to the multiset containing just the product of a and b and the program being finished:

$$\vdash \Phi \Rightarrow \Box(Y \to \Diamond((M = \mathbf{cons}(a * b, \mathbf{empty\_ms})) \wedge (P = \text{'finished'}))).$$

The final part of the proof is putting these results together: this part is simple temporal logic.

7.8.1 *A proof in temporal logic*

The most complicated of these parts is the first. We will look more closely at it. For each natural number $n \geq 2$ define:

$$\beta_n = \Phi \Rightarrow \Box(X_n \Rightarrow \Diamond X_2).$$

We establish β_{n_0} using the induction rule after first showing $\vdash \beta_2$ (by a straightforward use of the proof rules for simple temporal logic) and that $\vdash \beta_{n+1} \Rightarrow \beta_n$ for any $n \geq 2$.

We show $\vdash \beta_2$ by a straightforward use of the proof rules for simple temporal logic. By STLref we get $\vdash \Box(\neg\chi) \Rightarrow (\neg\chi)$. By propositional logic (effectively taking the contrapositive) we easily get $\vdash \chi \Rightarrow \neg\Box(\neg\chi)$ which is abbreviated by $\vdash \chi \Rightarrow \Diamond\chi$. The rule STLgen and propositional logic (with TLA formulae substituted for atoms) immediately give us β_2.

To show that $\beta_{n+1} \Rightarrow \beta_n$ we just need to use the fact that $\vdash \Phi \Rightarrow \Box(X_{n+1} \Rightarrow \Diamond X_n)$ (which we prove in the next subsection) along with a simple temporal logic proof via

$$\vdash \Box(a \Rightarrow \Diamond b) \wedge \Box(b \Rightarrow \Diamond c) \Rightarrow \Box(a \Rightarrow \Diamond c).$$

This latter temporal tautology follows from the transitivity rule for $\Box$ (STLtrans) along with the distributivity of $\Box$ over $\Rightarrow$ (STLdistr). It also needs the rule of temporal generalization (STLgen) as well as plenty of propositional logic and *modus ponens.*

We do not present this proof because, although it is a good example of straightforward propositional temporal logic, it is quite long and technical. Complete proof systems for propositional temporal logic and their uses are well known and well discussed. See, for example, Volume 1.

7.8.2 *A proof in TLA*

So our proof now hinges on showing

$$\vdash \Phi \Rightarrow \Box(X_{n+1} \Rightarrow \Diamond X_n)$$

for any $n \geq 2$. To do this we can most conveniently use the new TLA rule VWF1. If we prove the four premises:

1. $\vdash X_{n+1} \wedge [\mathcal{S}]_{(P,M)} \Rightarrow (X'_{n+1} \vee X'_n)$
2. $\vdash X_{n+1} \wedge \langle \mathcal{S}_1 \rangle_{(P,M)} \Rightarrow X'_n$
3. $\vdash X_{n+1} \wedge \text{Enabled}(\mathcal{S}_1 \wedge (P = P') \wedge (M = M')) \Rightarrow X_n$
4. $\vdash X_{n+1} \Rightarrow \text{Enabled}(\mathcal{S}_1)$

then we will have exactly the desired conclusion.

The four separate arguments needed are similar in some respects with the only novelty being the translation of Enabled($\mathcal{S}_1$) into

$$\exists q, k.\ \Gamma(P, M, q, k)$$

and a similar translation in the third rule.

We will only illustrate the second argument in detail. Translating away all the TLA abbreviation leaves us with the task of showing $\vdash \alpha \Rightarrow \beta$ where

$$\begin{aligned}\alpha = \textit{largest\_two}(a, b, M) \wedge \mathbf{size}(M, n+1) \wedge (P = q \circ p)\\ \wedge\, \Gamma(P, M, P', M') \wedge ((P \neq P') \vee (M \neq M'))\end{aligned}$$

and

$$\beta = \textit{largest\_two}(a, b, M') \wedge \mathbf{size}(M', n) \wedge (P' = q \circ p).$$

7.8.3 *A proof about* Γ

One use of the rule GAM along with propositional logic tells us that $\vdash \alpha \Rightarrow (\alpha_1 \vee \alpha_2 \vee \alpha_3)$ where

$$\begin{aligned}\alpha_1 = {} & \mathit{largest\_two}(a,b,M) \wedge \mathbf{size}(M,n+1) \wedge (P = q \circ p) \\ & \wedge \exists p'.\ (\Gamma(p,M,p',m') \wedge (P' = q \circ p')) \wedge (P \neq P');\end{aligned}$$

$$\begin{aligned}\alpha_2 = {} & \mathit{largest\_two}(a,b,M) \wedge \mathbf{size}(M,n+1) \wedge (P = q \circ p) \\ & \wedge (P \neq P') \wedge (M = M') \wedge (P' = q) \wedge \neg\exists p',k.\ [\Gamma(p,M,p',k)];\end{aligned}$$

$$\begin{aligned}\alpha_3 = {} & \mathit{largest\_two}(a,b,M) \wedge \mathbf{size}(M,n+1) \wedge (P = q \circ p) \\ & \wedge \exists p'.\ (\Gamma(p,M,p',M') \wedge (P' = q \circ p')) \wedge (M \neq M').\end{aligned}$$

The argument proceeds by showing that α_1 and α_2 are contradictory but that $\vdash \alpha_3 \Rightarrow \beta$. Each of these three arguments uses GAM again before becoming an argument about multiset sizes. We just concentrate on showing $\vdash \alpha_3 \Rightarrow \beta$.

Using the axiom **rule**(p) along with GAM we establish $\vdash \alpha_3 \Rightarrow \alpha_4$ where

$$\begin{aligned}\alpha_4 = {} & \mathit{largest\_two}(a,b,M) \wedge \mathbf{size}(M,n+1) \\ & \wedge \mathbf{rewrite}(p,M,M') \wedge (P' = q \circ p) \wedge (M \neq M').\end{aligned}$$

7.8.4 *A proof about rewrites and multiset sizes*

Finally we must show that $\vdash \alpha_4 \Rightarrow \beta$. Now it is time to bring in the definition of **rewrite**(p,m,m') in terms of the basic multiset operations and orderings on data objects. We have given this definition as an example in Section 7.7.2. A straightforward argument using the formal representations establishes

$$\begin{aligned}\vdash \quad & (x \geq z) \wedge (y \geq z) \\ & \wedge \mathbf{remove}(\mathbf{cons}(x, \mathbf{cons}(y, \mathbf{cons}(z, \mathbf{empty\_ms}))), M, k) \\ & \wedge \mathbf{add}(\mathbf{cons}(x, \mathbf{cons}(y, \mathbf{empty\_ms})), k, M') \\ & \wedge \mathbf{size}(M, n+1) \wedge \mathit{largest\_two}(a,b,M) \\ \Rightarrow \quad & (\mathbf{size}(M', n) \wedge \mathit{largest\_two}(a,b,M')),\end{aligned}$$

i.e. a p rewrite reduces the size of a multiset by one but keeps its two largest elements the same.

This completes a formal proof of the total correctness of the program.

8

INTERVALS AND PLANNING

8.1 Introduction

As described in [van Benthem, 1995], there are many and various motivations for using an interval temporal logic, which include philosophical considerations of time and events, natural language, processes in computations and planning problems. In this chapter we will motivate the study of interval temporal logics by glancing at philosophical arguments, seeing some of the interesting mathematics which underlies questions of expressiveness, decidability and axiomatic completeness for such logics and considering their application to planning problems. We will also see reasons why interval temporal logics have not become very popular: these include undecidability and a lack of expressive completeness.

The basic idea of using an interval temporal logic is that certain activities can only meaningfully be said to occur over an extended period of time. Examples of such activities, sometimes called *accomplishments*, include obtaining a doctorate, lifting a block onto a table, and crossing a road. To say that someone obtains a doctorate over the period i is not the same as saying that at each moment of i, that person was obtaining a doctorate. To understand some of the complications in trying to force this accomplishment into a point-based semantics, one need only consider the (admittedly rare) case of a student obtaining one doctorate in the period January 1997 to June 1999 and another in the period December 1998 to December 2002.

As mentioned above, there are problems with reasoning about the complications of overlapping intervals etc. Fortunately, for constraint problems in planning it may be sufficient just to consider networks of intervals with each pair related by some subset of the 13 possible basic relations which were introduced in [Allen, 1981] and which we will meet below. However, for more sophisticated reasoning about intervals an interval temporal logic such as that in [Halpern and Shoham, 1986] is better suited. We turn to the full logic first.

8.2 Interval logics

Although we expect to find truth of propositions and formulae evaluated on extended periods of time in this chapter, there are many possible variations on this theme. As usual, a temporal structure will consist of a set (here, a set of intervals or periods), relations between them and a valuation of the propositional atoms on them. We will see that there are very basic decisions to make about what constitutes a set of intervals, about what are the natural relations between intervals and about any restrictions there should be on the valuation of atoms.

The motivations for the actual decisions made can stem from philosophical considerations on the nature of time, from linguistic considerations on the nature of periods of time over which events take place, or from a consideration of what is convenient, natural or appropriate in any particular application from computer science or AI.

First consider the nature of intervals. It is becoming usual in computer science or AI to assume that intervals are just closed intervals defined by a pair of points from some underlying linear flow of time points. This is the approach we will take. However, it is far from ubiquitous. In both philosophical studies and some earlier computer science approaches, discussion on the nature of events often led to the suggestion of periods of time as the basic ontological unit rather than points. Points could either be defined as very special intervals (satisfying certain properties) or be outlawed as imaginary objects.

Next consider what would count as natural relations between intervals in an analogous way to the relation of precedence between points. Of course, it is possible to propose precedence between intervals as the natural relation. But it is also possible to propose a large set of other possible relations as important. For example:

- K is in between I and J iff J follows K which follows I;
- I meets J iff J follows I but there is no interval K in between I and J;
- I starts J iff there are intervals K, A and B such that A meets both I and J, I meets K, both K and J meet B.

There is a useful discussion of alternative relations in [van Benthem, 1983]. There, the relations of precedence and inclusion are chosen as basic.

The choice of natural relations between intervals is vital for following the intervals as basic objects approach. Once relations are chosen then the followers of this approach will need to give an account of what counts as an interval structure in terms of axioms or restrictions on the chosen relations. See [van Benthem, 1983] for an example of this.

If, on the other hand, intervals are taken to be objects defined in terms of points in a point structure, then the definition of an interval structure reduces to declaring the nature of the underlying point structure and the definition of which sets of points count as intervals. Then it should be possible to define many useful relations between intervals in terms of their constituent points. In this chapter we suppose that we are dealing with an underlying linear order of points. This may be an uneccessary restriction but it helps with presenting the material. Notice that we have still remained quite general and allow flows of time which are discrete, dense, Dedekind complete or not.

We have said that we will define intervals to be given by pairs of points from the underlying flow. We allow the pair (s, t) provided that $s \leq t$. We will take the pair (s, t) to represent the closed interval $(s, t) = \{x \in T | s \leq x \leq t\}$. Variations on the logic arise by allowing only open intervals (e.g. then there are no points) or allowing any convex set of points (e.g. in the rationals there are then intervals with no starting point). Many of these choices are discussed in [van Benthem, 1983] and [Marx and Venema, 1997].

In the rest of the chapter we suppose that $(T, <)$ is a linear order of points. We define the set of intervals on $(T, <)$ as $\mathcal{I}(T, <) = \{(s, t) \in T \times T | s \leq t\}$. For now, we can regard this as an interval frame by assuming implicitly that we can determine the start and end points of any given intervals in $\mathcal{I}(T, <)$ as well as the precedence ordering between the points. It is well known and proved in [Allen, 1981] and [van Benthem, 1983] that there are 13 different relations which represent various ways in which two intervals of time can relate to each other in a linear flow of time:

(a, b) equals (c, d) iff $a = c$ and $b = d$

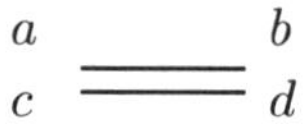

(a, b) precedes (c, d) iff $b < c$

(a, b) meets (c, d) iff $b = c$

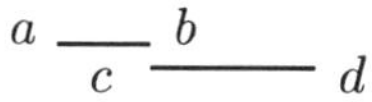

(a, b) overlaps (c, d) iff $a < c < b < d$

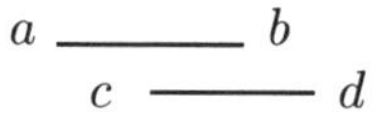

(a, b) is during (c, d) iff $c < a \leq b < d$

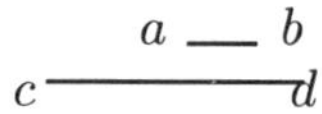

(a, b) starts (c, d) iff $c = a \leq b < d$

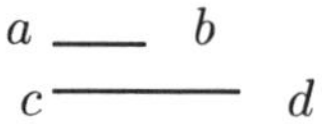

(a, b) ends (c, d) iff $c < a \leq b = d$

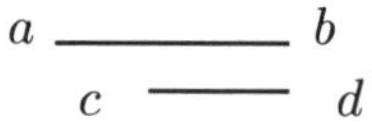

as well as the converses of the last six; for example, (a, b) is met by (c, d) iff $a = d$.

In planning problems it is sometimes sufficient just to reason with these relations, using Allen's interval algebra. However, for more general reasoning we need a modal logic. Thus we now turn to the question of valuations.

Given a set of propositional atoms, the most general approach—and the one which we will use—is to allow any map V from $\mathcal{I}(T, <)$ into the power set of atoms as a valuation. In certain applications it is natural to impose restrictions. A structure $(\mathcal{I}(T, <), V)$ is called *homogeneous* iff propositions hold on intervals exactly when they hold at all subintervals. Sometimes this is a natural assumption—for example, if the propositions relate to state information—but sometimes it is an unhelpful assumption—for example, if propositions relate to achievements such as Ruth walking to work.

The assumption of *solidity* on a proposition, i.e. that it does not hold of any two overlapping intervals, is also useful in certain planning and database applications. See also [Pratt and Francez, 1997] for a use in natural language semantics.

Another possible restriction is that of *flatness* or *locality*. This is that for each proposition, its truth agrees on any two intervals which share a common start point. We will see below that this assumption is important for some technical results about interval logics.

We now turn to consider a temporal logic of intervals. We have already met several interesting relations which can hold between intervals in an interval structure. Each of these gives rise to a modal diamond in the usual way. Different logics arise by choosing different diamonds as the basic modalities; see, for example, the logics in [Humberstone, 1979], [van Benthem, 1983] and [Allen and Hayes, 1985].

We shall look more closely at the logic, which we call HS logic, in [Halpern and Shoham, 1986] whose modalities include:

$\langle A \rangle$: at some interval beginning immediately after the end of the current one;

$\langle B \rangle$: at some interval during the current one, beginning when the current one begins; and

$\langle E \rangle$: at some interval during the current one, ending when the current one ends;

and their converses $\langle A^{\smile} \rangle, \langle B^{\smile} \rangle$ and $\langle E^{\smile} \rangle$.

The formal definitions for an interval structure $X = (\mathcal{I}(T, <), V)$ are as follows:

$$X, (s,t) \models \langle A \rangle \phi \text{ iff there is } u \in T \text{ such that } s \leq t \leq u \text{ and } X, (t,u) \models \phi;$$
$$X, (s,t) \models \langle B \rangle \phi \text{ iff there is } u \in T \text{ such that } s \leq u < t \text{ and } X, (s,u) \models \phi;$$
$$X, (s,t) \models \langle E \rangle \phi \text{ iff there is } u \in T \text{ such that } s < u \leq t \text{ and } X, (u,t) \models \phi.$$

We can express such things as:

- 'p is true at all subintervals of the current interval' as $[E][B]p$;
- 'the current interval is an instant' as $[E][B]\bot$.

8.3 Intervals and a compass logic

Here we compare interval logics with a two-dimensional temporal logic. The most straightforward two-dimensional temporal logic is the compass logic introduced by Venema in [Venema, 1990] and [Venema, 1992]. The language contains four interrelated modal diamonds: $\Diamond_{\uparrow}$, $\Diamond_{\downarrow}$, $\Diamond_{\rightarrow}$ and $\Diamond_{\leftarrow}$. Structures for this language consist of two linear orders $(T_1, <_1)$ and $(T_2, <_2)$; we shall call such a pair a *rectangular frame.* Two-dimensional valuations for atoms are made at ordered pairs from $T_1 \times T_2$. We can think of $(T_1, <_1)$ as lying vertically and $(T_2, <_2)$ as lying horizontally on a Cartesian grid. The truth of formulae is also defined at ordered pairs in a natural way. For example:

- $(T_1, <_1, T_2, <_2, g), t_1, t_2 \models p$ iff $p \in g(t_1, t_2)$;
- $(T_1, <_1, T_2, <_2, g), t_1, t_2 \models \Diamond_{\uparrow} A$ iff there is $s_1 \in T_1$ such that $t_1 <_1 s_1$ and $(T_1, <_1, T_2, <_2, g), s_1, t_2 \models A$.

It is useful to define some abbreviations including the corresponding universal modalities and some Boolean combinations of the basic modalities. For example:

$$\Box_{\uparrow} A \equiv \neg \Diamond_{\uparrow} \neg A$$
$$\Box A \equiv A \wedge \Box_{\rightarrow} A \wedge \Box_{\leftarrow} A \wedge \Box_{\uparrow} (A \wedge \Box_{\rightarrow} A \wedge \Box_{\leftarrow} A) \wedge \Box_{\downarrow} (A \wedge \Box_{\rightarrow} A \wedge \Box_{\leftarrow} A)$$
$$\Diamond_{\leftrightarrow} A \equiv \Diamond_{\leftarrow} A \vee A \vee \Diamond_{\rightarrow} A.$$

Notice that this notation extends the appealingly intuitive geographical analogy suggested by the notation for a modal two-dimensional logic in [Segerberg, 1973]. The intuition further suggests another possible application of this logic to the field of spatial reasoning. Although there are modal logics for spatial reasoning (such as the logic of convex hulls in [Bennett, 1996]), we know of no investigation of the use of modalities for compass directions in this field.

An example of the kind of statement one can make in the logic is

$$\Diamond_{\uparrow} \Box_{\rightarrow} A \rightarrow \Box_{\rightarrow} \Diamond_{\uparrow} A,$$

which is actually a validity, where by a validity we mean here a formula which is true at every ordered pair in every rectangular structure. A *satisfiable* formula, on the other hand, is a formula ϕ for which there exists some rectangular structure $\mathcal{T} = (T_1, <_1, T_2, <_2, g)$ and some pair (t_1, t_2) such that

$$\mathcal{T}, t_1, t_2 \models \phi.$$

Just to be clear, we also define semantic consequence by

$$\Gamma \models A$$

iff $\mathcal{T}, t_1, t_2 \models A$ whenever we have $\mathcal{T}, t_1, t_2 \models \gamma$ for all $\gamma \in \Gamma$.

As suggested in [Venema, 1990], it is rewarding to notice that interval temporal logics are closely related to two-dimensional temporal logics. We can use the compass language to describe interval structures. Then the interval logic is much the same as the compass logic with a diagonal but with half the points missing.

We turn $\mathcal{I}(T, <)$ into (half a) two-dimensional structure $(T, <, T, <, g)$ for the compass language with the diagonal constant δ by the following moves:

- truth of formulae and the valuation of atoms only takes place at pairs (s, t) with $s \leq t$;
- $p \in g(t_1, t_2)$ iff $t_1 \leq t_2$ and $p \in h(t_1, t_2)$, for atoms p;
- $(T, <, T, <, g), s, t \models \delta$ iff $s = t$.

Then interval properties can be expressed in compass logic:

- 'all subintervals satisfy p' is $\Box_{\rightarrow} \Box_{\downarrow} \, p$.

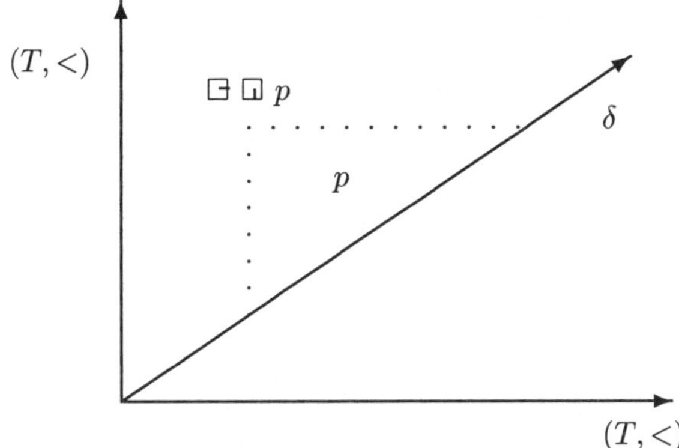

- 'any interval strictly in the future satisfies p' is $\Diamond_{\rightarrow} (\delta \wedge \Box_{\uparrow} \Box_{\rightarrow} (\neg \Diamond_{\leftarrow} \delta \rightarrow p))$.

In fact Venema shows that all the formulae of Halpern and Shoham's logic can be easily translated:

$\langle B \rangle p$ translates to $\Diamond_{\downarrow} \, p$;
$\langle \overline{B} \rangle p$ translates to $\Diamond_{\uparrow} \, p$;
$\langle E \rangle p$ translates to $\Diamond_{\rightarrow} \, p$;
$\langle \overline{E} \rangle p$ translates to $\Diamond_{\rightarrow} \, p$;
$\langle A \rangle p$ translates to $\Diamond_{\rightarrow} (\delta \wedge \Diamond_{\uparrow} \, p)$;
$\langle \overline{A} \rangle p$ translates to $\Diamond_{\downarrow} (\delta \wedge \Diamond_{\leftarrow} \, p)$.

More complicated formulae are translated recursively.

8.4 Expressive power

We may ask whether the HS logic is expressive enough for practical purposes. In [Venema, 1991], Venema describes a useful construct which is not able to be expressed in HS logic. This is the two-place *chop* operator C which is defined by

$$\begin{aligned} \mathcal{T}, t, u \models \phi C \psi \text{ iff } & \exists v \text{ such that } t \leq v \wedge v \leq u, \\ & \mathcal{T}, t, v \models \phi \text{ and } \mathcal{T}, v, u \models \psi. \end{aligned}$$

This operator would seem to have application in natural language semantics (as 'and–then'), program semantics (as sequential composition), planning and be a generally useful composition construct in two-dimensional logics.

Recall from Volume 1 that the way to answer general questions of expressive adequacy is to use the concept of expressive completeness. Roughly, this means that we can find an inverse translation to the usual one from temporal or modal languages to first-order ones. The situation here is slightly more complicated than in the case of point-based temporal logics because the natural first-order language to use as a yardstick is a language of points with the relation $<$. The complication is then that instead of the usual monadic language with $<$, we must use a dyadic language. For example, if the atom p, evaluated at (s, t), translates to the dyadic predicate $P(s, t)$, then the formula $\langle B \rangle p$ translates to

$$\exists x((s < x) \wedge (x < t) \wedge P(s, x)).$$

We would say that an interval logic is expressively (or functionally) complete iff for every formula $\phi(s, t)$ of the first-order language using dyadic predicates $<, P, Q, \ldots$, there is a formula in the language (using atoms $p, q, \ldots$) which translates to $\phi(s, t)$ (under the usual semantic-preserving translation). The truth, in the case of dense-time interval logics, is shown in [Venema, 1990] to be no: there is no expressively complete interval temporal language.

In Chapters 12 and 13 of Volume 1 we saw the close connection between expressive completeness results and various bounded pebble Ehrenfeucht–Fraïssé games. The non-expressive completeness result in [Venema, 1990] is another good example of this connection. Venema shows that allowing the use of extra pebbles always allows extra games to win. By the general connection, this effectively means that allowing extra connectives (with their own internal stock of bound variable symbols in dyadic translation) allows more structures to be distinguished.

In another interesting result of Venema's in [Venema, 1994], it is shown that it is possible to find an expressively complete set of two-dimensional modal operators if one restricts to *flat* evaluations. These are evaluations in which the truth of propositions only depends on the first coordinate, i.e. p will be true at (s, t) iff it is true at (s, u). Extending this result to the interval case—in which flatness means dependence on the beginning point—should be possible. For practical purposes, the question is whether flatness is a useful restriction for any applications of interval reasoning.

8.5 Undecidability of HS

We show that deciding validity in HS logic is undecidable. In fact we show that it is co-r.e.-hard.

With many two-dimensional temporal logics it is usually quite straightforward to show undecidability of validity using the domino or tiling technique. The technique was first applied to two-dimensional logic and other logics of programs in [Harel, 1983] where many different tiling techniques are used to establish various levels of undecidability of the logics. Other techniques such as coding of Turing machine runs (see [Halpern and Shoham, 1986] or [Halpern and Vardi, 1989]) can be used but the tiling approach is very natural in this context. In [Marx and Reynolds, 1997] it is shown that the compass logic itself is undecidable. The authors use tiling but the proof is not completely straightforward. A much better simple demonstration of tiling in action can be gained from considering the two-dimensional temporal logic X^2 with 'next' operators ⊖ , ⦶ , ⊝ and ⦷ as well as the compass operators used to describe structures where both dimensions of time are the natural numbers. This is very similar to the two-dimensional logic actually considered in [Harel, 1983].

We fix some denumerable set C of *colours*. *Tiles* are 4-tuples of colours and we define four projection maps so that each tile $\tau = (\text{left}(\tau), \text{right}(\tau), \text{up}(\tau), \text{down}(\tau))$. We say of a finite set T of tiles that T tiles $\mathbb{N} \times \mathbb{N}$ iff there is a map $\rho : \mathbb{N} \times \mathbb{N} \to T$ such that for each $i, j \in \mathbb{N}$,

- $\text{up}(\rho(i,j)) = \text{down}(\rho(i, j+1))$
- and $\text{right}(\rho(i,j)) = \text{left}(\rho(i+1, j))$.

The tiling problem for $\mathbb{N} \times \mathbb{N}$ is:

- given a finite set T of tiles, does T tile $\mathbb{N} \times \mathbb{N}$?

In [Robinson, 1971] it is shown that the tiling problem is co-r.e.-complete and hence undecidable.

It is now straightforward to use this result to show the undecidability of the logic X^2.

Given a finite set T of tiles we define (recursively) a formula ϕ_T of the logic such that the satisfiability of ϕ_T is equivalent to the tiling of $\mathbb{N} \times \mathbb{N}$ by T. This will be clear. Since deciding validity is just deciding satisfiability of negations, this shows that validity in X^2 is r.e.-hard.

The formula ϕ_T, which uses the elements of T as propositional atoms, is simply the conjunction of the following:

$$\begin{array}{l} \neg \ominus \top \wedge \neg \obar \top \\ \Box \bigvee_{\tau \in T} \tau \\ \Box \bigwedge_{\tau \neq \tau'} \neg(\tau \wedge \tau') \\ \Box \bigwedge_{\text{up}(\tau) \neq \text{down}(\tau')} \neg(\tau \wedge \obar\, \tau') \\ \Box \bigwedge_{\text{right}(\tau) \neq \text{left}(\tau')} \neg(\tau \wedge \ominus\, \tau'). \end{array}$$

Tiling proofs tend to be more complicated for other two-dimensional temporal logics. When the underlying flows are not necessarily the natural numbers then we must use the temporal logic to code in a discrete subflow. When the logic does not have next-time or *Until* operators then the coding gets more complicated again. See [Marx and Reynolds, 1997] for details.

Note that we could have easily followed Harel and modified this proof using a different tiling problem to show that the logic X^2 actually has a Π^1_1-hard validity problem. This implies that the logic is not recursively axiomatizable. In fact, when the underlying linear orders are restricted to be natural numbers, integers or reals then many of the two-dimensional and interval logics we have seen are non-axiomatizable. When the full class of linear orders is available then this modified tiling problem, involving the infinite repetition of a certain tile, is not able to be encoded in the two-dimensional logic and we can only prove undecidability.

Our main result in this section is the following:

Theorem 8.1. *Deciding validity in HS logic is r.e.-complete.*

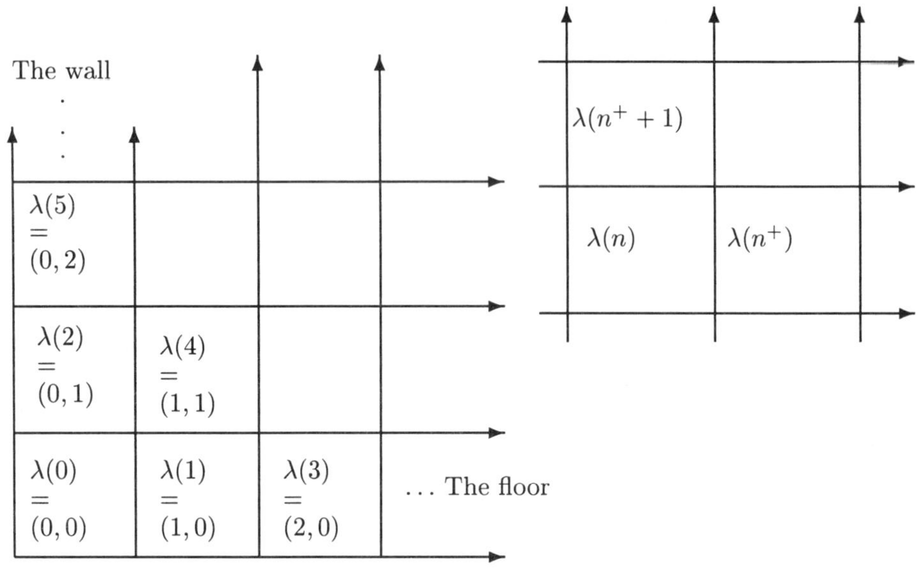

FIG. 8.1.

Given a finite set B of tiles, we will show how we can (recursively) find a formula ϕ so that the satisfiability of ϕ in HS logic is equivalent to the tiling of $\mathbb{N} \times \mathbb{N}$ by B.

8.5.1 *Some notation*

In what follows we will need an enumeration of $\mathbb{N} \times \mathbb{N}$.

Let $\lambda : \mathbb{N} \to (\mathbb{N} \times \mathbb{N})$ be the enumeration of $\mathbb{N} \times \mathbb{N}$ defined recursively by:

- $\lambda(0) = (0, 0)$;
- if $\lambda(n) = (0, j)$ then $\lambda(n + 1) = (j + 1, 0)$;
- otherwise, if $\lambda(n) = (i + 1, j)$ then $\lambda(n + 1) = (i, j + 1)$.

Let $\lambda^{-1} : (\mathbb{N} \times \mathbb{N}) \to \mathbb{N}$ be the inverse map.

Further, for any natural number n we define $n^+ = \lambda^{-1}(\lambda(n) + (1, 0))$, i.e. the number of the square to the right of square number n. For example, $0^+ = 1$ and $2^+ = 4$. Note that $\lambda(n^+ + 1)$ is the square above $\lambda(n)$ for any n.

We talk of the column of squares $\{(0, j)|j \geq 0\}$ as being the *wall* and say that $\lambda(n)$ is on the wall iff it is in this column. Similarly, we talk of the row of squares $\{(i, 0)|i \geq 0\}$ as being the *floor* and say that $\lambda(n)$ is on the floor iff it is in this row.

We will freely use all sorts of properties of this enumeration such as that $\lambda(n^+ + 1)$ is on the wall iff $\lambda(n)$ is.

The arrangement may be pictured as in Fig. 8.1.

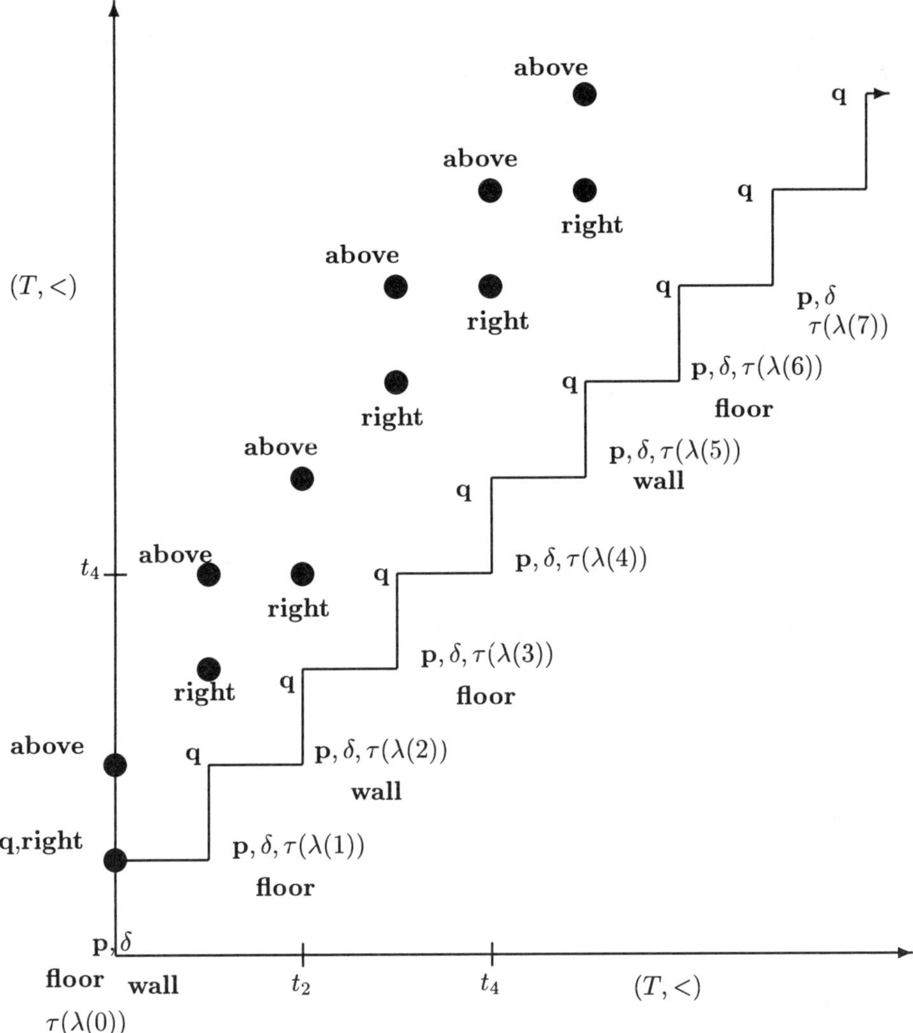

FIG. 8.2. The idea.

8.5.2 *The idea*

We construct ϕ using propositional atoms from $B \cup \{\mathbf{p}, \mathbf{q}, \mathbf{floor}, \mathbf{wall}, \mathbf{right}, \mathbf{above}\}$ (which is assumed to be a union of disjoint sets).

Corresponding to a tiling is a two-dimensional model which is pictured in Fig. 8.2.

To illustrate the coding used in the picture of this model we describe what corresponds to going rightwards from the tiling square $(0,1)$. Note that this square is $\lambda(2)$ and that under the tiling τ it will be tiled by $\tau(\lambda(2))$. Observe that the propositional atom $\tau(\lambda(2))$ is true at the point (t_2, t_2) in the model.

Notice that the only point of the form (t_2, x) which has the atom **right** true for any x is the point (t_2, t_4). This codes in the information that $\lambda(4)$ is the square to the right of $\lambda(2)$. Notice also that the atom $\tau(\lambda(4))$ is true at (t_4, t_4).

We will sometimes refer to the points of the form (x, y) where both x and y precede some t_i as the standard part of a model.

8.5.3 *The formula*

ϕ will be the conjunction of the following formulae which we will group according to their purposes.

We want some **p** and **q** steps:

C1 $\mathbf{p}$;
C2 $\Box(\mathbf{p} \rightarrow \Diamond_N\, \mathbf{q})$;
C3 $\Box(\mathbf{q} \rightarrow \Diamond_E\, \mathbf{p})$;
C4 $\Box(\mathbf{p} \rightarrow \delta)$.

We want **p** restricted horizontally to being on these steps:

C5 $\Box_W \Box_N (\neg\mathbf{p})$;
C6 $\Box(\mathbf{p} \rightarrow (\Box_N \neg\mathbf{p} \wedge \Box_S \neg\mathbf{p}))$;
C7 $\Box((\Diamond_W\, \mathbf{q} \wedge \Diamond_E\, \mathbf{p}) \rightarrow (\Box_N \neg(\mathbf{p} \vee \mathbf{q}) \wedge \Box_S \neg(\mathbf{p} \vee \mathbf{q})))$.

We want **p** restricted vertically to being on these steps:

C8 $\Box_S \Box_E \neg\mathbf{p}$;
C9 $\Box(\mathbf{p} \rightarrow (\Box_E \neg\mathbf{p} \wedge \Box_W \neg\mathbf{p}))$;
C10 $\Box((\Diamond_N\, \mathbf{q} \wedge \Diamond_S\, \mathbf{p}) \rightarrow (\Box_E \neg(\mathbf{p} \vee \mathbf{q}) \wedge \Box_W \neg(\mathbf{p} \vee \mathbf{q})))$.

We want **q** restricted to being on these steps:

C11 $\Box(\mathbf{q} \rightarrow \Diamond_S\, \mathbf{p})$;
C12 $\Box(\mathbf{q} \rightarrow (\Box_N \neg\mathbf{q} \wedge \Box_S \neg\mathbf{q}))$.

We want **right**, **above**, **floor**, **wall** restricted to being true only on grid points on and northwards of these steps:

C13 $\Box(\mathbf{right} \rightarrow \Diamond_S\, \mathbf{p})$.

We want **floor** and **wall** true at the origin:

C14 $\mathbf{floor} \wedge \mathbf{wall}$.

We want **wall** to be followed on the next step by **floor**:

C15 $\Box(\mathbf{wall} \rightarrow \Box_N (\mathbf{q} \rightarrow \Box_E (\mathbf{p} \rightarrow \mathbf{floor})))$.

We want **wall** to be restricted:

C16 $\Box_E \Box_N \neg(\mathbf{floor} \wedge \mathbf{wall})$;
C17 $\Box_E \Box_N (\mathbf{wall} \rightarrow \Diamond_W (\mathbf{above} \wedge \Diamond_S\, \mathbf{wall}))$;
C18 $\Box(\mathbf{above} \rightarrow \Box_W \neg\mathbf{above})$.

We want **right** to follow its own step pattern above the **p**/**q** steps:

C19 $\Box((\mathbf{p} \wedge \neg\mathbf{wall}) \rightarrow \Diamond_W\, \mathbf{right})$;

C20 $\Box(\mathbf{right} \to (\Box^{\uparrow}\, \neg\mathbf{right} \wedge \Box_{\downarrow}\, \neg\mathbf{right}))$;
C21 $\Box(\mathbf{p} \to \Diamond^{\uparrow}\, \mathbf{right})$;
C22 $\Box(\mathbf{right} \to \boxminus\, \neg\mathbf{right})$;
C23 $\Box(\mathbf{right} \to \boxminus\, \Box^{\uparrow}\, \neg\mathbf{right})$;
C24 $\Box(\mathbf{wall} \to \boxminus\, \neg\mathbf{right})$.

We want **above** to follow its own step pattern even further northwards:

C25 $\Box(\mathbf{right} \to \Diamond^{\uparrow}\, \mathbf{above})$;
C26 $\Box((\Diamond^{\downarrow}\, \mathbf{right} \wedge \Diamond^{\uparrow}\, \mathbf{above}) \to \Box_{\rightarrow}\, \neg\mathbf{p})$.

We want **wall** actually to be true every so often:

C27 $\Box(\mathbf{wall} \to \Diamond^{\uparrow}\, (\mathbf{above} \wedge \Diamond^{-}\, \mathbf{wall}))$;
C28 $\Box(\mathbf{above} \to (\Box^{\uparrow}\, \neg\mathbf{above} \wedge \Box_{\downarrow}\, \neg\mathbf{above}))$.

Finally, we want the tile propositions to be true at the right places:

C29 $\Box(\mathbf{p} \to \bigvee_{b \in B} \tau)$;
C30 $\Box(\bigwedge_{b \neq b'} \neg(b \wedge b'))$;
C31 $\bigwedge_{\mathrm{up}(b) \neq \mathrm{down}(b')} \Box\neg(b \wedge \Diamond^{\uparrow}\, (\mathbf{above} \wedge \Diamond^{-}\, b'))$;
C32 $\bigwedge_{\mathrm{right}(b) \neq \mathrm{left}(b')} \Box\neg(b \wedge \Diamond^{\uparrow}\, (\mathbf{right} \wedge \Diamond^{-}\, b'))$.

8.5.4 *Tiling implies satisfiability*

Suppose that B successfully tiles $\mathbb{N} \times \mathbb{N}$. Let $\rho : (\mathbb{N} \times \mathbb{N}) \to B$ be a tiling.

We will build a model of ϕ.

We define a valuation g for the atoms from $\{\mathbf{p}, \mathbf{q}, \mathbf{right}, \mathbf{above}, \mathbf{wall}, \mathbf{floor}\} \cup B$ on the frame $\mathcal{I}(\mathbb{N}, <)$:

1. $\mathbf{p} \in g(n, n)$ for each n;
2. $\mathbf{q} \in g(n, n+1)$ for each n;
3. $\mathbf{right} \in g(n, n^+)$ for each n;
4. $\mathbf{above} \in g(n, n^+ + 1)$ for each n;
5. $\mathbf{wall} \in g(n, n)$ iff $\lambda(n)$ is on the wall;
6. $\mathbf{floor} \in g(n, n)$ iff $\lambda(n)$ is on the floor;
7. $b \in g(n, n)$ iff $\rho(\lambda(n)) = b$, for each $b \in B$.

None of the atoms are true anywhere else.

It is straightforward to check that

$$(\mathcal{I}(\mathbb{N}, <), g), 0, 0 \models \phi.$$

8.5.5 *Satisfiability implies tiling*

Suppose that ϕ is satisfiable in HS logic. Say that $(T, <)$ is a linear order and $(\mathcal{I}(T, <), g), t_0, u_0 \models \phi$. In what follows we shall say 'by C14': for example, when we claim that something is a consequence of the fact that C14 holds in $(\mathcal{I}(T, <), g)$ at (t_0, u_0).

We will show that B tiles $\mathbb{N} \times \mathbb{N}$.

A simple induction using C1, C2, C3 and C4 gives us the following:

Lemma 8.2. *There is a sequence $t_0 < t_1 < t_2 < \ldots$ of elements of T such that for each $n \geq 0$:*

K102 $\mathbf{p} \in g(t_n, t_n)$;

and

K103 $\mathbf{q} \in g(t_n, t_{n+1})$.

We also have

K105 *if $\mathbf{p} \in g(t, u)$ and $u < t_n$ for some n then $t = t_m$ and $u = t_m$ for some $m < n$ (by C8, C9 and C10).*

We will use a function h to save notation. For $i, j \in \mathbb{N}$ we let $h(i,j) = g(t_i, t_j)$. We know from C14 that **floor** $\in h(0,0)$ and **wall** $\in h(0,0)$.

Lemma 8.3. *For all $n \geq 0$ we have:*

1. *if $\lambda(n^+)$ is on the floor then* **floor** $\in h(n^+, n^+)$;
2. $\neg$**wall** $\in h(n^+, n^+)$;
3. **right** $\in h(n, n^+)$;
4. **above** $\in h(n, n^+ + 1)$;
5. *if $\lambda(n^+ + 1)$ is on the wall then* **wall** $\in h(n^+ + 1, n^+ + 1)$.

Proof We proceed by induction on n. Suppose that these conditions are true for any m with $0 \leq m < n$.

1. Suppose that $\lambda(n^+)$ is on the floor. As $n^+ > 0$ we know that $\lambda(n^+ - 1)$ is on the wall.
   If $n = 0$ then $n^+ - 1 = 0$ so C14 tells us **wall** $\in h(n^+ - 1, n^+ - 1)$. Otherwise, $n^+ - 1 = (n-1)^+ + 1$ and the induction hypothesis, item 5, tells us **wall** $\in h(n^+ - 1, n^+ - 1)$.
   Then we just use C15, K102 and K103 to deduce that **floor** $\in h(n^+, n^+)$ as required.
2. If $\lambda(n^+)$ is on the floor then C16 gives us $\neg$**wall** $\in h(n^+, n^+)$ as required. Otherwise, we use C17 which is applicable as $n^+ > 0$ (in fact we must have $n > 1$). Suppose for contradiction that **wall** $\in h(n^+, n^+)$. By C17, there is some $x < t_{n^+}$ with **above** $\wedge \diamondsuit$ **wall** true at (x, t_{n^+}). But we already know **above** $\in h(n-1, (n-1)^+ + 1)$, and since $\lambda(n^+)$ is not on the floor, $(n-1)^+ + 1 = n^+$. Thus C18 tells us $x = t_{n-1}$. Thus $\diamondsuit$ **wall** is true at (t_{n-1}, t_{n^+}) and by K105 we must have **wall** $\in h(n-1, n-1)$.
   By our induction hypothesis (item 2) we cannot have $n - 1 = i^+$ for any i. Thus $\lambda(n-1)$ is on the wall. But then we have our contradiction as this implies $\lambda(n)$ and $\lambda(n^+)$ are on the floor.
3. Since we have $\neg$**wall** $\in h(n^+, n^+)$, by C19, there must be some $x < t_{n^+}$ with **right** $\in g(x, t_{n^+})$.
   By C13 there is $y < t_{n^+}$ such that $\mathbf{p} \in g(x, y)$. By K105, $x = t_i$ and $y = t_i$ for some i which must be less than n^+ (as $t_i < t_{n^+}$).

We will show that $i = n$ as required.

Suppose first for contradiction that $i < n$. Then our induction hypothesis tells us that **right** $\in h(i, i^+)$. By C20 we cannot have **right** $\in h(i, n^+)$.

Suppose next for contradiction that $n < i$. By C21, there is some $y > t_n$ such that **right** $\in g(t_n, y)$. There are three cases.

Case: $y < t_{n^+}$. Then, by K105, $y = t_j$ for some $j < n^+$ so **right** $\in h(n, j)$. Now $y > t_n$ so $j > 0$ and we may have $j = k^+$ for some $k < n$. But then the inductive hypothesis (item 3) tells us we have **right** $\in h(k, k^+)$ as well as **right** $\in h(n, k^+)$. This contradicts C20.

The other alternative (if j is not k^+ for any k) is that j is on the wall and then C24 tells us **right** $\notin h(n, j)$ to give us our contradiction.

Case: $t_{n^+} = y$. Thus **right** $\in h(n, n^+)$ and **right** $\in h(i, n^+)$. This contradicts C22.

Case: $t_{n^+} < y$. So $\diamondsuit$ **right** $\in g(t_n, t_{n^+})$ and $\diamondsuit$ **right** $\in g(t_n, t_{n^+})$. This contradicts C23.

4. So we have **right** $\in h(n, n^+)$. By C25, there is $y > t_{n^+}$ such that **above** $\in g(n, y)$. By C26 and K102, $y = t_{n^+ + 1}$ as required.
5. Finally suppose that $\lambda(n^+ + 1)$ is on the wall. This means that $\lambda(n)$ is also on the wall.

   If $n = 0$ then C14 tells us that **wall** $\in h(n, n)$. Otherwise, $n = k^+ + 1$ for some $k < n$ and the inductive hypothesis, item 5, tells us that **wall** $\in h(n, n)$.

   Now C27 tells us that there is $y > t_n$ such that **above** $\wedge \diamondsuit$ **wall** $\in g(t_n, y)$. By C28, and item 4, $y = t_{n^+ + 1}$.

   Thus there is $x > t_n$ such that **wall** $\in g(x, t_{n^+ + 1})$. By K102 and K105, $x = t_{n^+ + 1}$. Thus **wall** $\in h(n^+ + 1, n^+ + 1)$ as required.

□

To summarize the important results for each $n, m \in \mathbb{N}$:

- if $\lambda(m)$ is the square to the right of $\lambda(n)$ then **right** $\in h(n, m)$;
- if $\lambda(m)$ is the square above $\lambda(n)$ then **above** $\in h(n, m)$;
- **wall** is only true on the wall, i.e. **wall** $\in h(n, n)$ iff $\lambda(n)$ is on the wall.

Now we can finally use our tiling atoms. Define $\tau : (\mathbb{N} \times \mathbb{N}) \to B$ via $\tau(i, j)$ is the unique $b \in B$ such that $b \in h(n, n)$ where $n = \lambda^{-1}(i, j)$.

C29 shows that there is such a b, C30 shows that it is unique.

C31, C32 and the important results from the last induction show that:

- if $\tau(i, j) = b$ and $\tau(i, j + 1) = a$ then $\text{up}(b) = \text{down}(a)$;
- if $\tau(i, j) = b$ and $\tau(i + 1, j) = a$ then $\text{right}(b) = \text{left}(a)$.

So we have our result.

8.5.6 *Dedekind complete case*

In [Harel, 1986] it is shown that the following problem is Σ_1^1-complete and hence not recursively enumerable:

- Given a set $T = \{d_0, d_1, \ldots, d_n\}$ of tiles, can T tile $\mathbb{N} \times \mathbb{N}$ such that d_0 appears infinitely often in the first column?

We can use this result to show that HS logic is not recursively axiomatizable when we restrict the underlying linear order to being Dedekind complete.

Suppose that there is some $(T, <)$ containing an infinite ascending sequence. Suppose also that $(T, <)$ is Dedekind complete. Below we will show that the HS logic over $(T, <)$ is Σ_1^1-hard and so is not recursively axiomatizable. This will apply to the compass logics where both orders are the natural numbers, or integers or reals, for example.

We will use the formula ϕ from the earlier part of the paper but we will add some conjuncts to it. They are:

D1 $\Box(p \to \Diamond^{\uparrow} \Diamond^{\rightarrow} (d_0 \wedge \mathbf{wall}))$;

D2 $\Box^{\rightarrow} (\Diamond^{\updownarrow} (\Diamond^{\leftarrow} q \wedge \Diamond^{\rightarrow} p) \vee (\Box^{\uparrow} \neg p \wedge \Box^{\rightarrow} \Box^{\uparrow} \neg p) \vee \Diamond^{\updownarrow} (p \wedge \Diamond^{\leftarrow} q))$.

Let ψ be the conjunction of ϕ and these two formulae. We will show that the satisfiability of ψ in structures over U frames is equivalent to the tiling of $\mathbb{N} \times \mathbb{N}$ using the tiles of T with d_0 appearing infinitely often up the wall.

If we have a tiling with d_0 appearing infinitely often along the wall then it is clear that ψ has a U-model: just use the construction given above on one of the frames with ascending chains in both dimensions.

Conversely, if ψ has a model in which $((T, <), (T, <))$ is from U then this is also a model of ϕ. As above in this section we can find a tiling from this. Recall that we use the value of tiling atoms at a sequence of points $(t_0, t_0), (t_1, t_1), \ldots$. Note that the atom **wall** is true only at points (t_i, n_i) when $\lambda(i)$ is on the wall. If the sequence $t_0 < t_1 < t_2 < \ldots$ is unbounded then D1 (along with K104) tells us that d_0 appears infinitely often as a tile on the wall and we are done. Otherwise, as $(T, <)$ is Dedekind complete, there is some least upper bound t to the sequence.

Using C5, C6 and C7, we can show:

K104 if $\mathbf{p} \in g(t, u)$ and $t < t_n$ for some n then $t = t_m$ and $u = t_m$ for some $m < n$.

By D2 we have three cases as described below.

We may have $\Diamond^{\updownarrow} (\Diamond^{\leftarrow} q \wedge \Diamond^{\rightarrow} p)$ true at (t, t_0). In fact we will show that this does not arise. Say that $\Diamond^{\leftarrow} q \wedge \Diamond^{\rightarrow} p$ is true at (t, u) and that q is true at (t', u) for some $t' < t$. As t is the least upper bound of the sequence of t_is, we know there is some t_i such that $t' < t_i$. By C11 there is some $u' < u$ such that p is true at (t', u'). By K104, we know that $t' = t_j$ and $u' = t_j$ for some $j = 1, 2, \ldots$. By K103 and C12, $u = t_{j+1}$. By K102, p is true at (t_{j+1}, t_{j+1}) and, by C9, we cannot have $\Diamond^{\leftarrow} q \wedge \Diamond^{\rightarrow} p$ true at (t, u) unless $t < t_{j+1}$ which it is not.

We may have $\Box_{NS} \neg p \wedge \Box_{E} \Box_{NS} \neg p$ true at (t, t_0). Clearly there is no appearance of $d_0 \wedge$ **wall** outside the standard sequence. Thus we can conclude that $d_0 \wedge$ **wall** is true at infinitely many (t_i, t_i).

The third alternative is that $\Diamond_{NS} (p \wedge \Diamond_{W} q)$ is true at (t, t_0). Again we can show that this does not happen.

8.6 Axiomatization of HS logic

The set of validities of full HS logic is recursively enumerable. This follows immediately from the fact that the universal second-order theory of linear frames is r.e. (see Proposition 4.4.2 [Marx and Venema, 1997]). In fact, it is possible to give a finite axiom system explicitly. We borrow this from Marx and Venema in [Marx and Venema, 1997].

The system uses a non-orthodox (non-structural) irreflexivity rule in the style of those used in Chapter 6 of Volume 1. The rule involves the difference operator D which we can regard as an abbreviation:

$$D\phi = \Diamond_{N} \Diamond_{EW} \phi \vee \Diamond_{S} \Diamond_{EW} \phi \vee \Diamond_{W} \Diamond_{NS} \phi \vee \Diamond_{E} \Diamond_{NS} \phi.$$

Its irreflexivity rule (IRRD) is simply

$$\frac{(q \wedge \neg Dq) \rightarrow \phi}{\phi} \quad \text{provided the atom } q \text{ does not appear in } \phi.$$

The axiom system consists of the rules *modus ponens*, the various universal generalizations, substitution and IRRD along with:

0 all classical tautologies;
1 $\Box_{N} (p \rightarrow q) \rightarrow (\Box_{N} p \rightarrow \Box_{N} q)$
and the three duals of 1;
2 $p \rightarrow \Box_{S} \Diamond_{N} p$
and the three duals of 2;
3 $\Diamond_{N} \Diamond_{N} p \rightarrow \Diamond_{N} p$
and the eastern version of 3;
4 $\Diamond_{S} \Diamond_{N} p \rightarrow (\Diamond_{S} p \vee p \vee \Diamond_{N} p)$
and the east–west version of 4;
5 $\Diamond_{N} \Diamond_{S} p \rightarrow (\Diamond_{S} p \vee p \vee \Diamond_{N} p)$
and the east–west version;
6 $\Diamond_{N} \Diamond_{W} p \leftrightarrow \Diamond_{W} \Diamond_{N} p$;
7 $\Diamond_{E} \Diamond_{N} p \rightarrow \Diamond_{N} \Diamond_{E} p$;
8 $\Diamond_{S} \top \rightarrow \Diamond_{S} \Box_{S} \bot$
and the western version of 8;
9 $\Box_{S} \bot \leftrightarrow \Box_{E} \bot$.

Marx and Venema do not give a direct proof of the completeness of this system but, instead, give an interesting proof using several general techniques described in [Marx and Venema, 1997]. Essentially, they proceed as follows. They add an extra non-structural rule to the system above—the non-κ rule where κ is the modal formula $\Box\, p \to \Diamond\, p$. The extended system is shown to correspond in Sahlqvist manner to a set of first-order formulae which characterize the two-dimensional frames. The formula κ is used in a negative way to capture the requirement that $\Diamond$ and $\Diamond$ have disjoint accessibility relations. See [Venema, 1993] or [Hollenberg, 1994] for more details on the idea of negative definability, an extension of the usual Sahlqvist correspondence ideas. Thus a general completeness result (Theorem B.3.7 [Marx and Venema, 1997]) gives the completeness of the extended system. It is straightforward to show that the non-κ rule is a derived rule of the system presented above.

The usual additions of various axioms give us axiomatizations for other classes of frames: for example, for dense linear orders or discrete linear orders or the rational numbers.

It is an open problem, as with many other two-dimensional temporal logics, whether there is an axiomatization which only uses the traditional orthodox derivation rules.

8.7 Relation algebra approach

Important recent results concerned with reasoning with intervals have been obtained *via* the realization that intervals can be viewed from the perspective of algebras of binary relations. This connection was first examined in (earlier versions of) [Ladkin and Maddux, 1994]. There is a flourishing area of research on more general relation algebras but here we will just briefly describe the basic concepts and point out the connections with interval logics. More details can be found in [Hirsch, 1996] and [Hirsch and Hodkinson, 1997].

After the success in [Boole, 1948] of relating logic to a new algebra of unary relations, De Morgan [1864] was the first to attempt to do the same for binary relations. The extension to binary relations allows some extra operations on the relations and we use the signature with Boolean operators . (intersection) and $-$ (complementation), constants 0 (empty) and 1 (universe) as well as $1'$ (for the identity relation), unary function *converse* ($^\smile$) and binary function *composition* ;.

A *proper* relation algebra is an algebra $(A, 0, 1, ., -, ;, ^\smile, 1')$ where A consists of a set of sets of pairs from some set D which is closed under the operations and the operations have their intended interpretations. That is, the converse $R^\smile$ of a relation R on the domain set D is $(x, y) \in R^\smile$ iff $(y, x) \in R$ and the composition $R; S$ of relations R and S on D is

$$(x, y) \in R; S \text{ iff there is some } z \in D \text{ such that } (x, z) \in R \text{ and } (z, y) \in S.$$

In 1941, Tarski, in [Tarski, 1941], proposed a simple axiomatization for the class of relation algebras: an equivalent one, given in [Hirsch, 1993], is obtained by adding the following identities to those for a Boolean algebra:

1. $(a^\smile)^\smile = a$;

2. $1';a = a;1' = a$;
3. $(a;b);c = a;(b;c)$;
4. $a;(b+c) = (a;b) + (a;c)$;
5. $(a+b)^{\smile} = (a^{\smile}) + (b^{\smile})$;
6. $(-a)^{\smile} = -(a^{\smile})$
7. $(a;b)^{\smile} = (b^{\smile});(a^{\smile})$;
8. $(a^{\smile});(-(a;b)) \leq -b$;

where $a+b$ is an abbreviation for $-((-a).(-b))$ and $a \leq b$ is an abbreviation for $a+b=b$.

However, in [Lyndon, 1950], an abstract relation algebra, i.e. an algebra satisfying Tarski's axioms, was presented which is not proper, i.e. it does not have a representation as the algebra of a set of relations. In [Monk, 1964], it was shown that there is no finite axiomatization of the class of proper relation algebras. This is in marked contrast to Stone's theorem in the case with algebras of unary relations.

There are many interesting new results in the study of relation algebras and related algebraic logics. Some examples can be found in [Jónsson, 1991], [Maddux, 1991], [Marx and Venema, 1997] and [Hirsch and Hodkinson, 1997]. In the rest of this chapter we will just concentrate on the applications to reasoning with intervals and planning.

We shall give some examples of relation algebras below. We will present them by giving representations; we can do this as they will all be proper relation algebras. To present a relation algebra abstractly we need only define the Boolean algebra part with its *atoms*, i.e. minimal non-zero elements of the algebra, say which pairs are converse and then give the composition table for these atoms.

Three important relation algebras are those appearing in [Ladkin and Maddux, 1994].

The point algebra has eight elements including atoms $1'$, $<$ and $>$. It can be represented on the set of rational numbers. The representation of the atoms is as follows: $1'$ is the identity (which is self-converse); $<$ is the strictly less than relation and $>$ is its converse. The other elements of the algebra can be obtained by taking unions of these atoms. So, for example, there is an element $-1'$ which is $< + >$ and the element $\leq = < + 1' = - >$. The composition table includes such products as $(-1'); <= 1$ and $<; \leq = <$.

Allen's interval algebra is also a relation algebra. Its atoms are just the 13 relations between intervals defined in Section 8.2 above. This algebra can be represented in a very straightforward way as ordered pairs of rational numbers from the point relation $<$. We have such compositions as *during*; *meets* = *precedes* and *starts*$^{\smile}$; ends = *overlaps*$^{\smile}$.

The containment algebra has five atoms $1'$, c, $c^{\smile}$, n and a where $1'$, n and a are self-converse. This is a subalgebra of Allen's interval algebra. It can be represented by pairs of rationals from less than. As well as the identity relation, the atoms are represented by: c being 'contained in', $((x,y),(u,v)) \in c$ iff $u \leq$

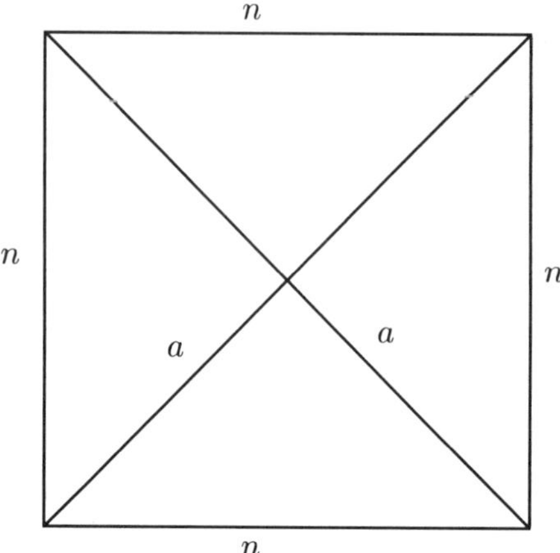

FIG. 8.3. A network in the containment algebra.

$x < y \leq v$; n being 'non-empty intersection', $((x, y), (u, v)) \in n$ iff $x < u \leq y \leq v$ or $u < x \leq v < y$; and a being 'apart', $((x, y), (u, v)) \in a$ iff $x < y < u < v$ or $u < v < x < y$.

The main use of these algebras in reasoning about intervals is in determining the consistency of networks of constraints. Often we must consider binary constraints. In the context of relation algebras this means that we have a finite complete directed labelled graph in which the edges are labelled with elements from a relation algebra. Such a *network* is consistent iff there is a representation of the algebra (as binary relations on a set of objects as defined above) and a map from the nodes of the network to objects in the set, such that the label on the edge between one node and another correctly describes a relation between the objects.

For example, from [Ladkin and Maddux, 1994] we have the network in the containment algebra shown in Fig. 8.3.

This is an interesting example, because, as shown in [Ladkin and Maddux, 1994], each triangle in the network is consistent but the whole network is not consistent. In [Hirsch, 1996], Hirsch gives some general conditions for determining whether network consistency in a relation algebra follows from consistency of triangles. This is true for Allen's interval algebra and the point algebra and makes checking consistency much easier. For Allen's algebra, the network consistency problem is NP-complete [Vilain and Kautz, 1986] and for the point algebra there is a cubic time algorithm.

8.8 Planning

In this section we will have a brief look at an area of AI in which interval-based temporal logics are much used. This is the task of *planning* which means deciding in what order to perform certain tasks in order to achieve a desired situation or *goal.*

Because these tasks or *actions* take time to accomplish, in many of the usual planning applications it is best to use an interval-based temporal logic as a foundation for the reasoning. The planning problem then becomes one of ordering these temporally extensive actions taking full account of the possibility that actions can overlap in their execution. Armed with an interval basis we can now formally define a planning problem. We follow the description in [Hirsch, 1993] of a modified version of Allen and Koomen's planner [Allen and Koomen, 1983]. Each particular problem has four components: an interval network, a planning environment, a goal state and a stock of available actions. The interval network just contains the few intervals mentioned in the planning environment and the goal state and details the relationships between them. That is, for every pair of intervals, we list those of the 13 relations which may hold between the pair.

The planning environment describes the fixed information about the problem. It is a list of interval names with associated propositions indicating that the propositions hold on that interval. For example, it might detail what is true on some initial interval.

The goal state is also such a list of intervals with propositions. It indicates which propositions we want to end up being true at each interval.

An action is formalized as follows. There is a small set $\{X_0, X_1, \ldots, X_n\}$ of variables for intervals involved in the execution of the action. The action will have a name, a set of preconditions describing which propositions must be true on each of the intervals X_i, a set of postconditions describing what will end up being true on the intervals (if we 'do' the action at time X_0) and a set of temporal constraints describing the possible relationships between the intervals.

For example, an action might be called brewing tea, have preconditions tea bag in cup at X_1, postconditions black tea in cup at X_2 and constraints X_0 during X_1 and $X_0 < X_2$.

A solution to a planning problem will be a set of intervals from a linear order with names including those mentioned in the problem, a list of which actions are done when and a list of which propositions are true at which times such that:

- the interval network is consistent with the original one;
- the preconditions and postconditions of each of the actions apply correctly;
- each proposition is true at an interval because either it was declared so in the planning environment or it is the postcondition of one of the actions.

It is important to realize that two things are going on in achieving a solution. One is the introduction of actions to explain the truth of propositions at various times but equally important is the realization of the interval network. This latter often includes deciding that two interval names name the same interval.

An algorithm for finding solutions to such planning problems is presented in [Allen and Koomen, 1983]. The basic approach of the Allen and Koomen planner is to work on the two tasks described above simultaneously. If the goal state requires proposition p to be true at interval i, the planning environment states that p is true at interval j and the equality relation is possible between i and j, then we say that p at i has a *possible causal explanation*—we simply make $i = j$. The planner keeps attempting the following two activities until every goal assertion has a possible causal explanation:

- *introducing* actions to explain goal assertions which do not have potential causal explanations; and
- *collapsing* intervals which do have potential causal explanations into the same interval.

This is just the outline of a procedure which can often involve back-tracking to change decisions which have subsequently lead to dead-ends.

Some interesting new work on this planner concerns the problem of collapsing several pairs of intervals at once. An example in [Hirsch, 1993] shows that even if each pair is separately consistently collapsible, together they may not be. Suppose that the planning environment states that p is true at i_0 and q at i_1 while the goal requires p at i_1 and q at i_2. Suppose that we have i_0 before or equal to i_1 and i_1 before or equal to i_2 but i_0 before i_2. Then we have one potential causal explanation for p at i_1 by collapsing i_0 and i_1 and a potential causal explanation for q at i_2 by collapsing i_1 and i_2. However, we cannot consistently collapse both pairs. A modified planning algorithm which can deal with this problem is presented in [Hirsch, 1993].

This *collapsing problem* is related to the persistence problem which infects many other approaches to planning. To make the algorithms more efficient or even just to ensure that a particular solution is found, various algorithms make tentative assumptions about the persistence of properties over time or, equivalently, the identity of intervals in which they are true. These assumptions may need to be retracted during back-tracking. Thus we find *stretching* in TMM in [Dean and McDermott, 1987] and the *persistence axiom* in the event calculus of [Kowalski and Sergot, 1986]. In all these approaches there is plenty of scope for developing better algorithms to solve the planning problem.

A recent introduction to planning problems and techniques can be found in [Allen *et al.*, 1991].

9

MANY-DIMENSIONAL SYSTEMS AND GENERALIZED QUANTIFIERS

9.1 Introduction

This chapter will study in more detail the properties of many-dimensional systems based on tree flows of time. As an application, we shall interpret generalized quantifiers in such systems.

9.2 The logic 1KY

Consider the propositional modal logic **K** for the modality $\Box$. This logic is complete for the Kripke semantics of all models of the form $(S, \prec, a, h)$, where S is the set of possible worlds, $\prec \subseteq S \times S, a \in S$ and h is the assignment, giving each $t \in S$ and each atomic q a value $h(t, q) \in \{0, 1\}$.

It is possible to assume that $(S, \prec, a)$ is a tree with root a. In other words, S can be taken as the set of all finite sequences of natural numbers, $a = \varnothing$, the empty sequence and $t \prec s$ iff for some $n, s = t * (n)$. Thus $\prec$ is a relation of immediate successorship on S and for every $t \in S$ there exists a unique sequence $t_0, t_1, \ldots, t_n$, with $t_i \in S$ such that $(a = t_0) \wedge (t = t_n) \wedge \bigwedge_{i=1}^{n} t_{i-1} \prec t_i$ holds.

Satisfaction is defined by the following clause for $\Box$:

- $t \vDash \Box A$ iff for all s such that $t \prec s$ we have $s \vDash A$.

Recall that **K** can be axiomatized by the following axioms and rules:

- all substitution instances of truth functional tautologies;
- $\vdash \Box(A \to B) \to (\Box A \to \Box B)$;
- $\vdash A$ implies $\vdash \Box A$.

Completeness is well known for the class of tree Kripke models, with the additional condition that each $t \in S$ has a unique predecessor.

Let us enrich the language of modal **K** with another modal operator Y (Y for yesterday), with the following truth definition:

- $t \vDash YA$ iff t does have an $\prec$ predecessor s and $s \vDash A$.

It is reasonable to assume that $(S, \prec)$ is such that every point has a predecessor. In this case $a \in S$ is not the 'first' point in any sense but only the 'distinguished' point.

Let **1KY** (the number 1 in the name is because it is one dimensional) be the system with $\Box$ and Y corresponding to the above semantics. We shall give precise definitions later with axioms and completeness theorems.

2KY is a two dimensional temporal logic which has $\Box$ and two unary operators Y_1 and Y_2.

A **2KY** model has the form $(S, \prec, a, h)$ where $(S, \prec, a)$ is a tree, and where h is an assignment giving for each $(t, s) \in S \times S$ and atomic q a value $h(t, s, q) \in \{0, 1\}$.

Evaluation is two-dimensional and is done as follows:[5]

- $(t, s) \vDash q$ iff $h(t, s, q) = 1$;
- $(t, s) \vDash Y_1 A$ iff $(t', s) \vDash A$, where t' is the predecessor of t;
- $(t, s) \vDash Y_2 A$ iff$(t, s') \vDash A$, where s' is the predecessor of s;
- $(t, s) \vDash \Box A$ iff for all t', s' such that $t \prec t' \wedge s \prec s'$ we have $(t', s') \vDash A$.

The truth definition for $\Box$ makes sense if we think of the one-dimensional case as the special case of $(t, t) \vDash A$.[6]

Thus:

- $(t, t) \vDash \Box A$ iff for all s such that tRs we have $(s, s) \vDash A$.

We can think of $(t, s) \vDash A$ as having t as the reference point and s as the evaluation point in the sense of Reichenbach [Reichenbach, 1947].

There is some connection between the two-dimensional propositional logic **2KY** and the one-dimensional monadic predicate logic **1QK**.

A monadic **K** model has the form $\mathbf{m} = (S, \prec, a, D, h)$ where $(S, \prec, a)$ is a tree and D is a domain. h is an interpretation giving for each monadic predicate P and each $t \in S$ a subset $h(t, P) \subseteq D$.

Satisfaction is defined in the traditional manner with

$$t \vDash P(x) \text{ iff } x \in h(t, P).$$

The connection with two-dimensional satisfaction is now clear. If we assume $D = S$, then in the two-dimensional case let:

- $h(t, P) = \{s \mid (t, s) \vDash p\}$;

or if $h(t, P)$ is given, we can have:

- $h(t, s, p) = 1$ iff $s \in h(t, P)$ iff $t \vDash P(s)$.

We need only one predecessor function Y and we have:

- $(t, s) \vDash Y_1 p$ iff $t \vDash YP(s)$;
- $(t, s) \vDash Y_2 p$ iff $t \vDash P(Ys)$;

where 'Ys' means the predecessor of s in $(S, \prec)$. That is, Y is used, with slight abuse of notation, also as a function symbol in the language.

To summarize, we have seen several motivations for two-dimensional modal **K**. We suggest to the reader that the best way to think of $(t, s) \vDash A$ is

According to the model named t, we have $s \vDash A$

[5] **2KY** is a strong two-dimensional logic, as defined in Definition 7.1.2 of Volume 1 [Gabbay *et al.*, 1994].

[6] This is called diagonal validity in Definition 7.1.3 of Volume 1 [Gabbay *et al.*, 1994].

(i.e. read it as $s \vDash_t A$).

Definition 9.1 (kKY Language and semantics).

1. The **kKY** language is a propositional language with atomic propositions, the classical connectives $\sim, \wedge, \vee, \rightarrow$ and the additional unary connectives $\boxminus, \diamondminus, Y_1, \ldots, Y_k, J_1, \ldots, J_k$. The notion of a wff is defined in the traditional manner.
2. Let $\mathbb{E} = \{e_1, e_2, e_3, \ldots\}$ be a set of distinct atoms and let $\mathbb{A} = \{a_1, a_2, a_3, \ldots\}$ be a set of distinct roots. Assume $\mathbb{A} \cap \mathbb{E} = \varnothing$. Consider the sets $(S_i, \prec_i)$ defined as follows: $S_i' = \{(a, e_1, \ldots, e_k) \mid k \geq 1, e_i \in \mathbb{E}, 1 \leq i \leq k\} \cup \{(a)\}$.
   If $k = 0$ we let $(a, e_1, \ldots, e_k)$ be (a).
   Define $\prec_i$ on S_i by letting $(a, e_1, \ldots, e_k) \prec_i (a, e_1', \ldots, e_r')$ iff $r = k+1$ and for all $i \leq k$, $e_i = e_i'$.
   These sets can be considered trees with roots a_i. Let $S = \bigcup_i S_i$. Consider a function $\mathbf{p}$, $\mathbf{p} : S \mapsto S$ defined as follows:
   - $\mathbf{p}((a_n)) = (a_{n+1})$, for $n \geq 1$;
   - $\mathbf{p}((a_i, e_1, \ldots, e_k)) = (a_i, e_1, \ldots e_{k-1})$ for $k \geq 2$;
   - $\mathbf{p}((a_i, e)) = (a_i)$.

   We can define $\prec$ on S by letting:
   - $x \prec y$ to mean $\mathbf{p}(y) = x$.

   Note that $\prec \restriction S_i = \prec_i$. We can think of $\mathbf{p}$ as a predecessor function.
   We can turn $(S, \prec, a)$, $a = (a_i), a_i \in \mathbb{A}$, into a possible world model. This we do next.
   Note that if $\mathbb{E}$ is countably infinite then for any $s \in S$, $(S, \prec, s)$ and $(S, \prec, a)$ are isomorphic.
3. A k-dimensional model for this language has the form $\mathbf{m} = (S, \prec, \bar{a}, h)$ where S is a set of worlds, $\prec$ is an irreflexive tree relation on S, as in (2) above, $\bar{a} = (a_1, \ldots, a_k)$ with $a_i \in S$, and h is an assignment giving for each $(t_1, \ldots, t_k) \in S^k$ and each atomic q a value $h(t_1, \ldots, t_k, q) \in \{0, 1\}$.
4. $\mathbf{m}$ satisfies the following. Let R be the reflexive and transitive closure of $\prec$. Let S_t, for $t \in S$, be $S_t = \{x \mid tRx\}$. Then:
   - $t \neq s \wedge \sim tRs \wedge \sim sRt \rightarrow S_t \cap S_s = \varnothing$;
   - $\forall t \exists ! s (sRt \wedge \forall y (sRy \wedge yRt \rightarrow s = y \vee t = y))$.
     We call this s the predecessor of t, i.e. $s = \mathbf{p}(t)$.
5. Let $\bar{t} = (t_1, \ldots, t_k)$ and $\bar{s} = (s_1, \ldots, s_k)$. Define $\bar{t} \prec \bar{s}$ iff $\bigwedge_{i=1}^{j} t_i \prec s_i$.
6. Satisfaction in models $\mathbf{m}$ is defined by the following clauses:
   - $\bar{t} = (t_1, \ldots, t_k) \vDash q$ iff $h(\bar{t}, q) = 1$;
   - the traditional definition for the classical connectives;
   - $\bar{t} \vDash \boxminus A$ iff for all $\bar{s}$ such that $\bar{t} \prec \bar{s}$ we have that $\bar{s} \vDash A$;
   - $(t_1, \ldots, t_k) \vDash Y_i A$ iff $(t_1, \ldots, t_{i-1}, s, t_{i+1}, \ldots, t_k) \vDash A$, where s is the predecessor of t_i;
   - $(t_1, \ldots, t_k) \vDash J_i A$ iff $(t_i, \ldots, t_i) \vDash A$;
   - A is said to hold in $\mathbf{m}$ iff $(a_1, \ldots, a_k) \vDash A$.

Definition 9.2. The following is a Hilbert system for **1KY**.[7]

1. All substitution instances of truth functional tautologies.
2. $\vdash \Box(A \to B) \to (\Box A \to \Box B)$.
3. $\vdash \sim YA \leftrightarrow Y \sim A$
   $\vdash Y(A \wedge B) \leftrightarrow YA \wedge YB$.
4. $\vdash A \to \Box YA$.
5. $\vdash Y\Box A \to A$.
6. $\vdash A$ implies $\vdash \Box A$ and $\vdash YA$.

Note that J_1 is redundant in our case.

Theorem 9.3 (Completeness theorem for 1KY). *Every consistent* **1KY** *theory has a model.*

Proof Let Δ be a consistent and complete theory. We construct a tree model $(S, \prec, 0, h)$ for Δ.

We define a sequence of trees $(S_n, \prec_n)$ and theories Δ_t, $t \in S_n$, by induction as follows.

Stage $n = 0$
Let $S_0 = \varnothing$ and $\Delta_0 = \Delta$.
Stage $n = 1$
For each wff $\Diamond\alpha \in \Delta_0$, let $\Delta^+_{(0,\alpha)} = \{\beta \mid \Box\beta \in \Delta_0\} \cup \{\alpha\}$. $\Delta^+_{(0,\alpha)}$ is consistent. Let $\Delta_{(0,\alpha)}$ be a complete theory containing $\Delta^+_{(0,\alpha)}$. It is easy to show that if $YA \in \Delta_{(0,\alpha)}$ then $A \in \Delta_0$. This is so because of the axioms

$$\vdash \sim A \leftrightarrow \Box Y \sim A$$
$$\vdash Y \sim A \leftrightarrow \sim YA.$$

Let $\Delta_{-1} = \{\alpha \mid Y\alpha \in \Delta_0\}$. This theory is complete and consistent and satisfies, for all β, if $\Box\beta \in \Delta_{-1}$ then $\beta \in \Delta_0$. This is because of the axiom $\vDash Y\Box\beta \to \beta$.

Let $S_1 = \{(-1), 0\} \cup S_0$.

Let $(-1) \prec_1 0 \prec_1 (0, \alpha)$.

Case $n + 1$
Assume $(S_n, \prec_n)$ has been defined. Assume the first element of this tree is $(-n)$. Further assume that the top (in the transitive closure of $\prec$) elements of S_n have the form $(-m, \alpha_1, \ldots, \alpha_{n-m})$, for $m \leq n$ (for $n = m$ we have $(-n)$ only), and that for each $i \leq i < n-m$, we have $t_i = (-m, \alpha_1, \ldots, \alpha_i) \in S_n$ and the following hold:

- $\Diamond\alpha_{i+1} \in \Delta_{t_i}$;
- for all $\beta, \Box\beta \in \Delta_{t_i}$ implies $\beta \in \Delta_{t_{i+1}}$;

[7] The temporal system $\mathbf{K}_t$ is defined in Section 3.2 of Volume 1. It contains the future connective G and the past connective H. If we extend $\mathbf{K}_t$ with axioms turning H into a yesterday operator (i.e. axiom 3 of our **1KY** for H) then the two systems become equal, with $\Box = G$ and $H = Y$.

- if $\Diamond\alpha \in \Delta_{t_i}$ then $t_i * (\alpha) \in S_n$.

Let t be a top point in S_n. Let $\Diamond\alpha \in \Delta_t$ and let $\Delta_{t*(\alpha)}$ be constructed as in case $n = 1$.

Let $S_t = \{t * (\alpha) \mid \Diamond\alpha \in \Delta_t\}$.

Let Δ_{-n-1} be constructed as in case $n = 1$.

Let $S_{n+1} = S_n \cup \{(-n-1)\} \cup \bigcup_{t \text{ top point}} S_t$.

Let $\prec_{n+1} = \prec \cup \{(-n-1), (-n)\} \cup \bigcup\{(t, t*(\alpha)) \mid t \text{ top point}\}$.

Then $(S_{n+1}, \prec_{n+1})$ satisfies the same assumptions.

Let $S = \bigcup_n S_n$ and $\prec = \bigcup_n \prec_n$. Let h be defined by $h(t, q) = 1$ iff $q \in \Delta_t$.

□

Lemma 9.4. *For any t and A, $t \vDash A$ iff $A \in \Delta_t$.*

Proof We check the cases of $\Box A$ and YA.

The case of $\Box$ is as in standard modal logic.

We check the case of YA. Assume $t = (-n, \alpha_1, \ldots, \alpha_k)$.

Let $t' = (-n, \alpha_1, \ldots, \alpha_{k-1})$ if $k \geq 2$, and let $t' = (-n)$ if $k = 1$, and finally if $t = (-n)$ let $t' = (-n-1)$. Clearly t' is the predecessor of t and it is easy to show that for all $A, A \in \Delta_{t'}$ iff $YA \in \Delta_t$.

This completes the proof of the lemma. □

Theorem 9.5. *Every formula of* **1KY** *is equivalent to a Boolean combination of pure formulae of the form $Y^m\alpha$, where α contains no occurrences of Y (i.e. α is a pure* **K** *wff) (referred to as normal form).*[8]

Proof By structural induction.

Clearly atomic wffs are in normal form and we do have closure under Boolean operations. Assume A has the form $\bigvee_i \bigwedge_j Y^{n_{i,j}}\alpha_{i,j}$, where $\alpha_{i,j}$ are formulae without Y.

We show that YA and $\Diamond A$ are also equivalent to wffs in normal form.

1. We have
$$YA \leftrightarrow \bigvee_i \bigwedge_j Y^{1+n_{i,j}}\alpha_{i,j}.$$

2. $\Diamond A$ is equivalent to
$$\bigvee_i \Diamond \bigwedge_j Y^{n_{i,j}}\alpha_{i,j}.$$

However, we have:

- $\Diamond(Y^n\alpha \wedge Y^m\beta) \leftrightarrow Y^{n-1}\alpha \wedge Y^{n-1}\beta$, if $m \geq 1$ and $n \geq 1$;
- $\Diamond(Y^n\alpha \wedge Y^m\beta) \leftrightarrow Y^{m-1}\beta \wedge \Diamond\alpha$, if $m \geq 1, n = 0$;
- $\Diamond(Y^n\alpha \wedge Y^m\beta) = \Diamond(\alpha \wedge \beta)$, if $n = m = 0$.

[8]Compare with Theorem 3.2.4 of Volume 1 [Gabbay *et al.*, 1994].

In any event, the above case analysis shows that

$$\Diamond \bigwedge_j Y^{m_{i,j}} \alpha_{i,j}$$

is equivalent to a conjunction of pure formulae. □

Example 9.6. Let us put in normal form the formula $\alpha = \Box\Diamond\Box(Y^2 p_1 \wedge Y p_2 \wedge p_1)$:

$$\begin{aligned}\alpha &\leftrightarrow \Box \sim \Box\Diamond(Y^2 \sim p_1 \vee Y \sim p_2 \vee \sim p_1)\\ &\leftrightarrow \Box \sim \Box(\Diamond Y^2 \sim p_1 \vee \Diamond Y \sim p_2 \vee \Diamond \sim p_1)\\ &\leftrightarrow \Box \sim \Box(Y \sim p_1 \vee \sim p_2 \vee \Diamond \sim p_1)\\ &\leftrightarrow \Box\Diamond(Y p_1 \wedge p_2 \wedge \Box p_1)\\ &\leftrightarrow \Box(p_1 \wedge \Diamond(p_2 \wedge \Box p_1))\\ &= \alpha'.\end{aligned}$$

Note the translation of α' back into $\mathbf{1Q}_0$ as given in Example 9.47.

Theorem 9.7 (Translation of 1KY into K). *There exists a faithful translation $\alpha \mapsto [\![\alpha]\!]$ from* **1KY** *into* **K** *(i.e. eliminating all occurrences of Y).*

Proof Let α be any wff of **1KY**. By the previous theorem α can be assumed to be in normal form

$$\alpha = \bigvee_i \bigwedge_j Y^{n_{i,j}} \alpha_{i,j}$$

where $\alpha_{i,j}$ contains no occurrences of Y, i.e. they are pure **K** formulae.

We shall rewrite each wff $\bigwedge_j Y^{n_{i,j}} \alpha_{i,j}$ as a different formula without occurrences of Y.

Consider a formula β of the above form:

$$\beta = \beta_k \wedge \bigwedge_{j=0}^{k-1} Y^{m_j} \beta_j$$

where β_j for $0 \leq j \leq k$ is a pure **K** formula.

β can be rewritten as β' where

$$\beta' = \beta_k \wedge Y^{n_{k-1}}(\beta_{k-1} \wedge Y^{n_{k-2}}(\beta_{k-2} \wedge \cdots \wedge Y^{n_0}\beta_0) \ldots).$$

Clearly β' is equivalent to β where $m_k = 0$ and $m_j = \Sigma_{i \geq j} n_i$, for $1 \leq j \leq k$.

To motivate our syntactical translation, let t_k be a point in a model with $t_k \vDash \beta$. Fig. 9.1 describes the state of affairs. We have

$$t_0 \prec x_{0,1} \prec \cdots \prec x_{0,n_1-1} \prec t_1 \prec \cdots \prec t_{k-1} \prec x_{k-1,1} \prec \cdots \prec x_{k-1,n_{k-1}-1} \prec t_k$$

with $t_j \vDash \beta_j$.

$$\begin{array}{cccccccc} \beta_0 & & \beta_1 & & \beta_{k-1} & & & \beta_k \\ \vdash\!\!\!-\!\!\!+\!\!\!- \ldots & \vdash\!\!\!-\!\!\!+\!\!\!- \ldots & & & -\!\!\!+\!\!\!-\!\!\!\dashv \ldots & -\!\!\!+\!\!\!-\!\!\!\dashv & & \\ t_0 & x_{0,1} & x_{0,n_1-1}\ t_1 & & t_{k-1} & x_{k-1,1} & x_{k-1,n_{k-1}-1} & t_k \end{array}$$

FIG. 9.1.

We therefore translate β to $[\![\beta]\!]$, where

$$[\![\beta]\!] = \beta_0 \wedge \Diamond^{n_1}(\beta_1 \wedge \Diamond^{n_2}(\ldots \beta_{k-1} \wedge \Diamond^{n_k}\beta_k)\ldots).$$

Equipped with this translation let

$$[\![\alpha]\!] = \bigvee_i [\![\bigwedge_j Y^{n_{i,j}}\alpha_{i,j}]\!].$$

□

Lemma 9.8. *α has a* **1KY** *model iff $[\![\alpha]\!]$ has a* **K** *model.*

Proof Assume α has a **1KY** model, $\mathbf{m} = (S, \prec, t, h)$. Then for some disjunct $\beta = \alpha_i, t \vDash \beta$. We can assume for convenience that β has the syntactical form above and we also let $t = t_k$. Since $t_k \vDash \beta$, we have points as in Fig. 9.1 with $t_i \vDash \beta_i, i = 0, \ldots, k$.

Then it is clear that $t_0 \vDash [\![\beta]\!]$. Therefore $\mathbf{m}_0(S, \prec, t_0, h) \vDash [\![\beta]\!]$. Hence $\mathbf{m}_0 \vDash [\![\alpha]\!]$.

Conversely, if some model $\mathbf{m} = (S, \prec, t_0, h)$ satisfies $[\![\alpha]\!]$, then it satisfies one disjunct of $[\![\alpha]\!]$, say $[\![\beta]\!]$, which we can assume is syntactically as above. Then there are points as in Fig. 9.1 and hence $t_k \vDash \beta$.

Thus, $\mathbf{m}' = (S, \prec, t_k, h)$ is a model of α. □

Remark 9.9. The previous theorem and especially its proof show that **1KY** is essentially the same as **K**. They have the same models and Y can be eliminated by using $\Diamond$ and by just shifting the actual world.

The reader should note that we cannot have a similar translation of $\mathbf{K}_t$ (Section 3.2 of Volume 1) into **K**. The reason is that the semantics for $\mathbf{K}_t$ allows for points x, y, z such that $x \prec y, z \prec y$ and x, z are not $\prec$ comparable.

The next step is to present **QK**, the quantified reflective modal **K**, with the locality predicate W.

9.3 The *k*-dimensional language kKY

This section offers an axiomatization of the logic **kKY** and gives a translation of the quantifier system $\mathbf{Q}_0$ into it.

The logic **kKY** is slightly unusual, in the sense that its semantics is k dimensional. Its set of possible worlds has the form S^k, where S is a set of points and S^k is the set of all k-tuples $\bar{t} = (t_1, \ldots, t_k), t_i \in S, i = 1, \ldots, k$. See Definition 9.1. We need to offer an axiom system for such semantics and possibly prove completeness by constructing a canonical model of complete and consistent theories.

This means that the complete and consistent theories Θ of our logic should be identified as theories of the form

$$\Theta = \Theta_{\bar{t}} = \{A \mid (t_1, \ldots, t_k) \vDash A\}$$

in some possible model $\mathbf{m} = (S, \prec, \bar{a}, h)$ as defined in Definition 9.1.

It is reasonable in this set-up to identify a point $t \in S$ with the tuple $(t, \ldots, t)$. Hence we must have enough axioms in our system to be able to single out the theories Θ which can be said to be of the form $\Theta = \Theta_{(t,\ldots,t)}$.

Thus we need 1 and 2 below:

1. If Θ is assumed to be of the form $\Theta_{\bar{t}}$, i.e. the set of all formulae holding at some $\bar{t} = (t_1, \ldots, t_k)$, then we must be able to 'extract' from Θ the points $t_i, i = 1, \ldots, k$, as some theories $\Theta_i = \Theta_{(t_i,\ldots,t_i)}$ derived (syntactically) from Θ.
2. Having identified all such theories Θ_i, we must be able to 'reconstruct' Θ from Θ_i. More generally, we must be able to present the set of all theories Θ as a product space S^k, from the set S of all theories of the form Θ_i.

Our answer to these two questions will yield both the axioms and the completeness proof for our logic. Recall that our language contains the operators J_i and Y_i, $i = 1, \ldots, k$. The semantical satisfaction conditions for these connectives are:

- $(t_1, \ldots, t_k) \vDash J_i A$ iff $(t_i, \ldots, t_i) \vDash A$;
- $(t_1, \ldots, t_k) \vDash Y_i A$ iff $(t_1, \ldots, t_{i-1}, \mathbf{p}(t_i), t_{i+1}, \ldots, t_k) \vDash A$;
- $(s, \ldots, s) \vDash Y_i J_i A$ iff $(\mathbf{p}(s), \ldots, \mathbf{p}(s)) \vDash A$.

If we view an arbitrary theory Θ as a $\Theta_{\bar{t}}, t = (t_1, \ldots, t_k)$, for some $t_i \in S, i = 1, \ldots, k$, then as we said, it is reasonable to take the 'points' s as $\Theta_{(s,\ldots,s)}$. Thus the 'point' t_i would be $\Theta_{(t_i,\ldots,t_i)}$, and we can identify S as the set of all theories of this form. Let $\Theta_i = \{\beta \mid J_i\beta \in \Theta\}$. Θ_i is a theory of such a form. Let us use the notation $\Theta_i = J_i\Theta$. Thus $S = \{J_i\Theta \mid \Theta$ an arbitrary complete and consistent theory of the logic and $i = 1, \ldots, k\}$.

Similarly, let $Y_i\Theta = \{\beta \mid Y_i\beta \in \Theta\}$.

We note that theories Γ of the form $\Gamma = J_i\Theta$ can be singled out because they have the property that for all β and all $j = 1, \ldots, k, (\beta \leftrightarrow J_j\beta) \in \Gamma$.

Such theories we shall call one-dimensional theories (D1 theories). They are going to be the building blocks of the canonical model for **kKY**.

The ordering between D1 theories is defined by

$$\Theta \prec_i \Theta' \text{ iff } \Theta = Y_i J_i \Theta'.$$

Our axioms will show that $\prec_i$ is independent of i.

For arbitrary theories we define

$$\Theta \prec \Theta' \text{ iff } \bigwedge_i J_i\Theta = J_iY_iJ_i\Theta',$$

which more or less means that

$$\Theta_{(t_1,\ldots,t_k)} \prec \Theta_{(t'_1,\ldots,t'_k)} \text{ iff } \bigwedge_i t_i \prec t'_i.$$

Our initial plan of building a canonical model and proving completeness can now be summarized as follows. We more or less take as our set of 'points' the set of all D1 theories. We construct from each sequence $(\Theta_1, \ldots, \Theta_k)$ of D1 theories a general theory $\Theta = \Theta_{(\Theta_1,\ldots\Theta_k)}$. We define a suitable $\prec$ on the set of theories and get ourselves a canonical model. Each Θ can be retrieved (we hope) from its sequence of D1 theories $(J_1\Theta, \ldots, J_k\Theta)$ as $\Theta = \Theta_{(J_1\Theta,\ldots,J_k\Theta)}$.

This is basically the naive idea of the completeness proof. It turns out, however, that matters are not that simple and that there are two problems which need to be overcome.

The first problem is that it is possible to have two theories $\Theta \neq \Gamma$ such that for all i, $J_i\Theta = J_i\Gamma$. Thus there are many theories that have the same S-coordinates in the above proposed naive canonical model of all theories. To overcome this difficulty, we need a careful construction that will choose one theory for each sequence $(\Theta_1, \ldots, \Theta_k)$ of D1 theories and will fit them together to form a model for our initial theory Θ. Thus the main non-traditional part of the proof is some sort of detailed inductive control and accounting of which theories go into the model and how they are to fit together.

The second problem is that we cannot naively take the D1 theories themselves as our points. In a real model we can have $t \neq s$ and yet $\Theta_t = \Theta_s$. If we take as points the theories Θ themselves, we will be limiting our expressive options. The solution to this problem is to take as points sequences of D1 theories of the form $(\Theta_1, \Theta_2, \ldots, \Theta_m)$, with $\Theta_1 \prec \Theta_2 \prec \cdots \prec \Theta_m$, and where Θ_m is the 'real' theory and $\Theta_1, \ldots, \Theta_{m-1}$ is the 'history' in the construction leading up to Θ_m. Thus a point in our model will be a k-sequence of theories of the above form.

Having explained our overall plan and our main difficulties, let us proceed with the axiomatization and the soundness and completeness proof. Our language is propositional, containing the classical connectives and the additional connectives $J_1, \ldots, J_k$ and $Y_1, \ldots, Y_k$, together with the modal necessity connectives $\boxminus$ and $\Diamond (\Diamond = \sim \boxminus \sim)$.[9]

Definition 9.10 (Hilbert system for kKY). Let the Hilbert system for **kKY** be defined by the following groups of axioms and rules.

1. All substitution instances of truth functional tautologies.
2. $\boxminus$ is a **K** modal operator:

[9]The reader should not confuse this notation with that of Section 8.3.

   - $\boxminus(A \to B) \to (\boxminus A \to \boxminus B)$;
   - $\vdash A$ implies $\vdash \boxminus A$.

3. Axioms for making J_i projection operators.
   (a) J_i is a 'next' modal operator:
   - $J_i(A \wedge B) \leftrightarrow J_i A \wedge J_i B$;
   - $J_i \sim A \leftrightarrow \ \sim J_i A$;
   - $\vdash A$ implies $\vdash J_i A$.

   (b) J_i are projections:
   - $J_j(\beta \leftrightarrow J_i\beta), i, j = 1, \ldots, k$;
   - $J_i J_i \beta \leftrightarrow J_i \beta, i = 1, \ldots, k$.

   (c) Interaction with $\boxminus, \diamondminus$:
   - $\bigwedge_{1 \le i \le k} \diamondminus J_i \beta_i \to \diamondminus \bigwedge_{1 \le i \le k} J_i \beta_i$;
   - $J_r(\boxminus J_i \beta \leftrightarrow \boxminus J_j \beta), 1 \le i, j, r \le k$;
   - $\diamondminus \top \to (\diamondminus J_i \beta \leftrightarrow J_i \diamondminus J_i \beta), 1 \le i \le k$.

   (d) J_i independence axioms: let π be a permutation of $\{1, \ldots, k\}$. Let $I_1 \cup I_2 = \{1, \ldots, k\}$, $I_1 \cap I_2 = \varnothing$ be a partition of $\{1, \ldots, k\}$. Then
   - $\dfrac{\vdash \bigwedge J_i \beta_i \to \bot}{\vdash \bigwedge J_{\pi i} \beta_i \to \bot}$.
   - $\dfrac{\vdash \bigwedge_{i \in I_1} J_i \alpha_i \to \sim \bigwedge_{j \in I_2} J_j \beta_j}{\vdash \bigwedge_{i \in I_1} J_i \alpha_i \to \boxminus \sim \bigwedge_{j \in I_2} J_j \beta_j}$.

4. Axioms for making Y_i 'yesterday' operators.
   (a) Y_i is a 'yesterday' modal operator:
   - $Y_i(A \wedge B) \leftrightarrow Y_i A \wedge Y_i B$;
   - $Y_i \sim A \leftrightarrow \ \sim Y_i A$;
   - $\vdash A$ implies $\vdash Y_i A$.

   (b) Y_i are independent operators:
   - $Y_i Y_j A \leftrightarrow Y_j Y_i A, 1 \le i \ne j \le k$;
   - $J_i(Y_j J_j \beta \leftrightarrow Y_r J_r \beta), 1 \le i, j, r \le k$;
   - $J_i Y_j J_j \beta \leftrightarrow Y_i J_i \beta$;
   - $J_j Y_1 \ldots Y_k(\beta \leftrightarrow J_i \beta), 1 \le i, j \le k$.

   (c) Interaction with $\boxminus$:
   - $J_i \alpha \to \boxminus J_i Y_i J_i \alpha$;
   - $\alpha \to \boxminus Y_1 \ldots Y_k \alpha$.

Theorem 9.11 (Soundness theorem for kKY). *If* $\vdash A$ *then* A *holds in all models.*

Proof We check the axioms one by one.

1. This case is obvious.

2. Obvious.
3. (a) Obvious.
   (b) Obvious.
   (c) We check each axiom.
   - $(t_1, \ldots, t_k) \models \bigwedge_i \diamondminus J_i\beta_i$ iff for some $s_1^i, \ldots, s_k^i$ such that $\bigwedge_j t_j \prec s_j^i$ we have $(s_1^i, \ldots, s_k^i) \models J_i\beta_i$.
     The latter means $(s_i^i, \ldots, s_i^i) \models \beta_i$, $i = 1, \ldots, k$, which implies $(s_1^1, \ldots, s_k^k) \models \bigwedge_i J_i\beta_i$ which implies $(t_1, \ldots, t_k) \models \diamondminus \bigwedge_i J_i\beta_i$.
   - $(t_r, \ldots, t_r) \models \boxminus J_i\beta$ iff for all $(s_1, \ldots, s_k)$ such that $\bigwedge_i t_r \prec s_i$ we have $(s_i, \ldots, s_i) \models \beta$, iff for all s such that $t_r \prec s$ we have $(s, \ldots, s) \models \beta$ iff $(t_r, \ldots, t_r) \models \boxminus J_j\beta$.
   - Assume that for some $(s_1, \ldots, s_k)$, $\bigwedge_i t_i \prec s_i$ hold. Then $(t_1, \ldots, t_k) \models \diamondminus J_i\beta$, iff for some $s_1, \ldots, s_k$ such that $\bigwedge_i t_i \prec s_i$ we have $(s_i, \ldots, s_i) \models \beta$ iff for some $s'_j, i = 1, \ldots, k, t_i \prec s'_j$ and $(s'_i, \ldots, s'_i) \models \beta$ iff $(t_i, \ldots, t_i) \models \diamondminus J_i\beta$.

   (d)
   - Assume for any $\bar{t}$ that it is not the case that $\bigwedge_i (t_i, \ldots, t_i) \models \beta_i$; then certainly there is no $\bar{s}$ such that
     $$\vdash \bigwedge_i (s_i, \ldots, s_i) \models \beta_{\pi i}.$$
   - Assume that for all $\bar{t}$ in all models if we have $\bigwedge_{i \in I_1} (t_i, \ldots, t_i) \models \alpha_i$ then we have that for some $j \in I_2, (t_j, \ldots, t_j) \models \beta_j$. Let there be a model with $\bar{s}$ such that $\bigwedge_{i \in I_1} (s_i, \ldots, s_i) \models \beta_i$, and an $\bar{s}'$ such that $\bar{s} \prec \bar{s}'$ and $\bigwedge_{j \in I_2} (s'_j, \ldots, s'_j) \models \beta_j$.
     Then we get a contradiction with the point $\bar{t}$, where
     $$t_i = \begin{cases} s_i, \ i \in I_1 \\ s'_i \ \ i \in I_2. \end{cases}$$

4. (a) Obvious.
   (b) Obvious.
   (c)
   - Assume $(t_1, \ldots, t_k) \models J_i\alpha$. Hence $(t_i, \ldots, t_i) \models \alpha$. Let $(s_1, \ldots, s_k)$ be such that $\bigwedge_i t_i \prec s_i$. Then $(s_1, \ldots, s_k) \models J_i Y_i J_i\alpha$ iff $(s_i, \ldots, s_i) \models Y_i J_i\alpha$ iff $(s_i, \ldots, t_i, \ldots, s_i) \models J_i\alpha$ (where t_i replaces s_i in the ith place) iff $(t_i, \ldots, t_i) \models \alpha$, which we assumed to hold.
   - The soundness of $\alpha \to \boxminus Y_1 \ldots Y_k\alpha$ can be easily checked.

□

To prove completeness we need to embark on a chain of definitions and lemmas, which will provide the building blocks for our construction of a canonical model.

Definition 9.12.

1. A theory is said to be saturated if it is complete and consistent in the logic.
2. A saturated theory is said to be one dimensional (D1 theory) iff for all $i = 1, \ldots, k$ and all $\beta, (\beta \leftrightarrow J_i\beta) \in \Theta$.
3. Let Θ be a theory. Define:
   - $J_i\Theta = \{\beta \mid J_i\beta \in \Theta\}$
   - $Y_i\Theta = \{\beta \mid Y_i\beta \in \Theta\}$.
4. Let Θ and Θ' be two D1 theories. Define $\prec_i$ by $\Theta \prec_i \Theta'$ iff $\Theta = Y_iJ_i\Theta'$. Note that by axiom 4(b) $\prec_i = \prec_j$, for $1 \leq i, j \leq k$.
5. Let Θ, Θ' be two theories:
   - Define $\Theta \prec_0 \Theta'$ iff for all $\beta, \boxminus\beta \in \Theta$ implies $\beta \in \Theta'$.
   - Define $\Theta \prec \Theta'$ iff for all i,
   $$J_i\Theta = J_iY_iJ_i\Theta'.$$
   Note that $\Theta \prec \Theta'$ does not imply $\Theta \prec_0 \Theta'$.
6. Let S be the set of all D1 theories. Let $\Theta_1, \ldots, \Theta_k \in S$. Define $\langle\Theta_1, \ldots, \Theta_k\rangle$ to be the set of all saturated theories Θ such that for all $i = 1, \ldots, k, \Theta_i = J_i\Theta$. We need axioms to prove that this set is non-empty.

Lemma 9.13. *Let Θ be a saturated theory. Then:*

1. *$J_i\Theta$ is a saturated D1 theory.*
2. *$Y_i\Theta$ is a saturated theory.*
3. *If Θ is a D1 theory then $Y_jJ_j\Theta = Y_iJ_i\Theta, i \leq i, j \leq k$.*
4. *If Θ is a D1 theory then $Y_1 \ldots Y_k\Theta = Y_iJ_i\Theta$, and it is a saturated D1 theory.*
5. *If $\Theta \prec_0 \Theta'$ then for $i = 1, \ldots, k$ we have $J_i\Theta \prec_i J_i\Theta'$.*

Proof

1. Follows from axioms 3(a).
2. Follows from axioms 4(a).
3. Follows from axioms 4(b).
4. Follows from axioms 4(b).
5. Assume $\Theta \prec_0 \Theta'$. We want to show that $J_i\Theta = Y_iJ_i\Theta'$. let $\alpha \in J_i\Theta$. Then $J_i\alpha \in \Theta$. Hence by axiom 4(c) $\boxminus J_iY_iJ_i\alpha \in \Theta$ and so $J_iY_iJ_i\alpha \in \Theta'$ and so $\alpha \in Y_iJ_i\Theta'$.

□

Let us make a first attempt at the construction process of our proposed canonical model, in order to see what lemmas we need. We start with an initial saturated theory Θ, for which we want to give a model. Assume $\bar{a} = (a_1, \ldots, a_k)$ is the point in the model such that $\Theta = \Theta_{\bar{a}}$. We already know that we can identify $J_i\Theta$ as $J_i\Theta = \Theta_{(a_i, \ldots, a_i)}$.

The first lemma we need is to supply theories of the form Θ_A, for any $\diamondminus A \in \Theta$, such that $\Theta_A = \Theta_{(s_1,\ldots,s_k)}, A \in \Theta_A$, with $t_i \prec s_i, i = 1,\ldots,k$. This means that we need to have that $J_i\Theta \prec_i J_i\Theta_A$ for $i = 1,\ldots,k$ as well as $\Theta \prec_0 \Theta_A$ all hold. The next lemma ensures that.

Lemma 9.14. *Let Θ be a saturated theory and let $\diamondminus A \in \Theta$. Then there exists a saturated Θ_A such that $\Theta \prec \Theta_A$ and $\Theta \prec_0 \Theta_A$ and $A \in \Theta_A$.*

Proof We show that the following set is consistent:

$$\{A\} \cup \{\beta \mid \boxminus\beta \in \Theta\} \cup \bigcup \{J_i Y_i J_i \alpha \mid J_i\alpha \in \Theta\}.$$

Otherwise for some α_r, β_r we have

$$\vdash \bigwedge_r \beta_r \wedge \bigwedge_{i,r} J_i Y_i J_i \alpha_r^i \to\sim A.$$

Hence

$$\boxminus \bigwedge_{i,r} J_i Y_i J_i \alpha_r^i \to \boxminus \sim A \in \Theta.$$

However,

$$\vdash J_i\alpha \to \boxminus J_i Y_i J_i \alpha.$$

Hence $\boxminus \sim A \in \Theta$, a contradiction.

We can extend the above set to a saturated theory Θ_A. □

Having ensured that for each $\diamondminus A \in \Theta$, there exists a Θ_A, we have another problem. Suppose we consider k such theories, say $\Theta_{A_r}, r = 1,\ldots,k$. This means that in our proposed future model we identified points s_i^r, $r = 1,\ldots,k, i = 1,\ldots,k$, such that $a_i \prec s_i^r$ and $J_i\Theta_{A_r} = \{\beta \mid (s_i^r,\ldots,s_i^r) \vDash \beta\}$.

Let $I_0 \cup \cdots \cup I_k = \{1,\ldots,k\}$ be a partition of $\{1,\ldots,k\}$ with pairwise disjoint sets I_j, $j = 0,\ldots,k$ (some of which may be $\varnothing$). We can create a new point $\bar{x} = (x_1,\ldots,x_k)$ by letting

$$x_j = \begin{cases} a_j \text{ if } j \in I_0 \\ s_j^r \text{ if } j \in I_r. \end{cases}$$

This point $\bar{x}$ is also in the proposed future model. We must be able to identify a theory $\Theta_{(x_1,\ldots,x_k)}$ from the components of Θ and Θ_{A_r}.

To achieve this we need two lemmas.

One for the case of $I_0 = \varnothing$ and one for the case of $I_0 \neq \varnothing$. The reason for these two cases is that for the case $I_0 = \varnothing$, we have $(a_1,\ldots,a_k) \prec (x_1,\ldots,x_k)$ and we therefore expect $\Theta \prec_0 \Theta_{(x_1,\ldots,x_k)}$. If $I_0 \neq \varnothing$, we do not expect the above relation to hold.

Lemma 9.15. *Let Θ, Θ_i be saturated theories such that $J_i\Theta \prec_i J_i\Theta_i$, $i = 1,\ldots,k$, and such that $\diamondminus\top \in \Theta$. Then the following theory is consistent:*

$$\{\beta \mid \boxminus\beta \in \Theta\} \cup \bigcup_{1\le i\le k} \{J_i\alpha \mid J_i\alpha \in \Theta_i\}$$

and can be extended to a saturated theory Θ'.

Proof Otherwise for some β_j such that $\boxminus\beta_j \in \Theta$ and some $J_i\alpha^i_r \in \Theta_i$ we have

$$\vdash \bigwedge_j \beta_j \to \ \sim \bigwedge_i \bigwedge_r J_i\alpha^i_r.$$

Hence

$$\vdash \bigwedge_j \boxminus\beta_j \to \boxminus \sim \bigwedge_i \bigwedge_r J_i\alpha^i_r$$

and

$$\boxminus \sim \bigwedge_i \bigwedge_r J_i\alpha^i_r \in \Theta.$$

Since $J_i\alpha^i_r \in \Theta_i$ we have $J_i\diamondminus J_i \bigwedge_r \alpha^i_r \in \Theta, i = 1, \dots, k$. By axiom 3(c), $\diamondminus J_i \bigwedge_r \alpha^i_r \in \Theta$, $i = 1, \dots, k$, and further from 3(c) we get $\diamondminus \bigwedge_i \bigwedge_r J_i\alpha^i_r \in \Theta$, a contradiction.

The above consistent set can be extended to a saturated theory Θ'. □

Lemma 9.16. *Let $\Theta, \Theta_i, i = 1, \dots, k$, be saturated theories such that $\Theta \prec_0 \Theta_i$ and let $\{1, \dots, k\}$ be decomposed into $k+1$ pairwise disjoint sets (some may be empty) $I_0 \cup \dots \cup I_k, I_0 \neq \varnothing$. Consider the theory $\Gamma = \bigcup_{j=0}^k \Gamma_j$, where*

$$\Gamma_j = \begin{cases} \{J_j\alpha \mid J_j\alpha \in \Theta\} \text{ if } j \in I_0 \\ \{J_j\alpha \mid J_j\alpha \in \Theta_i\} \text{ if } j \in I_i. \end{cases}$$

Then Γ is consistent and can be extended to a consistent and complete theory $\Theta_{I_0,\dots,I_k}$.

Proof Otherwise for some $J_j\alpha^r_j \in \Gamma_j$ we have

$$\vdash \bigwedge_{j\in I_0} \bigwedge_r J_j\alpha^r_j \to \ \sim \bigwedge_{i>0} \bigwedge_{j\in I_i} \bigwedge_r J_j\alpha^r_j.$$

Hence by axiom 3(d)

$$\vdash \bigwedge_{j\in I_0} \bigwedge_r J_j\alpha^r_j \to \boxminus \sim \bigwedge_{i>0} \bigwedge_{j\in I_i} \bigwedge_r J_j\alpha^r_j;$$

and hence

$$\boxminus \sim \bigwedge_{i>0} \bigwedge_{j\in I_i} \bigwedge_r J_j\alpha^r_j \in \Theta.$$

However, since $\Theta \prec_0 \Theta_i$ and for $i > 0$ and $j \in I_i$ $\bigwedge_r J_j\alpha^r_j \in \Theta_j$, we get that

$$\diamondminus \bigwedge_r J_j\alpha^r_j \in \Theta, i = 1, \dots, k,$$

and by axiom 3(c) we have $\diamondminus \bigwedge_{i>0} \bigwedge_{j\in I_i} \bigwedge_\gamma J_j\alpha^j_r \in \Theta$, which is a contradiction. □

Lemma 9.17. *Let* Θ *be a theory and let* π *be a permutation of* $\{1, \ldots, k\}$. *Then* $\bigcup_i \{J_i\alpha \mid J_{\pi i}\alpha \in \Theta\}$ *is consistent and can be extended to a complete and consistent theory* $\pi\Theta$.

Proof From axioms 3(d). □

We are now ready for the construction of the model. We construct a set S of points t in stages and define a partial function $\Delta_{\bar{t}}$ on S^k also in stages.

Definition 9.18.

1. Let $\mathbb{D}$ be the set of all D1 saturated theories. Let us denote the elements of $\mathbb{D}$ by letters $a, b, d, \ldots, \in \mathbb{D}$, to distinguish them from ordinary saturated theories Θ. Let $\mathbf{p}$ be a function symbol and define $\mathbf{p}(x) = Y_1 J_1 x$, for $x \in \mathbb{D}$. Note that since x is a D1 theory we also have that $Y_1 J_1 x = Y_i J_i x$ is a D1 saturated theory. Let $a_i, i = 1, \ldots, k$, be some fixed elements of $\mathbb{D}$.
2. Let S_i be the set of all sequences of the form $\bar{x} = (b, y_1, \ldots, y_m)$ such that the following hold:
   - if $m = 0$ then $\bar{x} = (b)$;
   - $b = \mathbf{p}(y_1)$;
   - $y_r = \mathbf{p}(y_{r+1}), 1 \leq r \leq m - 1$;
   - $b = \mathbf{p}^r(a_i)$ for some $r \geq 0$, where $\mathbf{p}^0(a_i) = a_i$ and $\mathbf{p}^{r+1}(a_i) = \mathbf{p}(\mathbf{p}^r(a_i))$.
3. Let $\prec_i$ be defined on S_i by the following two cases:
   - $(b, x_1, \ldots, x_n) \prec_i (b, x_1, \ldots, x_n, x_{n+1})$; and
   - $(\mathbf{p}^{r+1}(a_i)) \prec_i (\mathbf{p}^r(a_i))$.
4. Define a function Y on S_i as follows:
   - $Y((b, x_1, \ldots, x_{n+1})) = (b, x_1, \ldots, x_n)$;
   - $Y((b, x)) = (b)$;
   - $Y((b)) = (\mathbf{p}(b))$.

Given the sets $(S_i, \prec_i)$ we are ready to define our model. The points of the model are from S^k, where $S = \bigcup_{i=1}^k S_i$. $\prec$ is defined coordinatewise using $\prec = \bigcup_i \prec_i$. We give closure conditions and single out a set $\mathbb{S} \subseteq S^k$ such that for each $\bar{t} \in \mathbb{S}$ a saturated theory $\Delta_{\bar{t}}$ is defined.

Let us begin the construction.

Construction 9.19. Let Θ be a saturated theory. Let $a_i = J_i\Theta, i = 1, \ldots, k$. Define a set $\mathbb{S}$ of elements of $(S_1 \cup \cdots \cup S_k)^k$ and a function $\Delta_{\bar{t}}$ for $\bar{t} \in \mathbb{S}$, by closure rules as follows. $(\mathbb{S}, \Delta)$ is the smallest pair closed under the following rules:

Rule 1
Let $((a_1), \ldots, (a_k))$ be in $\mathbb{S}$ as well as $((a_i), \ldots, (a_i)), i = 1, \ldots, k$. Let $\Delta_{((a_i),\ldots(a_i))} = a_i$ and $\Delta_{((a_1),\ldots,(a_k))} = \Theta$.

Rule 2
Assume $\bar{t} = (t_1, \ldots, t_k) \in \mathbb{S}$ (note that each t_i has the form $t_i = (b_i, x_1, \ldots, x_n)$)

and that $\Delta_{\bar{t}}$ and $\Delta_{(t_i,\dots,t_i)}$ are defined. We assume that $\bar{t}, \Delta_{\bar{t}}$ satisfy the following (*) condition:

(*) $\Delta_{\bar{t}}$ is a saturated theory and $\Delta_{(t_i,\dots,t_i)}$ is a D1 theory and $\Delta_{(t_i,\dots,t_i)} = J_i\Delta_{\bar{t}}$.
Let A be a wff such that $\diamondsuit A \in \Delta_{\bar{t}}$. Consider Θ_A constructed as in Lemma 9.14. Let $t_i' = (t_i, J_i\Theta_A)$ (this notation means that $J_i\Theta_A$ is concatenated as last element of the sequence t_i). Let $\bar{t}' = (t_1', \dots, t_k')$. Then let $\bar{t}' \in \mathbb{S}$ and $(t_i, \dots, t_i) \in \mathbb{S}, i = 1, \dots, k$, with $\Delta_{\bar{t}} = \Theta_A$. $\Delta_{(t_i,\dots,t_i)} = J_i\Theta_A$. Clearly $\bar{t}', \Delta_{\bar{t}'}$ also satisfy the (*) condition.

Rule 3
Let $\bar{t} = (t_1, \dots, t_k), \Delta_{\bar{t}}, \bar{t} \in \mathbb{S}$, satisfy the (*) condition. Let $m_1, \dots, m_k$ be natural numbers ≥ 0; then $\bar{s} = (Y^{m_1}t_1, \dots, Y^{m_k}t_k)$ is also in $\mathbb{S}$ with $\Delta_{\bar{s}} = Y_1^{m_1} \dots Y_k^{m_k}\Delta_{\bar{t}}$ and $\Delta_{(Y^{m_i}t_i,\dots,Y^{m_i}t_i)} = J_i\Delta_{\bar{s}}$. Note that our axioms show that $J_i\Delta_{\bar{s}} = J_iY_i^{m_i}J_i\Delta_{\bar{t}}$ and the new pairs also satisfy the (*) condition.

Further note that if $\bar{t} \prec \bar{s}$ because $\bar{t} = (Y_1s_1, \dots, Y_ks_k)$ then the axioms tell us that $\Delta_{\bar{t}} \prec_0 \Delta_{\bar{s}}$.

Rule 4
Let $\bar{t}_r = (t_1^r, \dots, t_k^r)$ be in $\mathbb{S}$ for $r = 0, \dots, k$. Further assume that $\bar{t}_r$ and $\Delta_{\bar{t}_r}$ satisfy the (*) condition. Assume for each i that $t_i^0 \prec t_i^i$. Since $\Delta_{(t_i^0,\dots,t_i^0)} = J_i\Delta_{\bar{t}_0}$ and $\Delta_{(t_i^0,\dots,t_i^0)} \prec_i \Delta_{(t_i^i,\dots,t_i^i)} = J_i\Delta_{\bar{t}_i}$, it is clear that we are in the situation of Lemmas 9.15 and 9.16, as far as these theories are concerned. We can assume that $\Delta_{\bar{t}_0} \prec_0 \Delta_{\bar{t}_i}$ holds by induction on the construction.

Let $I_0, \dots, I_k$ be a decomposition of $\{1, \dots, k\}$ as in the discussion before Lemmas 9.15 and 9.16. Let $\bar{x}$ be the point as defined in that discussion. Then let $\bar{x}$ be in $\mathbb{S}$ and $\Delta_{\bar{x}}$ be the theory as defined in these lemmas. By the constructions in the lemmas the pair satisfies the (*) condition. In case $I_0 = \varnothing$, we have $\Delta_{\bar{t}_0} \prec_0 \Delta_{\bar{x}}$ by Lemma 9.15.

Rule 5
For any permutation π of $\{1, \dots, k\}$ and any $\bar{t} = (t_1, \dots, t_k)$ let $\pi\bar{t} = (t_{\pi 1}, \dots, t_{\pi k})$ be in $\mathbb{S}$ with $\Delta_{\pi\bar{t}} = \pi\Delta_{\bar{t}}$. Clearly the (*) condition continues to be satisfied.

We are now ready to define a model. Let

$$S = \bigcup S_i$$
$$\prec = \bigcup \prec_i .$$

For each $\bar{t} = (t_1, \dots, t_k) \in S^k$ and each atom q let $h(\bar{t}, q) = 1$ if $\bar{t} \in \mathbb{S}$ and $q \in \Delta_{\bar{t}}$ and $h(\bar{t}, q) = 0$ if $\bar{t} \in \mathbb{S}$ and $\sim q \in \Delta_{\bar{t}}$ and $h(\bar{t}, q)$ be arbitrary if $\bar{t} \notin \mathbb{S}$.

Lemma 9.20. *For all $\bar{t} \in \mathbb{S}$ and A we have*

$$A \in \Delta_{\bar{t}} \textit{ iff } \bar{t} \vDash A.$$

Proof By induction on A.

1. The case of q atomic follows from the definition.

2. The classical connectives present no difficulties.
3. The cases of J_i follow from the (*) property.
4. The case of Y_i follows from the axioms.
5. We check the case of $\diamondminus$. If $\diamondminus A \in \Delta_{\bar{t}}$ then by construction for some $\bar{s}$, $\bar{t} \prec \bar{s}$ and $A \in \Delta_{\bar{s}}$.
6. Assume $\boxminus A \in \Delta_{\bar{t}}$. Let $\bar{s}$ be such that $\bar{t} \prec \bar{s}$. We observe in the construction that points $\bar{s}$ such that $\bar{t} \prec \bar{s}$ all arise from closure rules 1, 3 and 4 and that in such cases we have $\Delta_{\bar{t}} \prec_0 \Delta_{\bar{s}}$.
   Hence we have $A \in \Delta_{\bar{s}}$.

□

Theorem 9.21 (Completeness). *Let Θ be a consistent and complete theory; then Θ has a model.*

Proof Proceed with Construction 9.19 and Lemma 9.20. This yields a model where Θ holds at $((a_1), \ldots, (a_k))$. □

9.4 The system QK

Definition 9.22 (The system QK). Let **QK** be the predicate modal system with the following axioms and rules.

1. All substitution instances of classical predicate logic theorems.
2. (a) $\Box(A \to B) \to (\Box A \to \Box B)$;
   (b) $\forall x \Box A(x) \to \Box \forall x A(x)$;
   (c) $\dfrac{\vdash A}{\vdash \Box A}$.
3. (a) $\vdash \exists x W(x)$;
   (b) $\vdash W(x) \to \Box^n \sim W(x)$;
   (c) $\Diamond^n W(x) \to \ \sim W(x)$;
   (d) $\Diamond(W(x) \wedge A) \wedge \Diamond(W(x) \wedge B) \to \Diamond(W(x) \wedge A \wedge B)$;
   (e) $W(x) \wedge W(y) \to \Box(W(x) \to W(y))$;
   (f) $\Diamond^n(W(x) \wedge W(y)) \to \Box^m(W(x) \to W(y))$.
4. $\Diamond^n(W(x) \wedge W(y)) \to \Box^m(A(x) \leftrightarrow A(y))$.
5. Let A be a formula with at most m nested occurrences of $\Box$. Then we adopt the following irreflexivity type rule (see [Gabbay, 1981a] and Chapter 6 of Volume 1 [Gabbay *et al.*, 1994]):
   $$\frac{\vdash A(x) \to \Box^{m+1} \sim W(x)}{\vdash \sim A(x)}.$$
   This rule says that if it is consistent to add $\exists x A(x)$ then it is consistent to add $\exists x(A(x) \wedge \Diamond^{m+1} W(x))$.[10]

[10]We need to write $\Box^{m+1} W(x)$ because $\Box$ is not transitive and hence $t \vDash A$ cannot have effect at worlds of distance more than m from t. If we do have transitivity then we write the rule as

Theorem 9.23 (Completeness theorem for QK). **QK** *is complete for the class of all reflective predicate modal* **K** *models of the form* $(S, \prec, a, h, g)$*, with locality predicate* W.

Proof

1. Soundness can be easily verified.
2. To show completeness, let Δ_W be the set of all instances of formulae of the form $\Box^n\varphi$, where φ is an instance of an axiom from groups 3–5). This is a consistent predicate **K** theory (with constant domains). It therefore has a constant domain tree model of the form $\mathbf{m} = (S, D, \prec, a, h, g)$, where D is the domain, and $(S, \prec)$ is a tree. We shall transform $\mathbf{m}$ into a desired reflective model by using the fact that $a \vDash \Delta_W$.

Using the axiom $\Box^n \exists x W(x)$, we know that for each world t, there exists an element $a_t \in D$ such that $t \vDash W(a_t)$. There may be more than one such element. We have enough axioms in Δ, however, to enable us to identify and merge all such elements into a single-element class. We also have enough axioms in Δ to make sure that each such $W(a_t)$ holds exactly at the world t.

We can now identify a_t with t and thus get a model with $S = D$. In this model $x = y$ can be defined as $W(x) \wedge W(y)$.

So let us prove all the above.

Consider a_t. By axiom 3(b) $W(a_t)$ cannot hold at any world above t and by 3(c) it cannot hold at any world below t. What about a constellation of worlds like $s \prec^m t \wedge s \prec^n t'$ and $W(a_t)$ holding at t and t'? This would contradict axiom 3(d). We need to assume that an A exists such that $t \vDash A$ and $t' \vDash \;\sim A$. Such A's can always be arranged during the construction of the original canonical Kripke model.

We thus have shown that $W(a_t)$ holds exactly at world t.

If $W(a_t) \wedge W(b)$ both hold at t, then by axioms 3(e), 3(f) and 4, a_t and b behave as if they are equal. So we can look at equivalence classes, and basically identify a_t and b.

Thus we can map S into D by $t \mapsto a_t$. The only problem which remains to be solved is that D may contain more elements (equivalence classes) than S. We need to ensure that for any $d \in D$, there exists a t somewhere such that $d = a_t$. Such an arrangement is best taken care of during the Henkin construction of the canonical model for Δ_W. Assume that we are at some theory Δ_s, which is supposed to be the theory of the world s, and that we have $\exists x A(x) \in \Delta_s$. We want to find or add a constant d such that $A(d) \in \Delta_s$. What we do at this stage is that we also add $\Diamond^{m+1} W(d)$ to Δ_s, for a suitable m. This can be done consistently because of axiom 5.

$$\frac{\vdash A(x) \to \Box \sim W(x)}{\vdash \sim A(x)}.$$

We believe axiom 5 is not needed but this is not our main concern in this chapter.

The above considerations prove the completeness theorem. □

9.5 Introducing generalized quantifiers

We introduce a variant of van Lambalgen's theory of generalized quantifiers. We introduce a generalized quantifier of the form $(Qx)A(x, u_1, \ldots, u_n)$, where the range of x depends on $(u_1, \ldots, u_n)$ taken as a sequence. We study the properties of this quantifier and compare it with the van Lambalgen generalized quantifier (dependent on sets). We also present a reduction of generalized quantifiers into certain many-dimensional propositional temporal logics, based on the modal logic **K**. In fact we can further translate these quantifiers into quantified modal predicate logic **QK** (based on modal **K**).

We assume a predicate language with the connectives $\{\sim, \wedge, \vee, \rightarrow\}$ and the quantifiers (x) and (Ex). This is the language of classical and intuitionistic predicate logics. We shall, however, give $(x)A(x)$ and $(Ex)A(x)$ a different, generalized quantifier, interpretation. Thus the traditional quantifiers (the classical or the intuitionistic) which we denote by $\forall$ and $\exists$ will not be available in the language.

To explain our results and their background, we first need to develop some notation.

Let $\mathbf{m} = (D, h)$ denote a classical model with domain D and interpretation h. h is a function giving for each n-place atomic predicate P a subset $h(P) \subseteq D^n$. If there are constants and function symbols in the language, then h gives to each k-place function symbol e of the language a function $h(e) : D^k \mapsto D$. A constant c is considered a zero-place function symbol. To define satisfaction we also need assignments. These are functions g assigning to each variable x of the language an element $g(x) \in D$.

Given a formula $A(u_1, \ldots, u_n)$ with the free variables $u_1, \ldots, u_n$, an assignment g with $g(u_i) = a_i$, $i = 1, \ldots, n$, and a model $\mathbf{m} = (D, h)$, we write $\mathbf{m} \vDash_g A(u_1, \ldots, u_n)$ or (where no misunderstanding can arise) $\mathbf{m} \vDash A(a_1, \ldots, a_n)$ exactly when $\mathbf{m}$ satisfies $A(u_1, \ldots, u_n)$ under the assignment g in the traditional sense.

Let $A(x, u_1, \ldots, u_n)$ be a formula with exactly the displayed free variables. The classical satisfaction clause for the quantifier $(x)A(x, u_1, \ldots, u_n)$ under g (with $g(u_i) = a_i$) is:

- $\mathbf{m} \vDash_g (x)A(x, u_1, \ldots, u_n)$ iff for all $d \in D, \mathbf{m} \vDash A(d, a_1, \ldots, a_n)$. (More precisely this means that for all g' such that $g(u_i) = g'(u_i)$ for $1 \leq i \leq n$, we have $\mathbf{m} \vDash_{g'} A(x, u_1, \ldots, u_n)$.)

Generalized quantifiers arise when the range of x is restricted to a set $V^g_{x,u_1,\ldots,u_n}$ dependent on the graph $\{(u_1, a_1), \ldots, (u_n, a_n)\}$ and on the variable x.

Let R be a relation between finite graphs $\{(u_i, a_i) \mid i = 1, \ldots, n\}$ and pairs (x, d), where $u_1, \ldots, u_n, x$ are pairwise distinct variables. Then the definition of satisfaction for the quantifier is:

- $\mathbf{m} \vDash_t (x)A(x, u_1, \ldots, u_n)$ iff for all d such that $\{(u_i, g(u_i)) \mid i = 1, \ldots, n\}$ $R(x, d)$ we have $\mathbf{m} \vDash A(d, a_1, \ldots, a_n)$ iff (equivalently) for all g' such that $g'(u_i) = g(u_i)$, $i = 1, \ldots, n$, and $\{(u_i, g'(u_i))\}R(x, g'(x))$ we have $\mathbf{m} \vDash_{g'} A(x, u_1, \ldots, u_n)$.

The reader is referred to the literature for details of possible quantifier options as well as for a historical discussion. See [Alechina, 1995; van Lambalgen, 1996; Andreka *et al.*, 1995; Németi, 1995; Alechina and van Lambalgen, 1996; van Benthem and Alechina, 1996] and especially [Alechina, 1995] and [van Benthem and Alechina, 1996].

The above clause for $(x)A(x)$ makes the quantifier actually dependent on the letter x (no change of bound variables is allowed). The systems most commonly studied in the literature make R a relation between finite sets of elements of D and elements of D (i.e. $R \subseteq D^* \times D$, where D^* is the set of all *finite* subsets of D).

In this case, (x) is not dependent on the letter x, and we have the following satisfaction clause:

- $\mathbf{m} \vDash (x)A(x, a_1, \ldots, a_n)$ iff for all $d \in D$ such that $\{a_1, \ldots, a_n\}Rd$ we have $\mathbf{m} \vDash A(d, a_1, \ldots, a_n)$.

Our approach is to make the relation R depend on finite sequences of D, i.e. $R \subseteq (\bigcup_n D^n) \times D$.

Let $\mathbf{Q}_0$ denote the basic system defined by the semantics with R dependent on sequences and let $\mathbf{Q}_1$ be the basic system corresponding to R dependent on sets. In this chapter we study both systems and their relationship.

The example below motivates our interest in the system $\mathbf{Q}_0$.

Consider the formula $B_0 = (u)A(x, y, c, u)$, where A is a four-place atomic predicate, c is an individual constant, and x, y are free in A.

The current notion of a generalized quantifier (system $\mathbf{Q}_1$) lets the range of (u) depend on the set of elements $\{c, x, y\}$ through a relation R. Thus evaluating B_0 in a model $\mathbf{m}$ we get (note the dependence on c!):

- $\mathbf{m} \vDash_g (u)A(x, y, c, u)$ iff for all d such that $\{h(c), g(x), g(y)\}Rd$, $\mathbf{m} \vDash_g A(x, y, c, d)$.

This makes (u) look like a modality, and, in fact, some authors do write the quantifier as

$$\Box_u A(x, y, c, u).$$

Our system $\mathbf{Q}_0$ is different in two main respects:

1. Since c is a constant in the formula and does not arise from quantifier instantiation, we have no dependence on c. The dependence is only on variable instantiations.
2. The dependence is also on the order of instantiation.

Thus we need to know whether x was substituted first or y first. Without this knowledge we cannot evaluate. Thus if y is substituted first and then x we will get:

- $\mathbf{m} \vDash_g (u)A(x, y, c, u)$ iff for all d such that $(g(y), g(x))Rd, \mathbf{m} \vDash_g A(x, y, c, d)$.

If x was substituted first and then y, we will get:

- $\mathbf{m} \vdash (u)A(x, y, c, u)$ iff for all d such that $(g(x), g(y))Rd, \vDash A(x, y, c, d)$.

We, of course, let R be a relation between sequences and points, and thus the two options will not be the same in general.

We can ask the question: why this choice of interpretation?

There are two main reasons for considering quantifiers dependent on sequences. Both reasons are technical. There is also some supporting evidence from some applications where such quantifiers seem more convenient to use. Such quantifiers also have some supporting evidence in natural language.

Our first observation is that in many applications the domain of discourse is itself structured. The structure of the domain may come from some additional considerations which are not directly formalized in the object level, but which nevertheless have direct influence on the deduction machinery of the systems. Here are some examples.

- The sequencing of elements in discourse representation structures. In DRS (see [Kamp and Reyle, 1993]) theory elements are introduced in nested boxes and elements from one box may or may not be accessible to predicates in other boxes. Consider a predicate $R(u_1, u_2, u_3)$ in some inner discourse box. The predicate slots are $1, 2, 3$ but the elements u_1, u_2, u_3 may have been introduced (in the DRS sense) in different boxes endowing these elements with a different order, say (u_2, u_3, u_1).
  This order may or may not have linguistic significance.
- In temporal labelled deductive systems or even in ordinary temporal Kripke models, there is a sequencing of elements according to their temporal 'coming into existence'.
- In Fine's theory of abstract objects, abstract elements have hierarchical dependence.
- In theoretical computer science, when evaluating an expression with variables through the so-called lazy evaluation, the result depends on the order of instantiation.

In all of the above systems, we have essentially two orderings (or more) of the variables of an atomic predicate $P(x_1, \ldots, x_n)$. One is the order of the slots of the predicate $P(x_1, \ldots, x_n)$. The other has to do with the extra structure on the elements $\{a_1, \ldots, a_n\}$ intended to be substituted as $x_1/a_2, \ldots, x_n/a_n$. This extra structure can be expressed through the order of their substitution into their respective slots in the predicate.

To the extent that a quantifier may depend on these elements, its dependence can be on the extra structure of the elements, namely on them as a sequence. The nature of the dependence may depend on the application area.

The second motivation for considering the quantifier $\mathbf{Q}_0$ and dependence on sequences comes from van Lambalgen's theory of the quantifier $\mathbf{Q}_1$ itself. The sequence dependence is already existing there! In fact, in the cause of technical purity and beauty of exposition, van Lambalgen himself could have stated his original theory, in our opinion, in terms of sequence dependence. The elegance of the sequence formulation becomes more apparent when we do our translation of the quantifiers into many-dimensional temporal logic and quantified modal logic **K**.

Let us say more about the implicit sequencing residing in van Lambalgen's theory itself.

Consider the closed formula

$$B(c) = (y)(x)(u)A(x, y, c, u).$$

This formula contains an ordering of quantifiers. If we want to make the inner quantifiers dependent on the sequence of external quantifiers governing them, then we should make the range of (u) depend on the (sequential) choices for (x) and (y). The constant c does not play a role here.

If we let $V_{\bar{u}}$ denote a range dependent on a sequence of variables $\bar{u} = (u_1, \ldots, u_n)$ (where the relation $\bar{u}Rx$ is the same as $x \in V_{\bar{u}}$), then we get that $B(c)$ can be translated into classical logic as follows ($\forall$ is the classical universal quantifier):

$$(\forall y \in V_{\varnothing})(\forall x \in V_y)(\forall u \in V_{(y,x)})A(x, y, c, u).$$

Clearly there is no dependence on c.

It is clear that we have hidden ordering on the variables y, x, u as witnessed by the sets $y \in V_{\varnothing}, x \in V_y, u \in V_{(y,x)}$. Note the formal resemblance between the indices $y, (y, x)$ and the set-theoretic notation of an ordered pair $\{y, \{y, x\}\}$.

The sceptic can, of course, turn this second point round and argue that we do not need to consider $\mathbf{Q}_0$ quantification (depending on sequences) since the van Lambalgen quantifier $\mathbf{Q}_1$ itself already has sequencing in it. This sceptic counter-argument can gain strength from Theorem 9.41 below which states that for any closed formula A, $\mathbf{Q}_0 \vdash A$ iff $\mathbf{Q}_1 \vdash A$.

We would reply to the sceptic by saying that $\mathbf{Q}_0$ has a neater theory at a very low generalization cost.

The quantifier system $\mathbf{Q}_0$ receives some supporting evidence from natural language, though, unfortunately, we could find nothing conclusive.

There are predicates in natural language where more than one ordering is relevant, as we shall now show. Consider the following two sentences:[11]

1. A priest and an undertaker visited every patient.
2. An undertaker and a priest visited every patient.

The range of patients depends on the set {priest, undertaker}. However, there is also a hidden temporal sequencing here. If the priest 'does his job' first and then the undertaker, the range would be that of dying patients, where the priest possibly takes the confession. If the undertaker comes first and then the priest,

[11] I am grateful to Ruth Kempson for these examples.

then the range would be of patients who have just died, where the priest conducts the burial ceremony.

Consider the predicate $R(x, y, z, c)$, meaning roughly that the plumber, electrician and joiner renovated the kitchen c. Assume that the work is done according to strict specifications. Here the order of slots in the predicate R determines who is the plumber, electrician and joiner. The experienced housewife/homebuilder will tell you that it makes a difference whether the plumber come first and then the joiner or whether the joiner does his bit first and then the plumber. This order can be expressed by the order of instantiation. Unfortunately, in English there seems to be no difference between the following two sentences:

Every joiner and every electrician and every plumber can do a perfect kitchen together.

Every plumber and every joiner and every electrician can do a perfect kitchen together.

There is some weak evidence for sequencing in other examples:

John and Mary were both at the party. All the guests thought he was much better dressed than her.

To say 'she was not as well dressed as him' sounds odd. Here we want a certain order of instantiation.

Note also that our logic is a sort of free logic as well as generalized quantifier logic (see [Bencivenga, 1986; Lambert and van Fraassen, 1972]); we do not have $\vDash B(c) \to (Ez)B(z)$, as we are not sure that the interpretation of c falls within the range $V_\varnothing$.

We now need the syntactical machinery to make our distinctions precise and the appropriate semantics for the language.

Definition 9.24.

1. Let **L** be the language of predicate logic with a stock of the following:
   (a) n-place atomic predicates of the form $P(x_1, \ldots, x_n)$;
   (b) the connectives $\{\neg, \wedge, \vee, \to, \top, \bot\}$ (which can be interpreted classically or intuitionistically, depending on our logic base);
   (c) the (generalized) quantifiers (x) and (Ex);
   (d) variables, constants and function symbols generating terms of the form $t(x_1, \ldots, x_k)$;
   (e) we also allow for a special slot substitution operator $\mathcal{S}$, which we consider as not part of the language but as a metalevel device for ordering variables.
2. The notion of a predicate wff with the set of free variables $\{x_1, \ldots, x_n\}$ is defined in the traditional manner. The symbol $\mathcal{S}$ is not used in this definition.
3. Let $A(x_1, \ldots, x_n)$ be a predicate wff with exactly the free variables $\bar{x} = (x_1, \ldots, x_n)$. Let $\bar{u} = (u_1, \ldots, u_n)$ be a sequence of variables and let π be a permutation of $\{x_1, \ldots, x_n\}$. The expression $\mathcal{S}^{\pi x_1, \ldots, \pi x_n}_{u_1, \ldots, u_n} A$ is called a

basic *slot formula* and represents the fact that we want to substitute u_i for $\pi x_i, i = 1, \ldots, n$, in the order indicated by $\mathcal{S}$.[12]
When no misunderstanding can arise we write $\mathcal{S}^{\bar{x}}_{\bar{u}} A$, assuming $\bar{x}$ is already ordered in the order of substitution. If $\bar{u} = \bar{x}$ we write $\mathcal{S}^{\bar{x}} A$.[13]

4. A classical model has the form $\mathbf{m} = (D, h)$, where h is the interpretation of the atomic predicates (i.e. for each n-place predicate P, $h(P) \subseteq D^n$), and of function symbols.
Let $\bar{u}$ be a sequence of variables and g, g' two assignments. We write $g =_{\bar{u}} g'$ to mean that for all x in $\bar{u}$ we have $g(x) = g'(x)$.
An assignment g is a function giving values in D to all variables.
5. It may be convenient for us to present a classical model as $\mathbf{m} = (D, h, g)$, i.e. as coming with an assignment already included, when this assignment does not change through the discussion. Sometimes we may write $\mathbf{m} = (D, h, \mathcal{G})$, where $\mathcal{G}$ is a family of assignments. We also abuse notation and write $A(a)$ to mean $A(x)$ with x free when the intention is that we are dealing with assignments g such that $g(x) = a$.

The substitution operator $\mathcal{S}$ is a device to indicate in what order the elements are substituted in a formula or a predicate. Thus if $P(x, y, z)$ is a three-place predicate and we want to substitute for y first, for x second and then for z the sequence of values (a, b, c), then we shall write $\mathcal{S}^{y,x,z}_{a,b,c} P(x, y, z)$. $\mathcal{S}$ is just a

[12]We said that the slot operator $\mathcal{S}$ is not part of the language but is a metalevel device for ordering free variables. We thus need not worry about questions like how $\mathcal{S}$ interacts with the connectives and quantifiers. However, we can allow $\mathcal{S}$ to be an operator in the object language. In this case we can give the following definition.

- Let $A(x_1, \ldots, x_n)$ be an $\mathcal{S}$-free formula of the language with the indicated exact free variables. Then $\mathcal{S}^{x_1,\ldots,x_n}_{u_1,\ldots,u_n} A$ is a basic slot formula. We say the formula is ready to accept quantification on the variable u_n, and that its free variable ordering is u_1 first, u_2 second, etc.
- A basic slot formula is a slot formula.
- If A and B are slot formulae whose free variable orderings are compatible (i.e. they have the same relative order on the common variables) then $A \wedge B$, $A \vee B$, $A \to B$ and $\sim A$ are also slot formulae with the combined ordering and the last variable in the ordering is the one to accept quantification.
- If A is a slot formula with free variables $u_1, \ldots, u_n$ with last variable u_n and next to last variable u_{n-1} and $\mathbf{Q}$ is any quantifier, then $(\mathbf{Q}u_n)A$ is a slot formula with free variables $u_1, \ldots, u_{n-1}$, ordered in the inherited ordering, and u_{n-1} is the last free variable ready for quantification.

The following reduction rules hold for the slot operator.

- $\mathcal{S}^{\bar{x}}_{\bar{u}}(A \wedge B) = \mathcal{S}^{\bar{x}}_{\bar{u}} A \wedge \mathcal{S}^{\bar{x}}_{\bar{u}} B$
$\mathcal{S}^{\bar{x}}_{\bar{u}} \sim A = \; \sim \mathcal{S}^{\bar{x}}_{\bar{u}} A$
$\mathcal{S}^{\bar{x}}_{\bar{u}}(\mathbf{Q}z)A(z, \bar{x}) = (\mathbf{Q}y)\mathcal{S}^{\bar{x},z}_{\bar{u},y} A(z, \bar{x})$
$\mathcal{S}^{\bar{y}}_{\bar{v}} \mathcal{S}^{\bar{x}}_{\bar{u}} A = \mathcal{S}^{\bar{y},\bar{x}}_{\bar{v},\bar{u}} A$
$\mathcal{S}^{\pi_1 \bar{x}}_{\bar{u}} \mathcal{S}^{\pi_2 \bar{y}}_{\bar{x}} A = \mathcal{S}^{\pi_1 \pi_2 \bar{y}}_{\bar{u}} A$, where π_1, π_2 are permutations.

[13]Consider a formula $A(x, y, z)$ with the three variables $\{x, y, z\}$ indicating slots and no order of substitution. $\mathcal{S}^{y,x,z}_{u_1,u_2,u_3} A$ indicates that we want u_1 substituted first in slot y etc. $\mathcal{S}^{y,x,z}_{y,x,z} A = \mathcal{S}^{y,x,z} A$ indicates that slot y should be filled first, x second and z third.

metalevel notation. $\mathcal{S}^{y,x,z}_{a,b,c}$ can be viewed as an external label to the formula $P(x,y,z)$, indicating order of variable substitution.

To see the usefulness of $\mathcal{S}$, let us look at another example.

Consider the formula $(y)(x)(Ez)P(x,y,z)$. According to standard evaluation procedures, the y place is substituted first, then the x place and then the z place. We get that the above formula holds iff the following:

- for all $a \in D$ such that $\varnothing Ra$ and for all $b \in D$ such that aRb there exists a $c \in D$ such that $(a,b)Rc$ and $P(b,a,c)$ hold.

Consider the last stage in the evaluation. We had $(Ez)P(b,a,z)$ to evaluate. The expression '$(Ez)P(b,a,z)$' is not sufficient to describe the evaluation state. We need to indicate that in this expression a was substituted first and then b. This fact determines the range of (Ez), being $\{d \mid (a,b)Rd\}$ as opposed to $\{d \mid (b,a)Rd\}$.

We can make use of the slot substitution operator $\mathcal{S}$ and present the intended meaning as

$$(Ez)\mathcal{S}^{y,x}_{a,b}P(x,y,z).$$

We should therefore present the evaluation process as follows:

$\mathbf{m} \vDash (y)(x)(Ez)P(x,y,z)$ iff for all $a \in D$ such that $\varnothing Ra$ we have $\mathbf{m} \vDash \mathcal{S}^{y}_{a}(x)$ $(Ez)P(x,y,z)$ iff for all $a,b \in D$ such that $\varnothing Ra \wedge aRb$ we have $\mathbf{m} \vDash \mathcal{S}^{y,x}_{a,b}(Ez)$ $P(x,y,z)$ iff for all $a,b \in D$, there exists a $c \in D$ such that $\varnothing Ra \wedge aRb \wedge (a,b)Rc$ and $\mathbf{m} \vDash \mathcal{S}^{y,x,z}_{a,b,c}P(x,y,z)$.

Therefore the semantic evaluation clause for expressions of the form $B = \mathcal{S}^{x_1,\ldots,x_n}_{a_1,\ldots,a_n}(y)A(y,x_1,\ldots,x_n)$ where $x,x_1,\ldots,x_n$ are all the free variables of A should be:

- $\mathbf{m} \vDash \mathcal{S}^{\bar{x}}_{\bar{a}}(y)A(y,\bar{x})$ iff for all $d \in D$ such that $\bar{a}Rd$ we have $\mathbf{m} \vDash \mathcal{S}^{\bar{x},y}_{\bar{a},d}A(y,\bar{x})$.

Definition 9.25 (Tree models and $\mathbf{Q}_0$ semantics).

1. Let $\mathbb{E} = \{e_1, e_2, e_3, \ldots\}$ be a set of distinct atomic points and let $a \notin \mathbb{E}$ be another point. a is called *the root* and will usually be taken to be the empty set $\varnothing$. We define a tree $\mathbb{S}_a$ as follows:

$$\mathbb{S}_a = \{(a, e_1, e_2, \ldots, e_k) \mid k \geq 1, e_i \in \mathbb{E}, \text{ for } 1 \leq i \leq k\} \cup \{(a)\}.$$

   Define $\prec_a$ on $\mathbb{S}_a$ by letting:

   - $(a, e_1, \ldots, e_k) \prec_a (a, e'_1, \ldots, e'_r)$ iff $r = k+1$ and $e_i = e'_i$, for all $1 \leq i \leq k$. We also allow $k = 0$ which gives $(a) \prec_a (a, e)$.

   Clearly if $\alpha, \beta \in \mathbb{S}_a$ then $\alpha \prec_a \beta$ iff $\beta = \alpha * (e)$ for some e and where $*$ denotes concatenation.

   Let $S \subseteq \mathbb{S}_a$. We say that S has no gaps iff $\beta \in S$, $\alpha \neq (a)$ and $\alpha \prec_a \beta$ imply $\alpha \in S$.

2. Let D be a set and assume that the empty set $\varnothing$ is not an element of D. We can present D as a tree set by using a one-to-one mapping $\mathbf{f} : D \mapsto \mathbb{S}_\varnothing - \{(\varnothing)\}$ such that $\mathbf{f}(D)$ has no gaps in $\mathbb{S}_\varnothing$.

We define the following decomposition of D. Let for $n \geq 1$,

$$D_n = \{d \in D \mid \mathbf{f}(d) \text{ has the form } (\varnothing, e_1, \ldots, e_n)\}.$$

Then clearly $\{D_n\}$ are pairwise disjoint sets, and $D = \bigcup_{n \geq 1} D_n$.
We can import $\prec_\varnothing$ from $\mathbb{S}_\varnothing$ as a relation $\prec$ on $D \cup \{\varnothing\}$ by letting:

- $\varnothing \prec d$ iff $d \in D_1$;
- $x \prec y, x, y \in D$ iff $\mathbf{f}(x) \prec_\varnothing \mathbf{f}(y)$.

We say that $(\mathbf{f}, \prec)$ is a tree decomposition of $D \cup \{\varnothing\}$.

3. Let $(\mathbf{f}, \prec)$ be a tree decomposition of $D \cup \{\varnothing\}$. Let R be a relation (of the form $(t_1, \ldots, t_k)Rt_{k+1}$) between finite sequences of elements of D and elements of D (i.e. $R \subseteq (\bigcup_{n \geq 0} D^n) \times D$, where $D^0 = \varnothing$ and $D^{n+1} = D^n \times D$).
   We say that $(\mathbf{f}, \prec)$ represents R faithfully iff the following hold:
   - $\varnothing Rd$ iff $\varnothing \prec d$;
   - $(t_1, \ldots, t_k)Rt_{k+1}$ iff $\varnothing \prec t_1 \wedge \bigwedge_{i=1}^{k} t_i \prec t_{i+1}$ iff $\varnothing Rt_1 \wedge \bigwedge_{i=1}^{k}(t_1, \ldots, t_i)Rt_{i+1}$.
4. A $\mathbf{Q}_0$ model has the form $\mathbf{m} = (D, R, h, g)$ where (D, h, g) is a classical model and R is a relation between finite sequences of elements of D and elements of D, i.e. $R \subseteq (\bigcup_n D^n) \times D$.
5. A $\mathbf{Q}_0$ model is said to be a tree model iff there exists a tree decomposition $(\mathbf{f}, \prec)$ of $D \cup \{\varnothing\}$ which represents R faithfully. This means that $D = \bigcup_n D_n$, where D_n are pairwise disjoint, and for each element $t \in D_n$, there exists a unique sequence of elements $t_1, \ldots, t_n$ such that $t_n = t, t_i \in D_i$, and $\varnothing Rt_1$ and for all $1 \leq i \leq n-1$ we have $(t_1, \ldots, t_i)Rt_{i+1}$.

We call $\prec$ the successor relation *derived* from R.

Definition 9.26 ($\mathbf{Q}_0$ satisfaction). Let $\mathbf{L}$ be the syntactical language of classical logic with the quantifiers (x) and (Ex). Let $\mathbf{m}$ be a $\mathbf{Q}_0$ model. We are going to define the notion of $\mathbf{m} \vDash A$ for a slot formula $\mathcal{S}^{\bar{u}}A$ for any formula A of $\mathbf{L}$. This definition of satisfaction will give the quantifiers the meaning of sequence generalized quantifiers and will be formulated using the syntactical help of the slot operator $\mathcal{S}$.

Let $A(x_1, \ldots, x_n)$ be a predicate logic formula with (all) the free variables set $\{x_1, \ldots, x_n\}$. These variables are not ordered. Let $\bar{u} = (u_1, \ldots, u_n)$ be a sequencing of the set of variables (i.e. $\{u_1, \ldots, u_n\} = \{x_1, \ldots, x_n\}$). We define satisfaction for expressions of the form $\alpha = \mathcal{S}^{\bar{u}}A$, which we refer to as slot formulae, by structural induction as follows.

The reader should bear in mind the obvious reduction rules of the substitution operator $\mathcal{S}$:

- $\mathbf{m} \vDash \mathcal{S}^{\bar{u}}P(x_1, \ldots, x_n)$ iff $(g(x_1), \ldots, g(x_n)) \in h(P)$;
- $\mathbf{m} \vDash \alpha \wedge \beta$ iff $\mathbf{m} \vDash \alpha$ and $\mathbf{m} \vDash \beta$;
- $\mathbf{m} \vDash \sim \alpha$ iff $\mathbf{m} \nvDash \alpha$;

- $\mathbf{m} \vDash \mathcal{S}^{\bar{u}}(y)A(y, x_1, \ldots, x_n)$ iff for all $g' =_{\bar{u}} g$ such that $g'(\bar{u})Rg'(y)$ holds we have, $\mathbf{m}' \vDash \mathcal{S}^{\bar{u},y}A(y, x_1, \ldots, x_n)$, where $\mathbf{m}' = (D, R, h, g')$.

Let $\mathcal{Q}_0$ be the class of all $\mathbf{Q}_0$ models. Let $\mathcal{S}^{\bar{x}}A$ be a slot formula. We write $\mathcal{Q}_0 \vDash \mathcal{S}^{\bar{x}}A$ or just $\vDash \mathcal{S}^{\bar{x}}A$ iff for all $\mathbf{m} \in \mathcal{Q}_0$ we have $\mathbf{m} \vDash \mathcal{S}^{\bar{x}}A$.

Note that the notion of satisfaction is defined for slot formulae. The tradition in classical logic is to evaluate an open formula $A(\bar{x})$ in $\mathbf{m}$ by evaluating essentially $A(g(\bar{x}))$. We cannot do the same here because we need the order of substitution of $g(\bar{x})$. So open formulae for us must be open slot formulae.

Example 9.27. Consider the wff $\alpha = (x)(y)(A(x) \wedge B(y))$ and the wff $\beta = (x)A(x) \wedge (y)B(y)$. In our sequence relation semantics as well as in the van Lambalgen set relation semantics, these two are not semantically equivalent.

We have

$$\mathbf{m} \vDash \alpha \text{ iff } (\forall x \in V_{\varnothing})(\forall y \in V_x), \mathbf{m} \vDash A(x) \wedge B(y)$$
$$\mathbf{m} \vDash \beta \text{ iff } (\forall u \in V_{\varnothing}), \mathbf{m} \vDash A(u) \wedge B(u).$$

This means movement of quantifiers across formulae must be done carefully.

Definition 9.28 ($\mathbf{Q}_1$ models and $\mathbf{Q}_1$ satisfaction). We similarly define the notions of $\mathbf{Q}_1$ models and $\mathbf{Q}_1$ satisfaction for the case where R is a relation between finite subsets and points of the domain. Of course in this case we can define satisfaction for open formulae since we need not have any order on the free variables.

Example 9.29. Let us now check the satisfaction of some well-known classical rules.

1. The rule of *modus ponens*, $\vDash A$ and $\vDash A \to B$ imply $\vDash B$:
   This rule cannot be stated without a detailed accounting of the order of variables. Consider $A(\bar{x}, \bar{y})$ and $B(\bar{x}, \bar{z})$. To evaluate A and B in a model we need order on the variables. There are several questions to ask when we check the validity of *modus ponens*:
   (a) Must $\bar{y}$ be empty?
   (b) Do we need to assume that $\bar{x}$ is ordered in the same way in A as in B?
   (c) Does $\bar{z}$ have to be ordered after $\bar{x}$ in B?
   Let us check *modus ponens* in the semantics and see what we need. Assume $\vDash A$ and $\vDash A \to B$ and that for some $\mathbf{m}, \mathbf{m} \nvDash B$. This means $\mathbf{m} \nvDash \mathcal{S}^{\bar{x},\bar{z}}_{g(\bar{x}),g(\bar{z})}B$.[14] However, $\mathbf{m} \vDash \mathcal{S}^{\bar{x},\bar{y}}_{g(\bar{x}),g(\bar{y})}A$ and $\mathbf{m} \vDash \mathcal{S}^{\bar{x},\bar{y},\bar{z}}_{g(\bar{x}),g(\bar{y}),g(\bar{z})}(A \to B)$. For '$\mathcal{S}^{\bar{x},\bar{y},\bar{z}}_{g(\bar{x}),g(\bar{y}),g(\bar{z})}A$' to match '$\mathcal{S}A$' in '$\mathcal{S}(A \to B)$', we must have that $\bar{x}$ is in the same order in A as in $A \to B$, and that $\bar{y} = \bar{z}$.

[14] $\mathcal{S}^{\bar{u}}A(x_1, \ldots, x_n)$ means that we want slot u_1 to be filled first etc. Let g be an assignment of a model $\mathbf{m}$, giving any variable u an element $g(u) \in D$. $\mathbf{m} \vDash \mathcal{S}^{\bar{u}}_{g(\bar{u})}A$ is a redundant abuse of notation allowing us visually to display the assignment g (of $\mathbf{m}$) involved in $\mathbf{m} \vDash \mathcal{S}^{\bar{u}}A$. See item 5 of Definition 9.25.

2. The rule of generalisation:
   Assume $\vDash A(\bar{x}, \bar{y}) \to B(\bar{x}, \bar{z})$, can we deduce $\vDash A \to (\bar{z})B$?
   Notice that we can quantify over $\bar{z}$ only when $\bar{z}$ is last in the order in B, as we indeed indicated.
   Let $\mathbf{m}$ be such that $\mathbf{m} \vDash A$ and $\mathbf{m} \nvDash (\bar{z})B$.
   Then for some $\bar{a}, \bar{b}$ and $\bar{c}$ we have $\mathbf{m} \vDash \mathcal{S}^{\bar{x},\bar{y}}_{\bar{a},\bar{b}} A$ and $\mathbf{m} \vDash \sim \mathcal{S}^{\bar{x},\bar{y},\bar{z}}_{\bar{a},\bar{b},\bar{c}}(A \to B)$.
   We get a contradiction only if $\bar{x}$ is ordered the same way in A and in B and $\bar{y} = \varnothing$.

We shall axiomatize $\mathcal{Q}_0, \mathcal{Q}_1$ in a later section. First we need to understand the R-semantics better. This is the task of the next section.

9.6 Properties of the semantics

This section develops properties of the semantics with relation R, for finite sets or for finite sequences.

Definition 9.30. A model $\mathbf{m} = (S, R, h, g)$ is said to be normalized iff R satisfies the following condition (which we call $SEQ(a_1, \ldots, a_{n+1})$) for all n and all $(a_1, \ldots, a_{n+1})$:

$$(a_1, \ldots, a_n)Ra_{n+1} \text{ iff } \varnothing Ra_1 \wedge \bigwedge_{i=1}^{n} (a_1, \ldots, a_i)Ra_{i+1}.$$

Theorem 9.31. *For any closed wff A, if A is false in some model then A is false in some normalized model.*

Proof The proof follows from the next lemma. □

Lemma 9.32. *Let $\mathbf{m} = (S, R, h, g)$ be a $\mathbf{Q}_0$ model. Let R_1 be defined by*

$$(u_1, \ldots, u_n)R_1 u_{n+1} \textit{ iff } \varnothing Ru_1 \wedge \bigwedge_{i=1}^{n} (u_1, \ldots, u_i)Ru_{i+1}.$$

Let $(a_1, \ldots, a_n)$ be such that $SEQ(a_1, \ldots, a_n)$ holds, i.e. $\varnothing Ra_1 \wedge \bigwedge_{i=1}^{n-1} (a_1, \ldots, a_i)$ Ra_{i+1}, and let $A(x_1, \ldots, x_n)$ be a formula with exactly the indicated free variables; then

$$\mathbf{m} \vDash A(a_1, \ldots, a_n) \textit{ iff } \mathbf{m}_1 \vDash A(a_1, \ldots, a_n)$$

where $\mathbf{m}_1 = (S, R_1, h, g)$.

Proof By induction on the structure of A.

The case of atomic A follows immediately. The classical connectives present no difficulties. We check the quantifier case. Note that by our assumption of $SEQ(a_1, \ldots, a_n)$, we have for arbitrary a_{n+1} that $(a_1, \ldots, a_n)Ra_{n+1}$ iff $(a_1, \ldots, a_n)R_1 a_{n+1}$. Hence, $\mathbf{m} \vDash (x)A(a_1, \ldots, a_n, x)$ iff for all a_{n+1} such that $(a_1, \ldots, a_n)Ra_{n+1}$ we have $\mathbf{m} \vDash A(a_1 \ldots, a_{n+1})$.

Since $SEQ(a_1, \ldots, a_n, a_{n+1})$ holds, we can use the induction hypothesis and continue: iff for all a_{n+1} such that $(a_1, \ldots, a_n)R_1 a_{n+1}$ we have $\mathbf{m}_1 \vDash A(a_1, \ldots, a_{n+1})$ iff $\mathbf{m}_1 \vDash (x)A(a_1, \ldots, a_n, x)$. □

Definition 9.33. Let $\mathbf{m} = (S, R, h, g)$ be a normalized $\mathbf{Q}_0$ model.

1. Define a sequence of sets $D'_n, n = 1, 2, \ldots$, as follows:
   Case 0
   Let $D'_0 = \{(0, a) \mid \varnothing R a\}$.
   Case $n + 1$
   Assume $D'_k, k \leq n$, have all been defined and that the elements of D'_k have the form $(k, (a_1, \ldots, a_k)), a_1, \ldots, a_k \in D$. Further assume that for $0 \leq k \leq n$ we have that for each $i \leq k, (i, (a_1, \ldots, a_i)) \in D'_i$ and that $(a_1, \ldots, a_{i-1})Ra_i$ hold.
   We define D'_{n+1}. For each $(n, (a_1, \ldots, a_n)) \in D'_n$ and each $d \in D$ such that $(a_1, \ldots, a_n)Rd$ let $(n + 1, (a_1, \ldots, a_n, d))$ be in D'_{n+1}.

   Let $D' = \bigcup_n D'_n$.
   Let R' be defined as

$$\{(\varnothing, (1, a)) \mid \varnothing R a\} \cup$$
$$\{(((1, a_1), \ldots, (n, (a_1, \ldots, a_n))),$$
$$(n + 1, (a_1, \ldots, a_{n+1}))) \mid \varnothing R a_1 \text{ and } \textstyle\bigwedge_{i=1}^{n} (a_1, \ldots, a_i) R a_{i+1}\}.$$

   For $t = (n, (a_1, \ldots, a_n))$ let $|t| = a_n$. Let h' be defined as follows:

$$h'(P) = \{(t_1, \ldots, t_k) \in (D')^k \mid (|t_1|, \ldots, |t_k|) \in h(P)\}.$$

   Let g' be any assignment such that for all variable x we have $|g'(x)| = g(x)$. Then the model $\mathbf{m}'$ is said to be a normalized tree model *for* $\mathbf{m}$.[15]

Theorem 9.34. *Let* $\mathbf{m}$ *be a normalized model and let* $\mathbf{m}'$ *be a normalized tree model for* $\mathbf{m}$. *Let* $A(x_1, \ldots, x_n)$ *be a wff with the free variables* $x_1, \ldots, x_n$. *Then for all* $t_1, \ldots, t_n \in D$ *such that* $SEQ(t_1, \ldots, t_n)$ *we have*

$$\mathbf{m}' \vDash \mathcal{S}^{x_1, \ldots, x_n}_{t_1, \ldots, t_n} A \textit{ iff } \mathbf{m} \vDash \mathcal{S}^{x_1, \ldots, x_n}_{|t_1|, \ldots, |t_n|} A.$$

Proof By structural induction on A.

1. For atomic A this holds by definition (of h').
2. The classical connectives present no difficulties.
3. Assume $\mathbf{m}' \vDash \mathcal{S}^{x_1, \ldots, x_n}_{t_1, \ldots, t_n}(y)A(y, x_1, \ldots, x_n)$. We show that $\mathbf{m} \vDash \mathcal{S}^{x_1, \ldots, x_n}_{|t_1|, \ldots, |t_n|}(y)A$.
   Let $d \in D$ be such that $(|t_1|, \ldots, |t_n|)Rd$. Since $\mathbf{m}$ is a normalized model, we have that $\varnothing R |t_1| \wedge \bigwedge_{i=1}^{n-1} (|t_1|, \ldots, |t_{i-1}|) R |t_i|$ holds. Let $t_{n+1} = (n +$

[15] We leave it to the reader to note that R' gives rise to a faithful tree decomposition for R in the sense of Definition 9.25.

$1, (|t_1|, \ldots, |t_n|, d))$. Then $(t_1, \ldots, t_n)R't_{n+1}$ holds and hence $\mathbf{m}' \vDash A(t_{n+1}, t_1, \ldots, t_n)$ and by the induction hypothesis $\mathbf{m} \vDash A(d, |t_1|, \ldots, |t_n|)$. Hence $\mathbf{m} \vDash (y)A(y, |t_1|, \ldots, |t_n|)$.

4. Assume $\mathbf{m} \vDash (y)A(y, |t_1|, \ldots, |t_n|)$. We show that $\mathbf{m}' \vDash (y)A(y, t_1, \ldots, t_n)$. Let t_{n+1} be such that $(t_1, \ldots, t_n)R't_{n+1}$.
Then certainly $(|t_1|, \ldots, |t_n|)R|t_{n+1}|$ and so since $\mathbf{m} \vDash A(|t_{n+1}|, |t_1|, \ldots, |t_n|)$ we get $\mathbf{m}' \vDash A(t_{n+1}, t_1, \ldots, t_n)$, and therefore $\mathbf{m}' \vDash (y)A(y, t_1, \ldots, t_n)$.

This proves the theorem. □

Theorem 9.35 (Finite model property). *Let β be a closed formula and assume β has a model then β has a finite model.*

Proof. We can assume that β has at most m nested quantifiers and that β has a normalized tree model $\mathbf{m} = (S, R, h, g)$.

We can assume that for some $a \in D$, $\varnothing Ra$ holds. Otherwise there is nothing to prove because all external outermost quantifier subformulae in β are true in the model if they are universal and false if they are existential, and β is reducible in this model to some propositional formula and hence has a finite model. We use selective filtration to build a finite model $\mathbf{n}$ which satisfies β.

β is built up from a finite number of predicates, $P_1, \ldots, P_k$.

Assume P_i is n_i place and let $n = n_1 + \cdots + n_k$. Consider the variables $x_1, \ldots, x_n$. Let Θ be the finite set of formulae closed under Boolean operations involving up to m nested quantifiers using free variables from $x_1, \ldots, x_n$ only.

We now construct a finite model $\mathbf{m}_\Theta$ as follows:

Stage 1

For any formula α of the form $(Eu)A(u)$ with maximal nested quantifiers at most m and no free variables such that $\mathbf{m} \vDash \alpha$, there must be an element $a_\alpha \in D$ such that $\varnothing Ra_\alpha$ holds and $\mathbf{m} \vDash A(a_\alpha)$.

Let $D_1 = \{a_\alpha\}, R_1 = \{(\varnothing, a_\alpha)\}$.

If no such α exists then let a be any element such that $\varnothing Ra$ and let $D_1 = \{a\}, R_1 = \{(\varnothing, a)\}$.

Stage $k + 1$

Assume $D_1, \ldots, D_k, R_1, \ldots, R_k$ have been defined. We now define D_{k+1}, R_{k+1}.

Let $\alpha \in \Theta$ be any formula of the form $(Eu)A(u, x_1, \ldots, x_k)$ with at most $m + 1 - k$ maximal nested quantifiers such that for some $a_i \in D_i, i = 1, \ldots, k$, we have

$$\mathbf{m} \vDash \mathcal{S}^{x_1, \ldots, x_k}_{a_1, \ldots, a_k}(Eu)A(u, x_1, \ldots, x_k).$$

Therefore there exists $a_\alpha \in D$ such that $(a_1, \ldots, a_k)Ra_\alpha$ and $\mathbf{m} \vDash \mathcal{S}^{x_1, \ldots, x_k, x_{k+1}}_{a_1, \ldots, a_k, a_\alpha} A(x_{k+1}, x_1, \ldots, x_k)$. Let $D_{k+1} = \{a_\alpha\}$ and let $R_{k+1} = \{(a_1, \ldots, a_k), a_\alpha\}$.

Obviously the process stops after $m + 1$ stages.

We can assume that D_i are all pairwise disjoint, since we are dealing with a tree model.

Let $D^* = D_1 \cup \cdots \cup D_{m+1}$.

Let $R^* = R_1 \cup \cdots \cup R_{m+1}$.

Let h^*, g^* be the restriction of h, g to the smaller model.

We prove the following:

Lemma 9.36. *For any $\alpha \in \Theta$ with at most $m+1-k$ nested quantifiers with $x_1, \ldots, x_k$ free and any $a_i \in D_i, i \leq k$, we have*

(*) $\mathbf{m}^* \vDash \mathcal{S}^{x_1,\ldots,x_k}_{a_1,\ldots,a_k}\alpha$ *iff* $\mathbf{m} \vDash \mathcal{S}^{x_1,\ldots,x_k}_{a_1,\ldots,a_k}\alpha$.

Proof

1. (*) holds for atomic α.
2. The classical connectives present no difficulties.
3. The case of existential quantifier follows from the construction.

This completes the proof of the theorem. □

Definition 9.37. Let (S, R, h, g) be a $\mathbf{Q}_1$ model. Define R_0 by $(a_1, \ldots, a_n)$ $R_0 a_{n+1}$ iff $\varnothing R a_1 \wedge \bigwedge_{i=1}^{n} \{a_1, \ldots, a_i\} R a_{i+1}$. Note that R_0 is normalized. Let $\mathbf{m}_0 = (S, R_0, h, g)$.

Lemma 9.38. *Let $A(x_1, \ldots, x_n)$ be a wff with exactly $x_1, \ldots, x_n$ free and let $(a_1, \ldots, a_n)$ be such that $\varnothing R a_1 \wedge \bigwedge_{i=1}^{n-1} \{a_1, \ldots, a_i\} R a_{i+1}$. Then*

$$\mathbf{m} \vDash A(a_1, \ldots, a_n) \textit{ iff } \mathbf{m}_0 \vDash \mathcal{S}^{x_1,\ldots,x_n}_{a_1,\ldots,a_n} A(x_1, \ldots, x_n).$$

Proof By induction on A. The atomic case and the case of the classical connectives present no difficulties.

Assume $\mathbf{m} \vDash (x)A(a_1, \ldots, a_n, x)$. We show $\mathbf{m}_0 \vDash \mathcal{S}^{\bar{x}}_{\bar{a}}(x)A$. Let a_{n+1} be such that $(a_1, \ldots, a_n) R_0 a_{n+1}$. Then certainly $\{a_1, \ldots, a_n\} R a_{n+1}$ and hence $\mathbf{m} \vDash A(a_1, \ldots, a_n, a_{n+1})$. Since $\varnothing R a_1 \wedge \bigwedge_{i=1}^{n} \{a_1, \ldots, a_i\} R a_{i+1}$ holds, then by the induction hypothesis, $\mathbf{m}_0 \vDash \mathcal{S}^{x_1,\ldots,x_{n+1}}_{a_1,\ldots,a_{n+1}}$ and therefore, since a_n was arbitrary, we have $\mathbf{m}_0 \vDash \mathcal{S}^{x_1,\ldots,x_n}_{a_1,\ldots,a_n}(x)A(x_1, \ldots, x_n, x)$.

Assume $\mathbf{m}_0 \vDash \mathcal{S}^{\bar{x}}_{\bar{a}}(x)A$. We show $\mathbf{m} \vDash (x)A$. Let a_{n+1} be such that $\{a_1, \ldots, a_n\}$ $R a_{n+1}$. Then by our assumptions $(a_1, \ldots, a_n) R_0 a_{n+1}$ holds and so $\mathbf{m}_0 \vDash \mathcal{S}^{\bar{x},x}_{\bar{a},a_{n+1}}$ $A(x_1, \ldots, x_n, x)$ and hence $\mathbf{m} \vDash A(a_1, \ldots, a_n, a_{n+1})$. Therefore, since a_{n+1} was arbitrary, we get $\mathbf{m} \vDash (x)A$. □

Definition 9.39. Let $\mathbf{m} = (S, R, h, g)$ be a normalized tree $\mathbf{Q}_0$ model. Let R_1 be defined by $\{t_1, \ldots, t_n\} R_1 t_{n+1}$ iff $(t_1, \ldots, t_n)$ is the unique sequence of elements such that $\varnothing R t_1 \wedge \bigwedge_{i=1}^{n} (t_1, \ldots, t_i) R t_{i+1}$ holds.

The model $\mathbf{m}_1 = (S, R_1, h, g)$ is the associated $\mathcal{Q}_1$ model of $\mathbf{m}$.

Lemma 9.40. *Let $\mathbf{m}$ and $\mathbf{m}_1$ be related as in the previous definition. Let $A(x_1, \ldots, x_n)$ be a wff with exactly the free variables indicated and let $(a_1, \ldots, a_n)$ be such that $\varnothing R a_1 \wedge \bigwedge_{i=1}^{n-1} (a_1, \ldots, a_i) R a_{i+1}$, then*

$$\mathbf{m} \vDash \mathcal{S}^{x_1,\ldots,x_n}_{a_1,\ldots,a_n} A \textit{ iff } \mathbf{m}_1 \vDash A(a_1, \ldots, a_n).$$

Proof By induction on A.

1. The atomic case and the case of the classical connectives present no difficulties.
2. Assume $\mathbf{m} \vDash \mathcal{S}^{\bar{x}}_{\bar{a}}(y)A(x_1, \ldots, x_n, y)$. We show $\mathbf{m}_1 \vDash (y)A(a_1, \ldots, a_n, y)$. Let a_{n+1} be arbitrary such that $\{a_1, \ldots, a_n\}R_1 a_{n+1}$. By definition of R_1 there exist unique $(b_1, \ldots, b_n)$ such that $\{b_1, \ldots, b_n\} = \{a_1, \ldots, a_n\}$ and $\varnothing R b_1 \wedge \bigwedge_{i=1}^{n-1}(b_1, \ldots, b_i)Rb_{i+1} \wedge (b_1, \ldots, b_n)Ra_{n+1}$.
   Since we are dealing with a tree model, we must have that $b_i = a_i, i = 1, \ldots, n$. Therefore we have $(a_1, \ldots, a_n)Ra_{n+1}$ and hence $\mathbf{m} \vDash \mathcal{S}^{x_1, \ldots, x_n, x_{n+1}}_{a_1, \ldots, a_n, a_{n+1}} A(x_1, \ldots, x_n, x_{n+1})$ and therefore $\mathbf{m}_1 \vDash A(a_1, \ldots, a_n, a_{n+1})$ by the induction hypothesis and hence $\mathbf{m}_1 \vDash (y)A(a_1, \ldots, a_n, y)$.
3. Assume $\mathbf{m}_1 \vDash (y)A(a_1, \ldots, a_n, y)$ and show $\mathbf{m} \vDash \mathcal{S}^{\bar{x}}_{\bar{a}}(y)A$. Let a_{n+1} be such that $\{a_1, \ldots, a_n\}R_1 a_{n+1}$ and so $\mathbf{m}_1 \vDash A(a_1, \ldots, a_n, a_{n+1})$ and by the induction hypothesis $\mathbf{m} \vDash \mathcal{S}^{\bar{x}, x_{n+1}}_{\bar{a}, a_{n+1}} A(x_1, \ldots, x_n, x_{n+1})$ and therefore $\mathbf{m} \vDash \mathcal{S}^{\bar{x}}_{\bar{a}}(y)A$.

This completes the proof. □

Theorem 9.41. *Let A be a closed formula; then*

$$\mathcal{Q}_0 \vDash A \text{ iff } \mathcal{Q}_1 \vDash A.$$

Proof Follows from Theorem 9.34, and Lemmas 9.38 and 9.40. □

The last theorem shows that, as far as closed formulae are concerned, the two semantics are the same. The difference is in how we handle formulae with free variables.

Consider the formula $B = (x)(y)(z)A$. In B the order of quantification is (x, y, z). This order can be simulated using set relation R_1 just as well as using a sequence relation R_0. We get

$$\vDash B \text{ iff } \vDash (\forall x \in V_\varnothing)(\forall y \in V_x)\forall z \in V_{x,y}A.$$

$V_{x,y}$ can depend on the set $\{x, y\}$. In the context of the evaluation of $\vDash B$ we know that x is instantiated before y because of the presence of V_x. In fact the set $\{V_x, V_{x,y}\}$ reminds us of the set-theoretic definition of an ordered pair $(x, y) = \{x, \{x, y\}\}$. What we cannot do in sequence language is to evaluate a formula of the form

$$\mathbf{m} \vDash (z)A(x, y, z)$$

without knowing the order of $\{x, y\}$. We cannot give $\{x, y\}$ simultaneous unordered substitutions; we must have an ordering on $\{x, y\}$.

Consider the formula $C(u, v) = A(u, v) \rightarrow (z)B(z)$ where the free variables are as indicated, and assume B is a tautology. We want to say $\vDash C(u, v)$. In $\mathbf{Q}_0$ we cannot say that because we need to give order on the variables. Thus both

$\mathcal{S}^{u,v}_{x,y}C$ and $\mathcal{S}^{v,u}_{y,x}C$ are theorems. The semantic evaluations of these formulae are, respectively,

$$\forall x, y \in D, \forall z \in V_{(x,y)}(A(x,y) \to B(z))$$

and

$$\forall x, y \in D, \forall z V_{(y,x)}(A(x,y) \to B(z)).$$

In comparison, the unordered substitution says

$$\forall x, y \in D, \forall z \in V_{\{x,y\}}(A(x,y) \to B(z)).$$

As another example, consider the formula $(x)(y)(y = x)$. This says in $\mathbf{Q}_0$ and in $\mathbf{Q}_1$ that $V_x = \{x\}$ for all $x \in V_\varnothing$.

We are now ready to axiomatize $\mathbf{Q}_0$, and proceed to the next section.

9.7 A completeness theorem for $\mathbf{Q}_0$

We now axiomatize $\mathbf{Q}_0$. We need to define the traditional notion of consistency and consequence.

Definition 9.42 (The system $\mathbf{Q}_0$). Let $\mathbf{Q}_0$ be the Hilbert system with the following axioms and rules:

1. All substitution instances of classical truth functional tautologies.
2. The rule of generalization:

$$\frac{\vdash \mathcal{S}^{\bar{x}}_{\bar{u}}A \to \mathcal{S}^{\bar{z},y}_{\bar{v},w}B}{\vdash \mathcal{S}^{\bar{x}}_{\bar{u}}A \to \mathcal{S}^{\bar{z}}_{\bar{v}}(y)B}, y \text{ not in } \mathcal{S}^{\bar{x}}_{\bar{u}}A.$$

3. $\vdash (y)\mathcal{S}^{\bar{x}}_{\bar{u}}(\alpha \to \beta) \to ((y)\mathcal{S}^{\bar{x}}_{\bar{u}}\alpha \to (y)\mathcal{S}^{\bar{x}}_{\bar{u}}\beta)$.
4. $\vdash \mathcal{S}^{\bar{x}}_{\bar{u}}(y)A(y,\bar{x}) \wedge \mathcal{S}^{\bar{x}}_{\bar{u}}(Ey)B(y,\bar{x}) \to \mathcal{S}^{\bar{x}}_{\bar{u}}(Ey)(A(y,\bar{x}) \wedge B(y,\bar{x}))$.
5. *Modus ponens*:
   $\vdash \mathcal{S}^{\bar{x}}_{\bar{u}}A$ and $\vdash \mathcal{S}^{\bar{x}}_{\bar{u}}(A \to B)$ imply $\vdash \mathcal{S}^{\bar{x}}_{\bar{u}}B$.

Definition 9.43 (Proof theory).

1. A theory Δ is a set of slot formulae.
2. We say $\Delta \vdash \alpha$ iff there exists a sequence of slot formulae $\alpha_1, \dots, \alpha_n = \alpha$ such that each element in the sequence is either in Δ or a theorem of $\mathbf{Q}_0$, or is obtained by *modus ponens* from the previous elements in the sequence. Note that the rule of generalization is not used in proofs from non-empty theories Δ.
3. Δ is consistent if $\Delta \nvdash \bot$.

Theorem 9.44 (Completeness theorem for $\mathbf{Q}_0$). *$\mathbf{Q}_0$ is complete for satisfaction in $\mathbf{Q}_0$ models.*

Proof
Soundness
This can be verified.

Completeness

Let β be a consistent $\mathbf{Q}_0$ formula. We want to construct a $\mathbf{Q}_0$ model $\mathbf{m}$ such that $\mathbf{m} \vDash \beta$. Let $\alpha_1, \alpha_2, \ldots$ be an enumeration of all slot formulae of the form $\mathcal{S}_{\bar{u}}^{\bar{x}} A$, where A is a wff of classical logic with the free variables $\{x_1, \ldots, x_n\}$. We want to enlarge $\{\beta\}$ to a saturated and complete theory Δ, giving rise to a model $\mathbf{m}_\Delta$. Recall that the semantical meaning of $\mathbf{m}_\Delta \vDash \mathcal{S}_{\bar{u}}^{\bar{x}} A$ involves not only that $\mathcal{S}_{\bar{u}}^{\bar{x}} A \in \Delta$ but also a relation R. Thus in defining Δ, a suitable R must also be constructed.

We define a sequence $(\Delta_n, R_n), \Delta_n, R_n$ finite, as follows:

CASE $n = 0$

We distinguish two subcases

1. β has the form $(Ey)\mathcal{S}_{\bar{u}}^{\bar{x}} A(y, \bar{x})$.

   Let u be a new variable not appearing in β. We claim $\{\beta, \mathcal{S}_{\bar{u},u}^{\bar{x},y} A\}$ is consistent. For otherwise
$$\vdash \beta \to \mathcal{S}_{\bar{u},u}^{\bar{x},y} \sim A$$
   and hence
$$\vdash \beta \to (y)\mathcal{S}_{\bar{u}}^{\bar{x}} \sim A$$
   a contradiction to the consistency of β.

   In this case let $\Delta_0 = \{\beta, \mathcal{S}_{\bar{u},u}^{\bar{x},y} A\}$ and let $R_0 = \{((u_1, \ldots, u_k), u)\}$.
2. β is not existential. In this case let $\Delta_0 = \{\beta\}$ and $R_0 = \varnothing$.

CASE $n + 1$

Assume Δ_n, R_n have been defined and Δ_n and R_n are consistent and finite. We now define Δ_{n+1}, R_{n+1}.

Consider α_n and $\sim \alpha_n$. One of these two is consistent with Δ_n. Denote by β_n the one which is consistent. We distinguish several subcases, according to the form of β_n.

1. Case $\beta_n = (Ey)\mathcal{S}_{\bar{u}}^{\bar{x}} A(y, \bar{x})$. We let u be a new variable not in Δ_n or R_n or β_n and let $\Delta_{n+1} = \Delta_n \cup \{\beta_n, \mathcal{S}_{\bar{u},u}^{\bar{x},y} A\} \cup \Gamma_n$ where
$$\Gamma_n = \{\mathcal{S}_{\bar{u},u}^{\bar{x},y} \varphi \mid \mathcal{S}_{\bar{u}}^{\bar{x}}(y)\varphi \in \Delta_n\} = \{\mathcal{S}_{\bar{u}}^{\bar{x}}(y)\varphi_j \mid j = 1, \ldots, r\}.$$
   We let $R_{n+1} = R_n \cup \{(\{u_1, \ldots, u_k\}, u)\}$.

   We claim Δ_{n+1} is consistent. Otherwise
$$\vdash \bigwedge \Delta_n \wedge \beta_n \to \left[\bigwedge_{j=1}^{r} (\mathcal{S}_{\bar{u},u}^{\bar{x},y} \varphi_j \to \mathcal{S}_{\bar{u},u}^{\bar{x},y} \sim A) \right]$$
   and hence
$$\vdash \bigwedge \Delta_n \wedge \beta_n \to (y) \left[\bigwedge_{j=1}^{r} (\mathcal{S}_{\bar{u}}^{\bar{x}} \varphi_j \to \mathcal{S}_{\bar{u}}^{\bar{x}} \sim A) \right].$$

Using the axiom

$$\vdash (y)(\mathcal{S}^{\bar{x}}_{\bar{u}}(\alpha \to \beta) \to ((y)\mathcal{S}^{\bar{x}}_{\bar{u}}\alpha \to (y)\mathcal{S}^{\bar{x}}_{\bar{u}}\beta))$$

we get a contradiction.

2. β_n has the form $\mathcal{S}^{\bar{x}}_{\bar{u}}(y)A(y,\bar{x})$. Let $\Delta_{n+1} = \Delta_n \cup \{\beta_n\} \cup \{\mathcal{S}^{\bar{x},y}_{\bar{u},u}A \mid \bar{u}R_n u\}$. Let $R_{n+1} = R_n$. We claim Δ_{n+1} is consistent. Otherwise for some $y_1, \ldots, y_r$ we have

$$\vdash \bigwedge \Delta_n \wedge \beta_n \to \quad \sim \bigwedge_{j=1}^{r} \mathcal{S}^{\bar{x},y}_{\bar{u},y_j} A.$$

But each such y_j was put into Δ_n in a previous stage of the construction because of some $\beta_j = \mathcal{S}^{\bar{x}}_{\bar{u}}(Ey)A_j$, and the wff $\mathcal{S}^{\bar{x},y}_{\bar{u},y_j}A_j$ was also put in at the same stage and hence is also present in Δ_n.
Let us regroup our formulae and get

$$\vdash \bigwedge \Delta'_n \wedge \mathcal{S}^{\bar{x}}_{\bar{u}}(y)A(y,\bar{u}) \wedge \bigwedge_{j=1}^{r} \mathcal{S}^{\bar{x}}_{\bar{u}}(Ey)A_j(y,\bar{x}) \to \\ \sim \bigwedge_{j=1}^{r}(\mathcal{S}^{\bar{x},y}_{\bar{u},y_j}A(y,\bar{x}) \wedge \mathcal{S}^{\bar{x},y}_{\bar{u},y_j}A_j(y,\bar{x}))$$

where $\bigwedge \Delta'_n$ does not contain any free $y_j, j = 1, \ldots, r$.
We can now universally quantify over $y_1, \ldots, y_r$.
This can be done as follows. Consider y_1. We have a provable formula of the form

$$\vdash \alpha \to \sim ((\mathcal{S}^{\bar{x},y}_{\bar{u},y_1}A \wedge \mathcal{S}^{\bar{x},y}_{\bar{u},y_1}A_1) \wedge \beta)$$

where y_1 is not free in α nor in β.
We can therefore use universal generalization on y_1 and get

$$\vdash \alpha \to \quad \sim ((Ey_1)(\mathcal{S}^{\bar{x},y}_{\bar{u},u_1}A \wedge \mathcal{S}^{\bar{x},y}_{\bar{u},y_1}A_1) \wedge \beta).$$

β contains similar expressions for $y_2, \ldots, y_r$. Doing the same again for y_2 and again for y_3 etc. we get:

- $\vdash \bigwedge \Delta'_n \wedge \mathcal{S}^{\bar{x}}_{\bar{u}}(y)A \wedge \bigwedge_{j=1}^{r} \mathcal{S}^{\bar{x}}_{\bar{u}}(Ey)A_j(y,\bar{x}) \to$
  $\sim \bigwedge_{j=1}^{r}(Ey_j)(\mathcal{S}^{\bar{x},y}_{\bar{u},y_j}A \wedge \mathcal{S}^{\bar{x},y}_{\bar{u},y_j}A_j(y_r,\bar{x}))$.

But, using the axiom

$$\vdash \mathcal{S}^{\bar{x}}_{\bar{u}}(y)A(y,\bar{x}) \wedge \mathcal{S}^{\bar{x}}_{\bar{u}}(Ey)A_j(y,\bar{x}) \to \mathcal{S}^{\bar{x}}_{\bar{u}}(Ey)(A \wedge A_j)$$

we get a contradiction.

3. β_n is neither existential nor universal. In this case let $\Delta_{n+1} = \Delta_n \cup \{\beta_n\}$ and $R_{n+1} = R_n$.

We are now ready to define the model. Let

$$\Delta = \bigcup_n \Delta_n, R = \bigcup_n R_n.$$

Let D_n be the set of all variables and constants appearing in Δ_n and R_n. Let $\mathbf{m}_n = \{D_n, R_n, h_n, g_n\}$ be the finite model, where D_n and R_n are as above and where g_n is the identity and h_n is defined by $h_n(P) = \{\bar{u} \mid P(\bar{u}) \in \Delta_n\}$, P atomic.

Let $\mathbf{m}_\Delta = (D, R, h, g)$, where D is the set of variables and constants of the language. R is defined above, g is the identity and for atomic P let $h(P) = \{\bar{u} \mid \mathcal{S}^{\bar{x}}_{\bar{u}} P(\bar{x}) \in \Delta\}$.

We claim

(*) $\mathbf{m}_\Delta \vDash A$ iff $A \in \Delta$, iff there exists a number n such that $\mathbf{m}_k \vDash A$ for all $k \geq n$.

The proof is by induction on the formula. The atomic case holds by definition. We check the quantifier case.

Assume $\mathcal{S}^{\bar{x}}_{\bar{u}}(y)A \in \Delta$ and let $\bar{u}Rd$ hold. We want to show $\mathcal{S}^{\bar{x},y}_{\bar{u},d} \in \Delta$. This holds because it was ensured by the construction.

Assume $\mathcal{S}^{\bar{x}}_{\bar{u}}(Ey)A \in \Delta$.

We are looking for a d such that $\bar{u}Rd$ holds and $\mathcal{S}^{\bar{x},y}_{\bar{u},d}A \in \Delta$. Again this was ensured by construction. □

Theorem 9.45 (Decidability). $\mathbf{Q}_0$ *is decidable.*

Proof We saw that $\mathbf{Q}_0$ has the finite model property, and since it is axiomatizable it is decidable. □

9.8 Translation into propositional temporal logic

This section shows how to translate $\mathbf{Q}_0$ and $\mathbf{Q}_1$ into many-dimensional propositional temporal logic based on modal logic **K**. The role of this section is to motivate and explain the temporal propositional logic used, and then to explain the translation informally. Later sections give all the formal machinery and theorems.

The lessons to be learned from the qualitative discussions of this section are mainly the following:

- generalized quantifiers can be viewed essentially as many-dimensional propositional temporal logics (based on modal **K**);
- generalized quantifiers with monadic predicates only, namely monadic $\mathbf{Q}_0$, is exactly equivalent to propositional modal **K**;
- the system $\mathbf{Q}_0$ with binary predicates only is equivalent to a version of monadic predicate modal logic **K**.

Consider the monadic fragment $\mathbf{1Q}_0$ of $\mathbf{Q}_0$. The number 1 stands for one-place predicates. Similarly $\mathbf{2Q}_0$ will stand for the fragment with binary relations etc. In a later section we shall show how $\mathbf{1Q}_0$ can be faithfully embedded in **1KY**. Meanwhile, the next example illustrates the idea.

Example 9.46. Consider the formula

$$B = (x)(Ey)(z)(P_1(x) \wedge P_2(y) \wedge P_1(z)).$$

Let $\mathbf{m} = (D, R, h, g)$ be a tree model. Then B holds in the tree model iff

$$(\forall x \in V_{\varnothing})(\exists y \in V_x)(\forall z \in V_{(x,y)})(P_1(x) \wedge P_2(y) \wedge P_1(z)).$$

Consider now the propositional language with $\Box$ and Y and the atomic propositions p_1 and p_2. We interpret $\mathbf{m} \vDash P(x)$ to mean $x \vDash p$ in the modal logic. Then the formula B can be translated into the formula

$$\alpha = \Box\Diamond\Box(YYp_1 \wedge Yp_2 \wedge p_1).$$

Let us evaluate α in the Kripke model $(S, \prec, \varnothing)$, where $\prec$ is the successor relation derived from R (see Definition 9.25). We have

$$\varnothing \vDash \alpha \text{ iff } \forall x(\varnothing \prec x \to \exists y(x \prec y \wedge \forall z(y \prec z \to x \vDash p_1 \text{ and } y \vDash p_2 \text{ and } z \vDash p_1))).$$

The two evaluations are the same.

It is also possible, as we shall see in a later section, to translate faithfully from **1KY** into $\mathbf{1Q}_0$. The next example illustrates the idea.

Example 9.47. Let $\alpha' = \Box(p_1 \wedge \Diamond(p_2 \wedge \Box p_1))$. The evaluation in the tree model $(S, \prec, \varnothing, h)$ is

$$\forall x(\varnothing \prec x \to x \vDash p_1 \wedge \exists y(x \prec y \wedge y \vDash p_2 \wedge \forall z(y \prec z \to z \vDash p_1)))$$

which is equivalent to

$$(\forall x \in V_{\varnothing})(P_1(x) \wedge (\exists y \in V_x)(P_2(y) \wedge (\forall z \in V_{(x,y)})P_1(z))).$$

The above is the semantic evaluation of

$$B' = (x)(P_1(x) \wedge (Ey)(P_2(y) \wedge (z)P_1(z))).$$

B' is equivalent to B of Example 9.46.

We shall give, in a later section, algorithms for translating any wff of $\mathbf{1Q}_0$ into a formula of **1KY** and vice versa.

The above considerations do not depend at all on the fact that we are dealing with the monadic fragment $\mathbf{1Q}_0$. We started with the monadic case to simplify matters. The real translation of $\mathbf{Q}_0$ is into many-dimensional propositional temporal logic [Gabbay *et al.*, 1994; Gabbay and Guenthner, 1982], which we have introduced.

Example 9.48. Consider the formula

$$C = (x)(Ey)(z)(R_1(x,z) \wedge R_2(y,x)).$$

Its translation into **2KY** is

$$\beta = \Box\Diamond\Box(Y_1Y_1r_1 \wedge Y_1Y_2Y_2r_2).$$

C can also be translated into monadic predicate **K**. Let $R_1(x,z)$ be translated into $x \vDash P_1(z)$ and similarly $R_2(y,x)$ into $y \vDash P_2(x)$; then C is translated into

$$\Box\Diamond\Box(\forall z(W(z) \to YYP_1(z)) \wedge \forall z(YYW(z) \to YP_2(x)))$$

where $W(z)$ is the 'locality' predicate satisfying

$$t \vDash W(z) \text{ iff } z = t.$$

Example 9.49. Consider the propositional constant E with the truth table:

- $(t, s) \vDash \text{E}$ iff $t = s$.

E can translate the equality predicate. Thus in the previous example, if '$R_2(y, x)$' in C were '$y = x$', then 'r_2' in β is replaced by 'E'.

In terms of the locality predicate W in the monadic predicate language we get

$$(t, s) \vDash \text{E iff } t = s \text{ iff } t \vDash W(s).$$

Remark 9.50. Note that $(x)(y)R(x, y)$ and $(y)(x)R(x, y)$ are translated the same way, as

$$\Box\Box\tau_1((x_1, x_2), R(x_1, x_2)) = \Box^2(Y_1 r)$$
$$\Box\Box\tau_1((x_1, x_2), R(x_2, x_1)) = \Box^2(Y_2 r).$$

So the commutativity of the quantifiers $(x)(y)A = (y)(x)A$ means $\vdash Y_1 r \leftrightarrow Y_2 r$.

Definition 9.51 (Translation τ_1). We present a translation τ_1 from $\mathbf{1Q_0}$ into $\mathbf{1KY}$.

Let $\bar{x} = (x_1, \ldots, x_n)$ be a sequence of distinct variables. Let A be a wff with its free variables from among $\{x_1, \ldots, x_n\}$. We define by induction a $\mathbf{1KY}$ formula $\tau_1(\bar{x}, A)$ as follows ($*$ denotes concatenation of sequences).

1. With each unary atom P_i of $\mathbf{Q}_0$, let us associate the propositional atom p_i of $\mathbf{1KY}$.
2. $\tau_1(\bar{x}, P(x_i)) = Y^{n-i}p$.
3. $\tau_1(\bar{x}, \sim A) = \ \sim \tau_1(\bar{x}, A)$.
4. $\tau_1(\bar{x}, A \wedge B) = \tau_1(\bar{x}, A) \wedge \tau_1(\bar{x}, B)$.
5. $\tau_1(\bar{x}, (y)A(y)) = \Box\tau_1(\bar{x} * (y), A)$, y a new distinct variable.
6. Let A be a closed wff of $\mathbf{1Q_0}$. Then its translation $\tau_1(A)$ is defined as $\tau_1(\varnothing, A)$.

Theorem 9.52. *For all closed A of $\mathbf{1Q_0}$, we have*

$$\mathbf{1Q_0} \vdash A \text{ iff } \mathbf{1KY} \vdash \tau_1(A).$$

Proof Let $\mathbf{m} = (S, \prec, a, h)$ be a tree model which can serve as a model of either $\mathbf{1Q_0}$ or $\mathbf{1KY}$.

Let $\bar{x} = (x_1, \ldots, x_n)$ be variables and let $g(x_i) = t_i \in S$ be an instantiation of these variables into S in such a way that $t_1 \prec t_2 \prec \cdots \prec t_n$ holds. Assume the correspondence of '$P(x)$' with '$x \vDash p$'. Consider $\tau_1(\bar{x}, A(\bar{x}))$, which is a wff of $\mathbf{1KY}$. This wff can be evaluated at x_n. We have for any such g:

- $x_n \vDash_g \tau_1((x_1, \ldots, x_n), A)$ iff $\mathbf{m} \vDash_g A(x_1, \ldots, x_n)$.

1. We check the atomic case:

$$\tau_1(\bar{x}, P(x_i)) = Y^{n-1}p.$$

   Hence

$$x_n \vDash_g Y^{n-1}p \text{ iff } x_i \vDash_g p \text{ iff } \mathbf{m} \vDash_g P(x_i).$$

2. The cases of $\sim$ and $\wedge$ present no difficulties.
3. $x_n \vDash_g \tau_1(\bar{x}, (y)A(\bar{x}, y))$ iff $x_n \vDash_g \Box\tau_1(\bar{x} * (y), A(\bar{x}, y))$ iff for all t such that $g(x_n) \prec t$ we have $t \vDash_g \tau_1(\bar{x} * (y), A(\bar{x}, y))$.
   Extend g to g' giving value to a new variable y with $g'(y) = t$; then we continue iff for all y such that $x_n \prec y$, $y \vDash_{g'} \tau_1(\bar{x} * (y), A(\bar{x}, y))$, and by the induction hypothesis, iff for all y such that $x_n \prec y$, $\mathbf{m} \vDash A(\bar{x}, y)$ iff $\mathbf{m} \vDash (y)A(\bar{x}, y)$.

□

Example 9.53. Recall formula B of Example 9.46:

$$B = (x)(Ex)(z)(P_1(x) \wedge P_2(y) \wedge P_1(z)).$$

Its translation into **1KY** was

$$\alpha = \Box\Diamond\Box(YYp_1 \wedge Yp_2 \wedge p_1).$$

α has a normal form (see Example 9.6) α':

$$\alpha' = \Box(p_1 \wedge \Diamond(p_2 \wedge \Box p_1)).$$

By the previous translation and theorem we have

$$\mathbf{1Q_0} \vdash B \text{ iff } \mathbf{K} \vdash \alpha'.$$

In fact, combining the two translations τ_1 and $[\![-]\!]$ and using the faithfulness of τ_1 we get, for all wffs A of $\mathbf{1Q_0}$,

$$\mathbf{1Q_0} \vdash A \text{ iff } \mathbf{K} \vdash [\![\tau_1(A)]\!].$$

Definition 9.54 (Translation τ_2). We present a translation τ_2 from **1KY** into $\mathbf{1Q_0}$.

Let $\bar{x}$ be a sequence of distinct variables. Let α be a formula of **K** with $\Box$ only. We define the formula $\tau_2(\bar{x}, \alpha)$ of $\mathbf{1Q_0}$ by induction. For atomic p_i of **1KY** associate the predicate P_i of $\mathbf{1Q_0}$.

1. $\tau_2((x_1, \ldots, x_n), p) = P(x_n)$.
2. $\tau_2(\bar{x}, \alpha \wedge \beta) = \tau_2(\bar{x}, \alpha) \wedge \tau_2(\bar{x}, \beta)$.

3. $\tau_2(\bar{x}, \sim \alpha) = \quad \sim \tau_2(\bar{x}, \alpha)$.
4. $\tau_2((x_1, \ldots, x_n)\Box\alpha) = (y)\tau_2((x_1, \ldots, x_n, y), \alpha)$ where y is a new variable.
5. Let $\tau_2(\alpha)$ be $\tau_2(\varnothing, \alpha)$.

The above define $\tau_1(\alpha)$ for α without Y. We want to define $\tau_2(\alpha)$ for an arbitrary formula of **1KY**. This we can do immediately since we have shown that **1KY** can be translated into **K**.

It is also instructive to translate directly from **1KY** into $\mathbf{1Q}_0$. The direct translation is the same as the one via **K** because it is based on the same idea (as seen from Figure 9.1).

By a previous theorem any such wff has a normal form

$$\alpha = \bigvee_i \bigwedge_j Y^{n_{i,j}} \alpha_{i,j}, \ \alpha_{i,j} \text{ are without } Y.$$

We now define $\tau_2(\alpha)$ for such an α.

Let $\alpha_i = \bigwedge_j Y^{n_{i,j}} \alpha_{i,j}$.

First we let $\tau_2(\alpha) = \bigvee_i \tau_2(\alpha_i)$, and worry about the translation $\tau_2(\alpha_i)$. This means we need to define τ_2 for a conjunction β of the form

$$\beta = \beta_k \wedge Y^{n_{k-1}}(\beta_{k-1} \wedge \cdots \wedge Y^{n_0}\beta_0) \ldots).$$

We translate β into the following formula:

$$(Et_0)(\tau_2(\beta_0) \wedge (Ex_{0,1}) \ldots (Ex_{0,n_1-1})(Et_1)(\tau_2(\beta_1) \wedge \\ \ldots (Et_{k-1})(\beta_{k-1} \wedge (Ex_{k-1,1}) \ldots (Ex_{k-1,n_{k-1}-1})(Et_k)\beta_k) \ldots).$$

Example 9.55. Consider the commutativity axiom

$$A = (x)(y)(P(x) \wedge Q(y)) \rightarrow (y)(x)((x) \wedge Q(y)).$$

The translation $\tau_1(A) = \alpha$, where

$$\alpha = \Box\Box(Yp \wedge q) \rightarrow \Box\Box(p \wedge Yq).$$

To translate back we need to reduce α to normal form. Let $\beta_1 = \Box\Box(Yp \wedge q), \beta_2 = \Box\Box(p \wedge Yq)$. We have

$$\beta_1 \leftrightarrow \Box(p \wedge \Box q)$$
$$\beta_2 \leftrightarrow \Box(q \wedge \Box p).$$

Hence $\tau_2(\alpha) = A'$, where

$$A' = (x)(P(x) \wedge (y)Q(y)) \rightarrow (y)(Q(y) \wedge (x)P(x)).$$

A and A' are equivalent.

Theorem 9.56 (Faithfulness of τ_2). *For τ_2 defined earlier, we have*

$$\mathbf{K} \vdash \alpha \textit{ iff } \mathbf{1Q}_0 \vdash \tau_2(\alpha).$$

Proof By structural induction on α.

We show that in any model $\mathbf{m} = (S, \prec, a, h)$ and any $(x_1, \ldots, x_n)$ such that $x_1 \prec x_2 \prec \cdots \prec x_n$ we have $x_n \vDash \alpha$ iff $\mathbf{m} \vDash \tau_2((x_1, \ldots, x_n), \alpha)$.

Of course we assume the basic correspondence of

$$x \vDash p \text{ iff } \mathbf{m} \vDash P(x).$$

The above we show by induction on α. We check the case of $\Box\alpha$.

$$x_n \vDash \tau_2((x_1, \ldots, x_n), \Box\alpha) \text{ iff } x_n \vDash (y)\tau_2((x_1, \ldots, x_n, y), \alpha) :$$

iff

$$\forall y(x_n \prec y \rightarrow y \vDash \tau_2((x_1, \ldots, x_n, y), \alpha))$$

iff (by the induction hypothesis)

$$\forall y(x_n \prec y \rightarrow \mathbf{m} \vDash \tau_2((x_1, \ldots, x_n, y), \alpha)$$

iff $\mathbf{m} \vDash (y)\tau_2((x_1, \ldots, x_n, y), \alpha)$. □

Corollary 9.57. $\mathbf{1Q_0}$ *is essentially equivalent to modal* $\mathbf{K}$.

Proof Follows from the translations. □

Definition 9.58 (Translation of $\mathbf{kQ_0}$ into kKY). The translation τ_1 is defined by the same clauses as in Definition 9.51 except for the atomic case. Clause (2) is to become the following:
(2, k): For atomic P, $\tau_1(\bar{x}, P(x_{i_1}, \ldots, x_{i_k})) = Y_1^{n-i_1}, \ldots, Y_k^{n-1_k} p$.

Theorem 9.59 (Faithfulness of τ_1). *For all closed A of* $\mathbf{kQ_0}$, *we have*

$$\mathbf{kQ_0} \vdash A \textit{ iff } \mathbf{kKY} \vdash \tau_1(A).$$

Proof Similar to the proof of Theorem 9.52. □

9.9 Translation into quantified modal logic

Since the generalized quantifiers have a modal meaning, it makes sense to translate them directly into quantified modal logic. Let $\alpha = (x)(y)P(x, y)$. When translating α into classical logic, we obtain $\forall x(\varnothing Rx \rightarrow \forall y(xRy \rightarrow P(x, y)))$. When we translate into quantified modal logic we obtain

$$\Box\Box\forall xy(W(x) \wedge \Diamond(W(y) \wedge P(x, y))),$$

where $W(z)$ is the locality predicate, satisfying

$$t \vDash W(z) \text{ iff } z = t.$$

We now motivate and explain this translation in three stages:

1. translate two-dimensional propositional logic **2KY** into one-dimensional monadic predicate temporal logic **Q1KY**;
2. translate $\mathbf{Q}_0$ into one-dimensional predicate temporal logic **QKY**;
3. translate $\mathbf{Q}_0$ into modal predicate logic **QK**.

Stage 1 is just a starter; 2 is a convenient intermediate step in order to get to 3, which is our main translation.

We now translate **2KY** into **Q1KY**, the monadic predicate theory of modal **1KY**. The idea of the translation is to look at models of the form $(S, D, \prec, 0, h)$ where the domain D is S itself. Such models are called *reflective*. In such a case we can consider $\{s \mid (t, s) \vDash p\}$ as the extension of a predicate P at world t. Thus $t \vDash P(s)$ iff $(t, s) \vDash p$. To translate Y_1 and Y_2 into monadic predicate logic we need to have in the language a symbol for the predecessor function. Let $\mathbf{e}(t)$ be the predecessor of t. Then we can translate $\mathbf{2Q}_0$ into monadic **1KY** as in Definition 9.58 with the atomic clause being

$$\tau_1(\bar{x}, P(x_{i_1}, x_{i_2})) = Y^{n-i_1} P(\mathbf{e}^{n-i_2}(x_n))$$

where x_n is the last point of the sequence

$$\bar{x} = (x_1, \ldots, x_n).$$

The above translation is not satisfactory because of the explicit use of the point x_n, which is only a parameter in the translation. x_n has to be eliminated. Fig. 9.2 explains the situation.

FIG. 9.2.

We are evaluating $x_n \vDash Y^{n-i_1} P(\mathbf{e}^{n-i_2}(x_n))$.

If we use the locality predicate $W(y)$, where $x_n \vDash W(y)$ iff $y = x_n$, then we can identify x_n using W, as well as eliminate $\mathbf{e}$.

Thus we let

$$\begin{aligned}\tau_1(\bar{x}, P(x_{i_1}, x_{i_2})) &= \forall y(W(y) \to Y^{n-i_1} P(\mathbf{e}^{n-i_2}(y))) \\ &= \forall z(Y^{n-i_2} W(z) \to Y^{n-i_1} P(z)).\end{aligned}$$

Definition 9.60 (Q1KY).

1. Let **Q1KY** be the reflective predicate language correspondng to the propositional logic **1KY**. This language contains the connectives $\Box$ and Y and a stock of n_i-place predicates $P_i^{n_i}(x_1, \ldots, x_{n_i}), i = 1, \ldots, n$. It also contains the locality predicate W and the classical quantifiers $\forall$ and $\exists$.

2. A model for this language has the form $(S, \prec, a, h, g)$, where $(S, \prec, a)$ is a tree model as before and h is an interpretation and g is an assignment, giving for each n-place predicate P and each $t \in S$, an extension $h(t, P) \subseteq S$ and each term x a $g(x) \in S$. The locality predicate is interpreted as before. Satisfaction is defined in the transitional modal logic manner.
3. We can define a translation $\tau_1(\bar{x}, A)$ from $\mathbf{Q}_0$ into **Q1KY** as follows. First associate with every $(n+1)$ place predicate $P(x_0, \ldots, x_n)$ of $\mathbf{Q}_0$ an $(n+1)$-place predicate $P_1(y_0, \ldots, y_n)$ of **Q1KY**.
   Let τ_1 be defined inductively as in Definition 9.58 except that the atomic clause is replaced by the following:

$$\tau_1(\bar{x}, P(x_{i_0}, \ldots, x_{i_n})) = \forall y_0, \ldots, y_n \left(\bigwedge_{j=0}^{n} Y^{n-i_j} W(y_j) \to P_1(y_0), \ldots, (y_n) \right).$$

Remark 9.61. As we have done in the case of the translation from $\mathbf{1Q}_0$ into **1KY**, the Y can be eliminated and $\mathbf{Q}_0$ can be translated into **QK**, quantified self-reflective predicate modal logic with $W(x)$.

Example 9.62 (Some translations). We translate from $\mathbf{Q}_0$ into **QKY** and then into **QK** some of the axioms from [van Lambalgen, 1996, p. 226].

1. Axiom (Q2): $\sim (x)(x \neq x)$
   The translation of $x = y$ is $W(x) \wedge W(y)$. Thus $x \neq x$ is $\sim W(x)$. Hence the translation of (Q2) is $\sim \Box\forall y(W(y) \to \ \sim W(y))$ which is $\sim \Box\forall y \sim W(y)$ which is a theorem.
2. Axiom (Q3): $(x)(y)(x \neq y)$
   $x = y$ must be translated as $W(x) \wedge W(y)$. Hence the translation of (Q3) is:

$$\Box\Box\forall x, y(YW(x) \wedge W(y) \to \ \sim W(x) \wedge W(y))$$

   which after rewriting is translated into **QK** as

$$\Box\forall x, y(W(x) \to \Box(W(y) \to \ \sim (W(x) \wedge W(y))))$$

   which again is a theorem.
3. Axiom (Q4) is clearly valid.
4. Axiom (Q5): $(x)A \wedge \forall x(A \to B) \to (x)B$
   This axiom is in a mixed language with the universal quantifier $\forall$, which we translate as itself. The translation of (Q5) into **QK** is

$$\Box\forall x(W(x) \to A(x)) \wedge \forall x(A(x) \to B(x)) \to \Box\forall x(W(x) \to B(x)).$$

   This is also obviously valid.
5. Axiom (Q6): $(x)A \wedge (x)B \to (x)(A \wedge B)$
   This is translated into

$$\Box\forall x(W(x) \to A(x)) \wedge \Box\forall x(W(x) \to B(x)) \to \Box\forall x(W(x) \to A(x) \wedge B(x)).$$

   This is also valid.

6. Axiom ($Q^{aa}7$): $(\bar{z})(x)(y)A \rightarrow (\bar{z})(y)(x)A$ where $\bar{z} = (z_1, \ldots, z_n)$
   The translation into **QKY** is

$$\Box^n\Box\Box\forall x, y(YW(x) \wedge W(y) \rightarrow A(x,y)) \rightarrow$$
$$\Box\Box\forall x, y(W(x) \wedge Y(W(y) \rightarrow A(x,y)))$$

   which transforms into **QK** as

$$\Box^n\Box\forall x\forall y(W(x) \rightarrow \Box(W(y) \rightarrow A(x,y))$$
$$\rightarrow \Box^n\Box\forall xy(W(y) \rightarrow \Box(W(x) \rightarrow A(x,y))).$$

   There is no obvious corresponding condition for this axiom in the tree semantics. However, if we relinquish the requirement for a tree base, a natural condition can be found.
   This kind of situation is familiar from propositional modal logic. Modal **K** is complete for strict tree semantics. Some extensions such as $\mathbf{K}+\Box A \rightarrow \Box^n A$ can still be complete for semantics derived from tree semantics. However there are axioms like $\Diamond^m\Box^k\Diamond^r A \rightarrow A$ which require a general relation satisfying some condition which cannot be 'based' on an underlying tree relation.
7. Axiom Q^u7: $(y)(\forall x)A \wedge (\forall y)(x)A \rightarrow (x)(\forall y)A$
   This corresponds to

$$\Box\forall y(W(y) \rightarrow \forall x A(x,y))$$
$$\wedge\forall y\Box\forall x(W(x) \rightarrow A(x,y))$$
$$\rightarrow \Box\forall x(W(x) \rightarrow \forall y A(x,y)).$$

9.10 Conclusion

We have seen many translations of generalized quantifiers into many-dimensional temporal logic. The most satisfactory among them are the translation of the monadic fragment of generalized quantifier logic $\mathbf{Q}_0$ into propositional modal **K**, and the translation of the full logic $\mathbf{Q}_0$ into predicate modal **K** with the locality predicate W.

These translations both keep the modal spirit of the generalized quantifier (which is translated as $\Box$) while still allowing for a transparent correspondence theory. This, we believe, is better than translating into classical logic, as has been done in the literature.

The translations into many-dimensional temporal logic rely on the tree property of the semantics and would not be effective for systems with semantics not directly derived from trees.

10

THE DECLARATIVE PAST AND IMPERATIVE FUTURE

10.1 Introduction

We propose a new paradigm in executable logic: that of the *declarative past and imperative future.* A future statement of temporal logic can be understood in two ways: the declarative way, that of describing the future as a temporal extension; and the imperative way, that of making sure that the future will happen the way we want. Since the future has not yet happened, we have a language which can be both declarative and imperative. We regard our theme as a natural meeting between the imperative and declarative paradigms.

More specifically, we describe a temporal logic with Since, Until and fixed point operators. The logic is based on the natural numbers as the flow of time and can be used for the specification and control of process behaviour in time. A specification formula of this logic can be automatically rewritten into an executable form. In an executable form it can be used as a program for controlling process behaviour. The executable form has the structure 'If A holds in the past then do B'. This structure shows that declarative and imperative programming can be integrated in a natural way.

Let $\mathcal{E}$ be an environment in which a program $\mathcal{P}$ is operating. The exact nature of the environment or the code and language of the program are not immediately relevant for our purposes. Suppose we make periodic checks at time $t_0, t_1, t_2, t_3, \ldots$ on what is going on in the environment and what the program is doing. These checks could be made after every unit execution of the program or at some key times. The time is not important to our discussion. What is important is that we check at each time the truth values of some propositions describing features of the environment and the program. We shall denote these propositions by $a_1, \ldots, a_m, b_1, \ldots, b_k$. These propositions, which we regard as units taking truth values $\top$ (true) or $\bot$ (false) at every checkpoint, need not be expressible in the language of the program $\mathcal{P}$, nor in the language used to describe the environment. The program may, however, in its course of execution, change the truth values of some of the propositions. Other propositions may be controlled only by the environment. Thus we assume that $a_1, \ldots, a_m$ are capable of being influenced by the program while $b_1, \ldots, b_k$ are influenced by the environment. We also assume that when at checktime t_n we want the program to be executed in such a way as to make the proposition a_i true, then it is possible to do so. We express this command by writing **exec** (a_i). For example, a_1 can be 'print the screen' and b_1 can be 'there is a read request from outside'; a_1 can be controlled by the program while b_1 cannot. **exec** (a_1) will make a_1 true.

To illustrate our idea further, we take one temporal sentence of the form

$$G[\bullet a \Rightarrow \bigcirc b]. \tag{10.1}$$

$\bullet$ is the 'yesterday' operator, $\bigcirc$ is the 'tomorrow' operator, and G is the 'always in the future' operator. One can view (10.1) as a wff of temporal logic which can be either true or false in a temporal model. One can use a temporal axiom system to check whether it is a temporal theorem etc. In other words, we treat it as a formula of logic.

There is another way of looking at it. Suppose we are at time n. In a real ticking forward temporal system, time $n+1$ (assume that time is measured in days) has not happened yet. We can find the truth value of $\bullet a$ by checking the past. We do not know yet the value of $\bigcirc b$ because tomorrow has not yet come. Checking the truth value of $\bullet a$ is a declarative reading of $\bullet a$. However, we need not read $\bigcirc b$ declaratively. We do not need to wait and see what happens to b tomorrow. Since tomorrow has not yet come, we can make b true tomorrow if we want to, and are able to. We are thus reading $\bigcirc b$ imperatively: 'make b true tomorrow'.

If we are committed to maintaining the truth of the specification (10.1) throughout time, then we can read (10.1) as: 'at any time t, if $\bullet a$ holds then execute $\bigcirc b$', or schematically, 'if declarative past then imperative future'. This is no different from the Pascal statement

```
if x<5 then x:=x+1
```

In our case we involve whole formulae of logic.

The above is our basic theme. This chapter makes it more precise. The rest of this introduction sets the scene for it, and the conclusion will describe existing implementations. Let us now give several examples:

Example 10.1 (Simplified Payroll). Mrs Smith is running a babysitter service. She has a list of reliable teenagers who can take on a babysitting job. A customer interested in a babysitter would call Mrs Smith and give the date on which the babysitter is needed. Mrs Smith calls a teenager employee of hers and arranges for the job. She may need to call several of her teenagers until she finds one who accepts. The customer pays Mrs Smith and Mrs Smith pays the teenager. The rate is £10 per night unless the job requires overtime (after midnight) in which case it jumps to £15.

Mrs Smith uses a program to handle her business. The predicates involved are the following:

$A(x)$	x is asked to babysit
$B(x)$	x does a babysitting job
$M(x)$	x works after midnight
$P(x,y)$	x is paid y pounds .

In this set-up, $B(x)$ and $M(x)$ are controlled mainly by the environment and $A(x)$ and $P(x,y)$ are controlled by the program.

We get a temporal model by recording the history of what happens with the above predicates. Mrs Smith laid out the following (partial) specification:

1. Babysitters are not allowed to take jobs three nights in a row, or two nights in a row if the first night involved overtime.
2. Priority in calling is given to babysitters who were not called before as many times as others.
3. Payment should be made the next day after a job is done.

Figure 10.1 is an example of a partial model of what has happened to a babysitter called Janet. This model may or may not satisfy the specification.

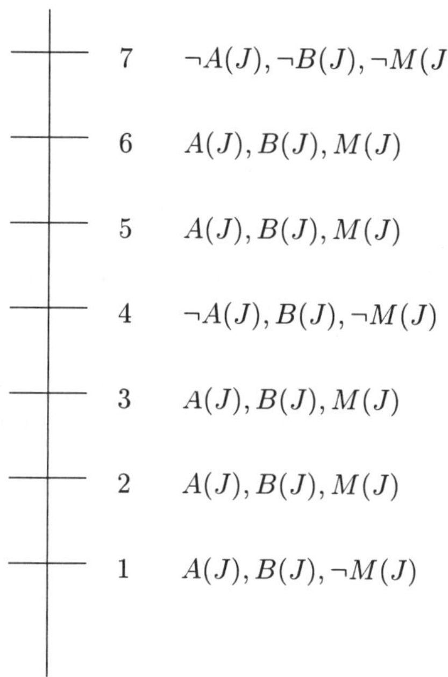

FIG. 10.1. A model for Janet.

We would like to be able to write down the specification in an intuitive temporal language (or even English) and have it automatically transformed into an executable program, telling us what to do day by day.

Example 10.2 (J. Darlington, L. While [1987]). Consider a simple program $\mathcal{P}$, written in a rewrite language, to merge two queues. There are two merge rules:

R1 Merge($a.x$, y) = a.**merge(x,y)**;

R2 Merge(x,$a.y$) = a.**merge(x,y)**.

That is, we have left or right merges. The environment E with which $\mathcal{P}$ interacts consists of the two queues x and y which get bigger and bigger over time. A real-life example is a policeman merging two queues of traffic. We use $t_0, t_1, t_2, \ldots$ as checktimes. The propositions we are interested in are:

A_1 The program uses the **R1** merge rule (left merge).
A_2 The program uses the **R2** merge rule (right merge).
B The left queue is longer than the right queue.

Notice that the proposition B is not under the complete control of the program. The environment supplies the queuing elements, though the merge process does take elements out of the queue. The propositions A_1 and A_2 can be made true by the program, though not in the framework of the rewrite language, since the evaluation is non-deterministic. The program may be modified (or annotated) to a program $\mathcal{P}_1$ which controls the choice of the merge rules. In the general case, for other possible programs and other possible languages, this may not be natural or even possible to do.

Example 10.3 (Loop checking in Prolog). Imagine a Prolog program $\mathcal{P}$ and imagine predicates $A_1, \ldots, A_m$ describing at each step of execution which rule is used and what is the current goal and other relevant data. Let B describe the history of the computation. This can be a list of states defined recursively. The loop checking can be done by ensuring that certain temporal properties hold throughout the computation. We can define in this set-up any loop-checking system we desire and change it during execution.

In the above examples, the propositions A_i, B_j change the truth value at each checktime t_k. We thus obtain a natural temporal model for these propositions (see Fig. 10.2).

$$\begin{array}{lll} \vdots & & \\ 3 & a_1 = \bot & b_2 = \top \\ 2 & a_1 = \bot & b_2 = \top \\ 1 & a_1 = \top & b_2 = \top \\ 0 & a_1 = \top & b_2 = \bot \end{array}$$

FIG. 10.2. An example temporal model.

In the above set-up the programmer is interested in influencing the execution of the program within the non-deterministic options available in the programming language. For example, in the merge case one may want to say that if the left queue is longer than the right queue then use the left merge next. In symbols

$$G[B \Rightarrow \bigcirc A1].$$

In the Prolog case, we may want to specify what the program should do in case of loops, i.e.

$$G[C \wedge \blacklozenge C \Rightarrow D],$$

where C is a complex proposition describing the state of the environment of interest to us ($\blacklozenge$ is the 'in the past' operator). $C \wedge \blacklozenge C$ indicate a loop and D says what is to be done. The controls may be very complex and can be made dependent on the data and to change as we go along.

Of course in many cases our additional controls of the execution of $\mathcal{P}$ may be synthesized and annotated in $\mathcal{P}$ to form a new program $\mathcal{P}*$. There are several reasons why the programmer may not want to do that:

1. The programming language may be such that it is impossible or not natural to synthesize the control in the program. We may lose in clarity and structure.
2. Changes in the control structure may be expensive once the program $\mathcal{P}*$ is defined and compiled.
3. It may be impossible to switch controlling features on and off during execution, i.e. have the control itself respond to the way the execution flows.
4. A properly defined temporal control module may be applicable as a package to many programming languages. It can give both practical and theoretical advantages.

In this chapter we follow option 4 above and develop an executable temporal logic for interactive systems. The reader will see that we are developing a logic here that on the one hand can be used for specification (of what we want the program to do) and on the other hand can be used for execution. (How to pass from the specification to the executable part requires some mathematical theorems.) Since logically the two formulations are equivalent, we will be able to use logic and proof theory to prove correctness.

This is what we have in mind:

1. We use a temporal language to specify the desirable behaviour of $\{a_i, b_j\}$ over time. Let $\mathcal{S}$ be the specification as expressed in the temporal language (e.g. $G[b_2 \Rightarrow \bigcirc a_1]$).
2. We rewrite automatically $\mathcal{S}$ into $\mathcal{E}$, $\mathcal{E}$ being an executable temporal module. The program $\mathcal{P}$ can communicate with $\mathcal{E}$ at each checktime t_i and get instructions on what to do.

We have to prove that:

- if $\mathcal{P}$ follows the instructions of $\mathcal{E}$ then any execution sequence satisfies $\mathcal{S}$, i.e. the resulting temporal model for $\{a_i, b_j\}$ satisfies the temporal formula $\mathcal{S}$;
- any execution sequence satisfying $\mathcal{S}$ is non-deterministically realizable using $\mathcal{P}$ and $\mathcal{E}$.

The proofs are tough!

Note that our discussion also applies to the case of shared resources. Given a resource to be shared by several processes, the temporal language can specify how to handle concurrent demands by more than one process. This point of view is dual to the previous one. Thus in the merge example, we can view the merge program as a black box which accepts items from two processes (queues), and the specification organizes how the box (program) is to handle that merge. We shall further observe that since the temporal language can serve as a metalanguage for the program $\mathcal{P}$ (controlling its execution), $\mathcal{P}$ can be completely subsumed in $\mathcal{E}$. Thus the temporal language itself can be used as an imperative language ($\mathcal{E}$) with an equivalent specification element $\mathcal{S}$. Ordinary Horn logic programming can be obtained as a special case of the above.

We can already see the importance of our logic from the following point of view. There are two competing approaches to programming:the declarative one as symbolized in logic programming and Prolog, and the imperative one as symbolized in many well-known languages. There are advantages to each approach and at first impression there seems to be a genuine conflict between the two. The executable temporal logic described in this chapter shows that these two approaches can truly complement each other in a natural way.

We start with a declarative specification $\mathcal{S}$, which is a formula of temporal logic. $\mathcal{S}$ is transformed into an executable form which is a conjunction of expressions of the form

`hold` C `in the past` $\Rightarrow$ `execute` B `now.`

At any moment of time, the past is given to us as a database; thus **hold**(C) can be evaluated as a goal from this database in a Prolog program. **execute**(B) can be performed imperatively. This creates more data for the database, as the present becomes past. Imperative languages have a little of this feature, e.g. **if x<5 let x:=x+1**. Here **hold**(C) equates to **x<5** and **execute**(B) equates to **x:=x+1**. The **x<5** is a very restricted form of a declarative test. On the other hand Prolog itself allows for imperative statements. Prolog clauses can have the form

`write(Term)` $\Rightarrow$ `b`

and the goal **write(Term)** is satisfied by printing. In fact, one can accomplish a string of imperative commands just by stringing goals together in a clever way.

We thus see our temporal language as a pointer in the direction of unifying in a logical way the declarative and imperative approaches. The temporal language can be used for planning. If we want to achieve B we try to execute B. The temporal logic will give several ways of satisfying **execute**(B) while **hold**(C) remains true. Any such successful way of executing B is a plan. In logical terms we are looking for a model for our atoms but since we are dealing with a temporal model and the atoms can have an imperative meaning we get a plan. We will try and investigate these points further.

10.2 The logic USF

This section describes the temporal system this chapter proposes for specification and execution. The proposed logic contains the temporal connectives *since*(S) and *until*(U) together with a fixed point operator φ. The formulae of USF are used for specifying temporal behaviour and these formulae will be syntactically transformed into an executable form. We begin with the definitions of the syntax of USF. There will be four types of well-formed formulae, pure future formulae (talking only about the strict future), pure past formulae (talking only about the strict past), pure present formulae (talking only about the present) and mixed formulae (talking about the entire flow of time).

Definition 10.4 (Syntax of USF for the propositional case). Let Q be a sufficiently large set of *atoms* (atomic propositions). Let $\wedge, \vee, \neg, \Rightarrow, \top, \bot$ be the usual classical connectives and let U and S be the temporal connectives and φ be the fixed point operator. We define by induction the notions of
wff (well- formed formula);
wff^+ (pure future wff);
wff^- (pure past wff);
wff^0 (pure present wff).

1. An atomic $q \in Q$ is a pure present wff and a wff. Its atoms are q.

2. Assume A and B are wffs with atoms $\{q_1, \ldots, q_n\}$ and $\{r_1, \ldots, r_m\}$ respectively. Then $A \wedge B$, $A \vee B$, $A \Rightarrow B$, $U(A, B)$ and $S(A, B)$ are wffs with atoms $\{q_1, \ldots, q_n, r_1, \ldots, r_m\}$.

   (a) If both A and B are in $wff^0 \cup wff^+$, then $U(A, B)$ is in wff^+.
   (b) If both A and B are in $wff^0 \cup wff^-$, then $S(A, B)$ is in wff^-.
   (c) If both A and B are in wff^*, then so are $A \wedge B$, $A \vee B$, $A \Rightarrow B$, where wff^* is one of wff^+, wff^- or wff^0.

3. $\neg A$ is also a wff and it is of the same type as A with the same atoms as A.

4. $\top$ (truth) and $\bot$ (falsity) are wffs in wff^0 with no atoms.

5. If A is a wff in wff^- (pure past) with atoms $\{q, q_1, \ldots, q_n\}$ then $(\varphi q)A$ is a pure past wff (i.e. in wff^-) with the atoms $\{q_1, \ldots, q_n\}$.

The intended model for the above propositional temporal language is the set of natural numbers $\mathbb{N} = 0, 1, 2, 3, \ldots$ with the 'smaller than' relation $<$ and variables $P, Q \subseteq \mathbb{N}$ ranging over subsets. We allow quantification $\forall x \exists y$ over elements of $\mathbb{N}$. So really we are dealing with the monadic language of the model $(\mathbb{N}, <, =, 0, P_i, Q_j \subseteq \mathbb{N})$. We refer to this model also as the *non-negative integers (flow of) time.* A formula of the monadic language will in general have free set variables, and these correspond to the atoms of temporal formulae. See Volume 1 for more details.

Definition 10.5 (Syntax of USF for the predicate case). Let $Q^* = \{Q^1_{n_1}, Q^2_{n_2}, \ldots\}$ be a set of predicate symbols. $Q^i_{n_i}$ is a symbol for an n_i-place predicate. Let $f^* = \{f^1_{n_1}, f^2_{n_2}, \ldots\}$ be a set of function symbols. $f^i_{n_i}$ is a function symbol for an n_i-place function. Let $V^* = \{v_1, v_2, \ldots\}$ be a set of variables. Let $\wedge$, $\vee$, $\neg$, $\Rightarrow$, $\top$, $\bot$, $\forall$, $\exists$ be the usual classical connectives and quantifiers and let U and S be the temporal connectives and φ be the fixed point operator. We define by induction the notions of:

$wff\{x_1,\dots,x_n\}$	wff with free variables $\{x_1,\dots,x_n\}$;
$wff^+\{x_1,\dots,x_n\}$	pure future wff with the indicated free variables;
$wff^-\{x_1,\dots,x_n\}$	pure past wff with the indicated free variables;
$wff^0\{x_1,\dots,x_n\}$	pure present wff with the indicated free variables;
$term\ \{x_1,\dots,x_n\}$	term with the indicated free variables.

1. x is a term in $term\ \{x\}$, where x is a variable.
2. If f is an n-place function symbol and $t_1,\dots,t_n$ are terms with variables $V_1^*,\dots,V_n^* \subseteq V^*$ respectively, then $f(t_1,\dots,t_n)$ is a term with variables $\bigcup_{i=1}^n V_i^*$.
3. If Q is an n-place atomic predicate symbol and we have $t_1,\dots,t_n$ as terms with variables $V_1^*,\dots,V_n^* \subseteq V^*$ respectively, then $Q(t_1,\dots,t_n)$ is an atomic formula with free variables, $\bigcup_{i=1}^n V_i^*$. This formula is pure present as well as a wff.
4. Assume A, B are formulae with free variables $\{x_1,\dots,x_n\}$ and $\{y_1,\dots,y_m\}$ respectively. Then $A \wedge B$, $A \vee B$, $A \Rightarrow B$, $U(A,B)$ and $S(A,B)$ are wffs with the free variables $\{x_1,\dots,x_n,y_1,\dots,y_m\}$.
   (a) If both A and B are in $wff^0 \cup wff^+$, then $U(A,B)$ is in wff^+.
   (b) If both A and B are in $wff^0 \cup wff^-$, then $S(A,B)$ is in wff^-.
   (c) If both A and B are in wff^*, then so are $A \wedge B$, $A \vee B$, $A \Rightarrow B$, where wff^* is one of wff^+, wff^- or wff^0.
5. $\neg A$ is also a wff and it is of the same type and has the same free variables as A.
6. $\top$ and $\bot$ are in wff^0 with no free variables.
7. If A is a formula in $wff^*\{x,y_1,\dots,y_m\}$ then $\forall x A$ and $\exists x A$ are wffs in $wff^*\{y_1,\dots,y_m\}$.
8. If $(\varphi q)A(q,q_1,\dots,q_n)$ is a pure past formula of propositional USF as defined in Definition 10.4, and if $B_i \in wff\ V_i^*, i=1,\dots,m$, as defined in the present Definition 10.5, then $A' = (\varphi q)A(q,B_1,\dots,B_m)$ is a wff in $wff\ \bigcup_{i=1}^m V_i^*$. If all of the B_i are pure past, then so is A'.
9. A wff A is said to be ***essentially propositional*** iff there exists a wff $B(q_1,\dots,q_n)$ of propositional USF and wffs $B_1,\dots,B_n$ of classical predicate logic such that $A = B(B_1,\dots,B_n)$.

Remark 10.6. Notice that the fixed point operator (φx) is used in propositional USF to define the new connectives, and after it is defined it is exported to the predicate USF. We can define a language HTL which will allow fixed-point operations on predicates as well; we will not discuss this here.

Definition 10.7. We define the semantic interpretation of propositional USF in the monadic theory of $(\mathbb{N},<,=,0)$. An assignment h is a function associating with each atom q_i of USF a subset $h(q_i)$ of $\mathbb{N}$ (sometimes denoted by Q_i). h can be extended to any wff of USF as follows:

$$h(A \wedge B) = h(A) \cap h(B);$$
$$h(A \vee B) = h(A) \cup h(B);$$

$$\begin{aligned}
h(\neg A) &= \mathbb{N} - h(A); \\
h(A \Rightarrow B) &= (\mathbb{N} - h(A)) \cup h(B); \\
h(U(A,B)) &= \{t | \exists s > t(s \in h(A) \text{ and } \forall y(t < y < s \Rightarrow y \in h(B)))\}; \\
h(S(A,B)) &= \{t | \exists s < t(s \in h(A) \text{ and } \forall y(s < y < t \Rightarrow y \in h(B)))\}.
\end{aligned}$$

The meaning of U and S just defined is the Until and Since of English, i.e. 'B is true until A becomes true' and 'B is true since A was true', as in Fig. 10.3 (notice the existential meaning in $U$and S).

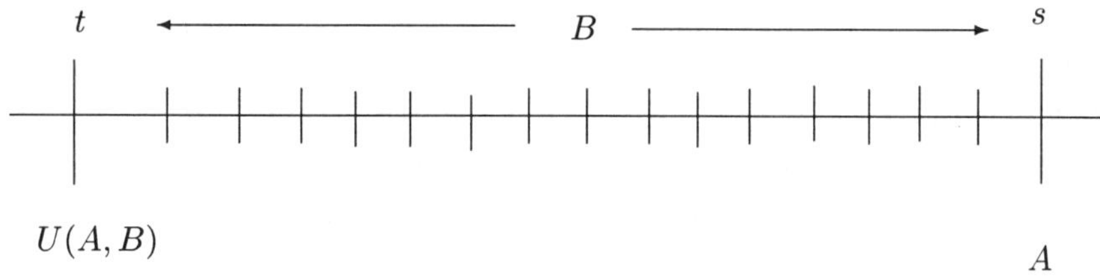

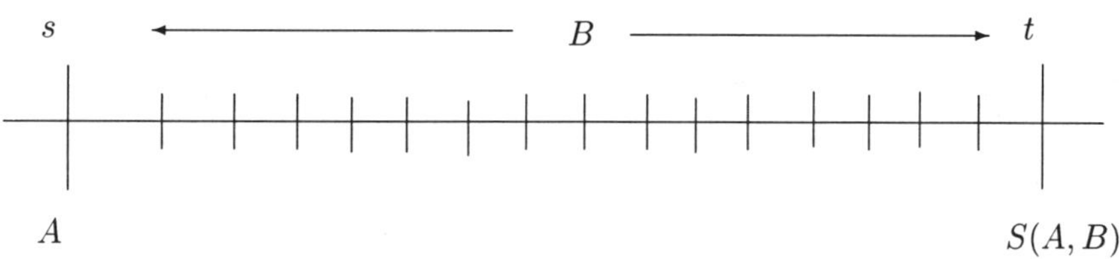

FIG. 10.3.

Finally we have the fixed point operator

$$h((\varphi q)A(q, q_i)) = \{n | n \in Q_n\}$$

where the sets $Q_n \subseteq \mathbb{N}$ are defined inductively by

$$\begin{aligned}
Q_0 &= h(A) \\
Q_{(n+1)} &= h_n(A(q, q_i))
\end{aligned}$$

where for $n \geq 0$

$$\begin{aligned}
h_n(r) = h(r) &\text{ for } r \neq q \\
Q_n \ &\text{ for } r = q.
\end{aligned}$$

This is an inductive definition. If n is a natural number, we assume inductively that we know the truth values of the formula $(\varphi q)A(q, q_i)$ at each $m < n$; then we obtain its value at n by first changing the assignment so that q has the same values as $(\varphi q)A$ for all $m < n$, and then taking the new value of $A(q, q_i)$ at n. Since A is pure past, the values of q at $m \geq n$ do not matter. Hence the definition is sound. So $(\varphi q)A$ is defined in terms of its own previous values.

This gives a fixed point semantics to formulae $(\varphi q)A(q, q_i)$, in the following sense. Suppose we have an assignment h. For any subset S of $\mathbb{N}$, let h_S be the assignment given by $h_S(r) = h(r)$ if $r \neq q$, and S if $r = q$. Then given A as above, we obtain a function $f : \wp\mathbb{N} \Rightarrow \wp\mathbb{N}$, given by $f(S) = h_S(A)$. f depends on h and A.

It is intuitively clear from the above that if $S = h((\varphi q)A)$ then $f(S) = S$, and that S is the unique solution of $f(x) = x$. So $h((\varphi q)A)$ is the unique fixed point of f. This is what we mean when we say that φ has a fixed point semantics. There are some details to be checked, in particular that the value at n of any past formula (even a complicated one involving φ) depends only on the values of its atoms at each $m < n$. For a full proof see [Hodkinson, 1989].

Definition 10.8 (Semantic definition of predicate USF). Let D be a non-empty set, called the domain, and g be a function assigning the following:

1. for each m-place function symbol f and each $n \in \mathbb{N}$ a function $g(n, f) : D^m \Rightarrow D$;
2. for each variable x and each $n \in \mathbb{N}$, an element $g(n, x) \in D$;
3. for each m-place predicate symbol Q and each $n \in \mathbb{N}$, a function $g(n, Q) : D^m \Rightarrow \{0, 1\}$.

The function g can be extended to a function $g(n, A)$, giving a value in $\{0, 1\}$ for each wff $A(x_1, \ldots, x_n)$ of the predicate USF as follows:

1. $g(n, f(t_1, \ldots, t_m)) = g(n, f)(g(n, t_1), \ldots, g(n, t_m))$;
2. $g(n, Q(t_1, \ldots, t_m)) = g(n, Q)(g(n, t_1), \ldots, g(n, t_m))$;
3. $g(n, A \wedge B) = 1$ iff $g(n, A) = 1$ and $g(n, B) = 1$;
4. $g(n, A \vee B) = 1$ iff either $g(n, A) = 1$ or $g(n, B) = 1$ or both;
5. $g(n, A \Rightarrow B) = 1$ iff either $g(n, A) = 0$ or $g(n, B) = 1$ or both;
6. $g(n, \neg A) = 1$ iff $g(n, A) = 0$;
7. $g(n, \top) = 1$ and $g(n, \bot) = 0$ for all n;
8. $g(n, U(A, B)) = 1$ iff for some $m > n, g(m, A) = 1$ and for all $n < k < m, g(k, B) = 1$;
9. $g(n, S(A, B)) = 1$ iff for some $m < n, g(m, A) = 1$ and for all $m < k < n, g(k, B) = 1$;
10. $g(n, \forall x A(x)) = 1$ for a variable x iff for all g' such that g' gives the same values as g to all function symbols and all predicate symbols and all variables different from x, we have $g'(n, A(x)) = 1$;
11. $g(n, \exists x A(x)) = 1$ for a variable x iff for some g' such that g' gives the same values as g to all function symbols and all predicate symbols and all variables different from x, we have $g'(n, A(x)) = 1$;
12. let $(\varphi q)A(q, q_1, \ldots, q_m)$ be a pure past formula of propositional USF, and let $B_i \in wffV_i^*$ for $i = 1, \ldots, m$. We want to define $g(n, A')$, where $A' =$

$(\varphi q)A(q, B_1, \ldots, B_m)$. First choose an assignment h such that $h(q_i) = \{n \in \mathbb{N} | g(n, B_i) = 1\}$. Then define $g(n, A') = 1$ iff $n \in h((\varphi q)A(q, q_1, \ldots, q_m))$.

Remark 10.9. If we let $h^*_g(A)$ be the set $\{n | g(n, A) = 1\}$ we get a function h^* like that of Definition 10.7.

Example 10.10. Let us evaluate $(\varphi x)A(x)$ for $A(x) = \blacksquare\neg x$, where $\blacksquare x = \neg S(\neg x, \top)$; see Example 10.11.1. We work out the value of $(\varphi x)A(x)$ at each n, by induction on n. If we know its values for all $m < n$, we assume that the atom x has the same value as $(\varphi x)A(x)$ for $m < n$. We then calculate the value of $A(x)$ at n. So, really, $(\varphi x)A(x)$ is a definition by recursion.

Since $\blacksquare\neg x$ is a pure past formula, its value at 0 is known and does not depend on x. Thus $A(x)$ is true at 0. Hence $(\varphi x)A(x)$ is true at 0.

Let us compute $A(x)$ at 1. Assume that x is true at 0. Since $A(x)$ is pure past, its value at 1 depends on the value of x at 0, which we know. It does not depend on the value of x at $n \geq 1$. Thus at 1, $(\varphi x)A(x) = A(x) = \blacksquare\neg\top = \bot$.

Assume inductively that we know the values of $(\varphi x)A(x)$ at $0, 1, \ldots, n$, and suppose that x also has these values at $m \leq n$. We compute $A(x)$ at $n+1$. This depends only on the values of x at points $m \leq n$, which we know. Hence $A(x)$ at $n+1$ can be computed; for our example we get $\bot$. So $(\varphi x)A(x)$ is false at $n+1$. Thus $(\varphi x)\blacksquare\neg x$ is (semantically) equivalent to $\blacksquare\bot$, because $\blacksquare\bot$ is true at 0 and nowhere else.

Another way to get the answer is to use the fixed point semantics directly. Let $f(S) = h(A)$, where $h(x) = S$, as above. Then by definition of f and g,

$$\begin{aligned} f(S) &= \{n \in \mathbb{N} | \neg\exists m < n(m \in S \wedge \forall k(m < k < n \Rightarrow k \in h(\top))\} \\ &= \{n \in \mathbb{N} | \forall m < n(m \notin S))\}. \end{aligned}$$

So $f(S) = S$ iff $S = \{0\}$. Hence the fixed point is $\{0\}$, as before.

Let us evaluate $(\varphi x)B(x)$ where $B(x) = S(S(x, a), \neg a)$. At time 0 the value of $B(x)$ is $\bot$. Let x be $\bot$ at 0. At time 1 the value of $B(x)$ is $S(S(\bot, a), \neg a) = S(\bot, \neg a) = \bot$. Let x be $\bot$ at 1 etc.... It is easy to see that $(\varphi x)B(x)$ is independent of a and is equal to $\bot$.

Example 10.11. We give examples of connectives definable in this system.

1. The basic temporal connectives are defined as follows:

Connective	Meaning	Definition
$\bullet q$	q was true 'yesterday'	$S(q, \bot)$
$\bigcirc q$	q will be true 'tomorrow'	$U(q, \bot)$
Gq	q 'will always' be true	$\neg U(\neg q, \top)$
$\Diamond q$	q 'will sometimes' be true	$U(q, \top)$
$\blacksquare q$	q 'was always' true	$\neg S(\neg q, \top)$
$\blacklozenge q$	q 'was sometimes' true	$S(q, \top)$

   Note that at 0, both $\bullet q$ and $\blacklozenge q$ are false.

2. The first time point (i.e. $n = 0$) can be identified as the point at which $\blacksquare\bot$ is true.

3. The fixed point operator allows us to define non-first-order definable subsets. For example, $e = (\varphi x)(\bullet\bullet x \vee \blacksquare\bot)$ is a constant true exactly at the even points $\{0, 2, 4, 6, \ldots\}$.
4. $S(A, B)$ can be defined from $\bullet$ using the fixed point operator.:

$$S(A, B) = (\varphi x)(\bullet A \vee \bullet(x \wedge B))'.$$

5. If we have

$$\mathit{block}(a, b) \triangleq (\varphi x)S(b \wedge S(a \wedge (x \vee \blacksquare\bot \vee \blacksquare\blacksquare\bot), a), b)$$

then $\mathit{block}(a, b)$ says that we have the sequence of the form

(block of bs)+(block of as)+...

recurring in the pure past, beginning yesterday with b and going into the past. In particular $\mathit{block}(a, b)$ is false at time 0 and time 1 because the smallest recurring block is (b, a) which requires two points in the past.

Definition 10.12 (Expressive power of USF). Let $\Psi(t, Q_1, \ldots, Q_n)$ be a formula in the monadic language of $(\mathbb{N}, <, =, 0, Q_1, \ldots, Q_n \subseteq \mathbb{N})$. Let $Q = \{t | \Psi(t, Q_i) \text{ is true}\}$. Then Q is said to be monadic first-order definable from Q_i.

Example 10.13. $\mathit{even} = \{0, 2, 4, \ldots\}$ is not monadic first-order definable from any family of finite or cofinite subsets. It is easy to see that every quantificational wff $\Psi(t, Q_i)$, with Q_i finite or cofinite subsets of $\mathbb{N}$, defines another finite or coinite subset. But *even* is definable in USF (Example 10.11.3 above). *even* is also definable in monadic second-order logic. In fact, given any formula $A(q_1, \ldots, q_n)$ of USF, we can construct a formula $A'(x, Y_1, \ldots, Y_n)$ of monadic second-order logic in the language with relations $\subseteq$ and $<$, such that for all h,

$$h(A) = \{m | A'(m, h(q_1), \ldots, h(q_n)) \text{ holds in } \mathbb{N}\}.$$

(See [Hodkinson, 1989].) Since this monadic logic is decidable, we get the following theorem.

Theorem 10.14. *Propositional USF is decidable. In other words, the set of wffs $\{A | h(A) = \mathbb{N}$ for all $h\}$ is recursive. See for example [Hodkinson, 1989].*

Theorem 10.15. *Many nested applications of the fixed point operator are no stronger than a single one. In fact, any pure past wff of USF is semantically equivalent to a positive Boolean combination (i.e. using $\wedge, \vee$ only) of wffs of the form $(\varphi x)A$, where A is built from atoms using only the Boolean connectives and $\bullet$ (as in Example 10.11.1,4). See [Hodkinson, 1989].*

Theorem 10.16 (Full expressiveness of S and U [Kamp, 1968]). *Let $Q_1, \ldots, Q_n \subseteq \mathbb{N}$ be n set variables and let $\Psi(t, Q_1, \ldots, Q_n)$ be a first-order monadic formula built up from $Q_1, \ldots, Q_n$ using $<, =$ and the quantifiers over elements of $\mathbb{N}$ and Boolean connectives. Then there exists a wff of USF, $A_\Psi(q_1,$*

$\ldots, q_n$), built up using S and U only (without the use of the fixed point operator φ) such that for all h and all Q_i the following holds:

$$\text{If } h(q_i) = Q_i \text{ then } h(A_\Psi) = \{t | \Psi(t, Q_i) \text{ holds in } \mathbb{N}\}.$$

Proof H. Kamp proved this theorem directly by constructing A_Ψ. Gabbay's proof was given in Volume 1. The significance of this theorem is that S and U alone are exactly as expressive (as a specification language) as first-order quantification $\forall$, $\exists$ over temporal points. The use of φ takes USF beyond first-order quantification. □

Theorem 10.17 (Separation theorem). *Let A be any wff of propositional USF built up using S and U only (no fixed point applications). Then A is effectively equivalent through a series of rewrite rules to a Boolean combination of wffs from $wff^+ \cup wff^- \cup wff^0$.*

Proof See [Gabbay *et al.*, 1994]. □

The above theorem is called the Separation Theorem because we separate the pure future, pure past and present. It is a true theorem for any Dedekind complete flow of time.

Theorem 10.18. *The full expressiveness of $\{S, U\}$ directly follows from Theorem 10.15.*

Proof See [Gabbay *et al.*, 1994]. □

Thus Theorems 10.17 and 10.18 provide an alternative proof for Theorem 10.16.

Theorem 10.19 (Well known). *Predicate USF without fixed point applications is not arithmetical ([Kamp, 1968]).*

10.3 USF as a specification language

The logic USF can be used as a specification language as follows. Let $A(a_1, \ldots, a_m, b_1, \ldots, b_k)$ be a wff of USF. Let h be an assignment to the atoms $\{a_i, b_j\}$. We say that h satisfies the specification A iff $h(A) = \mathbb{N}$.

In practice, the atoms b_j are controlled by the environment, and the atoms a_i are controlled by the program. Thus the truth values of b_j are determined as *events* and the truth values of a_i are determined by the program *execution*. As time moves forward and the program interacts with the environment, we get a function h, which may or may not satisfy the specification.

Let `event(`q`,` n`)` mean that the value of q at time n is truth, as determined by the environment and let `exec(`q`,` n`)` mean that q is executed at time n and therefore the truth value of q at time n is *true*. Thus out of `event` and `exec` we can get a full assignment $h =$ `event` $+$ `exec` by letting:

$h(q) = \{n | \texttt{event}(q,\ n) \text{ holds}\}$ for q controlled by the environment;
$h(q) = \{n | \texttt{exec}(q,\ n) \text{ holds}\}$ for q controlled by the program.

Of course our aim is to execute in such a way that the h obtained satisfies the specification.

We now explain how to execute any wff of our temporal language. Recall that the truth values of the atoms a_i come from the program *via* `exec(`a_i`,` m`)` and the truth values for the atoms b_j come from the environment via the function `event(`b_j`,` m`)`. We define a predicate `exec*(`A`,` m`)` for any wff A, which actually defines the value of A at time m.

For atoms of the form a_i, `exec*(`a_i`,` m`)` will be `exec(`a_i`,` m`)` and for atoms of the form b_j (i.e. controlled by the environment) `exec*(`b_j`,` m`)` will be `event(`b_j`,` m`)`. We `exec*` an atom controlled by the environment by 'agreeing' with the environment. For the case of A a pure past formula, `exec*(`A`,` m`)` is determined by past truth values. Thus `exec*` for pure past sentences is really a `hold` predicate, giving truth values determined already. For pure future sentences B, `exec*(`B`,` m`)` will have an operational meaning. For example, we will have

`exec*(`G `print,` m`) = exec*(print,` m`)` $\wedge$ `exec*(`G `print,` $m+1$`).`

We can assume that the wffs to be executed are pure formulae (pure past, pure present or pure future) such that all negations are pushed next to atoms. This can be done because of the following semantic equivalences.

1. $\neg U(a,b) = G\neg a \vee U(\neg b \wedge \neg a, \neg a)$;
2. $\neg S(a,b) = \blacksquare\neg a \vee S(\neg b \wedge \neg a, \neg a)$;
3. $\neg Ga = \Diamond\neg a$;
4. $\neg\blacksquare a = \blacklozenge\neg a$;
5. $\neg\Diamond a = G\neg a$;
6. $\neg\blacklozenge a = \blacksquare\neg a$;
7. $\neg[(\varphi x)A(x)] = (\varphi x)\neg A(\neg x)$.

See [Gabbay *et al.*, 1994] for 1 to 6. For 7, let h be an assignment and suppose that $h(\neg[(\varphi x)A(x)]) = S$. Because φ has fixed point semantics, to prove 7 it is enough to show that if h' is the assignment that agrees with h on all atoms except x, and $h'(x) = S$, then $h'(\neg(A\neg x)) = S$. Clearly, $h(\neg[(\varphi x)A(x)]) = \mathbb{N}\setminus S$. We may assume that $h(x) = \mathbb{N}\setminus S$. Then $h(A(x)) = \mathbb{N}\setminus S$. So $h(\neg A(x)) = S$ and $h'(\neg(A\neg x)) = S$ as required; 7 is also easy to see using the recursive approach.

Definition 10.20. We assume that `exec*(`A`,`m`)` is defined for the system for any m and any A which is atomic or the negation of an atom. For atomic b which is controlled by the environment, we assume `event(`b`,`m`)` is defined and `exec*(`b`,`m`)=event(`b`,`m`)`. For `exec*(`a`,`m`)`, for a controlled by the program, execution may be done by another program. It may be a graphical or a mathematical program. It certainly makes a difference what m is relative to now. If we want to `exec*(`a`,`m`)` for m in the past of now, then a has already been executed (or not) and so `exec*(`a`,`m`)` is a `hold` predicate. It agrees with what has been done. Otherwise (if $m \geq$ now) we do execute*.

1. `exec*(`$\top$`,`m`) =` $\top$.
2. `exec*(`$\bot$`,`m`) =` $\bot$.

3. exec*$(A \wedge B,\ m)$ = exec*$(A,m) \wedge exec*(B,m)$.
4. exec*$(A \vee B,\ m)$ = exec*$(A,m) \vee exec*(B,m)$.
5. exec*$(S(A,B),\ 0))$ = $\bot$.
6. exec*$(S(A,B),\ m+1)$ = exec*(A,m)
   $\vee$ [exec*(B,m) $\wedge$ exec*$(S(A,B),m)$].
7. exec*$(U(A,B),\ m)$ = exec*$(A,m+1)$
   $\vee$ [exec*$(B,m+1)\wedge$ exec*$(U(A,B),m+1)$].
8. exec*$((\varphi x)A(x),\ 0)$ = A_0, where A_0 is obtained from A by substituting $\top$ for any wff of the form $\blacksquare B$ and $\bot$ for any wff of the form $\Diamond B$ or $S(B_1, B_2)$.
9. exec*$((\varphi x)A(x),\ m+1)$ = exec*$(A(C),\ m+1)$, where C is a new atom defined for $n \leq m$ by exec*$(C,\ n)$ = exec*$((\varphi x)A(x),\ n)$. In other words exec*$((\varphi x)A(x),\ m+1)$ = exec*$(A((\varphi x)A(x)),\ m+1)$ and since in the execution of A at time $m+1$ we go down to executing A at time $n \leq m$, we will have to execute $(\varphi x)A(x)$ at $n \leq m$, which we assume by induction that we already know.
10. In the predicate case we can let
    exec*$(\forall y\ A(y))$ = $\forall y$ exec*$(A(y))$
    exec*$(\exists y\ A(y))$ = $\exists y$ exec*$(A(y))$.

We are now in a position to discuss how the execution of a specification is going to be carried out in practice. Start with a specification $\mathcal{S}$. For simplicity we assume that $\mathcal{S}$ is written in essentially propositional USF which means that $\mathcal{S}$ contains S, U and φ operators applied to pure past formulae, and is built up from atomic units which are wffs of classical logic. If we regard any fixed point wff $(\varphi x)D(x)$ as atomic, we can apply the separation theorem and rewrite $\mathcal{S}$ into an executable form $\mathcal{E}$, which is a conjunction of formulae such that

$$\Omega \equiv \bigwedge_k \left[\bigwedge_i C_{i,k} \Rightarrow \bigvee_j B_{j,k} \right]$$

where $C_{i,k}$ are pure past formulae (containing S only) and $B_{j,k}$ are either atomic or pure future formulae (containing U). However, since we regarded any (φx) formula as an atom, the $B_{j,k}$ can contain $(\varphi x)D(x)$ formulae in them. Thus $B_{j,k}$ can be for example $U(a, (\varphi x)[\bullet\neg x])$. We will assume that any such $(\varphi x)D(x)$ contains only atoms controlled by the environment; this is a restriction on $\mathcal{E}$. Again, this is because we have no separation theorem as yet for full propositional USF, but only for the fragment US of formulae not involving φ. We conjecture that—possibly in a strengthened version of USF that allows more fixed point formulae—any formula can be separated. This again remains to be done.

However, even without such a result we can still make progress. Although $(\varphi x)[\bullet\neg x]$ is a pure past formula within U, it is still an executable formula that only refers to environment atoms, and so we do not mind having it there. If program atoms were involved, we might have a formula equivalent to $\bigcirc \bullet$`print` (say), so that we would have to execute $\bullet$`print` tomorrow. This is not impossible: when tomorrow arrives we check whether we did in fact print yesterday, and return $\top$ or $\bot$ accordingly. But it is not a very intelligent way of executing the specification, since clearly we should have just printed in the first instance. This illustrates why we need to separate $\mathcal{S}$.

Recall the equation for executing $U(A, B)$:

$$\texttt{exec*}(U(A,B)) \equiv \bigcirc\texttt{exec*}(A) \vee (\bigcirc(\texttt{exec*}(B) \wedge \texttt{exec*}(U(A,B)).$$

If either A or B is of the form $(\varphi x)D(x)$, we know how to compute $\texttt{exec*}((\varphi x) D(x))$ by referring to past values. Thus $(\varphi x)D(x)$ can be regarded as atomic because we know how to execute it, in the same way as we know how to execute `write`.

Imagine now that we are at time n. We want to make sure the specification $\mathcal{E}$ remains true. To keep $\mathcal{E}$ true we must keep true each conjunct of $\mathcal{E}$. To keep true a conjunct of the form $C \Rightarrow B$ where C is past and B is future, we check whether C is true in the past. if it is true, then we have to make sure that B is true in the future. *Since the future has not happened yet, we can read B imperatively, and try to force the future to be true.* Thus the specification $C \Rightarrow B$ is read by us as

$$\texttt{hold}(C) \Rightarrow \texttt{exec*}(B).$$

Some future formulae cannot be executed immediately. We already saw that to execute $U(A, B)$ now we either execute A tomorrow or execute B tomorrow together with $U(A, B)$. Thus we have to pass a list of formulae to execute from today to tomorrow. Therefore at time $n+1$, we have a list of formulae to execute which we inherit from time n, in addition to the list of formulae to execute at time $n + 1$. We can thus summarize the situation at time $n + 1$ as follows:

1. Let $G_1, \ldots, G_m$ be a list of wffs we have to execute at time $n + 1$. Each G_i is a disjunction of formulae of the form atomic or negation of atomic or $\Diamond A$ or GA or $U(A, B)$.
2. In addition to the above, we are required to satisfy the specification $\mathcal{E}$, namely

$$\bigwedge_k \left[\bigwedge_i C_{i,k} \Rightarrow \bigvee_j B_{j,k} \right]$$

   for each k such that $\bigwedge_i C_{i,k}$ holds (in the past). We must execute the future (and present) formula $B_k = \bigvee_j B_{j,k}$ which is again a disjunction of the same form as in 1 above.

We know how to execute a formula; for example,

$$\texttt{exec*}(\Diamond A) = \bigcirc\texttt{exec*}A \vee \bigcirc\ \texttt{exec*}(\Diamond A).$$

$\Diamond A$ means 'A will be true'. To execute $\Diamond A$ we can either make A true tomorrow or make $\Diamond A$ true tomorrow. What we should be careful not to do is not to keep on executing $\Diamond A$ day after day because this way A will never become true. Clearly then we should try to execute A tomorrow and if we cannot, only then do we execute $\Diamond A$ by doing $\bigcirc$ `exec*(`$\Diamond A$`)`. We can thus read the disjunction `exec*(`$A \vee B$`)` as *first* try to `exec*`A and *then only if we fail* `exec*`B. This priority (left to right) is not a logical part of '$\vee$' but a procedural addition required for the correctness of the model. We can thus assume that the formulae given to execute at time n are written as disjunctions with the left disjuncts having priority in execution. Atomic sentences or their negations always have priority in execution (though this is not always the best practical policy).

Let $D = \bigvee_j D_j$ be any wff which has to be executed at time $n+1$, either because it is inherited from time n or because it has to be executed owing to the requirements of the specification at time $n+1$. To execute D, either we execute an atom and discharge our duty to execute, or we pass possibly several disjunctions to time $n+2$ to execute then (at $n+2$), and the passing of the disjunctions will discharge our obligation to execute D at time $n+1$. Formally we have

`exec*(`D`) = `$\bigvee_j$` exec*(`D_j`).`

Recall that we try to execute left to right. The atoms and their negations are supposed to be on the left. If we can execute any of them we are finished with D. If an atom is an environment atom, we check whether the environment gives it the right value. If the atom is under the program's control, we can execute it. However, the negation of the atom may appear in another formula D' to be executed and there may be a clash. See Examples 10.21 and 10.22 below. At any rate, should we choose to execute an atom or negation of an atom and succeed in doing so, then we are finished. Otherwise we can execute another disjunct of D of the form $D_j = U(A_j, B_j)$ or of the form GA_j or $\Diamond A_j$. We can pass the commitment to execute to the time $n+2$. Thus we get

`exec*(`D`) = `$\bigvee$` exec*(atoms of `D`) `$\vee$` exec*(future formulae of `D`).`

Thus if we cannot execute the atoms at time $n+1$, we pass to time $n+2$ a conjunction of disjunctions to be executed, ensuring that atoms and subformulae should be executed before formulae. We can write the disjunctions to reflect these priorities. Notice further that although, on first impression, the formulae to be executed seem to multiply, they actually do not.

At time $n = 0$ all there is to execute are heads of conditions in the specification. If we cannot execute a formula at time 0 then we pass execution to time 1. This means that at time 1 we inherit the execution of $A \vee (B \wedge U(A, B))$, where $U(A, B)$ is a disjunct in a head of the specification. This same $U(A, B)$ may be passed on to time 2, or some subformula of A or B may be passed. The number of such subformulae is limited and we will end up with a limited stock of formulae to be passed on. In practice this can be optimized. We have thus explained how to execute whatever is to be executed at time n. When we perform the execution sequence at times $n, n+1, n+2, \ldots$, we see that there are now two possibilities:

- We cannot go on because we cannot execute all the demands at the same time. In this case we stop. The specification cannot be satisfied either because it is a contradiction or because of a wrong execution choice (e.g. we should not have printed at time 1, as the specification does not allow anything to be done after printing).

- Another possibility is that we see after a while that the same formulae are passed for execution from time n to time $n+1$ to $n+2$ etc. This is a loop. Since we have given priority in execution to atoms and to the A in $U(A, B)$, such a loop means that it is not possible to make a change in execution, and therefore either the specification cannot be satisfied because of a contradiction or wrong choice of execution, or the execution is already satisfied by this loop.

Example 10.21. All atoms are controlled by the program. Let the specification be

$$Ga \wedge \Diamond \neg a.$$

Now the rules to execute the subformulae of this specification are

exec*(Ga) $\equiv$ exec*(a) $\wedge$ exec*(Ga)
exec*($\Diamond\neg a$) $\equiv$ exec*($\neg a$) $\vee$ exec*($\Diamond\neg a$).

To execute Ga we must execute a. Thus we are forced to discharge our execution duty of $\Diamond\neg a$ by passing $\Diamond\neg a$ to time $n+1$. Thus time $n+1$ will inherit from time n the need to execute $Ga \wedge \Diamond\neg a$. This is a loop. The specification is unsatisfiable.

Example 10.22. The specification is

$$b \vee Ga$$
$$\blacklozenge b \Rightarrow \Diamond\neg a \wedge Ga.$$

According to our priorities we execute b first at time 0. Thus we will have to execute $\Diamond\neg a \wedge Ga$ at time 1, which is impossible. Here we made the wrong execution choice. If we keep on executing $\neg b \wedge Ga$ we will behave as specified.

In practice, since we may have several choices in execution we may want to simulate the future a little to see if we are making the correct choice.

Having defined exec*, we need to add the concept of updating. Indeed, the viability of our notion of the declarative past and imperative future depends on adding information to our database. In this chapter we shall assume that every event that occurs in the environment, and every action exec-ed by our system, are recorded in the database. This is of course unnecessary, and in a future paper we shall present a more realistic method of updating.

10.4 The logic USF2

The fixed point operator that we have introduced in propositional USF has to do with the solution of the equation

$$x \leftrightarrow B(x, q_1, \ldots, q_m)$$

where B is a pure past formula. Such a solution always exists and is unique. The above equation defines a connective $A(q_1, \ldots, q_m)$ such that

$$A(q_1, \ldots, q_m) \leftrightarrow B(A(q_1, \ldots, q_m), q_1, \ldots, q_m).$$

Thus, for example, $S(p, q)$ is the solution of the equation

$$x \leftrightarrow \bullet p \vee \bullet (q \wedge x)$$

as we have $S(p, q) \leftrightarrow \bullet p \vee \bullet (q \wedge S(p, q))$. Notice that the connective to be defined ($x = S(p, q)$) appears as a unit in both sides of the equation.

To prove existence of a solution we proceed by induction. Suppose we know what x is at time $\{0, \ldots, n\}$. To find what x is supposed to be at time $n + 1$, we use the equation $x \leftrightarrow B(x, q_i)$. Since B is pure past, to compute B at time $n + 1$ we need to know $\{x, q_i\}$ at times $\leq n$, which we do know. This is the reason why we get a unique solution.

Let us now look at the following equation for a connective $Z(p, q)$. We want Z to satisfy the equation

$$Z(p, q) \leftrightarrow \bullet p \vee \bullet (q \wedge Z(\bullet p, q)).$$

Here we did not take $Z(p, q)$ as a unit in the equation, but substituted a value $\bullet p$ in the right-hand side, namely $Z(\bullet p, q)$. $\bullet p$ is a pure past formula. We can still get a unique solution because $Z(p, q)$ at time $n + 1$ still depends on the values of $Z(p, q)$ at earlier times, and certainly we can compute the values of $Z(\bullet p, q)$ at earlier times.

The general form of the new fixed point equation is as follows:

Definition 10.23 (Second-order fixed points). Let $Z(q_1, \ldots, q_m)$ be a candidate for a new connective to be defined. Let $B(x, q_1, \ldots, q_m)$ be a pure past formula and let $D_i(q_1, \ldots, q_m)$ for $i = 1 \ldots m$ be arbitrary formulae. Then we can define Z as the solution of the following equation:

$$Z(q_1, \ldots, q_m) \leftrightarrow B(Z[D_1(q_1, \ldots, q_m), \ldots, D_m(q_1, \ldots, q_m)], q_1, \ldots, q_m).$$

We call this definition of Z *second order*, because we can regard the equation as

$$Z \equiv \texttt{Application}(Z, D_i, q_j).$$

We define USF2 to be the logic obtained from USF by allowing nested applications of second-order fixed point equations. USF2 is more expressive than USF (Example 10.24).

Predicate USF2 is defined in a similar way to predicate USF.

Example 10.24. Let us see what we get for the connective $Z_1(p, q)$ defined by the equation

$$Z_1(p, q) \leftrightarrow \bullet p \vee \bullet (q \wedge Z_1(\bullet p, q)).$$

The connective $Z_1(p, q)$ says what is shown in Fig.10.4:

$Z_1(p, q)$ is true at n iff for some $m \leq n$, q is true at all points j with $m \leq j < n$, and p is true at the point $m_1 = m - (n - m + 1) = 2m - n - 1$. If we let $k = n - m$,

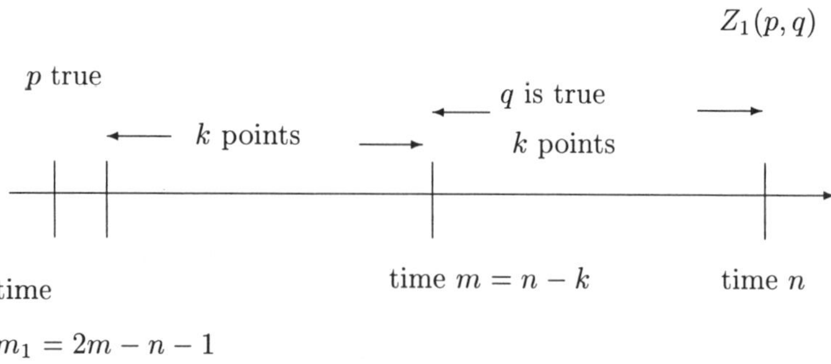

FIG. 10.4.

then we are saying that q is true k times into the past and before that p is true at a point which is $k+1$ times further into the past. This is not expressible with any pure past formula of USF; see [Hodkinson, 1989].

Let us see whether this connective satisfies the fixed point equation

$$Z_1(p,q) \leftrightarrow \bullet p \vee \bullet(q \wedge Z_1(\bullet p, q)).$$

If $\bullet p$ is true then $k = 0$ and the definition of $Z_1(p,q)$ is correct. If $\bullet(q \wedge Z_1(\bullet p, q))$ is true, than we have for some k the situation in Fig. 10.5:

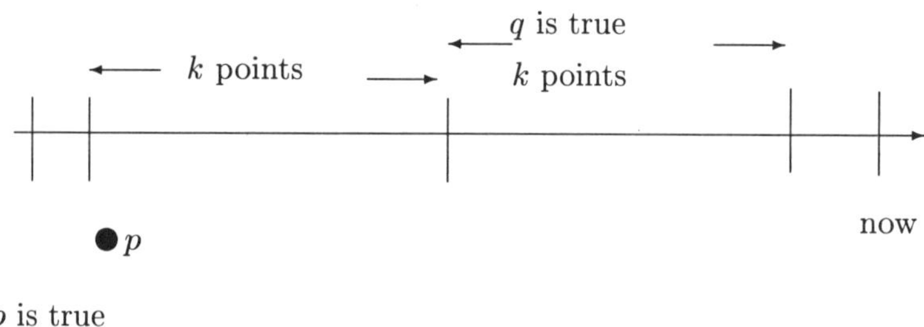

FIG. 10.5.

The definition of $Z_1(p,q)$ is satisfied for $k+1$.

Example 10.25 (Coding of dates). We can encode dates in the logic as follows:

1. The proposition $\neg\bullet\top$ is true exactly at time 0, since it says that there is no yesterday. Thus if we let

$$\mathbf{n} = \bot \text{ if } n \leq 0$$
$$0 = \neg\bullet\top$$

$$\mathbf{n} = \bullet(\mathbf{n} - \mathbf{1}).$$

then we have that $\mathbf{n}$ is true exactly at time n. This is a way of naming time n. In predicate temporal logic we can use elements to name time. Let $\mathtt{date}(x)$ be a predicate such that the following hold at all times n:

$\exists x\ \mathtt{date}(x)$
$\forall x(\mathtt{date}(x) \Rightarrow G\neg\ \mathtt{date}(x) \wedge \blacksquare\neg\ \mathtt{date}(x))$
$\forall x(\mathtt{date}(x) \vee \blacklozenge\ \mathtt{date}(x) \vee \lozenge\ \mathtt{date}(x))$.

These axioms simply say that each time n is identified by some element x in the domain that uniquely makes $\mathtt{date}(x)$ true, and every domain element corresponds to a time.

2. We can use this device to count in the model. Suppose we want to define a connective that counts how many times A was true in the past. We can represent the number m by the date formula $\mathbf{m}$, and define $\mathtt{count}(A, \mathbf{m})$ to be true at time n iff the number of times before n in which A was true is exactly $\mathbf{m}$. Thus in Fig. 10.6, $\mathtt{count}(A, \bullet\top \wedge \neg\bullet\bullet\top)$ is false at time 3, true at time 2, true at time 1 and false at time 0.

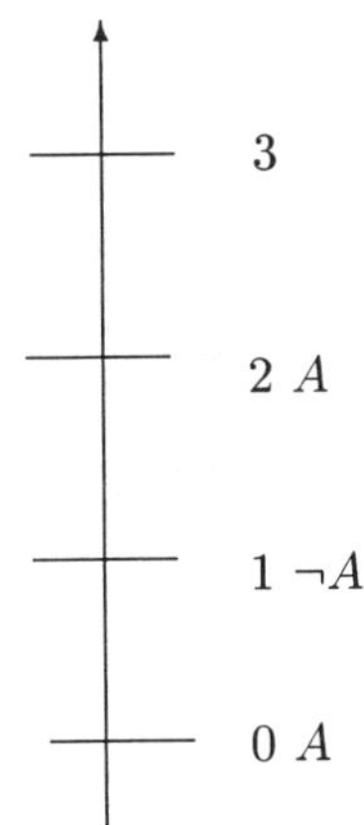

FIG. 10.6.

The connective count can be defined by recursion as follows:

$$\begin{aligned}\mathtt{count}(p, \mathbf{n}) \leftrightarrow\ & \bullet(\neg p \wedge \mathtt{count}(p, \mathbf{n})) \\ & \vee \bullet(p \wedge \mathtt{count}(p, \bigcirc\mathbf{n})) \\ & \vee (\neg\bullet\top \wedge \mathbf{n}).\end{aligned}$$

Note that $\bigcirc\mathbf{n}$ is equivalent to $\mathbf{n} - \mathbf{1}$. We have cheated in this example. For the formula $B(x, q_1, q_2)$ in the definition of second-order fixed points is here

$$\bullet(\neg q_1 \wedge x) \vee \bullet(q_1 \wedge x) \vee (\neg\bullet\top \wedge q_2).$$

This is not pure past, as q_2 occurs in the present tense. To deal with this we could define the notion of a formula $B(x, q_1, \ldots, q_m)$ being pure past in x. See [Hodkinson, 1989]. We could then amend the definition to allow any B that is pure past in x. This would cover the B here, as all xs in B occur under a $\bullet$. So the value of the connective at n still depends only on its values at $m \leq n$, which is all we need for there to be a fixed point solution. We do not do this formally here, as we can express count in standard USF2; see the next example.

Example 10.26. We can now define the connective $\mathtt{more}(A, B)$ reading 'A was true more times than B'.

$$\begin{aligned}\mathtt{more}(A,B) \leftrightarrow\ & \bullet(A \wedge \mathtt{more}(A,B)) \\ & \vee \bullet(\neg A \wedge \neg B \wedge \mathtt{more}(A,B)) \\ & \vee \bullet(\neg A \wedge \mathtt{more}((A \wedge \blacklozenge A), B)).\end{aligned}$$

(If $k > 0$, then at any n, $A \wedge \blacklozenge A$ has been true k times iff A has been true $k+1$ times.)

Note that for any $k > 0$, the formula $E_k = \neg \bullet^k \top$ is true exactly k times, at $0, 1, \ldots, k-1$. If we define

$$\mathtt{count*}(p, k) = \mathtt{more}(E_{k+1}, p) \wedge \neg\mathtt{more}(E_k, p),$$

then at any n, p has been true k times iff $\mathtt{count*}(p, k)$ holds. So we can do the previous example in standard USF2.

Theorem 10.27 (For propositional USF2). *Nested applications of the second-order fixed point operator are equivalent to one application. Any wff A of USF2 is equivalent to a wff B of USF2 built up using no nested applications of the second-order fixed point operator.*

10.5 Payroll example in detail

This section will consider in detail the execution procedures for the payroll example in Section 10.1.

First let us describe, in the temporal logic USF2, the specification required by Mrs Smith. We translate from the English in a natural way. This is important because we want our logical specification to be readable and have the same structure as in English.

Recall that the intended interpretation of the predicates to be used is

$A(x)$	x is asked to babysit
$B(x)$	x does a babysitting job
$M(x)$	x works after midnight
$P(x, y)$	x is paid y pounds.

'Babysitters are not allowed to take jobs three nights in a row, or two nights in a row if the first night involves overtime' is translated as

(a) $\forall x \neg[B(x) \wedge \bullet B(x) \wedge \bullet\bullet B(x)]$

(b) $\forall x \neg [B(x) \wedge \bullet (B(x) \wedge M(x))]$
(c) $\forall x [M(x) \Rightarrow B(x)]$.

Note that these wffs are not essentially propositional.

'Priority in calling is given to those who were not called before as many times as others' is translated as

(d) $\neg \exists x \exists y [\mathtt{more}(A(x), A(y)) \wedge A(x) \wedge \neg A(y) \wedge \neg \bullet M(y) \wedge \neg \bullet (B(y) \wedge \bullet B(y))]$.

'Payment should be made the next day after the job was done, with £15 for a job involving overtime, and £10 for a job not involving overtime' is translated as

(e) $\forall x [M(x) \Rightarrow \bigcirc P(x, 15)]$
(f) $\forall x [B(x) \wedge \neg M(x) \Rightarrow \bigcirc P(x, 10)]$
(g) $\forall x [\neg B(x) \Rightarrow \bigcirc \neg \exists y P(x, y)]$.

Besides the above we also have

(h) $\forall x [B(x) \Rightarrow A(x)]$.

Babysitters work only when they are called.

We have to rewrite the above into an executable form, namely

Past $\Rightarrow$ Present $\vee$ Future.

We transform the specification to the following:

(a′) $\forall x [\bullet B(x) \wedge \bullet \bullet B(x) \Rightarrow \neg B(x)]$
(b′) $\forall x [\bullet (B(x) \wedge M(x)) \Rightarrow \neg B(x)]$
(c′) $\forall x [\neg M(x) \vee B(x)]$.
(d′) $\forall x \forall y [\mathtt{more}(A(x), A(y)) \wedge \neg \bullet M(y) \wedge \neg \bullet (B(y) \wedge \bullet B(y)) \Rightarrow \neg A(x) \vee \neg A(y)]$
(e′) $\forall x [\neg M(x) \vee \bigcirc P(x, 15)]$
(f′) $\forall x [\neg B(x) \vee M(x) \vee \bigcirc P(x, 10)]$
(g′) $\forall x [B(x) \vee \bigcirc \forall y \neg P(x, y)]$
(h′) $\forall x [\neg B(x) \vee A(x)]$.

Note that **(e′)**, **(f′)** and **(h′)** can be rewritten in the following form using the $\bullet$ operator.

(e″) $\forall x [\bullet M(x) \Rightarrow P(x, 15)]$
(f″) $\forall x [\bullet (B(x) \wedge \neg M(x)) \Rightarrow P(x, 10)]$
(g″) $\forall x [\neg \bullet B(x) \Rightarrow \forall y \neg P(x, y)]$.

Our executable sentences become

(a*) $\mathtt{hold}(\bullet B(x) \wedge \bullet \bullet B(x)) \Rightarrow \mathtt{exec}(\neg B(x))$
(b*) $\mathtt{hold}(\bullet (B(x) \wedge M(x))) \Rightarrow \mathtt{exec}(\neg B(x))$
(c*) $\mathtt{exec}(\neg M(x) \vee B(x))$
(d*) $\mathtt{hold}(\mathtt{more}(A(x), A(y)) \wedge \neg \bullet M(y) \wedge \neg \bullet (B(y) \wedge \bullet B(y))) \Rightarrow \mathtt{exec}(\neg A(x) \vee \neg A(y))$
(e*) $\mathtt{exec}(\neg M(x) \vee \bigcirc P(x, 15))$
(f*) $\mathtt{exec}(\neg B(x) \vee M(x) \vee \bigcirc P(x, 10))$

(g*) $\texttt{exec}(B(x) \vee \bigcirc \forall y \neg P(x,y))$
(h*) $\texttt{exec}(\neg B(x) \vee A(x))$.

If we use **(e″)**, **(f″)**, **(g″)** the executable form will be

(e)** $\texttt{hold}(\bullet M(x)) \Rightarrow \texttt{exec}(P(x,15))$
(f)** $\texttt{hold}(\bullet (B(x) \wedge \neg M(x))) \Rightarrow \texttt{exec}(P(x,10))$
(g)** $\texttt{hold}(\neg \bullet B(x)) \Rightarrow \texttt{exec}(\forall y \neg P(x,y))$.

In practice there is no difference whether we use **(e**)** or **(e*)**. We execute $\bigcirc P$ by sending P to tomorrow for execution. If the specification is **(e**)**, we send nothing to tomorrow but we will find out tomorrow that we have to execute P.

11

THE METATEM PROGRAMMING LANGUAGE

METATEM is really a paradigm for programming languages rather than one particular language. There are three basic principles:

- programs should be expressed in a temporal language;
- programs should be able to be read declaratively;
- the operation of the program should be interpretative with individual program clauses operating according to the 'declarative past implies imperative future' idea.

The versions of METATEM described in [Barringer *et al.*, 1989], [Finger *et al.*, 1993], [Fisher and Barringer, 1991], [Fisher, 1993], [Reynolds, 1995], [Barringer *et al.*, 1996] and [Reynolds, 1997] use the temporal languages PML and FMO with Until and Since.

The basic idea of declarative languages is that a program should be able to be read as a specification of a problem in some formal language and that running the program should solve that problem. Thus we will see that a METATEM program can easily be read as a temporal sentence and that running the program *should* produce a model of that sentence. This is an ideal situation and we will see that the correctness question is not usually so trivial; owing to such behaviour as 'looping', sometimes the program does not construct a model.

There are related completeness issues for such a declarative programming language:

- can one specify all desired problems in that subset of the formal language used for the programs; and
- does the program guarantee to come up with a solution if one exists?

These are important questions which we will consider, but it is useful to remember that one can go too far in the quest for completeness. A very expressive language allows very difficult problems to be asked and relying on the program interpreter to find a solution might lead to inefficiencies—sometimes it is better to make the user rewrite the problem in a better way. However, although this restricts the language of programs, much work is going on to enable problems expressed in a more general way to be rewritten in the appropriate form. See the next section for a discussion of this.

The task of the METATEM program is to build a model satisfying the declared specification. This can sometimes be done by a machine following some arcane, highly complex procedure which eventually emerges with the description of the model. That would *not* be the METATEM approach. Because we are describing a programming language, transparency of control is crucial. It should be easy to follow and predict the program's behaviour and the contribution of the individual clauses must be straightforward.

Fortunately, these various disparate aims can be very nicely satisfied by the intuitively appealing 'declarative past implies imperative future' idea of [Gabbay, 1989]. The METATEM program *rule* is of the form $P \Rightarrow F$ where P is a strict past-time formula and F is a not necessarily strict future-time formula. The idea is that on the basis of the declarative truth of the past time P the program should go on to 'do' F.

A METATEM program is a list $\{P_i \to F_i \mid i = 1, \ldots, n\}$ of such rules and, at least in the propositional case, it represents the **PML** formula

$$\Box \bigwedge_{i=1}^{n} (P_i \to F_i).$$

The program is read declaratively as a specification: the execution mechanism should deliver a model of this formula. To do so it will indicate which propositions are true and which are false at time 0, then at time 1, then at time 2, etc.

To achieve this result one could follow the idea of [Pnueli and Rosner, 1989a] and compile the specification into a transducer which then builds a model. As we have said, METATEM uses an interpretive approach. The general idea is to go through the whole list of rules $P_i \to F_i$ at each successive stage and make sure that F_i gets made true whenever P_i is. This would seem to be straightforward as the P_i only depend on the past—so we know when they are true—and the F_i only depend on propositions in the present and future so we can make them true. However, we will see below that there are subtleties.

There are many versions of METATEM in existence. The main distinctions are:

- propositional versus first-order languages;
- closed systems versus reactive modules; and
- single-threaded execution versus concurrency.

Let us examine some of the main variants.

11.1 Propositional METATEM in a closed system

Here we restrict the language to PML so the program is designed to represent a formula of this language. But not just any formula is acceptable. In order to give a *declarative* temporal formula an *imperative* reading we limit ourselves to using rules of the form $P \Rightarrow F$ where P is a strict past-time formula and F is a not necessarily strict future-time formula.

As we have described above, a METATEM program is a list $\{P_i \to F_i \mid i = 1, \ldots, n\}$ of such rules representing the PML formula

$$\Box \bigwedge_{i=1}^{n} (P_i \to F_i),$$

and METATEM builds a model of this formula by going through the whole list of rules $P_i \to F_i$ at each successive stage and making sure that F_i gets made true whenever P_i is. Let us look at the procedure in more detail.

To build a model of $\Box \bigwedge_{i=1}^{n}(P_i \to F_i)$ we have to make each $P_i \to F_i$ true at every stage $n = 0, 1, \ldots$. This is done successively by stages. Suppose that we are now to decide on the truth values of propositions at stage $n \geq 0$.

The first step is to determine which rules *fire*, i.e. which P_i are true. Since the P_i are strict past, this just involves looking back at the behaviour of the process in the past. To do this we must have a database of these values. There are various concerns with deciding how to store and access this information efficiently and calculate truth values of complicated temporal formulae but we do not need to go into them here.

Say that $J \subseteq \{1, \ldots, n\}$ is the set of indices of rules which fired. Since the rules $P_i \to F_i$ for which P_i are false are already satisfied, we just have to make F_j true for every $j \in J$. To make F_j true might involve choosing truth values of atoms at time n but it may also involve *committing* ourselves to making some formula true in the future. Thus our procedure is actually slightly more complicated: at every stage, including now at stage n, we might have committed ourselves to making the formula χ_n true.

So we are now to make $\bigwedge_{j\in J} F_j \wedge \chi_n$ true. This can be done best by rewriting the formula equivalently in a standard form. It can be shown that every such formula is equivalent to a disjunction

$$\bigvee \psi_k$$

of conjunctions

$$\bigwedge_{m\in M_k} e_{km} \wedge \bigcirc\phi_k$$

where each e_{km} is a literal and ϕ_k is a not necessarily strict future-time formula. In order to do this one uses the important equivalence

$$\alpha U \beta \text{ iff } \bigcirc(\beta \vee (\alpha \wedge (\alpha U \beta)))$$

as well as the usual propositional equivalences. Recall that our U is a strict version of Until; as we will see below the program actually uses a non-strict version.

Now we see that there may be a choice: we must force one of the ψ_k to be true. We will discuss this non-determinism below but now suppose that ψ_k has been chosen. The procedure now is simple: we simply announce the truth of those propositions appearing positively in an e_{km}, the falsity of those appearing negatively and choose any truth value for the rest. These are the model's values for the state at time n. We can now update the database of history, carry over ϕ_k as the new commitment and move on to consider stage $n + 1$.

How then to choose ψ_k? A wrong choice would be to choose a disjunct ψ_k containing a conjunct $\bot$ or one containing a conjunct p and a conjunct $\neg p$. This is not too hard to avoid unless a wrong choice had beenmade previously in the past which now means that all disjuncts are contradictory. To recover from such a dead end we introduce *back-tracking* into the procedure and allow reversion of the execution procedure to the last choice point in the past and retraction of all anouncements of propositions truth values which were made since then.

Note that, although this is not usually stressed, the arbitrary choice of truth values of atoms mentioned above is also counted as a choice but most implementations make such propositions false. Another possible way of dealing with the choice could be to delay the choice until it is important in the hope of gaining some more information about what would be a sensible truth value to choose.

After back-tracking, the procedure, of course, makes a different choice to the ones it has previously tried at that point. If there are no more choices to try at this choice point we back-track to the previous choice point. When there are no more choice points to back-track to the execution fails. This should only happen if the specification is unsatisfiable.

One reason that, even without such a failure, we do not end up with a model of the specification is that the procedure may end up in a *loop* when the same sequence of states and unsatisfied commitments constantly recurs. To prevent this, in the propositional case, it is possible to implement a loop-checker which forces back-tracking whenever the procedure gets stuck in such a loop for long enough. There is an (admittedly very large) theoretical bound on the number of times a pattern of states and commitments can recur in the course of successfully building a model.

In order to avoid looping behaviour it is essential to make sensible decisions at choice points. An excellent strategy for achieving this is always to execute the disjunct which will satisfy the longest outstanding commitment. Furthermore, such a strategy is easy to implement using simply syntactic properties of the rewriting procedure above.

It has been shown in [Barringer *et al.*, 1989] that using such a strategy and with a loop-checker, the METATEM procedure outlined above will always eventually construct a model if the program is satisfiable.

11.2 The environment

The first complication we introduce to this situation is the very useful one of regarding a METATEM process as a *reactive module* (as defined in Chapter 3). This means that we immerse the process in an environment.

We could formalize this by supposing that the propositions in our language are partitioned into *component* propositions under the control of the METATEM process and *environment* propositions under the control of the environment.

We have seen that we can classify a reactive module situation as either *synchronous* or *asynchronous*. We will just consider the simpler synchronous situation in this subsection but see asynchronous modules in action when we examine concurrency.

One change to the METATEM procedure that we must make is to arrange for the environment's choices to be observed at each stage and to make sure that they are recorded in the history database.

The main change concerns back-tracking. In most applications of reactive modules back-tracking by the module is just not acceptable. Even in those rare examples when it is acceptable, the question arises as to whether the environment should also be allowed to retract its announcements when the module is back-tracking. Clearly to allow this is only fair. However, this would severely complicate the back-tracking mechanism as the procedure would have to record under which environment choices it has made its and to take this information into account when trying new choices.

So, although there might be some uses for back-tracking in a situation in which the environment's propositions are fixed, in general we will have to do without this ability in the reactive module version of METATEM. Obviously this has disastrous consequences for the completeness of the programming language and we can no longer expect a completeness result. However, the procedure can guarantee to build a model if there is one in the case of a *deterministic* program: i.e. a program in which there are no choices to make.

In this case study we will thus try always to employ deterministic programs. We will see that this cannot always be done. However, it can be seen that using rules with just one proposition—or a conjunction of propositions—as the future part along with the assumption that propositions are false unless explicitly made true will produce a deterministic program.

In the next section we will see that very useful work has already been done on converting any specification into a deterministic one.

If METATEM observed the usual division of propositions into component and environment ones as described in Chapter 3 then we would also want some explanation about finding an environment predicate in the future part of a rule. For example, if p is controlled by the environment in the little program $\bullet\, true \Rightarrow p$, where $\bullet\, q$ is defined as $\sim\ \bullet \sim q$, then the module has no way of guaranteeing that a model is built. In [Fisher, 1993] it is suggested that METATEM interprets such a rule as 'wait for p to be made true before proceeding'. This can be viewed as a means of synchronizing behaviour.

The existing version of METATEM (Michael Fisher's 1992 CMP) approaches the question of control of propositions in a very different way and so does not, in fact, need any such restrictions. Since we are going to use this language for an implementation we will examine the CMP approach.

In understanding this it is helpful to make a distinction between the propositions which appear in the module program and those which describe the interaction of the module and the environment. Say that P is the set of propositions which describe the interaction or *interface* of a module M. Let $P_I = \{p_I \mid p \in P\}$ be a set containing a unique new proposition p_I corresponding to each $p \in P$. Most of the propositions from P will also appear in the program. However, there may be other *internal* propositions appearing in the program as well. Let

$P_M = \{p_M \mid p$ appears in the program code$\}$. Assume that P_I and P_M are disjoint.

The basic idea is joint control by the environment and the module of all interface propositions: a proposition is made true exactly when either the module or the environment makes it true. The only restriction on this joint control is made in the *interface declaration*, a special part of the module program which also specifies how to relate these propositions to what is going on in the module. Some of the interface propositions are declared as being *listened to* by the module and some are declared as being able to be *announced* by the module. Call these sets L and $A \subseteq P$ for now. This is roughly an input/output idea. Note that these sets may overlap and do not have to mention all the propositions in the module.

Effectively the interface propositions not in A are under the sole control of the environment. They will not be announced by M. Note, however, that the interface definition is not primarily about control of the propositions.

Using the notation we now have we can explain the interaction of the module and the environment. Recall that we are dealing with the synchronous case at the moment. Let us see what happens at the instant of the completion of a step when the module and environment come to make their announcements. The module has decided by the process described above what truth values it would like to assign to the propositions appearing in the program. We regard these decisions as being about the corresponding propositions in P_M. If the module wants to make p true then p is recorded as being true for the purposes of the program and we will say that p_M is true. If the module decides to make p false then nothing gets recorded for the moment but we cannot yet decide whether p_M is true or false.

We have to deal with the interfacing. First the output. For all those p which the module wants to make true and which are in A, we simply have p announced. Thus p_I is true as well as p_M for those p.

Now the input. Also if the truth of a proposition $p \in P$ is announced by the environment we can regard p_I as being true. Furthermore, if $p \in L$ then we assume p_M is immediately made true and p is recorded as being true for the purposes of the program.

Thus for the purposes of the program, and its recording for future computations, p is true at that instant if either the program made it true or the environment made it true and p is in the listened-to set. The rest of the propositions are not recorded and so taken to be false.

For the purposes of the outside world p is true if either the environment made it true or M made it true and p is in the announced set.

One consequence of this procedure is that rules which purport to bring about the falsity of a proposition explicitly should not be taken too seriously if that proposition is in the listened-to set. The environment may make the proposition true anyway. For this reason we refrain from using negations in the future parts of rules in our case study. Given the 'false unless otherwise stated' approach to recording truths, negations are redundant there anyway.

One final question concerns the ability of METATEM programs to do anything apart from just announcing the truths of propositions. The answer is that the announcement of propositions can easily be arranged to have the desired side-effects. For example, propositions *clear-screen* and *turn-radio-on* will just need appropriate arrangements made within the computer and/or with external ports.

11.3 First-order METATEM

Moving to consider programming in the predicate temporal logic FML we allow much more expressiveness and, unfortunately, find a whole host of new difficulties. Below we first introduce a predicate METATEM with very restricted syntax to show how the basic mechanism works and then consider relaxing some of the restrictions. This will explain the motivation for the long list of complicated conditions with which we begin.

Again we want the general 'past implies future' format for rules but to keep tight control of the variables we require some more structure on each rule. In the *separated normal form* (SNF_f) of [Fisher, 1992] the program represents the formula

$$\Box \bigwedge_{i=1}^{n} \forall \overline{x_i}.(\forall \overline{y_i}.P_i(\overline{x_i}, \overline{y_i})) \rightarrow (\exists \overline{z_i}.F_i(\overline{x_i}, \overline{z_i}))$$

where each rule represents one of the following four types of formula:

An *initial* $\Box$ rule:

$$\forall \overline{x}. \left[\left(\forall \overline{y}. \bullet \bot \wedge \bigwedge_{b=1}^{k} l_b(\overline{x}, \overline{y}) \right) \rightarrow \exists \overline{z}. \bigvee_{j=1}^{r} m_j(\overline{x}, \overline{z}) \right].$$

An *initial* $\Diamond$ rule:

$$\forall \overline{x}. \left[\left(\forall \overline{y}. \bullet \bot \wedge \bigwedge_{b=1}^{k} l_b(\overline{x}, \overline{y}) \right) \rightarrow \exists \overline{z}. \Diamond l(\overline{x}, \overline{z}) \right].$$

A *global* $\Box$ rule:

$$\forall \overline{x}. \left[\left(\forall \overline{y}. \left(\odot \bigwedge_{i=1}^{n} k_i(\overline{x}, \overline{y}) \right) \wedge \bigwedge_{b=1}^{k} l_b(\overline{x}, \overline{y}) \right) \rightarrow \exists \overline{z}. \bigvee_{j=1}^{r} m_j(\overline{x}, \overline{z}) \right].$$

A *global* $\Diamond$ rule:

$$\forall \overline{x}. \left[\left(\forall \overline{y}. \left(\odot \bigwedge_{i=1}^{n} k_i(\overline{x}, \overline{y}) \right) \wedge \bigwedge_{b=1}^{k} l_b(\overline{x}, \overline{y}) \right) \rightarrow \exists \overline{z}. \Diamond l(\overline{x}, \overline{z}) \right].$$

where each k_i, l_i, m_j and l is a literal. It is shown in [Fisher, 1992] that every sentence can be transformed into this form.

Unfortunately the full generality of this form is very difficult to deal with so we will also suppose that

- the left-hand side of each rule is strict past, so $k = 0$;
- $\overline{y}$ and $\overline{z}$ are empty lists, i.e. all variables are one of the x_i;
- each variable symbol appears at least once in a positive literal on the left-hand side of the rule.

Thus our syntax looks more like

$$\bullet\bot \rightarrow \bigvee_{j=1}^{r} m_j(\overline{c})$$

$$\bullet\bot \rightarrow \Diamond l(\overline{c})$$

$$\forall\overline{x}. \left[\odot \bigwedge_{i=1}^{n} k_i(\overline{x}) \rightarrow \bigvee_{j=1}^{r} m_j(\overline{x}) \right]$$

$$\forall\overline{x}. \left[\odot \bigwedge_{i=1}^{n} k_i(\overline{x}) \rightarrow \Diamond l(\overline{x}) \right]$$

where each k_i, m_j and l is a literal and $\overline{c}$ is a tuple of constants.

Further assume that at any time the environment only announces a finite number of truths—as we have seen this is a sensible practical restriction. We will show that the process has only to announce a finite number of truths at any stage. This is true at the beginning as the only firing rules are those of the form $\bullet\bot \rightarrow \bigvee m$ or $\bullet\bot \rightarrow \Diamond l$ where m and l are ground literals. Negative literals do not need to be announced.

Let us now detail the basic recursive procedure.

We suppose that as well as announcing truths at each stage the process stores the finite list of truths—component and environment truths—for that stage. Remember that positive ground literals are assumed false unless otherwise recorded.

As in the propositional case, we have to determine at each stage, which rules fire. Here it is slightly more complicated because we want to determine, for each rule

$$\forall\overline{x}.(P(\overline{x}) \rightarrow F(\overline{x}))$$

which substitutions $[\overline{x} \mapsto \overline{d}]$ make $P(\overline{x})$ true now. It can be shown from our assumptions about the form of rules and the finite extent of past truths that there are only a finite number of such substitutions. For each of these we just need to make $F(\overline{d})$ true and this is done as in the propositional case as the component has control of announcing truths involving any predicates in F.

Once again back-tracking is an option but is not yet usually implemented. We saw that in the reactive module situation there are good reasons for not allowing back-tracking anyway.

Why do we have these restrictions and how could we lift them?

Having literals on the left-hand side of rules that are not nested under past operators makes the computational procedure more difficult and yet given all the other restrictions we are imposing we do not seem to lose any expressive power outlawing them. In most cases it seems possible to rewrite the program equivalently without them.

Now consider the restriction that each variable symbol in a rule must appear in at least one positive literal on the left-hand side. There are two sorts of examples where this does not hold. Consider first a rule like $\forall xy.[P(x) \rightarrow F(x,y)]$ in which a variable, here y, does not appear at all on the left-hand side. This means that if $P(a)$ happens to hold sometimes then the component will have to make $F(a,a), F(a,b), F(a,c)$, etc., all true. This is clearly a serious problem for our general procedure if the domain is infinite; we may have an infinite set of eventualities to satisfy. What is worse, e.g. if $F(x,y)$ is just a predicate, is the possibility that an infinite number of truths may need to be announced. Clearly this challenges our whole framework and has good reason not to be allowed.

Even if the domain is finite there is a not insurmountable problem here in that the interpreter does not know what the domain looks like. In the case when the domain is fixed for all uses of the program this can be easily solved by converting the program into a propositional one instead. If the domain is always finite but does vary from use to use of the program we might need some facility here for listing the domain.

Another example which breaks the restriction is a rule like $\forall xy. \bullet (p(x) \wedge \neg q(y)) \rightarrow R(x,y)$ in which y again does not appear in a positive literal on the left. Here, if $p(a)$ was true yesterday we need to find all the elements b of the domain for which $q(b)$ was not true yesterday. Again if the domain is finite and listed somewhere in the program this problem is solvable—we just look for those b for which $q(b)$ is *not* recorded as holding yesterday—but if the domain is infinite we have the same seemingly insurmountable problem of sometimes having to announce an infinite number of truths.

Now let us consider allowing $\overline{y}$ to be non-empty, i.e. allowing universally quantified variables within the left-hand side of the rule. In these rules, such as $\forall x.(\forall y. \bullet q(x,y)) \rightarrow F(x)$, we have the problem of not knowing which domain to contend with again.

Next let us consider allowing $\overline{z}$ to be non-empty i.e. allowing existentially quantified variables within the right-hand side of the rule. For example, if $P(a)$ holds and we have a rule $\forall x.P(x) \rightarrow (\exists z.F(x,z))$ then the process has to find an element d such that $F(a,d)$ is true. If environmentally controlled predicates are involved in F in a certain way it may be that the environment, will, in its own good time, supply the process with such a value—this could be a way of synchronously receiving a message. Similarly one could delay grounding the variable until a later stage when either the environment or component fixes its value. A different approach is to let the process choose a new constant symbol or functional term instead. This solution may not be acceptable in the finite domain case.

To describe a module interface in the first-order case we just use a straightforward generalization of that in the propositional case: predicates can be listened to and/or announced (or neither).

11.4 Concurrency

Once we have the ability to define a reactive module, it is not a great step further to generalize to the idea of a distributed system with several modules acting concurrently. This distributed system may be closed or it may itself be immersed in an environment. The modules within the system are just the ordinary reactive modules we have described above but the environment of a particular module is now the extra-system environment in combination with all the other modules. But the whole system is itself just an ordinary module as well. In fact the possibility of a great hierarchy of modules opens up here. At its depths are atomic modules possibly implemented by METATEM processes but at all other heights a module at one level turns out to be a distributed system of modules when we look within it.

Since we already know about implementing a module in METATEM, here we just need to describe the arrangements for communication between modules. In CMP, we use broadcast message passing as outlined in [Fisher and Barringer, 1991]. Essentially this means that all announcements are potentially available for every module in the system to hear. But, as usual, each module has an abstract interface specification indicating which predicates it listens to in its environment and which it may announce. Broadcast message passing contrasts with various channel systems such as that in [Hoare, 1985] in which messages pass from a specified sender to a specified receiver. The approaches are compared in [Fisher, 1993].

Internal predicates have an important use under such an arrangement as we can use the same predicate symbol in different modules to represent different predicates. So provided p is classed as an internal predicate in modules A and B then A could be making p true at a certain time, and recording that fact, while B is making its p false.

Once again we have the choice of a synchronous or asynchronous system. In the synchronous system, the overall controlling mechanism waits for all modules to be ready and then lets them make their announcements simultaneously. Then each module is told about the announcements which its interface classes as inputs and is left to get on with its next step.

This is not how distributed systems usually operate. We must turn now to consider the operation of a METATEM module in an asynchronous milieu. Here each module can make its announcements whenever it wants to. Each interested listener should be told then, but this means that the listener might be interrupted in the middle of the computation of the next step. So instead the CMP system arranges for a message queue to be built up for each module containing, in order, the announcements of interest to that module which have been made. Then, only when a module deigns to make an announcement—even an empty one—will it get access to its queue.

There are actually several options for the details of this access but an easily implementable arrangement may be for the process to read off the whole queue leaving it empty to collect messages announced during the next step. The module itself can then proceed, acting as if all the announcements contained within the queue happened yesterday. Note that the $\odot$ operator then loses its usual semantics. This also happens in different ways with the other options for reading through the queue.

One problem with this bunching arrangement is that if several identical messages arrive close together in the queue then they get collapsed into one message when bunched. We thus lose count of messages though it is easy to think of situations in which this is important. Consider, for example, a module which is required eventually to announce a q for each time it hears a p. If it is delivered a message queue containing five ps then a rule $p \rightarrow \Diamond q$ will only produce one q. In Michael Fisher's CMP this problem is avoided by the module only reading off the queue up until the first repetition of messages. Of course, then the operational meaning of $\bullet$ reflects even less the semantics of yesterday. This problem is discussed in [Reynolds, 1997].

11.5 Meta programming

As shown in [Barringer *et al.*, 1991], there is much potential for metalanguage capabilities to be useful as part of a MetateM language. See also [Brough *et al.*, 1996] for a discussion of what it means to be a metalanguage in this context. A metalanguage syntax allows formulae to be treated as objects for various purposes such as building interpreters and loop checking. A metalanguage operational semantics allows the program to interfere with its normal execution behaviour or even create a little simulation of itself for the purposes of planning ahead. Such facilities may be of great use in several activities. Examples include the manipulation of message queues, the creation of new modules and changing the interface of a module.

However, metalanguage handling may often be inefficient and as this facility is not yet implemented in MetateM we prefer to try to find *object-level* solutions to the problems we face in this case study.

11.6 Program syntax

In this subsection we introduce the syntax which we will use to implement a patient monitoring system (PMS) in the next chapter. Consider first the propositional case.

Strings consisting of lower case letters or a few symbols like - are used for proposition names. Call such a string an *atom*.

The logical operators are represented as follows:

PML:	$\top$	$\bot$	$\wedge$	$\vee$	$\neg$
CMP:	`true`	`false`	`&`	`|`	`~`

The temporal operators are

PML:	U^+	S	$\bigcirc$	$\odot$	$\bullet$	$\Diamond$	$\diamondsuit$ (filled)	$\Box$	$\blacksquare$	$\mathcal{W}$	$\mathcal{Z}$
CMP:	`U`	`S`	`N`	`Y`	`Q`	`F`	`P`	`G`	`H`	`W`	`Z`

It is straightforward to define the class $SPast$ of strictly past expressions and the class $NFut$ of not necessarily strict future expressions. A program rule looks like α=>β for some $\alpha \in SPast$ and $\beta \in NFut$.

A module program body looks like

$$\alpha{=}{>}\beta$$

or like

$$\begin{array}{c} \alpha_1{=}{>}\beta_1 \\ \cdot \\ \cdot \\ \cdot \\ \alpha_{n-1}{=}{>}\beta_{n-1} \\ \alpha_n{=}{>}\beta_n. \end{array}$$

The syntax of a module program must tell us about the control of propositions. The input or listened-to list is not present if it is empty but otherwise is represented by $(p_1,\ldots,p_n)$ where $p_1,\ldots,p_n$ are the input propositions. The output or announced list is also not present if it is empty but otherwises is represented by $[p_1,\ldots,p_n]$ where $p_1,\ldots,p_n$ are the output propositions.

In CMP we begin a module program with an atom being the module's name. The syntax is

< module name >< input list >< output list > : < program body > .

To define a system of METATEM modules CMP uses the syntactic arrangement of listing their programs. The list ends in a full stop after the last module program (which also ends in a full stop). There is an option for selecting asynchronous or synchronous running of the modules.

When we present an FML program to implement the PMS we will just use a straightforward generalization of the syntax for propositional CMP. The main differences will be that strings of lower case letters will stand for predicates (as well as propositions) and for constant symbols too (and of course module names). We will use the same upper case letters for temporal operators as before but as in Prolog a string of letters beginning with an upper case one will be used to represent a variable. This overuse of case conventions should not be ambiguous but any real first-order programming language will have to be more careful.

Variables will be assumed to be universally quantified across the whole rule (i.e. they are x_i types). In extensions of this language with less restrictions on the form of rules, some notation will have to be invented to distinguish other types of variables.

11.7 Synthesis and determinization

We have seen that even putting aside the question of whether a program achieves a model of its declarative reading or not, there is, because of all the syntactic restrictions we have had to impose, still a completeness/expressiveness question to ask. That is, what range of specifications can we describe in a METATEM program? Below we describe the rapidly growing body of work dedicated to answering such a question and showing how such a specification can be transformed into the required form.

11.7.1 *The propositional case*

To begin with, one may wonder whether the past implies future requirement restricts our ability to specify problems. In the propositional case this can be answered using the separation property of [Gabbay, 1989]. The separation property has been used to show that various temporal logics are equally as expressive as monadic first-order languages which describe time using the $<$ relation (see [Gabbay *et al.*, 1994]). It is shown in [Gabbay, 1989] and [Gabbay *et al.*, 1994], by much syntactic manipulation, that our PML language has this property over natural number's time. This means that every formula is equivalent to a Boolean combination of formulae each of which is strictly past, an atom or strictly future. By rearranging this combination we can produce an equivalent in the required form.

Similar syntactic rewriting methods are used in [Fisher, 1991] to show that every formula is equivalent to one in a very simple separated normal form.

In fact, rewriting can give us the best possible result of a *deterministic* METATEM program—that is one which never needs to back-track—to implement any implementable reactive module specification. See [Noël, 1992].

This is the METATEM version of the result by Pnueli and Rosner in [1989a] which shows how to build a finite state *transducer* to implement any such propositional specification. The transducer outputs the values for its propositions, inputs the environment's and changes its state at each stage.

Other important results are those concerning fixed point extensions of PML in [Hodkinson, 1995] and [Strulo *et al.*, 1993] which show how, in the absence of a reactive environment, we can even rewrite a specification from an expressive fixed point language so that it can be directly implemented in METATEM.

These general methods of implementation are seemingly quite inefficient so there is an incentive for trying to use any other bits of information we have on a situation. For example, we may be able to make some probabilistic descriptions about how the environment behaves. There is some very interesting work done on this in [Courcobetis and Yannakakis, 1990].

Another related efficiency question is deciding how to split up a specification for the best benefit of a distributed system. Yahknis and Yahknis have done some pioneering work on this topic in [Yakhnis and Yakhnis, 1990].

11.7.2 *The predicate case*

In marked contrast to the propositional case, there is still very much work to be done in the first-order case.

The only result here seems to be the rewriting algorithm in [Fisher, 1992] which shows constructively how to transform any FML specification into the SNF_f form we introduced in Section 11.3. Although the new formula may involve new predicate symbols, it has the desired properies that it is satisfiable iff the original one is and any model of the transformed formula is a model of the original one.

We will not need to use this transformation here as the specifications which we develop for each module will be immediately implementable because we are developing the specifications in tandem with the METATEM implementations. This is what prototyping is for. Nevertheless, there is much important work to be done in assessing the expressiveness of first-order METATEM languages and automatically producing implementations of first-order specifications—especially efficient ones. Indeed, after a successful session with the prototyping framework one would find such automatic generation very useful.

12

MetateM IN INTENSIVE CARE

12.1 Introduction

In these days with complex systems designed to cope with anything from exchanging stocks and shares to dispatching ambulances, it becomes increasingly more important to make sure that the design is adequate. Of course, it is not possible to be perfectly sure about the adequacy of a design. However, there are ways of attempting to be more certain and two of the most satisfactory are reasoned justification and testing. This chapter is about both.

We will see that the fact that such systems keep up an interaction with their environment means that formalisms based on temporal logic are relevant here. And we will see that the fact that the systems are complex and may involve separate communicating components will mean that considerations of modularity will also play a big part.

The work described here is intended to fit into a system development programme which involves the following aspects. A client has certain ideas about what he or she requires of a system and wants this system to be built. It is assumed that a formal description or *specification* of the requirements is made. This might be used for legal reasons but more importantly it greatly assists in a division of labour in building the system and in a formal justification of its correctness. It is also very useful in assisting the client to clarify his or her expectations. Of course, it is crucial that the specification really does capture the client's requirements.

We have mentioned that modularity and division of labour are important aspects here. As in [Barringer, 1987] we expect the building of the system to be carried out in a modular manner. Thus, the top-level task, the overall specification, is broken up into smaller tasks to be given to different teams to work on. They may in turn break up their tasks again etc. Here, too, specifications play their part as we describe the way in which the separate components built by the separate teams are put together, and how each component is supposed to behave. The idea is to specify this break-up so well that the separate teams do not have to waste time communicating with each other. If the specifications are adequate then it should be easy to justify that the overall system behaves the way it should by showing that each component behaves the way it should and that they are put together properly.

In this chapter we will describe a framework in which a *prototype* can be built. This means a simplified, probably inefficient, version of the system which can nevertheless be seen working (hopefully). A prototype has two very important roles to play in our programme. In one it can be used to demonstrate to the client roughly how things will work. This may help justify that the formal specification does indeed capture the client's expectations. The other purpose is for the top-level implementer to test and try different methods of breaking up the task into smaller modular ones and become clearer in deciding what is required from each component.

Specification needs a formal language. Some early attempts at specifying systems included the idea in [Lamport, 1977] of listing *liveness* properties and *soundness* properties of the system. Liveness properties are characterized as stating that something should happen while soundness properties state that something should always hold. These distinctions are still important but as we will see in this chapter a much more intricate description of the expected behaviour of the system is desired. For such purposes *temporal logic* is a very widely used and successful formalism (see, e.g. [Manna and Pnueli, 1992]). In temporal logic we use a surprisingly simple language capable of expressing and reasoning about processes changing over time.

More recently, temporal logic is also being used as a formalism in which to write actual programs. We can see many examples of *executable temporal logic* from the initial developments in [Moszkowski, 1986] to the very complex METATEM framework we describe below and the temporal logic programming in [Abadi and Manna, 1989]. The idea is simply that the program read as a sentence of a temporal language describes how it behaves when it runs. Thus executable temporal logics are *declarative* programming languages.

METATEM began with the simple idea of 'declarative past implies imperative future' idea in [Gabbay, 1989]. A complex expressive programming language took shape in [Barringer *et al.*, 1989] and an implementation of it was described in [Fisher and Owens, 1992]. Recently we have seen it applied to modelling rail networks in [Finger *et al.*, 1993] and developing into the concurrent programming language CMP in [Fisher, 1993].

As a higher level language it seems very appropriate for prototyping while lower level, more efficient—and less expressive—temporal languages like TEMPURA [Moszkowski, 1986] might be appropriate for final implementations.

There are great advantages in using the unified approach of the same temporal formalism for specification, implementation and formal justification. We will see in this chapter just how well the various steps in the design programme mesh together when we use such a framework.

In this chapter we want to demonstrate the success of METATEM as such a prototyping language and, seeing it in action, note any flaws. METATEM, especially the concurrent CMP version, is still in its early stages of development so we can expect many shortcomings. The use of this study will come in learning details about what aspects particularly need working on, perhaps with suggestions for improvements, and whether there are any more serious problems.

To this end we choose a system development problem from the software specification and design literature and try using the METATEM framework to take it through the stages of specification, implementation and justification. The example we choose is a patient monitoring system for an intensive care ward of a hospital from [Stevens *et al.*, 1974]. We describe the problem in Section 12.2.

In Section 12.3 we describe our temporal logic formalism. In Section 12.4 we use this formally to specify the patient monitoring system problem. In Section 12.5 we introduce the METATEM programming language. In Section 12.6 we consider theoretical problems of expressiveness of the language. In Section 12.7 we implement a very simple version of the hospital system. In Section 12.8 we describe a more realistic prototype implementation which we would want to be able to build when METATEM is better developed. In Section 12.9 we attempt some formal justification of the correctness of our prototype.

In Section 12.10 we compare the programme with other approaches to the problem including a quite detailed study of exactly the same problem done using a different formalism in [Castro and Kramer, 1990b].

All the while we have been noting lessons for METATEM which we present in Section 12.11.

In Section 12.12 we conclude and summarize our progress.

12.2 Informal description

In this chapter, we will apply METATEM to a patient monitoring system (PMS) for use in the intensive care wards of hospitals. This problem was initially discussed in [Stevens *et al.*, 1974] but a fuller description appears in [Castro and Kramer, 1990b] which analyses it in terms of what they call the temporal–causal formalization for specifications. Later we compare their approach with ours.

In the PMS there are two types of components: nurse components (NC's) and patient components (PC's). In any application of the system to a ward there will be a certain finite number of PC's but only one NC. In Fig. 12.1 we assume there are only two patients. Each PC is attached to a particular patient in the ward and automatically monitors various physiological measurements. For example, in this study we assume that blood pressure, pulse rate and temperature are measured. The NC is designed for use by the supervising nurse: the nurse can use the NC to find information about a particular patient and the NC also displays an alarm if any of the measurements of any of the patients strays from a certain predefined range. Although this is not mentioned in the previous literature, we make things a little more interesting by assuming that only one set of measurements can be displayed on the NC at a time and that the nurse can indicate to the NC when he or she has seen the information and finished with it.

At the risk of making early decisions about the actual implementation of the problem rather than only about its specification, we must say that the descriptions mentioned above do imply some form of distributed system with the separate components acting independently and communicating with each other rather than their being one centralized process. In that case the essential activities of the components are as follows:

PC: • constantly monitor, the three measurements of the attached patient,

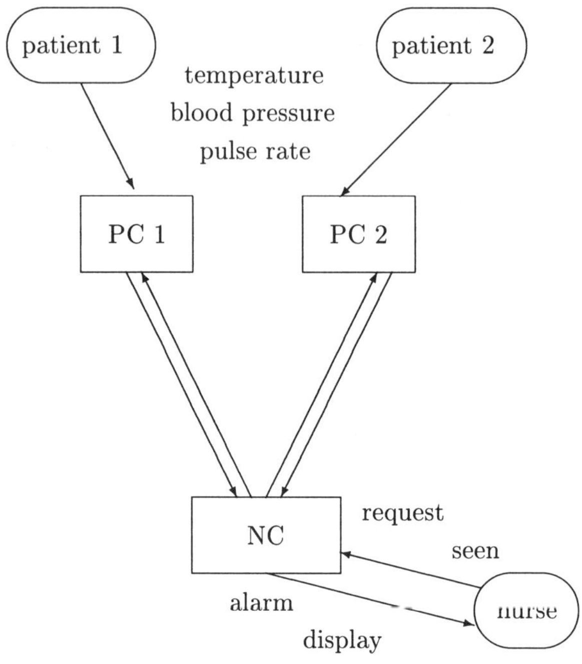

FIG. 12.1.

- if any of the values strays from its predefined range then send an alarm to the NC;
- if the NC requests measurements from the PC then send the latest measurements.

NC:

- if the NC receives an alarm from a particular PC then it is to display the alarm with an indication of which patient it comes from;
- if the nurse asks the NC for details on a particular patient then the NC should request them from the appropriate PC;
- when details arrive from a PC in response to a request then the NC should display them.

In any case, no matter how we implement the solution we would want the following conditions to be satisfied:

if the nurse requests information about a particular patient then it should be displayed; and

if any of the values of a patient strays from the safe ranges then the alarm should be activated as soon as possible.

There is the potential to make this arrangement more complicated by allowing patients to be admitted and discharged. We will consider this later.

In this chapter we are really testing METATEM to see how well it works as a prototyping framework or an implementation language. We would expect to be able to specify the problem formally, implement a solution and observe, prove (or otherwise test) the implementation against the specification and our informal expectations of the behaviour. This is an ideal pattern of development and, in fact, we expect feedback at every stage to allow reconsideration of preceding stages. For example, when we come to check the implementation against the specification we may see that despite being correct, the implementation exhibits some unattractive unforseen behaviour which means we have to refine the specification.

It is hard to present the full story of this development with feedback in this chapter so it might sometimes look as though we anticipated all the problems as we went through the three stages above. However, we will try to indicate where feedback was used.

On the other hand, we have had to make many allowances for METATEM's early stages of development, and the testing stage tells us more about what improvements we need to make to METATEM than about the PMS specification.

12.3 Specification

The task of a specification is unambiguously to state the problem to be solved without prejudging in any way the means of solving it. This ideal cannot often be achieved but we show that by using temporal logic we can almost do so.

Below we will show how we can regard a 'run' of a program as a temporal structure. Thus we will be able to use our formal temporal language to describe runs. The essence of a specification is the sentence in the formal language which distinguishes an acceptable run from an unacceptable one. In this way we formalize the desired behaviour of our PMS.

Note that this temporal logic specification sentence will just be one—probably very big—sentence but that it will be more intuitive to present it as a conjunction of several properties. For the sake of clarity of exposition we will also divide up those conjuncts according to the traditional safety/liveness dichotomy. See, for example, [Lamport, 1977] for details.

Below we will also see that there are many aspects of the framework for this formalization which have to be fixed too before we have truly delimited the behaviour. These include questions of the formal model of time, whether we use predicates or just propositions, the object domain, the nature of control of the predicates or propositions and the form of communication between the loci of control.

12.3.1 *Basics of the formalization*

To use a temporal language to describe a process, the first steps should be to decide on some predicates or propositions which can be usefully used to describe the state of the process at each time. Here it is clear that predicates *temp*(P,T), *pulse*(P,R) and *blood*(P,B) holding at a particular time can mean that patient P's temperature is then $T\circ$ C, pulse rate is R beats per minute and blood pressure is B units. We also might want *req*(P) to mean that the nurse is requesting the details on patient P, *alarm*(P), to mean that the alarm for patient P is being displayed, *display*(P,T,B,R) to mean that patient P's temperature, pulse rate and blood pressure are being displayed as T, R and B respectively, and finally *seen* will be true when the nurse indicates that he or she no longer needs to have the current information displayed.

It will also be noted that we will need some constants to describe the acceptable range of the physiological measurements. Let us say that *mint* is the minimum acceptable temperature and *maxt* the maximum acceptable temperature. Similarly we define *minp*, *maxp*, *minb* and *maxb*. One facility which we need here is $<$ comparison of values in the language. For now let us assume that the physiological measuments are positive integers, that the set of positive integers is a subset of the object domain and that $=$ and $<$ are defined on this subset as the usual relations.

The object domain can also be assumed to contain the patients. This is a problem if we want to admit or discharge patients but for now we suppose that there is some fixed set of patients.

It is not clear which model of time is appropriate for the PMS situation. With physiological measurements varying continuously and nurses making requests when they want to the real numbers suggest themselves at first. However, implementational considerations and the idea of a computer stepping its way through various states makes the natural numbers also appealing. Being the model on which, as we will see, METATEM is based means that this choice makes the implementation feasible: it is hard to see how a computer could make full use of a real-time model. But on the other hand, we are going to be dealing with a distributed system of several components stepping not necessarily in synchronization and a dense order like the reals or rationals becomes attractive. For now it is sensible to limit ourselves just to a linear flow, maybe with a first point, and consider further restrictions as appropriate.

12.3.2 *Liveness*

Given these decisions, we can now formally state our two overall requirements on the system. First a little helpful notation. Let us call a time when a patient's measurements stray from the given ranges a *life-threatening* situation. Let $\phi_{\text{temp}}(P)$ be the temporal formula $\exists T(\mathit{temp}(P,T) \wedge ((T < \mathit{mint}) \vee (\mathit{maxt} < T)))$, and similarly for ϕ_{blood} and $\phi_{\text{pulse}}(P)$. Also let $\phi = \phi_{\text{temp}}(P) \vee \phi_{\text{blood}}(P) \vee \phi_{\text{pulse}}(P)$ so that $\phi(P)$ is true exactly when P is in a life-threatening situation.

The alarm condition might be represented as

$$\text{LA:} \qquad \Box \forall P(\phi(P) \rightarrow \Diamond \mathit{alarm}(P)).$$

One problem with this is that it requires an alarm to be sounded after the measurement strays from its range for any instant no matter how brief the excursion is. This may be what the medical considerations call for but it is probably beyond the power of implementation by any piece of equipment. It is certainly not the kind of specification that step-by-step procedures like METATEM can handle directly. Fortunately this is not a serious problem and we can suppose that the measurements are made at discrete intervals and ideally made very frequently.

Another problem is the use of $\diamondsuit$ where, in the English description previously we used—or meant—'soon'. Our formal specification would seem to introduce the unwelcome possibility of an alarm which goes off two weeks late. One way to avoid this problem is to move to a metric temporal logic in which a time limit is specified. This would introduce many complications into the specification and in fact they are unnecessary. It will be seen that the METATEM programming approach interprets $\diamondsuit$ as 'as soon as possible', allowing for satisfying conflicting demands, being fair to competing demands and, of course, the mechanics of the computer(s) and system devices. This we would hope is soon enough for an alarm but it should be tested.

A third problem concerns the sounding of the alarm. A life-threatening situation will 'soon' result in *alarm*(P) being true. According to our semantics this means that P's alarm will be displayed for an instant during which the nurse may not notice it. One may consider leaving the alarm on until the nurse switches it off, but for this case study let us suppose that *alarm*(P) being true means that the nurse gets sufficient notification.

In a similar way we can formalize our display requirement as

$$
\begin{array}{l}
LD: \\
\qquad \Box \forall P\ (\ req(P) \quad \rightarrow \\
\qquad\qquad (\exists T \exists B \exists R \\
\qquad\qquad \diamondsuit(\qquad temp(P,T) \wedge blood(P,B) \wedge pulse(P,R) \\
\qquad\qquad\qquad\quad \wedge \diamondsuit(display(P,T,B,R)\,\mathcal{U}\,seen) \qquad))).
\end{array}
$$

Here we have some different ideas on the use of $\diamondsuit$. In one, we note that we require the measurements to be made *after* the request. This is arguably too strict a rendering: given the assumption that measurements are being made very frequently, perhaps it would have been acceptable for the current or even a very recent value to be displayed. Without notions of measured time it is quite hard to formalize 'recently' so we choose to stay with our over restrictive formalization on this point.

After the measurement is made we then want the results to be displayed. Using $\diamondsuit$ instead of an implicit soon seems acceptable here. However we have introduced a complication into the PMS situation by assuming the very real possibility that only one set of values can be displayed at one time. Thus this $\diamondsuit$ is to be read as 'as soon as possible allowing for conflicting demands'.

12.3.3 *Soundness*

It is very easy to build a model of the simple specification we have above: we simply keep sounding all the alarms and keep displaying measurements. Obviously the formalization does not capture our expectations about the PMS. We need to add some 'soundness' criteria: we need to require that we do not get unwarranted alarms and uncalled-for displays. We also need to formalize our requirement that only one set of values is displayed at a time.

First consider the unwarranted alarms. Recall that $\phi(P)$ represents a life-threatening situation for P. Then we may try the formula $\forall P\, \Box(alarm(P) \rightarrow \blacklozenge\, \phi(P))$. However, it is not hard to see that this is inadequate in preventing unwarranted alarms. For example, a $\phi(P)$ can be followed by any number of alarms ringing on into the distant future.

In fact it is very hard to express this requirement in FML—it is impossible. This problem is much the same as that of characterizing a (special case of a) buffer which was shown in [Koymans, 1987] to be impossible in this sort of temporal language.

One attempt at a solution may be to state our requirements slightly differently. We could require that the alarm is only sounded if there has been no alarm since a life-threatening situation occurred. Although this sounds reasonable and is able to be formalized easily, this requirement turns out to be quite difficult to implement in our distributed system. Another problem is that a PC may have to judge the relative timings of these two sorts of events.

Another way of seemingly avoiding this problem is just to keep sounding alarms until the nurse switches them off. But this has similar problems, as do other solutions specifying more closely the steps involved in notifying alarms in a distributed system.

So, instead of trying to change our requirements to suit our formal specification language, it is perhaps better to admit the deficiency in the language and try to improve the language so that it can handle what is a common specification problem. Fortunately several solutions in this vein exist—see [Manna and Pnueli, 1992]—and we can use one of the usual solutions by moving to the more expressive temporal language QML with quantified predicates. In this case we can require that there is a way of labelling life-threatening situations and alarms so that there is a one-to-one correspondence between them. This, along with the requirement that alarms sound soon after life-threatening situations, will give us our result. In fact, since an alarm may be caused by two distinct life-threatening situations, all we require is that every alarm is labelled to correspond with at least one life-threating situation. We write

$$
\begin{array}{ll}
SA: & \exists^1_Q od\ \exists^1_Q id\ \forall m \forall n\ \Box \\
& \quad (alarm(P) \wedge od(m) \rightarrow \blacklozenge\, (\phi(P) \wedge id(m))) \\
& \wedge\ (alarm(P) \wedge od(m) \rightarrow \neg\Diamond(alarm(P) \wedge od(m)) \\
& \wedge\ (alarm(P) \rightarrow \exists k.\ od(k)) \\
& \wedge\ (id(m) \wedge id(n) \rightarrow (m = n)) \\
& \wedge\ (\phi(P) \rightarrow \exists m.id(m) \wedge \Diamond(alarm(P) \wedge od(m))).
\end{array}
$$

For this to be satisfied there must be a way of defining two unary predicates *od* and *id* such that for any elements m and n from the object domain:

- if P's alarm goes off when $od(m)$ holds then there was a life-threatening situation for P when $id(m)$ held;
- it never happens twice (or more) that P's alarm goes off when $od(m)$ holds;
- if P's alarm goes off then there is some element k such that $od(k)$ holds;
- only one element k makes $id(k)$ true at each time;
- and each life-threatening situation is followed by an alarm with the corresponding label.

It is clear that the conjunction of these conditions is sufficient to rule out excess alarms. Note that it implies the liveness of alarms' condition (LA).

Because of the existential quantification on *id* and *od*, we do not actually have to implement them. We just have to construct a model so that they can be defined if one wants to (as indeed those responsible for proving correctness may have to do). However, in this case, we note that if one is averse to using the stronger QML language then one could follow the idea in [Hodkinson, 1995] and simply rewrite the specification without the predicate quantifications. The resulting FML specification would certainly be enough for our purposes but would involve the implementer in a little extra work to make sure the predicates *id* and *od* were built. Thus, technically, it is not the specification we are looking for.

To specify the soundness of displays—that the system displays values only in response to a request—we use a similar construct to arrive with a sentence (SD).

Finally to specify that we only get one set of values displayed at a time we need only write

$$
\begin{array}{ll}
UD: & \Box\ \forall P_1 T_1 B_1 R_1 \forall P_2 T_2 B_2 R_2 \\
& \quad (\ (display(P_1, T_1, B_1, R_1) \wedge display(P_2, T_2, B_2, R_2)) \\
& \rightarrow ((P_1 = P_2) \wedge (T_1 = T_2) \wedge (B_1 = B_2) \wedge (R_1 = R_2))).
\end{array}
$$

12.3.4 *As a reactive module*

As written, even with the soundness requirements, the formal specification is very easy to satisfy. We simply make *temp*, *pulse*, *blood* and *req* always empty predicates: $temp(P, T)$ etc. is false at every time point. This undesirable reading of the specification brings us to consider the nature of control in the framework here. This *does* need to be stated before the problem is fully described. Here we see that the process is itself responsible for determining the extensions of predicates *alarm* and *display* while the environment—that is everything else around including the nurse and all the patients—is responsible for the rest of the predicates *temp*, *pulse*, *blood* and *req*. It is beyond the power of the process to satisfy the specification in the way described above.

It is clear that our process is what is called *a reactive module*. The basic idea of a reactive module, as described, for example, in [Pnueli, 1986] or [Harel and Pnueli, 1985], is an object whose relationship with its environment changes over time in such a way that part of the relationship is under the control of the module and part under the control of the environment. The usual temporal logic approach to reactive modules is to posit a strict division between the predicates or propositions of the language. There are 'component predicates' under the control of the module such as the first group of predicates *alarm* etc. mentioned above and 'environment predicates' such as *temp* and the other group. The special case in which all the predicates are under the control of a module is often called a *closed system*.

With this distinction we can now be clearer about what we require from the implementation. As we said above, it does not seem adequate just to construct any model of the specification sentence. Instead, an implementation of a specification as a reactive module should give an effective strategy for, at each time t, choosing the extensions of the component predicates depending only on the extensions of predicates in the past of t in such a way that if the strategy is used in combination with any environmental behaviour, the temporal structure eventually (i.e. at time ω) constructed will be a model of the specification.

In the synchronous case—we will look at the asynchronous case below—this idea can be well illustrated by a game played between two players: the environment and the module. At each stage (i.e. time) $0, 1, 2, \ldots$ both players choose the extensions of their predicates. After ω moves together they will have fully described a temporal structure. If σ is a sentence in FML then we say that the module wins the play of the σ-game if the structure so constructed is a model of σ at 0. A strategy for the module in the σ-game is a prescription for the choice of extensions for the module at each stage which, of course, may only refer to the past choices of the module and environment. A winning strategy for the module for the σ-game is a strategy which, if followed by the module, will guarantee that it wins no matter what choices the environment makes. We say that ϕ is implementable (in the synchronous case) if and only if the module has a winning strategy for the σ-game.

In [Pnueli and Rosner, 1989a], Pnuelli and Rosner show that, at least under certain further constraints, this condition is equivalent to the validity of a certain branching time formula. To see the implementable reactive module in action we just make the module use the strategy to decide what to do at each stage and see what the environment does too.

There are further subtleties. In the case of predicate logic—as opposed to the propositional case—and in particular when the object domain is infinite, it may not be realistic just to have a procedure for determing the extensions of predicates. How could such an infinite amount of information be really necessary to determine the state of the relationship between the module and the environment? At first sight it seems that even a finite propositional model would always be adequate. However, we see in the PMS example at hand that it is convenient for us to use a predicate $temp(P, T)$ whose domain for the variable T is potentially infinite (apart from physiological restraints). We probably could limit ourselves here to a finite range but it seems a little inconvenient. Instead, for the convenience of the user, we will allow a first-order language but stipulate that all propositions are false and all predicates empty except as specifically announced.

Recall that a positive ground literal is a sentence of the form $p(\overline{a})$ where p is an n-ary predicate (a proposition being a 0-ary predicate) and $\overline{a}$ an n-tuple of terms in which no variable symbols appear. Say that a positive ground literal $p(\overline{a})$ is under the control of a module if and only if p is. Thus, we require that at a particular time t the module and the environment must just *announce* the true positive ground literals which are under their respective control. It will be assumed that there are only a finite number of such *truths* at any time. This assumption does restrict the number of specifications which are implementable.

As we will see below we will find it convenient not to keep the strict division of predicates. Some predicates will have joint controllers. In that case, a positive ground literal will be deemed to be true at a particular time if and only if any locus of control announces its truth.

Another complication which also concerns the communication between the module and its environment is the communication protocol. This can be either *synchronous*, when at each time both the module and the environment announce their truths, or *asynchronous*, when there is some interleaving of announcements. The asynchronous situation is common when the environment and the module go about there separate businesses at their own rates and make broadcasts of truths whenever they feel like it.

The choice between the two protocols is sometimes suggested by the situation—so it should be part of the specification—but sometimes it can be left as an implementation choice. However, the choice does once again decide whether particular specifications are implementable or not. In our PMS case it seems that the asynchronous choice is best.

Thus our specification for the PMS as a reactive module includes the control information (as summarized in Figure 12.2), the fact that the communication is asynchronous and the temporal sentences as before.

However, now that our specification includes these limitations on control it is easy to see that it is not, as it stands, implementable. The system cannot guarantee to display eventually requested information in the case when the nurse requests information on several patients but does not press 'seen'.

This is a common situation in the specification of reactive modules and it immediately suggests the idea of imposing obligations on the environment. This can be done by adding a further complication to our specification framework in

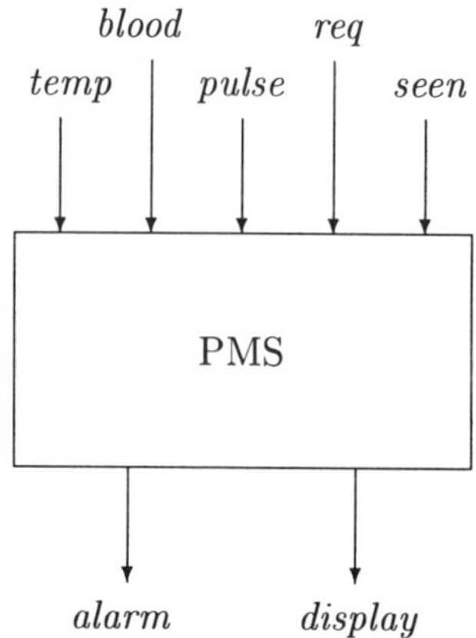

FIG. 12.2.

the form of explicit environment obligations. However, it is also straightforward to change the specification on the module to incorporate this idea without adding extra aspects to the framework. Basically by appropriate use of an implication sign we can allow the module to be relieved of some of its obligations in the case when the environment is not helpful.

Another assumption that we are making about the environment and which becomes evident while the proofs of correctness of the implementation are being done is that three measurements of the patients are taken and announced at the same time and that this actually happens from time to time. In fact, what we need is just that the following formula $\tau(P)$ holds for each patient P:

$$\Box\Diamond(\exists T \exists B \exists R(\mathit{temp}(P,T) \wedge \mathit{blood}(P,B) \wedge \mathit{pulse}(P,R))).$$

In the PMS example we want the alarm liveness condition LA and the soundness conditions SA and SD and UD to still hold no matter what the nurse does. However, in situations like that described above we want the liveness of response to requests to be weakened. The following should be adequate:

$$\begin{array}{lll} LD' : & & \\ \Box\forall P\ (& & \mathit{req}(P) \wedge \Box\Diamond \mathit{seen} \wedge \tau(P) \\ \quad \to (\exists T \exists B \exists R & & \\ \qquad \Diamond(& & \mathit{temp}(P,T) \wedge \mathit{blood}(P,B) \wedge \mathit{pulse}(P,R) \\ & & \wedge\ \Diamond(\mathit{display}(P,T,B,R)\,\mathcal{U}\,\mathit{seen}) \qquad)). \end{array}$$

12.3.5 *As a distributed system*

As we mentioned in the informal description of the problem, a distributed approach is required. The suggestion is to consider the NC and PCs as separate communicating processes.

This can be seen as a generalization of the reactive module approach as we can consider each process as a reactive module with, from its point of view, all the other processes and the extra-system environment together making up its environment. In specifying the behaviour of the totality of the system we may need to consider some new predicates describing the communication between modules. We can then write a sentence in the new language specifying each (sub)module's behaviour.

As well as the predicates *temp*, *blood*, *pulse*, *req*, *seen* and *alarm* already introduced we now bring in unary *pat–al* and *pat–req* and 4-ary *pat–val*. The intended meaning is as follows:

pat–al(P) being true indicates that the PC belonging to patient P is sending an alarm signal to the NC;

pat–req(P) signifies a request for information about patient P from the NC to the appropriate PC; and

pat–val(P, T, B, R) holds when P's PC is informing the NC that P's temperature is T, blood pressure is B and pulse rate is R.

We note that in a realistic situation the separate components will probably go about their tasks at their own rates and so we can suppose that the system will work in an asynchronous way.

In order to specify the behaviour of the various components (the NC and the several PC s) as reactive modules we need to decide on the control of the predicates. This can be done as shown in Fig. 12.3. Here we have a problem though: the extension of *pat–al*, for example, should be under the joint contol of the PCs. For each patient P, we want his or her PC to determine whether *pat–al*(P) is true or not. This conflicts with our stipulation that there should be a strict division of control of predicates. From the point of view of PC1, *pat–al* is controlled both by itself and by its environment which includes PC2.

Fortunately the solution to this problem is not too difficult given our notion of *truths* described in the last subsection. We will allow combined control of particular predicates or propositions as described there.

Thus *temp*, *blood*, *pulse*, *seen* and *req* are still totally under the control of the environment, *alarm*, *display* and *pat–req* are under the control of the NC, while *pat–al* and *pat–val* are each under joint control by the various PC s.

Obviously we still want the overall system to conform to the specification laid down in the last subsection:

$$LA \wedge LD' \wedge SA \wedge SD \wedge UD.$$

We will see in Section 12.9 that we can go about guaranteeing this in one of two ways:

- either implement the whole system and check the specification is satisfied directly;
- or find sentences to specify the behaviour of each of the modules and show how these sentences fit together to make the overall specification hold.

In fact we will use both approaches to good effect.

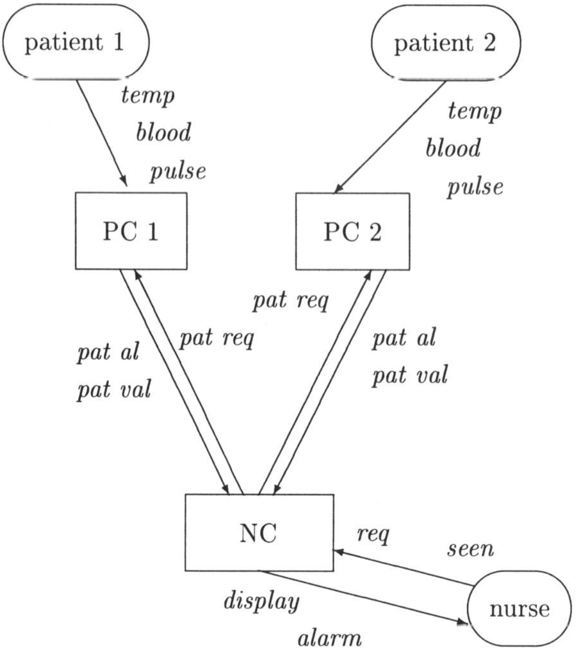

Fig. 12.3.

Recall that one of our tasks in this case study is to develop a formal specification of the PMS as a distributed system. This means that we are to decide how the different components are to interact to bring about the desired overall behaviour. We will have to give a specification for each component so that the separate components can then be implemented independently.

It turns out that one of the very important uses of a prototyping framework like METATEM is to enable us to implement quickly all the parts of the overall distributed system and hence debug the usually quite complicated mode of interaction and communication between the components. So for now we are going to depart from the task of specification and jump straight into implementing the system with components as detailed above and with the aim of producing the overall desired behaviour. Then, when we are sufficiently content with the behaviour of the system, we will try to specify formally what we need of each component to guarantee that it will behave in much the same way as our prototype component and, more importantly, in such a way that it will contribute, together with other components, to the overall system satisfying our requirements.

12.4 Propositional implementation of patient monitoring

At the moment the only working version of a METATEM interpreter is Michael Fisher's propositional CMP version of 1992. So, to see METATEM's PMS in action we will have to recast the full specification in propositional terms. Also, because we have no way of interfacing the system with an environment—not even simulated patients and nurses, let alone real ones (luckily)—we will have to make a closed distributed system in which there are METATEM modules acting as patients and nurses. This is how we would expect to use a prototyping framework anyway.

So let us whip up an appropriate specification. Suppose that we have two patients Bahir and Doris who give out good values or bad values every so often. The nurse can request to see the current value of either of the patients, indicate to the NC that he or she has seen the value or act on an alarm. We want the NC to sound an alarm at the appropriate time and deal with requests for information in sensible way. Each of the two PCs are to notify the NC of dangerous situations and satisfy the requests for information.

We can use the following list of modules and *their* propositions:

Doris: controls *value-D-gd* and *value-D-bd*;

Bahir: controls *value-B-gd* and *value-B-bd*;

PC-Doris: controls *patal-D*, *patval-D-gd* and *patval-D-bd*;

PC-Bahir: controls *patal-B*, *patval-B-gd* and *patval-B-bd*;

NC: controls *patreq-D*, *patreq-B*, *alarm*, *dis-D-gd*, *dis-D-bd*, *dis-B-gd* and *dis-B-bd*;

Nurse: controls *req-D*, *req-B*, *seen* and *act*.

The intended meanings of these propositions are clear. We can also describe the arrangements for communication and control as indicated in Fig. 12.4.

It seems that it is most realistic to regard this system as an *asynchronous* one because the several components would most likely act in their own good times. There is no reason why a patient should deliver his or her values at exactly the same rate as the nurse makes his or her actions or the non-human parts of the system step from stage to stage.

12.4.1 *The program*

Let us develop a specification of and a program for the system component by component.

Doris: It is a specification problem to decide whether the patient Doris delivers her measurements in response to a request from her PC or whether she just keeps on delivering measurements at her own rate. Let us suppose that it is the latter arrangement.

Thus we just want Doris to announce either *value-D-gd* or *value-D-bd* (but not both) at each stage. It is easy to get her to cycle continuously through a couple of *value-D-gd*s before a *value-D-bd*. We just need some new internal propositions p and q. Here is a module program:

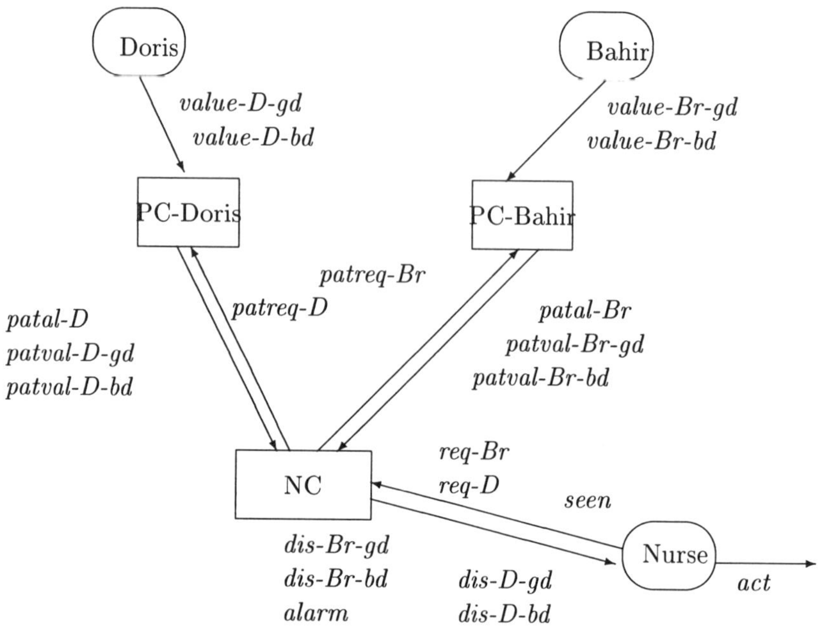

FIG. 12.4. Components of propositional implementation.

```
doris[value-doris-gd, value-doris-bd]:

  Q(p & q) => value-doris-gd & ~p & ~q;
  Y(~p & ~q) => value-doris-gd & p & ~q;
  Y(p & ~q) => value-doris-bd & p & q.
```

Bahir: This module is similar but for the sake of variety we will use a cycle of period 4:

```
bahir[value-bahir-gd, value-bahir-bd]:

  Q(p & q) => value-bahir-gd & ~p & ~q;
  Y(~p & ~q) => value-bahir-gd & p & ~q;
  Y(p & ~q) => value-bahir-gd & ~p & ~q;
  Y( ~p & q) => value-bahir-bd & p & q.
```

Nurse: The nurse does three things concurrently. He or she:

- acts on an alarm;
- indicates when displayed information is seen; and
- makes regular checks on each of the patients.

Let us suppose that the regular checks are done in a cyclical manner. One possibility is the four-stage cycle as indicated in the code below.

One improvement we could make is for the nurse to suspend other activities while acting on an alarm but this is not really worth the effort here.

```
nurse(alarm, dis-bahir-gd, dis-bahir-bd,
dis-doris-gd, dis-doris-bd )
[ act, seen, req-bahir, req-doris ]:

  Yalarm => act;
  Y( dis-bahir-gd | dis-bahir-bd |
                  dis-doris-gd | dis-doris-bd ) =>
             seen;
  Q(p & q) =>  ~p & ~q;
  Y(~p & ~q) => req-bahir & p & ~q;
  Y(p & ~q) => ~p & q;
   Y( ~p & q) => req-doris & p & q.
```

PC-Doris: The most important thing this module does is to send off a warning signal to the NC when it detects a bad value. If no value has arrived since a bad one then we may want to keep the alarm sounding, but in the specification of this problem we have assumed that even a brief burst of alarm is enough to alert the nurse.

When the PC gets a request it wants to send off the latest value. No value might have arrived yesterday so we use S to find out the most recent value that was received. Near the beginning of time there might have been no values received at all. In that case we will pretend the patient is all right and use $\mathcal{Z}$ to send off a good report.

We will see later in the case study that instead we should have waited for the next value to be received from the patient. This can be implemented just as easily.

One problem shown up by early experiments with implementation is the following. Although *value-D-gd* and *value-D-bd* are never announced at the same time, sometimes they arrive together in the message queue, so unless we are careful, we might send off two different values to the NC at once. Since this is unacceptable we make the arbitrary but fixed choice of a good report in such a case. We discuss this problem of *non-determinacy* later.

```
pcdoris( value-doris-gd, value-doris-bd, patreq-doris)
[patal-doris, patval-doris-gd, patval-doris-bd]:

  Y value-doris-bd => patal-doris;
  (~value-doris-bd)Z(value-doris-gd) & Y( patreq-doris)
          => patval-doris-gd;
  (~value-doris-gd)S(value-doris-bd &~value-doris-gd)
            & Y( patreq-doris)
          => patval-doris-bd.
```

PC-Bahir: As for PC-Doris.

```
pcbahir( value-bahir-gd, value-bahir-bd, patreq-bahir)
[patal-bahir, patval-bahir-gd, patval-bahir-bd]:

  Y value-bahir-bd => patal-bahir;
  (~value-bahir-bd)Z(value-bahir-gd) & Y( patreq-bahir)
           => patval-bahir-gd;
  (~value-bahir-gd)S(value-bahir-bd & ~value-bahir-gd)
               & Y( patreq-bahir)
           => patval-bahir-bd.
```

NC: Code for the alarm is straightforward. As we have said, there is no need to keep it ringing.

In an early version of this implementation the NC just sent on requests when they were received from the nurse and displayed information when it came in from a PC. This resulted in chaos with answers coming back in the wrong order and several arriving at once—even good and bad values from the same patient. Our restriction on the display was obviously not met.

In order to solve this problem we decided to institute a regime in which the NC passes successively through various states of waiting for information, displaying information or dealing with a new request. So that all requests are dealt with the NC alternates chances for requests to the two patients to get processed.

Notice that if several requests for information are put in before the NC gets around to dealing with them then they just result in one set of information eventually being displayed. This is adequate to meet the specification.

```
nurcomp(patal-doris, patal-bahir, patval-bahir-gd,
patval-bahir-bd, patval-doris-gd,patval-doris-bd,
seen ,req-bahir, req-doris)
[ patreq-bahir, patreq-doris,
 alarm, dis-doris-gd, dis-doris-bd,
           dis-bahir-gd, dis-bahir-bd]:

  Y(patal-doris | patal-bahir ) => alarm;

  Qfalse => chance-doris;

  Y chance-doris & (~shown-doris)S(req-doris)
         => wait-doris & patreq-doris;
  Y(wait-doris
        & ~(patval-doris-gd | patval-doris-bd))
         => wait-doris;
  Y((wait-doris & patval-doris-gd)
```

```
                        | dis-doris-gd & ~seen)
           => dis-doris-gd;
      Y((wait-doris
           & ~patval-doris-gd & patval-doris-bd)
                  | dis-doris-bd & ~seen)
           => dis-doris-bd;
      Y((dis-doris-gd | dis-doris-bd) & seen )
           => shown-doris & chance-bahir;

      Y chance-doris & (~req-doris)Z(shown-doris)
           => chance-bahir;

      Y chance-bahir & (~shown-bahir)S(req-bahir)
           => wait-bahir & patreq-bahir;
      Y(wait-bahir
           & ~(patval-bahir-gd | patval-bahir-bd))
           => wait-bahir;
      Y((wait-bahir
           & patval-bahir-gd) | dis-bahir-gd & ~seen)
           => dis-bahir-gd;
      Y((wait-bahir
              & ~patval-bahir-gd & patval-bahir-bd)
                 | dis-bahir-bd & ~seen)
           => dis-bahir-bd;
      Y((dis-bahir-gd | dis-bahir-bd) & seen )
           => shown-bahir & chance-doris;

      Y chance-bahir & (~req-bahir)Z(shown-bahir)
           => chance-doris.
```

12.4.2 *An example run*

```
CMP: executing asynchronously for 20 states.
pcdoris: OUTPUT { }
pcdoris: OUTPUT { }
nurse: OUTPUT { }
pcbahir: OUTPUT { }
pcdoris: OUTPUT { }
nurcomp: OUTPUT { }
pcbahir: OUTPUT { }
nurcomp: OUTPUT { }
bahir: OUTPUT { value-bahir-gd }
bahir: OUTPUT { value-bahir-gd }
bahir: OUTPUT { value-bahir-gd }
```

```
doris: OUTPUT { value-doris-gd }
nurse: OUTPUT { req-bahir }
nurse: OUTPUT { }
pcbahir: OUTPUT { }
nurse: OUTPUT { req-doris }
doris: OUTPUT { value-doris-gd }
doris: OUTPUT { value-doris-bd }
pcbahir: OUTPUT { }
bahir: OUTPUT { value-bahir-bd }
doris: OUTPUT { value-doris-gd }
nurse: OUTPUT { }
doris: OUTPUT { value-doris-gd }
nurcomp: OUTPUT { patreq-bahir }
pcdoris: OUTPUT { }
nurcomp: OUTPUT { }
pcdoris: OUTPUT { patal-doris }
pcdoris: OUTPUT { }
pcdoris: OUTPUT { }
nurcomp: OUTPUT { alarm }
pcdoris: OUTPUT { }
nurse: OUTPUT { req-bahir, act }
bahir: OUTPUT { value-bahir-gd }
doris: OUTPUT { value-doris-bd }
pcbahir: OUTPUT { patal-bahir, patval-bahir-gd }
pcbahir: OUTPUT { }
pcdoris: OUTPUT { patal-doris }
nurcomp: OUTPUT { alarm, dis-bahir-gd }
pcbahir: OUTPUT { }
nurse: OUTPUT { act, seen }
nurse: OUTPUT { req-doris }
pcdoris: OUTPUT { }
pcdoris: OUTPUT { }
bahir: OUTPUT { value-bahir-gd }
nurse: OUTPUT { }
doris: OUTPUT { value-doris-gd }
pcdoris: OUTPUT { }
nurcomp: OUTPUT { }
nurcomp: OUTPUT { patreq-doris }
bahir: OUTPUT { value-bahir-gd }
nurse: OUTPUT { req-bahir }
nurse: OUTPUT { }
bahir: OUTPUT { value-bahir-bd }
bahir: OUTPUT { value-bahir-gd }
nurse: OUTPUT { req-doris }
```

```
pcbahir: OUTPUT { }
pcbahir: OUTPUT { patal-bahir }
nurcomp: OUTPUT { alarm }
bahir: OUTPUT { value-bahir-gd }
doris: OUTPUT { value-doris-gd }
nurcomp: OUTPUT { }
nurse: OUTPUT { act }
nurse: OUTPUT { req-bahir }
nurse: OUTPUT { }
doris: OUTPUT { value-doris-bd }
pcbahir: OUTPUT { }
bahir: OUTPUT { value-bahir-gd }
doris: OUTPUT { value-doris-gd }
nurcomp: OUTPUT { }
nurse: OUTPUT { req-doris }
nurcomp: OUTPUT { }
pcbahir: OUTPUT { }
nurse: OUTPUT { }
nurse: OUTPUT { req-bahir }
pcdoris: OUTPUT { patal-doris, patval-doris-gd }
bahir: OUTPUT { value-bahir-bd }
bahir: OUTPUT { value-bahir-gd }
nurcomp: OUTPUT { alarm, dis-doris-gd }
pcdoris: OUTPUT { }
pcbahir: OUTPUT { patal-bahir }
pcdoris: OUTPUT { }
doris: OUTPUT { value-doris-gd }
doris: OUTPUT { value-doris-bd }
pcdoris: OUTPUT { patal-doris }
doris: OUTPUT { value-doris-gd }
nurse: OUTPUT { act, seen }
pcbahir: OUTPUT { }
nurse: OUTPUT { req-doris }
pcdoris: OUTPUT { }
pcdoris: OUTPUT { }
pcdoris: OUTPUT { }
doris: OUTPUT { value-doris-gd }
pcbahir: OUTPUT { }
nurcomp: OUTPUT { alarm }
nurcomp: OUTPUT { patreq-bahir }
nurcomp: OUTPUT { }
pcdoris: OUTPUT { }
nurcomp: OUTPUT { }
pcbahir: OUTPUT { patval-bahir-gd }
```

```
pcbahir: OUTPUT { }
pcbahir: OUTPUT { }
pcbahir: OUTPUT { }
doris: OUTPUT { value-doris-bd }
bahir: OUTPUT { value-bahir-gd }
nurcomp: OUTPUT { dis-bahir-gd }
doris: OUTPUT { value-doris-gd }
nurcomp: OUTPUT { dis-bahir-gd }
doris: OUTPUT { value-doris-gd }
bahir: OUTPUT { value-bahir-gd }
bahir: OUTPUT { value-bahir-bd }
nurcomp: OUTPUT { dis-bahir-gd }
bahir: OUTPUT { value-bahir-gd }
pcbahir: OUTPUT { }
bahir: OUTPUT { value-bahir-gd }
bahir: OUTPUT { value-bahir-gd }
doris: OUTPUT { value-doris-bd }
doris: OUTPUT { value-doris-gd }
pcbahir: OUTPUT { patal-bahir }
doris: OUTPUT { value-doris-gd }
bahir: OUTPUT { value-bahir-bd }

EXITING\ldots
```

12.4.3 *Lessons learnt*

Much was learnt in the process of constructing this implementation and observing the behaviour of it and several less satisfactory preliminary versions.

One thing to notice is that we always have the situation in which all of the components make their steps at about the same rate. Admittedly their 20 steps are distributed pleasingly randomly throughout the period, but the restriction that they get through exactly 20 steps in this time is both unrealistic and not in the interests of a prototyping system in which many and varied experiments should be able to be undertaken.

In fact, a serious problem with the implementation is disguised by this restriction. In some previous longer experiments, in which the random scheduling of steps was different, it was seen that a PC was very slow in responding to both life-threatening situations and requests for information. This turned out to be due to the fact that its message queue had been previously swamped with good values from its patient. Under the strict bunching rules in force for the processing of the queue, this meant that the PC had to work through these good values one at a time even though more important information was waiting at the other end of the queue. If we had had the facility to be able to change the rate of steps of the different processes this problem might have shown up even more clearly. We discuss this problem later.

Another problem observed in longer runs and also possibly more likely to have been visible with variable scheduling rates was the following violation of the specification. We noted that the PC sends off the latest measurements in response to a request. It was noticed that sometimes a measurement comes into a PC and then there are no more announcements from the patient for a while. During this time, the nurse might have made a request and this might have been passed on to the PC. In this case the PC will send back old information to be measured before the nurse's request. This violates the propositional version of our liveness of the display requirement. In the next section we take this discovery into account in writing a first-order program.

Another problem we face in the first-order program which was hinted at here is the non-determinancy problem mentioned in the discussion on PC-**Doris**. There we had to choose arbitrarily to send off a good report if both good and bad measurements arrive together. We will see in the next section that the problem is far more serious in the first-order case.

Finally, we note some problems which were discovered in observing incorrect implementations. These tell us about the lack of robustness of some of our module programs. For example, it was seen that the NC program is very dependent on the good behaviour of the PC s. If a PC breaks down the NC may become completely stuck waiting for a value to be returned and if a PC mistakenly sends in two reports of measurements when only asked for one', or sends in an unrequested report, then the NC can easily display two sets of measurements at once. Robustness is a very important consideration and such behaviour should be eliminated from any final implementation.

12.5 Possible real program

In this section we dream. Ignoring the fact that METATEM is in its early stages of development we suppose that we have every reasonable facility that a METATEM prototyping system should have and present a possible program for implementing our PMS distributed system. Here we just note the new facilities that we are assuming of METATEM and we discuss them in more detail in Section 12.11. Recall that our task is to build a prototype which will establish a feasible specification for the separate modules.

Of course we can suppose that the running and debugging of the implementation is done in a nice package environment but we will not go into these aspects here as there are more fundamental decisions to be made about the operation of the framework first. However, we might consider interfering interactively in the running of the program by allowing the user to play at being the nurse. This seems possible to engineer but for now we will suppose that once again we want the system to simulate the patients and nurse as well as run system components. Assuming two patients again, we have six components in this closed system.

We saw in the last section that the bunching algorithm for the reading of message queues was not satisfactory. In the program below we will assume that this algorithm is changed in exactly the way we want: that is the whole of a component's queue is emptied at each broadcast by the component.

We will also assume that the package allows the user to make his or her own choice about the relative firing rates of the components. Here, for example, we might want the patients' measurements to be made very often compared to the actions of the nurse. If such control is not given to the user then we could build in our own desired time control mechanism into the program in several ways. One could be to introduce a new process called a clock which sends out 'ticks'. Then each process could be limited to broadcast only at certain multiples of 'ticks'. But this is all a lot of extra work and it seems a reasonable request to ask for user-controlled broadcast rates. Of course, this whole discussion shows that we again desire asynchronous behaviour by the different modules.

Enough then of the setting for the program. Now let us consider each process in turn. Recall that first-order METATEM has, at the moment, severe restrictions on the form of rules. A lot of the difference between this program and the one in the last section will be because we try not to go beyond this syntax unless it seems absolutely necessary to introduce new constructs.

12.5.1 *Program*

The predicates used are 2-ary *temp*, 2-ary *blood*, 2-ary *pulse* under shared control of the various patients, 0-ary *seen*, 0-ary *act* and 1-ary *req* under the control of the nurse, 1-ary *alarm*, 4-ary *display* and 1-ary *pat–req* under the control of the NC, and 1-ary *pat–al* and 4-ary *pat–val* under shared control of the various PCs.

Doris: To allow the patient to broadcast a realistic range of measurements it seems best to suppose that some random-number generator is available. Without going into the exact form of a simulator for fluctuations in the temperature (etc.) of a critically ill patient—but, nevertheless, noting that some research might need to be done by the implementer and some heavy-duty mathematical functions might need to be provided in the METATEM language—we assume here that a program construct $RND(\delta)$ exists as a term where δ is the description of a probabilistic distribution. We suppose each time a formula containing the term is made true that the term gets substituted by a real or integer value according to the distribution δ.
Recall that the three measurements are announced simultaneously.

```
doris[temp, blood, pulse]:
Y true => temp( RND(temp-distribution) );
Y true => blood( RND(blood-distribution) );
Y true => pulse( RND(pulse-distribution) ).
```

Bahir: As for Doris.

Nurse: We assume that the nurse acts on alarms and sees displays immediately. He or she will cycle through patients in some order requesting their information. It is likely that the nurse will not make a request at each step, but

using a random-integer generator $RNDI(0,9)$ with a similar behaviour to RND above, we can make him or her do so about every 10 steps.
To produce the cycling in the order Doris, Bahir, Doris, Bahir, ... we assume that 1-ary *begin* and 2-ary *next* are some internal predicates. $begin(P)$ will only ever be true at the first moment of time and then only with with $P = doris$. After that the nurse waits to request (wtr) information on Doris until he or she does so when the right random number (here 0) turns up. Because of the constant extension of *next* the nurse keeps her requests alternating between information on each of the two patients.
Note that the nurse will never request information about a non-existent patient, a problem which we discuss in Section 8.2.

```
nurse(alarm, display)[act, seen, req]:
Y alarm(P) => act;
Y display(P,V) => seen;

Y begin(P) => wtr(P);
Y ( wtr(P) & ~req(P) ) => wtr(P);
Y( wtr(P) & req(P) & next(P,Q) ) => wtr(Q);

Y ( wtr(P) & ( RNDI( 0,9) =0 ) )
        => req(P);

Q false => begin(doris);
Y true => next(doris, bahir);
Y true => next(bahir, doris).
```

PC-Doris: To detect a life-threatening situation and send off a *pat–al* is straightforward. Note that here we need < comparison on values.
Answering requests is more of a problem for two reasons. First, our specification requires that when a request is received we send off values which are measured subsequently. To ensure that we do not send off old values we have to delay answering the request for at least a step and then send off the next values to arrive to make sure that they are 'fresh'. Hence the use of a proposition *reqd* which is made true by a request and stays true until values arrive and are sent off.
As we will see when we consider the NC component, it is crucial for our implementation that PC s only send off values when they are asked and only send off one set of values. This brings us to our next problem: it is possible that two (or more) sets of values arrive while the PC is making one step. Thus, if it has a request to answer, the PC has to choose one set to send. It seems that the restricted METATEM language for the first-order case is not adequate to implement this requirement.
As a possible solution to this problem we introduce a new three-place auxiliary function NDC which enables such *non-deterministic choice*. The

idea is as follows. Suppose that $\phi(\overline{y}, x)$ is a formula with free variables only from the set $\{x, y_1, \ldots, y_n\}$. In our program ϕ is only ever a positive literal but perhaps this need not always be so. Suppose further that $\overline{a}$ is an n-tuple of ground terms. Then the construct $NDC(X, \phi(\overline{a}, X), Y)$ can appear as a literal in the program for any variable symbols X and Y. To define its semantics suppose that at some time t, B is the set of all the ground terms which make $\phi(\overline{a}, b)$ true now. Then if B is empty we define $NDC(X, \phi(\overline{a}, X), d)$ to be false for all d at time t. On the other hand if B is not empty then the computer must choose one value $b \in B$ and at time t, $NDC(X, \phi(\overline{a}, X), d)$ is true if and only if $d = b$.

So in our program, if $\textit{temp}(\textit{doris}, 37.9)$ and $\textit{temp}(\textit{doris}, 43)$ are both assumed true at the same moment by PC-**Doris**—no doubt because it is running much slower than the thermometer—then it is up to the program to pick one of 37.9 or 43, say 37.9, and make $NDC(T, \textit{temp}(\textit{doris}, T), 37.9)$ true and $NDC(T, \textit{temp}(\textit{doris}, T), 43)$ false.

We discuss this construct and other approaches in Section 12.7.

Notice that the PC assumes that the three measured values come in simultaneously.

Finally, consider the question of keeping *reqd* true until a set of values comes in. We want to say that this happens if yesterday *reqd* was true and it was not true that there is a value which came in. We see that we can use NDC to express this negated existential quantifier.

```
pc-doris(temp, blood, pulse,patreq)[patal, patval]:

Y( temp(doris,T) & (T< mint) ) => patal(doris);
Y( temp(doris,T) & (T> maxt) ) => patal(doris);
Y( blood(doris,B) & (B< minb) ) => patal(doris);
Y( blood(doris,B) & (B> maxb) ) => patal(doris);
Y( pulse(doris,R) & (R< minr) ) => patal(doris);
Y( pulse(doris,R) & (R> maxr) ) => patal(doris);

Y( patreq( doris) ) => reqd;

Y( reqd & ~NDC(T, temp(doris,T), T) ) => reqd;

Y( reqd & NDC(T, temp(doris,T), T))
  & NDC(B, blood(doris,B), B))
  & NDC(R, pulse(doris,R), R))
    => patval(doris,T,B,R).
```

PC-Bahir: As for PC-**Doris**.

NC: Here we use the *begin*, *next* pattern as in the nurse's program to allow the NC to go through the list of patients satisfying requests for information. This is because we only want to send out one *pat–req* at a time and be waiting for one reply. Meanwhile the nurse might be making many varied requests. These are remembered in a simple yes/no fashion by setting *reqd*(P) to be true when a request for P comes in and keeping it true until the NC sends out a *pat–req*(P). Then it makes *wait*(P) true and keeps that true until a set of values comes back. Then that is displayed until the nurse has *seen* it which signals the NC to start checking through other requests. Notice that we are having to use the new predicate *reqd* where, in the propositional program, we were able to use a since expression. In order to satisfy the specification we may only cancel a request when we send off a *pat–req*. If we ignored requests for P coming in during a wait for P we may end up displaying 'old' information gathered before the request but displayed after it. To avoid this we have to process such a subsequent request at a later time.

Notice also that instead of using a negated NDC to express $\neg\exists$ as in the program for PC-**Doris** we bring about the end of a waiting period by using the *stop-wait* proposition. This is an acceptable method of expressing $\neg\exists$ when we can delay the effect for one step.

In Section 12.7 we discuss a completely different way of arranging for the NC to deal eventually with requests.

```
nurcomp(patal,patval,seen,req)[patreq,alarm,display]:

Y patal(P) => alarm(P);

Y begin(P) => chance(P);

Y( req(P) ) => reqd(P);
Y( reqd(P) & ~chance(P) ) => reqd(P);

Y( chance(P) & reqd(P) ) => wait(P);
Y( chance(P) & reqd(P) ) => patreq(P);

Y( wait(P) & ~ stop-wait(P) ) => wait(P);
Y( wait(P) & patval(P, T,R,B) ) => display(P, T,R,B);
Y( display(P,T,R,B) & ~ seen ) => display(P, T,R,B);
Y( patval(P,T,R,B) ) => stop-wait(P);

Y( display(P,T,R,B) & seen & next(P,Q) ) => chance(Q);

Q false => begin(doris);
Y true => next(doris, bahir);
Y true => next(bahir, doris).
```

We must also remember to finish off the whole program with a full stop.

12.5.2 *Extensions*

There are several ways in which we could extend this program and make it more realistic. Several concern making it more forgiving to the mistakes and malfunctions of the environment or breakdowns within the system. For example, it is very dangerous for the NC to keep waiting for a reply from a PC. Perhaps some kind of indication of long waiting and a method of cancelling a request could be incorporated. There is a similar need for notification of faulty measuring devices.

Another problem is that although the NC will not attempt anything silly, like sending out a *pat–req* for a non-existent patient, it will not indicate to the nurse that he or she has asked for information on such a patient. This should be easy to add but in the final PMS to be supplied to hospitals it seems that there must be a very user-friendly way of checking through the patients.

This brings us to the problem of admission and discharge of patients. We note that the NC is quite amenable to changes in the list of patients but that we do not jave here the facilities for adding or removing modules dynamically. In fact it is probably not true that we must be able to change the system and interface definitions of existing modules during its operation in order to implement such a program. The system could probably be set up with a finite fixed group of PCs which could be reused again and again with different patients and sit quietly when they are not being used. No interfaces seem to need to be flexible.

12.6 Proving correctness

In this section we are going to examine two tasks concerned with the justification that our implementation is correct. We will try to identify exactly what it is about the behaviour of our various METATEM modules which causes the overall system to live up to the requirements for a PMS. Thus, in the process of the justification we would hope to produce the detailed modular specification required. Unfortunately there is quite a lot of work to be done to achieve this and there is still much of METATEM's theoretical support that needs development before such a task can be properly completed.

This justification would seem to be a matter of using formal reasoning techniques to show that the program implements the formal specification. However, as pointed out in [Barringer, 1987], there may be a gap between the client's original expectations and our formal specification. Obviously there is no way of completely closing this gap but one of the overall purposes of the prototyping procedure is to make this gap smaller. The client can see the prototype in action and again supply feedback so the specification and the implementation can be refined or changed.

In developing the PMS implementation we have already seen provisional attempts at a formal specification being shown to be incorrect. Some of the refinements which we have taken on board in the course of undertaking this case study are:

- outlawing uncaused alarms; and

- the various assumptions on the behaviour of the environment.

We have also seen questions arise which will need to be answered by the client, i.e. the informal description was inadequate. These include:

- details about what to do if PC breaks down;
- more details about the working of the NC as far as the ability of the display to show numbers of sets of measurements and details about how the nurse requests information;
- more information on how the measuring devices work; and
- the requirements for admitting and discharging patients.

Now, suppose that our prototyping and other considerations have convinced us that the formal specification has adequately captured the client's expectations. Let us now consider the question of whether our implementation reifies the formal specification: we have to show that every possible behaviour of the program is a model of the sentence $LA \wedge LD' \wedge SA \wedge SD \wedge UD$.

This may be attempted directly by using a formal temporal semantics for the METATEM distributed system and by doing some reasoning in an appropriate temporal logic. However, as we have suggested above, there are important reasons for not proceeding in this way. One of our tasks here is to develop a modular specification of a distributed system solution to the PMS problem. Thus we want formal specifications for each of the components so that no matter how a particular component is implemented to satisfy its specification, the components acting together are guaranteed to satisfy the overall specification.

Unfortunately we will only be able to outline how this process of modular specification might be accomplished. This is because the theoretical basis underpinning the work is only in its early days of development. There are three separate aspects to this basis, each of them an important area needing much investigation.

The first is the task of composing specifications. In Section 12.6.1 we will show how specifications of different modules can be composed in this way and say something about their combined operation. We follow the basic ideas recently formalized in [Manna and Pnueli, 1992].

Since we also want to establish that we have, in fact, implemented the specification, we will also have to show that each component does satisfy its modular specification. To do this we will need to use a formal semantics for the behaviour of METATEM modules, something which, alas, is only just being developed (see [Fisher, 1993]). Later, in Section 12.6.2, we give an outline of this development.

Note that in both the compositional reasoning and the reasoning about a single component we will be using temporal logic. We will see what assumptions we have to make about this logic but try and use it in a semantic way. With further work on automated proof techniques for such logics—the third area needing much work—we may one day be able to give complete syntactic renditions of such proofs, but for now we just have to hope to be rigorous in taking shortcuts. See [Barringer, 1987] for a discussion about having to take shortcuts in the formal reasoning.

In the rest of the section we will consider the overall specification as its five conjuncts, and, considering each in turn, we will see what is needed from each separate component in order to achieve a model of the conjunct. When the machinery for doing this reasoning is established, this strategy will allow us to present a complete modular specification of our PMS system.

12.6.1 *Modular proofs*

The idea of a compositional proof is to relate the specification of individual modules to the specification of a system of which they form a part. The important contribution described in [Manna and Pnueli, 1992] is the formalization of the ideas of control of communication channels—what we use predicates for and what [Manna and Pnueli, 1992] use variables for. This is crucial for composing modules.

Without defining exactly what a module is we can define a *module specification* as a pair consisting of:

- an *interface specification*, being details on the control of predicates, i.e. which truths are within the power of the module to announce and which are within the power of the environment; and
- a temporal logic formula representing the desired behaviour of the module in its environment.

Because we use predicates here for communication, the details of control correspond to the *interface* in [Manna and Pnueli, 1992].

When putting modules together in one system it is important to realize that sometimes this is not possible to do because of their respective interface specifications. We say that a set of module specifications are *interface compatible* if and only if the details on the control of predicates contained in the various interface specifications are not contradictory.

In case we have several modules $M_1, \ldots, M_n$ acting together in a system (perhaps with its own extra-system environment) we write $M_1 \| \ldots \| M_n$ to represent the system. Again we do not want to try to formalize this here but below we will reason about the the behaviour of such a system using the usual interleaving idea. See [Manna and Pnueli, 1992] for details.

As in [Manna and Pnueli, 1992] we say that a temporal sentence ϕ is *modularly valid* for a module M if and only if for any module E which is interface compatible with M, any behaviour of $M \| E$ is a model of ϕ.

In [Manna and Pnueli, 1992] we have the very useful claim that

> If formulas ϕ_1 and ϕ_2 are modularly valid over modules M_1 and M_2, respectively, then $\phi_1 \wedge \phi_2$ is modularly valid over $M_1 \| M_2$.

In the proofs below we will make great use of this rule but we must note that there is still much work to be done here. Not the least of this is in defining exactly what we want a module to be in the METATEM setting, how to formalize the idea of the interface specification and how we formally present the totality of a system of modules.

12.6.2 *Semantics for CMP*

In order to show that our METATEM modules live up to their modular specifications we will need to use a formal semantics for the language. An outline of such a semantics for a CMP program—not just an individual module—is given in [Fisher, 1993]. Unfortunately this is very incomplete and uses an interval-based approach which does not seem appropriate here, so we will present an outline of a slightly different semantics for use in the proofs.

We consider the behaviour of a CMP module M with its message queue in an environment with broadcast message passing. We shall suppose that at each broadcast the queue is completely read, bunched and emptied. As in [Fisher, 1993] the semantics will describe the message passing and the behaviour of M.

We will introduce a new proposition tick_M which is true at exactly the instants when M makes its broadcasts. These times might be called *ticks* of M. The predicates p of the language will represent themselves in an obvious way, i.e. $p(\overline{a})$ will be true at exactly those times at which $p(\overline{a})$ is broadcast by some module in the system (or the environment).

We will also need to allow for the fact that M pretends that certain truths happen at different times. To do this we introduce a new n-ary predicate symbol p_M for each n-ary predicate symbol which appears in M's program.

For those predicates p which are read by M from the environment we can thus describe the reading of the message queue by

$$\Box\forall\overline{x}.(p(\overline{x}) \to ((\mathrm{tick}_M \wedge p_M(\overline{x})) \vee (\neg\mathrm{tick}_M)U(\mathrm{tick}_M \wedge p_M(\overline{x})))).$$

The two conjuncts in the consequent here represent:

- the possibility that $p(\overline{a})$ is announced by M itself and so $p_M(\overline{a})$ and $tick_M$ are true; and
- the possibility that $p(\overline{a})$ is announced by another module and so gets into the queue making $p_M(\overline{a})$ true at the next tick of M.

Notice that this is more specific than the implication with an eventuality appearing in [Fisher, 1993]; for some of the soundness proofs below we need to have much more detail about the timing of these events.

For those predicates p which M broadcasts, we can describe the act of broadcasting by

$$\Box\forall\overline{x}.(p_M(\overline{x}) \to p(\overline{x})).$$

Finally, we consider the internal behaviour of M itself.

The semantics of the program line `Q false => p(` $c_1,\ldots,c_n$ `)` is just

$$\Box((\mathrm{tick}_M \wedge \blacksquare(\neg\mathrm{tick}_M)) \to p_M(\overline{c})).$$

To define the semantics of the program line `Y(` C_1 `&` ... `&` C_m `) => p(` $u_1,\ldots,u_n$`)` for some literals C_i and some terms u_j using the variable symbols

$\texttt{X}_1, \ldots, \texttt{X}_k$, we first convert the conjunction of the C_i into a formula $\phi(\overline{x})$ in an obvious way. Then we replace each of the predicates q in ϕ by q_M to get ϕ_M. The semantics for the line is then

$$\Box\forall\overline{x}.(\text{tick}_M \wedge \phi_M \rightarrow (\neg\text{tick}_M)U(\text{tick}_M \wedge p_M(\overline{u}))).$$

It is clear how this idea is generalized to other types of program rules.

We will also need conditions forcing the falsity of positive literals not explicitly made true by the module or its environment. Something like the Clark completion for logic programs is needed here. See [Clark, 1978].

To find the semantics for the module M, we just take the conjunction of all these rules although a complete account may also need extra properties like disallowance of convergence of ticks $\Box(\neg\text{tick}_M)S(\text{tick}_M)$.

There may also need to be conditions on the overall behaviour of the system, such as the need to disallow pirate broadcasters. To formalize this suppose that each predicate p has a set C_p of modules which share in its control. Then we can specify

$$\Box\forall\overline{x}. \left[p(\overline{x}) \leftrightarrow \left(\bigvee_{M \in C_p} p_M(\overline{x})\right)\right].$$

There is still much work to be done on such a semantics, particularly in a compositional setting.

12.6.3 *Liveness of alarm*

Let us look at the easiest required property to prove our PMS implementation: LA. We want to prove that

$$\Box\forall P(\phi(P) \rightarrow \Diamond \mathit{alarm}(P))$$

will be true of any behaviour of the system regardless of what the environment does.

This is an easy requirement to modularize: it is clear that if

$$\Box(\phi(P) \rightarrow \Diamond \mathit{pat\text{-}al}(P))$$

is modularly valid for PC-P and

$$\Box(\mathit{pat\text{-}al}(P) \rightarrow \Diamond \mathit{alarm}(P))$$

is modularly valid for NC then

$$\Box(\phi(P) \rightarrow \Diamond \mathit{alarm}(P))$$

follows from the conjunction and so is modularly valid for any system containing PC-P and NC.

To prove our result from here involves, as well as the Barcan axiom, a short argument based on the fact that for any object P capable of making $\phi(P)$ true we have a module PC-P in the system. This might be best formalized in a *many-sorted* logic with the variable P being assumed to be of sort *patient*, but it will also need a formalization of the set of modules in the system.

Now let us go back to our modular requirements. First, we show that

$$\Box(\phi(doris) \to \Diamond \mathit{pat\text{-}al}(doris))$$

is modularly valid for PC-**Doris**.

So suppose that at some time t there is a $T_{\mathrm{bad}} > maxt$ such that $temp(doris, T_{\mathrm{bad}})$ holds.

Note that, using the rigidity of the $>$ comparison on numbers, we know that $T_{\mathrm{bad}} > maxt$ always.

We need to assume that PC-**Doris** completes a step after time t. This is a form of what is called *process justice* in [Manna and Pnueli, 1992]. In our semantics for CMP we establish this and take care of the liveness of messages passing into a module with such axioms as

$$\Box\forall PT.(temp(P,T) \to (\neg \mathrm{tick}_{\mathrm{pc-doris}})U(\mathrm{tick}_{\mathrm{pc-doris}} \wedge temp_{\mathrm{pc-doris}}(\overline{x}))).$$

From this we deduce that $\mathrm{tick}_{\mathrm{pc-doris}} \wedge temp_{\mathrm{pc-doris}}(doris, T_{\mathrm{bad}})$ holds at some time after t.

But one of the conjuncts of the semantics of PC-**Doris** is just

$$\begin{array}{ll} \Box\forall T & [(\mathrm{tick}_{\mathrm{pc-doris}} \wedge temp_{\mathrm{pc-doris}}(doris, T) \wedge (T > maxt)) \\ \to & (\neg \mathrm{tick}_{\mathrm{pc-doris}})U(\mathrm{tick}_{\mathrm{pc-doris}} \wedge patal_{\mathrm{pc-doris}}(doris))]. \end{array}$$

Thus we deduce that $patal_{\mathrm{pc-doris}}(doris)$ holds at some further time.

Now we need to use the axioms establishing liveness of message broadcasting. We have

$$\Box[patal_{\mathrm{pc-doris}}(doris) \to \Diamond \mathit{pat\text{-}al}(doris)]$$

from which we deduce that at a yet later time $\mathit{pat\text{-}al}(doris)$ holds as required.

Going through the same sorts of arguments for NC gives us the other half of the modular specification.

12.6.4 *The rest of the specification*

LA is the most straightforward property to prove. In the previous subsection we saw the process of generating requirements on modules in order to prove the overall properties of the system. As will be gathered from the discussion below, this is much more difficult to do in the case of the other properties. The machinery is not well enough developed yet to complete the task here but let us look quickly at some of these properties and see what sorts of axioms they might require of the semantics.

Consider first the property LD':

$$\begin{array}{ll} \Box \forall P \; (\; req(P) \wedge \Box \Diamond seen \wedge \tau(P) & \rightarrow \\ \quad (\exists T \exists B \exists R & \\ \quad \Diamond(temp(P,T) \wedge blood(P,B) \wedge pulse(P,R) & \\ \quad \wedge \Diamond(display(P,T,B,R) \mathcal{U} seen) &)) \end{array}$$

in which we recall that $\tau(P)$ says that the three measurements for P arrive simultaneously.

To show that the nurse's requests eventually lead to the desired information being displayed, we must do a very complex compositional proof.

For each PC we need to know that:

- it will eventually respond to a request;
- it answers with only one set of values;
- the values sent to NC will be measured after the PC receives the request; and
- no module other than PC-P_0 sends in values of the form *pat–val*(P_0, T, B, R).

To prove these we need axioms concerning such wide-ranging matters as the properties of NDC, the details of the reading of the message queues and the control details about the predicate *pat–val*.

For the NC we will need to prove that $\phi_1 \rightarrow \phi_2$ is modularly valid for it, where ϕ_1 includes assumptions about the nurse repeatedly making *seen* true, that $\tau(P)$ holds for each P and that the properties listed above hold for *each* PC. Under these assumptions we can prove ϕ_2, which says that a request of P's information will eventually lead to a subsequent measurement being displayed.

For the rest of the specification we need much the same varied group of axioms but the proofs are a little easier. The soundness of alarms follows from requiring soundness of *pat–al*(P) with respect to life-threatening situations for P of PC-P and soundness of *alarm*(P) with respect to *pat–al*(P) of the NC. We also need to know that no other modules broadcast alarms. The soundness proofs can then be done by constructing appropriate labelling of causes and effects.

12.7 Comparison with other approaches

12.7.1 *Other methods of implementing msbjles*

There are many ways of implementing modules as part of a distributed system. A popular general approach is that of regarding the separate modules as distinct *objects* having their own behaviour and communicating with other objects by message passing. Within this general approach we find CMP as well as the more developed *actor* model of [Agha, 1986].

As pointed out in [Fisher, 1993], the actor model differs fundamentally from CMP in that the separate modules—or actors—only exhibit any behaviour at all in response to receiving a message. So although this model exists as a direct implementation in [Lieberman, 1986] and is used as the method of introducing concurrency into such languages as ABCL [Yonezawa, 1990] and concurrent prolog in [Prasad, 1991] and [Kahn *et al.*, 1987], it is not suited to our purposes in the PMS where, for example, constant monitoring of life-threatening situations needs to be undertaken.

Another difference between actor models and CMP is that actors only receive and process one message at a time. This means that slow actors may suffer from the problem of being left behind, as do slow modules in the unsatisfactory existing implementation of CMP. It is important to remember that we are assuming the final version of CMP will not exhibit this behaviour.

A third difference is that actors only send messages addressed to another specific actor—an example of what is called point-to-point message passing. This is in contrast to the broadcast message passing of CMP. Some discussion of the relative merits of these two approaches can be found in [Fisher, 1993] but here we note that in our prototype implementation of the PMS the extra machinery needed in an implementation to define the routes and recipients of messages would have been an unnecessary extra distraction from the basic task of determining the temporal properties of the communication arrangements. Once we have settled these issues we need to look at where messages are going to and then implement communication channels if that is indeed more efficient.

As another type of object-based system mention must also be made of the Linda parallel programming paradigm of [Gelerter *et al.*, 1985]. With its central tuple space this provides another mechanism for communication between modules—here called processors. The processors themselves can be programmed in any of a variety of languages but only interact with the tuple space through special Linda operators. This can be seen as a broadcast message-passing arrangement.

12.7.2 *Executable temporal logic*

Given this wide variety of arrangements for implementing the system and the individual modules within it, we must examine the benefits of presenting yet another approach. First, recall that we are using METATEM in its CMP form as an implementation tool for preparing a prototype of a distributed system which allows us to investigate and correct:

1. the overall specification; and
2. the modular specification of the system.

Assuming that temporal logic is our language of specification we make the claim that METATEM is a good tool because there is a very intuitive and straightforward relationship between the temporal specification of a module and its program. Although the formal semantics for CMP is in its early stages of definition this claim seems very plausible because of the care which we have taken in relating the operational definition of the program constructs with their declarative reading. The ease with which we were able naturally to translate informal correctness arguments about the program into formal equivalents supports this claim.

None of the alternative methods of implementing systems surveyed above offers this cosy relationship between the specification formalism and the implementation language.

Perhaps some hope for a rival comes from another very well-developed executable temporal logic in the form of TEMPURA of [Moszkowski, 1986]. In fact in [Hale, 1987] there is a very detailed description of the process of building a TEMPURA implementation of the temporal logic specification for a multiple lift system given in [Barringer, 1985]. The main differences to note here are that TEMPURA is a much lower level, less directly expressive language and that it has not been developed to fit into a distributed system environment. These would cause problems in our case study. We also note that TEMPURA is designed to fit into a specification framework using an interval temporal logic (as opposed to the point-based FML) but the PMS could, no doubt, have been specified in terms of intervals.

12.7.3 *Process algebras*

This brings us to an examination of the possibility of using formalisms other than temporal logic in which to do this prototyping development. A major approach to modelling concurrent programs is the process algebra development seen for example in [Hoare, 1985] and [Milner, 1989]. Although concurrent programs are not quite the same as distributed systems, this distinction is not important here as a distributed system can be thought of as just a particular type of concurrent program—especially for simulation purposes. Perhaps we could do all the work in a process algebra environment instead.

The problem here is that these algebras are designed to model concurrent programs and so are really about implementation instead of specification. In fact, in some recent work [Zhou, 1987] a process algebra (CSP) is used towards implementing the internal behaviour of modules specified in temporal logic. Such programming languages as OCCAM [INMOS Ltd., 1984] show process algebras being used for implementation.

From informal beginnings in, for example, the discussions of specification in [Hoare, 1985], much work is now being done on providing a specification framework for process algebras. One way is to use non-determinacy and a partial ordering on processes to make some attempt at capturing the specification/implementation dichotomy as in [Hennessy, 1988]. Another approach is by using a modal logic as in the modal mu-calculus (see e.g. [Stirling and Walker, 1989]).

Some useful work on providing machine assistance for developing and verifying implementations from specifications is described in [Cleaveland *et al.*, 1989].

Nevertheless there are still reasons to criticize some of these approaches to specification on the grounds that they sometimes distinguish between processes on the basis on internal structure rather than purely on the basis of observable external behaviour.

12.7.4 *Temporal–causal logic*

The same applies to the Petri net approach (see [Reisig, 1985]). Despite this, recent work in the series by Castro and Kramer [Castro and Kramer, 1990a; Castro and Kramer, 1990b]; [Castro, 1990; Castro, 1991] has focused on using a Petri net approach in specifying a PMS. An interesting aspect of their work is that recognizing the inability of Petri nets to state liveness properties and their tendency to become unmanageably complex quickly, and giving the now outdated argument that temporal logic cannot provide a modular approach to specification, the authors have defined a pluralistic logic based on Petri nets but also incorporating temporal aspects. This is called temporal-causal logic.

It must be pointed out that this work is also in its early stages and is simply a specification language without any closely associated prototyping programming language. However, a quite detailed attempt at specifying the PMS is attempted in [Castro and Kramer, 1990b]. From this we can make some comparisons with our temporal logic approach.

The first problem is that the temporal–causal logic is less well established than temporal logic. No formal semantics is given for the logic and this is a serious problem since some of the axioms given in the proof system look unintuitive. For example, in one axiom (axiom 9 in [Castro, 1990]) we are told that for any state property of a system, and any event which can occur, either the event causes that property to hold or the event causes its negation to hold. Since events (like the arrival of a request for information from a patient at a PC) can occur many times this seems to prevent concurrency within and between modules. Without a formal semantics, a proof system cannot be justified.

Another problem with the lack of a formal semantics is that it is not clear that adding causality in this way to a temporal logic allows us to say anything more. It is quite easy in temporal logics to say that the occurrence of a certain event a always brings about b holding for a while: $\Box(a \rightarrow (bU\top))$.

Perhaps the most serious problem with a lack of semantics is that it will be hard to give a formal semantics of an implementation in this language.

As we have said, and as can be seen from the following Petri net specification diagram of a PC (Fig. 12.5), the main problem with this approach is the almost implementation-like detail introduced by the Petri net basis. This shows a lack of abstraction as we concentrate on internal details of the module rather than just on its externally observable behaviour. For example, without going into the details of the semantics, we see that separate cycles are proposed for the alarm part and the answering requests part of the module. Can this concurrency be observed externally?

12.8 Lessons for METATEM

During the course of this case study we discovered several aspects of the existing METATEM language which did not seem correctly defined for our purposes and

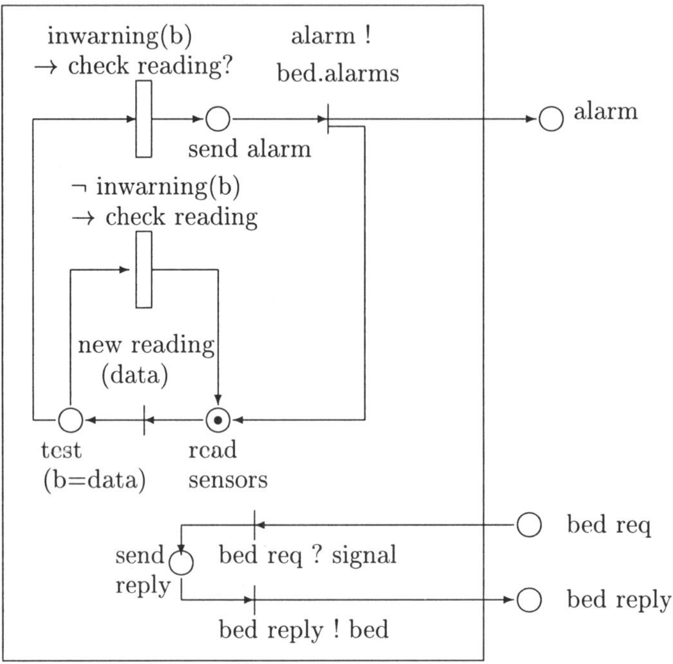

FIG. 12.5. Petri net description of a PC.

many examples of features which would need to be added to form a realistic prototyping framework. Let us look in some detail at these points.

We saw that we needed a few built-in arithmetic functions and relations. These should not be too difficult to supply along with all sorts of other built-in predicates. Although we consider some important differences in Section 12.8.2 below, the similarities between METATEM and Prolog syntax suggest we could probably use a lot of the usual Prolog library procedures—see [Filipič, 1988]. Perhaps, in METATEM, we would want to add a wider range of random-number generators which are useful for simulations.

Again, considering the similarities with Prolog syntax, it may not be worth the trouble but there are some arguments for introducing *typing* of objects in the object domain and so types of variable symbols. All the usual points about this aiding robustness of intermodule communication and of changing programs apply. On the other hand the module declaration parts will become more complicated. We saw in Section 12.3 how typing of modules might play a role in proving correctness of systems.

12.8.1 *The package*

An essential part of any package like METATEM is a comfortable environment in which to program, observe and debug. In considering the PMS example we have seen a need for the observing stage to be carried out by a realistic simulator with different processes being observed in different windows. Considering that an interactive program is an exemplar of a reactive module, we also see that there is potential for doing this well. Unlike a traditional programming language oriented to the transformational paradigm, it seems that METATEM might very naturally be able to handle a convenient collection of interactive input and output procedures for screen control and keyboard input. In our example it would have been very useful to cast the NC module as an interactive module taking instructions from the keyboard and displaying information and alarms on its little bit of screen.

We saw that part of creating an environment for realistic simulations is to allow the user to control the scheduling rate between the various asynchronous processes. To achieve this would seem to take only some usual sort of random element in the system controlling program.

12.8.2 *Rigid predicates*

We saw in our implementation that we sometimes want to be able to define certain predicates to have constant extension. In the NC, for example, it would have been convenient to declare *next* to be such a *rigid* predicate. In the implementation we achieved the rigidity of *next* (except at the first step when it did not matter) by having its extension computed at every step. This seems rather wasteful.

We can easily imagine other situations in which such a predicate has a very large extension or even an infinite one, such as that of list membership. It would be stupid to try and calculate this each time.

Although none appeared in our example, there is a related class of predicates which are also worth considering here, namely *derived* predicates. This is a classification which, as we show below, can be introduced to capture the idea that sometimes there is a 'rigid' logical relationship between various flexible predicates which remains fixed as they evolve through time. For the sake of efficiency, or even just implementability, it may make sense to exploit this relationship.

We will introduce a modification to METATEM which will allow us to do this by classifying predicates into *basic* and *derived* and show that rigid predicates can be efficiently implemented within this framework as just a special type of derived predicate.

For now just suppose that our METATEM rules have single predicates as their heads. We syntactically divide the predicates up into derived ones and basic ones and we divide the module program up into a derived part which looks like a Prolog program and a basic part which uses our restricted METATEM syntax. In the basic part we bar derived predicates from the heads of rules and in the derived part we bar basic predicates from the heads of clauses.

We only record the truth of basic positive ground literals in the historical database. To determine the truth of basic ground literals at a certain time we look up the database. To determine the truth of derived literals we use the usual Prolog resolution procedure in the derived part of the program.

For example, consider the program line

```
Y( display(P,T,R,B) & seen & next(P,Q) )=> chance(Q);
```

when *next* is declared derived and the rest of the predicates basic. If we find that *display*(*doris*, 38, 70, 120) and *seen* were true yesterday, say, we come to trying to find Q such that *next*(*doris*, Q) was true. Instead of looking in the database for yesterday's truths we go to the derived part of the program, namely

```
next( doris, bahir).
next(bahir, doris).
```

and using Prolog's resolution strategy quickly find $Q = bahir$ is a solution. In keeping with our semantics we would have to try for other solutions too, but finding none the NC can just announce *chance*(*bahir*).

As another example consider an implementation of the PMS which allows us to admit and discharge patients. Suppose that we have a basic predicate *patlist* which is true of the list of patients each time. Say *patlist*([*bahir*, *jim*, *auntie*]) was true yesterday. Now the derived program may, in part, look like

```
next( X, Y) :- patlist( [H |T ]), nextl(H, [H|T], X, Y).

nextl( F, [X| [Y|T] ], X, Y).
nextl( F, [X], X, F).
```

When asked about *next*(*jim*, Q) being true yesterday, this should tell us that $Q = auntie$ is the only solution.

Notice that we could have used a more METATEM-like syntax here and written

```
patlist( [H|T] ) & nextl( H ,[H|T], X, Y)
 => next( X, Y);
```

etc.

There are questions to be answered about allowing temporal operators into derived part of the program but even without allowing this we see that we have been able provide a not too messy means of increasing expressiveness and efficiency.

It may be thought that allowing other modules to have access to the 'definitions' contained in the derived part of a module's program might save some repetition of complicated code. This is true but detracts from the realism of the simulation of a distributed system in which all information should be passed through the proper channels.

12.8.3 *Basic expressiveness*

In the implementation we saw that we were able to express a remarkable variety of temporal relationships with just a very restricted syntax of program rule. Generally our rules were of the form $\bullet \bot \rightarrow p(\overline{c})$ or $\forall \overline{x}.[\, \odot(C) \rightarrow p(\overline{u})]$ where p is a predicate, $\overline{c}$ a tuple of constants, C a conjunction of literals and $\overline{u}$ a tuple of terms. Thus the program was almost totally deterministic.

The only real deviation from this pattern was our need to introduce the non-deterministic choice operator NDC. Although this has a reasonable, if not completely simple, semantics, it might be worth considering some other solutions to the problem. Recall that the problem is one of choosing just one from a whole host of values which might have made a predicate true yesterday. In the next subsection we see that waiting to synchronize with an incoming message presents another solution while in the following subsection we recall that changing the message queue bunching algorithm provides a yet different solution. Now let us look at some variations on the theme of non-deterministic choice.

One way which provides a more low-level control of the operation of NDC is to introduce set operations into the basic repertoire of the program and allow the non-deterministic choice of elements from a set. For example, allow a basic relation *in* to be the element-of relation between objects and sets and introduce $NDCS(element, set)$ as the operator. Our PC-**Doris** program can now contain the line

```
Y( temp(doris, T) ) => T in Tempset
```

etc. to collect values in sets and the line

```
Y( reqd & NDCS(T,Tempset)
       & NDCS(B, Bloodset)
     & NDCS(R, Pulseset) )
 => patval( doris,T,B,R)
```

to select values and send them off. This construction seems eminently implementable if we promise to stick to finite sets.

We saw in the implementation for PC-**Doris** that the NDC operator was useful in expressing a negated existential quantifier within the body of a rule. Although our implementation for the NC showed that this was not necessary, we note that the alternative method used there—via the *stop-wait* proposition—required an extra step to stop the waiting. It is conceivable that an extra step may sometimes be crucial and so we can conclude that in the absence of any type of non-deterministic choice, some other immediate 'not exists' operator might be called for.

There were other situations where non-determinism might have been useful. These include the scheduling of answers to requests by the NC. It would have been possible to to write something like

```
Y( req(P) ) => F patreq(P)
```

along with conditions to stop more than one request going out at once. Thus we could use METATEM's memory for obligations on the future to ensure that the *pat–req* is eventually made. This saves us cycling through the patients and may involve some use of a non-deterministic choice by the METATEM interpreter itself when it comes to try to satisfy several incompatible eventualities.

Another place where eventualities would have been useful is in arranging for a PC to answer a request. We discuss this in the next section.

12.8.4 *Synchronizing communications*

Many distributed systems rely to some extent on synchronous message passing—even in an asynchronous mileu. We need to supply the CMP language with some such facility. There is some discussion of this in [Fisher, 1993] where it is suggested that allowing environment-controlled predicates in the heads of rules might be a way of achieving this. If p is controlled by the environment, and the module finds in the head of a firing rule that it must make $p(X)$ true, for example, then the module must wait until it hears $p(a)$ from the environment for some a and immediately the module goes ahead with its step and broadcasts.

There seems to be several serious problems with this approach. One is that there is no mention of a means for letting the environment know that by broadcasting $p(a)$ it is synchronizing with the module. The environment may jump in with such a broadcast before the module has even realized that it needs to wait for a $p(X)$. This does not seem to reflect the usually cooperative aspect of synchronous message passing.

Another problem is that it is not clear what to do when the module is waiting for two different broadcasts at once but receives them at different times.

Still, it seems to be worthwhile investigating this approach further. Alternatives might include built-in metainstructions to wait and synchronize with a specific other module. A related communications question concerns the use of channels for message passing between specific modules as opposed to broadcast message passing. This is discussed in [Fisher, 1993] but it seems that there is much work still to be done on providing CMP with a realistic range of these facilities.

One use we could find for the method of synchronization described above is a little trick so that we do not need non-deterministic choice to implement PC-**Doris**. If we use

```
Y( patreq(doris) ) => temp(doris, T)
& blood(doris,B) & pulse(doris,R) & patval(doris, T,B,R)
```

to effect the answering of requests then, after a request, PC-**Doris** will wait for the very next broadcast of values before immediately sending off a *patval*. As Doris only broadcasts one set of values at a time, this will allow us to satisfy the problematical requirement that PC-**Doris** only send one answer to a request. Given the problems with this approach and the fact that the synchronization is only used here incidentally, we decided not to use this code.

A superficially similar approach to dealing with the request-answering problem is that implied in the following code:

```
Y( patreq(doris) ) =>
   F(   temp(doris, T)
      & blood(doris,B)
     & pulse(doris,R)
 & F( patval(doris,T,B,R)).
```

This is self-explanatory because it is very close to our specification. The code also has a straightforward procedural interpretation which, it will be noted, involves no use of synchronicity. However, it is not possible to write such code in the restricted syntax which we were using. Furthermore, if such code was possible to use we would have to be very careful about the bunching of messages to avoid running into the same old non-determinacy problems. This brings us to an examination of bunching.

12.8.5 *The message queue*

From the simple propositional implementation of the PMS, one of the main lessons to be learnt was that the existing algorithm for the bunching of messages in the message queue was unsuitable for our simulation purposes. Slower modules would tend to get left further and further behind dealing with messages from the distant past and leaving a longer and longer queue still to be dealt with. Here we will examine the alternative algorithms for message reading and see if we can suggest a version which will be generally applicable.

There is a whole range of possible message-bunching algorithms. At one extreme is the 'whole queue' algorithm which we assumed for developing the predicate version of an implementation: when a module broadcasts it is given the whole of its message queue to process in the next step. At the other extreme is a 'one at a time' bunching algorithm in which the module is only given the first message from the queue (i.e. the oldest) to deal with. In between are many other algorithms including the one used in the existing CMP implementation which reads up until the first repeated message or takes the whole queue if there are no repeated messages.

We argue that any algorithm apart from the 'whole queue' one suffers from the serious problem described above. Having slow modules being left behind is unrealistic and also seem to present a problem for the formal temporal semantics of CMP. One obvious problem is that the yesterday or last-time operator completely loses its usual semantics. Instead we suggest that the 'whole queue' algorithm is used and if it is desired that a module should work slowly through its message queue then the module program should say this explicitly.

Now let us consider the disadvantages of this choice and if there are ways around them. There seems to be two main reasons for wanting to use a different algorithm. They are:

- the inability to *count* the number of separate broadcasts of identical messages which arrive in one bunch; and
- the inability to discern the temporal *ordering* of any of the messages which arrive in one bunch.

It seems that both of these problems come closer to solution if at each broadcast we deliver to the module the whole queue in the form of a list of messages in order rather than as a set. The problem then becomes one of providing the module program with enough machinery to deal adequately with such a list.

One unsatisfactory way of doing this if so desired for a particular module, is to use meta-language facilities in order to lead the module through the queue as if the messages were arriving one at a time or in small packets defined in some other way. Thus can we simulate other bunching algorithms by using metalanguage. This is unsatisfactory because the metalanguage is in general much less efficient as a programming language and the formal semantics is also much more complex.

It is much better to provide the necessary machinery at the object level. This needs much more work but as the message list is just a piece of temporal information—the list in order of broadcasts between the last two module ticks—it must not be too difficult to express properties of the list within the bodies of rules. In fact, by using an appropriate predicate and assuming the natural numbers are in the object domain, it is even possible to count occurrences of the same message.

One suggestion is to step back from the very restricted syntax written as $\mathtt{Y}(B_1\& \ldots \& B_n)$ involving a (supposed) yesterday operator and allow a more semantically justifiable use of S in combination with a special 'tick' proposition.

12.8.6 *Control of predicates*

We have seen that the issue of control of predicates was very important in several aspects of our use of CMP. These included:

- the procedure for the operation of broadcast message passing;
- the question of implementability of specifications; and
- modular proofs of correctness of programs with respect to specifications.

While it is essential for the operation of the module that some information is declared in the syntax of the module for the message-passing procedure, we argue that there are also good reasons to include some explicit declaration of control for the purpose of correctness. There are cases in which the sort of information to be declared is slightly different for the two purposes.

A module can sometimes guarantee to satisfy a specification only if there are restrictions on the control of predicates. Thus in the interests of robustness and modular construction of distributed systems it is important that this information is declared explicitly. When such a framework is available a construction team can then implement a module and provide it with an interface definition adequate to ensure the satisfaction of the specification. When modules are put together to form a system some kind of syntactical correctness check can be made on the interfaces to ensure that they are all compatible.

Consider a simple example of a module which must guarantee that p is announced soon after the environment announces q and at no other time. For the purposes of message passing we just need to know that q is under the control of the environment and p is under the control of the module. However, for the purposes of modular correctness we also need to guarantee that p is under the control of no other module, i.e. only this module is allowed to announce p.

We also saw in Section 12.7 that in some cases, for the purposes of correctness of the overall system, we might want to declare general information about control of a predicate. For example, we might want to say that *pat–val* is only controlled by modules of type PC. This type of information more properly belongs to the system as a whole rather than a particular module. It certainly becomes important if the system of modules is treated as a module itself.

Some work needs to be done on the syntax for declaring all this type of information. This includes the cases where joint control of predicates can be more specifically defined in a syntactical way. We saw an example of this where *pat–val*(P_0, T, B, R) is only controlled by module PC-P_0.

12.9 Conclusion

On one level of understanding, the task of this case study was to use the METATEM framework to prototype a PMS in the hope of developing an overall specification for it and a specification of it as a distributed system. We obviously failed at this task although we did end up with an overall specification and the rudiments of a distributed one. If one does not count the very simplified propositional implementation then we were not able to present a working prototype of the system. Thus the formal specification which we ended up with—no matter how well argued for—suffers from being untested against those intuitive expectations which rely for satisfaction on a working demonstration.

We must be even less confident about the separate specifications of the various modules in the system. Even those parts which were able to be formalized were neither tested on a prototype nor even very well argued for.

Obviously the fault lies with the fact that METATEM and its formal semantical environment is only in the early stages of development. In fact, in this study we were able to identify many particular aspects of this framework which would need to be developed to allow us to do a better job of prototyping and suggest in what way the development might proceed.

Now on a more positive note, we saw very good reason throughout this study to assume that given those particular improvements identified, the whole process of specification, implementation and justification would have proceeded remarkably well. We saw how in contrast to those other approaches mentioned in Section 12.7, METATEM would provide a very comfortable unified framework for each of the stages in the process.

Furthermore, in Sections 12.6 and 12.8 we saw very good reasons why one would expect many of these improvements to be achieved without too much difficulty. None of them seemed to imply any great technical difficulty. Thus, in a few years, a similar case study should be able to present a very different, much more successful account of the development of the PMS.

And this brings us to what was, after all, the real purpose of this case study—to determine what aspects of METATEM need development and learn some lessons about METATEM by attempting to use it in a realistic prototyping situation. By this criterion this case study must be regarded as very successful. In Section 12.9 we have identified many particular aspects of the framework which need development and in Section 12.6 we made a start at sorting out some kind of detailed formal semantics for the framework.

Some of the problems identified in Section 12.11 seem very straightforward problems which must have been solved many times before in other contexts. These include supplying METATEM with a comfortable programming and observing environment, adding Prolog-like rigid predicates, and adding some kind of non-deterministic choice operator.

Some of the problems which seem likely to require a little more thought or policy decisions include allowing for synchronous communications, waiting for an environment input in other ways, handling the message queue and tidying up the interface definitions.

In the more distant future some thought will have to be applied in allowing for a hierarchy of modules and dynamic changes to the system. This study did not have much to say about these problems.

On the theoretical side we saw that it was very important for the purposes of proving correctness of our modules to have an adequate account of the formal temporal semantics of the behaviour of a METATEM module. We saw that this was not yet available. There is much work still to be done on developing such a semantics and this should be done in tandem with some of the other improvements suggested above.

In short, we have identified many directions in which work must be done but have seen that a very successful development is promised and does not seem too far away.

13

NON-MONOTONIC CODING OF THE DECLARATIVE PAST

13.1 Example of a temporal database

13.1.1 *The babysitter specification*

Mrs Smith is running a babysitter service. She has a list of reliable teenagers who can take on a babysitting job. A customer interested in a babysitter would call Mrs Smith and give the date on which the babysitter is needed. Mrs Smith then calls a teenager employee of hers and arranges the job. She may need to call several of her teenagers until she finds one who accepts. The customer pays Mrs Smith and the service pays the teenager. The rate is £10 per night unless the job requires overtime (after midnight) in which case it jumps to £15.

Mrs Smith uses a program to handle her business. The predicates involved are the following:

$A(x)$:	x Actually does a babysitting job,
$M(x)$:	x works overtime (More),
$C(x)$:	x is Called and asked to babysit,
$P(x,y)$:	x is Paid y pounds.

In this set-up, $A(x)$ and $M(x)$ are controlled mainly by the environment and $C(x)$ and $P(x,y)$ are controlled by the program. We get a temporal model by recording the history of what happens with the above predicates.

Mrs Smith laid out the following specifications:

1. babysitters are not allowed to take jobs three nights in a row, or two nights in a row if the first night involved overtime;
2. priority in calling is given to babysitters who were not called before as many times as others;
3. payment should be made the next day after a job is done.

Figure 13.1 is an example of a partial model of what has happened to the babysitter Janet. P is only written if it occurs. This model does not satisfy the specification, since Janet was not paid.

Mrs Smith would like to be able to write down the specification in an intuitive temporal language (or even English) and have it automatically transformed into an executable program, telling her what to do day by day.

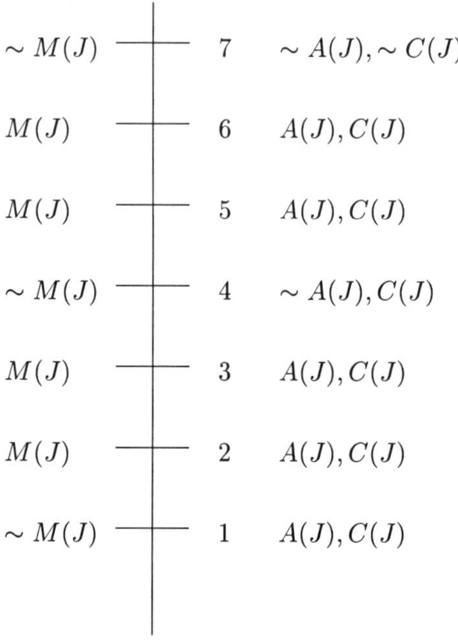

FIG. 13.1.

13.1.2 *Specifying the system*

We formally describe the specification required by Mrs Smith. We translate from the English in a natural way. This is important because we want our logical specification to be readable and have the same structure as in English.

Specification: (1) Babysitters are not allowed to take jobs three nights in a row, or two nights in a row if the first night involves overtime. Translation: (a) $G\forall x \sim [A(x) \wedge YA(x) \wedge YYA(x)]$; (b) $G\forall x \sim [A(x) \wedge YM(x)]$.

Specification: (2) Priority in calling is given to those who were not called before as many times as others. Translation: Let $More(A, B)$ be the connective 'A was true in the past more times than B' (see Example 10.26 and [Gabbay, 1989], where More was recursively defined using second-order fixed points; it cannot be defined in USF). More can obviously be easily implemented. The specification becomes: (c) $G \sim \exists x \exists y[More(C(x), C(y)) \wedge C(x) \wedge \sim C(y) \wedge \sim YM(y) \wedge \sim YA(y) \wedge YA(y)]$.

Specification: (3) Payment should be made the next day after the job was done. Translation: (d) $G\forall x[M(x) \rightarrow TP(x, 15)]$; (e) $G\forall x[A(x) \wedge \sim M(x) \rightarrow TP(x, 10)]$; (f) $G\forall x[\sim A(x) \rightarrow T \sim \exists y P(x, y)]$.

Besides the above specification we also have that babysitters work only when they are called, and only work overtime if they actually work. Translation: (g)

$G\forall x(A(x) \to C(x))$; (h) $G\forall x(M(x) \to A(x))$.

13.1.3 *Executing the specification*

We have to rewrite the above into an executable form, namely Past $\to$ Present $\vee$ Future. See Chapters 10 and 11.

Transformations:

(a') $G\forall x[YA(x) \wedge YYA(x) \to\sim A(x)]$

(b') $G\forall x[Y(A(x) \wedge M(x)) \to\sim A(x)]$

(c') $G\forall x\forall y[MT(C(x), C(y)) \wedge \sim YM(y) \wedge \sim Y(A(y) \wedge YA(y)) \to\sim C(x) \vee C(y)]$

(d') $G\forall x[\sim M(x) \vee TP(x, 15)]$

(e') $G\forall x[\sim A(x) \vee M(x) \vee TP(x, 10)]$

(f') $G\forall x[A(x) \vee T\forall y \sim P(x, y)]$

(g') $G\forall x[\sim A(x) \vee C(x)]$

(h') $G\forall x[\sim M(x) \vee A(x)]$.

Note that (d'), (e') and (f') can be rewritten in the following form using the Y operator:

(d'') $G\forall x[YM(x) \to P(x, 15)]$

(e'') $G\forall x[Y(A(x) \wedge \sim M(x)) \to P(x, 10)]$

(f'') $G\forall x[\sim YA(x) \to \forall y \sim P(x, y)]$.

Our executable sentences become

$(a*)$ $\text{IST}(Y(A(x) \wedge YA(x))) \to Exec(\sim A(x))$

$(b*)$ $\text{IST}(Y(A(x) \wedge M(x))) \to Exec(\sim A(x))$

$(c*)$ $\text{IST}(More(C(x), C(y)) \wedge \sim YM(y) \wedge \sim Y(A(y) \wedge YA(y))) \to Exec(\sim C(x) \vee C(y))$

$(d*)$ $Exec(\sim M(x) \vee Tp(x, 15))$

$(e*)$ $Exec(\sim A(x) \vee M(x) \vee Tp(x, 10))$

$(f*)$ $Exec(A(x) \vee T\forall y \sim P(x, y))$

$(g*)$ $Exec(\sim A(x) \vee C(x))$

$(h*)$ $Exec(A(x) \vee \sim M(x))$,

If we use (d''), (e''), (f'') the executable form will be

$(d**)$ $\text{IST}(YM(x)) \to Exec(P(x, 15))$

$(e**)$ $\text{IST}(Y(A(x) \wedge \sim M(x))) \to Exec(P(x, 10))$

$(f**)$ $\text{IST}(\sim YA(x)) \to Exec(\forall y \sim P(x, y))$.

IST stands for is-true, as in Section 13.3. Here, $\text{IST}(\phi)$ is an instruction to find out if ϕ is true, either by using the decoding rules on the database, or by directly querying the coded database. It returns the value true if ϕ is true, and false otherwise. We will discuss its implementation below.

In practice there is no difference whether we use $(e**)$ or $(e*)$. We execute TP by sending P to tomorrow for execution. If the specification is $(e**)$, we send nothing to tomorrow but we will find out tomorrow that we have to execute P.

We have used the babysitting agency as an example of how to execute a temporal formula. In the next section we study the possibility of coding or compressing the information we keep on the past history. Then in Section 13.3 below we will return to the babysitter example in order to illustrate the method.

13.2 Querying and coding the past

13.2.1 *Introduction*

Coding the past is essentially compressing a database. We want to record a temporal model. Rather than simply put into our database the truth or falsehood of every atomic statement at every point in time, we might want to condense or code the data in some way. In this section we give two ways of doing this. We show how to recover the full temporal model from the coded data. We then discuss how to evaluate queries directly on the coded data, rather than first reconstructing the temporal model. Why do we want to code the past at all? Well, this depends on the application area. It may be in order to save time and space. When a book is borrowed from a library, the librarian may make up a card saying so, destroying it when the book is returned. One finds out whether a book is on loan by seeing if there is a card for it. This is a simple but useful coding of the past. Without it, the librarian would have to record the whole temporal model, which could mean writing out a new card each day for every book in the library, saying whether it is on loan or not. There are other considerations in coding the past. If you are designing an evolving system where you know what kinds of questions you are going to ask, then you can write a direct imperative program (based on the relationship between temporal connectives and finite automata) to give you the answers you want at any time. If you don't know what is going to be asked you can still do this (see Section 13.3), or you may want to record everything. However, in application areas where the majority of the data is rules, you are better off with the deductive inferential theory—the coding. The examples we give are quite simple and we may not need to code the past but may use instead a full record of what happened. Indeed, in practical database systems some coding may be done but not the extensive logical coding we do here to illustrate the method. But it is always good to have the ability to code logically, should we want to.

13.2.2 *Coding the past using persistence*

Imagine that we are giving a course of lectures (Fig. 13.2). John started attending the lectures only in lecture 4 and at that lecture he sat next to Mary. If we record the past properly, as a mathematical model, we have to record fully, for every lecture, who attended and who sat next to whom. This may be expensive. In case of attendance, there is some measure of persistence and we can therefore just record the beginnings and ends of periods of attendance. Thus we may record only the following three items. We ignore the attendance details about Mary in this discussion. $R(x, y)$ means that x sat next to y.

Lecture 4	$A(J), R(J, M)$
Lecture 3	
Lecture 2	
Lecture 1	
Lecture 0	$\sim A(J)$

FIG. 13.2.

We are coding the past. We record only part of the temporal model in our database, but we do so in such a way that we can always recover the whole model.

How does our coding work? How do we find out whether John attended at time 3? We pick out the most recent entry in the database (here, $\sim A(J)$ at time 0), and decide on the answer. Since we only have entries when the situation changed, the last (most recent) entry persists and we decide that John did not attend lecture 3. We can express this informally and roughly with the connective S: John attended at time t if 'S (there is an entry in the database saying that John attended, the database has no entry about John's attendance)' is true at t. We will do this more formally in Section 13.3.

We can further economize on the entries by realizing that if we record $R(J, M)$ (John sat next to Mary) then certainly John attended. Thus we do not need the entry $A(J)$ at time 4.

Our database can thus be encoded, using the predicate logic-like approach, as $\sim A^*(0, J), R^*(4, J, M)$. We have the following three rules valid at all times:

1. ϕ is true at time t if there is an entry in the coded database of the form $\phi^*(t)$ (for ϕ equals $A(x)$ or $R(x, y)$; x, y can be any students);
2. $A(x)$ is true at time t if there is an entry in the coded database of the form $R^*(t, x, y)$ or of the form $R^*(t, y, x)$ (i.e. x attended if it is recorded that x sat next to someone);
3. $A(x)$ is true at time t if there is an entry in the database of the form $A^*(s, x), s < t$, and there is no entry in the database of the form $\sim A^*(k, x)$ for $s < kEt$.

We expect to be able to state our rules in some suitable logic, perhaps predicate or temporal logic or a mixed logic. We first reduce the value $\text{IST}(T, t, A)$ to the atomic level. The computation at the atomic level is sent to the second database D. D is a non-monotonic coded database, using rules 1–3 and reading the coded entries. We thus have a metabase in the sense of Section 13.6 below.

Rule 3 describes persistence. Note that it is important to apply rule 2 before applying rule 3, and to stop if it succeeds. At time 4, we first apply rule 2 to add to the database $A^*(4, J)$ and then apply rule 3. If we applied rule 3 first we would (wrongly) get that John did not attend at time 4. Applying rule 2 would then give a contradiction. We are thus in the realm of non-monotonic defeasible reasoning. To describe what we are doing we need to talk about entries, rules and their priorities, and about putting entries in and out of databases. We need a special language for that, which is capable of metalanguage features. Such a language could be the language of hereditarily finite predicates (HFPs) of Volume 1 or any other suitable language. In Section 13.3 we give a method of dealing with this using the ordinary connectives.

We are taking the database here as a given 'black box'. We apply our rules to its output, in order to decide whether John attended at a particular time. All that we ask of the black box is that it can answer certain queries in such a way that our rules then give the correct answers. But in practice, inside the box there may be a whole new system. Perhaps it is a straightforward coded database, as we have imagined. Or maybe it itself contains another coded database together with its own internal decoding rules, which it uses to decide whether there is a certain entry when asked to do so by 1–3 above. Here we are concatenating the decoding process. Or it may be a real-time system which, when asked if John attended at time 2, prints a note asking the lecturer. The lecturer may have highly non-monotonic rules for deciding whether John was there, such as 'he was there if I remember him being there, or if he signed the attendance sheet, unless his signature is fuzzy (in which case I suspect that a friend of his signed the sheet for him)'. (What if I remember him but his signature is fuzzy?) In these more complex cases the idea of the temporal model seems less central, as we are really dealing with an involved system of rules without obvious semantics.

13.2.3 *Coding the past by recording changes*

We have analysed coding of the past using persistence. The opposite situation arises where the changes are due to impulse. Such an example is the 'Block World'. The domain D consists of wooden blocks which are placed at various points. The relations say what block is positioned where. Changes are made in time by taking one block and putting it somewhere else. More formally, the state of affairs at time n is represented by a first order model M_n. Each model is the domain D together with extensions for the relations of the language. We pass from M_n to M_{n+1} by changing (inverting) a small number of relations from holding to non-holding, and vice versa. The situation calculus is an example of impulse coding.

The important point is that to describe M_n, it is most natural to describe the sequence of (impulse) changes done to M_0. So the natural representation of the past is in terms of changes or events, rather than codings of temporal patterns. The block world is such an example. Changes are made by taking one block and putting it somewhere else. So we are making at most two changes at any one time.

Let us try this for our attendance example. In persistence coding we recorded, roughly speaking, the end points of intervals. So if John attends first at time 4, we record $\sim A(J)$ at 0, $A(J)$ at 4, and infer $\sim A(J)$ at 1, 2, 3 by persistence.

What do we do with impulse coding? We record $\sim A(J)$ at time 0; we must record all of M_0, or agree on a default convention. Then we record something like 'Change(A, J)!' at time 4. We still use persistence to infer $\sim A(J)$ at 2, 3—the whole idea of impulse coding relies on persistence. The difference is that we do not record A or $\sim A$, but only that A has changed its value. The advantage is a saving of one bit of space, the equivalent of 1 bit. The disadvantage is that we must refer all the way back to time 0 to get the actual meaning. This means that there is a problem with updating. Suppose that John attends at time 5. How do we enter this? In persistence coding there are two options:

1. just write $A(J)$ at 5; and
2. observe that 1 is not needed, as the most recent entry was $A(J)$ (at time 4), so we need do nothing.

But in impulse coding, we must refer back to all prior times, to know what to do. The context must always be considered: option 1 is not available. A consequence is that an impulse-coded database is always maximally compressed. No information that it holds is ever redundant, and there is a one–one correspondence between databases and models. (This does not necessarily mean that impulse coding is the most efficient possible way of coding the model as, perhaps, some of the data could still be replaced by decoding rules.) In practice a mixture of persistence and impulse coding can be used.

The frame problem, as we understand it, arises when we have a past history represented as a list of changes to the original model at time 0. It is a computational problem involving changing databases. Suppose we know that a database started initially as D_0 and has undergone a series of discrete changes leading to D_n. Suppose we know that the database satisfies some integrity constraints and hence that these changes probably affected other parts of the database. The frame problem is to find a systematic way of checking, for a given entry in D_0, whether it is affected in D_n.

13.2.4 *Query evaluation directly on the coded past*

We have seen the following schematic dependence in Fig. 13.3:

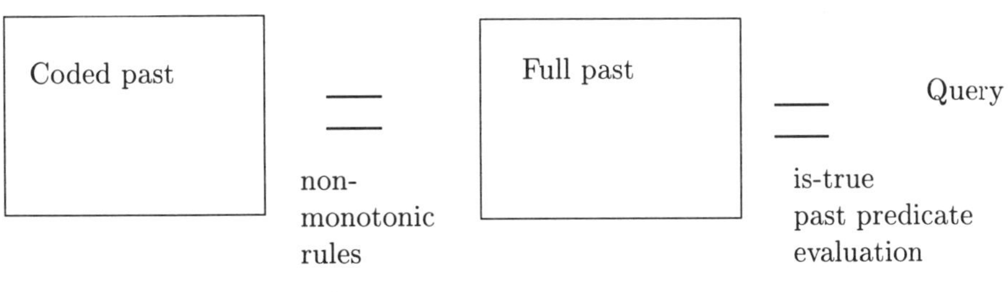

FIG. 13.3.

Using the coded past we first build up using (in general) non-monotonic reasoning the full past database, which we then use to evaluate IST(ϕ) for any ϕ.

Let us make a mental jump! Since the value of IST(ϕ) ultimately depends on the coded past, we can write a program evaluating IST(ϕ) directly on the coded entries! This can certainly be done. There may be different ways of coding the past, different ways of using non-monotonic reasoning on the codes and different ways of writing a programming language for evaluating IST(ϕ). Another such programming language is event calculus. Whatever we do, we are in the area of deductive inferential databases. We consider one approach in the Appendix.

Which coding should we choose, given that we want to evaluate a certain class of queries? The impulse representation is natural for answering metaquestions about what has been done to the model. It is not as natural as persistence for answering queries about relational extensions. When we ask 'Did John attend at time 4?', we really have to compute M_4. So we apply the impulse changes to M_0 to get M_1, apply the next batch of changes to get M_2, and so on, until we get the model M4. In the case of persistence the representation (i.e. coding) is already in terms of truth extensions and we do not usually have to look so far back in time as M_0. Thus, when we are not combining it with non-monotonic rules, the persistence case does not have a frame problem by its very nature, because its natural representation is already extensional.

13.3 Coding the past in the babysitter example

13.3.1 *The choice of coding*

The first step is to decide how to code the past in the example. We could use the following logical relationships:

$$\begin{array}{ll} M(x) & \text{if } TP(x, 15) \\ A(x) & \text{if } TP(x, 10) \\ A(x) & \text{if } M(x) \\ C(x) & \text{if } A(x). \end{array}$$

However, from the practical point of view it is better not to rely on the pay predicate for coding as it is easy to have non-payment occur, contrary to specification. The other predicates $C(x)$, $M(x)$ and $A(x)$ seem less likely to go wrong in practice. Consider for example the following history; see Fig. 13.4.

The data comes in groups of chains of events taking one to three days. We can have any of the following seven figures for these chains of events, for days $(n-2)$, $(n-1)$ and n; see Fig. 13.5.

In cases (a), (b) and (c) the babysitter John was not allowed to work on day n. In cases (d) and (f) John did not work on day n because he refused a call and in cases (e) and (g) he was not called at all on day n.

Any history of events can be decomposed into a sequence of grouped events of the above form, (a) ... (g). For our example in Fig. 13.4, we have the sequence of Fig. 13.6.

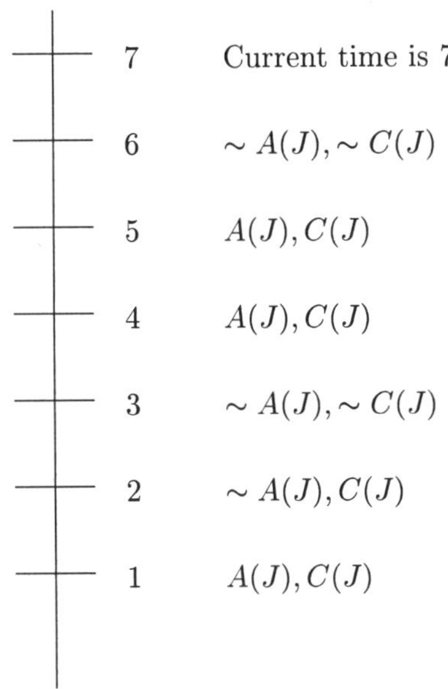

FIG. 13.4.

It is sufficient for our coding history to be able to code each of these sequences, (a) ... (g). Suppose we agree to record $C(J)$ only when the call was turned down (i.e. $\sim A(J)$). Also we record $\sim C(J)$ only when the call was not allowed on account of $[YA(J)$ and $YYA(J)]$ or $YM(J)$—the former if we record $A(J)$ at $n-1$, and the latter if not. In addition to that, we can record $A(J)$ whenever necessary and allow it to supplement the information not yet coded. Thus the cases (a) ... (g) are coded as in Fig. 13.7, so that, for example, sequence (a) is coded by entering $\sim C(J)$ at n and $A(J)$ at $n-1$.

The model of Fig. 13.4 thus becomes the table of entries of Fig. 13.8 below.

Remark 13.1. Notice that if the current time is n, we may have to record explicitly the full information of items $(n-1)$ and $(n-2)$. Imagine a situation as in the diagrams (a) or (c) of Fig. 13.9.

When time $n-1$ becomes history, i.e. when the current time moves on to be n so that we now know whether we had $M(J)$ at $n-1$ or not, we can code the diagrams (a) and (c) as in Fig. 13.10. (Note that (a′) has $n-1$ whilst (c′) has $n-2$.) Until then, we must code the information for time $n-2$ as it is.

We could further code the model by distinguishing between cases (a) and (c). For example, we know there is payment to be made. Thus we can code the situation in Fig. 13.11.

This is not such a good idea, because in practice we may not pay (even though we have to!) and so this coding is not compatible with non-monotonic principles

$$
\begin{array}{llll}
\text{(a):} & n & - & \sim A(J) \\
 & & - & \\
 & n-1 & - & A(J) \\
 & & - & \\
 & n-2 & - & A(J) \\
\text{(b):} & n & - & \sim A(J) \\
 & & - & \\
 & n-1 & - & M(J) \\
\text{(c):} & n & - & \sim A(J) \\
 & & - & \\
 & n-1 & - & M(J) \\
 & & - & \\
 & n-2 & - & A(J) \\
\text{(d):} & n & - & \sim A(J), C(J) \\
\text{(e):} & n & - & \sim A(J) \\
 & & - & \\
 & n-1 & - & A(J) \\
\text{(f):} & n & - & \sim A(J), C(J) \\
 & & - & \\
 & n-1 & - & A(J) \\
\text{(g):} & n & - & \sim A(J) \\
 & . & - & . \\
 & . & - & . \\
 & . & - & . \\
 & m & - & \sim A(J)
\end{array}
$$

FIG. 13.5.

$$
\begin{array}{ll}
- & \text{(c)} \\
- & \\
- & \text{(g)} \\
- & \\
- & \text{(f)}
\end{array}
$$

FIG. 13.6.

of reading from partial information.

(a):	n	— $\sim A(J)$	$\sim C(J)$
	$n-1$	— $A(J)$	$A(J)$
	$n-2$	— $A(J)$	
(b):	n	— $\sim A(J)$	$\sim C(J)$
	$n-1$	— $M(J)$	
(c):	n	— $\sim A(J)$	$\sim C(J)$
	$n-1$	— $M(J)$	
	$n-2$	— $A(J)$	$A(J)$
(d):	n	— $\sim A(J), C(J)$	$C(J)$
(e):	n	— $\sim A(J)$	$\sim A(J)$
	$n-1$	— $A(J)$	$A(J)$
(f):	n	— $\sim A(J), C(J)$	$C(J)$
	$n-1$	— $A(J)$	$A(J)$
(g):	n	— $\sim A(J)$	will be
	.	— .	incorporated
	.	— .	as a
	.	— .	default
	m	— $\sim A(J)$	

FIG. 13.7.

6: $\sim C(J)$
4: $A(J)$
2: $C(J)$
1: $A(J)$

FIG. 13.8.

13.3.2 *Recovering the model from the coding*

Given a table of entries, we take the following steps to build the full model. We assume that the entries are a coding of a real model and therefore we do not get into an inconsistency situation. Consider the following updating rules:

$$\begin{array}{lcl} n-1 & — & A(J), \sim M(J) \\ & — & \\ n-2 & — & A(J) \end{array} \qquad \begin{array}{lcl} n-1 & — & M(J) \\ & — & \\ n-2 & — & A(J) \end{array}$$

(a) (c)

FIG. 13.9.

$$\begin{array}{lcl} n & — & \sim C(J) \\ & — & \\ n-1 & — & A(J) \end{array} \qquad \begin{array}{lcl} n & — & \sim C(J) \\ & — & \\ n-2 & — & A(J) \end{array}$$

(a') (c')

FIG. 13.10.

(R0) Sequences (a), (b), (c) If '$n :\sim C(J)$' is in the table, add the entry '$n :\sim A(J)$'.

(R1) Sequence (a) (no overtime) If the following hold, '$n :\sim C(J)$' is in the table and '$(n-1) : A(J)$' is in the table then add the entry: '$(n-2) : A(J)$'.

(R2) Sequence (b), (c) (overtime) If the following hold, '$n :\sim C(J)$' is in the table and there is no entry '$(n-1) : A(J)$' in the table, then add the entries '$n :\sim M(J)$', '$(n-1) : M(J)$', '$(n-1) : A(J)$'.

(R3) Sequences (d), (f) If the entry '$n : C(J)$' is in the table, then add '$n :\sim A(J)$'.

These rules (in particular (R1) and (R2)) must be applied in the order given. Their effect is to fill in a block completely for A, and if M occurs in a block, to fill in the details, making M true at $n-1$ and false at n. If M does not occur, it will be made false by persistence ((R5) below). Having exhausted the use of rules (R0) ... (R3) we get a new table. We continue to apply the rules below to the new table in the order they are given.

(R4) Persistence of $A(J)$

(a) If '$\exists k < n$['$k : A(J)$' is an entry and $\forall m[k < m < n \to$ there is no entry of the form '$m :\sim A(J)$']], then add '$n : A(J)$'.

(b) If $\exists k < n$['$k :\sim A(J)$' is an entry and $\forall m[k < m < n \to$ there is no entry of the form '$m : A(J)$']] or $\forall k < n$, [neither '$k : A(J)$' nor '$k :\sim A(J)$' is an entry], then add '$n :\sim A(J)$'.

Note that as we do (R4a) first, (R4b) can be replaced by the simpler:

(b') if '$n : A(J)$' is not yet an entry, add '$n :\sim A(J)$'.

In fact, as (R0)–(R3) have already filled in all positive As, we could even use (b') only, without the persistence rule (R4a) at all. This would also allow deletion of (R0). There are similar possibilities for M. But we keep persistence for illustration.

(R5) Persistence of $M(J)$. Similar to (R4).

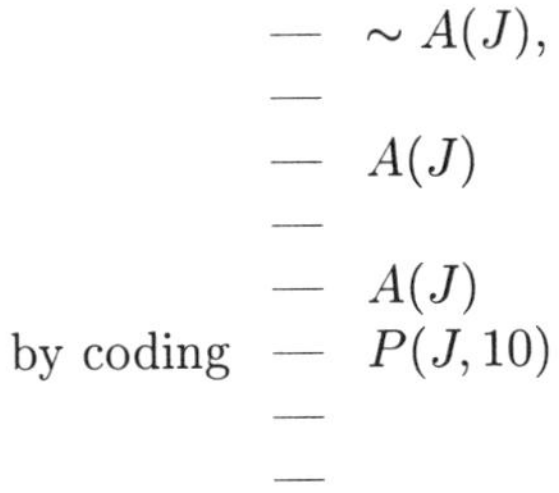

— $\sim A(J)$,
—
— $A(J)$
—
— $A(J)$
by coding — $P(J, 10)$
—
—

FIG. 13.11.

(R6) J works $\rightarrow$ J is called. If either '$n : A(J)$' or '$n : M(J)$' is in the table, then add '$n : C(J)$'.

(R7) Payment. If $(n+1)$ is still in the past ($n+1 <$ current time) then:
   (a) If '$n : M(J)$' is an entry, add '$(n+1) : P(J, 15)$'.
   (b) If '$n : A(J)$' is an entry but not '$n : M(J)$', add '$(n+1) : P(J, 10)$'.

(R8) If you cannot add '$C(n, J)$' by other rules then add '$\sim C(n, J)$'.

Let us use the above rules to construct the history in Fig. 13.4 from the table in Fig. 13.8. Assume the current time is 7. The table is shown in Fig. 13.12

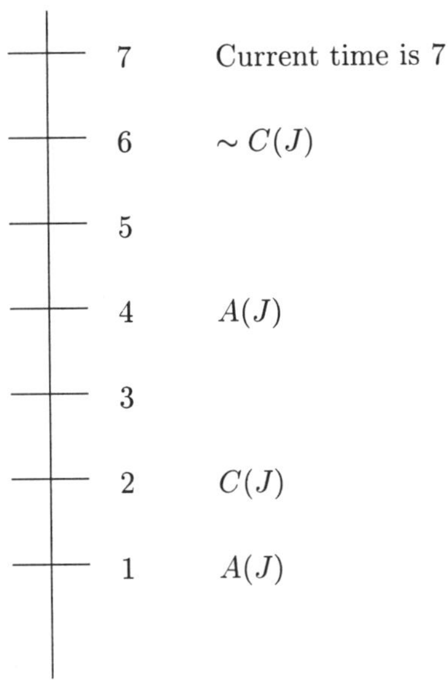

FIG. 13.12.

By following rules (R0) and (R2) we add '$6 :\sim A(J), \sim M(J)$' and '$5 : A(J), M(J)$'. By applying rule (R3) we add '$2 :\sim A(J)$' to the table. Thus we get the table in Fig. 13.13.

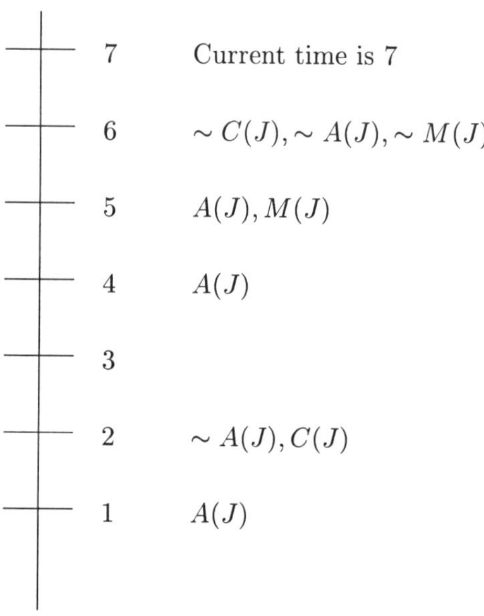

FIG. 13.13.

Now continuing the application of the rules (R4)–(R8) to the table of Fig. 13.13 will give us the full model.

Recall that we want to compute the queries of the form $?n : A(J)$ directly on the original table. We have rules (R0)–(R7) at our disposal. However, we have to apply them in a certain order. Take for example the query $?2 : A(J)$. If we apply (R4) first, on the table of Fig. 13.12, the answer will be 'yes'. However, if we apply rule (R3) first, then the answer should be 'no'.

13.3.3 *Updating the coded database*

Let us now discuss the problem of updating. At any current time n, we can have the following cases:

(u1) we call John and he accepts and either works overtime or not;

(u2) we do not call John either because we cannot or because we did not want to;

(u3) we call John and he does not accept.

The cases are illustrated as follows:

- Time n: execution possibilities
  Possibility 1: $A(J), \sim M(J)$.
  Possibility 2: $A(J), M(J)$.

Possibility 3: $\sim C(J)$, either because worked $M(J)$ yesterday or worked $A(J)$ the past two days.
Possibility 4: $\sim C(J), \sim A(J)$, for no special reason.
Possibility 5: $\sim A(J), C(J)$.

The updating will be different in each case.

- Time n: updating possibilities

Possibility 1: Add to the table of entries the entries '$n : A(J)$' and '$n :\sim M(J)$'.
Possibility 2: Add '$n : M(J)$'.
Possibility 3:
(a) If the entry '$(n-1) : M(J)$' is in the table then delete it and add '$n :\sim C(J)$'.
(b) If '$(n-1) : A(J)$' and '$(n-1) :\sim M(J)$' and '$(n-2) : A(J)$' are in the table then delete '$(n-1) :\sim M(J)$' and '$(n-2) : A(J)$' and add '$n :\sim C(J)$'.
Possibility 4: If the latest (i.e. the greatest $m < n$) entry (time-wise) is '$m :\sim A(J)$' or '$m :\sim C(J)$' $m < n$, then do nothing. If the latest entry is '$m : A(J)$' or '$m : M(J)$' or '$m : C(J)$' then add '$n :\sim A(J)$'.
Possibility 5: Add '$n : C(J)$'.

13.3.4 *The program*

We are now ready to describe an outline of a program to handle the babysitter problem. The general form of such a program is as follows:

Module 1: the coded database up to the current time. In our case it is Fig. 13.12.

Module 2: specification rules, in executable form.

Module 3: the program to evaluate whether an IST statement is true at the database. This is possibly a defeasible non-monotonic system (following Nute's terminology (see [Nute, 1986]).

Module 4: the set of executable commitments currently (time n) to be executed. This set is inherited from time $n-1$.

Module 5: the updating module containing a program to update all five other modules.

Module 6: auxiliary counters, among them a clock for the current time. These counters can carry special values from time $n-1$ to time n that help evaluate the IST predicate. The nature of this module changes with any new specification. In fact, for a given specification one can make do only with this module if enough information is put in. In this case only queries that arise from the specification can be answered.

Modules 1 and 2 have been discussed already.

13.3.4.1 *Module 3: a program for a defeasible shell* We have to decide how to represent the data and rules. The best language for that is N-Prolog, being a language which is its own metalanguage; however, since Prolog is widely available, we will use Prolog with some metalanguage features.

We record the table of entries as '$A(n,J)$' for '$n : A(J)$' and 'Neg$A(n,J)$' for '$n :\sim A(J)$'. The metapredicate table will record the table of entries. Thus we have *Table*$(A(n,J))$ and *Table*$(NegA(m,J))$ among the items in our Prolog database. We also have rules, $R_1,\ldots,R_m,R_{(m+1)},\ldots,R_{m'}$. In our example there were seven rules, and the first three had to be used first. If we give a Prolog formulation to the rules, there may be a total of m' rules, and the first m of them are to be used first. This is a simple system of non-monotonic defeat. In the general case the priorities among rules and data can be made more complex.

Given a proof in Prolog of an entry, e.g. $A(k,J)$, there may also be a proof of *NegA*(k,J). The proof we accept is the one which uses the first m rules first. Since our system of priorities in the babysitter example is very simple, we can implement it more efficiently using metapredicates. We do not need a fully fledged defeasible shell. Let *Table* be a predicate describing the data in the table of entries. We first apply the first m rules to obtain *Table*1 and only then apply the rest of the rules to *Table*1. We thus begin with the meta rules:

(R meta 1): *Table*1(X) if *Table*(X).

We now code rules (R0) to (R3).

(R0): *Table*1(*NegA*(n,J)) if *Table*(*NegC*(n,J)).

(R1): *Table*1$(A(n-2,J))$ if *Table*(*NegC*(n,J)) $\wedge$ *Table*$(A(n-1,J))$

(R2.1): *Table*1(*NegM*(n,J)) if *Table*(*NegC*(n,J)) $\wedge \neg$ *Table*$(A(n-1,J))$.
Here $\neg$ is Prolog negation by failure.

(R2.2): *Table*1$(M(n-1,J))$ if *Table*(*NegC*(n,J)) $\wedge \neg$ *Table*$(A(n-1,J))$.

(R2.3): *Table*1(*NegA*$(n-1,J)$) if *Table*(*NegC*(n,J)) $\wedge \neg$ *Table*$(A(n-1,J))$.

(R3): *Table*1(*NegA*(n,J)) if *Table*$(C(n,J))$.

(R4.1): $A(n,J)$ if *Table*1$(A(k,J)) \wedge k \leq n \wedge \neg[k \leq m \wedge m \leq n \wedge$ *Table*1(*NegA*(m,J))], where $\neg$ is negation by failure. Note that because m is not instantiated we have here the meaning of $\forall m \sim [\,]$.

(R4.2): *NegA*(n,J) if *Table*1(*NegA*(k,J)) $\wedge\, k \leq n \wedge \neg[m \leq n \wedge k \leq m \wedge$ *Table*1 $(A(m,J))]$.

(R5): Similar to (R4.1) and (R4.2) for 'M' replacing 'A'.

(R6): $C(n,J)$ if $A(n,J) \vee M(n,J)$.

(R7):

(a) $P(n,J,15)$ if $n <$ current time $\wedge\ M(n-1,J)$

(b) $P(n,J,10)$ if $n <$ current time $\wedge\ A(n-1,J) \wedge \neg M(n-1,J)$.

(R8): *NegC*(n,J) if $\neg C(n,J)$. Here $\neg$ is negation by failure and n,J must be instantiated under C.

We need the following additional meta rule:

(Rmeta 2): X if *Table*1(X).

Note that the rules are correct for the current time n provided that at the time $n-1$ (just before the current time) and time $n-2$ we make sure we put the complete information without coding in the table of entries. Thus if the current time is n and at $n-1, M(n-1, J)$ occurred, we also put in the data $A(n-1, J)$. At time $n+1$ we can put $\sim C(n, J)$ in the data and delete both $A(n-1, J)$ and $M(n, J)$. But we cannot do that before time ticks on to be $n+1$. We could improve our coding and put at time n only $M(n-1, J)$, but for such coding we have to add the additional rule $A(n, J)$ if $[n+1 = \text{Current time}] \wedge M(n, J)$. We choose not to do so, because in practice one finds out about overtime $(M(J))$ after one has assigned $(C(J) \wedge A(J))$ a babysitter. So first $A(n, J)$ goes in the data and then $M(n, J)$, and obviously it is not practical to delete $A(n, J)$ and use a rule to retrieve it just for today, when the next day (tomorrow) both $M(n, J)$ and $A(n, J)$ should be deleted anyway!

We have thus defined module 3 for our problem and we are now ready to execute the specification.

13.3.4.2 *Module 4: the set of executable commitments* These are obtained from the future parts of the specification. That is, if our specification has the form $\text{IST}(C) \rightarrow Exec(B)$ then the pure future parts of B give rise to commitments to the future, as in $Exec(U(A, B))$ iff $TExec(A) \vee TExec(B \wedge U(A, B))$. In our particular example the future parts are of the form $TP(J, 10)$ or of the form $TP(J, 15)$. Thus the elements of module 4 are 'pay' statements, to be executed immediately. The meaning of $Exec(P(x, y))$ is determined by the system. It can send a command to another program which finds the address of x and posts x a cheque y.

13.3.4.3 *Module 5: updating* We are now ready to deal with the updating module. We must make sure it is activated only after the execution module does its job. The way we update depends on what was executed. Thus we must record what was executed. We cannot update as part of the execution process because what we update depends on the totality of what was executed, and so we have to wait. The best way of handling it is first to record (separately) entries of what was executed, and then to use this new table of entries to compute the updates. We thus amend the general rule for execution to be $Exec*(atom) \leftrightarrow Exec(atom) \wedge ExecRecord(atom)$, where *atom* can be either an atomic sentence or its negation. *Exec* depends on the system and *Exec Record* puts an entry in a table call *Record*. The updating rules are given below. They correspond to the various possibilities of what was recorded. Earlier in this section we presented two tables, one for the execution possibilities and one for the updating possibilities. We need to state in an imperative language what to do for each possibility. For example, if possibility 1 occurs, namely the execution at time n was $A(J)$ and $\sim M(J)$, then *Record*$(A(n, J))$ and *Record*$(NegM(n, J))$ will be true and the update is (according to the updating possibilities table for time n): add to the database (i.e. module 1) the items *Table*$(A(n, J))$ and *Table*$(NegM(n, J))$.

This can be programmed imperatively. In fact, its nature is imperative. If we insist on doing it in Prolog, we will have to cheat and write something like update if *Record*($A(n,J)$) $\wedge$ *Record*($NegM(n,J)$) and Cut $\wedge$ *AddTable*($A(n,J)$) $\wedge$ *AddTable*($NegM(n,J)$). *Add* and Cut are Prolog commands; we write a rule like the above for each possibility. This is in the tradition of Prolog doing imperative commands. A common example is: a if Print b.

We write the updating rules imperatively:

(1) *AddTable*($A(n,J)$) and *Table*($NegM(n,J)$) if *Record*($A(n,J)$) and *Record*($NegM(n,J)$).

(2) *AddTable*($M(n,J)$) if *Record*($A(n,J)$) and *Record*($M(n,J)$).

(3a) Delete *Table*($M(n-1,J)$) and *AddTable*($NegC(n,J)$) if *Table*($M(n-1,J)$) and *Record*($NegC(n,J)$).

(3b) Delete *Table*($NegM(n-1,J)$) and Delete *Table*($A(n-2,J)$) and *AddTable*($NegC(n,J)$) if *Record*($NegC(n,J)$) and *Table*($A(n-1,J)$) and *Table*($NegM(n-1,J)$) and *Table*($A(n-2,J)$).

(4) *AddTable*($NegA(n,J)$) if *Record*($NegC(n,J)$) and *Record*($NegA(n,J)$) and the latest entry in the table is of the form *Table*($A(m,J)$) or *Table*($M(m,J)$).

(5) *AddTable*($C(n,J)$) if *Record*($NegA(n,J)$) and *Record*($C(n,J)$).

Note that only after all the updating has been done, including the preparation of module 3, do we then set the current time from n to $n+1$. The resetting of the time must also be executed.

13.3.4.4 *Module 6: auxiliary counters* We now come to deal with module 6. This contains counters of interest to us. If we think sensibly about the babysitter example and analyse what we really need to make it work, we find that at any time we need to know the following:

1. who we can call and in what order;
2. who we shall pay.

To that purpose we need only record the full information about the past two days and record how many times each individual has been called up to now. The latter will give us the priority who to call now. We will know who to pay from last day's information. If we do only that, we will be able to run our system to specification. We will not be able to answer an arbitrary query at time n, unless we use the other modules.

In fact, commercial databases probably keep full information about some recent days together with some useful parameters (module 5) and keep the distant past as a full (uncoded) database on disk in storage. In view of the simplicity and cost-effectiveness of this approach one may ask why adopt any different point of view? The answer to that lies in reasoning from partial information and change. The coded table of entries can be a table of partial information and our entire process (except possibly module 5) still applies. Further, we can change our specification as we go along to handle unforseen evolution of the system. We would like to stress that what is important is what is practical and works. We should keep an open mind in order to adopt the best method for the problem in hand.

14

A LOGICAL VIEW OF TEMPORAL DATABASES

Traditional approaches to temporal databases are either not formal or deal only with static features, considering mainly a single state of the database. The temporal nature of data in temporal databases, however, requires a dynamic treatment of several features which are absent from non-temporal systems.

In this chapter and the next one, a temporal logic framework is proposed to cope simultaneously with both the static and the dynamic aspects of temporal data. The framework consists of a two-dimensional temporal model. One-dimensional temporal models deal with static aspects such as temporal querying and data representation, and are presented in this chapter. The second temporal dimension is introduced as a natural extension that captures the dynamics of temporal updates, and is described in the next chapter. The resulting two-dimensional temporal model is used formally to characterize the differences between valid-time and transaction-time, to characterize the notion of 'now' in a temporal database, and to give a formal model for bitemporal databases.

14.1 Introduction

Dealing with temporal data is an activity that inherits the intrinsic complexity of modelling, querying and manipulating data in general. It also contains the extra difficulties of dealing explicitly with time. This extra hardship is not neglectable and deserves special attention.

It may seem trivial to add a temporal dimension to data simply by adding a time stamp to it. But immediately there appear new issues, absent from non-temporal data, which have to be addressed. For example, it is not clear what that time stamp is intended to represent. Is it the time when the data was true in the universe of discourse, or is it the time when the system learned about that data? Traditionally, the former has been called the valid time of data and the latter the transaction time of data. But if it is one or the other, how does that affect the manipulation of the time-stamped data?

There are other problems which were once purely metaphysical but in the context of temporal data need to have their practical consequences analysed. Does the nature of time affect the modelling and manipulation of temporal data? If time is considered a discrete set of time points or a dense set of temporal intervals, can we 'say' the same things? Can we represent all the things that we want to express about time? At what cost? All these questions bring up the need for a formal, precise discussion of the concepts involved.

The area of temporal databases has been addressing those problems for more than a decade. In this chapter we propose to treat these exclusively temporal issues under a unique framework, namely under a temporal logic perspective. Logic has always been present in relational databases as an underlying framework; the notion of a relation can be seen as an interpretation of a predicate in a logical (finite) model, and the relational calculus is a logical query language. When dealing with temporal databases, it is natural to extend the underlying framework to temporal logic. Unfortunately, apart from a few exceptions, temporal database presentations have tended to hide this underlying temporal logic basis. We hope to make it more explicit here.

What is particularly new about our approach is that we put the emphasis on the *dynamics of data*, differing from the traditional approach that focuses on the *statics* of a single state of the database. We intend to show that the stress put into the dynamics of data is fundamental to capture the evolving nature of temporal data.

The static aspects are not despised or relegated. Quite the opposite, we show how a temporal logic approach can be developed to deal with both aspects of data. That is done by presenting the several *temporal dimensions* that a temporal model may have. The static aspects, such as querying and data representation, are dealt with a single temporal dimension. The latter is indeed a non-issue in non-temporal databases, for it is obvious how to represent data; in temporal databases, however, there are uncountably many possible databases and only a countable set may ever be represented in a computer. So temporal data representation becomes an issue that requires special attention, to be dealt with one-dimensional temporal logics.

The second temporal dimension naturally appears when we try to describe some dynamic aspects. Suppose there is a one-dimensional temporal database that stores either only the transaction-time of data or only the valid-time of data, but we do not know which. There is no query that can be posed to the database such that, according to its answer, it is possible to tell whether it is a valid-time or a transaction-time database. Since queries reveal only the static aspect of data, this indicates that valid time versus transaction time is an implicitly dynamic issue. Indeed, the two-dimensional description of the database has several uses:

- It is used formally to present the evolution, i.e. the dynamic aspects, of a temporal database. We show that the two-dimensional aspects have to be taken into consideration even when a single (valid-time or transaction-time) temporal dimension is stored by the database.
- It is used to characterize the distinctions between valid-time and transaction-time in a formal, temporal logic framework. Such characterization is done via an axiomatization over the two-dimensional model.
- It is used to clarify the notion of 'now' in a temporal database. 'Now' is the place in time where the *database observer* is situated when accessing the database by querying or updating. It is shown how this notion can be captured as the diagonal of the two-dimensional model, and some properties of it are studied.

- Finally, it is shown how the two-dimensional model may be used as the formal basis for *bitemporal databases*, wherein both valid-time and transaction-time dimensions are stored.

The contents here can be perceived differently depending on the reader's background. For the logician, this chapter can be seen as a case of 'applied logic', where a temporal logic theory is applied to describe a temporal database. For the computer scientist, the database practitioner or anyone else interested in temporal information, this chapter provides a unified formal description of the principles underlying temporal databases.

The discussion in the following two chapters is organized in the following way. Sections 14.2, 14.3 and 14.4 deal with the static aspects. Section 14.2 presents the one-dimensional temporal finite model as a temporal database, defining a query language for it; Section 14.3 studies several possible data representations and their expressivity; Section 14.4 analyses the issue of safety with regards to temporal queries. Chapter 15, Section 15.1 starts to deal with the dynamic aspects, by introducing the two-dimensional model and using it to describe the evolution of a temporal database. Section 15.2 shows how to characterize the distinction between valid-time and transaction-time databases by an axiomatization over a two-dimensional model. Finally, Section 15.3 shows another application of the two-dimensional model, describing databases that store both valid-time and transaction-time.

14.2 One-dimensional temporal databases

We give first a formal presentation of one-dimensional temporal databases in a temporal logic perspective. A one-dimensional temporal database is one that associates a single time stamp with each database fact (also tuple, or atom). A time stamp is a set of time points; this notion will be further refined when we discuss data representation.

We start by analysing the language and semantics of first-order temporal logic, FOTL, over a finite signature. Finiteness is an important property to be taken into account in the database presentation, for databases are supposed to be finite repositories of information. That will motivate the definition of temporal queries as the restricted class of *safe* FOTL-formulae.

14.2.1 *The temporal query language*

We consider the signature of a two-sorted first-order language, without function symbols but including equality, where one sort is *domain* and one sort is *time*. A *database signature* (or, in accordance with the database terminology, a *database schema*) is a pair $\Sigma = (\mathcal{C}, \mathcal{P})$, where $\mathcal{C}$ is a countable set of constant symbols and $\mathcal{P}$ is a set of predicate symbols (or, in the database context, relation names), such that each predicate symbol is associated with an arity $r \geq 1$; this is the unnamed approach to describe the schema. In the named approach, each predicate name is associated with a finite set of *attribute names*, the size of which is the arity of the predicate. With the help of an order $<_{att}$ on the set of attribute names, both approaches are equivalent. The schema is finite if $\mathcal{P}$ is finite, and we only consider finite schemas here.

Let $(T, <)$ be a *flow of time*, where T is a set of time points and $<$ is a binary order over T. The objects we are dealing with are two sorted: the sort of the elements of $\mathcal{C}$ is domain and the sort of elements of T is time. Given an n-ary predicate $R \in \mathcal{P}$, each position (also attribute, or column) k, $1 \leq k \leq n$, is associated with a sort, $sort_k(R) \in \{\text{domain}, \text{time}\}$. If a position is associated with the time sort, this position is said to record *user-defined time*. User-defined time receives no special treatment from the system, and for all purposes is dealt with in the same manner as domain elements. The domain sort may be subdivided in several types, but this is not relevant to our discussion and we do not cover typed domains.

Let V_d and V_t be countably infinite sets of domain and time variables, respectively, and let V be the disjoint union of V_d and V_t; a *term* is either a constant or a variable; functional symbols are normally absent from databases. The *atomic formulae* of our language are of the form $a_1 = a_2$, for a_1 and a_2 of the same sort, and $R(a_1, \ldots, a_n)$, where R is a predicate symbol of arity $arity(R) = n$ and a_is are terms such that the sort of a_i is $sort_i(R)$.

As usual with temporal language presentations, there is always a choice between the *pure first-order approach* and the *temporal connectives* approach:

- In the pure first-order approach, each predicate of the schema is extended with an extra time attribute, say the last one, and the language is extended with a binary predicate, $<$, whose interpretation is fixed to be the order of the flow of time; $<$ is frequently an infinite relation and therefore not like any other database predicate.
- In the connectives approach, the language is extended with temporal connectives. We consider here the binary connective S (Since) and U (Until). Other unary connectives can be derived in terms of those, namely F (sometimes in the future), G (always in the future), P (sometimes in the past) and H (always in the past); over discrete flows of time, it is also possible to derive $\bigcirc$ (next time) and $\bullet$ (previous time).

$$\begin{array}{lll} FA =_{def} U(A, true) & GA =_{def} \neg F \neg A & \bigcirc A =_{def} U(false, A) \\ PA =_{def} S(A, true) & HA =_{def} \neg P \neg A & \bullet A =_{def} S(false, A). \end{array}$$

The language is also extended with the unary predicate $time(t)$, where the time points can be seen as part of the database domain (alternatively, we could have a two-sorted approach where t in $time(t)$ is of sort time).

Both approaches can be easily shown to be equivalent. It has been shown in [Abiteboul *et al.*, 1996] that, without the unary predicate $time(t)$, the first-order approach is strictly more expressive than the connective approach; in fact, it has been shown that a query of the form

Are there two distinct moments at which a relation (or predicate) has exactly the same values?

is impossible to express in the connective approach without the predicate $time(t)$, but can be done in the two-sorted first-order approach.

The insertion of $time(t)$ is very useful for practical uses of temporal databases and it is even traditional in other connective-based approaches to temporal query languages [Gabbay and McBrien, 1991]. Since there is no loss of expressivity in the connective-cum-$time(t)$ approach, it is our opinion that this approach provides a clearer separation between the temporal and non-temporal parts of the language; besides, temporal statements can be more concisely expressed using the connectives. So we choose the connective approach to develop the query language.

We consider the existential quantifier, $\exists$, as primitive and syntactically define the universal quantifier as $\forall \equiv \neg\exists\neg$; for reasons that will become clear when we discuss the safeness of queries, we will consider both $\wedge$ and $\vee$ Boolean connectives as primitive. The set of free variables of a formula A is represented by $free(A)$. The set of wffs of the language of FOTL over the schema $(\mathcal{C}, \mathcal{P})$ is defined as:

- every atomic formula is in it, the set of variables occurring in an atomic formula are all free;
- if A, B are in it, so are $\neg A$, $A \wedge B$, and $A \vee B$, with free variables $free(A)$, $free(A) \cup free(B)$ and $free(A) \cup free(B)$, respectively;
- if A, B are in it, so are $U(A, B)$ and $S(A, B)$, with free variables $free(A) \cup free(B)$;
- if A is in it, so is $\exists x A$, with free variables $free(A) - \{x\}$.

We write $A(x_1, \ldots, x_m)$ to indicate that $x_1, \ldots, x_m$ are all the free variables of A. A formula is *ground* if it contains no variables nor quantifiers. A subformula (or, for that matter, any symbol) occurs positively in a formula if it occurs within the scope of an even number of $\neg$ symbols; it occurs negatively if it occurs within the scope of an odd number of $\neg$ symbols.

A *temporal query* based on formula A is an expression of the form

$$Q_A = \{x_1, \ldots, x_n \mid A(x_1, \ldots, x_n)\},$$

such that A is an FOTL formula with only free variables $x_1, \ldots, x_n$.

Example 14.1. Consider the following finite schema $\Sigma = (\mathcal{C}, \mathcal{P})$:

$\mathcal{C} = \{\text{strings of characters}\}$
$\mathcal{P} = \{employee\}$

where $employee(Name, Salary, Department, Birthdate)$ is a four-place predicate symbol. *Birthdate* is a used-defined time attribute. The following are wffs over this schema:

$employee(\text{Peter}, \text{2K}, \text{Marketing}, 17/9/1962)$;
$\bullet\, \exists u\, employee(x, \text{2K}, \text{Marketing}, u) \wedge \neg(x = \text{Peter})$;
$P\exists y \exists z\, employee(x, y, z, u) \;\wedge\; \neg\exists y_1 \exists z_1\, employee(x, y_1, z_1, u)$;
$P\, employee(x, y, \text{R\&D}, u) \;\vee\; employee(x, y, \text{R\&D}, u)$.

The first formula is ground and atomic, x is free in the second, x and u in the third one, and x, y, u are free in the last one. In the third formula, the predicate symbol *employee* occurs both positively and negatively.

It is important to mention that a standardized language to query temporal databases has recently been generated [Snodgrass, 1995]. As would be expected, such a language is an extension of the standard query language SQL of traditional databases. It has also been shown that there is a fragment of TSQL2 that has equivalent querying power to temporal logic [Bölen *et al.*, 1996]. This fragment does not contain the aggregation functions that are usual in database query languages, and it also imposes that a tuple in a temporal relation cannot be associated to two different time stamps (the time stamps have to be *coalesced* in a single one); the queries also have to be *local*, a restriction that limits the occurrence or explicit references to time in the query.

If this is respected the temporal logic can be used as an equivalent query language, and a provably correct translation is proposed in [Bölen *et al.*, 1996]. The temporal logic involved is based on the connective approach but does not contain an explicit predicate $time(t)$. If $time(t)$ were added to the temporal logic, it is possible that the restrictions of queries to local ones could be relaxed, so that a larger fragment of TSQL2 could be expressively equivalent to the query language of connectives-cum-$time(t)$.

14.2.2 *Temporal database instances*

A ***first-order finite structure*** over the schema $(\mathcal{C}, \mathcal{P})$ is a pair (D, I), where D is a countably infinite set, called the *domain*, and I is a *finite interpretation* consisting of an interpretation of constants, $I(c) \in D$, and a finite interpretation of predicate symbols, $I(R) \subseteq_{finite} D^n$, where n is the arity of R.

The flow of time $(T, <)$ that underlies the database is assumed to be linear with no end points, i.e. the following conditions apply for every $t, t', t'' \in T$:

- *Transitivity*: if $t < t'$ and $t' < t''$ then $t < t''$.
- *Irreflexivity*: it is never the case that $t < t$.
- *Totality*: $t = t'$ or $t < t'$ or $t' < t$.
- *Unboundedness*: for every $t \in T$ there exists t_1 and t_2 such that $t_1 < t < t_2$.

We later examine further conditions such as *discreteness* and *denseness* which, with the help of some extra conditions, will help us obtain flows of time that are isomorphic, respectively, to the integers and to the rationals.

A *temporal database instance* (or just a *temporal database*) for schema $\Sigma = (\mathcal{C}, \mathcal{P})$ over the flow $(T, <)$ is a temporal structure $\mathcal{DB} = (T, <, g)$, where g is a *temporal valuation* that associates every time point $t \in T$ to a first-order finite structure $g(t) = (D_t, I_t)$. In the spirit of conventional databases, it is further required that $\mathcal{DB}$ respects the additional conditions of *constant domains* and *rigid constants*, i.e. for every $t, t' \in T$ and $c \in \mathcal{C}$ it must always be the case that

$$D_t = D_{t'} = D \text{ and}$$
$$I_t(c) = I_{t'}(c)$$

where the set D is called the *domain of the database*; only the interpretation of predicate symbols is *flexible*, i.e. may vary from one time point to another. The symbol '$=$' is not a database predicate, for it is an infinite relation and it has a rigid interpretation over time. It is usual and very convenient for databases to consider $D = \mathcal{C}$, i.e. the domain of interpretation is the set of constant symbols such that I is the identity over constants. Unless otherwise stated, we will assume such a simplification, known as the *Herbrand interpretation*.

The *active domain* with respect to a predicate symbol $R \in \mathcal{P}$ is the set $adom(R)$ of all the domain and time elements occurring in $I_t(R)$, for some t. The active domain of a formula A, $adom(A)$, is the union of the active domains of all the predicate symbols occurring in A and the set of constant symbols in A. The active domain of the database is $adom(\mathcal{DB})$ the union of the active domains for all predicate symbols. A temporal database $\mathcal{DB}$ is *domain finite* if $adom(\mathcal{DB})$ is finite. As a final restriction on a temporal database as a temporal model, we constrain all databases to be domain finite. Domain finiteness is one step in the direction of allowing infinite information to be stored in the database, for the temporal part of the information is allowed to be infinite. Such a restriction will be respected throughout the chapter; for a generic extension to finitely representable infinite databases, see [Kanelakis *et al.*, 1990].

The definition of a temporal database above has followed a temporal and modal logic tradition in its presentation. Another view of a database, certainly more in line with what the database community defines as a database, is the following. A (one-dimensional) *temporal tuple* over a relation $R \in \mathcal{P}$ with $arity(R) = n$ is a time-stamped n-tuple $\tau : \langle a_1, \ldots, a_n \rangle$ such that $\tau \subseteq T$ is the time stamp and $a_i \in \mathcal{C}$, $sort(a_i) = sort_i(R)$, for $1 \leq i \leq n$; $\langle a_1, \ldots, a_n \rangle$ is called an *atom* or a non-temporal tuple; the *truth value* of an atom is *true* at t if $t \in \tau$, otherwise it is *false*. A *temporal relation* or *temporal relation instance* r over R is a finite set of temporal tuples over R, but each tuple time stamp τ may be an infinite set. A temporal database $\mathcal{DB}$ is a set of temporal relations over all relation names in the schema. Note that in this perspective the database domain is automatically constant and identical to $\mathcal{C}$. Furthermore, the finite number of temporal tuples per relation guarantees the finiteness of the active domain. In fact, both perspectives can be clearly seen as equivalent, by making

$$\tau : \langle a_1, \ldots, a_n \rangle \in r \text{ iff } \langle a_1, \ldots, a_n \rangle \in I_t(R), \text{ for all } t \in \tau.$$

Both perspectives will be used interchangeably in the chapter.

For the semantics of FOTL-formulae in a database, let a *(rigid) variable assignment* v be a mapping that associates every domain variable $x \in V_d$ to a domain value $v(x) \in D$ and every time variable $u \in V_d$ to a time point $t \in T$; the assignment is time independent, so the variables are treated as rigid elements with respect to time. An assignment v' is an *x-variant* of an assignment v if they agree on the values of all variables in $V = V_d \cup V_t$ except, possibly, on the value of x.

Consider a temporal database instance $\mathcal{DB} = (T, <, g)$ and an assignment v. To define the semantics of the formulae of FOTL, it is convenient to extend the assignment over all terms by making $v(c) = I_t(c)$ for $c \in \mathcal{C} = D$. We define a formula A to be true in $\mathcal{DB}$ at time t under assignment v, writing $\mathcal{DB}, v, t \models A$, by induction on the structure of formulae.

$\mathcal{DB}, v, t \models R(a_1, \ldots, a_r)$	iff $\langle v(a_1), \ldots, v(a_r)\rangle \in I_t(R)$.
$\mathcal{DB}, v, t \models a_1 = a_2$	iff $v(a_1) = v(a_2)$.
$\mathcal{DB}, v, t \models time(u)$	iff $v(u) = t$.
$\mathcal{DB}, v, t \models \neg A$	iff it is not the case that $\mathcal{DB}, v, t \models A$.
$\mathcal{DB}, v, t \models A \wedge B$	iff $\mathcal{DB}, v, t \models A$ and $\mathcal{DB}, v, t \models B$.
$\mathcal{DB}, v, t \models A \vee B$	iff $\mathcal{DB}, v, t \models A$ or $\mathcal{DB}, v, t \models B$.
$\mathcal{DB}, v, t \models S(A, B)$	iff there exists a $t' \in T$ with $t' < t$ and $\mathcal{DB}, v, t' \models A$ and for every $t'' \in T$, whenever $t' < t'' < t$ then $\mathcal{DB}, v, t'' \models B$.
$\mathcal{DB}, v, t \models U(A, B)$	iff there exists a $t' \in T$ with $t < t'$ and $\mathcal{DB}, v, t' \models A$ and for every $t'' \in T$, whenever $t < t'' < t'$ then $\mathcal{DB}, v, t'' \models B$.
$\mathcal{DB}, v, t \models \exists x A$	iff there exists an assignment v' x-variant of v such that $\mathcal{DB}, v', t \models A$.

The answer of a query $Q_A = \{x_1, \ldots, x_n \mid A(x_1, \ldots, x_n)\}$ is the temporal relation it generates in the database,

$$Q_A(\mathcal{DB}) = \{\tau : \langle v(x_1), \ldots, v(x_n)\rangle \mid \mathcal{DB}, v, t \models A(x_1, \ldots, x_n) \text{ and } t \in \tau\}.$$

By insisting that the result of a query be a finite temporal relation, it turns out that not all FOTL-formulae are acceptable as queries, so we have to restrict queries to *safe* ones. This topic will be discussed in detail in Section 14.4. But first we have to determine how a temporal database is to be represented. We next study temporal data representations over the flows $\mathbb{Z}$ and $\mathbb{Q}$ and discuss their *data expressivity*.

Example 14.2. Consider a database instance $\mathcal{DB}$ for a schema containing a three-place relation name *employee(Name, Salary, Dept)*. We take the integer-like flow of time to be that of months, $\{\ldots, Jan92, Feb92, Mar92, \ldots\}$. Let *employee* be given by

$$\begin{array}{l}[Jan90, Dec92] : \langle \text{Peter 1K R\&D}\rangle \\ [Jan93, Apr93] : \langle \text{Peter 2K Marketing}\rangle \\ [Jan90, Dec92] : \langle \text{Paul 1K R\&D}\rangle \\ [Sep91, Apr93] : \langle \text{Mary 3K Finance}\rangle.\end{array}$$

If we consider a variable assignment v such that $v(x) = Peter$ and $v(y) = 1K$, then

$$\mathcal{DB}, v, Apr93 \not\models P\exists y \exists z\, employee(x, y, z) \;\wedge\; \neg\exists y_1 \exists z_1\, employee(x, y_1, z_1)$$

for Peter is currently an employee; furthermore, the query asking about past and present employees in the R&D department and their salary, posed at $Jan93$, is

$$\{x, y \mid (P\, employee(x, y, R\&D) \;\vee\; employee(x, y, R\&D)) \wedge time(Jan93)\}$$

and generates the following temporal relation:

$[Jan90, Jan93]$: ⟨Peter 1K⟩
$[Jan90, Dec92]$: ⟨Paul 1K⟩.

14.3 Temporal data representation

In non-temporal databases, the issue of data representation never arises because it is an obvious one. But as pointed out by [Kabanza *et al.*, 1990], there are uncountably many possible temporal databases and we are therefore limited to finitely representing just a few among those. In [Baudinet *et al.*, 1991] a formal distinction was made between temporal data and temporal query expressivity, the former being our concern in this section.

With the view that a temporal relation is a finite set of time-stamped tuples, the issue of data representation consists of deciding how to represent the set of time points that can be used as a tuple time stamp. By taking a time stamp to be a *set* of time points, we bypass most of the discussion in the temporal database literature of whether time stamps should be represented by points, intervals or temporal elements (to be defined below) [Tansel *et al.*, 1993], relegating that discussion to the physical representation of a relation in a temporal database system.

In general, given a linear flow of time $(T, <)$, a time stamp can be viewed as a set defined by $\{t \in T \mid \phi(t)\}$, where ϕ is a first-order formula in a language that contains, at least, a countable subset of the elements of T as constants, the special constant symbols $-\infty$ and $+\infty$, and the predicate symbols $=$ and $<$ ($\leq$ is the usual abbreviation). We examine next two forms of data representation, leading to *temporally bounded databases* and *periodic databases*.

14.3.1 *Temporally bounded databases*

Definition 14.3. A temporal database $\mathcal{DB} = (T, <, g)$ is *temporally bounded* if there are time points t_{max} and t_{min} such that:

(a) whenever $t < t_{min}$ then $g(t) = g(t_{min})$; and

(b) whenever $t > t_{max}$ then $g(t) = g(t_{max})$.

In other words, all the non-temporal tuples have their truth value 'persisting' to the past before t_{min} and to the future after t_{max} in temporally bounded databases. In the database of Example 14.2, all time stamps are temporal elements (actually, simple intervals), and note that $t_{min} = Dec89$ and $t_{max} = May93$. Temporal boundedness is a global property of the database. We show how temporal boundedness can be obtained through a constraint on the time stamps.

Let $\tau = \{t \in T \mid \phi(t)\}$ be a time stamp; τ is a *temporal element* if $\phi(t)$ is a formula with single free variable t which is a disjunction of formulae of the form $t = t_0$ and $t_1 < t < t_2$, with $t_0 \in T$ and $t_1, t_2 \in T \cup \{\pm\infty\}$. Clearly, temporal elements can be defined by the union of finitely many intervals, namely the intervals corresponding to the disjuncts in $\phi(t)$. The original definition of temporal elements over a discrete flow of time [Gadia and Vaishnav, 1985] presupposes a canonical format for temporal elements, namely a finite sequence of pairs, $[t_1, u_1], \ldots, [t_m, u_m]$, such that $t_i \leq u_i$ and $u_i < t_{i+1}$, $1 \leq i \leq m$. Clearly, every temporal element can be transformed into an equivalent canonical temporal element, and this transformation can be done in polynomial time in the number of intervals.

Proposition 14.4. *Let $\mathcal{DB} = (T, <, g)$ be a temporal database over a linear flow of time. If all time stamps in $\mathcal{DB}$ are temporal elements then $\mathcal{DB}$ is temporally bounded.*

Proof Let t_{min} be the minimal time point in T occurring in all the finitely many time stamps in $\mathcal{DB}$, and let t_{max} be the maximal one. So g cannot change its value outside the interval $[t_{min}, t_{max}]$ and $\mathcal{DB}$ is temporally bounded. □

The converse is true for $\mathbb{Z}$, i.e. all temporally bounded databases over $(\mathbb{Z}, <)$ can be represented by temporal elements. In fact, for $\mathbb{Z}$, there may be only finitely many intervals between t_{min} and t_{max}, all of which are representable by temporal elements. This fails to be true for other linear flows of time, such as $\mathbb{Q}$, over which there are temporally bounded databases that are not representable by temporal elements for the following reasons:

- for *dense* flows of time, there may be infinitely many intervals between t_{min} and t_{max}, which are not representable by temporal elements;
- in flows of time containing *gaps*, an interval may end or start in a gap, but since the gap does not belong to the flow of time, such an interval is not representable by temporal elements.

We say that temporal elements *completely represent* temporally bounded databases over $\mathbb{Z}$, but are an *incomplete representations* over flows that are dense or contain gaps. Complete data representation is also called the *data expressiveness* of a temporal database representation formalism in [Baudinet *et al.*, 1991].

The practical importance of temporal boundedness lies in the fact that the absolute majority of proposed temporal database systems and implementation prototypes described in the literature are in this class; for samples and pointers refer to [McKenzie and Snodgrass, 1991; Tansel *et al.*, 1993; Loucopoulos *et al.*, 1990]. The database of Example 14.2 is temporally bounded.

14.3.2 *Periodic databases*

Temporally bounded databases exclude the presence of atoms that occur repeatedly often in time, e.g. 'every Wednesday'. This is a kind of information that recurs infinitely often in the flow of time, either to the future or to the past, or both. However, the only kind of infinite information allowed by temporal elements is 'always into the future' or 'always into the past'. To extend data expressivity, the notion of a *periodic temporal database* is introduced. Only the integer flow of time will be considered for this discussion.

Definition 14.5. A temporal database $\mathcal{DB} = (\mathbb{Z}, <, g)$ is *ultimately periodic* if there are time points t_{max} and t_{min} and a non-null, positive $\pi \in \mathbb{Z}$ such that:

(a) whenever $t < t_{min}$ then $g(t) = g(t - \pi)$; and

(b) whenever $t > t_{max}$ then $g(t) = g(t + \pi)$.

We show how ultimately periodic databases are related to Pressburger definable temporal tuples. A Pressburger formula is a first-order formula over the language consisting of constant symbols 0, unary function symbol *succ*, binary functional symbol $+$, and binary predicate symbols $=$ and $<$. Pressburger formulae are standardly interpreted over $\mathbb{Z}$. A time stamp $\tau = \{t \in \mathbb{Z} \mid \phi(t)\}$ is *Pressburger definable* if $\phi(t)$ is a Pressburger formula with a single free variable t. A temporal tuple is Pressburger definable if its time stamp is.

Proposition 14.6. *Ultimately periodic temporal databases are completely defined by Pressburger definable temporal tuples.*

Proof We know that a unary Pressburger formula $\phi(t)$ is equivalent, by quantifier elimination, to a Boolean combination (say, in disjunctive normal form) of formulae of one of the four basic forms [Enderton, 1972; Boolos and Jeffrey, 1989]

$$t = c \quad t < c \quad c < t \quad t \equiv_m c$$

where the language was extended with binary predicates $<$ and countably many $\equiv_m$ representing equality modulo m, $m \geq 2$, and $c \in \mathbb{Z}$. The first three components are temporal element formulae, and therefore are used to determine $[t_{min}, t_{max}]$. The last one is responsible for the periodic nature of the data. It follows that every database with Pressburger definable time stamps is ultimately periodic.

To prove that every ultimately periodic database with period π is representable with Pressburger definable tuples, consider the database over the interval $[t_{max} + 1, +\infty)$. Every atom $p(a_1, \ldots, a_{arity(p)}) \in g(t')$ for $t' \in [t_{max} + 1, t_{max} + \pi]$ can be represented with Pressburger definable time stamp $(t > t_{max}) \wedge ((t - t_{max}) \equiv_\pi t')$. Similarly for the interval $(-\infty, t_{min} - 1]$. The interval $[t_{min}, t_{max}]$ is finite and the database over that period can be represented with temporal elements, which are Pressburger formulae. □

This result is similar to one obtained by Chomicki and Imieliński [Chomicki and Imieliński, 1988], in which it is proved that temporal deductive queries in language $\mathrm{Datalog}_{1S}$ over finite temporal data over $\mathbb{N}$ have data expressivity of ultimately periodic sets. Another way of defining time stamps for periodic databases, as in [Kabanza *et al.*, 1990], relies on *linear repeating points* (*lrps*), i.e. a set of the form

$$\{a * n + b \mid n \in \mathbb{Z}\} \quad (a, b \in \mathbb{Z})$$

as values for the temporal attributes of temporal relations together with linear constraints over these attributes. [Kabanza *et al.*, 1990] shows that, under certain conditions, the temporal tuples over lrps are Pressburger definable.

Periodic databases are usually found in connection with temporal deductive databases, where the power of recursion is coupled with temporal expressivity [Baudinet *et al.*, 1993; Abadi and Manna, 1989; Chomicki and Imieliński, 1988].

Example 14.7. Consider a periodic database with a binary relation *special(TypeOfPayment, Value)* to represent special payments periodically. An end-of-year bonus would then be represented as

$$(t \equiv_{12} Dec90) : \langle \text{Bonus},\ 50\text{K} \rangle.$$

This example is, admittedly, a gross simplification of 'real' periodic data. In most cases the period of repetition is not totally regular; for instance, if we want to express that 'the last Friday of a month is a pay day' this is an irregularly periodic data. To express such a property (and many others) in a clear and concise way, it is necessary for the system to support *calendar* information; this, however, remains outside the scope of this chapter and we will not discuss calendars here.

14.4 Safe queries

The fact that we want the result of temporal queries to be *finite* temporal relations forces us to restrict the format of the formulae that are acceptable as queries.

A *domain-independent* formula $A(x_1, \ldots, x_n)$ is one whose interpretation generates a finite n-place relation containing only domain elements in its active domain, i.e. for every time t the set of tuples of domain elements $\langle v(x_1), \ldots, v(x_n)\rangle$ such that $\mathcal{DB}, v, t \models A(x_1, \ldots, x_n)$ is finite and contains only elements that are in $adom(A)$. Clearly, $adom(A)$ is finite, so an acceptable query should ideally be a domain-independent formula, but unfortunately it is an undecidable problem to tell whether a formula is domain independent [Fagin, 1982; Ullman, 1988]. Thus we syntactically define the class of *safe* formulae below as an alternative sufficient condition to obtain domain independence. The basic idea behind safeness is that of 'limiting' all free variables that appear in disjunctions, negations and temporal subformulae.

For disjunctions $A \vee B$, it is simply required that A and B share the same free variables, for if we have the non-safe formula $A(x) \vee B(y)$ such that $A(x)$ holds for some domain element $x_0 \in D$, there are infinitely many pairs (x_0, d), $d \in D$, satisfying the query.

For negations, the idea of limiting a variable is similar to that of *range-restricted* clauses in logic programming [Lloyd, 1987]. Basically, all free variables inside a negation are required to occur positively outside the negation. For example, the formula $\neg A(x,y)$ will not be safe, but in $\neg A(x,y) \wedge B(x) \wedge C(y)$ the free variables of negated A are limited owing to their positive occurrence in B and C.

An extra safeness problem occurs with temporal formulae over discrete flows of time: the temporal formula $S(A(x), B(y))$ will not be safe, because if $A(x)$ is true at the previous moment for some $x_0 \in D$, there are infinitely many pairs (x_0, d), $d \in D$, satisfying the formula. The formula $S(A(x) \wedge B(y), B(y))$[16], however, does not present that problem and it is considered safe; that is how the semantics of the S operator is defined in the U, S-based temporal algebra in [Gabbay and McBrien, 1991] in terms of the semantics we present here. In general, for temporal formulae of the form $S(A, B)$ and $U(A, B)$ to be safe it is required that:

(a) free variables that are limited inside A are also considered limited in $S(A, B)$ and $U(A, B)$;

(b) all free variables occurring in B have to be limited outside B.

The variable y is not limited outside $B(y)$ in the non-safe formula $S(A(x), B(y))$, but it is limited in safe formulae such as $S(A(x) \wedge B(y), B(y))$ and $S(A(x), B(y)) \wedge C(y)$.

We follow Ullman's [1988] formal presentation of safe formulae for standard non-temporal databases; further discussions on safeness can be found in [Zaniolo, 1986; Ramakrishnan *et al.*, 1987; Abiteboul *et al.*, 1995]. But before we present the formal definition, we make just a small remark: a subformula X is a maximal conjunction in a formula A if X is a subformula that is not part of a conjunction. For example, in the formula $(\neg(A \wedge B \wedge C) \vee S(A, B)) \wedge E$ there are six maximal conjunction subformulae, namely $A \wedge B \wedge C$, $\neg(A \wedge B \wedge C)$, $S(A, B)$, A, B and the whole formula.

Definition 14.8. Safe queries A formula is safe when:

1. If it contains a subformula that is the disjunction of B_1 and B_2, then B_1 and B_2 share the same free variables, i.e. the subformula is of the form $B_1(x_1, \ldots, x_n) \vee B_2(x_1, \ldots, x_n)$.
2. If it contains a subformula that is a maximal conjunction $B_1 \wedge \cdots \wedge B_m$, then all the free variables appearing in the B_i must be limited in the following sense:
    - a variable is limited if it occurs in some B_i, where B_i is not an equality nor is it negated nor temporal;
    - if B_i is of the form $x = c$ or $c = x$, where c is a constant, then x is limited;

[16]This is equivalent to defining the semantics of $S(A, B)$ as

$$\mathcal{DB}, v, t \models S(A, B) \text{ iff } \exists s < t \text{ and } \mathcal{DB}, v, s \models A \text{ and } \forall s',\ s \leq s' < t \text{ implies } \mathcal{DB}, v, s' \models B$$

and similarly for $U(A, B)$.

- if B_i is of the form $x = y$ or $y = x$, where y is a limited variable, then x is limited;
- if B_i is of the form $S(A, C)$ or $U(A, C)$, then, recursively, all free variables that are limited in A are limited in B_i.

3. It contains a negated subformula of the form $\neg B(x_1, \ldots, x_k)$ only in the terms of (b), i.e. $\neg B(x_1, \ldots, x_k)$ must be part of a conjunction or temporal formula such that all x_i are limited.
4. It contains a subformula of the form $S(A, B)$ or $U(A, B)$ only in the terms of (b), i.e. $S(A, B)$ or $U(A, B)$ must be part of a maximal conjunction such that all the free variables are limited (this rule applies only to temporal databases over discrete flows of time).

A *(safe) query* Q is then represented by $Q_A = \{x_1, \ldots, x_m \mid A(x_1, \ldots, x_m)\}$, where $A(x_1, \ldots, x_m)$ is a safe formula with free variables $x_1, \ldots, x_m$.

Unless otherwise mentioned, all queries will be assumed to be safe. We can also define a *snapshot query*

$$Q_{A_t} = \{t : x_1, \ldots, x_m \mid A(x_1, \ldots, x_m)\}$$

as the temporal query $Q_{A_t} = \{x_1, \ldots, x_m \mid time(t) \wedge A(x_1, \ldots, x_m)\}$. The resulting *snapshot relation* is $\{t : \langle v(x_1), \ldots, v(x_m)\rangle \mid \mathcal{DB}, v, t \models A(x_1, \ldots, x_m)\}$. Each tuple in the generated relation is said to *satisfy* the query Q at time t.

Proposition 14.9. *Safe queries are domain independent for:*

1. *temporally bounded databases over a discrete or a dense flow of time; and*
2. *ultimately periodic databases over a discrete flow of time.*

Proof The proof is by induction on the number of nested temporal operations in a formula. We know from [Ullman, 1988] that every non-temporal subformula with limited variables generates only finite relations over their active domain, and so do safe disjunctions; thus this is the case for any time t. In fact, this may be derived from Codd's original result on the equivalence of the (finitely based) relational algebra and the relational calculus [Codd, 1970; 1972]. With regards to safe temporal subformulae of the form $S(A, B)$ and $U(A, B)$, by the induction hypothesis A is domain independent so it can only generate finite relations over $adom(A)$ at any time t, and B has all its free variables limited. We then examine the following cases:

1. If the database is temporally bounded, only finitely many tuples of domain elements (corresponding to the free variables) can satisfy $S(A, B)$ and $U(A, B)$ at every time point and safe temporal formulae are domain independent. Note that if the flow of time is dense, we do not even need to bound B's variable to obtain this result;
2. If data is periodic, we have to face the possibility of A being true infinitely often in the future (or past), such that the union of all tuples satisfying A at all times is infinite, thus violating the domain independence of FA (PA).

Note, however, that by the induction hypothesis A is domain independent, $adom(A)$ is finite and the union of all tuples satisfying A at all times is made up of elements of $adom(A)$, and hence finite. So again only finitely many tuples of domain elements can satisfy $S(A, B)$ and $U(A, B)$ at every time point and the result is proved. □

Safe queries provide a large enough class of queries that are domain independent and membership of that class can be efficiently computed. But in temporal databases, domain independence is not everything we want from a query language. If a database is represented by a class of time stamps, the temporal relations generated by queries should also be expressed by the same class of time stamps. A temporal query language is *closed form* if it satisfies that property, i.e. if queries over data represented by a certain class of time stamps (e.g. temporal elements or Pressburger definable time stamps) generate temporal relations representable in that same class of time stamps. Note that closed form queries are always domain independent.

Proposition 14.10. *Safe queries are closed form for:*

1. *temporal elements over a discrete or a dense flow of time; and*
2. *Pressburger-defined time stamps over a discrete flow of time.*

Proof By induction on the structure of the formula. Atomic queries are obviously closed form. We analyse the other cases separately:

1. Suppose the data is time stamped by temporal elements. Then for conjunction, disjunction, negation and existential quantification, it suffices to notice that temporal elements are closed under, respectively, intersection, union, complementation and projection. For the temporal cases of the form $S(A, B)$ and $U(A, B)$, by the induction hypothesis A and B generate a temporal relation that is representable by temporal elements, so $S(A, B)$ and $U(A, B)$ can only generate intervals.
2. Suppose the data is represented by Pressburger-defined time stamps. Similarly, note that for conjunction, disjunction, negation and existential quantification it suffices to notice that the class of Pressburger-defined time stamps is closed under, respectively, intersection, union, complementation and projection. For the temporal cases of the form $S(A, B)$ and $U(A, B)$, by the induction hypothesis A and B generate a temporal relation that is representable by Pressburger-defined time stamps, so $S(A, B)$ and $U(A, B)$ can only generate recurring intervals which are obviously Pressburger definable. □

Note that we may have non-safe queries that are closed form, domain independent. For instance, the formula

$$A(x, y, z) \wedge \neg(FB(x, y) \vee C(y, z))$$

is non-safe, but it generates only closed form finite relations over $adom(\mathcal{DB})$ because it is logically equivalent to the safe formula

$$A(x, y, z) \wedge \neg FB(x, y) \wedge \neg C(y, z).$$

In fact, the temporal formula $GA(x)$, defined as $\neg S(\neg A(x), true)$, is not safe, but it is equivalent over $\mathbb{Z}$ to the safe formula $\bigcirc(A(x) \wedge GA(x))$.

Example 14.11. In the database of Example 14.2, if we want to know the names of employees that were sacked in the past, as of *Apr*93, we can pose the following safe snapshot query:

$$\{Apr93 : x \mid P\exists y \exists z\, employee(x, y, z) \;\; \wedge \;\; \neg\exists w \exists v\, employee(x, w, v)\}$$

which generates the one-place unary relation $\{Apr93 : \langle\text{Paul}\rangle\}$. Note that the time stamp indicates the time the data is valid, *not* the time of the sacking of the employee. If we want to know the name and salaries of employees that have ever worked in the R&D department, as of *Apr*93, we pose the following safe query to the database:

$$\{x, y \mid P\, employee(x, y, \text{R\&D}) \;\; \vee \;\; employee(x, y, \text{R\&D})\}$$

which generates the relation $\{[Jan90, Dec92] : \langle\text{Peter 1K}\rangle,\ [Jan90, Dec92] : \langle\text{Paul 1K}\rangle\}$.

A brief comment on complexity issues is made here. Even though a logic is undecidable, it can still be used efficiently to compute safe queries. For example, first-order logic is undecidable but safe first-order queries are computed in polynomial time. Full first-order temporal logic cannot even be finitely axiomatized over $\mathbb{Z}$, but safe temporal queries over temporally bounded databases can be computed in polynomial time too; see [Gabbay and McBrien, 1991; Tuzhilin and Clifford, 1990; McBrien, 1993]. For dense flows of time, queries over temporal databases represented with temporal elements can also be computed in polynomial time and closed form [Kanelakis *et al.*, 1990]. For periodic databases over $\mathbb{Z}$, queries without negation can be computed in PTIME, but with negation the problem becomes NP-hard [Kabanza *et al.*, 1990]; all these results are for data complexity measure, i.e. it is assumed that the size of the temporal database is much larger than that of the query.

So much for the static aspects of temporal databases. In the next chapter we proceed with the formal study of temporal databases, investigating their dynamic aspects.

15

A LOGICAL VIEW OF TEMPORAL DATABASE DYNAMICS

In traditional databases, the update of a piece of data means the replacement of the current value of the data by a new one. In temporal databases we are presented with the extra possibility of changing the past, the present and the future. In a nutshell, a temporal update is a 'change in history'. Note the double reference to time in such an expression: *change* relates to temporal evolution, while *history* refers to temporal record. These two notions of time are independent and coexistent. In analysing updates in temporal databases we have to be able to cope simultaneously with those two notions of time. For that, we present next a *two-dimensional temporal logic*, a formalism that will allow for the simultaneous handling of two references of time.

The discussion develops as follows. We first present a propositional version of two-dimensional temporal logic in Section 15.1.1, then in Section 15.1.2 we show that such a propositional view suffices for the analysis of temporal updates of atomic information, and finally we present the semantics of temporal database evolution based on the two-dimensional model. The remaining sections of this chapter further investigate the applications of the two-dimensional model to characterize the distinction between transaction time and valid time, and to serve as the logical foundation of bitemporal databases.

15.1 Temporal updates

We start our discussion on temporal updates by defining the formal framework that will serve as the basis for the rest of this presentation.

15.1.1 *Propositional two-dimensional temporal logic*

There are several modal and temporal logic systems in the literature which are called *two-dimensional*; all of them provide some sort of double reference to an underlying modal or temporal structure. More systematically, two-dimensional systems have been studied as the result of combining two one-dimensional logic systems [Finger and Gabbay, 1992; 1996]. In [Finger and Gabbay, 1996] two criteria were presented to classify a logical system as two dimensional (see also Chapter 6):

- *the connective approach:* a temporal logic system is two dimensional if it contains two sets of connectives, each set referring to a distinct flow of time;
- *the semantic approach:* a temporal logic system is two dimensional if the truth value of a formula is evaluated with respect to two time points.

The two criteria are independent and there are examples of systems satisfying each criterion alone, or both. For the purposes of this work, the two-dimensional temporal logic satisfies both criteria, and is thus a *broadly two-dimensional logic*. We present a system with limited expressivity; one of the flows of time (the one related to data evolution) is assumed to be discrete ($\mathbb{Z}$) and the operators over it can only refer to the previous and next times. However, it will be shown that such a system is strong enough for our modelling purposes.

So let $\mathcal{L}$ be a countable set of propositional atoms. Besides the Boolean connectives, we consider two sets of temporal operator. The *horizontal operators* are the usual 'Since' (S) and 'Until' (U) two-place operators, together with all the usual derived operators; the horizontal dimension will be used to represent *valid time* temporal information. The *vertical dimension* is assumed to be a $\mathbb{Z}$-like flow and the only operators over such dimensions are the one-place operators 'next vertical' ($\bar{\bigcirc}$) and the 'previous vertical' ($\bar{\bullet}$); in general, we use barred symbols when they refer to the vertical dimension. The vertical dimension will be used to represent *transaction-time* information. Two-dimensional formulae are inductively defined as:

- every propositional atom is a two-dimensional formula;
- if A and B are two-dimensional formulae, so are $\neg A$ and $A \wedge B$;
- if A and B are two-dimensional formulae, so are $S(A, B)$ and $U(A, B)$;
- if A is a two-dimensional formula, so are $\bar{\bigcirc} A$ and $\bar{\bullet} A$.

On the semantic side, we consider two flows of time: the horizontal one $(T, <)$ and the vertical one $(\overline{T}, \overline{<})$; the vertical flow is actually always assumed to be $(\mathbb{Z}, <_{\mathbb{Z}})$. Two-dimensional formulae are evaluated with respect to two dimensions, typically a time point $t \in T$ and a time point $\bar{t} \in \overline{T}$, so that a *two-dimensional plane model* is a structure based on two flows of time $\mathcal{M} = (T, <, \overline{T}, \overline{<}, g)$. The *two-dimensional assignment* g maps every pair of time points $t, \bar{t}$ into a set of atoms $g(t, \bar{t}) \subseteq \mathcal{L}$, namely the set of atoms that are true at that point in the two-dimensional plane. The model structure can be seen as a two-dimensional plane, where every point is identified by a pair of coordinates, one for each flow of time (there are other, non-standard models of two-dimensional logics which are not planar; see [Finger and Gabbay, 1996]).

The fact that a formula A is true in the two-dimensional plane model $\mathcal{M}$ at point $(t, \bar{t})$ is represented by $\mathcal{M}, t, \bar{t} \models A$ and is defined inductively as:

$\mathcal{M}, t, \bar{t} \models p$	iff $p \in g(t, \bar{t})$ for $p \in \mathcal{L}$.
$\mathcal{M}, t, \bar{t} \models \neg A$	iff it is not the case that $\mathcal{M}, t, \bar{t} \models A$.
$\mathcal{M}, t, \bar{t} \models A \wedge B$	iff $\mathcal{M}, t, \bar{t} \models A$ and $\mathcal{M}, t, \bar{t} \models B$.
$\mathcal{M}, t, \bar{t} \models S(A, B)$	iff there exists a $t' \in T$ with $t' < t$ and $\mathcal{M}, t', \bar{t} \models A$ and for every $t'' \in T$, whenever $t' < t'' < t$ then $\mathcal{M}, t'', \bar{t} \models B$.

$\mathcal{M}, t, \bar{t} \models U(A, B)$ iff there exists a $t' \in T$ with $t < t'$ and $\mathcal{M}, t', \bar{t} \models A$ and for every $t'' \in T$, whenever $t < t'' < t'$ then $\mathcal{M}, t', \bar{t} \models B$.

$\mathcal{M}, t, \bar{t} \models \bar{\bigcirc} A$ iff $\mathcal{M}, t, \bar{t}+1 \models A$.

$\mathcal{M}, t, \bar{t} \models \bullet A$ iff $\mathcal{M}, t, \bar{t}-1 \models A$.

Note that the semantics of the horizontal and vertical operators are totally independent of each other, i.e. the horizontal operators have no effect on the vertical dimension and similarly for the vertical operators. If we consider the formula without the vertical operators, we have a one-dimensional US-temporal logic which is called the (one-dimensional) horizontal temporal logic; similarly for the vertical temporal logic. If $\mathcal{M}$ is a two-dimensional model and A is a two-dimensional formula, we say that A ***holds over*** $\mathcal{M}$, and write $\mathcal{M} \models A$ if for every $t \in T$ and every $\bar{t} \in \mathbb{Z}$, $M, t, \bar{t} \models A$ (see Chapter 6).

The two-dimensional logic above has very nice properties 'inherited' from the horizontal logic. For example, as presented in [Finger and Gabbay, 1996; Finger, 1994], the two-dimensional logic is a conservative extension of the horizontal logic and it is completely axiomatizable and decidable if that is the case for the horizontal temporal logic.

We now examine some properties of the diagonal in two-dimensional plane models. The diagonal is a privileged line in the two-dimensional model intended to represent the sequence of time points we call 'now', i.e. the time points on which an historical observer is expected to traverse. The observer is, therefore, on the diagonal when he or she poses a query (i.e. evaluates the truth value of a formula) on a two-dimensional model. We have already fixed the vertical (transaction-time) flow of time to be $\mathbb{Z}$. In this context, the notion of a diagonal as the sequence of 'now' time points makes sense only in the horizontal (valid-time) flow of time $(T, <)$ such that $\mathbb{Z} \subseteq T$, in which which case we can compare the elements of both flows of time with respect to $<$ and $=$. Such an assumption is held for the rest of the chapter.

So let δ be a special atom that denotes the points of the diagonal, which is characterized by the following property: for every $t \in T$ and every $\bar{t} \in \mathbb{Z}$

$$\mathcal{M}, t, \bar{t} \models \delta \quad \text{iff} \quad t = \bar{t}.$$

The diagonal is illustrated in Fig. 15.1.

Some properties of the diagonal are illustrated below. The following formula hold over all two-dimensional plane models:

D₁ $P\delta \vee \delta \vee F\delta$

D₂ $\delta \to (G\neg\delta \wedge H\neg\delta)$

D₃ $\delta \to \bar{\bigcirc} F\delta$.

Indeed, the three formulae above can be said to characterize the diagonal, owing to the following (cf. Proposition 6.24).

Proposition 15.1. *Let $\mathcal{M} = (T, <, \overline{T}, \overline{<}, g)$ be a (generic) two-dimensional plane such that $\mathcal{M} \models \mathbf{D_1} \wedge \mathbf{D_2} \wedge \mathbf{D_3}$. Then the relation*

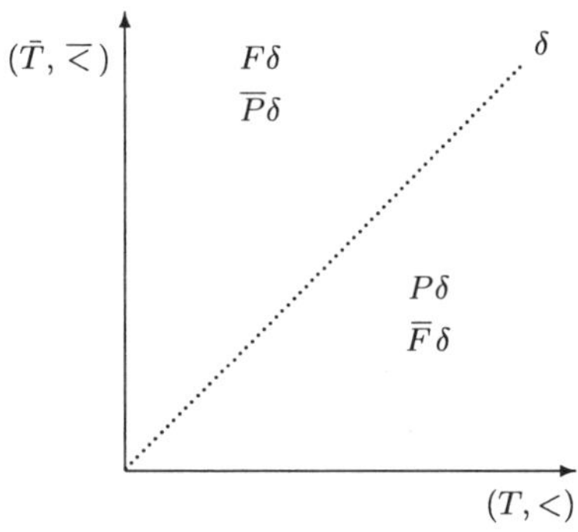

FIG. 15.1. The two-dimensional diagonal

$$i = \{(\bar{t}, t) \in \overline{T} \times T \mid \mathcal{M}, t, \bar{t} \models \delta\}$$

is a homomorphism that maps $(\overline{T}, \overline{<})$ *into* $(T, <)$.

Proof From $\mathbf{D_1}$ it follows that i is defined at all $\bar{t} \in \overline{T}$ and from $\mathbf{D_2}$ it follows that i is indeed a function. $\mathbf{D_3}$ gives us that $i(\bar{t}+1) > i(\bar{t})$, so i is order preserving. □

This result reinforces the claim, namely that a diagonal with properties $\mathbf{D_1} \wedge \mathbf{D_2} \wedge \mathbf{D_3}$ only makes sense when the time points of the vertical flow of time can be compared with the horizontal ones. The embedding $\overline{T} \subseteq T$ adopted earlier is just one way of guaranteeing such comparability.

The diagonal is interpreted as the sequence of time points we call 'now'. The diagonal divides the two-dimensional plane into two semiplanes. The semiplane that is to the (horizontal) left of the diagonal is 'the past', and the formula $F\delta$ holds over all points of this semiplane. Similarly, the semiplane that is to the (horizontal) right of the diagonal is 'the future', and the formula $P\delta$ holds over all points of this semiplane. Fig. 15.1 displays this fact.

Before we see how the two-dimensional plane model can be used to describe atomic updates in temporal databases, it is necessary to show how the propositional approach of this section can be used with the first-order treatment so far.

15.1.2 *Propositional abstractions*

The propositional approach of the previous section differs from the first-order treatment of temporal features in the previous chapter. To reconcile these two different approaches this section presents a propositional abstraction of database queries.

Let $\mathcal{DB}$ be a temporal database over schema $\Sigma = (\mathcal{C}, \mathcal{P})$ and let $adom(\mathcal{DB})$ be its (finite) active domain. We define the propositional schema $\mathcal{L}$ abstracting from Σ as consisting of the following propositional atoms:

- $[\![a = b]\!]$, for each $a, b \in \mathcal{C}$;

- $[\![time(t)]\!]$, for each $t \in T$;
- $[\![R(a_1, \ldots, a_{ar(R)})]\!]$, for each $R \in \mathcal{P}$ and each $a_i \in \mathcal{C}$.

It is no coincidence that we choose first-order $[\![]\!]$-enclosed atoms to represent propositions: the first item above allows us to equate two constant symbols (this latter equality when relating to two distinct constant symbols will actually generate propositions that are always false, in the same way that equating two identical symbols will generate propositions that are always true); the second item above generates a proposition for each possible time (these atoms will be true at a single time point and false at all other time points); and the third item generates a proposition for each possible ground predicate. The generalization of this notation will give us a propositional abstraction of FOTL-formulae. If v is an assignment, let B^v be a *v-grounded formula* obtained by substituting every free variable x occurring in B by $v(x)$. For every safe FOTL-formula A, we define its *propositional abstraction* with respect to v by taking $B = A^v$ and generalizing the propositional notation above over v-grounded formulae, denoted by $[\![B]\!]$, as:

- $[\![\neg B]\!] = \neg[\![B]\!]$;
- $[\![B_1 \wedge B_2]\!] = [\![B_1]\!] \wedge [\![B_2]\!]$;
- $[\![S(B_1, B_2)]\!] = S([\![B_1]\!], [\![B_2]\!])$;
- $[\![U(B_1, B_2)]\!] = U([\![B_1]\!], [\![B_2]\!])$;
- $[\![\exists x B]\!] = \bigvee_{d \in adom(\mathcal{DB})} [\![B(x := d)]\!]$;

where $B(x := d)$ is the formula obtained by substituting all free occurrences of x in B by d; the last item above is a well-defined propositional formula because $adom(\mathcal{DB})$ is a finite set.

Definition 15.2. Given a database $\mathcal{DB} = (\mathbb{Z}, <, g)$ over a schema $\Sigma = (\mathcal{C}, \mathcal{P})$, we say that a propositional model $\mathcal{M} = (\mathbb{Z}, <, h)$ over $\mathcal{P}$ *abstracts from $\mathcal{DB}$* if for every assignment v extended over constant symbols, and every $t \in \mathbb{Z}$, it is the case that for every atomic formula of the form $R(a_1, \ldots, a_r)$,

$$[\![R(a_1, \ldots, a_r)]\!] \in h(t) \quad \text{iff} \quad \langle v(a_1), \ldots, v(a_r)\rangle \in I_t(R);$$

for every atomic formula of the form $time(u)$

$$[\![time(u)]\!] \in h(t) \quad \text{iff} \quad v(u) = t;$$

and for every atomic formula of the form $a = b$

$$[\![a = b]\!] \in h(t) \quad \text{iff} \quad v(a) = v(b).$$

A straightforward induction on the structure of formulae then shows that

$$\mathcal{DB}, v, t \models B(x_1, \ldots, x_m) \quad \text{iff} \quad \mathcal{M}, t \models [\![B^v]\!].$$

The relation generated by the safe formula $B(x_1, \ldots, x_m)$ can be expressed as

$$\{\tau : \langle v(x_1), \dots, v(x_m)\rangle | \text{ there exists } t \in \tau \text{ such that } \mathcal{M}, t \models [\![B(x_1 := v(x_1), \dots, x_m := v(x_m))]\!]\}.$$

The propositional abstraction as defined above has nothing especially 'temporal' about it, the whole purpose of it being the elimination of variables and quantifiers from safe formulae. However, it is important for sending us back to the propositional framework. From now on, we refer to the contents of a database by its propositional abstraction $\mathcal{M}$; moreover, we can refer to the update of ground atomic information in a database as the update of propositional atoms. We assume that the countably infinite domain D of the database remains the same after the update, and so does the database schema.

Example 15.3. Consider the database $\mathcal{DB}$ from Example 14.2; let v be an assignment such that $v(x) = Peter$ and $v(y) = 10\text{K}\neg \in adom(\mathcal{DB})$. A few of the properties of the database propositional abstraction model $\mathcal{M} = (\mathbb{Z}, <, h)$ are:

- $[\![employee(\text{Peter}, 1\text{K}, \text{R\&D})]\!] \in h(t)$ iff $Jan90 \leq t \leq Dec92$;
- $[\![employee(\text{Peter}, \text{y}, \text{R\&D})^v]\!]\neg \in h(t)$, for every $t \in \mathbb{Z}$.

Consider the following safe formula about sacked employees

$$A(x) = P\exists y \exists z\, employee(x, y, z) \ \wedge\ \neg\exists y \exists z\, employee(x, y, z),$$

and its propositional abstraction under v:

$$[\![A(x)^v]\!] = P\left(\bigvee_{c \in adom(\mathcal{DB})} \bigvee_{d \in adom(\mathcal{DB})} [\![employee(Peter, c, d)]\!]\right) \wedge \neg\left(\bigvee_{c \in adom(\mathcal{DB})} \bigvee_{d \in adom(\mathcal{DB})} [\![employee(Peter, c, d)]\!]\right).$$

It follows that, for all $t \in \mathbb{Z}$, $\mathcal{DB}, v, t \models A(x)$ iff $\mathcal{M}, t \models P[\![A(x)^v]\!]$.

Notation: We may sometimes abuse the notation and represent the propositional abstraction of a predicate formula by the formula itself. We do this when no ambiguity is implied, mainly when we refer to atomic formulae with no free variables.

15.1.3 *Temporal database evolution*

In describing the evolution of a temporal database, we have to distinguish the database evolution from the evolution of the world it describes. The 'world', also called the *universe of discourse*, is understood to be any particular set of objects in a certain environment that we may wish to describe. The database, in its turn, contains a description of the world. Conceptually, we have to bear in mind two distinct types of evolution:

- The *evolution of the modelled world* is the result of changes in the world that occur independently of the database.
- A temporal database contains a description of the history of the modelled world that is also constantly changing due to database updates, generating

a sequence of database states. This *evolution of the temporal description* does not depend only on what is happening at the present; changes in the way the past is viewed also alter this historical description. Moreover, changes in expectations about the future, if those expectations are recorded in the database, also generate an alteration of the historical description. This process is also called *historical revision*.

These two distinct concepts of evolution are reflected by a distinction between two kinds of flows of time, whether their time points refer to a moment in the history of the world, or whether they are associated with a moment in time at which a historical description is in the database.

Several different names are found in the literature for these two time concepts. The former is called *evaluation time* [Kamp, 1971; Gabbay *et al.*, 1994], *historical time* [Finger, 1992], *valid time* [Snodgrass and Ahn, 1985] and *event time* [McKenzie and Snodgrass, 1991], among other names. The latter time concept is called *utterance time* [Kamp, 1971], *reference time* [Gabbay *et al.*, 1994], *transaction time* [Finger, 1992; Snodgrass and Ahn, 1985] and *belief time* [Sripada, 1990]. In this presentation we chose to follow a glossary of temporal database concepts proposed in [Jensen *et al.*, 1992], calling the former valid time, which is associated with the horizontal dimension in our two-dimensional model, and calling the latter transaction time, which is associated with the vertical dimension.

So we use the two-dimensional plane model to cope simultaneously with the two notions of time in the description of the evolution of a temporal database, as illustrated in Fig. 15.2.

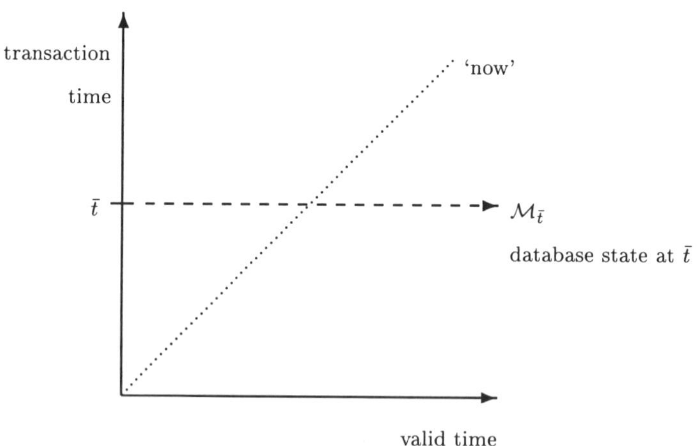

FIG. 15.2. Two-dimensional database evolution

Let $\mathcal{M} = (T, <, \overline{T}, \overline{<}, g)$ be a two-dimensional plane model; its *horizontal projection* with respect to the vertical point $\bar{t} \in \overline{T}$ is the one-dimensional temporal model

$$\mathcal{M}_{\bar{t}} = (T, <, g_{\bar{t}}),$$

such that, for every propositional atom q, time points $t \in T$ and $\bar{t} \in \overline{T}$,

$$q \in g_{\bar{t}}(t) \quad \text{iff} \quad q \in g(t, \bar{t}).$$

It follows that for every horizontal US-formula A and for every $t \in T$ and $\bar{t} \in \overline{T}$,

$$\mathcal{M}_{\bar{t}}, t \models A \quad \text{iff} \quad \mathcal{M}, t, \bar{t} \models A.$$

The horizontal projection represents a state of a temporal database. Since we are interested in describing the evolution of database states as a linear and discrete process, we fix $\overline{T} = \mathbb{Z}$.

Updating temporal databases requires that, besides specifying the atom to be inserted or deleted, we specify the time where the atom is to be inserted or deleted. For that reason, it is convenient to use the notation of time-stamped formulae to represent the data being inserted and deleted; in most cases, we will restrict our attention to time-stamped atoms only. The restriction on the time stamp (i.e. if it is a temporal element or a Pressburger definable set) is assumed to be the same as that of the underlying database. As a result, infinite updates are possible as long as such update is representable in the database.

An *update pair* (θ_+, θ_-) consists of two finite disjoint sets of time-stamped atoms, where θ_+ is the *insertion set* and θ_- is the *deletion set*; by disjoint sets, in the context of data representation, it is meant that it is not the case that $\tau_+ : p \in \theta_+$ and $\tau_- : p \in \theta_-$ such that $\tau_+ \cap \tau_- \neq \emptyset$. We say that an update pair determines or characterizes a *database update* $\Theta_{\bar{t}}$ occurring at transaction time $\bar{t} \in \mathbb{Z}$ if the application of the update function $\Theta_{\bar{t}}$ to the database state $\mathcal{M}_{\bar{t}} = (T, <, g_{\bar{t}})$ generates a database state $\Theta_{\bar{t}}(\mathcal{M}_{\bar{t}}) = (T, <, \Theta_{\bar{t}}(g_{\bar{t}}))$ satisfying, for every propositional atom q and every time point $t \in T$:

- if $\tau : q \in \theta_+$, then $q \in \Theta_{\bar{t}}(g_{\bar{t}})(t)$ for every $t \in \tau$;
- if $\tau : q \in \theta_-$, then $q\neg \in \Theta_{\bar{t}}(g_{\bar{t}})(t)$ for every $t \in \tau$;
- if neither $t : q \in \theta_+$ nor $t : q \in \theta_-$, then $q \in \Theta_{\bar{t}}(g_{\bar{t}})(t) \quad \text{iff} \quad q \in g_{\bar{t}}(t)$.

The first item corresponds to the insertion of atomic information, the second one corresponds to the deletion of atomic information, and the third one corresponds to the persistence of the unaffected atoms in the database. Note that the disjoint sets θ_+ and θ_- are represented in the same way as the underlying database, so that we can represent a temporal database update schematically as

$$\Theta_{\bar{t}}(\mathcal{M}_{\bar{t}}) = \mathcal{M}_{\bar{t}} \cup \theta_+ - \theta_-$$

because there are only finitely many atoms in the database. When the sets θ_+ and θ_- are not disjoint, i.e. the update is trying to insert and remove the same information, the situation is undetermined; typically this would mean that the transaction in which the update was generated should be rolled back, but we do not deal with the notion of transactions here. The update $\Theta_{\bar{t}}$ is a database state transformation function. An update may be empty ($\theta_+ = \theta_- = \emptyset$), in which case the transformation function is just the identity and the database state remains the same.

Let $\{\Theta_{\bar{t}}\}_{\bar{t}\in\mathbb{Z}}$ be a sequence of database updates; we say that such a sequence is representable by temporal elements (resp. Pressburger definable) time stamps if each update is characterized by sets θ_+ and θ_- that are representable by temporal elements (resp. Pressburger definable) time stamps. We say that a two-dimensional plane model $\mathcal{M}$ *represents the evolution in time of a temporal database* through the update sequence $\{\Theta_{\bar{t}}\}_{\bar{t}\in\mathbb{Z}}$ if, for every $\bar{t} \in \mathbb{Z}$, $\Theta_{\bar{t}}(\mathcal{M}_{\bar{t}}) = \mathcal{M}_{\bar{t}+1}$.

Proposition 15.4. *Let $\mathcal{M}$ be a two-dimensional model representing the evolution of a temporal database through the update sequence $\{\Theta_{\bar{t}}\}_{\bar{t}\in\mathbb{Z}}$ with initial time $\bar{t}_0$, such that $\mathcal{M}_{\bar{t}_0}$ is representable by temporal elements (resp. Pressburger definable) time stamps. Then, for every $\bar{t} \geq \bar{t}_0$, $\mathcal{M}_{\bar{t}}$ is representable by temporal elements (resp. Pressburger definable) time stamps iff $\{\Theta_{\bar{t}}\}_{\bar{t}\in\mathbb{Z}}$ is so.*

Proof The two directions of the iff-condition are proved separately.

($\Leftarrow$) Follows directly from the fact that temporal elements and Pressburger definable sets are closed under set union and set difference.

($\Rightarrow$) Suppose $\mathcal{M}_{\bar{t}}$ and $\Theta_{\bar{t}}(\mathcal{M}_{\bar{t}}) = \mathcal{M}_{\bar{t}+1}$ are both representable by temporal elements. Then either there is only a finite amount of time-stamped formulae $\tau : q$ changing its value from one state to the next, in which case $\Theta_{\bar{t}}$ is bounded, or there are atoms q_i such that their value has changed in finitely many times. In this last case, since both $\mathcal{M}_{\bar{t}}$ and $\mathcal{M}_{\bar{t}+1}$ are temporally bounded, there must be times t' and t'' after which and before which, respectively, $t : q$ was always inserted or always deleted. Therefore $\Theta_{\bar{t}}$ is representable by temporal elements.

Suppose $\mathcal{M}_{\bar{t}}$ and $\mathcal{M}_{\bar{t}+1}$ are both representable by Pressburger definable time stamps; we only consider periodic data, for the other cases have been covered by temporal elements. Clearly, there has been only the addition and deletion of finitely many periodic tuples between $\mathcal{M}_{\bar{t}}$ and $\mathcal{M}_{\bar{t}+1}$, so $\Theta_{\bar{t}}$ is representable by Pressburger definable time stamps. □

The proposition above shows us that updates representable in the same way as the database are the kind of update for which the two-dimensional models describe the evolution.

Example 15.5. Consider the monthly evolution of the database of Example 14.2. Suppose that at *Apr*93 we decide to increase retroactively Mary's monthly salary to 5K for the whole year, which is illustrated in Fig. 15.3.

This situation is described by an update $\Theta_{Apr93} = (\theta_+, \theta_-)$, where

$$\begin{aligned} \theta_+ &= \{[Jan93, Dec93] : employee(\text{Mary}, \text{5K}, \text{Finance})\} \\ \theta_- &= \{[Jan93, Dec93] : employee(\text{Mary}, \text{3K}, \text{Finance})\}. \end{aligned}$$

Note that we specified the deletion of Mary's old salary until *Dec*93; the same effect would be achieved had we only specified the deletions until *Apr*93, i.e. the same horizontal projection would have been generated for $\mathcal{M}_{May93}$. Note also that we have changed not only the past, but also Mary's salary at the present, *Apr*93, and also its expectation for the future.

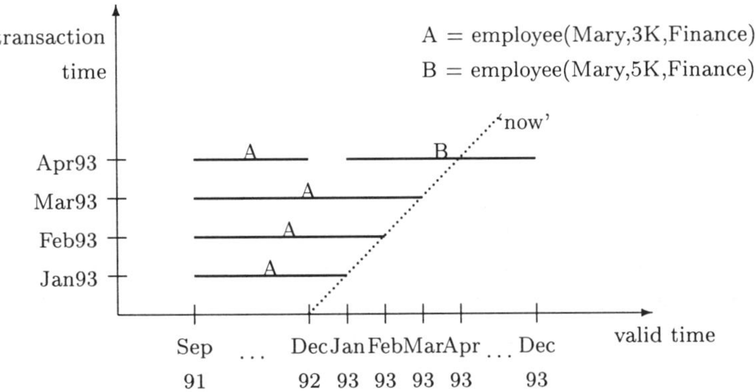

FIG. 15.3. Two-dimensional diagram of database evolution.

If we wanted to increase Mary's salary indefinitely to the future, we could have used the following infinite update $\Theta_{Apr93} = (\theta_+, \theta_-)$, where

$$\begin{aligned} \theta_+ &= \{[Jan93, +\infty) : employee(\text{Mary}, 5\text{K}, \text{Finance})\} \\ \theta_- &= \{[Jan93, Dec93] : employee(\text{Mary}, 3\text{K}, \text{Finance})\}. \end{aligned}$$

This illustrates the result of Proposition 15.4, where time boundedness is preserved by this infinite update.

15.2 Valid-time and transaction-time databases

In the previous section we pointed out a conceptual difference between the valid-time and transaction-time flows of time, and treated them separately but simultaneously in the two-dimensional model of the evolution of a temporal database. For that, we assumed that the database records the valid-time flow of time. However, it need not be so.

In their taxonomy of time in databases, Snodgrass and Ahn [1985] distinguish a *historical database* (called here a *valid-time database*), a temporal database that records only valid time, from a *rollback database* (called here a *transaction-time database*), a database that records only transaction time. The temporal data in a valid-time database is supposed to describe the evolution of the modelled part of the world that occurs independently from the database. On the other hand, the temporal information associated with any piece of data recorded in a transaction-time database is generated automatically and is supposed to represent the times when that information is held in the database; therefore, transaction time has no existence independent of the database. For example, suppose that at (transaction time) *Sep*91 the piece of data *employee*(Mary, 3K, Finance) is inserted in the transaction-time database; this information will persist for all times t, from t starting at $Sep91$, until the time that data is deleted from the database. Suppose the deletion happens at (transaction time) $Apr93$, so that in the database model we have

$$[Sep91, Apr93] : employee(\text{Mary, 3K, Finance}).$$

Note that the temporal information was generated automatically at the times of insertion and deletion, without the need to supply them externally. If we want then, at $Apr93$, to increase Mary's salary retroactively to the beginning of the year, we cannot record this fact in the transaction-time database. The transaction-time database records only a sequence of (non-temporal) database states, attributing to each state a (transaction) time stamp, thus allowing us to reconstruct a previous state but not to modify it.

No present and past query that may be posed to a temporal database can tell whether it is a valid-time or a transaction-time database. In fact, the query language and the notion of a correct answer (the generated relation) for a query are exactly the same in both cases. Even if we do not have in the database any information concerning a time later than the current time, based on the fact that transaction-time databases record only the present and past states of the database, we cannot guarantee that the given temporal database is a transaction-time one, for it is perfectly legal for a valid-time database to store only data about the present and past.

The difference between valid-time and transaction-time databases lies in their dynamic behaviour, not in their static query answering. We still consider the state of any temporal database to be the horizontal projection of a two-dimensional plane model with respect to some vertical (hence transaction-time) point. A *tt-update* (transaction-time update) is a state transformation function $\Theta_{\bar{t}}$, where $\bar{t}$ is the current (transaction) time, determined by the update pair (θ_+, θ_-) satisfying the following conditions:

(a) the formulae of θ_+ and θ_- are of the form $\bar{t'} : q$ where $\bar{t'} \geq \bar{t}$, i.e. there are no updates in the past;

(b) $\bar{t} : q \in \theta_+$ (resp. $\bar{t} : q \in \theta_-$) iff for all $\bar{t'} \geq \bar{t}$, $\bar{t'} : q \in \theta_+$ (resp. $\bar{t'} : q \in \theta_-$), i.e. updates persist to the future.

It is then possible to use the two-dimensional plane model to describe the evolution of a transaction-time database so that we may characterize it by the properties of the two-dimensional plane model that describe its evolution. The basic distinction between valid-time and transaction-time databases is that in transaction-time databases we cannot change the past, whereas in valid-time databases any change is allowed. To characterize this impossibility to change the past, we will have first to characterize the existence of a 'now' in the two-dimensional plane model.

Recall that in Section 15.1.1 we used a special propositional symbol δ to characterize the 'diagonal' of a two-dimensional model $\mathcal{M}$. The points where δ

holds are exactly those where the valid and transaction times coincide, so we use those diagonal points as the ones where 'now' holds.

To characterize the persistence of present data towards the future, we make the one-dimensional formula

Persist $(\delta \wedge q) \rightarrow Gq$

hold over a two-dimensional model $\mathcal{M}$ for every literal q, where a literal is an atom or the negation of an atom.

The 'no change in the past' feature of transaction-time databases is characterized as the persistence of atomic information of the 'now', i.e. on the diagonal, towards the vertical future. Therefore, the following formula must hold over two-dimensional plane models that represent the evolution of a temporal database:

Roll $((\delta \vee F\delta) \wedge q) \rightarrow \bar{\bigcirc} q$,

where q is any literal. The subformula $(\delta \vee F\delta)$ represents the fact that we are in the present or past, so whatever information we have then will persist to the next vertical moment, when it will certainly be part of the (horizontal, therefore the database's) past. By making such a formula hold over the two-dimensional plane model, we guarantee the persistence of the information about the past in all states of the database. The following property generalizes **Roll** for a larger class of formulae; a proof is provided in [Finger, 1994].

Lemma 15.6. *If* **Roll** *holds over a two-dimensional $\mathcal{M}$ for any literal q, it also holds over $\mathcal{M}$ for any one-dimensional US-formula that does not contain future operators, i.e. does not contain U and its derived operators.*

The characterization of transaction-time databases as 'no updates in the past' is given by the following.

Theorem 15.7. *Let $\mathcal{M}$ be a two-dimensional model representing the evolution of a temporal database through the update sequence $\{\Theta_{\bar{t}}\}_{\bar{t} \in \mathbb{Z}}$ such that* **Persist** *holds over $\mathcal{M}$. Then the following are equivalent:*

(a) every update $\Theta_{\bar{t}}$ is a tt-update;

(b) **Roll** *holds over $\mathcal{M}$.*

Proof (a $\Rightarrow$ b) A *tt*-update does not update the past, so for $t \leq \bar{t}$, by the semantics of two-dimensional updates, the atomic information in $\mathcal{M}$ at t, t persists into the vertical, transaction-time future. So **Roll** holds over $\mathcal{M}$.

(b $\Rightarrow$ a) Assume that (b) holds and suppose that at transaction time $\bar{t}$ there was an update in the past time $t \leq \bar{t}$. Let q be a literal affected by the update such that $\mathcal{M}, t, \bar{t} \models \neg q$ and $\mathcal{M}, t, \bar{t}+1 \models q$. Because $t < \bar{t}$ we know that $\mathcal{M}, t, \bar{t} \models \neg\delta \vee F\delta$, but also $\mathcal{M}, t, \bar{t} \models \neg q \wedge \bar{\bigcirc} q$, contradicting **Roll**. So the update in the past might not have happened. □

The formulae **Persist** and **Roll** are not axioms in the usual sense. In fact, they are second-order constraints on a two-dimensional model, for they require that some property hold 'for all literals'.

It follows that a transaction-time database is one whose evolution is described by a two-dimensional model $\mathcal{M}$ such that **Persist** and **Roll** hold over $\mathcal{M}$. The persistence of information on a transaction-time database is illustrated in Fig. 15.4. It is no coincidence that there is a diagonal symmetry in Fig. 15.4, where it can be seen that a literal q that holds at the diagonal (i.e. at some current time) persists into the horizontal future in the current database state, and also persists into the vertical future throughout all the future database states, when it will be part of the unmodifiable past.

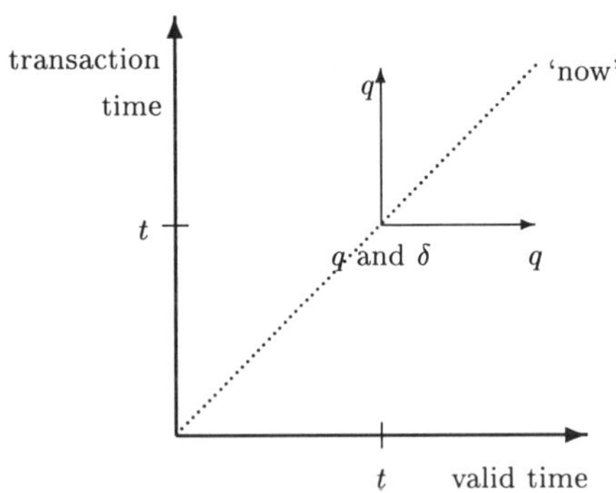

FIG. 15.4. Persistence of atomic data in transaction time databases

A temporal database is a valid-time one if it is possible that a two-dimensional model describing its evolution does not satisfy the three second-order axioms above. That does not mean that all two-dimensional models describing the evolution of a valid-time database invalidate all those metalevel axioms, because it is possible for a valid-time database to behave like a transaction-time one, i.e. the valid-time database can simulate a transaction-time one.

15.3 Bitemporal databases

Bitemporal database systems store both the valid time and the transaction time associated with each piece of data. It is just natural to use the two-dimensional plane model as a formal basis for such systems. Or else, we may think of a bitemporal database as a representation of a temporal two-dimensional plane model $\mathcal{M} = (T, <, \overline{T}, \overline{<}, g)$. The fact that a tuple $\langle a_1, \ldots, a_n \rangle$ is in relation r at valid time t and transaction time $\bar{t}$ can be represented as

$$\langle t, \bar{t} \rangle : r(a_1, \ldots, a_n).$$

A state of a bitemporal database is no longer the horizontal projection of the model, as in the one-dimensional case. The state of a bitemporal database at transaction time $\bar{t}_0$ is the semiplane generated by the equation $\bar{t} \leq \bar{t}_0$, as illustrated in Fig. 15.5.

A query language for bitemporal databases can be obtained by extending the one-dimensional query language with transaction-time operators $\overline{S}$ and $\overline{U}$ and

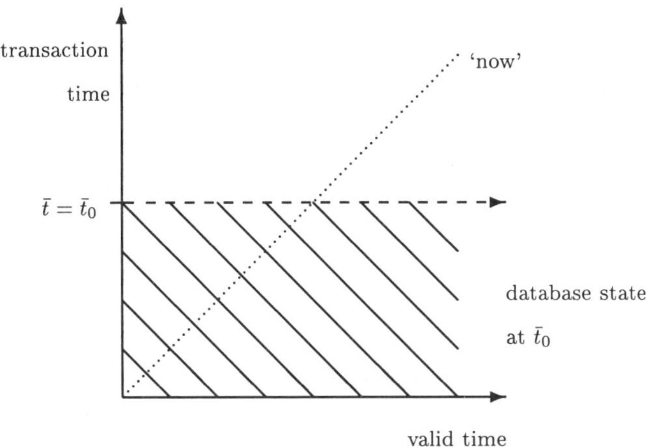

FIG. 15.5. The state of a bitemporal database.

their standard derived operators, plus an extra predicate $ttime(t)$, where the sort of t is time and $ttime(t)$ is true only at current time t. The usual $time(t)$ predicate is interpreted as referring to valid-time.

All the choices we have for one-dimensional temporal databases appear twofold in bitemporal databases. For each of the two flows of time, it has to be independently decided what is their nature (integer-like, $\mathbb{Q}$-like, etc.. The issues of data representation also appear for each dimension, e.g. we may decide to have periodic data in the valid-time dimension but only temporally bounded data in the transaction-time dimension.

As a natural extension of what was previously explained, to represent the evolution of a bitemporal database, one could expect it to be necessary to use a three-dimensional representation. However, since one of the dimensions represents transaction-time, no updates in the future are allowed in it, so that the evolution of a bitemporal database is monotonic. As a consequence, from the state of the database at a given transaction-time we can obtain the states at all previous transaction-times.

Many other temporal and non-temporal dimensions can be added to the database (at least, in theory). For example, one can think of a different kind of transaction-time, called *decision time*, where each piece of data is associated to the time when that data was decided to be true (not when it was inserted in the temporal database); note that the value of decision time is always smaller than the corresponding transaction-time value, for the decision to insert data always precedes the insertion. Other variations on transaction-time and valid-time can certainly be devised. Non-temporal dimensions appear in geographical databases, where coordinates, curves, areas and volumes may be associated to data. However, with respect to serious implementation, more-than-one-dimensional databases still have to prove to be feasible.

15.4 The granularity of transaction-time and valid-time

The valid-time flow and the transaction-time flow may be different in their ontological nature. But even when they are both linear and discrete flows of time, their *granularity* need not be the same. In fact, in all the previous examples we have used a *month* as the granularity of valid time. If the same granularity is applied to the transaction time, it would mean that an update made in the current state of the database will only take effect next month! Admittedly, that is far from a desirable situation.

So the granularity of the transaction flow of time is allowed to be distinct from that of the valid flow of time, and in general the transaction flow will have a finer granularity. But this statement can be misleading in the following sense. We know that *seconds* is a finer measure of time than *minutes*, but there always is the same number of seconds within a minute. A constant behaviour like that, however, is not observable between the transaction and valid flows of time. This is due to the fact that while valid time models the time in the real world, transaction time is *constructed* by the applications running on the database. Let us explain this view of a constructed transaction time more carefully. In the absence of any updates, i.e. no application is changing the database, transaction time and valid time evolve hand in hand, as illustrated below in Fig. 15.6.

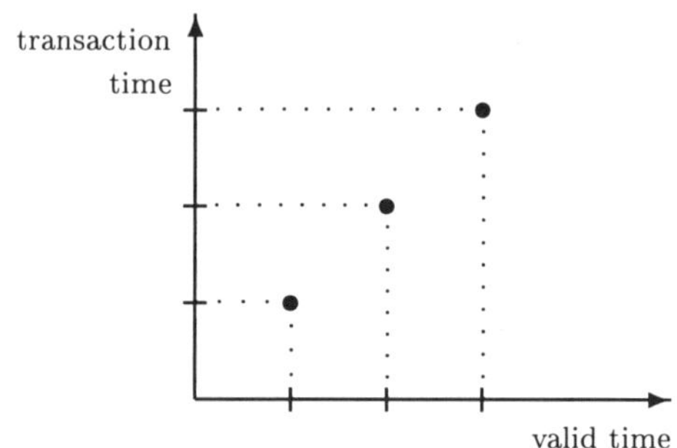

FIG. 15.6. No updates applied to the database.

Once there is at least one transaction updating the database, a new scenario emerges. What determines the 'real-time duration' between two transaction times is, according to our semantics, the speed of processing an update by the database system. This temporal interval is usually much smaller than the 'real-time duration' between two valid-time points. A single transaction running on the database may operate a chain of updates, all of which will be done at the same valid time, but at consecutive transaction times. A similar behaviour is produced by several transactions running concurrently in the database. This is illustrated at Fig. 15.7.

Figure 15.7 actually shows that the result of this imbalance between the granularity of the transaction and valid flows of time is to change the shape of the two-dimensional diagonal. The diagonal is, in this case, composed of two

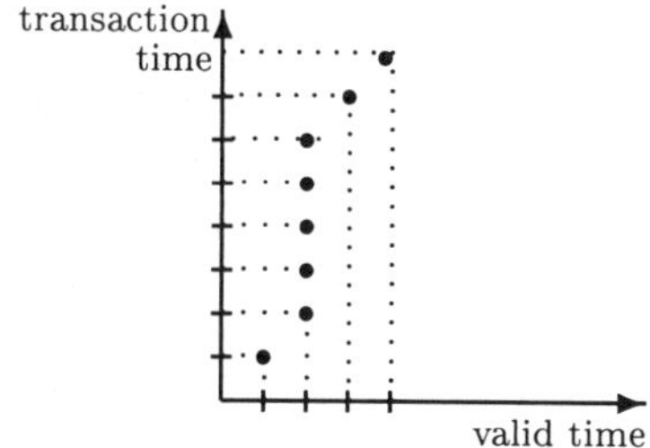

FIG. 15.7. Database with updating transactions.

types of segments, namely 45° segments, that indicate that no changes occur in the database, and 90° segments, parallel to the transaction-time axis, which indicate a chain of updates applied to the database during a single valid time. Note that two 90° segments need not be over the same number of transaction-time points, indicating a distinct number of updates occurring within a single valid time. This is illustrated in Figure 15.8.

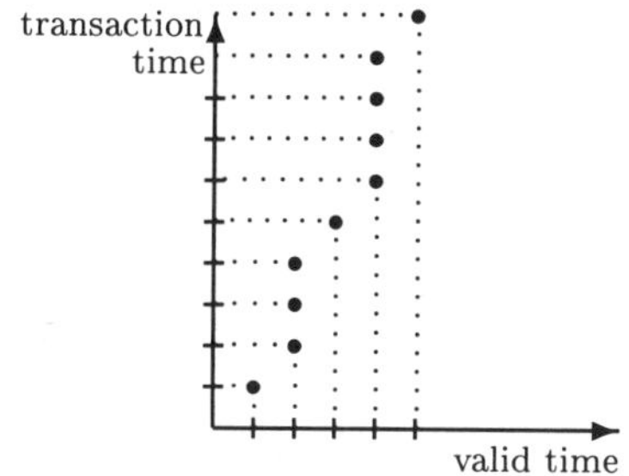

FIG. 15.8. Updates with distinct length.

If the atomicity of transactions is taken into consideration, Fig. 15.8 can be interpreted in the following way: each 90° dislocation as the update resulting of a full transaction, and the several 90° dislocations as representing several transactions running concurrently and being serialized in the given order. Note that, as a consequence, the value of 'now' remains constant during a transaction; that is, a transaction does not see the time passing. A temporal database that adopts such a strategy is said to support *perceivedly instantaneous* transactions. This is indeed a desirable effect, for a varying 'now' may lead to unexpected behaviour; temporal transaction, however, remains outside the scope of this presentation. A discussion on the full range of possibilities for the semantics of 'now' within a transaction in a valid-time database can be found in [Finger and McBrien, 1996].

Formally, the difference in granularity between transaction time and valid time in the *perceivedly instantaneous* strategy is modelled by a function f that maps elements of the transaction-time flow $(\bar{T}, \bar{<})$ into the valid-time flow $(T, <)$, $f : \bar{T} \to T$, such that

$$f(\bar{t}+1) = \begin{cases} f(\bar{t}), & \text{indicating a } 90^\circ \text{ segment} \\ f(\bar{t})+1, & \text{indicating a } 45^\circ \text{ segment.} \end{cases}$$

We can look at f as partitioning $(\bar{T}, \bar{<})$ at disjoint, non-empty intervals over a discrete linear flow, each interval containing a finite number of points. Since the intervals are disjoint, the order $\bar{<}$ can be applied to the set of intervals, and the resulting order on intervals, $(f(\bar{T}), \bar{<})$, is isomorphic to the valid-time flow $(T, <)$. In such a setting, a diagonal point has the format $(\bar{t}, f(\bar{t}))$. The set of diagonal points remains monotonically increasing, so it still divides the two-dimensional plane into two areas that can be called 'the past' and 'the future'. The diagonal remains the set of two-dimensional points called 'now'.

15.5 Conclusion

In this chapter we have shown how both static and dynamic aspects of temporal databases can be modelled within a single temporal logic framework by using a multidimensional temporal approach. Several issues dealt with here are not confined to the domain of temporal databases, but are applicable to any system dealing with temporal reasoning and temporal data.

With respect to temporal databases, our presentation abstracted completely from any details of physical representation. This includes the fact of whether temporal element time stamps ought to be stored as the union of several intervals in a single tuple or as several tuples time stamped by a single interval. We firmly believe that considerations as such must not influence the design of the query language, and must be relegated to the physical level, hidden from any database user or application.

One aspect of the physical level of temporal databases remains for further investigation and that is whether temporal transactions should have a special treatment or whether they can be dealt in the same way as non-temporal ones.

16

TEMPORAL CONCEPTUAL-LEVEL DATABASES

M. Finger and P. McBrien

In this chapter we describe the temporal database concepts from a higher abstraction level. Such an abstraction level is called the conceptual level and it is from this level that the process of designing databases and database applications is normally initiated.

Our aim with this chapter is to show that all the database concepts do exist from this higher abstraction level. This means that the temporal concepts presented in earlier chapters relate to data in a general way, and it is not restricted to its presentation in a relational, fixed format. Indeed, we show that we can manipulate the database exclusively from this higher level. Such a manipulation includes a generation of a conceptual-level schema, semantics, querying, updates and even the inclusion of active rules in the database.

The starting point of the presentation of the conceptual level is the creation of a structured schema which is claimed to be better suited to model the universe of discourse (UoD) of real applications. Such a structured schema is presented in the form of an *ERT diagram* (an Entity–Relationship diagram extended with valid-Time concepts). In particular, we focus on the several temporal marks that are included in ERT to cope with the nuances of the representation of temporal features.

The structural placement of the elements of the ERT diagram imposes certain implicit constraints on the semantics of *instances* of the diagram. Those constraints are also reflected on what updates should be allowed and how updates are interpreted. Querying an instantiated diagram is done with the ERL temporal conceptual query language. ERL also has capabilities of representing triggered active rules, allowing the modelling of complex data constraints as well as creating a conceptual level temporal active database.

This rich collection of features allows the development of real applications from the conceptual level, as described in [Persson and Wohed, 1991; Wangler, 1993]. To enable the transformation of conceptual-level applications into running systems we need a translation mechanism that maps a conceptual-level database to an existing commercial relational system.

A translation mechanism is sketched in Section 16.1 without, however, presenting the full details which are dependent on the target (commercial) database system. We can then proceed to describe in detail the elements used in the (valid-time) temporal modelling of a UoD using ERT diagrams; this will be done in Section 16.2, where the emphasis will be on the modelling of temporal features and the semantics of ERT instances. Section 16.3 then shows how such a diagram can be formally queried. The transformation of a conceptual-level database from a passive repository to an active one is finally described in Section 16.4, where the full ERL rule language, together with its active semantics, will be presented.

16.1 Motivation

Designing a database, be it temporal or not, is far from a trivial task. The activity of designing a database schema involves dealing with abstractions on a conceptual level that is higher than the level of relations. Relations are a flat, homogeneous way to represent information, but the the UoD, or the part of the real world that is being modelled by the system, turns out to be a much less homogeneous collection of interacting elements.

Therefore, in order to better model the UoD we need structures that are better abstractions than simple, flat relations. It was with this goal in mind that Peter Chen proposed an entity–relationship diagram as a higher level modelling abstraction [Chen, 1976]. The activity of constructing those higher-level abstractions of data became known as *conceptual modelling*, which is normally performed by a category of professionals known as systems analysts.

It is important to stress that the fact that systems analysts are not normally associated with formal, rigorous mathematicians does not imply that the models that they build cannot be seen as rigorous, formal models of the UoD. Quite the opposite; it is only by considering those conceptual models as a mathematical object with a formal definition and semantics that they can become useful tools for the design of database schema and applications, with the support of a *provably correct* translation into a lower level runnable system. Indeed, since we have a higher level view of data from the conceptual level, why not access and manipulate the data from this level alone? Let us see how to do that.

From a formal point of view, a conceptual-level diagram will be nothing but a structured database schema, as opposed to the 'flat' schema one finds in relational databases. At the conceptual level, we can consider that we have a conceptual database with a (structured) schema, an instantiation, and that queries can be posed to it. In this way, we can talk about translating the conceptual database into a relational database in such a way that the translation is property preserving.

What properties do we want to preserve? The basic guideline is that we want the conceptual-level database to be *observationally equivalent* to the database obtained after the translation. More specifically, we want to translate the schema, instance and queries in the conceptual level into a schema, instance and queries in the relational level that are observably equivalent, as illustrated in Fig. 16.1.

The observations required to be equivalent are the following: any query posed to the conceptual-level representation is translated to a relational-level query that generates an *equivalent* answer. So query answers are the only observations we are considering worth preserving. But note that we stated this requirement in terms which demand the answers to be only equivalent, not identical; this

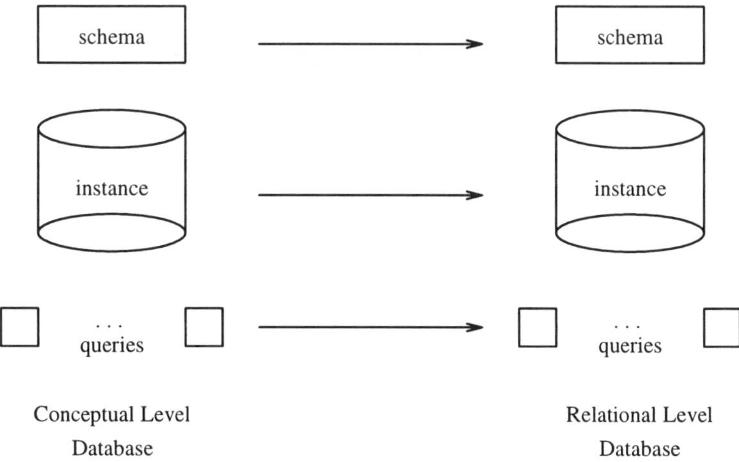

FIG. 16.1. Translation from conceptual to relational.

means that once the answer is obtained at the relational level there is one more translation step from the relational answer back to the conceptual level before we can talk about equality of answers.

In these four translations, the conceptual- to the relational-level cases need schema, instance and query translation and the relational- to the conceptual-level case needs answer translation.

The notion of observable equivalence can be summarized by the commutativity of the diagram in Fig. 16.2. On the conceptual level, we have a conceptual database CDB and a conceptual query q generating an answer $q(CDB)$. Suppose α is a translation, such that at the relational level we have database CDB_α, a query q_α generating an answer $q_\alpha(CDB_\alpha)$. Now, to compare the result of the query obtained in the relational with that of the conceptual level, we need a reverse translation, α^{-1}, so that we can bring both results together, $q(CDB) = \alpha^{-1}(q_\alpha(CDB_\alpha))$. Indeed, if we can construct translations α and α^{-1} we can simulate a conceptual-level querying using those translations and a relational-level database.

This same notion of query-observability can be used to incorporate the dynamics of the database in the translating process. Updates to the conceptual level are translated into updates in the relational level such that, after the update, the results of any queries to the conceptual level remain equivalent to their corresponding queries in the relational level.

Going one step further, we can have active rules in the conceptual level too, but then we have to include external actions as part of the observation set, so that, for example, if a conceptual rule causes a program to be executed in a given situation, its relational equivalent should cause the same program to be executed in the equivalent (i.e. the translated) situation.

The basic idea of these translations is the following: we can pretend that we

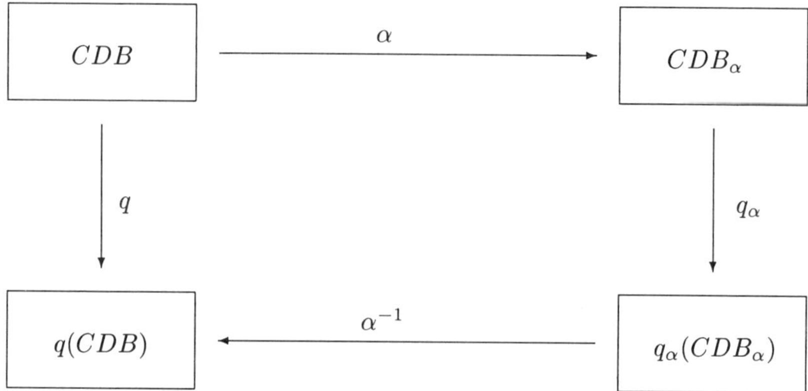

FIG. 16.2. Observable equivalence.

have a conceptual-level database, and that we access and manipulate data *only from the conceptual level*; the translations take care of implementing the conceptual access and manipulation of data using, for example, an existing commercial database at the relational end of the diagram in Fig. 16.2.

So far the issue of time has been totally absent from our discussion on conceptual-level databases and their implementation. Let us bring it into the picture now. One of the advantadges of having a flat and homogeneous model at the relational level is that we can have a homogeneous treatment of time. Indeed, for a relation to become temporal, all we had to do was to add a times tamp to each of its tuples. But once the conceptual model contains extra 'structure', one has to see how the notion of time can be accommodated into the structure. A first approach would simply be to insert a time stamp in all the structure elements, and this seems easily achievable. However, this temporal extension is not sufficient. It fails to capture all the temporal subtleties of possible relationships between temporal entities. Certain types of temporal relationships do impose some constraints on the range of values the time stamps can assume, a constraint that does not exist in the relational model. So the temporal semantics of the elements of the diagram has to deal with the temporal constraints that their instances have to obey.

We now have to fill in the details of this generic discussion. We should now proceed by introducing the elements of the diagram we want to consider. The presentation of an actual target relational model and the detailed translation remain outside the scope of this work.

16.2 ERT diagrams

We start by showing how to compose the conceptual-level schema of a conceptual database, which will be represented by an *ERT diagram*. The ERT diagram was developed by the TEMPORA project [Loucopoulos *et al.*, 1990] as a tool for capturing temporal aspects of the UoD, having in mind its possible applications for temporal databases.

The original form and presentation of the ERT diagram can be found in [McBrien *et al.*, 1992; Wangler, 1992], which we slightly adapt here to bring it closer to more standard ER (Entity–Relationship)notation, such as that used by [Batini *et al.*, 1992]. The ERT semantics developed in [Wangler, 1992] is not a formal one, and we take the task of presenting in this chapter a formal semantics for both ERT and its query language ERL, in the style of semantics found throughout this book. The application of ERL in the modelling of medium to large applications (i.e. non-toy-sized case studies) is described in [Persson and Wohed, 1991; Wangler, 1993].

Another ER model extended to deal with time has been developed by Elmasri *et al.* [1993], which extended the extended entity relationship (EER) model with temporal capabilities, generating a temporal EER model, TEER. Like ERT, it is used to model valid-time information, and it comes with a query language (GORDAS) allowing the manipulation of a conceptual-level database. However, ERT differs from it in several points concerning the modelling of time:

- ERT distinguishes between temporal and non-temporal (eternal) entities, while in TEER all entities are, in principle, temporal;
- ERT has a much richer way of capturing the temporality of relationships given by the several temporal markings given below;
- ERT allows one to define temporal restrictions in hierachies.

These all witness the preoccupation of ERT designers in generating a modelling tool capable of capturing the several distinct but important temporal subtleties found in modelling real applications, aiming at being more than just a temporal extension of an existing modelling tool. We proceed now by describing the elements of the ERT.

16.2.1 *Entity classes*

The basic element of the diagram is the *entity class* (sometimes, for brevity, we call it just entity), which in ERT comes in two formats represented as

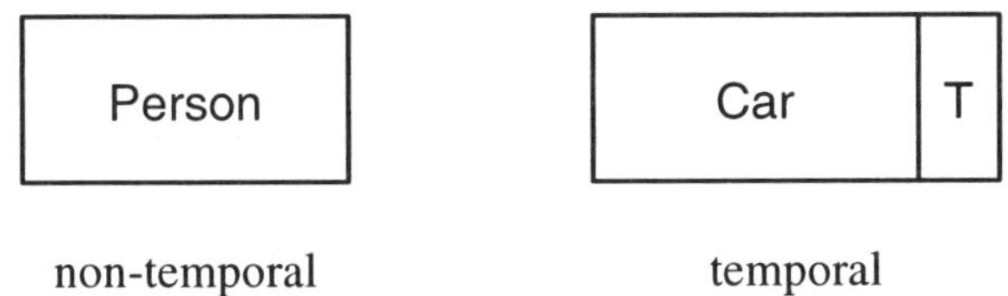

A non-temporal entity class such as Person above is actually a perennial entity. It means that in the model we are building we are not interested in when people are born or die—all we want to know is whether a particular person ever existed or not. On the other hand, the entity class Car has a T-mark, which means that cars may exist at some times in the UoD and not in others. For example, we may be interested in the dates of manufacture and destruction of a car.

Since our model is a formal one, we have to present the semantical counterpart of each element we introduce. Each element of the ERT diagram will be semantically instantiated. If $\mathcal{I}$ is an instance of the whole diagram and e is an entity class in it, $\mathcal{I}(e)$ represents the instance of that entity class. To define $\mathcal{I}(e)$, consider a flow of time $(T, <)$ and a countable set of unique identifiers, $Id = \{id_0, id_1, \ldots\}$. An *instance* of an entity class $\mathcal{I}(e)$ is a finite set of time-stamped identifiers, i.e.

$\mathcal{I}(e)$ is a finite set of pairs $\langle id, \tau\rangle$, with $id \in ID$ and $\tau \subseteq T$.

For perennial entity classes (non T-marked), all time stamps $\tau = T$, i.e. all identifiers are eternal. Typically the time stamp τ will be a temporal element or an interval over a discrete flow of time. Each identifier in an instance identifies a single entity in the model (hence the diagram element above being appropriately called an entity class for it can contain several entity identifiers). Furthermore, each entity class will be named with a unique name in the diagram.

Entity classes are associated to attributes

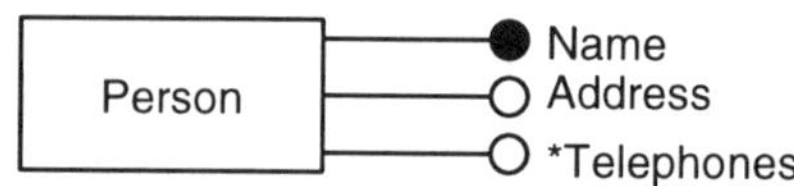

Each attribute *att* refers to a domain $dom(att)$. In the example above, we have that $dom(\mathsf{Name}) = dom(\mathsf{Address}) = string$ and $dom(\mathsf{Telephones}) = 2^{dom(\mathsf{Telephone})}$; Telephones is a multivalued attribute, as indicated by the *-symbol. Semantically speaking, in a non-temporal environment an instance of attribute is a function from the set of identifiers to its domain, $\mathcal{I}(att) : Id \to dom(att)$. This function is partial, for it is defined only for the identifiers present in the entity e to which the attribute is connected.

In a temporal environment there are non-temporal and temporal attributes (carrying T-marks). Non-temporal attribute instances remain functions from the identifiers in the range of $\mathcal{I}(e)$ to the attribute domain. In the temporal case, the instance of an attribute contains a set of pairs value–time stamp for the different values the attribute may assume during the existence of the entity. Formally, we cope with T-marked attribute semantics in the following way. Let *att* be an attribute connected to entity class e. Then $\mathcal{I}(att)$ is a function from the identifiers in range of $\mathcal{I}(e)$ to the attribute domain such that

$$\mathcal{I}(e) = \langle id, \tau\rangle \Rightarrow \mathcal{I}(att)(id) = \{\langle a_1, \tau_1\rangle, \ldots, \langle a_n, \tau_n\rangle\}$$

$$\text{where} \begin{cases} a_i \in dom(att) \text{ for } 0 \leq i \leq n \\ \tau_i \cap \tau_j = \emptyset \qquad \text{for } i \neq j,\, 0 \leq i, j \leq n \\ \displaystyle\bigcup_{i=1}^{n} \tau_i \subseteq \tau. \end{cases}$$

The reason for allowing $\bigcup_{i=1}^{n} \tau_i \subseteq \tau$ instead of $\bigcup_{i=1}^{n} \tau_i = \tau$ is that some attributes may be optional and not have a value at all times; see the end of this section for notation for optional attributes. In the unmarked case the constraint is simplified to $\mathcal{I}(e) = \langle id, \tau \rangle \Rightarrow \mathcal{I}(att)(id) = \langle a, \tau \rangle$. By refining the example above under this temporal view, we can obtain

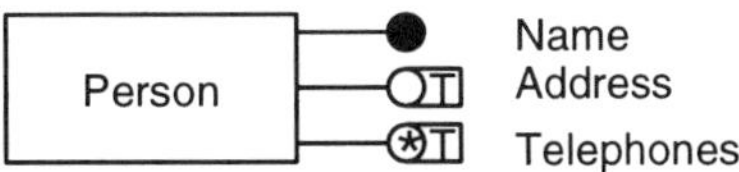

where a person can now have different addresses at different times, and similarly with telephone numbers, but the name of the person remains the same throughout time. Two time stamped identifiers that agree on all attributes on all times *need not be* identical.

The black-marked attribute above represents a *determiner attribute* and it imposes a functional dependency on identifiers; that means, if *att* is a determiner attribute, $\mathcal{I}(att)$ is a bijection. In the example above, each name determines the value of an identifier and, by consequence, its address and telephones at different times. Two entities with the same values for the determiner are necessarily identical. It is also possible to have compound determines which include more than on attribute, as illustrated in (a):

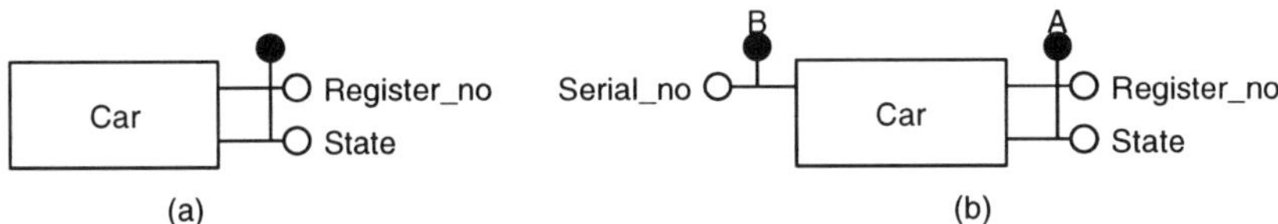

In this case, a car is determined both by its registration number and the state in which it was registered. Semantically this means that there is a bijection between car identifiers and pairs of registration numbers and state names. All the attibutes involved in a determiner, simple or compound, cannot be temporal. It is also possible to have alternative determiners, as illustrated in (b); both the serial number and the pair $\langle$ registration number, state $\rangle$ determine the car, and thus determine each other.

An entity class without a determiner is called *weak*. Weak entity classes have either external determiners (e.g. another entity connected to it by a relationship) or a hybrid internal–external composed determiner that combine some of its attibutes with an external entity. Since there is nothing temporal in this discussion we will not pursue it here; for a detailed account of determiners and their representation, refer to [Batini *et al.*, 1992].

It is common to omit all or several of the attributes of an entity class in a complex ERT diagram to avoid cluttering. Since entity class names are unique in the diagram, the full set of attributes can be presented separately for each entity class.

16.2.2 *Relationship classes*

A *relationship class* (or a relationship, for short) is an association between two entity classes which present us with particular interesting semantical choices for its temporal behaviour. The ownership relationship between Person and a Car is illustrated as

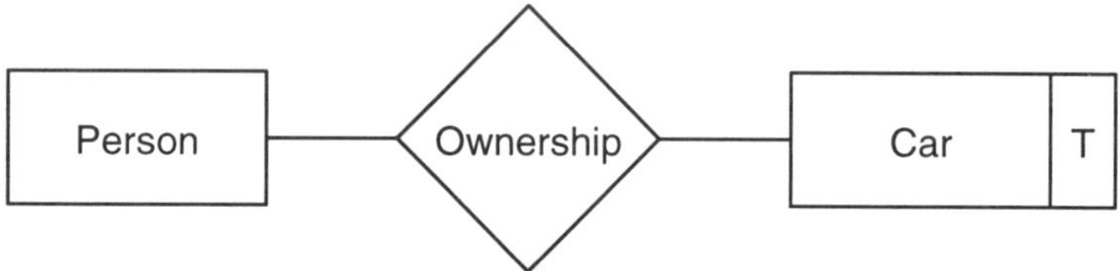

More frequently, however, instead of representing a relationship class by a single name we will create aliases for it and use the aliases as labels on the arcs connecting the relationship to each entity class. In this way we use the 'directional' aliases owns and belonging_to for the ownership relationship class; this will be quite helpful in the definition of a 'close to natural language' query expression later on

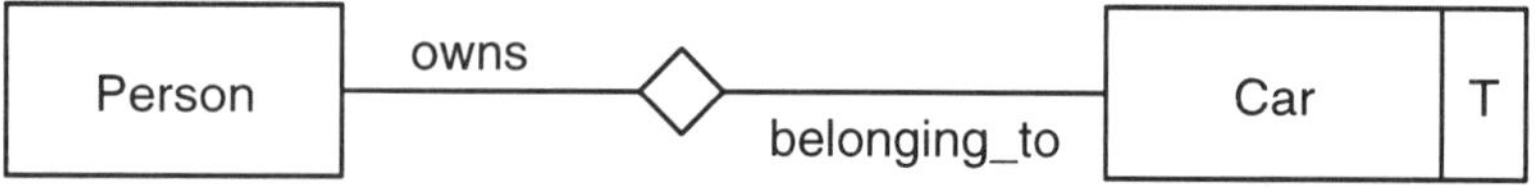

We can express the ownership relationship in the diagram above in the following way: (i) from left to right, 'a person owns a car'; (ii) from right to left, 'a car belonging to a person'. In (i) the reading generated a sentence, while in (ii) it generated a noun phrase. Actually in (ii) we are focusing on the entity and using the relationship to restrict it; the same could be done with (i) if we read it as 'a person *that* owns a car'.

In a non-temporal setting, an instance of a relationship class connecting two entity classes contains a subset of the possible pairs built from the instances of the entity classes. But in our setting time has to be taken into account, and there are several choices.

Before we consider the addition of T-marks to relationships, in the manner that was done previously to entity classes and attributes, let us first consider what an unmarked relationship should mean. The instances of the entity classes participating in the relationship are now time stamped. It makes no sense to throw away the time stamps and declare that the relationship is eternal when the entities involved in that relationship are not. So unmarked relationships between temporal entities do not, in general, have eternal instances; they must have a time stamp and that time stamp must be obtained from the time stamps of the involved entities. Intuitively, a standard (unmarked) relationship between two entites should last as long as the entities involved coexist. Formally, if r is an unmarked relationship between entities e_1 and e_2, and $\mathcal{I}$ is an instance of the ERT diagram, then the instance $\mathcal{I}(r)$ of the relationship r is a set of triples such that

$$\langle id_1, id_2, \tau\rangle \in \mathcal{I}(r) \Rightarrow \begin{cases} \langle id_1, \tau_1\rangle \in \mathcal{I}(e_1) \\ \langle id_2, \tau_2\rangle \in \mathcal{I}(e_2) \\ \tau = \tau_1 \cap \tau_2. \end{cases}$$

By defining the time stamp of an instance of an unmarked relationship as the intersection of the time stamps of the entities involved, our model remains a true extension of a non-temporal model. For in a non-temporal model, all elements are unmarked, so entities are eternal and their relationships are time stamped by the intersection of two eternities, and hence they are eternal too.

However, looking back at our example of the ownership of the car, we see that it must be incorrect to leave the relationsip unmarked, for it prevents the buying and selling of cars. As it stands, the ownership of a car lasts as much as the life of the car (intersected with the eternity of a person). To allow for a better representation, we introduce T-marked relationships:

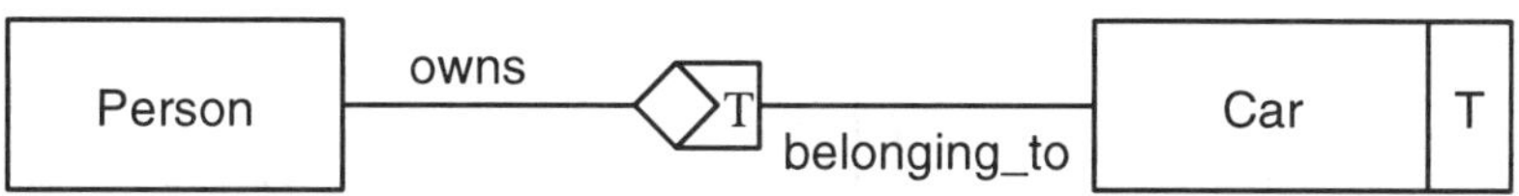

A T-marked relationship allows a relationship to hold for a subset of the time that both entities involved coexist. So while a person can only own a car while it exists, the ownership can terminate without the destruction of the car (i.e. by selling the car), and it can start after the car's manufacture (by buying a used car). A genuine unmarked relationship can be found, for instance, between a car and its manufacturer:

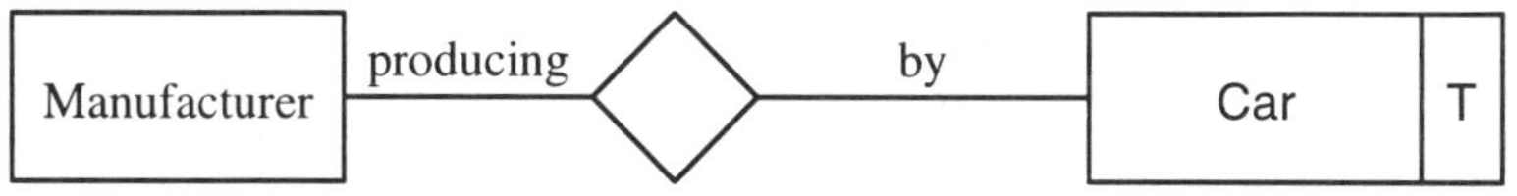

The T-mark, however, does not solve all the problems with temporal relationships, for there may exist a genuine relationship between two entities that never coexist. The typical case is the relationship grandparent_of that may hold between two persons in a UoD where people's lifespans are relevant. For those cases we introduce H-marked relationships:

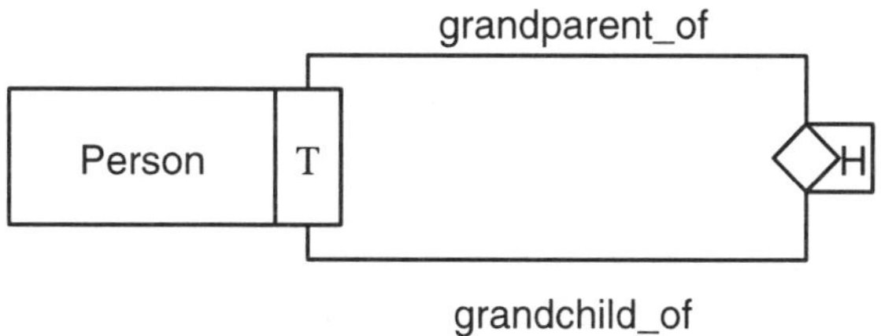

An H-marked relationship holds over the convex union of the time stamps of the two entities involved, defined by $Convex(\tau_1, \tau_2) = \{t \in T \mid \exists t_1 \in \tau_1, t_2 \in \tau_2, t_1 \leq t \leq t_2\}$. For example, the relationship grandparent_of/grandchild_of starts holding when the grandparent is born; if he or she dies before the grandchild is born, the relationship still holds during that gap—it only finishes when the grandchild ceases to exist. Such a relationship is not inserted in the model when the grandfather is born—that would be fortunetelling. More realistically, the relationship may be inserted when the grandchild is born, retroactively changing the past from that time back to the birth of the grandparent; it persists until the grandchild dies. Admittedly, modelling the grandparent_of/grandchild_of relationship in such a way may not be the most appropriate in all cases; it is possible that we want it to hold just while the grandchild is alive and other restrictions may be imposed in similar situations. To cope with those problems, we allow for the double HT-marking, meaning that the relationship holds over some subset of the convex union of the time stamps of the two entities involved:

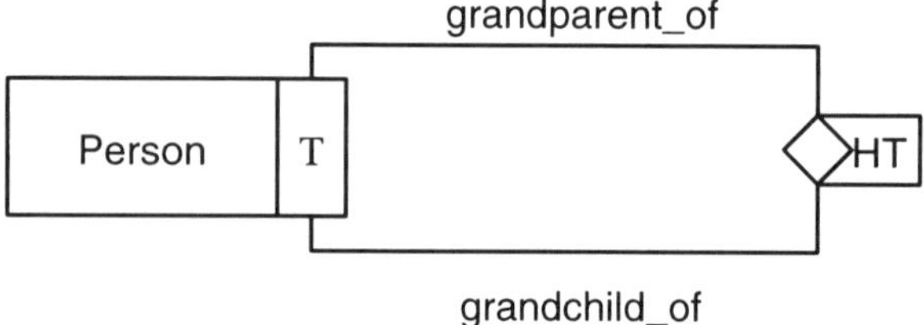

The following summarizes the temporal semantics for the four kinds of temporal relationships. Let r be a relationship between entities e_1 and e_2, and let $\mathcal{I}$ be an instance of the ERT diagram; then the instance $\mathcal{I}(r)$ of the relationship r is a finite set of triples such that

$$\langle id_1, id_2, \tau\rangle \in \mathcal{I}(r) \Rightarrow \begin{cases} \langle id_1, \tau_1\rangle \in \mathcal{I}(e_1) \\ \langle id_2, \tau_2\rangle \in \mathcal{I}(e_2) \\ \tau \text{ is such that } \begin{cases} \tau = \tau_1 \cup \tau_2 & \text{if } r \text{ is unmarked} \\ \tau \subseteq \tau_1 \cup \tau_2 & \text{if } r \text{ is T-marked} \\ \tau = Convex(\tau_1, \tau_2) & \text{if } r \text{ is H-marked} \\ \tau \subseteq Convex(\tau_1, \tau_2) & \text{if } r \text{ is HT-marked.} \end{cases} \end{cases}$$

The semantics of relationships is normally further restricted by cardinality constraints. In our notation, a cardinality constraint is a positive integer interval annotated on the arcs of the relationship. For example, consider the following cardinality constraints:

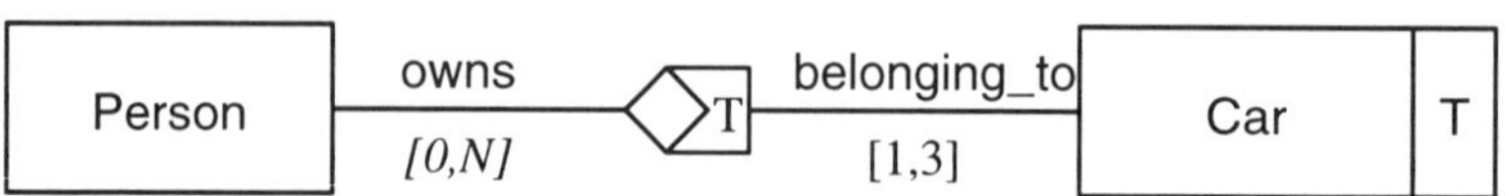

Its interpretation is the following. *At each moment of time*, a person can own any number of cars—from no cars to an unbound number N of cars; conversely, a car must belong to at least one person, and at most to three persons. So every relationship r between entities e_1 and e_2 is subjected to two cardinality constraints, namely $c(\langle e_1\ r\ e_2\rangle)$ and $c(\langle e_2\ r\ e_1\rangle)$. In the example above, we have $c(\text{Person owns Car}) = [0, N]$ and $c(\text{Car belonging\_to Person}) = [1, 3]$.

Formally this may be expressed in the following way. Let $c(\langle e_1\ r\ e_2\rangle) = [a, b]$ be a cardinality constraint. For each element $\langle id_1, \tau_1\rangle \in \mathcal{I}(e_1)$ and for each $t \in \tau_1$, then

$$a \leq |\{x \mid \langle id_1, x, \tau\rangle \in \mathcal{I}(r) \text{ and } t \in \tau\}| \leq b.$$

We can extend the cardinality notation to attributes as well, as in the following:

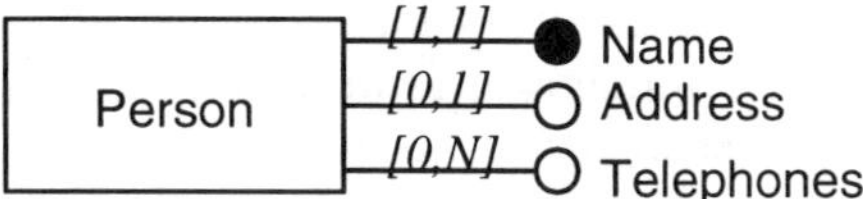

The cardinality shows us that Address is an optional attibute, with minimum 0 and maximum 1 value. The cardinality $[0, N]$ for telephones substitutes the *-notation, and tells us that Telephones maps a person to a possibly empty (0), finite but unbound (N) set. The cardinality $[1, 1]$ on Name determines that it is a truly functional attribute, with exactly one value for each person that is mandatory; in fact, whenever a cardinality is not explicitly mentioned for an attribute, we will assume it to be $[1, 1]$ by default.

Similarly, we could label the attribute arcs in the same way we did for relationships, but instead of doing it explicitly, we will assume attribute arcs are implicitly named by the preposition with, so we can read an instance of the diagram above as 'a person with name 'Joseph', with address '12 Nowhere Place', with telephones '{123-4567, 987-6543}' '.

Finally, in Chen's original ER model it was allowed for relationships to involve more than two entities, but our model will be restricted to binary relationships. We are also excluding attributes from relationships. Those exclusions do not diminish the formal expressivity of the model, for those constructions can be replaced with the constructions we do allow. For example, it is well known that an n-ary relationship can be replaced with an entity class plus n binary relationships:

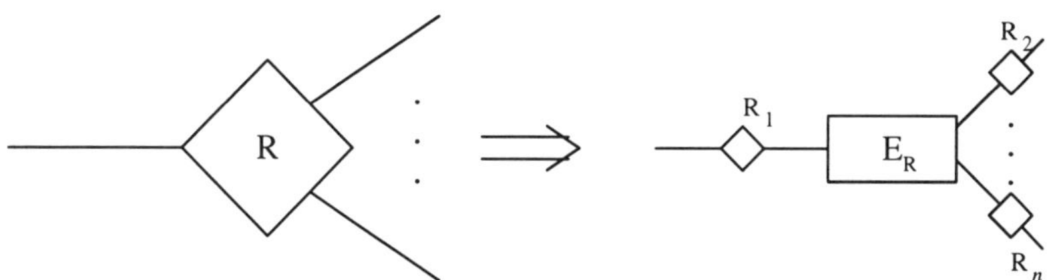

As a matter of fact, n-ary relationships are consistently reported as 'hard to understand' and we believe that the alternative formulation is at least as clear;

furthermore, the querying language is much simplified and stays much closer to 'natural language' if non-binary relationships are kept out. A relationship that demands an attribute can be transformed into an entity class with two relationships, and the attributes attached to the new entity class:

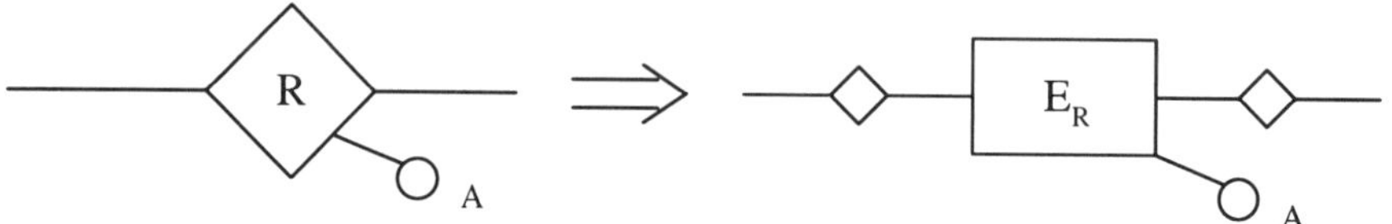

Our claim for keeping this feature out is much less strong, basically to keep the simplicity of the model. If the need for relationship attributes increases, they can be reintroduced in the diagram (and also in the query language) with the expected semantics in a straightforward way.

16.2.3 *Hierarchies*

Chen's model [Chen, 1976] included only entities and relationships, but a now largely adopted extension of the model is isa-hierarchies. Hierarchies embody the specialization/generalization data abstraction.

Hierarchies of entity classes capture the possible specialization of a given entity class with the inheritance of attributes and the addition of new ones. For instance, an entity class such as Person in our representation of the world can be refined into an entity class that includes drivers:

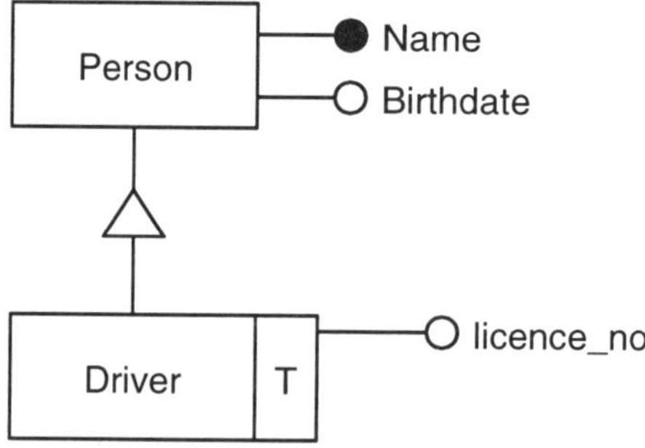

So a Driver is a Person and thus inherits all its attributes: a Driver also has a name and a Birthdate, even though it is not explicitly shown in the drawing; it is determined by name as well. Furthermore a Driver has an exclusive attribute licence_no, meaning that not all persons have a licence number, but those who are drivers certainly do. In the hierarchy above, we say that the entity class Person is the *parent* and that Driver is the *child*. Besides inheriting the attributes of its parent, a child entity class also inherits the relationships in which the parent participates. So if a Person can own a car, so can a Driver. Also note that there are persons that may not be drivers.

We are of course more interested in the role that T-marks assume in a hierarchy. The convention we adopt is the following:

(a) an unmarked child inherits the T-mark and the time stamp from its parent;

(b) a T-marked child restricts the duration (time stamp) of its parent.

Let us focus on (b) first. In the hierarchy above, instances of Person are eternal, which means we are only interested if a person exists or not. But a Driver is a specialization of a Person and we are interested in when a person becomes a driver—which would be a *migration* of an identifier from the parent class to its child; we are going to see in the semantics that the identifier does not disapear from the parent, but appears in the child with a shorter life. We may also be interested when the person stops being a driver, e.g. when the licence expires.

On the other hand, a Ferrari is a car, and since Car is a T-marked entity, from item (a) above we would have *(i)* below:

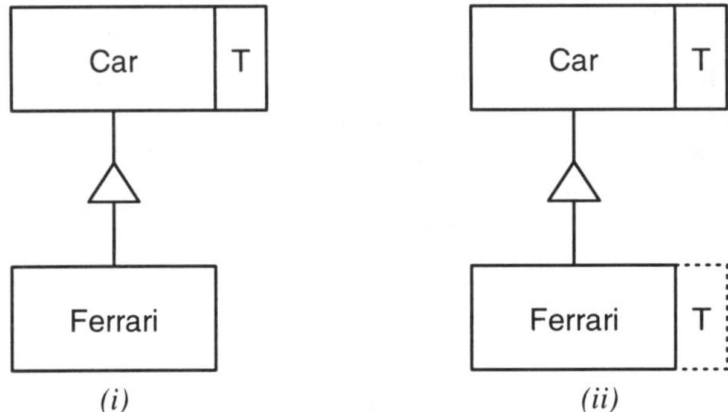

meaning that a Ferrari is a Car. But such a representation is misleading, for Ferrari looks like an eternal entity class. So we adopt the notation of *(ii)*, where entity classes that inherit their parent's temporality appear with a dotted T-mark.

Hierarchies also permit us to capture finer-grained details of the UoD. For example, the hierarchies below shows that both Ferraris and Ladas are cars, but that the class of Ferraris is disjoint from the class of Ladas.

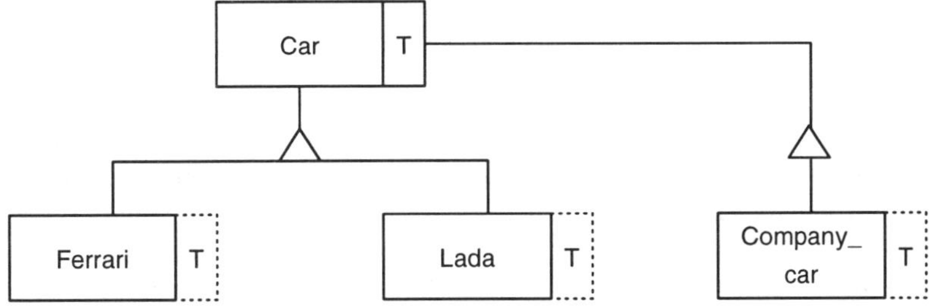

However, company cars are also cars, and they are represented in a different class, meaning that they can be Ferraris, Ladas or neither.

All hierarchies shown so far are *partial*, in the sense that an instance of a child entity class is an instance of the parent entity class, but an instance of the parent entity class needs not be in one of its children. So cars can be Ferraris or Ladas, but not necessarily remain an unspeciali'zed car. On the other hand, we can have *total hierarchies*; for example, a Person must be either a Man or a Woman:

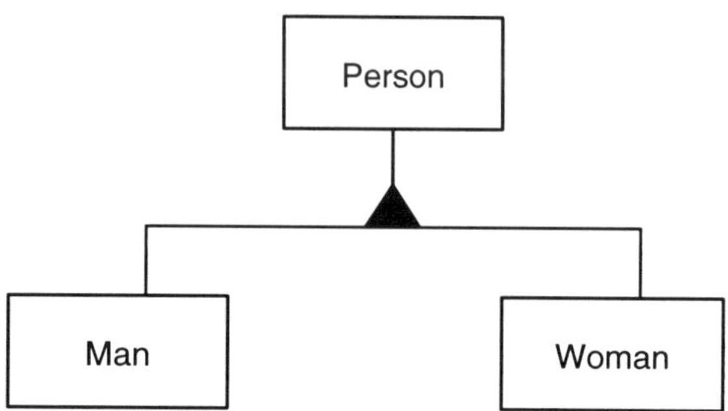

Total hierarchies actually partition the parent entity class into disjoint classes; obviously, in a total hierarchy, the parent must have more than one child. Total hierarchies can also be used to represent the *generalizations* of several entity classes. For example, the diagram

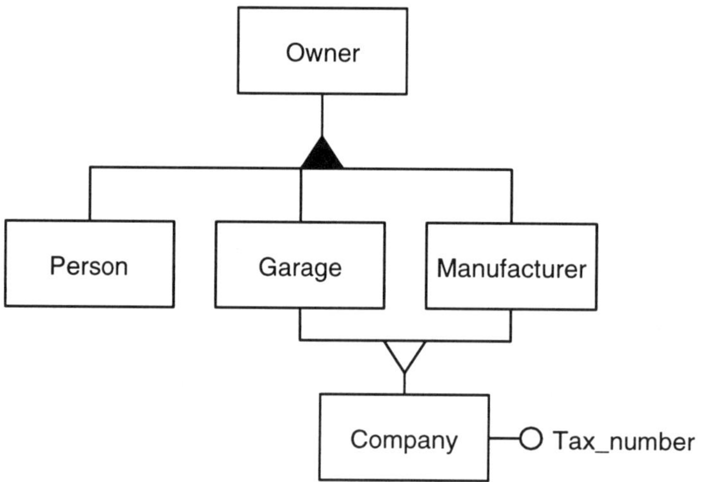

intends to model the fact that a (car) owner can be a person, or a garage or a (car) manufacturer; the owner is therefore a generalization of three other independent entity classes. The diagram also shows the possibility of having multiple inheritance: that is, Garage and Manufacturer are both owners, inheriting from Owner its relationships (e.g. owning a car, not represented in the diagram); they are also both companies, inheriting from Company its attributes (e.g. the tax number).

We allow multiple inheritance for an entity class, i.e. an entity may be the child of more than one parent, provided the parent entity classes involved do not have attributes and a relationship with identical names that would cause a conflict in inheritance. However, we must note that multiple inheritance is not always a good modelling tool; in our experience we find that multiple inheritance is sometimes difficult to understand and hard to handle.

Smaller hierarchies can be composed to form larger ones. An obvious restriction for large hierarchies is that they must not be cyclic.

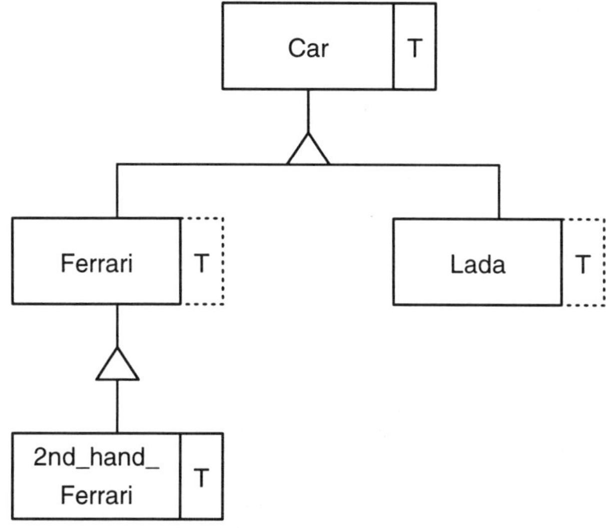

Just note in the hierarchy above that Ferrari inherits its (dotted) T-mark from Car, but 2nd_hand_Ferrari has a solid T-mark, indicating its instances have a more restricted lifespan than the instances of Ferrari. In other words, a car remains a second-hand Ferrari for a shorter period than the period that it remains a Ferrari (which is, by the diagram, its whole life). This is in accordance with principle (b) above.

Formally, we create an isa_0 relation between a child entity class and its parent, e.g. Ferrari isa_0 Car and 2nd_hand_Ferrari isa_0 Ferrari. The isa relation is the transitive closure of isa_0, $isa = isa_0^*$, e.g. 2nd_hand_Ferrari isa Car. For the semantics, suppose we have a hierarchy that is either partial or total, as in

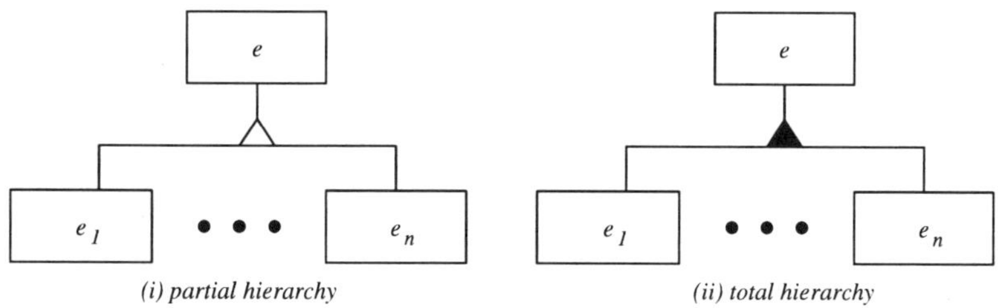

(i) partial hierarchy *(ii) total hierarchy*

Let e be the parent node and let e_i be one of its children. Suppose $\mathcal{I}$ is an instance of the diagram. With respect to the hierarchy, we have the following constraint:

$$\langle id_i, \tau_i \rangle \in \mathcal{I}(e_i) \Rightarrow \begin{cases} \exists \tau, \langle id_i, \tau \rangle \in \mathcal{I}(e) \\ \text{such that} \begin{cases} \tau_i = \tau, \text{ if } e_i \text{ is not T-marked} \\ \tau_i \subseteq \tau, \text{ if } e_i \text{ is T-marked.} \end{cases} \end{cases}$$

Furthermore, for sibling entity classes we have

$$\mathcal{I}(e_i) \cap \mathcal{I}(e_j) = \emptyset \text{ for } i \neq j.$$

It follows that, for partial hierarchies, the instance of the parent e contains that of the children:

$$\mathcal{I}(e) \supseteq \bigcup_{i=1}^{n} \mathcal{I}(e_i);$$

but for total hierarchy $\mathcal{I}(e)$ is partitioned into the children, resulting in

$$\mathcal{I}(e) = \bigcup_{i=1}^{n} \mathcal{I}(e_i).$$

16.2.4 *A full ERT diagram*

In Fig. 16.3 a more complete ERT diagram is presented containing the (part of a) model of a car ownership as viewed from a car registration authority. This example is taken from a situation proposed as a trial case for different modelling tools in [ISO, 1982]. A somewhat more refined and detailed ERT modelling of the same problem can be found in [Wangler, 1993].

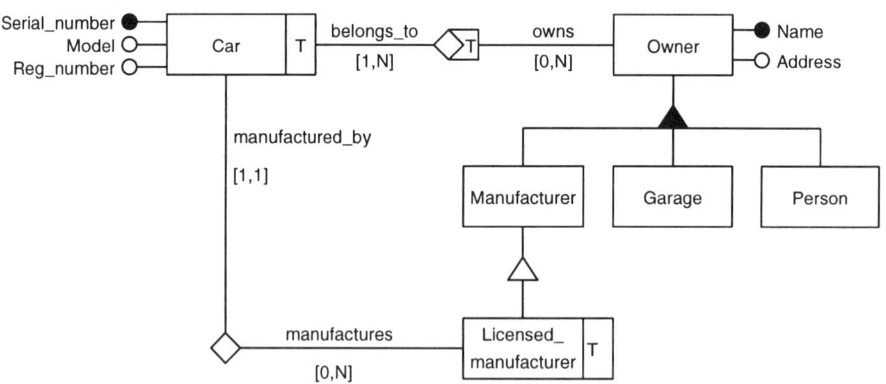

FIG. 16.3. Car registration model.

The important aspects that we try to capture with Fig. 16.3 are the following:

- a manufacturer, a garage, a person, or sets of them (due to $[0, N]$-cardinality constraint) can be a registered car owner;

- the name and address of each car owner must be known;
- a manufacturer can start production only after it becomes licensed to do so, and it is important to know the date the licence takes effect (this is handled implicitly by the T-mark in Licensed_manufacturer);
- a car is uniquely identified by its serial number. It also has a registration number given by the registration authority and a model.

Some features which are *not* modelled by the diagram, and if one wants to include them, should be formulate using integrity constranits (see below):

- if a car is owned by a set of owners, they must all be persons;
- manufacturers cannot own used cars;
- a garage cannot sell a car to another garage.

So there are some limitations to the expressivity of ERT diagrams.

We formally present an ERT diagram as a graph-like structure represented by a 6-tuple $G = \langle E, R, A, H, Mark, Conf \rangle$, where:

- E is a set of entity classes;
- R is a set of relationship classes;
- A is a set of attributes;
- H is a set of hierarchies;
- $Mark$ represents the temporal marking of the elements of the diagram; it is a function that takes an element of E, R or A and returns a value representing the fact that the element is either unmarked or T-marked, or in the case of an element of R, H-marked or HT-marked;
- $Conf$ is a function that gives the configuration of the diagram in the following way:
    * for each entity class $e \in E$, $Conf(e) = \langle A_d, A_{nd} \rangle$, where $A_d \subseteq A$ is the set of determinant attributes of e, and $A_{nd} \subseteq A$ is the set of non-determinant attributes of e;
    * for each relationship $r \in R$, $Conf(r) = \langle e_1, e_2, alias_{12}, c_{12}, alias_{21}, c_{21} \rangle$, where $alias_{12}$ and $alias_{21}$ are the directional aliases for the relationship and c_{12} and c_{21} are the corresponding cardinality constraints;
    * for each hierarchy $h \in H$, $Conf(h) = \langle e, \{e_i\}, type \rangle$, where $e \in E$ is the parent of the hierarchy, $\{e_i\} \subseteq E$ is the set of children and $type$ indicates whether the hierarchy is partial or total.

16.3 The ERL query language

We present in this section the ERL query language, which is capable of extracting information from (an instance of) a conceptual-level database. Unlike what it may apparently look like, ERL *does not* stand for entity–relationship language but is an acronym for extended rule language (the extension refers to its treatment of time).

That means ERL also has capabilities of updating the database and of representing rules that transform the conceptual-level database from a passive repository of data into an active one, in the same way that was described in Chapter 17. So ERL allows us to construct a conceptual level temporal active database.

The ERL query fragment is presented here in two phases. First, we introduce the *canonical language*. This part of the ERL looks very much like a relational calculus query language, which is in many senses similar to the query language we have been using so far (with some extra 'syntactic sugar'); a semantics for queries is also presented.

In the second phase, several abbreviations are introduced to the language, making the query language appear better suited for querying graph-like structures such as ERT diagrams. The abbreviations, however, are no more than that. They maintain the semantics of the formulae for which they provide an abbreviation.

We leave for the next section the discussion of the other uses of ERL, namely updates, constraints and active rules.

16.3.1 *Canonical ERL*

We start with the definition of the basic concepts which are, at the lower level, different enough from the usual logical languages to deserve special attention.

Atomic formulae A *term* θ in ERL, as usual in databases, is either a constant or a variable; variables in ERL always start with a capital letter, following a Prolog-like notation. Terms are sorted and the possible sorts in an ERL associated to an ERT diagram are the 'identifier' sort, the 'time' sort and the sorts of the domains associated with the attributes in the diagrams.

Given an ERT diagram $G = \langle E, R, A, H, Mark, Conf \rangle$, a *qualified term* is an expression of the form $\mathsf{name}\ \theta$, such that:

- name is either an entity name in E or an attribute name in A;
- both name and term θ have to be of the same sort; the sort of an entity name is 'identifier' and the sort of an attribute name is its domain.

The set of ERL atomic formulae based on an ERT $G = \langle E, R, A, H, Mark, Conf \rangle$ consists of:

- qualified terms, $\mathsf{name}\ \theta$, where $\mathsf{name} \in E$ is an entity name;
- $\theta_1\ op\ \theta_2$, where $op \in \{=, <, >, \leq, \geq, \neq\}$ and θ_1 and θ_2 are terms of the same sort;
- $\mathsf{name}_1\ \theta_1\ \mathsf{rel}\ \mathsf{name}_2\ \theta_2$, where $\mathsf{name}_1\ \theta_1$ and $\mathsf{name}_2\ \theta_2$ are qualified terms and rel is a *relationship allowed* between name_1 and name_2, which means:
  - ∗ name_1 is an entity name and name_2 is an attribute name, and name_1 either possesses or inherits name_2, in which case $\mathsf{rel} \equiv \mathsf{with}$;
  - ∗ both name_1 and name_2 are entity names and there are name'_1 and name'_2 such that name_1 isa name'_1 and name_2 isa name'_2 and there is a directional alias named rel from name_1 to name_2. In other words,

name_1 and name_2 either are directly linked by rel or inherit such a link.

For example, given the ERT diagram in Fig. 16.3, we can construct the following well-formed atomic formulae:

- Qualified terms: car C, name 'Joseph', person P, etc.
- Comparisons: C = P, X < 26, X ≠ Y, etc.
- Relationships: person P with name 'Joseph', garage G owns car C, car C manufactured_by licensed_manufacturer L, etc.

Compound formulae The formulae of the ERL query language are of the form:

- atomic formulae;
- A and B, A or B, not A, where A and B are ERL formulae;
- exists X for_which A, forall X such_that A, where A is an ERL formula;
- A since B, A until B, where A and B are ERL formulae.

As can be clearly seen, the ERL query language is obviously a first-order modal–temporal language adapted to a structured signature. With respect to the temporal modalities, one-place modalities can be defined in terms of since and until in the usual way:

sometime_in_past $A =_{def}$ true since A
sometime_in_future $A =_{def}$ true until A
always_in_past $A =_{def}$ not sometime_in_past not A
always_in_future $A =_{def}$ not sometime_in_future not A
previous_time $A =_{def}$ A since true
next_time $A =_{def}$ A until true.

Also, in order to obtain the same expressivity for the temporal–modal language as we get for pure first-order temporal language with a sort for time, we introduce a qualified term time X, via the back door, where X is a term of the sort 'time'.

ERL semantics Let $\forall$ be a variable assignment mapping each variable into an element of its corresponding sorted domain. $\forall$ is extended to all terms in the usual way. Let $\mathcal{I}$ be an instantiation of an ERT diagram $G = \langle E, R, A, H, Mark, Conf\rangle$ and let $(T, <)$ be a $\mathbb{Z}$-like flow of time. The fact that a formula A holds in the instantiation $\mathcal{I}$ at time t under assignment $\forall$ is represented by $\mathcal{I}, t, \forall \models A$.

We start to define $\mathcal{I}, t, \forall \models A$ in the case of atomic formulae. This first case simply states that a qualified term is true if it is interpreted as an instance of the corresponding entity at a time which includes the evaluation time. The evaluation of time and comparison are immediate. The last case involves the truth of relational expressions; if it is a real relationship, then the entities involved must be related in the relationship instance at the time of evaluation, otherwise it is a pseudo-relationship with relating an entity to one of its attributes, and the attribute value must hold at the time of evaluation.

Formally, this gets the following formal definition:

$\mathcal{I}, t, \forall \models \mathsf{name}\theta$	iff $\mathsf{name} \in E$, $\mathcal{I}(\mathsf{name}) = \langle id, \tau \rangle$ such that $\forall(\theta) = id$ and $t \in \tau$.
$\mathcal{I}, t, \forall \models \mathsf{time}\ \theta$	iff $\forall(\theta) = t$.
$\mathcal{I}, t, \forall \models \theta_1\ op\ \theta_2$	iff $\forall(\theta_1)\ op\ \forall(\theta_2)$
$\mathcal{I}, t, \forall \models \mathsf{n}_1\ \theta_1\ \mathsf{rel}\ \mathsf{n}_2\ \theta_2$	iff either $\mathsf{rel} \in R$, $\mathsf{n}_1, \mathsf{n}_2 \in E$, in which case $\mathcal{I}(\mathsf{n}_1) = \langle \forall(\theta_1), \tau_1 \rangle$ and $\mathcal{I}(\mathsf{n}_2) = \langle \forall(\theta_2), \tau_2 \rangle$ such that $\langle \forall(\theta_1), \forall(\theta_2), \tau \rangle \in \mathcal{I}(\mathsf{rel})$ and $t \in \tau$; or $\mathsf{rel} = \mathsf{with}$, $\mathsf{n}_1 \in E$, $\mathsf{n}_2 \in A$, in which case $\mathcal{I}(\mathsf{n}_1) = \langle \forall(\theta_1), \tau_1 \rangle$ so $\langle \forall(\theta_2), \tau \rangle \in \mathcal{I}(\mathsf{n}_2)(\forall(\theta_1))$ and $t \in \tau$.

The inductive cases of the semantics are pretty much the standard ones:

$\mathcal{I}, t, \forall \models$ not A	iff	it is not the case that $\mathcal{I}, t, \forall \models A$.
$\mathcal{I}, t, \forall \models A$ and B	iff	$\mathcal{I}, t, \forall \models A$ and $\mathcal{I}, t, \forall \models B$.
$\mathcal{I}, t, \forall \models A$ or B	iff	$\mathcal{I}, t, \forall \models A$ or $\mathcal{I}, t, \forall \models B$.
$\mathcal{I}, t, \forall \models$ exists X such_that A	iff	there exists an $\forall'$, X-variant of $\forall$, such that $\mathcal{I}, t, \forall' \models A$.
$\mathcal{I}, t, \forall \models$ forall X for_which A	iff	for all $\forall'$, X-variant of $\forall$, $\mathcal{I}, t, \forall' \models A$.
$\mathcal{I}, t, \forall \models A$ since B	iff	there exists an $s < t$ such that $\mathcal{I}, s, \forall \models B$ and for every u, $s \leq u < t \rightarrow \mathcal{I}, u, \forall \models A$.
$\mathcal{I}, t, \forall \models A$ until B	iff	there exists an $s > t$ such that $\mathcal{I}, s, \forall \models B$ and for every u, $t < u \leq s \rightarrow \mathcal{I}, u, \forall \models A$.

16.3.2 *Abbreviations of ERL*

Even though we have full functional expressivity in ERL, this does not mean that it cannot be improved. First-order languages, which ERL deeply resembles, are developed over flat schemas (signatures), but ERL is to be used in connection with graph-like ERT diagrams, so it has to be adapted to conform to paths, multiple branches and similar constructions. We develop such adaptations as abbreviations to canonical ERL to highlight the fact that no extra expressivity is gained by that.

Relationship chains In connection with the ERT diagram illustrated in Fig. 16.3, there is the following (inherent or inherited) path:

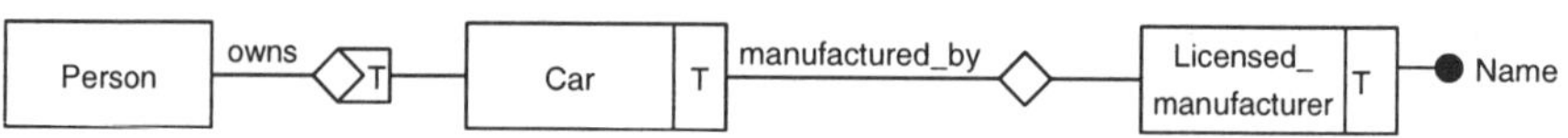

and we may want to express about it the fact that a person owns a car from Fiat, by writing

person P owns car C and
car C manufactured_by licensed_manufacturer L and
licensed_manufacturer L with name 'Fiat'.

But we may want to abbreviate such a long expression with several repeated qualified terms. For that, we simply follow the path, writing instead

person P owns car C manufactured_by licensed_manufacturer L with name 'Fiat'.

Note that the conjuction is now implicit and is omitted by contracting identical qualified terms. This is all there is to path terms.

Relationship branching One entity in an ERT diagram may be linked to several other elements, namely attributes and other entities (via relationships). To specify an entity one frequently has to provide a value to several of its links at once. For example, in this subgraph extracted from Fig. 16.3.

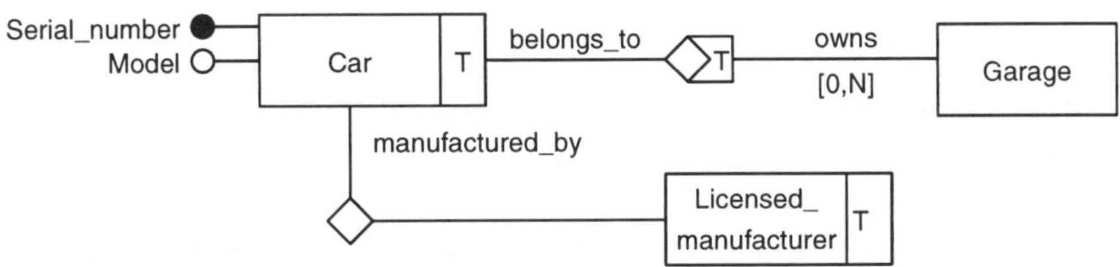

we may want to define a particular car (or set of cars) by giving the following constraints in an ERL formula:

car C belongs_to garage G and
car C with serial_number 'ABC1234' and
car C manufactured_by licensed_manufacturer L and
car C with model '2000'.

But the common qualified term can be factored out based on the structure of the diagram:

car C [belongs_to garage G,
with serial_number 'ABC1234',
manufactured_by licensed_manufacturer L,
with model '2000'].

The path and branching abbreviations can be combined; for example,

car C [belongs_to garage G owns car C_1,
with serial_number 'ABC1234',
manufactured_by licensed_manufacturer L [with name 'Ferrari',
with address A],
with model '2000'].

It is important to note that in the abbreviations we can freely use attributes and relationships that are inherited by an entity in exactly the same way that we can refer to one attribute or relationship that is phisically indicated in the diagram. The paths in the exerpts of diagrams above do not exist physically in Fig. 16.3, but hold through the isa relation through the entities in the graph.

Implicit quantification If we return to our first path example, in which we want to state just that a person owns a Fiat, without caring for the identities of the person or the car, then in that case we would existentially quantify the unwanted information:

exists C such_that (
exists L such_that (
person P owns car C manufactured_by licensed_manufacturer L with name 'Fiat')).

However, this makes the formula quite 'heavy' and awkward to read. So we note that implicit existential quantification is also possible, provided the quantified variable is used only once in the formula. By omitting the once-occurring variable and the quantifier from a qualified term, the formula above becomes simply

person P owns car manufactured_by licensed_manufacturer with name 'Fiat'.

This implicit existential quantification works just as well with branching:

person P owns car [manufactured_by licensed_manufacturer with name 'Fiat',
with model '147'].

However, one has to be careful with negation, for 'not exists' is not the same as 'exists not'. Since the existential quantification is always as external as possible, 'not exists' has to be stated explicitly, while 'exists not' is dealt with correctly.

This last abbreviation also allows us to rewrite name X and X *op* Y as name *op* Y, for any of the allowed comparison operators, so if we had a number_of_doors attribute for cars we could legally write the formula car with number_of_doors > 2.

We would like to draw attention to a point on the good design of ERT diagrams to improve readability of ERL formulae. This involves a practice that we *have not* followed so far, at least not systematically. Compare the following formulae representing a car owned by a garage that ones another car:

car C belongs_to garage owns car C_1

and

car C belonging_to garage owning car C_1.

The second formulation is much closer to what one could consider a 'natural' way of expressing the sentence. This naturality is provided by the use of gerunds to label the arcs of the diagram, instead of verbs in the present, third person. Furthermore, gerunds provide a 'relative tense' that adapts better with the use of the temporal operators, as in

car C belonging_to garage G and sometime_in_past (garage G owning car C_1)

as opposed to a more awkward car C belongs_to garage G and sometime_in_past (garage G owns car C_1). So the general rule is to use non-finite verbs (gerunds and participles) to label branches.

16.3.3 *Conceptual queries*

We are now in a position to define the notion of a query over a conceptual-level database instance $\mathcal{I}$. In the same way that a query in the relational model generates a new relation (schema plus instance), we want a conceptual query to generate a new diagram and an instantiation of it.

So a conceptual-level query must have two parts: a new schema (diagram) and an instance over the new schema. Formally, a *temporal query* to a database instance $\mathcal{I}$ over an ERT diagram $G = \langle E, R, A, H, Mark, Conf \rangle$ consists of a temporal query expression and an extension to the ERT diagram G; such an extension need not be (graph-)connected to the existing ERT diagram.

The extension to the ERT diagram generated by a query consists of a new T-marked node e_{new} and m new attributes att_{x_i}, where m is the number of free variables in the query. The *query expression* has the form

$$\{x_1, \ldots, x_m \mid A(x_1, \ldots, x_m)\}$$

where $A(x_1, \ldots, x_m)$ is a temporal ERL formula with free variables $x_1, \ldots, x_m$. The temporal relation generated by the query consists of the set of $(m+1)$-tuples

$$\{\forall(x_1), \ldots, \forall(x_m), \tau \mid t \in \tau \text{ and } \mathcal{I}, t, \forall \models A(x_1, \ldots, x_m)\}.$$

The instantiation of the extension of the ERT diagram is carried out as follows. For each tuple in the temporal relation generated by the query, a new identifier id_j is created. The instance of the new entity $\mathcal{I}(e_{new})$ receives the pair of time-stamped identifiers $\langle id_j, \tau_j \rangle \in \mathcal{I}(e_{new})$. Finally, for every instance $\langle id_j, \tau_j \rangle$ of the entity e_{new} there is an instance of the corresponding attribute $\mathcal{I}(att_{x_i})(id_j) = \{\langle \forall(x_i), \tau \rangle\}$.

Let us give an example. Suppose we wanted to pose the following query to the ERT diagram in Fig. 16.3: who are the persons and garages such that the person bought his or her car from the garage? This can be expressed in ERL by asking who the persons are that own a car that in the previous moment belonged to a garage. In ERL, this looks like

```
{ PName, GName | Person [with name PName,
                         owns car C]  and
                previous_time (
                Garage [ with name GName,
                         owns car C])}.
```

The temporal relation generated contains time stamped two-place tuples (i.e. triples), such that the time stamp is the time the person (first position) bought the car from the garage (second position). Also note that the variable C is needed to refer to the same car at different times. The extension to the ERT diagram of Fig. 16.3 is the following:

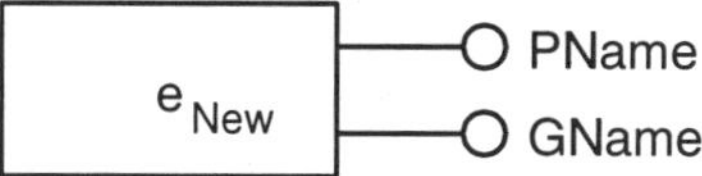

Note that this extension is totally disconnected from the diagram of Fig. 16.3. But we may want to know who are the buyers from garages at a particular moment. This may be expressed by a snapshot query.

The *snapshot query expression* has the form

$$\{x_1, \ldots, x_m; t \mid A(x_1, \ldots, x_m)\}.$$

The snapshot relation generated by the query consists of the set of $(m+1)$-tuples

$$\{\forall(x_1), \ldots, \forall(x_m), \{t\} \mid DB, t, \forall \models A(x_1, \ldots, x_m)\}.$$

The current-time snapshot query asking for the persons and garages such that the person is a car owner now and has remained as such since buying the car from the garage is expressed as

```
{ PName, GName; now | (Person [with name PName,
                               owns car C ])
                      since(
                      Garage [ with name GName,
                               owns car C ])}.
```

The use of the special value now is necessary to refer the query to the exact moment when it is posed. In such a context, now becomes the intersection of the notions of valid and transaction time. For a detailed discussion on the semantics of now in temporal query languages, please refer to Chapter 15.

Exactly the same concepts of *domain independence* and *safe temporal query* that apply to temporal relational queries apply at the conceptual level too. We will not duplicate their definition here, for it is totally analogous to that described in Chapter 14. We will always assume that queries are safe, as are those in the examples above.

16.4 Conceptual level updates, actions and rules

The full data manipulation capability is present at the conceptual level as well. A conceptual-level database state (i.e. instance) can not only be queried but also be changed. So far, we have considered that at the conceptual level only valid time is stored, but in discussing updates the notion of transaction time will be present, even if not stored. This is because the dichotomy between valid time and transaction time pertains to *temporal data* at any level of abstraction, and it is not restricted just to the flat relational model. If this is true when we move 'upwards' in the abstraction ladder, it is even more so when we move 'downwards', when we approach the physical levels of data storage.

Updates Coming back to the conceptual level, we see that the notion of an update is indeed totally analogous to that in the conceptual level (no small surprise, then, that we can do full data manipulation at the conceptual level).

An update is seen as a function $\Theta_{\bar{t}}$ that, applied to an instance $\mathcal{I}$ of a (conceptual) database at transaction time $\bar{t}$, generates another instance $\Theta(\mathcal{I})$ such that both instances $\mathcal{I}$ and $\Theta(\mathcal{I})$ refer to the same ERT diagram. That means that updates are seen at all levels of abstraction as database transformation functions that leave the schema invariant and potentially change only the instances.

An update $\Theta(\mathcal{I})$ is defined by the disjoint sets $I_{\bar{t}}$ of insertions and $D_{\bar{t}}$ of deletions to be applied simultaneously (and thus simulating eventual modifications) on an instance $\mathcal{I}_{\bar{t}}$ at transaction time $\bar{t}$. Both $I_{\bar{t}}$ and $D_{\bar{t}}$ are represented as partial instances of the database schema. The update semantics is the usual: whatever data that is inserted will be present in the updated state; whatever is deleted will be absent; and that not referenced by the update will persist from one instant to the next one.

The major difference between conceptual updates and their lower level counterparts is that the conceptual instance comes together with implicit constraints, namely:

(a) An identitifier can take part in an instance of a relationship class only if it belongs to the instance of the corresponding entity class taking part in the relationship; furthermore, the temporal restrictions imposed by the H- and T-marks have to be respected.

(b) An identifier can take part in the instance of an entity class only if it takes part in the instances of all its ancestors in a hierachy; furthermore, its associated time stamp is limited by that of all its ancestors and their eventual T-marks.

(c) Cardinality constraints have to be respected by all instances of relationship classes.

(d) Cardinality constraints also have to be respected for attributes.

These problems are treated by the following automatic actions in the system:

(a_1) The insertion of a time-stamped pair $\langle id_1, id_2, \tau\rangle$ in a relationship class instance $\mathcal{I}(r_{1,2})$ forces the insertion of time-stamped identifiers $\langle id_1, \tau\rangle$ in $\mathcal{I}(e_1)$ and $\langle id_2, \tau\rangle$ in $\mathcal{I}(e_2)$. This is only possible if $r_{1,2}$ is either unmarked or T-marked. In the case of an H- or HT-marked relationship, if there are no occurrences of $\langle id_1, \tau\rangle$ in $\mathcal{I}(e_1)$ and $\langle id_2, \tau\rangle$ in $\mathcal{I}(e_2)$ we do not know which time stamp to insert in $\mathcal{I}(e_1)$ and $\mathcal{I}(e_2)$, so the update is rejected (aborted).

(a_2) The deletion of an identifier id from an entity class instance forces the removal of all time-stamped pairs from the relationship classes in which it occurs. Also, all the attributes referring to id are deleted.

(a_3) If the removal of an identifier id is not total, but just for a period of time, all attributes and T-marked and unmarked relationship class instances involving id have to be updated accordingly.

(b_1) The insertion of a time stamped identifier $\langle id, \tau\rangle$ in an entity class forces its insertion in all the instances of that entity's ancestors. That is, insertions propagate upwards in a hierarchy.

(b_2) Likewise, the removal of a time stamped identifier $\langle id, \tau\rangle$ from an entity class forces its removal from all the instances of that entity's descendants. That is, deletions propagate downwards in a hierarchy.

(c) All updates that violate the cardinality constraints of relationship classes are rejected.

(d) An insertion of an attribute that allows at most one value is treated as a modification, i.e. it forces the deletion of the old value. All other cardinality constraint violations are rejected.

The insertion and deletion sets I_x and D_x that characterize an update can be seen as if closed under the rules above. If the update is rejected, those sets are emptied. This last consideration is an oversimplification, for a transaction may contain several updates and decisions on rejection for constraint violation may be left until the end of transaction execution. Issues involving database transactions—which may exist in the conceptual level as well as in the lower levels—remain outside the scope of this work.

ERL rules The conceptual-level database can also be transformed from a passive repository into an active one. The process is, again, totally analogous to that of the relational level and includes the addition of rules and a rule execution mechanism to the database. In the conceptual level, this is achieved with the inclusion of ERL rules.

An ERL rule has the following format:

when *trigger*
if *query*
then *action*.

The *trigger* behaves just like ordinary triggers (see Chapter 17), which in a conceptual-level setting means that a trigger is a non-persisting entity, while in Chapter 17 a trigger was a non-persisting relation. Parameters that a trigger may carry are treated as attributes, so a trigger with a series of parameters has the following format:

trigger_name [with parameter$_1$ X_1, ..., with parameter$_n$ X_n].

Since the trigger attributes are ofter unnamed, the format above can be further abbreviated to a relation-like trigger_name($X_1, \ldots, X_n$). The non-persistency of triggers means that once the trigger is inserted it exists only for one single transaction-time unit, disappearing after it. In the conceptual level, this can be taken literally: that is, even if a sequence of triggers with identical parameters is inserted at consecutive transaction times, at each moment the parameters are associated to distinct identifiers, and an identifier is never reused.

The actions that are allowed are *deterministic*, which in this case is defined as follows:

- every qualified term entity_name θ is a deterministic action, meaning the insertion of a new entity instance;

- every relationship expression is a deterministic action, meaning the insertion of a relationsip class or an attribute value;
- every trigger expression is a deterministic action, meaning the firing of the trigger;
- if A and B are deterministic actions, so is A and B;
- if A, B are deterministic actions, so are the following temporal expressions: next_time A, previous_time A, always_in_future A, always_in_past A and also the expressions of the form A until (time θ and B) and A since (time θ and B).

It is also possible to add external procedure calls to the deterministic actions; provided they do not occur within past temporal operators. Note that there are no quantifiers in actions. As usual, the free variables occurring in an action must also occur in the query part.

The semantics of action execution is the usual two-dimensional one. A rule of the form

when *trigger*
if *query*
then *action*

is interpreted as the two-dimensional formula

(δ and *trigger* and *query*) $\rightarrow$ *action*

to be forced to hold at all transaction times. Since we have δ as the antecendent, this means that the rule has to be verified only at the current time, i.e. on the two-dimensional diagonal. All the problems of *time paradoxes* (see Chapter 17) occur at this level owing to the conflicts between rule semantics and temporal updates.

17

TEMPORAL ACTIVE DATABASES: A BACKGROUND FOR CREATING AND DETECTING TIME PARADOXES

In this chapter we show how to create logical links between information associated to possibly distinct times in history. We can then discuss the effects that changes in history may have upon those temporal links.

For the purpose of establishing these links, we extend a valid-time database with *temporal rules*, and we provide those rules with an execution semantics; the resulting combined system is called an *active valid-time database*.

Temporal rules have the general form

$$Condition \Rightarrow Action$$

which, intuitively speaking, imply that whenever *Condition* is verified in the database, then *Action* is forced to hold. On the formal side, we will see that rules are constraints posed on the two-dimensional evolution of the valid-time database.

This two-dimensional view of temporal rules gives us an operational way to see how a valid-time database evolves. Those rules can be recursively applied in this way and so they may lead to an infinite evolution. We need to study ways in which these infinite executions can be avoided, studying the properties of *termination* and *confluence* of a set of temporal rules; in this process, we aim to establish a *declarative semantics* for temporal rules, and compare it with the two-dimensional execution semantics that is more operational in nature.

For efficiency and practical reasons, it is usual to extend rules with *triggers*. Triggered rules are represented as

$$Trigger\text{: } Condition \Rightarrow Action$$

with the intuition that both *Trigger* and *Condition* have to be satisfied for the *Action* to be executed. We will see that the basic differences between a *Condition* and a *Trigger* lie in their *persistence properties*, i.e. triggers are very short-lived entities which become false immediately after they are made true. We will see that triggers can help us control the serial execution of rules, therefore improving the efficiency of rule execution.

We have then created the background in which to analyse the following problem:

What are the effects of changes in history?

The setting in which such a question can meaningfully be posed and answered is, of course, an active valid-time database. We have already seen that valid-time databases allow us to change information about any time in history. So we know the meaning of 'changing the history' and we know how to do it. Furthermore, with the extension of temporal databases with active rules, we know how to create logical links between temporal information associated to any points in history. Therefore we know how one time in history may 'affect' any other time. To answer the question above, it remains to be defined what are the 'effects of changing history'.

The goal of the final part of this chapter is to study precisely what the effects are of changing the history and how to detect them, whenever possible.

For that we define a declarative *valid-time interpretation* of the temporal rules in an active database. We show that, under the execution semantics, the occurrence of updates at any time may cause an invalidation of the valid-time interpretation of rules at some time in the database state, generating a *time paradox*. In the same way that database updates were interpreted as *changes in history*, these time paradoxes are interpreted as *the effects of changing history* or *how changes in history affect other times*. We classify these time paradoxes and, in order to detect their occurrences, the notions of *temporal and syntactical dependences* of a rule are studied. These notions are then used to develop algorithms that perform the detection of time paradoxes. All algorithms developed in this chapter are collected and presented in an appendix (Section 17.10) at the end of the chapter.

17.1 Introduction

Databases in general, and valid-time databases as presented in the previous chapter, are passive repositories of data wherein any data is, in principle, updatable; no interaction takes place between the data either to generate or remove other data, or to execute a program in the database environment. When such an interaction is allowed, the database becomes an *active* (or reactive) repository, and the basic mechanism that enables such behaviour is the addition of *logical rules* to a database.

In the case of valid-time databases, the interaction is introduced by means of the inclusion of *temporal rules* in the database. The presence of rules in a database does not contradict the view of the database as a model that we have followed so far. Differently from the traditional perspective in logic programming [Lloyd, 1987] and temporal logic programming [Barringer *et al.*, 1989], rules will not be seen as a theory from which inferences can be drawn. In our approach, rules are seen as constraints that force the two-dimensional evolution of the database to occur according to the execution semantics of the rules. As a consequence, we will be providing the database with logical links whereby the existence of some data will force the insertion, deletion or modification of some other data.

The imperative future is not adequate for valid-time databases Temporal rules were first introduced by [Gabbay, 1989] under the paradigm of 'imperative future', which was then applied and further developed in several papers [Barringer *et al.*, 1989; 1991; 1996; Loucopoulos *et al.*, 1990; Manning and Torsun, 1989; Finger *et al.*, 1991; 1993] an extensive discussion on the imperative future can be found in other chapters of this book.

The imperative future approach was initially developed having in mind an underlying transaction-time flow, in which the past is never updated. It should not be a surprise that some incompatibilities exist between such a view and a valid-time dynamic view of data evolution and temporal rules.

Gabbay's temporal rules were temporal formulae of the form

$$\Box(Condition \rightarrow Action);$$

where *Condition* is a Boolean combination of pure present and pure past formulae and *Action* is a pure future formula; therefore, *Condition* was restricted to querying the past and present, whilst Action could only influence the future. The propositional atoms are of two kinds, *controllable* and *environment* atoms, such that environment atoms can only appear in *Condition* but not in *Action*. The intended meaning of such rules is that, whenever *Condition* holds against the past and present data in history, *Action* is executed imperatively by making it hold in the future; hence the name *imperative future*.

While that semantic interpretation of temporal rules does provide the database with links between data associated to distinct time points, and despite its intuitive appeal, the restriction to the format 'past and present implies future' has a few problems.

With respect to the format of *Condition*, we note that valid-time databases can also store information about the future, which is then interpreted as an expectation about what is going to happen. It is reasonable to base the execution of actions not only on the data recorded about the past and present, but also on the expectations on the future. For example, if one expects to attend a meeting abroad in a couple of weeks' time, it is reasonable to book tickets for the flight now, which may indeed be performed automatically if the information is present in the database; if share prices are expected to fall, it is reasonable to try to sell them. So in a valid-time database, forcing *Condition* to refer to the present and past only is too restrictive.

On the other hand, accepting any pure future formulae as actions brings the problem of deciding which atomic information is to be inserted in the database and at which time, a problem due to the existence of indeterminate actions of the form $A \vee B$ and FA. It may also become intractable to decide whether a set of rules containing such actions is always consistent. So this aspect of the format of *Actions* is too liberal. And yet, to confine actions to be pure future formulae is too restrictive, for we have the possibility of updating data associated with any time in a valid-time database.

As a consequence, we will improve on the imperative future restrictions on the format of rules. First, conditions will be able to refer to any point of time. Second, actions will be able to change any point in time. And third, actions will have to be *deterministic*, i.e. we rule out disjunctive actions as well as actions whose execution time is not determined. A formal definition of deterministic action is given below.

Actions can do more than updates Allowing actions to perform temporal updates in the database is surely a way of providing a logical link between temporal data. But we also want the database to interact with its environment. For that, we need to have the ability to start a program in the database environment.

So program calls will be added as a basic block in the construction of actions. They will be predicates whose name is linked not to a database relation, but to a program invocation. Their arguments will be associated with parameters passed for the program invocation. We do that by extending the basic action vocabulary with *external actions*, e.g. the automatic flight booking may be one of such programs, and the share-dealing program may be another. It is clear that external actions can be performed only at the current time, i.e. on the two-dimensional diagonal; they cannot be executed retroactively, for we can only change the stored past, not the execution of the past. External actions can be scheduled for a future time, but they are only executed when that future time becomes the present.

We therefore modify the format of temporal rules to cope with the problems mentioned above. We distinguish between two kinds of atomic actions. *Database actions* are atomic updates represented by negated and non-negated atoms (i.e. by literals). *External actions* are non-negated atoms associated to programs such that, whenever an atomic external action a is executed, a is inserted at the current time in the database and its associated program π_a is executed in the database environment. Since we are dealing with a first-order database, every external predicate action is associated to a program and the predicate arguments are seen as parameters passed to the program; all variables in an external action atom must be bound to a value at execution time, so that the propositional abstraction can be extended naturally. For example, the formula $\neg employee(\text{Peter}, 1\text{K}, \text{R\&D})$ is a database deletion action and $employee(\text{Mary}, 5\text{K}, \text{Finance})$ is a database insertion action; the formula $pay(x, y)$ is an external action associated to a program π that takes as parameters a person name, x, and an integer y, and prints a \$$y$ payment cheque to x.

17.2 Temporal rules

We can now formalize all the items discussed so far with respect to the format of temporal rules in a valid-time database.

Definition 17.1. A *temporal rule* is a formula of the form

$$Condition(x_1, \ldots, x_m, y_1, \ldots, y_n) \Rightarrow Action(x_1, \ldots, x_m)$$

where *Condition* is any temporal query formula, i.e. a safe formula, with free variables $x_1, \ldots, x_m, y_1, \ldots, y_n$. The notation indicates that the free variables of $Action(x_1, \ldots, x_m)$ are necessarily a subset of those occurring in *Condition*.

Action$(x_1, \ldots, x_m)$ is a *deterministic action*, which is defined as follows in two steps. We start with the above-defined distinction beween database actions and external actions. First define *update actions*:

- every database action is an update action: a positive literal $p(x_1, \ldots, x_r)$ is an insertion and a negative literal $\neg p(x_1, \ldots, x_r)$ is a deletion;
- if A and B are update actions, so are $\bigcirc A$, $\bullet A$, GA, HA, and $U(time(t) \wedge A, B)$, $S(time(t) \wedge A, B)$ and $A \wedge B$.

Then define *deterministic action*:

- every update action and every external action is a deterministic action;
- if A and B are deterministic actions, so is $A \wedge B$.

Note that all free variables occurring in the *Action* part of the rule must occur in the *Condition* part. Since *Condition* is a safe formula, this guarantees that there will only be a finite number of actions to be executed and all its arguments will then be bound to a value. We can now state why deterministic actions deserve such a name.

Lemma 17.2. *Let Δ be a one-dimensional temporal model and let*

$$\mathcal{V}_t = \{v | \Delta, t, v \models Condition(x_1, \ldots, x_m, y_1, \ldots, y_n)\}.$$

Then, for a given $t \in T$, exactly one of the following holds:

(a) there is no Δ' such that, for each $v \in \mathcal{V}_t$,

$$\Delta', t, v \models Action(x_1, \ldots, x_m);$$

(b) there are unique minimal Δ^+ and Δ^- such that, for each $v \in \mathcal{V}_t$,

$$(\Delta \cup \Delta^+) - \Delta^-, t, v \models Action(x_1, \ldots, x_m).$$

Proof We start by defining, for every deterministic action A, valuation v and time points $a, b \in T$, $a < b$, the set $\Delta^v_{[a,b]}(A)$ recursively on the structure of deterministic actions:

$$\begin{aligned}
&\Delta^v_{[a,b]}(\ell) = \{t : \ell | t \in [a,b]\}, \ell \text{ is a positive or negative literal}\\
&\Delta^v_{[a,b]}(\bigcirc A) = \Delta^v_{[a+1,b+1]}(A)\\
&\Delta^v_{[a,b]}(\bullet A) = \Delta^v_{[a-1,b-1]}(A)\\
&\Delta^v_{[a,b]}(GA) = \Delta^v_{[a+1,+\infty]}(A)\\
&\Delta^v_{[a,b]}(HA) = \Delta^v_{[-\infty,b-1]}(A)\\
&\Delta^v_{[a,b]}(U(time(x) \wedge A, B)) = \begin{cases} \Delta^v_{[v(x),v(x)]}(A) \cup \Delta^v_{[a+1,v(x)-1]}(B), & \text{if } a < v(x)\\ \{\bot\}, & \text{otherwise}\end{cases}\\
&\Delta^v_{[a,b]}(S(time(x) \wedge A, B)) = \begin{cases} \Delta^v_{[v(x),v(x)]}(A) \cup \Delta^v_{[a+1,v(x)-1]}(B), & \text{if } a < v(x)\\ \{\bot\}, & \text{otherwise.}\end{cases}
\end{aligned}$$

For $a \geq b$ we make $\Delta^v_{[a,b]}(A) = \emptyset$. Now construct the set

$$\Gamma = \bigcup_{v \in \mathcal{V}_t} \Delta^v_{[t,t]}(Action(x_1, \ldots, x_m)).$$

It is easy to see that if $\perp \in \Gamma$, then it is impossible to satisfy *Action*, for it demands setting a time (t) in the past (future), with t in the present or future (past), satisfying condition (a) above.

So suppose $\perp \notin \Gamma$. The elements of Γ are of the form $t : \ell$, where ℓ is a literal; so let us construct

$$\Gamma^+ = \{t : \ell \in \Gamma | \ell \text{ is a positive literal}\}$$
$$\Gamma^- = \{t : p | t : \neg p \in \Gamma\}$$

It follows from the definition of the semantics of formulae that to satisfy *Action* a model must include Γ^+ as a submodel and must be disjoint from Γ^-. So if $\Gamma^+ \cap \Gamma^- \neq \emptyset$ then it is impossible to have a model satisfying *Action*, and we are in condition (a) above.

Finally suppose $\Gamma^+ \cap \Gamma^- = \emptyset$. We construct the sets Δ^+ and Δ^- as follows:

$$\Delta^+ = \Gamma^+ - \Gamma$$
$$\Delta^- = \Gamma^- \cap \Gamma.$$

Since all models satisfying *Action* have to include Δ^+ and have to be disjoint from Δ^-, the definition above guarantees that these sets are the minimal ones satisfying, for each $v \in \mathcal{V}_t$

$$(\Delta \cup \Delta^+) - \Delta^-, t, v \models Action(x_1, \ldots, x_m).$$

Their uniqueness follows straight from their minimality, satisfying (b). □

Before we can give a formal semantics of rule execution, a brief discussion of action execution is opportune. First note that the format of update actions allows negation only when applied to atoms. A positive update atom predicate intends the insertion of a tuple or set of tuples for that predicate in the database; a negated atom predicate intends the deletion of a tuple or set of tuples. Note that neither disjunctions nor the temporal operators F and P are allowed in the construction of complex update actions. If the positive or negative atom occurs inside the scope of a temporal operator, this is to indicate the times where that information is to be inserted or deleted, always taken with respect to the current time. It is trivial to extend the action language to allow for constructs such as

literal@$TimeInterval$

representing the insertion/deletion of the literal for the defined time interval.

The format of deterministic actions does not allow the occurrence of an external action inside any temporal operators. As a result, external actions can be executed only at the current time. This avoids the problem of retroactive external actions, but it also prevents external actions to be scheduled for the future. The latter problem can be avoided in two ways: either we extend the definition of deterministic actions to allow external actions to occur inside the scope of future operator $\bigcirc$, or we may use auxiliary predicates to help us schedule external actions in the future. The 'auxiliary predicate' will be a trigger, discussed below. In any case, both solutions are satisfiable and equivalent, so we will stick with the second one and defer its presentation until we discuss triggers.

As a final remark on external actions, note that they appear in an *Action* formula in the format of a predicate. It is their execution semantics that differentiates them from update actions. For example, we could have an external action *delete_file(FileName)* associated with a program at the database environment that removes files from a disk. So the execution of the deterministic action

$$\neg image(Title, Storage) \wedge G\neg image(Title, Storage) \wedge delete\_file(Storage)$$

deletes records of an *image* two-place predicate at the current time and for all future time points and activates a program that should delete a file (whether this program succeeds or not is no concern of the action; all it cares about is that the program is activated with the correct parameters). Furthermore, we assume that the execution of external actions is recorded in the database, so for each file deleted an instance of predicate *delete_file(FileName)* is inserted in the database at the time corresponding to the start of its execution in the environment.

Notation: In the following, we represent both *Condition*- and *Action*-parts of rules as FOTL-formulae, since this is the natural way to express rules in a real database. There will be no conflict here, for we define the semantics of rule execution in terms of its propositional abstraction. Also, to avoid pedantism and unnatural notation, we use first-order formulae instead of their propositional abstractions when no ambiguity is implied, i.e. we abuse notation by writing '*Condition*' instead of '$[\![Condition^v]\!]$, for some valuation v', and similarly with *Action*.

17.3 Semantics of rule execution

We turn our attention to rule execution. The double arrow ($\Rightarrow$) was used in rule definition instead the single arrow ($\rightarrow$) to highlight the fact that its semantical value is not that of $\rightarrow$. We are going to present two semantical interpretations of temporal rules $Condition \Rightarrow Action$, namely the two-dimensional execution semantics of rules (in this section) and the one-dimensional valid-time interpretation of rules (in Section 17.6). In none of them is $\Rightarrow$ identical to $\rightarrow$. As a consequence, both *Condition* and *Action* are safe temporal formulae belonging to the FOTL language defined in Section 14.2, but not the rule itself.

Owing to the new format of temporal rule in Definition 17.1 which does not satisfy the imperative future restrictions, the intuitive notion of 'past and present implies future' no longer holds over the valid-time flow; however, it will be recovered over the transaction-time flow of time by the two-dimensional execution semantics of temporal rules given below.

Definition 17.3 (Execution semantics). Let $\mathcal{M}$ be a two-dimensional model representing the evolution of an active valid-time database containing the set of temporal rules

$$Condition_i \Rightarrow Action_i.$$

Consider the finite set

$$act(t) = \{[\![Action_i^v]\!] \mid \text{there exists } i, v \text{ such that } \mathcal{M}, t, t \models [\![Condition_i^v]\!]\}$$

of actions fired at the diagonal point t. The *semantics of rule execution* says that if $act(t)$ is satisfiable, then $\bar{\bigcirc}(\bigwedge act(t))$ must hold at the next transaction time, i.e.

$$\mathcal{M}, t, t \models Condition_i \quad \text{implies} \quad \mathcal{M}, t, t+1 \models Action_i.$$

Note that this semantics is equivalent to reading the temporal rule as the two-dimensional formula $(\delta \wedge Condition_i) \rightarrow \bar{\bigcirc} Action_i$ holding over $\mathcal{M}$, which satisfies the format 'present implies future' over the transaction-time flow. The elements of $act(t)$ are called *executed actions at time t*. For every external atomic action in $act(t)$ its associated program is executed against the database environment. Rules are always executed at present time, i.e. on the diagonal of the two-dimensional plane; in this sense, those programs will always be executed at some current time.

If the set $act(t)$ is unsatisfiable the situation is undefined. Typically this would mean that the database transaction that has caused the generation of the invalid state will be rolled back so as to restore a satisfiable state of the database. Note that since we have deterministic actions, the unsatisfiability check can be done efficiently in the following way: at each valid time affected by the action execution, just check whether there is an identical piece of information that is both inserted and deleted. It is also possible to have an external action *Rollback* that forces the abortion of the transaction. The treatment of transactions remains outside the scope of this work.

One important note here concerns the granularity of the transaction time. We are assuming here the *perceivedly instantaneous* model for temporal evolution described in Section 15.4, so that the value of 'now' perceived at transaction time $t+1$ is the same as the one at time t; the diagonal has some chunks at 45° and some chunks at 90° (namely, those that represent the steps of the execution of a recursively triggered set of rules). Hence, a sequence of recursively triggered rules executes with the same value of 'now'.

A temporal active database can then be defined.

Definition 17.4 (Active database). An *active valid-time database* is a valid-time database enhanced with a finite set of temporal rules such that, if $\mathcal{M}$ is a two-dimensional representation of the evolution of the database, then $\mathcal{M}$ satisfies the execution semantics of Definition 17.3.

This definition accommodates the view of the database as a model with the presence of rules as part of the database.

Concerning the ways an active database can interact with its environment, the presence of rules adds a bidirectionality to that interaction that does not exist in non-active databases. Both rules and updates have an effect on the evolution of the database, each taking part as one side of a two-way interaction between the database and its environment:

- environment $\rightarrow$ database: the environment acts upon the database by updating it;
- database $\rightarrow$ environment: the rules react to the data in the database, changing the data in it and, possibly, executing a program in the environment.

Example 17.5. Suppose that we add the following rule to the database of Example 14.2, so that an employee is to be paid every month the amount corresponding to the previous month's salary,

$$\bullet employee(Person, Salary, Dept) \Rightarrow pay(Person, Salary).$$

Action $pay(Person, Salary)$ is associated to a program that prints a cheque of amount $Salary$ to $Person$. Suppose we have the following information in the database at transaction time $Apr93$:

- $[Jan90, Dec92]$: $employee(\text{Peter}, \text{1K}, \text{R\&D})$
- $[Jan93, Apr93]$: $employee(\text{Peter}, \text{2K}, \text{Marketing})$
- $[Jan90, Dec92]$: $employee(\text{Paul}, \text{1K}, \text{R\&D})$
- $[Sep91, Apr93]$: $employee(\text{Mary}, \text{3K}, \text{Finance})$.

If that information had been added monthly to the database since $Jan90$, owing to the execution of the rules the following would also be in the database:

- $[Feb90, Jan93]$: $pay(\text{Peter}, \text{1K})$
- $[Feb93, Apr93]$: $pay(\text{Peter}, \text{2K})$
- $[Feb90, Jan93]$: $pay(\text{Paul}, \text{1K})$
- $[Oct91, Apr93]$: $pay(\text{Mary}, \text{3K})$.

17.4 Properties of the rule execution semantics

There are basically two main properties of a rule execution semantics that are considered very important and which apply here [Widom and Ceri, 1996; Aiken *et al.*, 1992]:

Termination deals with the problem of whether it is possible to guarantee if a set of rules will terminate, i.e. if the database will reach a state where no rules are triggered.

Confluence is the property of a set of rules that guarantees that, no matter in which order the rules are chosen, at termination (if this is at all reached) the database state reached will be the same independent of rule order choice.

We will now analyse these properties with regard to the semantics proposed above.

17.4.1 *Termination*

In non-temporal databases, the termination of a set of rules are normally guaranteed with the join combination of two properties:

(a) the syntactic restrictions on the format of rules is such that the rules impose a database transition function τ that is *inflationary*, i.e. for each database state Δ, $\Delta \subset \tau(\Delta)$;

(b) for a given initial database and a set of rules, there are only finitely many constants in its active domain. So there are only finitely many database states reachable with recursively triggered rules.

It was shown, e.g. [Kolaitis and Papadimitriou, 1991], that this is enough to guarantee that successive applications of τ reach a fixed point in finitely many steps, i.e. $\tau^{n+1}(\Delta) = \tau^n(\Delta)$, and therefore the recursive triggering is halted. Fixed points are the basis for a *declarative* semantics of rules in which the final state of rule processing is described without actually referring to the intermediate steps. By contrast, our semantics is more of the *operational* type, for the process from moving from one state to the other is specified.

Unfortunately, both conditions above fail for the temporal rules in Definition 17.1. First, our rules allow the conditions to contain negated formulae, and that does not guarantee that the associated transition function will be inflationary. Second, differently from non-temporal databases, there are infinitely many finite temporal databases for a finite active domain, because we are dealing with labelled atomic data. In fact, in the case of temporally bounded databases, there are countably many distinct temporal databases for a given active domain. For example, consider the deterministic rule

$$F(A \wedge time(t)) \Rightarrow F(time(t) \wedge \bigcirc A),$$

stating that if A is true sometime in the future, it will be made true in the next moment. This rule is enough to guarantee an infinite chain of recursive activations, each step adding a new moment where A is true. However, the infinite behaviour of the recursive activation of this rule is to make A true at all moments after t, which is expressible as a single rule:

$$F(A \wedge time(t)) \Rightarrow F(time(t) \wedge GA).$$

So the infinite behaviour of some rules may be obtained from the finite behaviour of a modified set of rules. However, another problem is presented here, which was not present in the non-temporal case. Consider the rule

$$F(A \wedge time(t)) \Rightarrow F(time(t) \wedge \bigcirc\bigcirc A),$$

stating that if A is true sometime in the future, it will be true at the second next time in the future. The infinite behaviour of this self-activating rule is to make A true at every other time, starting at t. However, if we start with a temporally bounded database, this state reached after infinitely many activations of this rule is *not* representable in a temporally bounded database.

The way several practical implementations solve this problem is by imposing a limit (which can be a time-out or a counter) on the number of activations allowed to a set of rules. If this limit is reached, the rule processing is aborted and returned to its initial state.

Restricted temporal rules If we want to guarantee termination, the format of temporal rules has to be restricted even more, so that the problems listed above can be avoided. For that purpose, we introduce the class of *restricted temporal rules*, of the form

$$Condition(x_1, \ldots, x_m, y_1, \ldots, y_n) \Rightarrow Action(x_1, \ldots, x_m)$$

subjected to the following condition:

- *Actions* are restricted so that only the following temporal operators are allowed: $\bigcirc$, $\bullet$, G and H.

The restriction on *Action* aims at reducing the number of temporal database states reachable in the presence of a given set of rules to a finite number. Note that the two examples above of non-terminating, infinite, temporal rule processing are not allowed owing to the syntactical restrictions on actions.

The transformation function $\tau_{\mathcal{R}}$ induced by a set of rules $\mathcal{R}$ is built in the following way. For each rule $R_i \in \mathcal{R}$, consider the transformation τ_i that, when applied to a database state Δ, generates two sets of labelled formula, Δ_i^+ and Δ_i^-, given by the determinism lemma 17.2. The transformation $\tau_{\mathcal{R}}$ creates a set of temporally labelled annotated data; the annotations are $+$, for insertion, and $-$, for deletion. These annotations are introduced to guarantee that $\tau_{\mathcal{R}}$ remains an inflationary function. So $\tau_{\mathcal{R}}(\Delta)$ is defined as

$$\begin{aligned}\tau_{\mathcal{R}}(\Delta) = \Delta & \\ & \cup \bigcup_{R_i \in \mathcal{R}} \{t : +\ell | t : \ell \in \; Delta_i^+\} \\ & \cup \bigcup_{R_i \in \mathcal{R}} \{t : -\ell | t : \ell \in \; Delta_i^-\}.\end{aligned}$$

It is obvious from this definition that $\tau_{\mathcal{R}}$ is inflationary. This comes at the expense of forcing us to deal with annotations $+$ and $-$ and broadening the notion of *conflict*, which is taken by the PARK rule semantics of [Gottlob *et al.*, 1996] to be any pair of insert/delete of the same literal throughout the rule processing, and not only at a single execution step (remember that we are trying to eliminate the notion of 'execution step' to be able to provide a declarative semantics for rules).

A *conflict* occurs if there are t and ℓ such that both $t : +\ell$ and $t : -\ell$ are in $\tau_{\mathcal{R}}(\Delta)$. Given set Δ, $\Delta^\bullet$ is obtained by deleting all its elements annotated with

+ and −. For a conflict-free set Δ, its *closure* $\overline{\Delta}$ is constructed by inserting the + elements into $\Delta^\bullet$ and deleting the − elements from it. In a recursive activation of the set of rules $\mathcal{R}$, i.e. in constructing the composition

$$\tau_{\mathcal{R}}^{n+1}(\Delta) = \tau_{\mathcal{R}}(\tau_{\mathcal{R}}^{n}(\Delta)),$$

the *Condition*-part of the rules is tested against $\overline{\tau_{\mathcal{R}}^{n}(\Delta)}$. As a consequence, a composition is only defined if $\tau_{\mathcal{R}}^{n}(\Delta)$ is conflict free (and hence it is conflict free for all $i < n$).

We say that a database state Δ' is *reachable* from Δ by rules $\mathcal{R}$ if $\Delta' = \overline{\tau_{\mathcal{R}}^{n}(\Delta)}$ for some n.

Lemma 17.6. *Let Δ be a temporal database state and let $\mathcal{R}$ be a set of restricted temporal rules. Then there are only finitely many database states reachable from $\mathcal{R}$ by recursively processing $\mathcal{R}$.*

Proof Let n be the maximal number of nested $\bigcirc$s and $\bullet$s occurring in an action in $\mathcal{R}$. Let t be the current time. Outside the interval $[t-n, t+n]$, the actions in $\mathcal{R}$ can only insert a value for all times or delete it for all times (using G and H). Since the active domain is finite, there are only finitely many such changes possible. Inside $[t-n, t+n]$, also due to the finiteness of the active domain, there are only finitely many possibly reachable database states. So there are only finitely many states reachable. □

A set of rules $\mathcal{R}$ is conflict free for database state Δ if $\tau_{\mathcal{R}}^{n}(\Delta)$ is conflict free for all n.

Lemma 17.7. *Let Δ be a temporal database state and let $\mathcal{R}$ be a set of restricted temporal rules conflict-free for Δ. Then there exists a finitely reachable fixed point of $\tau_{\mathcal{R}}$.*

Proof It follows directly from its definition that $\tau_{\mathcal{R}}$ is inflationary. If $\mathcal{R}$ is conflict free for Δ, then $\tau_{\mathcal{R}}^{n}$ is defined. Then by Lemma 17.6 we know that there are only finitely many reachable states, so there must be a fixed point for $\tau_{\mathcal{R}}$. □

This lemma also tells us that restricted rules will not take a temporally bounded database and transform it into a periodic one.

Now we must decide what to do when a state is reached in which there is a conflict; in this case we will abort a transaction. To extend the fixed point result for this situation, we will define a *correction function* κ such that

$$\kappa(\Delta) = \begin{cases} \Delta, & \text{if } \Delta \text{ does not contain a conflict} \\ \Delta^\bullet \cup \{\mathit{abort}\}, & \text{otherwise.} \end{cases}$$

The new constant *abort* is added to the vocabulary, and we have to create a function $\tau'_{\mathcal{R}}$ extending $\tau_{\mathcal{R}}$ to deal with it in the following way:

$$\tau'_{\mathcal{R}}(\Delta) = \begin{cases} \tau'_{\mathcal{R}}(\Delta), & \mathit{abort} \notin \Delta \\ \Delta, & \text{otherwise.} \end{cases}$$

The idea is that we should apply the correction κ after every application of the transformation $\tau'_{\mathcal{R}}$. Then we should find the fixed point of the compound function obtained by composing κ with $\tau'_{\mathcal{R}}$, $\kappa \circ \tau'_{\mathcal{R}}$.

Theorem 17.8. *Let Δ be a temporal database state and let $\mathcal{R}$ be a restricted set of temporal rules. Then there exists a finitely reachable fixed point of $\kappa \circ \tau'_{\mathcal{R}}$.*

Proof Straightforward form the definition of κ and Lemma 17.7. □

The fixed point of $\kappa \circ \tau'_{\mathcal{R}}$ is the *declarative semantics* of restricted temporal rules. For a database Δ such that the set of rules $\mathcal{R}$ is conflict free for Δ, it is easy to see that this fixed point can be reached using the execution semantics of Section 17.3. However, the reverse is not true, i.e. there are cases in which $\mathcal{R}$ is not conflict free for Δ, but the execution semantics of Section 17.3 still continues executing without aborting.

There are conflict correction methods much more sophisticated than the κ function above (see [Gottlob *et al.*, 1996]), but none can solve this problem. In fact, we can have the set of restricted rules

$$p \Rightarrow \neg p$$
$$\neg p \Rightarrow p$$

that are never conflict free, so that the restricted declarative semantics will always cause an abortion; but the execution semantics of Section 17.3 will create an infinite alternate sequence of ps and $\neg p$s. However, after every cycle of rule execution it is easy to check whether a conflict has arisen. The conclusion on the termination property of a set of rules is the following:

- To guarantee termination, one must use restricted temporal rules; the two-dimensional execution semantics can then be used, provided some extra checking is made to force abortions when conflicts are detected. This small change in the execution semantics of Section 17.3 guarantees that rule executions will always terminate.
- If infinite executions are not considered a problem, full deterministic actions can be used in conjunction with the two-dimensional execution semantics.

In the following, we will assume that the rules of execution always terminate.

17.4.2 *Confluence*

The semantics of temporal rules given in the previous section has important operational repercussions. A rule of the form

$$\mathit{Condition} \Rightarrow \mathit{Action}$$

is seen as forcing, at every current time (t, t) in the two-dimensional evolution model $\mathcal{M}$, that

$$\mathcal{M}, t, t \models \mathit{Condition} \rightarrow \bar{\bigcirc} \mathit{Action}.$$

This implies that, if there are two rules $\mathit{Condition}_1 \Rightarrow \mathit{Action}_1$ and $\mathit{Condition}_2 \Rightarrow \mathit{Action}_2$ such that at a given current time both antecedents are satisfied, then both actions will have to be enforced at the next transaction time, keeping the valid time fixed.

Operationally this means that every current time the *Condition*-part of *all rules* has to be queried into a database. For each tuple thus obtained an action is generated. From the deterministic action lemma 17.2, we know how to compute such action when it exists. At the next transaction time, *all actions are enforced*. This operational behaviour obviously satisfies the declarative two-dimensional semantics. Furthermore, the next state is deterministically fixed, and thus, assuming termination, *convergence is guaranteed* for any set of temporal rules executing in any initial database state.

This semantics is totally homogeneous with respect to rules in the database in the sense that:

(a) there are no preferences among rules—no rule is seen as 'privileged' or prioritized with respect to the others, and an inconsistent set of enforced actions causes an abortion;

(b) there are no instantial preferences—if a condition is verified for more than a single tuple, all tuples satisfying the condition are considered for action generation.

Those two semantical choices are not the norm among the several proposed rule execution semantics existing in the literature. For a description of the myriad of choices one faces in choosing a semantics for rule execution refer to [Widom and Ceri, 1996, Chapter 1]. We are interested in analysing here the semantics that violates one or both of the 'no-preferences' policies for rules and instantiations.

Rule priority Suppose that instead of posing the *Condition*-part of all rules as queries to the database at every current time, we were to choose, query and execute one rule at a time. Thus we would have a cycle of choice, execution and new choice, until no rule can be satisfied or chosen.

The immediate problem we are faced with is the rule choice mechanism. It must provide as a basic property a *fair choice* among rules, i.e. that all rules will eventually be chosen. Given fairness, the possibilities still abound; we can have a round-robin selection; rules can be chosen randomly; the least recently used rule may be tried; rules can be dynamically assigned a priority, in such a way that the highest-priority rule is always chosen. In the last case, the priority assignment mechanism that guarantees fairness still remains a problem. In some applications it is also possible to include domain knowledge in the choice mechanism.

The second problem of this operational semantics is rule confluence. Confluence requires that the final state that will be reached will be the same no matter what choice mechanism is used. If confluence is guaranteed, the system may try to choose the smallest path that leads to the final state, increasing the system's efficiency. The problem is that confluence is much harder to guarantee than fairness, even if the choice mechanisms being considered are restricted. Confluence depends not only on the choice mechanism, but on the rules themselves. It may be very difficult to predict the way rules interact with each other. In our no-choice rule execution mechanism (which is indeed a choose-all mechanism), confluence is not an issue, for the final result of rule application is deterministic.

The possible lack of confluence in a choice mechanism indicates that it does not conform to the two-dimensional rule execution semantics.

In normal (non-temporal) active databases, the majority of the systems have some way of choosing rules; this mechanism is not known to the rule designer, so it behaves as if it is random. However, it is theoretically provable that the no-choice mechanism—a non-temporal execution policy consisting of the repetition of a cycle of querying all rules and then executing all actions until no rules are satisfied—can simulate all choice mechanisms and therefore it is as expressive as all the others.

Instantial priority On top of the choice of rule, one can add a mechanism that chooses only one instance among the many that can satisfy the chosen rule, and execute the corresponding action. This seems particularly suitable to database systems that provide query answers one at a time (unlike standard relational databases, which treat data as a set), for it is more in the spirit of imperative programming languages. This is the mechanism underlying expert system programming languages such as OPS5 [Brownston *et al.*, 1985].

All the problems due to fairness and confluence are inherited here.

Still, it is possible to have a considerably more efficient rule execution mechanism that does not check all rules at all times, and yet conforms to two-dimensional execution semantics. This can be achieved with the addition of rule triggers.

17.5 Triggers

In this section temporal rules are enhanced with *triggers*. Informally, a trigger is a mechanism that enables the processing of a rule. Only when a rule is enabled (i.e. when its trigger is fired) is its antecedent checked against the temporal database and the corresponding actions are executed; otherwise, when the trigger of a rule is not fired, the rule remains disabled and is ignored.

A trigger will be represented by a guard (i.e. a label) placed in front of the rule:

$$\mathit{trigger} : \mathsf{Condition} \Rightarrow \mathsf{Action}$$

and it is read as

when *trigger* is fired **if** Condition holds **then** execute Action.

A rule without a trigger can be thought of as being labelled by truth, $\top$, so it is always enabled.

One of the main reasons for including triggers in rules is to increase the system's efficiency. Without triggers, *all* rules have their antecedents checked against the database at every rule evaluation cycle. When triggers are present, only the subset of enabled rules has their antecedents evaluated. We also say that the rule is *fired* when its trigger is.

Another good reason for including triggers is that they may be treated as channels through which the outside world communicates with the database, in the sense of [Milner, 1989], for instance in the rule sketched below:

$$alarm\_ringing : \mathsf{door\_open} \Rightarrow \mathsf{close\_door}.$$

The fact that the alarm is ringing is *external* to the system. This rule is activated only when it is communicated to the system that the alarm is ringing; therefore the trigger is called *external*. Only then is it checked whether the door is open.

Triggers may also be used in connection with an event *internal* to the database, such as the insertion (+) or deletion (−) of a fact. For example, when a door is recorded closed, we may wish to issue several warnings:

$$+door\_closed : \mathsf{emergency\_mode} \Rightarrow (\ \mathsf{lock\_door} \wedge trigger\_lights\ ).$$

As also seen in the example above, the Action-part of the rule can also fire an *internal* trigger (*trigger_lights*), which on its turn will activate another rule at the next transaction-time. Therefore a *chain of control* of rule execution may be created through the rules.

Finally, there may be triggers related to events associated to a resource managed by the system, such as the system clock, file access, etc. For example

$$time\_is(8am) : \mathsf{user\_logged\_in(User)} \Rightarrow \mathsf{good\_morning\_to(User)}.$$

As we see, triggers can also carry parameters and have the same 'appearance' as a normal database predicate.

There must be a mechanism for linking external and system triggers to their corresponding external and system events, but this will not be discussed here.

We will use the two-dimensional view of temporal database evolution to give a formal semantics to triggers. *Internal, external* and *system* triggers are seen as *non-persistent predicates*. We have to distinguish between trigger predicates and data predicates. When we defined the update semantics of data predicates in Section 15.1, whatever data was neither inserted nor deleted had its validity persisting into the following database state at the following transaction-time.

However, trigger predicates have a 'fixed duration' of a single transaction-time unit. The insertion of a trigger atomic predicate in the database corresponds to the *firing of a trigger* with the respective set of parameters. A trigger fired (i.e. inserted) at transaction-time t will hold at the next transaction-time, $t+1$. It will only hold at transaction-time $t+2$ if has been refired (i.e. reinserted) at $t+1$; otherwise it is removed from the database.

With this dynamic semantics for trigger predicates, we can construct *trigger expressions* by combining trigger predicates with Boolean operators. A guarded rule of the form

$$trigger\_expression : \mathsf{Condition} \Rightarrow \mathsf{Action}$$

is then evaluated simply as

$$(trigger\_expression \wedge \mathsf{Condition}) \Rightarrow \mathsf{Action}.$$

The difference between *trigger_expression* and Condition remains solely in the dynamic behaviour of its components.

The operational semantics now tells us that the *Condition*-part of all triggered rules have to be queried into the database and all the corresponding actions have to be executed in the next transaction time. This strategy remains deterministic, so rule confluence remains a non-issue.

Since it is normal that at any time only a small number of triggers is fired, then only a small number of queries has to be posed to the database during rule evaluation. This is the basic assumption underlying our claim of improved efficiency. Also a cycle of trigger firing, condition checking and action execution is terminated when a no-fired triggers is reached, a much cheaper test than the no *Condition* satisfied in the untriggered case.

17.6 The valid-time interpretation of rules

Let us first consider a temporal active scenario distinct from the one we have so far. Suppose we have in the database only rules of the form 'past implies present', without triggers; also suppose that the past is never changed by an update so that, once a rule is executed in the database, both the condition part that holds in the database (we call it the rule's *support* in the database) and the executed action that is consequently recorded in the database remain fpr ever in the database and are never changed.

In this scenario, one may try to confront the rules of an active valid time database against the database state, i.e. testing for the execution of rules at every valid time in that state. If one does such a confrontation expecting to find that, at every valid-time, the support and the recorded actions of a rule must either both hold or fail, we say one is using a valid-time interpretation of the rules.

In the presence of arbitrary updates in the past or with rules with a more liberal format, like those adopted in the previous sections, the valid-time interpretation of rules may not hold. This (perhaps intuitive) static view of rules does not follow, in the general case, directly from the dynamic semantics of rule execution and it may indeed become invalid owing to the occurrence of database updates, generating a *time paradox*. This section discusses the kinds of time paradoxes that may appear from the conflict between execution semantics and the valid-time interpretation in the presence of generic updates.

We can express these ideas more precisely. In a *valid-time interpretation of rules* every rule support holding at a database state should imply that its corresponding action was executed and recorded in the database. Furthermore, to avoid actions being executed without any support, in a kind of 'spontaneous generation of actions', a recorded action should hold in the database only if accompanied by its corresponding support; that would be equivalent to reading the '$\Rightarrow$' symbol in rules as '$\leftrightarrow$' and not simply as '$\rightarrow$'. In other words, under the valid-time interpretation, the rule

$$Condition \Rightarrow Action$$

is understood as the formula

$$\Box(Condition \leftrightarrow Action)$$

holding over the database state. Such a view is the one originally presented as the declarative past and imperative future view of a temporal database; there, however, the issue of updating the past was not raised, so no conflict was generated.[17]

There are several update-generated ways of invalidating the declarative valid-time interpretation of rules, thus generating a time paradox. Updates, either coming from the environment or resulting from the execution of an action, can affect both the support and the record of an executed action with respect to some past time t, falsifying the valid-time interpretation; such falsification can occur both in the case the rule was once executed in the past at t and in the case it was not, i.e. in the case the support of the rule did not hold at the diagonal point t.

The importance of those time paradoxes comes from the fact that they are interpreted as being the problems that arise from changes in history. In the absence of the temporal links given by temporal rules, no such anomaly could exist. Within the framework of an active valid-time database we can therefore study the problems of changing the past—or, indeed, the problems of changing any time in history. Next we discuss and classify such time paradoxes. Their detection in a database is the subject of the remaining sections of this chapter.

17.7 Time paradoxes

There are four ways in which the one-dimensional valid-time interpretation of rules can be invalidated by the two-dimensional execution of temporal rules. Of course, rules continue to be executed according to their two-dimensional semantics. It is just their 'intuitive' (but false) one-dimensional reading that gets violated.

17.7.1 *Non-supported actions*

After the execution of a rule an update can falsify its support, leaving a recorded action in a database state in which there is no apparent justification for its existence. Under the valid-time interpretation of rules, this is a contradiction, for *Action* holds but not *Condition*.

Formally, let $\mathcal{M}$ be a model describing the database evolution; we say that $[\![Action^v]\!]$ becomes *non-supported* at transaction time x and valid time t if it was executed in the past moment $t < x$, i.e. $\mathcal{M}, t, t \models [\![Condition^v]\!]$, such that until transaction time $x-1$ the support and recorded actions of the rule persist, i.e. for $t < y \leq x-1$, $\mathcal{M}, t, y \models [\![(Condition \wedge Action)^v]\!]$, but after an update Θ_{x-1},

[17] The presence of ***triggers*** in the rules is irrelevant for the present discussion. All we need to know is whether the rule is *supported* or not. In the case of triggered rules, this is equivalent to considering the trigger as part of *Condition*.

$\mathcal{M}, t, x \models \neg[\![Condition^v]\!] \wedge [\![Action^v]\!]$. Typically, an update in the past is the cause for the appearance of a non-supported action in the database, but since the support of a rule is not restricted to the past only, any update can cause it.

Example 17.9. In the temporal active database of Example 17.5, the action *pay*(Mary, 3K) was executed from *Sep*91 until *Apr*93. Consider the situation illustrated in Example 15.5 where, at *Apr*93, Mary's salary is increased to 5K retroactively to the beginning of the 1993 year. This would leave the database with action *pay*(Mary, 3K) non-supported from *Feb*93 until *Apr*93. In other words, at *Apr*93 there is no longer a justification in the database for having paid Mary only 3K from *Feb*93 until *Apr*93. This situation is illustrated in Fig. 17.1

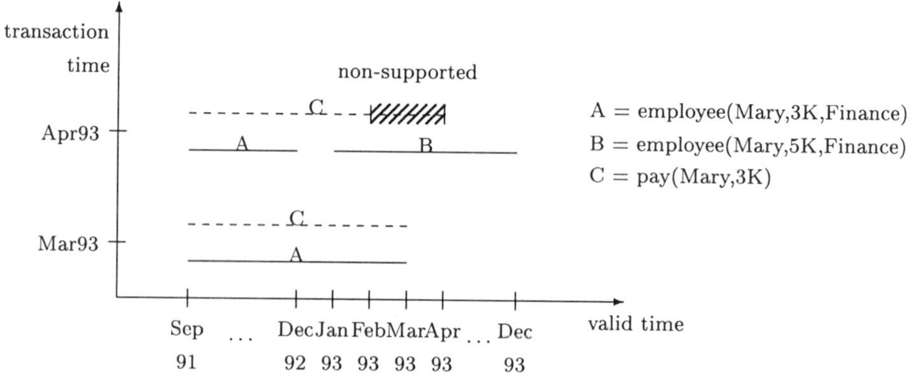

FIG. 17.1. Non-supported payment.

Note that the rule we are dealing with does respect the imperative future format restriction of 'past implies present or future', but a non-supported actions may still occur. The only way to avoid such a time paradox would be to forbid updates in the past, transforming the valid-time database into a transaction-time one.

17.7.2 *Retroactive actions*

An update, mainly one of the past, may leave the database in a state where there exists support at a past time t for an action to have been executed in the past, but the action was not executed at t. It is obvious that, if an employee is not in the database, they will not be paid a salary; salaries are only paid to employees that exist in the database *at the time of the payment*. But if an update inserts a new employee retroactively two months in the past, this employee will not have been paid two months salary because that information was not present at the time of payment. In this case, the payment of their salary is said to be a *retroactive action.*

Recall that, according to the semantics of rule execution, rules may only be executed at time points corresponding to the two-dimensional diagonal, i.e. at the current time, so no rule can be executed in the past. As a consequence, a retroactively fired rule and its corresponding retroactive action will never be executed.

The retroactive firing of an action can be seen as dual to the appearance of non-supported action. In the case of non-supported actions, both *Condition* and *Action* are true before the update, but after it *Condition* is falsified and *Action* still holds; in the case of retroactive actions, on the other hand, both *Condition* and *Action* are false before the update, yet *Condition* holds after the update but not *Action*.

Formally, if $\mathcal{M}$ is a model describing the evolution of an active database containing the rule $Condition \Rightarrow Action$, we say that an action or rule is *retroactively fired* at valid time t according to transaction time $x > t$ if $\mathcal{M}, t, t \not\models [\![Condition^v]\!]$ for some valuation v, so that the rule is not fired and $[\![Action^v]\!]$ is not executed, and this situation persists until transaction time $x - 1$, i.e. for $t \leq y \leq x - 1$, $\mathcal{M}, t, y \not\models [\![Condition^v]\!]$ and $\mathcal{M}, t, y \not\models [\![Action^v]\!]$, but after an update Θ_{x-1}, at the database state at transaction time x, $\mathcal{M}, t, x \models [\![(Condition \wedge \neg Action)^v]\!]$. The formula $\Box(Condition \leftrightarrow Action)$ does not hold over the database state $\mathcal{M}_x$.

As in the previous case, an update in the past is typically the cause for its appearance; however, since the support of a rule is not restricted to the past, any update can in principle cause it.

Example 17.10. Continuing Example 17.9 on Mary's retroactive salary increase from 3K to 5K, we see that the action *pay*(Mary, 5K) was retroactively fired from *Feb*93 until *Apr*93, i.e. it has a justification in the current state of the database for not having been executed, but the database state reflects that the execution never happened. This situation is illustrated in Fig. 17.2

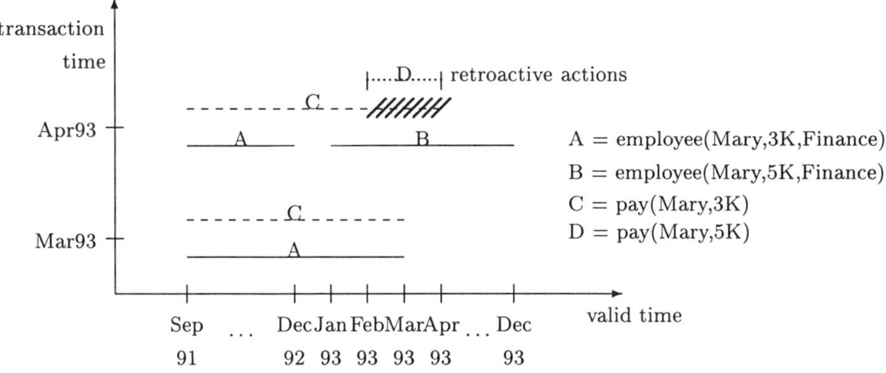

FIG. 17.2. Retroactive firing of payment action.

Note that there is no obvious 'corrective action' to the time paradox generated by a retroactively fired action. In Mary's case, the payment of the outstanding amount may depend on the company's policy. The payment may be made through an extra cheque at the current month, or its value may be added to next month's payment, or it may be converted into shares of the company, etc.

The simple correction of the violation of the valid-time interpretation is no solution to the time paradox, because, in this case, it would mean adding to the database the payment of a 5K cheque to Mary, which never occurred, or the deletion of her new salary, undoing her retroactive increase for the sake of a paradox-free database; clearly, those are not adequate corrective actions. The situation is no better if we try to 'correct' non-supported actions. The specification of such corrective action, if done formally so as to be executed automatically, may have to take into account the dynamic nature of retroactive actions and therefore it would require the expressivity of two-dimensional temporal logic. We do not explore these 'corrective actions' in this work, concentrating only on the detection of the occurrence of problematic situations.

Examples 17.9 and 17.10 show the simultaneous occurrence of non-supported and retroactive actions due to a modification update. When this happens, we call them *connected actions*. Note, however, that there can exist non-supported actions without a connected retroactive one, e.g. the removal of all past records of an employee leaving all past payments non-supported. There can also exist retroactive actions without a connected non-supported one, e.g. the retroactive hiring of an employee, retroactive firing and payments that never occurred.

17.7.3 *Rule violation and faked execution*

After a rule was triggered and executed at valid time t, an update may remove the recorded actions of the rule, leaving the database in a state such that, at t, the support of a rule is in the database but not its corresponding recorded actions. Therefore, it is said that the execution of the rule was violated.

Let $\mathcal{M}$ be a two-dimensional model representing the evolution of an active valid-time database and let $Condition \Rightarrow Action$ be a rule in it. We say that there is a *rule violation* in $\mathcal{M}$ at valid time t and transaction time x if

$$\mathcal{M}, t, x \models Condition \wedge \neg Action \wedge \overline{S}\,(\delta \wedge Condition, Condition \wedge Action).$$

It excludes the case where both the recorded actions and the support of a rule are simultaneously deleted, for then there is no invalidation of the valid-time interpretation.

The dual of rule violation occurs when an update in the past inserts in the database the recorded actions of a rule that was not fired at that past time, therefore faking its execution.

We say that a *faked execution* is introduced in $\mathcal{M}$ at valid time t and transaction time x if

$$\mathcal{M}, t, x \models Action \wedge \overline{S}\,(\delta \wedge \neg Condition, \neg Action).$$

Note that this definition excludes the case where the action was executed, then a later update removed both its recorded actions and support, and then only the recorded actions are restored to the database; in this case, the execution is not considered faked because it actually happened, although it does violate the valid-time interpretation. The execution is considered faked even if the support of the rule is simultaneously inserted in the database with its recorded actions, even though there is no violation of the valid-time interpretation in this case; programs associated with an external action included in the faked recorded actions will not be executed, for programs are executed only at a present time (note that it follows from the definition that $t < x$), so the execution is still considered faked.

Faked actions are not considered a serious problem that deserves having its occurrences always detected. In fact, a faked execution may be even part of the 'corrective action' taken in the case of a retroactively fired rule, in which case the faked action is no longer seen as a paradox.

17.7.4 *Summary*

Table 17.1 contains a summary of the possible effects of temporal updates in active valid-time databases. The first four rows present the update-generated time paradoxes arising from conflicts between the execution semantics and the valid-time interpretation of rules. The other three rows present the cases where no invalidation of the valid-time interpretation occurs, namely when neither support nor recorded actions of the rule are changed, and when both support and recorded actions are simultaneously removed from the database, therefore keeping the validity of the valid-time interpretation.

The fact that an action was executed or not at a certain time t cannot be detected by looking at the state of a one-dimensional valid-time database. If the support of a rule is in the database at time t but not its recorded actions, either it may be the case that it was there at execution time, but an update later removed its recorded actions, or it may be the case that the support did not hold at t, but was later introduced by an update. The detection of a past execution is, in fact, two dimensional and cannot be extracted from the database state.

Table 17.1 *Effects of temporal updates in active databases.*

	Executed	*Support*	*Recorded action*
Non-supported action	Yes	False	True
Retroactive action	No	True	False
Rule violation	Yes	True	False
Faked execution	No	True/False	True
No change	Yes	True	True
	No	False	False
Simultaneous deletion	Yes	False	False

In the following we concentrate on a method for detecting the occurrence of such time paradoxes.

17.8 Syntactic and temporal dependences

We describe here a method for detecting the appearance of non-supported actions in the database and we show how this method can be used to detect their connected retroactive actions. The method produces, as a side-effect, a way of detecting rule violation, but we do not attempt to detect faked executions. The detection of loss of support is particularly interesting in the case of non-supported external actions, for then a program has no justification in the database for its past execution against the database environment and, since we cannot change the past state of the environment, those actions are of particular interest for detection. In the following, whenever we refer to a non-supported or retroactive action, unless otherwise specified, we mean an external action.

The naive method for detecting loss of support consists of rechecking the database for rule support of executed actions at every past time point after every update. Of course, this solution just uses brute force and is computationally very expensive; therefore it is unacceptable.

The first thing deserving notice is that it is not necessary to check every rule at every time point in the past if we keep a log of executed rules and their execution times; we postpone an exact description of this log until later, but note that this log may increase indefinitely, so we concentrate on the detection of loss of support in a 'recent past', i.e. we may fix, a priori, the time interval we will be searching in the past. The log allows us to recheck only the rules that were once executed.

However, even with the log there are still too many checks to be done, for an update usually does not affect all the data at all times in the database. Ideally, we should only recheck the support of executed rules whose support was affected by an update. To deal with this idea of 'affected support' we introduce the notions of *syntactic dependence* and *temporal dependence* of a query.

The syntactic dependence of a first-order formula consists of the set of the predicate symbols occuring in the formula. The *positive syntactic dependence* of a formula is the set of predicate symbols occurring within the scope of an even number of $\neg$-symbols; the *negative syntactic dependence* of a formula is the set of predicate symbols occurring within the scope of an odd number of $\neg$-symbols. The positive/negative dependences of a rule are those of its *Condition*-part.

The temporal dependences of a formula A at a valid time t consist of sets of time points at which the truth or falsity of atomic formulae in the database gives support to the truth of A at t and, as in the syntactic case, there are positive and negative temporal dependences. It follows that the temporal dependence of a formula actually depends on the database state, i.e. it relies on the semantics of the formula. In order to define the temporal semantics formally, we make use of an extended representation of temporally labelled formulae and we define the semantics of such labelled formulae where, if t is a time point, d_+ and d_- are sets of time points and A is a temporal formula, $(t, d_+, d_-) : A$ is a well-formed temporally labelled formula. Let x be the current time and let $\mathcal{M}_x = (\mathbb{Z}, <, h)$ be a temporal model representing the database state at x. The expression

$$\mathcal{M}_x \models (t, d_+, d_-) : A$$

means that the formula A is true in $\mathcal{M}_x$ at time t with positive temporal dependence d_+ and negative temporal dependence d_-. Table 17.2 contains the extended definition of the semantics of temporally labelled formulae with temporal dependences.

Note that $\mathcal{M}_x \not\models (t, d_+, d_-) : A$ does not imply $\mathcal{M}_x \models (t, d_+, d_-) : \neg A$; the former means that A is not true at t with temporal dependences d_+ and d_-, but it may well be true at t with other sets of dependences; the latter means that $\neg A$ is true at t with the dependences d_+ and d_-. Table 17.2 therefore treats separately each case of negation. Note that it is also possible to have several distinct temporal dependences for the same formula at the same time, e.g. if $s < u < t$ and q holds at s and u then both $\mathcal{M}_x \models (t, \{s\}, \emptyset) : Pq$ and $\mathcal{M}_x \models (t, \{u\}, \emptyset) : Pq$. As in the syntactic case, the positive/negative dependences of a rule at time t are those of its *Condition*-part.

Lemma 17.11. *There exist temporal dependence sets d_+ and d_- such that $\mathcal{M}_x \models (t, d_+, d_-) : A$ if and only if $\mathcal{M}_x, t \models A$.*

Proof The proof is by induction on Table 17.2; the 'if and only if' is part of the induction hypothesis. For the basic cases, note that $p \in h(t)$ iff $\mathcal{M}_x, t \models p$ iff $\mathcal{M}_x \models (t, \{t\}, \emptyset) : p$ and $p \notin h(t)$ iff $\mathcal{M}_x, t \models \neg p$ iff $\mathcal{M}_x \models (t, \emptyset, \{t\}) : \neg p$. The non-negated cases in Table 17.2 are straightforward to prove and are therefore omitted; double negation and negation of conjunction are also straightforward and omitted.

The interesting parts of Table 17.2 are those concerning the negation of the temporal operators over $\mathbb{Z}$. We discuss here the case for the S-operator; for the U-operator the situation is analogous. Recall the semantics of $S(A, B)$ where it holds at a point t iff

(a) A holds somewhere to the past of t, at s; and

(b) B holds at all points u between s and t.

The negation of the formula $S(A, B)$ is satisfied if either of those cases is not. The first one is not satisfied if there is no such s at the past where A holds, so $H\neg A$ holds at t; by the induction hypothesis, $\mathcal{M}_x \models (s, d^s_+, d^s_+) : \neg A$ for all $s < t$ and by Table 17.2, $\mathcal{M}_x \models (t, \bigcup_{s<t} d^s_+, \bigcup_{s<t} d^s_+) : \neg S(A, B)$. The second one fails to hold over an integer-like flow of time if, going towards the past, we reach $\neg B$ before we reach A; over a $\mathbb{Z}$-like flow of time, this means that $\neg B \wedge \neg A$ is satisfied in the past and since then $\neg A$ holds, which can be expressed as the formula $S(\neg B \wedge \neg A, \neg A)$ holding at t. Therefore, by the induction hypothesis, there exists $s < t$, $\mathcal{M}_x \models (s, d^s_+, d^s_-) : \neg B \wedge \neg A$, and for all u, $s < u < t$, $\mathcal{M}_x \models (s, d^u_+, d^u_-) : \neg A$, and by Table 17.2, $\mathcal{M}_x \models (t, \bigcup_{s \le v < t} d^v_+, \bigcup_{s \le v < t} d^v_-) : \neg S(A, B)$. For dense flows of time and for flows that allow 'gaps',[18], there are

[18] A flow contains a gap if it contains an infinite ascending/descending sequence but does not contain the least upper/greatest lower bound of such a sequence.

other possibilities to falsify the second case which we need not take into account here. On the other hand, if $\mathcal{M}_x \models (t, d_+, d_-) : \neg S(A, B)$, then by Table 17.2 either of the above two cases is unsatisfied, so the induction hypothesis gives us $\mathcal{M}_x, t \models \neg S(A, B)$. This concludes the proof. □

When there is a generic update in the database, several time points are affected. Let Θ_x be an update determined by the pair (θ_+, θ_-). The set of time points positively affected by the update, $Aff_+(\Theta_x)$, is defined as

$$Aff_+(\Theta_x) = \{t \mid t : q \in \theta_+\}$$

and the set of time points negatively affected by the update, $Aff_-(\Theta_x)$, as

$$Aff_-(\Theta_x) = \{t \mid t : q \in \theta_-\}.$$

The support of formulae is preserved after an update under the following case.

Lemma 17.12. *For every transaction time x and every valid time t and every formula A such that $\mathcal{M}_{x-1} \models (t, d_+, d_-) : A$, if $Aff_+(\Theta_{x-1}) \cap d_- = Aff_-(\Theta_{x-1}) \cap d_+ = \emptyset$ then $\mathcal{M}_x \models (t, d_+, d_-) : A$*

Proof By induction on Table 17.2. For the base cases, if $A = q$ then $d_+ = \{t\}$ and $d_- = \emptyset$; it follows that $t : q \notin \theta_-$, so by update semantics we have $\mathcal{M}_x, t \models q$ and $\mathcal{M}_x \models (t, d_+, d_-) : q$. If $A = \neg q$ then $d_- = \{t\}$ and $d_+ = \emptyset$; it follows that $t : q \notin \theta_+$, so $\mathcal{M}_x \models (t, d_+, d_-) : \neg q$. This finishes the base cases.

For the inductive cases, we only examine the cases involving the temporal operator S; the other cases are either analogous or straightforward. If $\mathcal{M}_{x-1} \models (t, d_+, d_-) : S(A, B)$ then there exists $s < t$, $\mathcal{M}_{x-1} \models (s, d_+^s, d_-^s) : A$, and for all u, $s < u < t$, $\mathcal{M}_{x-1} \models (u, d_+^u, d_-^u) : A$, where $d_+ = \bigcup_{s \leq v < t} d_+^v$ and $d_- = \bigcup_{s \leq v < t} d_-^v$. By the induction hypothesis, there exists $s < t$, $\mathcal{M}_x \models (s, d_+^s, d_-^s) : A$, and for all u, $s < u < t$, $\mathcal{M}_x \models (u, d_+^u, d_-^u) : A$, so $\mathcal{M}_x \models (t, d_+, d_-) : S(A, B)$.

If $\mathcal{M}_{x-1} \models (t, d_+, d_-) : \neg S(A, B)$ we have to examine two cases. Suppose for all $s < t$, $\mathcal{M}_{x-1} \models (s, d_+^s, d_-^s) : \neg A$, where $d_+ = \bigcup_{s<t} d_+^s$ and $d_- = \bigcup_{s<t} d_-^s$; in this case, by the induction hypothesis, for all $s < t$, $\mathcal{M}_x \models (s, d_+^s, d_-^s) : \neg A$ and $\mathcal{M}_x \models (t, d_+, d_-) : \neg S(A, B)$. For the second case, suppose that there exists $s < t$, $\mathcal{M}_{x-1} \models (s, d_+^s, d_-^s) : \neg B$, and for all u, $s < u < t$, $\mathcal{M}_{x-1} \models (u, d_+^u, d_-^u) : \neg A$ where $d_+ = \bigcup_{s \leq v < t} d_+^v$ and $d_- = \bigcup_{s \leq v < t} d_-^v$. Then, by the induction hypothesis, $\mathcal{M}_x \models (s, d_+^s, d_-^s) : \neg B$, and for all u, $s < u < t$, $\mathcal{M}_x \models (u, d_+^u, d_-^u) : \neg A$; it follows that $\mathcal{M}_x \models (t, d_+, d_-) : \neg S(A, B)$, which concludes the proof. □

By combining temporal dependences and temporal affectedness we get the following necessary condition for an action to become non-supported.

Theorem 17.13 (Non-supported actions). *Let $\mathcal{M}$ describe the evolution of an active temporal database containing the rule Condition $\Rightarrow$ Action such that*

$\mathcal{M}_{x-1} \models (t, d_+, d_-) : Condition$. *A necessary condition for an executed Action to become non-supported at transaction time* x *and valid time* t *is*

$$[Aff_+(\Theta_{x-1}) \cap d_-] \cup [Aff_-(\Theta_{x-1}) \cap d_+] \neq \emptyset.$$

Proof Suppose $\mathcal{M}_{x-1} \models (t, d_+, d_-) : Condition$ so, by Lemma 17.11, $\mathcal{M}_{x-1}, t \models Condition$. If we assume that $Aff_+(\Theta_{x-1}) \cap d_- = Aff_-(\Theta_{x-1}) \cap d_+ = \emptyset$, by Lemma 17.12 it follows that $\mathcal{M}_x \models (t, d_+, d_-) : Condition$ and, by Lemma 17.11, $\mathcal{M}_x, t \models Condition$, which contradicts the fact that *Action* becomes non-supported at transaction time x and valid time t. □

A similar result can be obtained for the syntactic dependences on the first-order view of a database. Let Θ_x be an update determined by (θ_+, θ_-). Let $Pred(\theta_+)$ be the set of predicate symbols whose $[\![]\!]$-abstraction occurs in θ_+, and similarly for $Pred(\theta_-)$ with respect to θ_-.

Theorem 17.14 (Syntactic Dependences). *Let* $\mathcal{M}$ *describe the evolution of an active temporal database containing the rule* $Condition \Rightarrow Action$ *such that* s_+ *and* s_- *are, respectively, its positive and negative syntactic dependences and* $\mathcal{M}_{x-1}, t \models Condition$. *A necessary condition for an executed Action to become non-supported at transaction time* x *and valid time* t *such that* Θ_x *is an update determined by* (θ_+, θ_-) *is*

$$[Pred(\theta_+) \cap s_-] \cup [Pred(\theta_-) \cap s_+] \neq \emptyset.$$

Proof Suppose $\mathcal{M}_{x-1}, t \models Condition^v$ and $\mathcal{M}_x, t \models \neg Condition^v$ for some v. A simple induction on the structure of *Condition* shows us that $[Pred(\theta_+) \cap s_-] \cup [Pred(\theta_-) \cap s_+] \neq \emptyset$. For the base cases, if $Condition = p(x)$ then $s_+ = \{p\}$ $s_- = \emptyset$; then *Condition* can only be falsified if $t : [\![p(x)^v]\!]$ is deleted, so $p \in Pred(\theta_-)$. If $Condition = \neg p(x)$ then $s_- = \{p\}$ and $s_+ = \emptyset$; then *Condition* can only be falsified if $t : [\![p(x)^v]\!]$ is inserted, so $p \in Pred(\theta_+)$. The inductive cases are all straightforwardly proved and we omit the details. □

Since the notion of syntactic dependence is primarily a first-order one, it is reasonable to ask how the definition of temporal dependences translates from the propositional abstraction into the first-order case. For that, we combine the definition of temporal dependences from Table 17.2 and the propositional abstraction from Section 15.1.2. Recall that $R_\mathcal{D}$ is the set of all relevant domain elements of the database $\mathcal{D}$; furthermore we apply the three-place labels to safe first-order temporal formulae. The quantifier-free cases are basically those of Table 17.2; for the quantified cases we obtain:

$\mathcal{D}, v \models (t, d_+, d_-) : \exists x A$ iff there exists v' an x-variant of v such that $v'(x) \in R_\mathcal{D}$ and $\mathcal{D}, v' \models (t, d_+, d_-) : A$.

$\mathcal{D}, v \models (t, d_+, d_-) : \neg\exists x A$ iff for every v' an x-variant of v such that $v'(x) \in R_\mathcal{D}$, $\mathcal{D}, v' \models (t, d_+^{v'(x)}, d_-^{v'(x)}) : \neg A$, where $d_+ = \bigcup_{c \in R_\mathcal{D}} d_+^c$ and $d_- = \bigcup_{c \in R_\mathcal{D}} d_-^c$.

It follows from the safeness of formula A that both Lemmas 17.11 and 17.12 and Theorem 17.13 are extended to the first-order case with the definition above complementing Table 17.2. Note that a formula of the form $\exists x A$ may have several distinct temporal dependences.

We now examine the feasibility of computing the temporal dependences during query evaluation. Table 17.2 can be seen as a reasonable means for actually computing the temporal dependences during a query evaluation. In fact, real databases containing large tables demand better optimizations to achieve acceptable response times for queries, but for the sake of showing the feasibility of computing temporal dependences for quantifier-free formulae, we consider Table 17.2 satisfactory; the issue of incorporating the computation of temporal dependences to the optimization of queries is outside the scope of this work.

The extension dealing with quantifiers, however, does present us with a computational problem. Although the positive version of $\exists x A$ does not pose any problem, the negative case of $\neg\exists x A$ seems to demand that the temporal dependences for $\neg A(x)$ be calculated for each element of $R_{\mathcal{D}}$, which places a great computational burden; note that it leads to considering domain elements that are not even relevant to A. One possible improvement is to consider only the domain elements that are relevant to A, but this would still place a considerable burden on the system. We propose here a more straightforward computation of the dependences in that case; such a simplification will not affect the correctness of Algorithm 17.20 for the detection of non-supported actions.

Consider the labelled formula $(t, d_+, d_-) : \neg\exists x A$. The computation of (d_+, d_-) is done in the following (syntactical) way. We start with $d_+ = d_- = \emptyset$ and let $q(x)$ represent an atom containing the variable x. Then:

- if there is an atom $q(x)$ occurring positively in A, but not within the scope of any temporal operator, then $d_- := \{t\}$;
- if there is an atom $q(x)$ occurring negatively in A, but not within the scope of any temporal operator, then $d_+ := \{t\}$;
- if there is an atom $q(x)$ in A occurring positively inside the scope of a past operator, but not within the scope of any future operator, then $d_- := d_- \cup \{s \mid s < t\}$;
- if there is an atom $q(x)$ in A occurring negatively inside the scope of a past operator, but not within the scope of any future operator, then $d_+ := d_+ \cup \{s \mid s < t\}$;
- if there is an atom $q(x)$ in A occurring positively inside the scope of a future operator, but not within the scope of any past operator, then $d_- := d_- \cup \{s \mid s > t\}$;
- if there is an atom $q(x)$ in A occurring negatively inside the scope of a future operator, but not within the scope of any past operator, then $d_+ := d_+ \cup \{s \mid s > t\}$;
- if there is an atom $q(x)$ in A occurring positively inside the scope of both a future and a past operator, then $d_- := \mathbb{Z}$;

- if there is an atom $q(x)$ in A occurring negatively inside the scope of both a future and a past operator, then $d_+ := \mathbb{Z}$.

This leaves us with the following result.

Proposition 17.15. *There is a polynomial time algorithm that calculates the temporal dependences of a formula against a temporal database state.*

We conclude that it is effective to compute the temporal dependences of queries, which will then allow us to detect some of the time paradoxes described in Section 17.6.

17.9 Detection of time paradoxes

The results of the previous section allow us to discuss ways of detecting some of the time paradoxes arising from the conflicts between the execution semantic of temporal rules and the valid-time interpretation of rules. We concentrate basically on the algorithm for the detection of non-supported external actions, i.e. actions that caused a program to be executed against the environment; eventually other kinds of time paradoxes will be detectable following this path. The presentation of the algorithms will be done in a pseudo-programming language. All algorithms are collected and presented in the Appendix (Section 17.10).

We represent the data structures by means of Prolog-style structures, i.e. functional terms of first-order logic. We start by indicating the representation of temporal dependences as a list of pairs of integer numbers

$$[(s_1, e_1), \ldots, (s_n, e_n)]$$

such that, for every i, $1 \leq i \leq n$, s_i and e_i are integer numbers, $s_i \leq e_i$ and $s_{i+1} > e_i + 1$. The pairs (s_i, e_i) are intended to represent the interval $\{x \mid s_i \leq x \leq e_i\}$ and the list is supposed to represent the disjoint union of those intervals; eventually, s_1 may be equal to '*', where the pair $(*, e_1)$ represents the set $\{x \mid x \leq e_1\}$, and similarly e_n may be equal to '*' so that $(s_n, *)$ represents the set $\{x \mid x \geq s_n\}$. According to that definition, the list $[(*, 5), (6, 9), (12, 12)]$ is not acceptable as a representation of temporal dependences because $s_2 = 6$ is not greater than $e_1 + 1 = 5 + 1$. If we coalesce the two initial pairs, generating the list $[(*, 9), (12, 12)]$, we obtain an acceptable representation. This process of generating acceptable representations from generic lists of pairs can be done automatically; a description of such a *coalesce function* can be found in [McBrien, 1993], together with a description of algorithms on how to insert or delete time points to this representation of sets of time points.

We assume that every rule in the active database has a unique identification. This information will be used to indicate which rules will have to be rechecked for the purpose of confirming its support.

Two auxiliary tables are defined in the form of prolog predicates. The first table is a static one, i.e. it can be generated taking as input only the set of rules in the database and does not depend on the contents of the database at any time. The *table of syntactic dependences* has the form

$$syntactic\_dependences(Pred\_Name, Pos\_dep\_list, Neg\_dep\_list)$$

where *Pred_Name* is the name of a predicate in the database, *Pos_dep_list* is a list of rule identifiers such that each correspondent rule has *Pred_Name* as a positive syntactic dependence; similarly for *Neg_dep_list* with respect to negative syntactic dependence. In order to construct such a table we do the following. For each rule, we construct the parsing tree for its *Condition*-part; predicate names will be at the leaves of the tree. We then assign either '+' or '-' to every node in the tree in the following way. The root node receives '+'. If a node contains the $\neg$-symbol, its children receive the opposite sign as the node itself received; otherwise, the children receive the same sign as the father. For each predicate name on a leaf we include the rule identifier on its list of positive dependences if it is assigned a '+'; otherwise we include it in the list of negative dependences. Figure 17.3 shows the parsing tree for the formula $\exists x \exists y \neg(\neg p(x) \wedge (p(y) \wedge q(x)))$, in which p occurs both positively and negatively, and q occurs negatively.

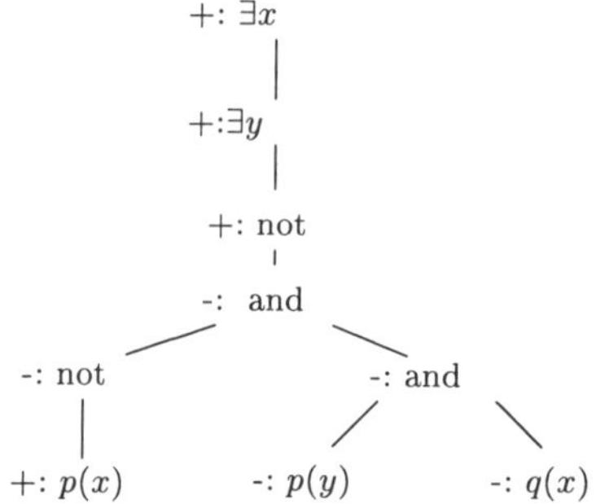

FIG. 17.3. Parsing tree for $\exists x \exists y \neg(\neg p(x) \wedge (p(y) \wedge q(x)))$.

Pred_Name is supposed to be the key of this table, so there are no two entries in the table with the same value for *Pred_Name*. For convenience, we use the following two functions to access the *syntactic_dependences*. All algorithms are presented in an informal pseudo-programming language, in the fashion algorithms are presented in basic books of relational database theory such as [Maier, 1983] and [Ullman, 1988]. An SQL-like notation is used to access the tables.

Algorithm 17.16 (Positive syntactic dependences).

```
Input: a predicate name
Output: a list of rule identifiers
POSSYNT(PredName)
BEGIN
Select Pos_dep_list from syntactic_dependences
   where Pred_Name = PredName;
return Pos_dep_list;
END
```

Algorithm 17.17 (Negative syntactic dependences).

```
Input: a predicate name
Output: a list of rule identifiers
NEGSYNT(PredName)
BEGIN
Select Neg_dep_list from syntactic_dependences
   where Pred_Name = PredName;
return Neg_dep_list;
END
```

The second table is a dynamic one, i.e. it is constructed as the database runs, and contains information about the external actions that were executed in the past by the system. The table *executed_action*(*Action*, *Rule_id*, *Parms*, *Time*, d_+, d_-) consists of the following:

Action the action name and the parameters passed to the external action at execution time are stored in the format of a Prolog term, e.g. *pay*(Peter, 50);

Rule_id the identifier of the rule whose firing caused the action to be executed;

Parms the list of parameters $[x_1, \ldots, x_m]$ that $Condition(x_1, \ldots, x_m)$ is true of at the time of execution;

Time the time at which the action was executed;

d_+, d_- lists representing positive and negative temporal dependences of the rule at *Time*.

The insertion of information in the table occurs at execution time. We assume that the temporal dependences are generated together with the query evaluation according to the description for first-order formulae in the previous section. The key for this table is composed of the *Action* name and parameters, and the *Time* of its execution. This table implements the previously cited 'log of executed actions' and it is a substitute for storing all the information about the transaction dimension.

For convenience, we provide two functions to access its table. Both take as input a rule identifier and a set of time points and both return a set of rows from the *executed_action* table such that, in the first one, the row contains the input rule identifier and a positive d_+ intersection with the input set of time points is non-empty; the second one does a similar thing with respect to d_-.

Algorithm 17.18 (Positive intersection rows).

```
Input: a rule ID and a list of time points.
Output: a set of rows from executed_action table.
POSROWS(RID, TIMES)
BEGIN
ROWS := ∅;
For every row of the table given by
   Select Action, Rule_id, Parms, Time, d+, d-
   from executed_action
```

```
   where RID = Rule_id and d+ ∩ TIMES ≠ ∅
   do     ROWS := ROWS ∪ {row(Action, Rule_id, Parms, Time, d+, d-)}
END
```

Algorithm 17.19 (Negative intersection rows).

```
Input: a rule ID and a list of time points.
Output: a set of rows from executed_action table.
NEGROWS(RID, TIMES)
BEGIN
ROWS := ∅;
For every row of the table given by
   Select Action, Rule_id, Parms, Time, d+, d-
   from executed_action
   where RID = Rule_id and d- ∩ TIMES ≠ ∅
   do     ROWS := ROWS ∪ {row(Action, Rule_id, Parms, Time, d+, d-)}
END
```

The detection of the appearance of non-supported actions takes as input the pair (θ_+, θ_-) that characterize an update. Each set is represented as a finite list of labelled elements of the form

$$times : atom$$

where *times* is a Prolog-style list representing a set of time points as described previously and *atom* is a ground atomic predicate in the database. We also assume there is a function PREDNAME that takes as input an atom and returns its predicate name, and a function COND that takes as input a rule identifier and returns the condition part of that rule. We then have the following algorithm to detect the occurrence of non-supported actions.

Algorithm 17.20 (Detection of non-supported actions).

```
Input: update sets θ+ and θ-.
Output: A set of time-labelled non-supported actions

DETECT_NONSUP(θ+, θ-)
BEGIN
NONSUP := ∅;

/* The first part of the algorithm deals with insertions */
For every times : atom in θ+
BEGIN
   For every rule id RID in table NEGSYNT( PREDNAME( atom ) )
   BEGIN
      For each row(Action, RID, Parms, Time, d+, d-)
```

```
        in NEGROWS( RID, times )
        BEGIN
           If the query COND( RID ) is unsatisfied by Parms at Time
           and Time : Action holds in the database
           then
              NONSUP := NONSUP ∪ {Time : Action};
              delete the row from the executed_action table.
              /* OBS */
           else if COND( RID ) is satisfied with new temporal
              dependences d'+ ≠ d+ or d'- ≠ d-,
              then modify the row in the executed_action table
              with dependences d'+ and d'-.
        END
     END
  END


  /* The second part of the algorithm deals with deletions */
   For every times : atom in θ-
  BEGIN
     For every rule id RID in table POSSYNT( PREDNAME( atom ) )
     BEGIN
        For each row(Action, RID, Parms, Time, d+, d-)
        in POSROWS( RID, times )
        BEGIN
           If the query COND( RID ) is unsatisfied by Parms at Time
           and Time : Action holds in the database
           then
              NONSUP := NONSUP ∪ {Time : Action};
              delete the row from the executed_action table. /* OBS */
           else if COND( RID ) is satisfied with new temporal
              dependences d'+ ≠ d+ or d'- ≠ d-,
              then modify the row in the executed_action table
              with dependences d'+ and d'-.
        END
     END
  END
  return( NONSUP )
  END
```

The parts of the algorithm with the comments `/* OBS */` will be used later when we extend it to detect connected retroactive actions. We first prove the correctness of the algorithm as it is.

Theorem 17.21 (Correctness of Algorithm 17.20). *Algorithm 17.20 detects all actions that become non-supported.*

Proof Algorithm 17.20 selects a (possibly empty) subset of rules to have its *Condition*-part rechecked. This selection occurs in two stages for both insertion and deletion. The first stage consists of selecting the rules whose syntactic dependences are affected by the update; by Theorem 17.14, no relevant rule is missed out. With this initial selection as input, a second selection is done based on the temporal dependences. In the non-quantified case, the correctness of this second selection follows directly from Theorem 17.13; the quantified case follows from the discussion at the end of last section, where we generate a big enough interval for the temporal dependences. As a consequence, we are guaranteed to recheck every rule that loses its support after the update.

Obviously, we assume that the tables *syntactic_dependences* and *executed_action* are generated and maintained correctly as described previously. In fact, the correct maintenance of the latter affects the correctness of the algorithm, as we will see next.

When the condition part is rechecked, three possibilities may occur. The first is that the rule is satisfied with the same temporal dependences as those stored in *executed_action*, in which case there is nothing to be done. The second possibility is when the query is satisfied, but with temporal dependences distinct from those stored at the table; this corresponds to the fact that the update may change the support of a rule (i.e. the evaluation of the condition now visits different time points) without leaving it unsatisfied. Theorem 17.13 uses the value of the temporal dependences just before the update, not at evaluation time, so in order to use it correctly to select the set of rules to recheck, we have to update table *executed_action* with the current version of d_+ and d_-, which is what the algorithm does. Finally, if the recheck fails and $Time : Action$ still holds then, by definition, *Action* has become non-supported at the current transaction time, at the valid time $Time$ when the rule was executed in the past. □

A slight alteration to Algorithm 17.20 can do better than just detect non-supported actions. In the case when the *Condition*-part of a rule is not satisfied by the parameters $Parms$ in the *executed_action* table, it is possible that it is satisfied by a different set of parameters that did not satisfy it at the execution time. This characterizes the appearance of a *retroactive action* connected to the non-supported action just detected. This, however, does not guarantee that all retroactive actions are detected, just those connected ones. In order to change Algorithm 17.20 to detect all the connected retroactive actions, we add to the initialization the variable `RETRO` initially set to $\emptyset$ and that is also returned at the end. Then, in the places where the comment `/* OBS */` occurs in Algorithm 17.20, we add the following piece.

Algorithm 17.22 (Detection of connected retroactive actions).

```
/* OBS: A non-supported action was detected at
```

$row(Action, \texttt{RID}, Parms, Time, d_+, d_-)$
*/
If the query COND(RID) is satisfied at $Time$ by $Parms' \neq Parms$
and there is no row in table *executed_action* such that
it contains $Action(Parms')$, RID, $Parms'$ and $Time$
then
RETRO := RETRO $\cup$ $\{Time : Action(Parms')\}$

In the algorithm above, $Action(Parms')$ represents the action we obtain by suitably substituting the new parameters in *Action*. We could execute Algorithm 17.22 after every recheck instead, so that we may even find some retroactive actions not connected to any non-supported action. Although this would generate only correct answers, i.e. every action detected has been retroactively fired, there are still no guarantees that all retroactive actions would be detected. The problem of finding all the retroactive actions comes from the fact that no information is stored about the rules and parameters that were unsatisfied at execution time. This must clearly be so, for there are infinitely many ways a rule might be unsatisfied (given a countably infinite domain). Nor can we afford to recheck every rule at every time point in the past after every update. So the detection of just connected retroactive actions seems a good compromise.

A complete version of Algorithm 17.20, including two occurrences of Algorithm 17.22 and the correct initialization of RETRO, is presented in the Appendix.

Note that in Algorithm 17.20 the detection of one non-supported action is independent of all the others. We may decide that it is worthwhile to apply the detection just to a certain group of rules and with the same algorithm; that would decrease the size of both auxiliary tables and consequently the time of rechecking, making the whole process more efficient. The algorithm will then correctly select the rules from the restricted set that must be rechecked.

The propagation of loss of support is one issue not discussed yet. Suppose we have the following rule:

$$\bullet^7 wednesday \Rightarrow wednesday$$

stating that if seven days ago was Wednesday, then today is Wednesday again. Suppose we have been executing this rule for several months and the database is populated with several *wednesday*s, when we delete the first occurrence of a *wednesday*. The second occurrence of a *wednesday* will become non-supported, but in fact it is reasonable to expect that all the subsequent Wednesdays be pointed as non-supported through propagation. This corresponds to considering a detected non-supported action as an automatically deleted action. In this case, it is enough to reapply the algorithm above after every automatic deletion to detect, recursively, all propagated non-supported actions. The deletion of a detected non-supported action is, however, a 'corrective action' to the detection of a time paradox; we have already decided to leave the user to decide when and how to apply corrective actions, and the process of detecting non-supported actions, even propagated ones, should not change the state of the database. A solution for that would be to fake the deletion of the detected non-supported action during the propagation, i.e. consider the detection and propagation of loss of support as a database process—a transaction—that temporarily deletes from the database the detected non-supported action and call Algorithm 17.20 until it returns an empty set, at which point we have reached the end of the propagating process and the temporarily deleted information may be restored. This process may be enriched with the temporary insertion of retroactive actions eventually detected.

To finalize the detection of time paradoxes, since we have already stated that we are not interested in detecting faked executions, all we have to do is show how to detect rule violation. The process is very simple if we restrict ourselves to the detection of violation of executed external actions, for then we may use the information stored in table *executed_action*.

Algorithm 17.23 (Detection of rule violation for external actions).

```
Input: update set θ_.
Output: A set of time-labelled actions
DETECT_VIOLATION(θ_)
BEGIN
VIOLATE := ∅
   For all t : atom in θ_ do
      If PRED(atom) is an external action and
      atom occurs in executed_action(Action, Rule_id, Parms, Time, d+, d-)
         with Time = t
         and Action = atom
         and COND(Rule_id) is satisfied by Parms
      then VIOLATE := VIOLATE ∪ {t : atom};
return(VIOLATE);
END
```

If we assume that the table *executed_action* contains all executed external actions, the algorithm clearly detects all rule violations caused by the deletion of an executed external action. In practice, it is reasonable to expect neither the database nor the auxiliary tables to retain all the information about the past, but just that a 'recent' past be kept, while the 'distant' past may be periodicaly transferred to tapes and only brought back to the database for special applications; the definition of what 'recent' means is clearly a database design decision. As long as the auxiliary tables used here cover the same period of time towards the past as the database itself, the algorithms presented in this section will remain correct; this periodical removal of information from the database into tapes is considered as a huge update that will cause the table *executed_action* to be maintained appropriately, but we may wish to disable the detection of non-supported actions at that time.

All the main algorithms described are to be executed after the database has been updated. They can be executed immediately after the update or after a transaction has committed, in which case the detection of non-supported actions, connected retroactive actions and rule violation can be considered as an independent transaction. This has the advantage that, except for the computation and storage of temporal dependences, no extra overhead is placed on the original transaction due to the detection of time paradoxes.

With respect to the worst-case complexity of the algorithm for the detection of non-supported actions, we note that if every rule is fired at all times and with temporal dependences equal to the whole set of time points, the algorithm ends up rechecking every rule at every time point after every update, and therefore degenerating into the naive method for the detection of loss of support. That extreme case, however, appears very unlikely to occur in practice, in which case the described algorithm should perform well. A more detailed analysis of the complexity of the presented algorithms should take into consideration the complexity of the accesses to the database itself, and therefore remains outside the scope of this thesis. The number of rules and the average number of actions executed at each transaction time should also play a role in determining this complexity. A good evaluator of the efficiency 'of the detection of loss of support would be the ratio r between the number of non-supported actions detected and the number of rules rechecked; clearly, $0 \leq r \leq 1$, and the closer r is to 1, the smaller is the number of useless rechecks the algorithm performed. To evaluate r, however, would require having the algorithm implemented in a database system running a real application, which is also outside the scope of this presentation.

This finishes our presentation of the detection of update-generated time paradoxes arising from conflicts between the semantics of rule execution and the static valid-time interpretation of rules in temporal active databases.

17.10 Appendix: collected algorithms

This appendix presents all the algorithms developed in this chapter.

17.10.1 *Auxiliary algorithms*

We present here the algorithms that manipulate the data structures

$$syntactic\_dependences(Pred\_Name, Pos\_dep\_list, Neg\_dep\_list)$$
$$executed\_action(Action, Rule\_id, Parms, Time, d_+, d_-).$$

The first two algorithms manipulate the first data structure for the retrieval of the rules affected by a given positive/negative syntactic dependence.

Algorithm 17.24 (Positive syntactic dependences (equivalent to Algorithm 17.16)).

```
Input: a predicate name
Output: a list of rule identifiers
   in which the predicate appears as
   a positive syntactic dependence
```

```
POSSYNT(PredName)
BEGIN
Select Pos_dep_list from syntactic_dependences
   where Pred_Name = PredName;
return Pos_dep_list;
END
```

Algorithm 17.25 (Negative syntactical dependencies (equivalent to Algorithm 17.17)).

```
Input: a predicate name
Output: a list of rule identifiers
   in which the predicate appears as
   a negative syntactic dependence

NEGSYNT(PredName)
BEGIN
Select Neg_dep_list from syntactic_dependences
   where Pred_Name = PredName;
return Neg_dep_list;
END
```

The following algorithms manipulate the second data structure for the retrival of executed actions that may have been affected by an update.

Algorithm 17.26 (Positive intersection rows (equivalent to Algorithm 17.18)).

```
Input: a rule ID and a list of time points.
Output: a set of rows from executed_action table
   with the same rule ID, RID,
   and positive dependence overlapping TIMES.
POSROWS(RID, TIMES)
BEGIN
ROWS := ∅;
For every row of the table given by
   Select Action, Rule_id, Parms, Time, d+, d-
   from executed_action
   where RID = Rule_id and d+ ∩ TIMES ≠ ∅
   do    ROWS := ROWS ∪ {row(Action, Rule_id, Parms, Time, d+, d-)}
return ROWS;
END
```

Algorithm 17.27 (Negative intersection rows (equivalent to Algorithm 17.19)).

```
Input: a rule ID and a list of time points.
Output: a set of rows from executed_action table
   with the same rule ID, RID,
   and negative dependence overlapping TIMES.

NEGROWS(RID, TIMES)
BEGIN
ROWS := ∅;
For every row of the table given by
   Select Action, Rule_id, Parms, Time, d+, d-
   from executed_action
   where RID = Rule_id and d- ∩ TIMES ≠ ∅
   do    ROWS := ROWS ∪ {row(Action, Rule_id, Parms, Time, d+, d-)}
return ROWS;
END
```

17.10.2 *Main algorithms*

We present here a combination of the main algorithms to detect non-supported actions, Algorithm 17.20, and its extension to detect connected retroactive actions, Algorithm 17.22. Note that the output now is a pair of sets, namely the set of detected non-supported actions and the set of connected retroactive actions, generated by a recent update.

Algorithm 17.28 (Detection of non-supported and connected retroactive actions).

```
Input: update sets θ+ and θ-.
Output: A pair containing a set of time-labelled non-supported
   actions and a set of retroactive actions.

DETECT_NONSUP(θ+, θ-)
BEGIN
NONSUP := ∅;
RETRO := ∅;

/* The first part of the algorithm deals with insertions */
For every times : atom in θ+
BEGIN
   For every rule id RID in NEGSYNT( PREDNAME( atom ) )
   BEGIN
      For each row(Action, RID, Parms, Time, d+, d-)
      in NEGROWS( RID, times )
      BEGIN
         If the query COND( RID ) is not satisfied by Parms at
         Time and Time : Action holds in the database
```

```
        then
        BEGIN
           NONSUP := NONSUP ∪ {Timc : Action};
           delete the row from the executed_action table.
           /* OBS: The detection of retroactive actions */
           /* is inserted here (Algorithm 17.22)*/
           If the query COND( RID ) is satisfied at Time for Parms'
              ≠ Parms and there is no row in table executed_action such
              that it contains Action(Parms'), ttRID, Parms' and Time
           then
              RETRO := RETRO ∪ {Time : Action(Parms')};
        END
        else if COND( RID ) is satisfied with new temporal
           dependences d'_+ ≠ d_+ or d'_- ≠ d_-,
           then modify the row in the executed_action table
           with dependences d'_+ and d'_-.
     END
  END
END


/* The second part of the algorithm deals with deletions */
For every times : atom in θ_-
BEGIN
  For every rule id RID in POSSYNT( PREDNAME( atom ) )
  BEGIN
     For each row(Action, ttRID, Parms, Time, d_+, d_-)
     in POSROWS( RID, times )
     BEGIN
        If the query COND( RID ) is not satisfied by Parms at
        Time and Time : Action holds in the database
        then
        BEGIN
           NONSUP := NONSUP ∪ {Time : Action};
           delete the row from the executed_action table.
           /* OBS: The detection of retroactive actions */
           /* is inserted again here (Algorithm 17.22) */
           If the query COND( RID ) is satisfied at Time for
              Parms' ≠ Parms and there is no row in table
              executed_action such that it contains
              Action(Parms'), ttRID, Parms' and Time
           then
              RETRO := RETRO ∪ {Time : Action(Parms')};
        END
```

```
        else if COND( RID ) is satisfied with new temporal
           dependences d'_+ ≠ d_+ or d'_- ≠ d_-,
           then modify the entry in the table
           with dependences d'_+ and d'_-.
    END
  END
END
return( (NONSUP, RETRO) )
END
```

The final algorithm deals with the detection of rule violations.

Algorithm 17.29 (Equivalent to 17.23: Detection of rule violation for external actions).

Input: update set θ_-.
Output: A set of time-labelled actions

```
DETECT_VIOLATION(θ_)
BEGIN
VIOLATE := ∅;
  For all t : atom in θ_ do
    If PRED(atom) is an external action and
    atom occurs in executed_action(Action, Rule_id, Parms, Time, d+, d-)
      with Time = t
      and Action = atom
      and COND(Rule_id) is satisfied Parms
    then VIOLATE := VIOLATE ∪ {t : atom};
return(VIOLATE);
END
```

Table 17.2 *Temporal dependences.*

$\mathcal{M}_x \models (t, \{t\}, \emptyset) : p$	iff	$p \in h(t)$.
$\mathcal{M}_x \models (t, \emptyset, \{t\}) : \neg p$	iff	$p \notin h(t)$.
$\mathcal{M}_x \models (t, d_+, d_-) : \neg\neg A$	iff	$\mathcal{M}_x \models (t, d_+, d_-) : A$.
$\mathcal{M}_x \models (t, d_+, d_-) : A \wedge B$	iff	$\mathcal{M}_x \models (t, d'_+, d'_-) : A$ and $\mathcal{M}_x \models (t, d''_+, d''_-) : B$ where $d_+ = d'_+ \cup d''_+$ and $d_- = d'_- \cup d''_-$.
$\mathcal{M}_x \models (t, d_+, d_-) : \neg(A \wedge B)$	iff	$\mathcal{M}_x \models (t, d_+, d_-) : \neg A$ or $\mathcal{M}_x \models (t, d_+, d_-) : \neg B$.
$\mathcal{M}_x \models (t, d_+, d_-) : S(A, B)$	iff	there exists $s < t$, $\mathcal{M}_x \models (s, d^s_+, d^s_-) : A$, and for all u, $s < u < t$, $\mathcal{M}_x \models (u, d^u_+, d^u_-) : B$, where $d_+ = \bigcup_{s \leq v < t} d^v_+$ and $d_- = \bigcup_{s \leq v < t} d^v_-$.
$\mathcal{M}_x \models (t, d_+, d_-) : \neg S(A, B)$	iff	for all $s < t$, $\mathcal{M}_x \models (s, d^s_+, d^s_-) : \neg A$, where $d_+ = \bigcup_{s<t} d^s_+$ and $d_- = \bigcup_{s<t} d^s_-$, or there exists $s < t$, $\mathcal{M}_x \models (s, d^s_+, d^s_-) : \neg B \wedge \neg A$, and for all u, $s < u < t$, $\mathcal{M}_x \models (s, d^u_+, d^u_-) : \neg A$ where $d_+ = \bigcup_{s \leq v < t} d^v_+$ and $d_- = \bigcup_{s \leq v < t} d^v_-$.
$\mathcal{M}_x \models (t, d_+, d_-) : U(A, B)$	iff	there exists $s > t$, $\mathcal{M}_x \models (s, d^s_+, d^s_-) : A$, and for all u, $t < u < s$, $\mathcal{M}_x \models (u, d^u_+, d^u_-) : B$, where $d_+ = \bigcup_{t < u \leq s} d^u_+$ and $d_- = \bigcup_{t < u \leq s} d^u_-$.
$\mathcal{M}_x \models (t, d_+, d_-) : \neg U(A, B)$	iff	for all $s > t$, $\mathcal{M}_x \models (s, d^s_+, d^s_-) : \neg A$ where $d_+ = \bigcup_{t<s} d^s_+$ and $d_- = \bigcup_{t<s} d^s_-$, or there exists $s > t$, $\mathcal{M}_x \models (s, d^s_+, d^s_-) : \neg B \wedge \neg A$ and for all u, $t < u < s$, $\mathcal{M}_x \models (s, d^u_+, d^u_-) : \neg A$ where $d_+ = \bigcup_{t \leq v < s} d^v_+$, and $d_- = \bigcup_{t \leq v < s} d^v_-$.

18

CALENDAR LOGIC

Hans Jürgen Ohlbach

18.1 Introduction

In this chapter we link temporal logic to real-time calendar systems. Instead of the very general operators like 'sometime in the future' or 'always in the future', we introduce relativized operators of the kind 'sometime next week' or 'always in office hours'. Calendar logic is developed in three stages. We start with a system for specifying everyday temporal notions like 'hours', 'weeks', 'weekend', 'office hour', 'holiday' within a given calendar system. In this system one can check whether a given point in time lies within an interval specified by a time term, or whether a subsumption relation holds between two time terms.

In the next stage these temporal notions are integrated into a propositional modal logic with two parameterized operators 'sometimes within τ' and 'always within τ' where τ may be one of the previously defined temporal notions. An example for a statement in this logic is

$$[2000, year] : \langle June(x_{year})\rangle election.$$

It expresses that some time in June in the year 2000 there is an election. $[2000, year]$ denotes the time region corresponding to the year 2000. $June(x_{year})$ is a time term. It denotes a function that takes a year-coordinate as argument and returns a month-coordinate. Temporal expressions like these are not built in, but they can be defined with some primitive constructors.

A characteristic feature of this logic is that the time regions involved are always finite. There is no 'sometime in the future' operator which refers to an unbounded time region. Instead of this, one can say for example 'sometime within the next million years'. This refers to a finite time region, which makes an important difference. The restriction to finite time regions is actually the reason that calendar logic can be translated into classical propositional logic, although at the expense of an exponential increase of the formulae's size. Nevertheless, this result means that satisfiability of formulae in calendar logic is decidable. It can be decided with any propositional decision procedure. Because of the exponential increase in the size of the formulae this may not be the most efficient method for deciding satisfiability. As an alternative we therefore provide a tableau algorithm for calendar logic. The tableau method has more possibilities for guiding and optimizing the proof search.

In the last stage, calendar logic will be combined with other (multi)modal logics, e.g. modal logics of knowledge and belief. A simple statement in this logic is

$$[1998, year] : [believe\_I]\langle x_{year} + 1\rangle\langle June(y_{year})\rangle election.$$

(In 1998 I believe that at sometime in June the year after (i.e. 1999) there is an election.) A formula in a combination of calendar logic and an $\mathcal{ALC}$-like description logic [Schmidt-Schauß and Smolka, 1991] is

$$\langle 1996\rangle(car \wedge \langle \text{sold}\rangle woman).$$

It denotes the set of cars sold to women in the year 1996.

The time regions related to the temporal operators are again finite. Therefore it is possible to eliminate the temporal operators and translate the formulae of the combined system into formulae of the basic multimodal logic. Consequently, satisfiability in the combined system is decidable if satisfiability in the multimodal logic itself is decidable. Because the elimination of the temporal operators blows up the formulae exponentially, inference systems which avoid this exponential explosion are to be preferred. A combination of the tableau systems for calendar logic and the given multimodal logic $\mathcal{L}$ might be possible in each case, but it is very difficult to describe it without making assumptions about the tableau system for $\mathcal{L}$.

An alternative for a calculus operating directly on the formulae of the logic is the translation approach. Satisfiability of a formula is checked by translating the formula into another logic and checking the satisfiability of the translated formula using inference systems of the other logic. We show how a particular translation method, namely *functional translation* [Ohlbach, 1991], can be used to get a decision method for a combination of calendar logic with a large class of multimodal logics. The result of the functional translation are formulae in a language very similar to first-order predicate logic. In principle one can therefore use any predicate logic inference system for the translated formulae. A better performance, however, can be achieved by tailoring the inference rules to the datatypes involved, which in our case are finite temporal intervals. We present a resolution-like inference system with special mechanisms for dealing with the temporal intervals. This (theory) resolution system is sound and complete, and it terminates if the corresponding resolution system for the multimodal logic component alone terminates without particular strategies. This is the case for all modal logics with the finite model property and an upper bound for the number of worlds in the model. For simple logics like K, KD, T, B, S5, however, much tighter limits are known to enforce the termination.

18.2 The language of time terms

Many times I have tried to phone a colleague in another country, maybe on another continent; I look into my database, get his or her office phone number, dial it, and get no answer. Maybe I forgot the time difference between time zones, and it is actually the middle of the night over there, or in this country there is a public holiday, or it is lunch time, or we have moved to daylight saving time, and everything is shifted by one hour compared with last week, when I phoned him or her at the same time of the day. A clever database would know all about this, and if I asked for the phone number for Mr. X, it would give me the actually valid number, or at least warn me that it is not very likely to get somebody at the phone at this time of day or year.

One can implement of course a notion like 'office hour' with a special algorithm that takes into account all the phenomena of real calendar systems (cf. Dershowitz and Reingold's excellent book on calendrical calculations [Dershowitz and Reingold, 1997]). Many commercial software products offer some kind of calendar manipulations, but in general they are limited to one calendar system and do not convert between different systems. Much more convenient and flexible, however, would be a simple abstract specification language for these kinds of temporal notions, which reduces the actual computation tasks to some standard algorithms. Such a specification language must be based on a mathematical model of calendar systems with all their specialities.

Let us start with the western calendar system, the Gregorian calendar. We have years, months, weeks, days, hours, minutes, seconds, milliseconds, etc. Although at first glance the system looks quite simple, there are many phenomena which make a formal model difficult. First of all, there are different time zones. A year in Europe is not the same as, for example, a year in America. Not all years have the same length. In leap years a year is a day longer. Not all days have the same length. In many countries one day in the spring is an hour shorter and one day in the autumn is an hour longer than usual. Even minutes might not all have the same length if some 'time authorities' decide to insert some seconds into a minute to recalibrate their reference clocks.

Some time measures are exactly in phase with each other. For example, a new year starts exactly at the same time point when a new month, day, hour, minute, second starts. But it does not start at the same time point as a new week. Nevertheless, weeks are quite often counted as week 1 in a year, week 2, etc. Many businesses allocate their tasks to particular week numbers in a year. If you go into a furniture shop and buy a new sofa, they might well tell you that the sofa will be delivered in week 25, and you have to look in your calendar to find out when on earth week 25 is.

Weeks are not in phase with months and years, but they are in phase with hours and minutes and seconds, which themselves are in phase with months and years. Quite confusing, isn't it? Things would be much easier if calendar systems were decimal systems, with one fixed unit of time as basis, and all other units as fractions or multiples of this basis. Since this is not the case, we need to model the time units in a different way. Therefore in the next section we begin with the definition of time units as independent partitionings of a universal time axis. Based on this, a specification language for temporal notions denoting sets of time points will be presented. Its semantics is such that we obtain algorithms for deciding whether or not a given point in time is within the set of time points denoted by a given time term (e.g. whether it is currently an office hour, say, in Recife, Brazil).

18.2.1 *Reference Time and Time Units*

There is no simple way of choosing a popular unit such as a year as the basis for a mathematical model, and defining everything else relative to this unit. Years have different lengths, and they depend on the time zone. Fortunately there is a well-established time standard, which is available on most computer systems and in most programming languages. The origin of time in this standard is 1.1.1970, Greenwich Mean Time (GMT). There are functions which give you the number of seconds (or even milliseconds) elapsed from the origin of time until right now. Other functions can map this number back to a common date format (year, month, day, hour, minute, second), either in GMT or in local time.

In our reference time standard we therefore assume a linear time axis and a time structure which is isomorphic to the set of real numbers. In a particular implementation we are free to move point 0 in this time axis to some point in history, 1.1.1970 in Unix systems, or 1.1.0 relative to the Gregorian calendar, or even at the time of the Big Bang. This is only a linear transposition by a fixed number.

Definition 18.1 (Reference time line). Let $\mathcal{T}$ be an isomorphic copy of the set of real numbers. $\mathcal{T}$ is the reference time line. The time point 0 is called the *reference origin of time.*

Unfortunately it is not the case that the common time units—minutes, hours, days, weeks, months, years—can be defined as multiples of a smallest unit. Nevertheless, all of them define a partitioning of the time axis into sequences of time intervals. That means that we can count for example years as 'year 0 after the origin, year 1 after origin', etc. The only difficulty is that different years may have different lengths (measured in the reference system). To specify a time unit U, e.g. 'GMTyear', as a sequence of intervals of different lengths, we need three things. First of all, these time units have their own coordinate system which is isomorphic to the integers. This means that we can identify each time interval measured by the given time unit by a particular integer. Moreover, we know that if n is the coordinate of a given interval, then $n+1$ is the coordinate of the next interval in this sequence.

Second we need for each time unit U a function $U_{\mathcal{N}}$ that maps reference time points to the time unit's own coordinates. For example, if the reference time axis is the GMT time measured in seconds from 1.1.1970, then for U = GMTday, $U_{\mathcal{N}}$ might map all the points in the half-open interval[19] $[0, 86400[$ to the GMTday coordinate 0 (if we count 1.1.1970 as day 0). In another time zone Z with, say,

[19]The notation $[b, e[$ for half-open intervals denotes the set of all points t with $b \leq t < e$. We could have chosen half-open intervals $]b, e]$. This does not matter. We cannot choose closed intervals because in this case subsequent intervals are not disjoint.

one hour difference to GMT (earlier), we would then have a unit $U' = $ Zday and a mapping $U'_{\mathcal{N}}$ which maps the interval $[-3600, 82800[$ to day 0.

Finally we need for each unit U a mapping $U_{[[}$ which maps the unit's own coordinates back to the reference time axis. $U_{[[}(n)$ actually computes the beginning and the end of U's time interval n. For $U = $ GMTday, we would for example have $U_{[[}(0) = [0, 86400[$ and for U' we would have $U'_{[[}(0) = [-3600, 82800[$.

It is important to notice that at this stage of our model, we assume a separate definition of $U_{[[}$ and $U_{\mathcal{N}}$ for each time unit U we are interested in. For the standard time units like years, months, days, hours, minutes, seconds, these functions are usually available in one form or another in a programming language, both for the GMT time zone, and for the local time zone and, via the Internet, for all other time zones. All the information about the particular calendar system, leap years, leap seconds, daylight saving time, etc., must be encoded in $U_{[[}$ and $U_{\mathcal{N}}$. Therefore these algorithms are usually quite complex.

The correlations between different time units so far are indirect. From a coordinate n in U's system, we can use $U_{[[}$ to go back to the reference system, and from there with $U'_{\mathcal{N}}$ to the coordinate system of some other unit U'. But this way we can correlate all units with each other, regardless of the calendar system, time zone, or granularity. For example, we can map the Gregorian calendar GMT-minute with coordinate n to, say, the corresponding hour in the Chinese calendar system provided the two functions GMT-minute$_{[[}$ and Chinese-hour$_{\mathcal{N}}$ for the two time units are available.

Definition 18.2 (Time units). Let $\mathcal{U}$ be a set of *time unit symbols.*

If $U \in \mathcal{U}$ is a time unit symbol, an *interpretation* $\Im_T(U)$ is defined as the triple $(\mathcal{N}_U, U_{\mathcal{N}}, U_{[[})$ where':

- $\mathcal{N}_U$ is a U-coloured isomorphic copy of the set $\mathbb{Z}$ of integers.[20] $\mathcal{N}_U$ is the time unit's own coordinate system.
- $U_{\mathcal{N}} : \mathcal{T} \mapsto \mathcal{N}_U$ is a function mapping the reference time axis $\mathcal{T}$ to the U-coordinates.
- $U_{[[} : \mathcal{N}_U \mapsto Int(\mathcal{T})$ is a function mapping the elements n of the U-coordinates to the half-open interval $[b, e[$ in the reference time line which corresponds to the U-coordinate n. $Int(\mathcal{T})$ is the set of all half-open intervals over the reference time line.

  We call a finite set $i \in Int(\mathcal{T})$ of half-open intervals a *time region.*

If $U_{[[}(n) = [b, e[$ then $U^b_{[[}(n) = b$ denotes the beginning of the interval and $U^e_{[[}(n) = e$ denotes the end of the interval.

We require:

- $U^e_{[[}(n) = U^b_{[[}(n+1)$ for all $n \in \mathcal{N}_U$
  (there are no gaps between U-coordinates);

[20]A formal definition of $\mathcal{N}_U$ would be $\mathcal{N}_U = \{(i, U) \mid i \in \mathbb{Z}\}$ with the usual structure of integer numbers imposed on $\mathcal{N}_U$.

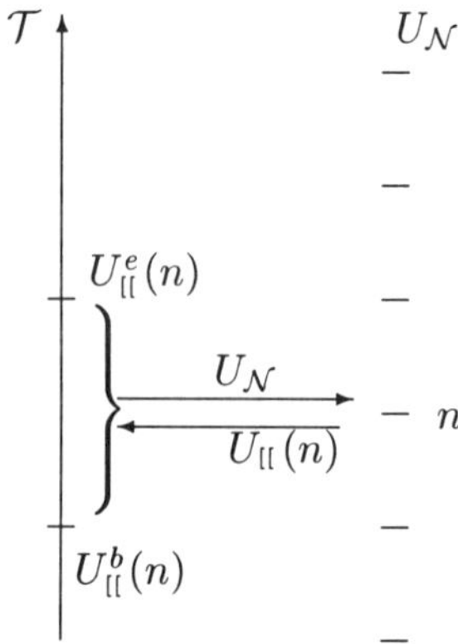

FIG. 18.1. Relation between the reference time system and a time unit U.

- $\forall t \in \mathcal{T}: \ t \in U_{[[}(U_{\mathcal{N}}(t))$ and $\forall n \in \mathcal{N}_U : \ U_{\mathcal{N}}(U_{[[}(n)) = n$ ($U_{\mathcal{N}}$ and $U_{[[}$ are inverse to each other).

It is important to notice that sets of half-open intervals which are open all at the same side, in our case the upper side, are closed under union and intersection. See Figure 18.1 for a typical configuration.

18.2.2 *Temporal notions*

With the functions $U_{\mathcal{N}}$ and $U_{[[}$ we can convert back and forth between a reference time system and an arbitrary time unit U. Unfortunately, time units are not the only irregular structures in a calendar where a fully fledged programming language is necessary to compute the details. Some of the Christian holidays also have a quite complex temporal cycle. The date for Easter, for example, is defined to be the first weekend after the first full moon in spring. Since in our context it makes no sense to define a language powerful enough to compute the moon cycle (this should be done in C or any other suitable programming language), we assume that certain functions, e.g. *easter-time*, are available. *easter-time*, for example, would be a function that takes a year-coordinate and gives back a day-coordinate (Esophaster Sunday).

Given $U_{\mathcal{N}}$ and $U_{[[}$ and special functions like *easter-time*, we need only a few further constructs to specify quite complex temporal notions and to do the corresponding computations.

18.2.2.1 *The U_within_V constructor* Let us start with something very familiar, the notion of the month 'February'. 'February' denotes a set of time points. There are, however, different possibilities. If I say 'February', do I mean a particular February, or maybe even the set of all Februaries in the history of humankind, or more sophisticatedly, a function that gives me for a particular year the particular February in this year? The last interpretation is the most general one, because with this we can reconstruct the other two interpretations. Moreover, it is a quite natural one. If you ask somebody to define 'February', he would say something like 'February is the second month in a year'. What he has in mind is therefore a notion of 'ith month in a year', which is a function taking a year y and an integer i and yielding a month. If $i = 2$, we get a February, if $i = 3$, we get a March, etc.

Both 'month' and 'year' are time units. Therefore we can generalize the notion 'ith month in a year' from the two time units 'month' and 'year' to two arbitrary time units U and V. For each pair of time units U and V we introduce a function *U_within_V* with two arguments, a V-coordinate and an integer. The value of $U\_within\_V(v, i)$ would be a U-coordinate.

Returning to the 'February' example, we have U = month and V = year. Stating

$$\textit{February}(y) \stackrel{\text{def}}{=} \text{month\_}\textit{within}\text{\_year}(y, 2) \tag{18.1}$$

we get a definition of a function *February* that takes a year-coordinate y and yields a month-coordinate (as the second month in a year), provided we have a suitable definition of the function month_*within*_year. In this case, we can use the *February*-function to determine whether a given point t in the temporal reference system is in a February. We compute $y = \text{year}_{\mathcal{N}}(t)$ and get the year-coordinate. $m = \textit{February}(y)$ gives us the corresponding month-coordinate of the February in the year y. Now we check whether t lies in the interval $\text{month}_{[[}(m)$. If this is so, then t is in a February; if not, it is not in a February. Since all the regular and irregular things about the time units 'month' and 'year' are built into the functions $\text{year}_{\mathcal{N}}$ and $\text{month}_{[[}$, and, by the definition of *U_within_V* below, also into this function, *February*(y) gives us a precise definition of the notion of 'February'.

Before we come to a general definition of the *U_within_V* function, let us see whether there are other useful examples of the application of this function. The date 'February 20th' denotes a particular day in a year. Following the same ideas as above, we would like to encode this as a function that maps the coordinate of a year to the coordinate of a day, the 20th February. We have got the function *February* which maps year-coordinates to month-coordinates. In a very similar fashion, 20th denotes a function which maps a month-coordinate m to a day-coordinate. Therefore we can define something like

$$20\textit{th-in-month}(m) \stackrel{\text{def}}{=} \text{day\_}\textit{within}\text{\_month}(m, 20).$$

Composing this with the *February*-function, we get

$$20\textit{th-in-February}(y) = \text{day\_}\textit{within}\text{\_month}(\textit{February}(y), 20)$$

which maps a year-coordinate y to a day-coordinate. The usual date notation year:month:day:hour:minute:second can in fact be encoded by using the *U_within_V* function for the corresponding time units, and function composition.

As another example,

$$Monday(w) \stackrel{\text{def}}{=} \text{day\_}within\text{\_week}(w, 1) \tag{18.2}$$

defines Monday to be the first day in a week, by mapping a week-coordinate w to a day-coordinate. Although we have not yet given a formal definition of the *U_within_V* function, its meaning seems to be quite obvious. In all cases, where the two units U and V are in phase, i.e. the beginning of a V unit (i.e. year) coincides with the beginning of a U unit (i.e. month), it is in fact obvious.

But sometimes there is a subtlety we have to consider. For example, the beginning of a year does not coincide with the beginning of a week (years and weeks are not in phase). It is therefore not clear whether week_*within*_year should count the first week overlapping with a new year as week number 1 (inclusive interpretation), or only the first week which lies completely within the new year (exclusive interpretation). One can avoid making a choice on this simply by providing both versions: $U\_within^e\_V$ is the exclusive reading and $U\_within^i\_V$ is the inclusive reading. For all those pairs of time units which are in phase with each other, both versions coincide.

Let us consider the definition of *U_within_V* for these cases first.

month_*within*_year(y, i) is an example. In order to get the month coordinate for the year y and integer i, we first need to compute with $k \stackrel{\text{def}}{=} \text{year}^b_{[[}(y)$ the reference time of the beginning of the year. Since years and months are in phase, $\text{month}_{\mathcal{N}}(k)$ yields the month-coordinate of the corresponding January in the year y. If this is counted as the first month, we need to add $i - 1$ to get the month-coordinate of the ith month in the year y. The definition of $U\_within\_V(n, i)$ in the general case is therefore just

$$U\_within\_V(v, i) \stackrel{\text{def}}{=} U_{\mathcal{N}}(V^b_{[[}(v)) + i - 1.$$

If the two time units are not in phase, we add either $i - 1$ or i in the exclusive reading, depending on whether there is an overlap at the beginning of the interval or not. These definitions compute answers even for terms like month_*within*_year(1980, 100) (the 100th month in the year 1980) or year_*within*_month(200000, 100) (the 100th year in the month 200000.) It depends on the application whether this might be useful or not.

18.2.2.2 *The begin_U and end_U constructors* Suppose we have a built-in function $springbegin(y)$ which maps a year-coordinate to a day-coordinate (giving the first day in spring). If we want to define 'the Monday in the first week in spring', we cannot use the 'Monday'-function above, because this needs as an input a week-coordinate. But given a day-coordinate d, we can of course get the week-coordinate w of the week containing the given day. This is just $\text{week}_{\mathcal{N}}(\text{day}^b_{[[}(d))$, i.e. from d we compute the reference time of the beginning of the day d, and from there the week-coordinate corresponding to this reference time.

In this example, we went from the finer day-coordinate to the coarser week-coordinate. We might also want to go the other way round, say from a coarser year-coordinate to the finer day-coordinate, either to the beginning of the year (which could be achieved with the *U_within_V* constructor), or to the end of the year, which could only be achieved with the *U_within_V* constructor if the number of days within a year were fixed.

To do this kind of type casting, we introduce the *begin*_U and *end*_U constructors. For a given time unit U and a given V-coordinate v of another time unit V, *begin*_U(v) yields the U-coordinate of the reference time corresponding to the *beginning* of the v-interval, and *end*_U(v) yields the U-coordinate of the reference time corresponding to the *end* of the v-interval. Both yield the same result if V is coarser than U. A useful example for the application of the *end*_U constructor is

$$\textit{new-years-eve}(y) \stackrel{\text{def}}{=} \textit{end}\_\text{day}(y).$$

This function maps a year-coordinate y to the day-coordinate of its last day, new year's eve.

18.2.2.3 *Set constructors* With the constructors we have got so far, we can select single coordinate units within bigger time units. This is still too limited for many purposes. For example, a weekend usually consists of two days, which can only be specified by joining the two days into a set. Therefore we introduce the constructors for manipulating sets of coordinates.

First of all we introduce the familiar set constructors $\{\ldots\}, \cup, \cap, \setminus$. $\{n\}$ turns the single element n into a singleton set $\{n\}$. $\cup, \cap, \setminus$ denote union, intersection and difference of sets. Now we can define for example

$$\textit{weekend}(w) \stackrel{\text{def}}{=} \{\textit{Saturday}(w)\} \cup \{\textit{Sunday}(w)\}$$

as a function mapping week-coordinates to the union of the two day-coordinates corresponding to the Saturday and Sunday in a week. In order to check whether a given reference time t is actually in a weekend, we compute $\{d_1, d_2\} = \textit{weekend}$ $(\text{week}_{\mathcal{N}}(t))$ and check whether t is within the interval $\text{day}_{[[}(d_1)$ or $\text{day}_{[[}(d_2)$.

Some precautions must be taken with the binary set constructors in order not to mix coordinates of different time units. Although something like $\textit{weekend}(w) \cup$ hour_*within*_week$(w, 36)$ which computes the union of days and hours makes sense set-theoretically, it does not make sense in our setting. Therefore in the actual language we need a type mechanism to prevent mixing coordinates of different time units.

18.2.2.4 *The eecomposition constructor U_s* Besides the singleton set constructor $\{\ldots\}$, there is another constructor, which is quite useful for turning individual coordinates into sets. This is the *U*_s constructor. If for example w is a week-coordinate, then day_s(w) denotes the set of all day-coordinates within the week w. Now, for example, we can define

$$\textit{working-days}(w) \stackrel{\text{def}}{=} \text{day\_s}(w) \setminus \textit{weekend}(w)$$

as a function mapping week-coordinates w to the set of day-coordinates corresponding to the complement of the weekend-coordinates.

For time units which are not in phase with each other, e.g. weeks and years, we need a refined U_s constructor. For computing all the weeks within a year, we have to decide whether to include the weeks which overlap partially with the year (inclusive reading) or only the weeks which are completely included in the year (exclusive reading). Therefore we introduce two versions of the U_s constructor again, the U^e_s (exclusive) and the U^i_s (inclusive) version. The U_s constructor is particularly useful for converting between different time zones or calendar systems. If we have, for example, a time unit A-day (day in time zone A) and B-hour (hour in time zone B) and an A-day coordinate d then B-hour_s$^e(d)$ yields the set of B-hours contained completely within the A-day d, and B-hour_s$^i(d)$ yields the set of B-hours overlapping with the A-day d.

18.2.2.5 *Individuals and sets* Each function ϕ producing a set of U-coordinates for a time unit U can become, in a canonical way, an argument of a function $\tau(n)$ accepting a single U-coordinate. We just define $\tau(\phi) \stackrel{\text{def}}{=} \{\tau(n) \mid n \in \phi\}$. For example,

$$\textit{all-Mondays}(y) \stackrel{\text{def}}{=} \textit{Monday}(\text{week\_s}(y))$$

denotes the set of all Mondays in the year y. week_s(y) yields the set of all week-coordinates for the year y, and the definition of *Monday* is applied to all of them separately. The resulting day-coordinates are collected in a set.

We finish this section with a more complicated example. We want a realistic definition of the notion of 'office hour'. There are different ways to define this. We might start with the definition of lunch time per day:

$$\textit{lunch-time}(d) \stackrel{\text{def}}{=} \text{hour\_}\textit{within}\text{\_day}(d, 12).$$

(Lunch time is between 12 and 1 o'clock.) Next we define $ohd(d)$ (office hours per day):

$$\textit{ohd}(d) \stackrel{\text{def}}{=} \text{hour\_}\textit{within}\text{\_day}(d, [9..17]) \setminus \textit{lunch-time}(d).$$

(We use hour_*within*_day$(d, [9..17])$ as an abbreviation for $\{\text{hour\_}\textit{within}\text{\_day}(d, 9)\} \cup \ldots \cup \{\text{hour\_}\textit{within}\text{\_day}(d, 17)\}$. In the next step we define office-hours per week $ohw(w)$ and take the weekends out:

$$\textit{ohw}(w) \stackrel{\text{def}}{=} \textit{ohd}(\text{day\_s}(w) \setminus \textit{weekend}(w)).$$

day_s$(w) \setminus \textit{weekend}(w)$ gives us all day-coordinates of the week w except for the weekend-coordinates. *ohd*(day_s(w)) is applied to each day-coordinate and the union of the resulting sets is formed. Let us further assume that we have a definition of $\textit{holidays}(y)$ which computes the set of all holidays in the year y. Now we can finally define office hours per year:

$$\textit{ohy}(y) \stackrel{\text{def}}{=} \textit{ohw}(\text{week\_s}(y)) \setminus \text{hour\_s}(\textit{holiday}(y)).$$

In an actual implementation, one would of course not compute these coordinates as sets, but maybe represent them as sets of intervals, and do the corresponding set operations on the intervals.

18.2.2.6 *Periods of time* Time units are not only coordinates for fixed intervals in the reference time line. They are also used for specifying other periods of time. If somebody says for example 'for the next two years ... ', he usually means 'for a period of time beginning now and lasting for two years'. Unfortunately notions like 'for two years' are not as precise as they seem to be. If next year is a leap year, does 'for two years' mean 365 + 365 days or 366 + 365 days? When does this period start and end? Does it start exactly at this moment and end at some point at some day in two years' time, or does it include today as a whole day and the last day after two years' time as a whole day?

I propose a flexible encoding of time periods where the user can choose the precise meaning. The constructor is *U_period_V*. For example, day_*period*_year$(d, 2)$ denotes the set of day-coordinates from the day d up to a day after two years' time. This includes the begin and the end day. If the two years' time period is to be located more precisely, say, at a resolution of hours, one can choose hour_*period*_year$(h, 2)$ which yields the set of hour-coordinates from hour h up to an hour after two years' time. (One can also, of course, define an exclusive reading where the coordinates at the beginning and at the end are not included.)

The problem that a period of two years' time itself is also not precisely defined, can be solved, or at least be approximated, in the following way: consider again day_*period*_year$(d, 2)$. Day d defines an interval $[d^b, d^e[$ in the reference time line. We compute the middle $t_d = (d^b + d^e)/2$ of this interval, and from this the current year-coordinate y. y itself corresponds to an interval $[y^b, y^e[$, and $t_d \in [y^b, y^e[$. t_d therefore divides $[y^b, y^e[$ into a fraction $f = (t_d - y^b)/(y^e - y^b)$. f is the fraction of the year y elapsed since y^b. Now we go two years further to $y' = y + 2$ and get the time point $t'_d = y'^b + f \cdot (y'^e - y'^b)$ which corresponds to the same fraction f. From t'_d we get the day-coordinate of the corresponding day in two years' time.

18.2.2.7 *Arithmetic* The results of all these functions that have been defined so far are either single coordinates or sets of coordinates, and a coordinate is nothing but an integer. Therefore nothing stops us from embedding the time language into an integer arithmetic language. This way we can, for example, define

$$tomorrow(d) \stackrel{\text{def}}{=} d + 1$$

as a function that returns for a given day-coordinate d the coordinate of the following day. Even conditional terms with arithmetic comparisons are possible. Borrowing the $c?a : b$ construct from the programming language C (if c holds then a, otherwise b) we can define the 'first Monday after the beginning of springtime' by

$$\begin{aligned}&Monday(begin\_\text{week}(springbegin(y))) < springbegin(y)\ ?\\ &\quad Monday(begin\_\text{week}(springbegin(y)) + 1)\ \ :\\ &\quad Monday(begin\_\text{week}(springbegin(y)))\end{aligned}$$

(if the beginning of springtime lies within a week such that the Monday is actually before the beginning of springtime, then we choose the Monday of the next week, otherwise we take this Monday.)

18.2.3 *The time term language*

We now turn these informal ideas into a formal specification language. Its semantics is purely functional and can be defined in terms of a kind of interpreter which can actually execute the computations indicated above. A specification S of temporal notions is a list of equations $f(u_1,\ldots,u_n) \stackrel{\text{def}}{=} \varphi[u_1,\ldots,u_n]$ where on the left-hand side a new function symbol f is introduced, and defined by φ. S should be such that all defined function symbols can be eliminated in a finite sequence of rewrite steps.

In order not to mix coordinates of different time units (all of them are essentially integers), we need a sorted (typed) language where for each function the sort of the arguments and the sort of the resulting values are specified [Schmidt-Schauß, 1989]. The sort structure consists of two parts. The 'single value part' contains the sort symbol $\mathbb{Z}$ (for the integers) and the time unit symbols $\mathcal{U}$ used as further sort symbols denoting the time units' own coordinates. In the 'set value part' we have for each time unit sort symbol $U \in \mathcal{U}$ a symbol U^* denoting sets of U-coordinates. In addition there is a sort *Bool* for arithmetic comparisons.

Most of the function symbols can have 'overloaded' sort declarations where the sort of the function value depends on the sort of its arguments.

Definition 18.3 (The time term language). We define a first-order (many-sorted) time term language $\mathcal{T}_{\mathcal{U}}$ as follows:

- The set of (element) sort symbols consists of the (finite) set $\mathcal{U}$, the special symbol $\mathbb{Z}$ for integers, and the sort *Bool* for Boolean values. In addition we have for each $U \in \mathcal{U}$ a (set) sort symbol U^* (denoting sets of U-elements).
- For each $U \in \mathcal{U}$, each $u \in \mathcal{N}_U$ is a *constant symbol* of sort U. Each integer is a constant of sort $\mathbb{Z}$.[21]
- For each sort $U \in \mathcal{U}$ we have an unlimited supply of variable symbols of that sort (there are no 'set variables', no integer variables and no Boolean variables).
- We can have an arbitrary number of built-in functions of sort $U_1 \times \cdots \times U_n \to U$. (*eastertime*(*year*) would be a typical built-in function.)[22]
- Depending on the set $\mathcal{U}$ of time units, there is a certain set of derived function symbols.

[21] With these constants we have a name for each coordinate. For example, the constant 1997 of sort 'year' (actually the pair (1997,year)) denotes the year 1997.

[22] In the sort declarations $U_1 \times \cdots \times U_n \to U$ for function symbols, the U_k are the *domain-sorts*, and the U is the *range-sort*.

  * For $U, V \in \mathcal{U}$ we have function symbols $U\_within^e\_V$ and $U\_within^i\_V$, both of sort $V \times \mathbb{Z} \to U$ and $V^* \times \mathbb{Z} \to U^*$ (a single V-coordinate as input yields a single U-coordinate as output, whereas a set of V-coordinates yields a set of U-coordinates as output).
  * For each $U \in \mathcal{U}$ we have the 'begin' and 'end' constructors $begin\_U$ and end_U, both of sort $V \to U$ and $V^* \to U^*$ for all sorts $V \in \mathcal{U}$.
  * For each $U \in \mathcal{U}$ we have the set constructor U_s of sort $V \to U^*$ and $V^* \to U^*$ for all sorts $V \in \mathcal{U}$.
  * For $U, V \in \mathcal{U}$ we have the function symbol U_period_V of sort $V \times \mathbb{Z} \to U^*$ and $V^* \times \mathbb{Z} \to U^*$.
  * The 'singleton set' operator $\{\ldots\}$ of sort $U \to U^*$ for each sort $U \in \mathcal{U}$ maps single elements to singleton sets.
  * The set connectives $\cap, \cup, \setminus$, all of sort $U^* \times U^* \to U^*$, for all $U \in \mathcal{U}$ denote the usual set operations intersection, union and set difference.
  * For each time unit $U \in \mathcal{U}$ there are two 'decomposition function' symbols U^e_s and U^i_s with sort declarations $V \to U^*$ for each sort (time unit) $V \in \mathcal{U}$.

- We include the standard integer arithmetic symbols $+, -, *$, all of sort $U \times U \to U^*$, $U^* \times U \to U$ and $U \times U^* \to U^*$ for all $U \in \mathcal{U}$.
- There is the special Boolean conditional operator $\ldots ? \ldots : \ldots$ of sort $Bool \times U \times U \to U$ for all $U \in \mathcal{U}$.
- The standard integer comparison operators $=, <, >, \leq, \geq$ of sort $U \times U \to Bool$ for all $U \in \mathcal{U}$ can be used in the conditional part of the conditional operator.
- Other function symbols can only be used as abbreviations if they have suitable non-recursive definitions of the form $f(u_1, \ldots, u_n) \stackrel{\text{def}}{=} \tau[u_1, \ldots, u_n]$.[23] One can always discard these function symbols by using their defining equation as rewrite rule (this is actually the main requirement on these definitions).

The sort declarations for the function symbols guarantee that they do not compute coordinates of mixed time units (sorts).

Notice that the complement operator is not included in the time term language. This is deliberate, because for most of the algorithms working with the time terms the assumption that the time regions are finite is absolutely essential. The complement of a finite time region is infinite, and infinite time regions require very different treatment.

Proposition 18.4 (Unique range-sort). *Since the terms of our language $\mathcal{T}_\mathcal{U}$ are standard first-order terms with overloaded sort declarations, given the sorts of the free variables in the term they always have a unique range-sort.*

[23]The notation $\tau[u_1, \ldots, u_n]$ means that τ is a term containing the variables $u_1, \ldots, u_n$ at some places. $u_1, \ldots, u_n$ are τ's *free variables*.

We define the semantics of $\mathcal{T}_\mathcal{U}$-terms by giving the definitions of all the built-in functions. From a logical point of view, the semantics of the language $\mathcal{T}_\mathcal{U}$ is based on a single model only.

Definition 18.5 (Semantics of the time term language). Given the interpretation $\Im_T$ for the time unit symbols $\mathcal{U}$ (Definition 18.2) we define an interpretation $\Im = (\Im_T, \mathcal{S}, \mathcal{V}, \mathcal{F})$ for the time term language $\mathcal{T}_\mathcal{U}$:

- $\mathcal{S}$ maps (element) sort symbols U to $\mathcal{N}_U$ and (set) sort symbols U^* to the set of all subsets of $\mathcal{N}_U$.
- $\mathcal{V}$ is the variable assignment. For each variable u of sort U we have $\mathcal{V}(u) \in \mathcal{N}_U$.
- The function interpretation $\mathcal{F}$ maps constant symbols to themselves. It also contains the interpretation of the built-in functions (such as *eastertime*).
- The other functions are interpreted by $\mathcal{F}$ as follows:
  * If $v \in \mathcal{N}_V$ and $n \in \mathbb{Z}$ then $\mathcal{F}(U\_within^i\_V)(v,n) = U_\mathcal{N}(V^b_{[[}(v)) + n - 1$.
  * $\mathcal{F}(U\_within^e\_V)(v,n) = \begin{cases} U_\mathcal{N}(V^b_{[[}(v)) + n - 1 & \text{if } U^b_{[[}(U_\mathcal{N}(V^b_{[[}(v))) < V^b_{[[}(v) \\ U_\mathcal{N}(V^b_{[[}(v)) + n & \text{otherwise.} \end{cases}$
  * If $v \in \mathcal{N}_V$ then
    $\mathcal{F}(U^i\_s)(v) = \{u \in \mathcal{N}_U \mid V_\mathcal{N}(U^b_{[[}(u)) = v \text{ or } V_\mathcal{N}(U^e_{[[}(u)) = v\}$.
    $\mathcal{F}(U^e\_s)(v) = \{u \in \mathcal{N}_U \mid V_\mathcal{N}(U^b_{[[}(u)) = v \text{ and } V_\mathcal{N}(U^e_{[[}(u)) = v\}$.
  * $\mathcal{F}(begin\_\mathrm{U})(v) = U_\mathcal{N}(V^b_{[[}(v))$ if v is of sort $V \in \mathcal{U}$.
    $\mathcal{F}(end\_\mathrm{U})(v) = U_\mathcal{N}(V^e_{[[}(v))$ if v is of sort $V \in \mathcal{U}$.
  * $\mathcal{F}(U\_period\_V)(u,n) = \{u, \ldots, w\}$ where w is determined as follows:

    Let $t \stackrel{\text{def}}{=} (U^e_{[[}(u) + U^b_{[[}(u))/2$ the middle of the u-interval.
    Let $v \stackrel{\text{def}}{=} V_\mathcal{N}(t)$ the V-coordinate of the middle.
    Let $f \stackrel{\text{def}}{=} \frac{t - V^b_{[[}(v)}{V^e_{[[}(v) - V^b_{[[}(v)}$ the fraction elapsed since the beginning.
    Let $v' \stackrel{\text{def}}{=} v + n$ the new V-coordinate.
    Let $t' \stackrel{\text{def}}{=} V^b_{[[}(v') + f \cdot (V^e_{[[}(v') - V^b_{[[}(v'))$ the end time point.
    Let $w \stackrel{\text{def}}{=} U_\mathcal{N}(t')$ the end U-coordinate.
  * If the argument to one of the functions is a set instead of a single coordinate then the function is applied to each set and the results are collected in a set.
  * The conditional ...?... : ..., the set operators and the arithmetic terms are interpreted in the standard way. If an argument to one of the arithmetic functions is a set, then the function is applied to each element and the results are collected in a set.
- Arbitrary time terms $\tau[u_1, \ldots, u_m]$ with free variables $\{u_1, \ldots, u_n\}$ with $sort(u_i) = U_i$ for $1 \le i \le n$, and range-sort U, can now be interpreted in two ways. The first way is the standard homomorphic extension of the interpretation $\Im$. If $\tau = u$ is a variable then $\Im(u) \stackrel{\text{def}}{=} \mathcal{V}(u)$. If $\tau = f(t_1 \ldots, t_n)$

is a complex term then $\Im(\tau) \stackrel{\text{def}}{=} \mathcal{F}(f)(\Im(t_1), \ldots, \Im(t_n))$. The binding of the variables u_i to some U_i-coordinates is contained in the variable assignment $\mathcal{V}$. Therefore $\Im(\tau)$ is well defined and yields a set of U-coordinates.

- Much more interesting and useful is the interpretation of time terms as functions mapping a reference time point t to some U-coordinates. If for a variable v of sort V and a V-coordinate n we define $\Im[v/n]$ to be the interpretation which is like $\Im$, but the variable assignment maps v to n, then we can define an $\Im(\tau)(t)$ as a function mapping a reference time point t to some U-coordinates.
  If $\tau = u$ is a variable, then
  $$\Im(u)(t) \stackrel{\text{def}}{=} \Im[u/U_{\mathcal{N}}(t)](u).$$
  If $\tau[u_1, \ldots, u_m]$ is a complex term, then
  $$\Im(\tau)(t) \stackrel{\text{def}}{=} \Im[u_1/U_{1\mathcal{N}}(t), \ldots, u_m/U_{m\mathcal{N}}(t)](\tau).$$
  Since the time point t completely determines the interpretation of τ, we usually write $\tau(t)$ instead of $\Im(\tau)(t)$.
  If i is an interval or a set of intervals, we define
  $$\tau(i) \stackrel{\text{def}}{=} \bigcup_{t \in i} \tau(t).$$
  Furthermore let
  $$\tau_{[[}(t) \stackrel{\text{def}}{=} \bigcup_{u \in \tau(t)} U_{[[}(u).$$

$\Im(\tau[u_1, \ldots, u_m])(t)$ computes the value of $\tau[u_1, \ldots, u_m]$ at a given reference time point t by first computing the corresponding U_i-coordinates $U_{i\mathcal{N}}(t)$, binding the variables u_i to these coordinates and then evaluating them in the usual way. The result is either a single U-coordinate or a set of U-coordinates.

The $\tau_{[[}$ function goes one step further and turns the computed U-coordinates back into intervals in the reference time system.

Using the interpreter $\Im$ we can now check the *instance* relation between a given reference time point and a $\mathcal{T}_{\mathcal{U}}$ term (e.g. whether the current moment in time lies in the office hours of my colleague in Brazil).

Notice that $\tau_{[[}(i)$ may be an empty time region. A natural example where this may occur is the definition of *leap-day* as a function mapping year-coordinates to day-coordinates:

$$\begin{aligned} \mathit{leap\text{-}day}(x_{year}) \stackrel{\text{def}}{=} \mathit{end}\_\mathrm{day}(\mathit{February}(x_{year})) &= \\ \mathrm{day}\_\mathit{within}\_\mathrm{month}(\mathit{February}(x_{year}), 29)&? \\ \mathit{end}\_\mathrm{day}(\mathit{February}(x_{year})) &: \{\}. \end{aligned}$$

(If the last day of a February in a year is the 29th then the leap day is the 29th, otherwise the leap day is empty.) $(\mathit{leap\text{-}day}(x_{year}))_{[[}$ is empty for all non-leap years.

Definition 18.6 (Instance relation). A time point t in the reference system is an instance of the term τ with range-sort U (we write $t \in \tau$), iff $t \in \tau_{[[}(t)$ or, which is equivalent, iff $U_{\mathcal{N}}(t) \in \tau(t)$.

According to this definition, we compute the U-coordinate corresponding to t and check whether this is one of the coordinates computed from τ.

A more complicated relation is the subsumption relation. An example might be the problem to figure out whether my colleague's morning office hours in Brazil always lie within my local afternoon. (Then I could always phone him in the afternoon without first checking it in the database.) Unfortunately this general subsumption relation is usually not decidable. This is due to the fact that we treat the functions $U_{\mathcal{N}}$ and $U_{[[}$ as a kind of black box where the most peculiar things may be encoded, e.g. a transition to a completely new calendar system in the year 2100.

For temporal relationships from everyday life, this general subsumption relationship, however, is really too general. I certainly won't be interested in knowing whether in the year 2100 my colleague's morning office hours in Brazil lie within my local afternoon. For me it is completely sufficient to know this until his or my own time of retirement.

Definition 18.7 (Subsumption relation). A term τ_1 is subsumed by a term τ_2 ($\tau_1 \subseteq \tau_2$) iff for all time points t in the reference system, $\tau_{1[[}(t) \subseteq \tau_{2[[}(t)$.

A *restricted subsumption* relation ($\tau_1 \subseteq_i \tau_2$) for the time region i holds between τ_1 and τ_2 iff for all time points $t \in i$, $\tau_{1[[}(t) \subseteq \tau_{2[[}(t)$.

Since the restricted subsumption relation still quantifies over an infinite number of time points, this definition is not yet a suitable basis for an algorithm. In order to decide restricted subsumption, we exploit the fact that the time terms depend on the coordinates of the time units, and they are therefore constant over the period in the reference time line corresponding to the same coordinates of the time unit. For example, the time term *February*(y) (Definition 18.1) depends on the year-coordinate. It is therefore constant over a whole year, i.e. over all the points t in the reference time line with $\text{year}_{\mathcal{N}}(t) = y$.

Exploiting this observation, each time term $\tau[u_1, \ldots u_n]$ can be used to partition each finite interval in the reference time line into the finitely many subintervals where $\tau[u_1, \ldots u_n]$ represents a constant function.

As a simple example, consider i to be week 53 in the year 1970. i overlaps partially with 1970 and partially with 1971. *February*(x_{year})(i) therefore yields the month-coordinates for the February in 1970 and 1971. This is computed by first partitioning i into the part of the week lying in the year 1970, and the part of the week lying in 1971. For each part, *February*(x_{year}) is then evaluated separately.

This partitioning can then be used to trigger finite case analysis for checking whether τ_1 subsumes τ_2 over a time region i. i is partitioned into the set $\{i_1, \ldots, i_n\}$ of subregions over which τ_1 and τ_2 are constant, and then, for $1 \leq k \leq n$, $\Im(\tau_1)(t_k) \subseteq \Im(\tau_2)(t_k)$ is checked for some arbitrary chosen $t_k \in i_k$.

$1 \leq k \leq n$, $\Im(\tau_1)(t_k) \subseteq \Im(\tau_2)(t_k)$ is checked for some arbitrary chosen $t_k \in i_k$.

Definition 18.8 (Partitioning algorithm). For a non-empty interval $i = [i^b, i^e[\subseteq \mathcal{T}$ and a time term $\tau[u_1, \ldots, u_n]$ with $U_k \stackrel{\text{def}}{=} sort(u_k)$, $k = 1, \ldots, n$, we define the *partitioning of i with respect to τ* as

$$\delta(i, \tau) = \{[i^b, t[\} \cup \delta([t, i^e[, \tau)$$

where

$$t \stackrel{\text{def}}{=} \begin{cases} min(\{U_k{}^b_{[[}(U_{k\mathcal{N}}(i^b) + 1) \in i \mid 1 \leq k \leq n\}) \\ i^e \text{ if this set is empty.} \end{cases}$$

If i is a set of intervals, let $\delta(i, \tau) \stackrel{\text{def}}{=} \bigcup_{j \in i} \delta(j, \tau)$.

Since $\delta(i, \tau)$ only depends on the free variables $\{u_1, \ldots, u_n\}$ in τ, we can write $\delta(i, \{u_1, \ldots, u_n\})$ for $\delta(i, \tau)$.

This version can be generalized to sets $\{\tau_1, \ldots, \tau_m\}$ of time terms. If $\{u_1, \ldots, u_n\}$ is the set of all free variables occurring in $\{\tau_1, \ldots, \tau_m\}$, we get

$$\delta(i, \{\tau_1, \ldots, \tau_m\}) \stackrel{\text{def}}{=} \delta(i, \{u_1, \ldots, u_n\})$$

and this is computed as indicated above.

The partitioning algorithm decomposes i by computing the U_k-coordinate $U_{k\mathcal{N}}(i^b)$ corresponding to the beginning of the interval i and then checking whether the beginning of the next interval corresponding to the next U_k-coordinate still lies in i or not. If it still lies in i, this is a candidate for the lower border of the next subinterval. The real border is determined by the time unit U_k which gives the smallest such border. (A properly implemented algorithm would, of course, try the most fine-grained time unit U_k first, e.g. seconds before minutes before hours, etc.)

Notice that the partitioning function only depends on the time units corresponding to τ's free variables, not on τ itself.

Proposition 18.9 (Soundness of the partitioning algorithm). *For each interval $j \in \delta(i, \tau)$ (Definition 18.8) and for each $\{t_1, t_2\} \subseteq j$, $\tau_{[[}(t_1) = \tau_{[[}(t_2)$.*

Proof By induction on the number of intervals in $\delta(i, \tau)$. In the base case, $\delta(i, \tau) = i$, i.e. i itself is the only component, and for $1 \leq k \leq n$, $U_k{}^b_{[[}(U_{k\mathcal{N}}(i^b) + 1) > i^e$. Therefore for all $t \in i$, $U_{k\mathcal{N}}(t)$ yields the same U_k-coordinate. Since $\tau[u_1, \ldots, u_n]$ depends on the U_k-coordinates only, $\tau_{[[}(t_1) = \tau_{[[}(t_2)$ for all $\{t_1, t_2\} \subseteq i$.

In the induction step, $\delta(i, \tau)$ consists of more than one interval and $[i^b, s[$ is the leftmost subinterval of the decomposition. According to the definition of δ, t is the smallest value such that $t = U_k{}^b_{[[}(U_{k\mathcal{N}}(i^b) + 1)$ lies still in i. That means for all $k \in \{1, \ldots, n\}$: $U_{k\mathcal{N}}(t_1) = U_{k\mathcal{N}}(t_2)$ for all $t_1 \in [i^b, t[$ and $t_2 \in [i^b, t[$. Thus, $\tau_{[[}(t_1) = \tau_{[[}(t_2)$ for the first interval in $\delta(i, \tau)$. The induction hypothesis directly applies to the remaining intervals. □

Proposition 18.10 (Evaluation of time terms over intervals). *For a $\mathcal{T}_{\mathcal{U}}$-formula τ and a finite half-open interval $i \subseteq \mathcal{T}$:*

$$\tau(i) = \bigcup_{j \in \delta(i,\tau)} \tau(j^b).$$

Proof $\delta(i,\tau)$ partitions i into the finitely many subintervals where τ is a constant function. Therefore for each such subinterval j we can just choose an arbitrary element s, in particular the first one, $s = j^b$, to compute $\tau(s)$ and join these intervals together. □

Proposition 18.11 (Decidability).

- *The instance relation is decidable. $\tau_{[[}(t)$ yields only finitely many intervals. Therefore just check $t \in \tau_{[[}(t)$, which amounts to finding some $[i^b, i^e[\ \in \tau_{[[}(t)$ with $i^b \leq t < i^i$.*
- *Without special assumptions about the functions $U_{\mathcal{N}}$ and $U_{[[}$, the subsumption relation is, in general, not decidable.*
- *The restricted subsumption relation $\tau_1 \subseteq_i \tau_2$ over an interval $i = [t_1, t_2[$ is decidable. Using the partitioning algorithm of Definition 18.8, partition this interval into the finitely many subintervals where the two time terms are constant functions, choose for each subinterval j some point $t \in j$, and check the subsumption relation $\tau_{1[[}(t) \subseteq \tau_{2[[}(t)$ for this point. To do this, check for each $[i^b, i^e[\ \in \tau_{1[[}(t)$ whether there is some $[j^b, j^e[\ \in \tau_{2[[}(t)$ with $[i^b, i^e[\ \subseteq [j^b, j^e[$.[24]*

18.2.3.1 *Top-down interpretation* The interpretation $\Im(\tau)$ of a time term τ works bottom up in the usual recursive style. There is a recursive descent into the term down to the variable level, and then the actual computation is done on the way back to the top level of the term. This can be quite expensive if large sets of coordinates are generated. For example, the interpretation of the term seconds_s(y) where y is a year-variable causes the generation of the set of all second-coordinates within the year y (a set with about 31 million elements).

For the instance check $t \in$ seconds_s(y) (Definition 18.6) this would be much too expensive. A *top-down interpretation* is much more efficient in this case. How could this work? For checking $t \in$ seconds_s(y), we exploit the fact that the term seconds_s(y) has range-sort 'second'. Therefore one can compute the second-coordinate $s = second_{\mathcal{N}}(t)$, and then compute the set of all year-coordinates y such that $s =$ seconds_s(y). There is only one y in this case, and this can be computed quite easily, using the semantics of the seconds_s-constructor. If $t \in year_{[[}(y)$ then the instance relation holds, otherwise not.

To do this kind of top-down interpretation, one needs for each constructor $f(u_1, \ldots, u_n)$ in the language, an algorithm for computing for a given u the set

[24] $\tau_{[[}(t)$ should be computed in such a way that the sequences of intervals without gaps is constrained into one single interval.

actually produce this result. A top-down interpreter can then descend into a time term and pass the potential arguments as wanted results to the functions at the deeper level of the term. This has not yet been investigated in detail, but for the constructors in the time term language this seems to be quite feasible.

There are, however, examples where this strategy also causes the computation of large sets of coordinates. Therefore a mixed bottom-up, top-down strategy seems to be the most efficient way to decide the instance relation.

18.3 Pure calendar logic

The pure calendar logic $\mathcal{CL}$ is the next stage in our system. Its syntax is almost like the syntax of propositional multimodal logic K_m [Chellas, 1980; Chargov and Zakharayaschev, 1997]. There are the parameterized modal operators $\langle\tau\rangle$ and $[\tau]$ where τ is a time term in the language $\mathcal{T_U}$. The intuitive meaning for a formula like

$$[\mathit{February}(y_{year})]\mathit{sunshine}$$

is 'there was sunshine throughout this year's February'.

This formula is interpreted in a given reference time point t as follows: from t one gets $y_{year} = year_{\mathcal{N}}(t)$, from that the interval corresponding to $\mathit{February}(y_{year})$, and in *all* points (semantics of the box operator) of this interval, $\mathit{sunshine}$ must be true.

Whereas the $[\tau]$-operator is a universal quantification over the points corresponding to $\tau(s)$, $\langle\tau\rangle$ is the dual operator and works as an existential quantification. The formula

$$\langle x_{day} + 1\rangle \mathit{go\text{-}to\text{-}cinema}$$

expresses 'tomorrow I'll go to the cinema'. $\langle x_{day} + 1\rangle$ denotes some particular time point tomorrow, at which *go-to-cinema* is supposed to be true. One can also express 'tomorrow I'll go to the cinema' together with the extra fact 'and I'll stay there for two hours' by

$$\langle x_{day} + 1\rangle(\mathit{go\text{-}to\text{-}cinema} \wedge [\{y_{hour}, y_{hour} + 1\}]\mathit{in\text{-}cinema}).$$

This formula restricts the duration of the two hours to the interval corresponding to the beginning of the first hour until the end of the second hour. With the *U_period_V* constructor in the time term language it is also possible to refer to an interval of two hours' length starting at an arbitrary point in time.

Notice that in calendar logic there are no classical temporal logic operators like 'always in the future'. Instead of this, however, we may for example have an 'always in the next century' operator.

There is an extra feature of calendar logic which has no correspondence in modal logic: we allow for time terms at formula positions. For example,

$$[x_{day} + 1](\mathit{Monday}(w_{week}) \rightarrow \neg \mathit{work})$$

expresses 'if tomorrow ($x_{day} + 1$) is Monday ($\mathit{Monday}(w_{week})$), I don't need to work'. $\mathit{Monday}(w_{week})$ is a $\mathcal{T_U}$-term at a formula position. It is true at a reference time point if this time point lies in the interval specified by $\mathit{Monday}(w_{week})$.

If t is again a reference time point, then the interpretation of this formula works as follows: $x_{day} \stackrel{\text{def}}{=} day_{\mathcal{N}}(t)$ yields the current day-coordinate. $x_{day}+1$ is tomorrow's day-coordinate. For *each* (semantics of $[\tau]$) reference time point t' corresponding to $x_{day}+1$ one must check whether $Monday(w_{week}) \to \neg work$ holds at t'. With $w_{week} \stackrel{\text{def}}{=} week_{\mathcal{N}}(t')$ we get the week-coordinate of t'. Using (18.2), $d \stackrel{\text{def}}{=} Monday(w_{week})$ yields a day-coordinate. If $t' \in day_{[[}(d)$ then $Monday(w_{week})$ is true for t', and we have to check whether $\neg work$ holds as well at t'.

We may also have subsumption relationships of time terms at formula positions. For example,

$$\langle x_{day}+1\rangle(noon(x_{day}) \subseteq lunch\text{-}time(x_{day}) \to ring\text{-}home)$$

expresses 'if tomorrow's lunch time is at noon, I'll ring home'. $lunch\text{-}time(x_{day})$ and $noon(x_{day})$ are both time terms. The subsumption relationship is true at a time point t if the time region $noon(x_{day})_{[[}(t)$ is a subset of $lunch\text{-}time(x_{day})_{[[}(t)$.

Subsumption relations in both directions, i.e. $\tau_1 \subseteq \tau_2 \wedge \tau_2 \subseteq \tau_1$, can be abbreviated using the equality symbol: $\tau_1 = \tau_2$. Below, (18.3) is an example of the use of equations between time terms.

Definition 18.12 (Syntax of calendar logic). If $\mathcal{T_U}$ is a language of time terms (Definition 18.3) and $\mathbb{P}$ a set of predicate symbols then the formulae of calendar logic $\mathcal{CL}_{\mathcal{T_U}}$ are built according to the following grammar:

$$\begin{aligned}\mathcal{CL}_{\mathcal{T_U}} := \mathbb{P} \mid \mathcal{T_U} \mid \mathcal{T_U} \subseteq \mathcal{T_U} \mid \neg\mathcal{CL}_{\mathcal{T_U}} \mid \mathcal{CL}_{\mathcal{T_U}} \wedge \mathcal{CL}_{\mathcal{T_U}} \mid \mathcal{CL}_{\mathcal{T_U}} \vee \mathcal{CL}_{\mathcal{T_U}} \mid \\ \langle\mathcal{T_U}\rangle\mathcal{CL}_{\mathcal{T_U}} \mid [\mathcal{T_U}]\mathcal{CL}_{\mathcal{T_U}}.\end{aligned}$$

$\tau_1 = \tau_2$ abbreviates $\tau_1 \subseteq \tau_2 \wedge \tau_2 \subseteq \tau_1$.

A *statement* of calendar logic is either $\langle n, U\rangle : \varphi$ or $[n, U] : \varphi$ where U is a time unit, n is one of U's coordinates and φ is a calendar logic formula.

A calendar logic formula is interpreted relative to some time point in the reference time line. For expressing a fact about a very concrete point in time, extra syntax is necessary. One could, for example, express a fact as a tuple $t : \varphi$ where t is a point in the reference time line, i.e. a real number. These real numbers, however, will never be explicitly accessible. They are usually deeply hidden in some timer devices. Moreover, for associating a particular event with the time when it happened, points in the reference time line are in general much too sharp. In most cases, the time when an event happened can be located only within a certain interval. Therefore we choose the syntax $\langle n, U\rangle : \varphi$ or $[n, U] : \varphi$ for expressing statements about events happening at some point in time. U is a time unit (second, minute, hour, etc.), and n is a concrete coordinate of this time unit. $\langle n, U\rangle : \varphi$ means φ is true at *some* point in the interval corresponding to the U-coordinate n, and $[n, U] : \varphi$ means φ is true at *all* points in the interval corresponding to the U-coordinate n. For example, the event 'Jack was born in 1980' would be expressed as $\langle 10, year\rangle : born(Jack)$ where 10 is the tenth year in the Unix reference time line starting with 1970. As another example, 'in 1996 Jack lived in London' could be expressed as $[26, year] : lives\text{-}in(London, Jack)$

1980' would be expressed as $\langle 10, year\rangle : born(Jack)$ where 10 is the tenth year in the Unix reference time line starting with 1970. As another example, 'in 1996 Jack lived in London' could be expressed as $[26, year] : lives\text{-}in(London, Jack)$ where the box operator expresses that Jack lived in London all the time in the year 1996.

Finally we will see how our introductory example 'yesterday I worked for eight hours with one hour lunch break at noon' could be expressed in calendar logic. Let d be today's day-coordinate. A first formalization of this statement is

$$\langle d, day\rangle : \langle x_{day} - 1\rangle(([\{y_{hour}, \ldots, y_{hour} + 8\} \setminus noon(z_{day})]work) \\ \wedge [noon(z_{day})]lunch)$$

with

$$noon(z_{day}) \stackrel{\text{def}}{=} \text{hour\_}within\text{\_day}(z_{day}, 12).$$

This statement would be true even if I started working yesterday at 11p.m. To insist that the whole period of work was yesterday, one could write

$$\langle d, day\rangle : \langle x_{day} - 1\rangle(([\{y_{hour}, \ldots, y_{hour} + 8\} \setminus noon(z_{day})]work) \\ \wedge\ day\_s(y_{hour}) = day\_s(y_{hour} + 8) \wedge [noon(z_{day})]lunch) \tag{18.3}$$

$day\_s(y_{hour}) = day\_s(y_{hour} + 8)$ restricts the *hour*-coordinate y_{hour} where I began to work, such that eight hours later is still the same day.

Definition 18.13 (Semantics of calendar logic). An interpretation $\Im \stackrel{\text{def}}{=} (\Im_{\mathcal{T}_\mathcal{U}}, \mathcal{P})$ for the formulae of calendar logic $\mathcal{CL}_{\mathcal{T}_\mathcal{U}}$ consists of a fixed $\mathcal{T}_\mathcal{U}$-interpretation $\Im_{\mathcal{T}_\mathcal{U}}$ (Definition 18.5) for the time terms $\mathcal{T}_\mathcal{U}$, and an assignment $\mathcal{P} : \mathbb{P} \mapsto 2^{\mathcal{T}}$ which assigns to each predicate symbol $p \in \mathbb{P}$ the set of reference time points where p is true.

For a reference time point $t \in \mathcal{T}$ we define:

$$\begin{array}{lll}
\Im, t \models p & \text{iff } t \in \mathcal{P}(p) & \text{where } p \in \mathbb{P} \\
\Im, t \models \tau & \text{iff } t \in \tau_{[[}(t) & \text{where } \tau \in \mathcal{T}_\mathcal{U} \\
\Im, t \models \tau_1 \subseteq \tau_2 & \text{iff } \tau_{1[[}(t) \subseteq \tau_{2[[}(s) & \\
\Im, t \models \langle\tau\rangle\varphi & \text{iff } \Im, t' \models \varphi \text{ for some } t' \in \tau_{[[}(t) & \\
\Im, t \models [\tau]\varphi & \text{iff } \Im, t' \models \varphi \text{ for all } t' \in \tau_{[[}(t) &
\end{array}$$

$$\begin{array}{ll}
\Im \models \langle n, U\rangle : \varphi & \text{iff } \Im, t \models \varphi \text{ for some } t \in U_{[[}(n) \\
\Im \models [n, U] : \varphi & \text{iff } \Im, t \models \varphi \text{ for all } t \in U_{[[}(n).
\end{array}$$

Algorithms operating on logical formulae are usually more efficient and easier to define if the formulae are in some kind of normal form. The *negation normal form* where all negation symbols are moved inside as far as possible simplifies the treatment of negation considerably.

Definition 18.14 (Negation normal form). There is a *negation normal form* for $\mathcal{CL}_{\mathcal{T}_\mathcal{U}}$-formulae where all negation symbols are moved in front of the predicate symbols or the time terms. An arbitrary $\mathcal{CL}_{\mathcal{T_U}}$-formula or statement can be brought into negation normal form using the following rewrite system:

$$\begin{aligned}
\neg\neg\varphi &\to \varphi \\
\neg(\varphi_1 \wedge \varphi_2) &\to \neg\varphi_1 \vee \neg\varphi_2 \\
\neg(\varphi_1 \vee \varphi_2) &\to \neg\varphi_1 \wedge \neg\varphi_2 \\
\neg\langle\tau\rangle\varphi &\to [\tau]\neg\varphi \\
\neg[\tau]\varphi &\to \langle\tau\rangle\neg\varphi \\
\neg\langle n, U\rangle : \varphi &\to [n, U] : \neg\varphi \\
\neg[n, U] : \varphi &\to \langle n, U\rangle : \neg\varphi.
\end{aligned}$$

It is easy to check that the negation normal form algorithm is an equivalence preserving transformation on $\mathcal{CL}_{\mathcal{T_U}}$-formulae and statements. Putting formulae into negation normal form is only a critical operation if many equivalences are present. $\varphi \Leftrightarrow \psi$ is turned into $(\neg\varphi \vee \psi) \wedge (\neg\psi \vee \varphi)$. This is an exponential transformation if φ or ψ themselves contain equivalences. By introducing predicate symbols as abbreviations for φ and ψ, one can avoid this—a well-known trick in predicate logic theorem proving.

In first-order predicate logic there is a particularly useful normal form, the *conjunctive normal form*, or *clause form*. It consists of conjunctions of clauses, and each clause is a disjunction of literals. Literals are positive or negative *atoms* $p(t_1, \ldots, t_n)$. This simple normal form is not possible for $\mathcal{CL}_{\mathcal{T_U}}$-formulae because the temporal operators do not distribute over conjunction and disjunction. Nevertheless, there is a similar normal form, which we also call *clause normal form*. In this clause form the conjunctions are moved to the top level as far as possible, and the disjunctions are moved down as far as possible. The clause normal form is the basis for a further transformation which eliminates the temporal operators altogether.

Definition 18.15 (Clause normal form). The equivalences

$$\varphi_1 \vee (\varphi_2 \wedge \varphi_3) \Leftrightarrow (\varphi_1 \vee \varphi_2) \wedge (\varphi_1 \vee \varphi_3) \quad \text{(distributivity)} \tag{18.4}$$

$$\varphi_1 \wedge (\varphi_2 \vee \varphi_3) \Leftrightarrow (\varphi_1 \wedge \varphi_2) \vee (\varphi_1 \wedge \varphi_3) \quad \text{(distributivity)} \tag{18.5}$$

$$[\tau](\varphi_1 \wedge \varphi_2) \Leftrightarrow [\tau]\varphi_1 \wedge [\tau]\varphi_2 \tag{18.6}$$

$$\langle\tau\rangle(\varphi_1 \vee \varphi_2) \Leftrightarrow \langle\tau\rangle\varphi_1 \vee \langle\tau\rangle\varphi_2 \tag{18.7}$$

can be used as rewrite rules to put all $\mathcal{CL}_{\mathcal{T_U}}$-formulae and statements from their negation normal form into a *clause normal form*.

The *disjunctive clauses* and *conjunctive clauses* of the clause normal form for $\mathcal{CL}_{\mathcal{T_U}}$-formulae look like

$$\bigvee_k [\tau_k]\varphi_k \vee \bigvee_l \langle\tau_l\rangle\psi_l \vee \bigvee_m \tau_m \vee \bigvee_n (\tau_{n1} \subseteq \tau_{n2}) \vee \varphi \tag{18.8}$$

$$\bigwedge_k [\tau_k]\varphi_k \wedge \bigwedge_l \langle\tau_l\rangle\psi_l \wedge \bigwedge_m \tau_m \wedge \bigwedge_n (\tau_{n1} \subseteq \tau_{n2}) \wedge \varphi \tag{18.9}$$

where the φ_k are disjunctive (conjunctive) clauses (due to (18.6)) and the ψ_l are conjunctive clauses (due to (18.7)); the τ_m, the τ_{n1} and the τ_{n2} are $\mathcal{T_U}$-terms. φ is a disjunction (conjunction) of positive or negated predicate symbols.

From now on we shall assume that all formulae and statements are put into clause normal form.

18.3.1 *Translation of $\mathcal{CL}_{\mathcal{T_U}}$ into propositional logic*

$\mathcal{CL}_{\mathcal{T_U}}$-statements $[n, U]\varphi$ and $\langle n, U\rangle\varphi$ specify a finite time region $i = U_{[[}(n)$ within which φ is supposed to be valid. φ itself may contain temporal operators $[\tau]$ and $\langle\tau\rangle$ specifying other finite time ranges $\tau_{[[}(i)$ within which the corresponding subformulae are supposed to hold. Since $\mathcal{CL}_{\mathcal{T_U}}$-formulae are finite, there are only finitely many time regions involved, and they can all be computed as sets of half-open intervals in the reference time line. This observation is exploited to define a sequence of further transformations which finally eliminates all the temporal operators and leaves us with propositional formulae.

This means that satisfiability in $\mathcal{CL}_{\mathcal{T_U}}$ is decidable by any decision procedure for propositional logic. The bad news is that the translation into propositional logic may blow up the formulae exponentially. The worst-case time complexity of this method for deciding satisfiability is therefore double exponential.

The translation into propositional logic consists of four different steps. For the first step two further *time region* operators $[i]$ and $\langle i\rangle$ are introduced. The parameter i is a concrete time range. $\langle[10, 100[, [200, 300[\rangle\varphi$ for example means that φ is true at some time either between 10 and 100 or between 200 and 300 in the reference time line. $[i]\varphi$ means that φ is true in *all* points $t \in i$ and $\langle i\rangle\varphi$ means that φ is true in *some* point $t \in i$.

The translation $\rightarrow_I$ defined below turns the temporal operators $[\tau]$ and $\langle\tau\rangle$ into time region operators $[i]$ and $\langle i\rangle$.

Example 18.16 (For turning temporal into time region operators). In these examples we assume the reference time line is counted in days, starting from the year 2000. The numbers are approximate.

Example 1: $[2000, year] : [January(x_{year})]election$. (in January in the year 2000 there is an election.) $[2000, year]$ denotes a time region $[0, 366[$ and the January in the year 2000 ranges from day 0 to the end of day 31. Therefore the translation yields $[[0, 366[][[0, 32[]holiday$.

Notice that the first box operator is no longer necessary. $[[0, 32[]holiday$ contains the same information as the formula with the two box operators. Since the semantics of the time region operators does not depend on the actual time point, this is no coincidence.

Example 2: $[2000, year] : \langle January(x_{year})\rangle(sunshine \rightarrow \langle weekend(y_{week})\rangle$ $visit)$. (I will visit you at some weekend in January 2000, if the sun is shining at that time.) The first two translation steps are the same as above. The result is $[[0, 366[]\langle[0, 32[\rangle(sunshine \rightarrow \langle weekend(y_{week})\rangle)visit$. The last translation step must take into account that there are different weekends in January 2000, and we do not know which of them the diamond operator refers to. A translation into

$$[[0,366[]\langle[0,32[\rangle(sunshine \rightarrow \langle[5,7[,[12,14[,[19,21[,[26,29[\rangle visit)$$

would be wrong. This formula can be satisfied in a situation where sunshine is at the 1st January and the visit would be in the last weekend of January. The meaning of the original formula, however, is more restrictive. The implication $sunshine \rightarrow \langle weekend(y_{week})\rangle visit$ enforces that the weekend of the week in which *sunshine* is true is meant, and not any weekend in January.

The original meaning of the implication can be preserved, if the translation inserts a case analysis at a higher level. To this end the time region corresponding to $January(x_{year})$ is split into the subranges where $weekend(y_{week})$ is constant. Since there are four weeks in January, we get four subranges. The correct translation is then

$$\begin{aligned}[[0,366[]\ (&\langle[0,7[\rangle\ \ (sunshine \rightarrow \langle[5,7[\rangle\ \ visit) \vee \\ &\langle[7,14[\rangle\ (sunshine \rightarrow \langle[12,14[\rangle visit) \vee \\ &\langle[14,21[\rangle(sunshine \rightarrow \langle[19,21[\rangle visit) \vee \\ &\langle[21,29[\rangle(sunshine \rightarrow \langle[26,29[\rangle visit)).\end{aligned}$$

This formula is false if the sunshine is in the first week and the visit is in the last week of January. It can only be made true if the visit is in the same week as the sunshine.

The temporal operators in temporal statements like $[n,U]:\varphi$ are translated recursively into time region operators. We start with turning the $[n,U]$-operator into $[U_{[[}(n)]$. Now the time region for which φ is supposed to hold is made explicit, namely $U_{[[}(n)$. In the recursive part of the translation, we deal with formulae like $[i]\langle\tau\rangle\varphi$ where i is a time region and τ a time term. The time region i is split into the subregions $\{i_1,\ldots,i_h\} = \delta(i,\tau)$ where τ is constant. $[i]\langle\tau\rangle\varphi$ is first turned into $[i_1]\langle\tau\rangle\varphi \wedge \cdots \wedge [i_h]\langle\tau\rangle\varphi$ and then into $[i_1]\langle\tau_{[[}(i_1)\rangle\varphi \wedge \cdots \wedge [i_h]\langle\tau_{[[}(i_h)\rangle\varphi$. The $\tau_{[[}(i_k)$ are now time regions, and the recursive translation can go down into φ.

Definition 18.17 ($\mathcal{CL}^I_{\mathcal{T}_\mathcal{U}}$). $\mathcal{CL}^I_{\mathcal{T}_\mathcal{U}}$ is defined as an extension of calendar logic with two further *time region* operators $[i]$ and $\langle i\rangle$ where i is not a time term, but a time region. The semantics of the operators is

$$\begin{aligned}&\Im,t \models [i]\varphi \text{ iff } \Im,t' \models \varphi \text{ for all } t' \in i\\ &\Im,t \models \langle i\rangle\varphi \text{ iff } \Im,t' \models \varphi \text{ for some } t' \in i.\end{aligned}$$

Notice that the semantics of these operators does not depend on the actual time point s. Therefore instead of the notation $\Im,t \models [i]\ldots$ or $\Im,t \models \langle i\rangle\ldots$ the irrelevant t can be dropped by simply writing $\Im \models [i]\ldots$ or $\Im \models \langle i\rangle\ldots$.

Definition 18.18 (Translation of $\mathcal{CL}_{\mathcal{T}_\mathcal{U}}$ into $\mathcal{CL}^I_{\mathcal{T}_\mathcal{U}}$). We define rewrite rules for translating the $\mathcal{CL}_{\mathcal{T}_\mathcal{U}}$-operators into time region operators.

The rewrite rules for $\mathcal{CL}_{\mathcal{T_U}}$-statements are

$$[n, U] : \varphi \rightarrow_I [U_{[[}(n)]\varphi$$
$$\langle n, U\rangle : \varphi \rightarrow_I \langle U_{[[}(n)\rangle\varphi.$$

The rewrite rules for $\mathcal{CL}_{\mathcal{T_U}}$-formulae are defined for the clause normal form (Definition 18.15):

$$[i](\bigvee_k [\tau_k]\varphi_k \vee \bigvee_l \langle\tau_l\rangle\varphi_l \vee \varphi)$$
$$\rightarrow_I$$
$$\bigwedge_{r=1}^h [i_r](\bigvee_k [\tau_{k[[}(i_r)]\varphi_k \vee \bigvee_l \langle\tau_{l[[}(i_r)\rangle\varphi_l \vee \varphi)$$

$$\langle i\rangle(\bigwedge_k [\tau_k]\varphi_k \wedge \bigwedge_l \langle\tau_l\rangle\varphi_l \wedge \varphi)$$
$$\rightarrow_I$$
$$\bigvee_{r=1}^h \langle i_r\rangle(\bigwedge_k [\tau_{k[[}(i_r)]\varphi_k \wedge \bigwedge_l \langle\tau_{l[[}(i_r)\rangle\varphi_l \wedge \varphi)$$

where

- $[i]$ and $\langle i\rangle$ are time region operators;
- φ is a disjunction (conjunction) of temporal statements τ, subsumption relations $\tau_1 \subseteq \tau_2$ and propositional literals;
- $\{i_1, \ldots, i_h\} \stackrel{\text{def}}{=} \delta(i, \{\tau_1, \ldots\})$ and $\{\tau_1, \ldots\}$ are all the time terms occurring at the top level in the clauses and in φ.

Proposition 18.19. *The translation $\rightarrow_I$ is an equivalence-preserving transformation.*

Proof For the $\mathcal{CL}_{\mathcal{T_U}}$-statements:

$\Im \models [n, U]\varphi$ iff $\Im, t \models \varphi$ for all $t \in U_{[[}(n)$
iff $\Im, t \models [U_{[[}(n)]\varphi$ for all t
$\Im \models \langle n, U\rangle\varphi$ iff $\Im, t \models \varphi$ for some $t \in U_{[[}(n)$
iff $\Im, t \models \langle U_{[[}(n)\rangle\varphi$ for all t.

For $\mathcal{CL}_{\mathcal{T_U}}$-formulae:

$\Im \models [i](\bigvee_k [\tau_k]\varphi_k \vee \bigvee_l \langle\tau_l\rangle\varphi_l \vee \varphi$
iff $\Im, t \models (\bigvee_k [\tau_k]\varphi_k \vee \bigvee_l \langle\tau_l\rangle\varphi_l \vee \varphi)$
for all $t \in i$
iff $\Im, t \models (\bigvee_k [\tau_k]\varphi_k \vee \bigvee_l \langle\tau_l\rangle\varphi_l \vee \varphi)$
for all $i_r \in \delta(i, \{\tau_1, \ldots\})$ and for all $t \in i_r$
iff $\bigvee_k(\forall t' \in \tau_{k[[}(i_r) : \Im, t' \models \varphi_k)$ or $\bigvee_l(\exists t' \in \tau_{l[[}(i_r) : \Im, t' \models \varphi_l)$ or $\Im, t \models \varphi$
for all $i_r \in \delta(i, \{\tau_1, \ldots\})$ and for all $t \in i_r$
(this is because the time terms τ are constant over i_r)
iff $\bigvee_k(\Im \models [\tau_{k[[}(i_r)]\varphi_k)$ or $\bigvee_l(\Im \models \langle\tau_{l[[}(i_r)\rangle\varphi_l)$ or $\Im, t \models \varphi$
for all $i_r \in \delta(i, \{\tau_1, \ldots\})$ and for all $t \in i_r$
iff $\Im \models \bigwedge_{r=1}^h [i_r](\bigvee_k [\tau_{k[[}(i_r)]\varphi_k \vee \bigvee_l \langle\tau_{l[[}(i_r)\rangle\varphi_l \vee \varphi)$.

The proof for the rule for the diamond operator is dual to the proof for the box operator. □

It turns out that the time terms τ at formula positions and the subsumption relations $\tau_1 \subseteq \tau_2$ can be eliminated completely after this translation step. The next examples illustrate the idea.

Example 18.20 (For eliminating temporal formulae). We use the same reference time line as in Example 18.16. Consider

$$[2000, year] : \langle January(x_{year})\rangle(\mathrm{Friday}(y_{week}) \wedge election).$$

(Some Friday in January 2000 there is an election.) The first two translation steps are as in Example 18.16: $[[0,366[]\langle[0,32[\rangle(\mathrm{Friday}(y_{week}) \to election)$. In the next step, $\langle[0,32[\rangle$ has to be decomposed into the subranges where $\mathrm{Friday}(y_{week})$ is constant. These are the four weeks in January. Thus, we get

$$\begin{array}{ll} [[0,366[](\langle[0,7[\rangle & (\mathrm{Friday}(y_{week}) \to election)) \vee \\ \quad (\langle[7,14[\rangle & (\mathrm{Friday}(y_{week}) \to election)) \vee \\ \quad (\langle[14,21[\rangle & (\mathrm{Friday}(y_{week}) \to election)) \vee \\ \quad (\langle[21,29[\rangle & (\mathrm{Friday}(y_{week}) \to election)). \end{array}$$

Now the time regions in the diamond operator can be reduced to the time regions where Friday is true. The result is

$$[[0,366[](\langle[5,6[\rangle election \vee \langle[12,13[\rangle election \vee \langle[19,20[\rangle election \vee \langle[26,27[\rangle election)$$

(This is in fact equivalent to $[[0,366[]\langle[5,6[,[12,13[,[19,20[,[26,27[\rangle election.$) The example is continued in Example 18.28.

Lemma 18.21 (Elimination of temporal formulae). *If the time terms τ, τ_1 and τ_2 are constant over a time region i then the following equivalences hold:*

$$\begin{array}{lll} \langle i\rangle(\tau \wedge \varphi) & \Leftrightarrow \langle i \cap \tau_{[[}(i)\rangle\varphi & \\ \langle i\rangle(\neg\tau \wedge \varphi) & \Leftrightarrow \langle i \setminus \tau_{[[}(i)\rangle\varphi & \\ \langle i\rangle(\tau_1 = \tau_2 \wedge \varphi) & \Leftrightarrow \bot & \textit{if } \tau_{1[[}(i) \neq \tau_{2[[}(i) \\ \langle i\rangle(\tau_1 = \tau_2 \wedge \varphi) & \Leftrightarrow \langle i\rangle\varphi & \textit{if } \tau_{1[[}(i) = \tau_{2[[}(i) \\ \langle i\rangle(\tau_1 \neq \tau_2 \wedge \varphi) & \Leftrightarrow \bot & \textit{if } \tau_{1[[}(i) = \tau_{2[[}(i) \\ \langle i\rangle(\tau_1 \neq \tau_2 \wedge \varphi) & \Leftrightarrow \langle i\rangle\varphi & \textit{if } \tau_{1[[}(i) \neq \tau_{2[[}(i) \\ [i](\tau \vee \varphi) & \Leftrightarrow [i \setminus \tau_{[[}(i)]\varphi & \\ [i](\neg\tau \vee \varphi) & \Leftrightarrow [i \cap \tau_{[[}(i)]\varphi & \\ [i](\tau_1 = \tau_2 \vee \varphi) & \Leftrightarrow \top & \textit{if } \tau_{1[[}(i) = \tau_{2[[}(i) \\ [i](\tau_1 = \tau_2 \vee \varphi) & \Leftrightarrow [i]\varphi & \textit{if } \tau_{1[[}(i) \neq \tau_{2[[}(i) \\ [i](\tau_1 \neq \tau_2 \vee \varphi) & \Leftrightarrow \top & \textit{if } \tau_{1[[}(i) \neq \tau_{2[[}(i) \\ [i](\tau_1 \neq \tau_2 \vee \varphi) & \Leftrightarrow [i]\varphi & \textit{if } \tau_{1[[}(i) = \tau_{2[[}(i). \end{array}$$

The equivalences follow straightforwardly from the semantics of the operators and the temporal formulae.

Lemma 18.22 (Simplification rules). *The following equivalences hold in $\mathcal{CL}^I$ for all formulae φ and ψ:*

$$\begin{array}{llll}
[i][j]\varphi \Leftrightarrow [j]\varphi & [i](\varphi \wedge [j]\psi) \Leftrightarrow ([i]\varphi \wedge [j]\psi) & [i](\varphi \vee [j]\psi) \Leftrightarrow ([i]\varphi \vee [j]\psi) \\
[i]\langle j\rangle\varphi \Leftrightarrow \langle j\rangle\varphi & [i](\varphi \wedge \langle j\rangle\psi) \Leftrightarrow ([i]\varphi \wedge \langle j\rangle\psi) & [i](\varphi \vee \langle j\rangle\psi) \Leftrightarrow ([i]\varphi \vee \langle j\rangle\psi) \\
\langle i\rangle[j]\varphi \Leftrightarrow [j]\varphi & \langle i\rangle(\varphi \wedge [j]\psi) \Leftrightarrow (\langle i\rangle\varphi \wedge [j]\psi) & \langle i\rangle(\varphi \vee [j]\psi) \Leftrightarrow (\langle i\rangle\varphi \vee [j]\psi) \\
\langle i\rangle\langle j\rangle\varphi \Leftrightarrow \langle j\rangle\varphi & \langle i\rangle(\varphi \wedge \langle j\rangle\psi) \Leftrightarrow (\langle i\rangle\varphi \wedge \langle j\rangle\psi) & \langle i\rangle(\varphi \vee \langle j\rangle\psi) \Leftrightarrow (\langle i\rangle\varphi \vee \langle j\rangle\psi)
\end{array}$$

and

$$\langle\emptyset\rangle\varphi \Leftrightarrow \bot \qquad [\emptyset]\varphi \Leftrightarrow \top.$$

The proof is straightforward. It uses the fact that the interpretation of the time region operators does not depend on the actual time point.

These equivalences can be used as rewrite rules from left to right. The result of rewriting a $\mathcal{CL}^I$-formula with these rules is a modal degree 1 normal form. There are no longer any nested temporal operators.

The different time regions in the modal operators can interact with each other. For example, $[[0,100[]p$ and $\langle[50,60[\rangle\neg p$ are contradictory because $[50,60[$ is a subset of $[0,100[$. $[[0,100[]p$ and $\langle[50,150[\rangle\neg p$ are not contradictory because $\neg p$ might be true between 100 and 150.

There are two options for dealing with these operators. One can define special inference rules that deal with the intervals in a suitable way, or one can make the different time regions independent. For example, $[[0,100[]p$ and $\langle[50,150[\rangle\neg p$ are equivalent to $[[0,50[]p \wedge [[50,100[]p$ and $\langle[50,100[\rangle\neg p \vee \langle[100,150[\rangle\neg p$. The second version is obtained by decomposing the time regions into their atomic components. Now the internal structure of the decomposed time regions does not matter any more. This is a well-known trick from Boolean algebra theory. A finite set $\{i_1,\ldots,i_n\}$ of sets generates a finite atomic Boolean algebra by closing the set under intersection, union and complement. The atoms of this Boolean algebra consist of the 2^n intersections $\pm i_1 \cap\cdots\cap \pm i_n$ (where $\pm i$ means either i or its complement). For example, the set $\{[0,100[,[50,150[\}$ has the four atoms $[0,50[$, $[50,100[$, $[100,150[$ together with the complement of $[0,200[$. In the next definition we formalize this idea.

Definition 18.23 (Atomic decomposition for $\mathcal{CL}^I$). Let φ be a $\mathcal{CL}^I_{\mathcal{T}u}$-formula with $M \stackrel{\text{def}}{=} \{i_1,\ldots,i_k\}$ being the time regions occurring in the time operators contained in φ. The closure of M under intersection, union and complement forms a finite atomic Boolean algebra A. Let $atoms(A)$ be the atoms of A, and let $atoms(i) \stackrel{\text{def}}{=} \{a \in atoms(A) \mid a \subseteq i\}$ be the atomic decomposition of i.

We define the following rewrite system:

$$\begin{array}{l}
[i]\varphi \rightarrow_A \bigwedge_{a\in atoms(i)} [a]\varphi \\
\langle i\rangle\varphi \rightarrow_A \bigvee_{a\in atoms(i)} \langle a\rangle\varphi.
\end{array}$$

The result of applying $\rightarrow_A$ exhaustively to φ is called the *atomic decomposition* of φ.

Proposition 18.24. *$\rightarrow_A$ is an equivalence-preserving transformation.*

a can be translated into a unary predicate $a(s)$. It is not necessary that the truth set of the predicate a be a time region. It can be an arbitrary set in an arbitrary domain. The only restriction is that for different time regions a and b, the disjointness condition

$$\forall x\ a(x) \Leftrightarrow \neg b(x)$$

is guaranteed. The translation rules for the box and diamond operator are:

$[a](p_1 \vee \cdots \vee p_n)$ is translated into $\forall s\ a(s) \rightarrow (p_1(s) \vee \ldots \vee p_n(s))$.
$\langle a\rangle(p_1 \wedge \cdots \wedge p_n)$ is translated into $\exists s\ a(s) \wedge (p_1(s) \vee \ldots \vee p_n(s))$.

Because there are no nested time region operators in the decomposed clauses, Skolemization of the existential quantifier yields a constant symbol. [25]Therefore the clauses are translated into the function free fragment of predicate logic. Satisfiability of formulae is decidable in this fragment (because the Herbrand universe is finite). Thus, this is actually a proof for the decidability of calendar logic.

We will not go into the technical details of this translation, because one can even do better. The decomposed clauses can in fact be translated into classical propositional logic. Each formula $[a]\varphi$ is replaced with a new predicate symbol $b_{a\varphi}$ and each formula $\langle a\rangle\psi$ is replaced with a new predicate symbol $d_{a\psi}$. The new predicate symbols are not independent of each other and of the old predicate symbols. Therefore some extra axioms are needed to correlate these different variable sets.

Truth of $b_{a\varphi}$ indicates that $[a]\varphi$ is true, and therefore φ must be true for all time points in a. This is made explicit by adding an implication (cf. (18.12) below)

$$b_{a\varphi} \rightarrow R_a(\varphi).$$

$R_a(\varphi)$ consistently renames all predicate symbols p occurring in φ to p_a. (p_a indicates that p is true for all time points in a.)

Truth of $d_{a\psi}$ indicates that $\langle a\rangle\psi$ is true, and therefore ψ must be true for *some* time point in a. This is made explicit by adding an implication (cf. (18.11) below)

$$d_{a\psi} \rightarrow R_{a\psi}(\psi).$$

$R_{a\psi}(\psi)$ consistently renames all predicate symbols p occurring in ψ to $p_{a\psi}$. ($p_{a\psi}$ indicates that p is true for some particular time point in a. This point may be different for each occurrence $\langle a\rangle\psi$ in different subformulae. Therefore $p_{a\psi}$ must be different for each ψ.)

[25] *Skolemization* is the process of replacing an existentially quantified variable with a constant or function symbol. For example, $\exists x\ p(x)$ can be skolemized to $p(a)$ where a is a Skolem constant. $\forall y\ \exists x\ q(x, y)$ can be skolemized to $\forall y\ q(f(y), y)$. f is a Skolem function. In predicate logic, skolemization is a satisfiability-preserving transformation, not an equivalence transformation.

Truth of $d_{a\psi}$ indicates that $\langle a\rangle\psi$ is true, and therefore ψ must be true for *some* time point in a. This is made explicit by adding an implication (cf. (18.11) below)

$$d_{a\psi} \to R_{a\psi}(\psi).$$

$R_{a\psi}(\psi)$ consistently renames all predicate symbols p occurring in ψ to $p_{a\psi}$. ($p_{a\psi}$ indicates that p is true for some particular time point in a. This point may be different for each occurrence $\langle a\rangle\psi$ in different subformulae. Therefore $p_{a\psi}$ must be different for each ψ.)

Truth of $[a]\varphi$ further implies that φ is true in the particular time points introduced by the diamond operators. Therefore for each $b_{a\varphi}$ and $d_{a\psi}$ we add (cf. (18.14) below)

$$b_{a\varphi} \to R_{a\psi}(\varphi).$$

The soundness and completeness of this translation relies on the fact that the atoms a of the atomic decomposition of the original time intervals are disjoint time regions. Box and diamond operators for different (atomic) time regions a never interact with each other.

Definition 18.25 (Translation into propositional logic). We define a translation $PL0$ of decomposed formulae into classical propositional logic.

Let $R_a(\varphi)$ be a function that renames φ such that every predicate symbol p occurring in φ is replaced with a new predicate symbol p_a.

Let $R_{a\psi}(\varphi)$ be a function that renames φ such that every predicate symbol p occurring in φ is replaced with a new predicate symbol $p_{a\psi}$.

The translation function $PL0'$ introduces for each formula $[a]\varphi$ a new predicate symbol $b_{a\varphi}$ and for each formula $\langle a\rangle\psi$ a new predicate symbol $d_{a\psi}$:

$$[a]\varphi \to_{PL0} b_{a\varphi} \tag{18.10}$$

$$\langle a\rangle\psi \to_{PL0} d_{a\psi}. \tag{18.11}$$

Let $PL0'(\eta)$ be the result of applying the rewrite relation $\to_{PL0}$ exhaustively to η.

The translation function $PL0$ is defined

$$\begin{aligned} PL0(\eta) \stackrel{\text{def}}{=} & PL0'(\eta) \\ & \wedge \textstyle\bigwedge_{b_{a\varphi}} b_{a\varphi} \Rightarrow R_a(\varphi) && (18.12) \\ & \wedge \textstyle\bigwedge_{d_{a\psi}} d_{a\psi} \Rightarrow R_{a\psi}(\psi) && (18.13) \\ & \wedge \textstyle\bigwedge_{b_{a\varphi}, d_{a\psi}} b_{a\varphi} \Rightarrow R_{a\psi}(\varphi). && (18.14) \end{aligned}$$

Lemma 18.26 (Soundness and completeness of the *PL0* translation). *A decomposed formula η is satisfiable in $\mathcal{CL}_{\mathcal{T}_\mathcal{U}}$ if and only if the translated formula $PL0(\eta)$ is satisfiable in propositional logic.*

Proof ($\to$) Suppose $\Im$ satisfies η. We define a propositional interpretation $\Im'$ and show that it satisfies the translated formula.

If $\Im \models [a]\varphi$ then $\Im'(b_{a\varphi}) \stackrel{\text{def}}{=} true$, otherwise $false$. (18.15)
If $\Im \models \langle a\rangle\psi$ then $\Im'(d_{a\psi}) \stackrel{\text{def}}{=} true$, otherwise $false$. (18.16)
If $\Im \models [a]\varphi$ then $\Im, t \models \varphi$ for all $t \in a$. We choose one $t_0 \in a$ and define for each propositional variable p occurring in φ:
if $\Im, t_0 \models p$ then $\Im'(p_a) = true$, otherwise $false$. (18.17)
If $\Im \models \langle a\rangle\psi$ then $\Im, t \models \psi$ for some $t \in a$. We define for each propositional variable p occurring in $\{\psi, \varphi_1, \ldots, \varphi_m\}$ where $\{\varphi_1, \ldots, \varphi_m\} \stackrel{\text{def}}{=} \{\varphi \mid \Im \models [a]\varphi\}$:
if $\Im, t \models p$ then $\Im'(p_{a\psi}) = true$, otherwise $false$. (18.18)

This guarantees that $\Im'$ gives the same truth values to the variables $b_{a\varphi}$ and $d_{a\psi}$ as $\Im$ does to $[a]\varphi$ and $\langle a\rangle\psi$. Therefore $PL0'(\eta)$ is satisfied by $\Im'$; (18.17) guarantees that if $\Im'(b_{a\varphi}) = true$ then $\Im'$ satisfies $R_a(\varphi)$, i.e. $\Im'$ satisfies (18.12); and (18.18) guarantees that

- if $\Im'(d_{a\psi}) = true$ then $\Im'$ satisfies $R_{a\psi}(\psi)$, i.e. $\Im'$ satisfies (18.13) and
- if $\Im'(b_{a\varphi}) = true$ then $\Im'$ satisfies $R_{a\psi}(\varphi)$, i.e. $\Im'$ satisfies (18.14).

Thus, $\Im'$ satisfies $PL0(\eta)$.

($\Leftarrow$) The proof of this direction exploits the fact that the time regions occurring in the decomposed clauses are all disjoint, and that the reference time line is a dense time structure. In integer-like time structures it would not work because there might not be enough time points in a time region to satisfy all relevant diamond operators.

Suppose a propositional interpretation $\Im'$ satisfies $PL0(\eta)$. We construct a temporal interpretation $\Im$ for η.

Suppose $\Im'(b_{a\varphi}) = true$. In this case $\Im'(R_a(\varphi)) = true$ by (18.12). For each $p_a \in R_a(\varphi)$ we define: if $\Im'(p_a) = true$ then $\Im, t \models p$ for each $t \in a$. (This is possible without getting conflicts for some $t \in a$ because all the time regions a are disjoint.) This definition together with $\Im'(R_a(\varphi)) = true$ ensures $\Im, t \models \varphi$ for all $t \in a$. Thus, $\Im \models [a]\varphi_k$.

Suppose $\Im'(d_{a\psi}) = true$. In this case $\Im'(R_{a\psi}(\psi)) = true$ by (18.13), and for all $b_{a\varphi}$ where $\Im'(b_{a\varphi}) = true$ we have $\Im'(R_{a\psi}(\varphi)) = true$ as well, by (18.14). That means $\Im'(R_{a\psi}(\psi) \wedge R_{a\psi}(\varphi_1) \wedge \cdots \wedge R_{a\psi}(\varphi_m)) = true$ where $\{\varphi_1, \ldots, \varphi_m\} = \{\varphi \mid \Im'(b_{a\varphi}) = true\}$.

For each $d_{a\psi}$ we choose a different time point $t_{a\psi} \in a$ as the time point where ψ is to be made true. This is possible because the reference time line is a dense structure. Therefore there are always enough time points to choose from. When constructing the interpretation of ψ in $t_{a\psi}$, we exploit (18.14) to make sure that for all formulae $[a]\varphi$ which are satisfied in $\Im$ by the above construction, φ is satisfied in $t_{a\psi}$. For each $p_{a\psi}$ occurring in $R_{a\psi}(\psi) \wedge R_{a\psi}(\varphi_1) \wedge \cdots \wedge R_{a\psi}(\varphi_m)$ we define: if $\Im'(p_{a\psi}) = true$ then $\Im, t_{a\psi} \models p$. This, together with $\Im'(R_{a\psi}(\psi) \wedge R_{a\psi}(\varphi_1) \wedge \cdots \wedge R_{a\psi}(\varphi_m)) = true$ guarantees $\Im, t_{a\psi} \models \psi \wedge \varphi_1 \wedge \ldots \wedge \varphi_m$. Thus, $\Im \models \langle a\rangle\psi$. This means $b_{a\varphi}$ and $d_{a\psi}$ have the same truth value in $\Im'$ as $[a]\varphi$ and $\langle a\rangle\psi$ have in $\Im$. Thus, $\Im \models \eta$. □

Theorem 18.27. *Satisfiability of $\mathcal{CL}_{\mathcal{T}_\mathcal{U}}$-formulae is decidable.*

was reduced to

$$[[0,366[](\langle[5,6[\rangle election \vee \langle[12,13[\rangle election \vee \langle[19,20[\rangle election \vee \langle[26,27[\rangle election).$$

The outermost box operator is superfluous (Lemma 18.22) and can be dropped. The remaining time regions are all disjoint. Therefore they themselves are the atoms in the atomic decomposition. The propositional version is therefore simply

$$\begin{array}{l}(d_{[5,6[election} \vee d_{[12,13[election} \vee d_{[19,20[election} \vee d_{[26,27[election}) \wedge \\ d_{[5,6[election} \rightarrow election_{[5,6[} \wedge \\ d_{[12,13[election} \rightarrow election_{[12,13[} \wedge \\ d_{[19,20[election} \rightarrow election_{[19,20[} \wedge \\ d_{[26,27[election} \rightarrow election_{[26,27[}.\end{array}$$

18.3.2 *A tableau system for calendar logic*

The translation of $\mathcal{CL}_{\mathcal{T_U}}$-formulae into propositional logic may increase the size of the formulae exponentially. Therefore it could be more efficient for a calculus to work on the original and not on the translated syntax. As an alternative to the translation method we provide a tableau system for deciding the satisfiability of $\mathcal{CL}_{\mathcal{T_U}}$-formulae.

A tableau system makes implicit information explicit and does a systematic case analysis for disjunctive information. Starting with a formula or a list of formulae, the tableau system operates by choosing a tableau rule and some formulae to which the rule is applicable and modifies the formulae in the way the rule specifies. Disjunctive rules split the whole set of formulae into different branches corresponding to the different cases to be investigated. The tableau rules can operate on different branches simultaneously. A branch is *closed* as soon as a contradiction is found. This is checked with some extra rules, the *closing conditions*. A branch which is not closed, but to which no rule is applicable any more, is an *open branch*. An open branch describes a model for the original formula. If all branches close then the initial formula set is not satisfiable.

Tableau entries are usually formulae of the given logic labelled with some extra information which is related to the semantics of the logic. In our case the labels are time regions, or *reference time symbols*. An entry $i : \varphi$ where i is an interval expresses that the formula φ is true at all points in the interval. An entry $t : \varphi$ where t is a reference time symbol expresses that φ is true at some point in time, and t is the name of this point (t is not a number). In all cases these reference time symbols t are constrained by an entry $t \in i$ where i is an interval.

A fourth kind of tableau entries are tuples $(i : \varphi_1 \vee \cdots \vee \varphi_n \mid t_1, \ldots, t_k)$. They are used to control the processing of disjunctions.

Definition 18.29 (Tableau entries). A *tableau entry* can be

(i) $t : \varphi$ (point statement or point entry) where t is a symbol denoting a reference time point (a *reference time symbol*) and φ is a $\mathcal{CL}_{\mathcal{T_U}}$-formula; or

entry $t : \varphi$ where t is a reference time symbol expresses that φ is true at some point in time, and t is the name of this point (t is not a number). In all cases these reference time symbols t are constrained by an entry $t \in i$ where i is an interval.

A fourth kind of tableau entries are tuples $(i : \varphi_1 \vee \cdots \vee \varphi_n \mid t_1, \ldots, t_k)$. They are used to control the processing of disjunctions.

Definition 18.29 (Tableau entries). A *tableau entry* can be

(i) $t : \varphi$ (point statement or point entry) where t is a symbol denoting a reference time point (a *reference time symbol*) and φ is a $\mathcal{CL}_{\mathcal{T_U}}$-formula; or
(ii) $i : \varphi$ (interval statement or interval entry) where i is a time region and φ is a $\mathcal{CL}_{\mathcal{T_U}}$-formula; or
(iii) $t \in i$ (membership statement or constraint for t) where t is a reference time symbol and i is a time region; or
(iv) $(i : \varphi_1 \vee \cdots \vee \varphi_n \mid t_1, \ldots, t_k)$ ($\vee$-control entry) where $i : \varphi_1 \vee, \cdots \vee \varphi_n$ is an interval statement and $t_1, \ldots, t_k$ are reference time symbols.

We shall always use the letters t and i to distinguish reference time symbols from time regions.

Definition 18.30 (Semantics of tableau entries). A *tableau interpretation* $\mathcal{E} = (\Im, \mathbb{T})$ consists of a $\mathcal{CL}_{\mathcal{T_U}}$-interpretation $\Im = (\Im_{\mathcal{T_U}}, \mathcal{P})$ (Definition 18.13) and a component $\mathbb{T}$ mapping reference time symbols to reference time points.

$$\begin{array}{ll} \mathcal{E} \models t : \varphi & \text{iff } \Im, \mathbb{T}(t) \models \varphi \\ \mathcal{E} \models i : \varphi & \text{iff } \Im, s \models \varphi \text{ for all } s \in i \\ \mathcal{E} \models t \in i & \text{iff } \mathbb{T}(t) \in i \end{array}$$

$$\begin{array}{l} \mathcal{E} \models [n, U] : \varphi \text{ iff } \Im, s \models \varphi \text{ for all } s \in U_{[[}(n) \\ \mathcal{E} \models \langle n, U \rangle : \varphi \text{ iff } \Im, s \models \varphi \text{ for some } s \in U_{[[}(n). \end{array}$$

If $\mathcal{E} \models \varphi$, we say $\mathcal{E}$ satisfies (is a model of) φ.

Definition 18.31 ($\mathcal{CL}_{\mathcal{T_U}}$-statements $\to$ tableau entries). The following replacement rules turn a set of $\mathcal{CL}_{\mathcal{T_U}}$-statements (Definition 18.12) into tableau entries:

$$\langle n, U \rangle : \varphi \to \begin{array}{c} t : \varphi \\ t \in U_{[[}(n) \end{array}$$

$$[n, U] : \varphi \to U_{[[}(n) : \varphi.$$

It is straightforward to prove that a set of $\mathcal{CL}_{\mathcal{T_U}}$-statements has a model if and only if the corresponding transformed set of tableau entries has a model.

The tableau rules below exploit the semantics of the $\mathcal{CL}_{\mathcal{T_U}}$-connectives in most cases in a quite natural way. There are, however, a few tricky parts. The connective which is most difficult to handle turned out to be the disjunction in connection with intervals. A tableau entry $i : \varphi_1 \vee \cdots \vee \varphi_n$ expresses that at each point in the interval i one of the φ_k is true. Since intervals consist of infinitely many points, an exhaustive case analysis for all points is not possible. Instead of this, these entries are processed by triggering a computation which figures out the subintervals $i_k \subseteq i$ for which it is consistent to assume that φ_k is true. If it turns out that the union of all these subintervals is not i itself, then a closing condition applies because points in i have been detected for which none of the φ_k can be true.

Definition 18.32 (Tableau rules). The tableau rules for checking the consistency of a set of $\mathcal{CL}_{\mathcal{T}_\mathcal{U}}$-statements consists of *extension rules*, *transformation rules* and *closing conditions*. The extension rules just add new facts or new branches in the usual tableau style. The transformation rules operate destructively at certain parts of the tableau. The closing conditions are to be checked whenever something has been changed in the tableau. As soon as they detect an inconsistency they declare a branch to be closed.

The tableau rules below are self-explanatory. $left \rightarrow right$ indicates a transformation rule which replaces the $left$ part with the $right$ part. $left \rightarrow right_1 \mid \cdots \mid right_n$ means deleting $left$ and splitting the current branch into n new branches with $right_k$ in the kth branch instead of $left$. $left \rightarrow \emptyset$ means simply deleting $left$.

Extension rules are displayed as

$$\frac{\begin{array}{c}\varphi_1\\ \vdots \\ \varphi_n\end{array}}{\psi_1 \mid \cdots \mid \psi_k.}$$

This means that if there are terms matching $\varphi_1 \ldots \varphi_n$ in the branch then split the branch into k sub-branches (k may of course be 1) and add ψ_i to the ith branch. Most of the rules come with additional restrictions for their applicability.

- **Inference rules**

$(\wedge_\exists)$ $t : \varphi_1 \wedge \varphi_2 \quad \rightarrow_{\wedge_\exists} \quad \begin{array}{l} t : \varphi_1 \\ t : \varphi_2. \end{array}$ (t is a reference time symbol).

$(\wedge_\forall)$ $i : \varphi_1 \wedge \varphi_2 \quad \rightarrow_{\wedge_\forall} \quad \begin{array}{l} i : \varphi_1 \\ i : \varphi_2. \end{array}$ (i is a time region).

$(\vee_\exists)$ $t : \varphi_1 \vee \varphi_2 \quad \rightarrow_{\vee_\exists} \quad t : \varphi_1 \mid t : \varphi_2.$

$(\forall_\vee)$

$$\frac{\begin{array}{c} i : \varphi_1 \vee \varphi_2 \\ t \in j \end{array}}{\begin{array}{c|c|c} t \in (i \cap j) & t \in (i \cap j) & t \in (j \setminus i) \\ t : \varphi_1 & t : \varphi_2 & \end{array}}$$

if $i \cap j \neq \emptyset$ and $t : \varphi_1$ and $t : \varphi_2$ are not in the branch.

$(\vee^1_\forall)$ $i : \varphi_1 \vee \cdots \vee \varphi_n \rightarrow_{\vee^1_\forall} \begin{array}{c|l} i : \varphi_1 \vee \cdots \vee \varphi_n & i : \varphi_2 \vee \cdots \vee \varphi_n \\ (i : \varphi_2 \vee \cdots \vee \varphi_n \mid t) & \\ t : \varphi_1 & \\ t \in i & \end{array}$

if (i) none of the φ_i are time terms, (ii) none of the φ_i is itself a disjunction, (iii) the $(\vee^k_\forall)$-rule is not applicable, and (iv) this rule has not previously been applied to $i : \varphi_1 \vee \cdots \vee \varphi_n$ in the branch. t is a fresh new reference time symbol.[26]

$(\vee^k_\forall)$ $\quad \dfrac{i : \varphi_1 \vee \cdots \vee \varphi_n}{(i : \varphi_k \vee \cdots \vee \varphi_n \mid t_1, \ldots, t_l)} \to_{\vee^k_\forall}$

$$\begin{array}{c|l} i : \varphi_1 \vee \cdots \vee \varphi_n & i : \varphi_1 \vee \cdots \varphi_k \vee \varphi_{k+2} \ldots \vee \varphi_n \\ (i : \varphi_{k+1} \vee \ldots \vee \varphi_n \mid t_1, \ldots, t_l, t) & \\ t : \varphi_{k+1} & \\ t \in i & \end{array}$$

if this rule has not previously been applied to $(i : \varphi_k \vee \cdots \vee \varphi_n \mid t_1, \ldots, t_l)$ and if no other rule except $(\vee^1_\forall)$ is applicable. t is again a fresh, new reference time symbol. The right branch is superfluous if $k = n$.[27]

$(\Diamond_\exists)$ $\quad \begin{array}{l} t \in i \\ t : \langle\tau\rangle\varphi \end{array} \to_{\Diamond_\exists} \begin{array}{c|c|c} t \in i_1 & \ldots & t \in i_n \\ t_1 : \varphi & \ldots & t_n : \varphi \\ t_1 \in \tau_{[[}(i_1) & \ldots & t_n \in \tau_{[[}(i_n) \end{array}$

where $\{i_1, \ldots, i_n\} = \{j \in \delta(i, \tau) \mid \tau_{[[}(j) \neq \emptyset\}$ and the t_k are fresh,new reference time symbols.

$(\Diamond_\forall)$ $\quad i : \langle\tau\rangle\varphi \to_{\Diamond_\forall} \begin{array}{c} t_1 : \varphi \\ t_1 \in \tau_{[[}(i_1) \\ \vdots \\ t_n : \varphi \\ t_n \in \tau_{[[}(i_n) \end{array}$

where $\{i_1, \ldots, i_n\} = \{j \in \delta(i, \tau) \mid \tau_{[[}(j) \neq \emptyset\}$ and the t_k are fresh, new reference time symbols.

$(\Box_\exists)$ $\quad \begin{array}{l} t \in i \\ t : [\tau]\varphi \end{array} \to_{\Box_\exists} \begin{array}{c|c|c} t \in i_1 & \ldots & t \in i_n \\ \tau_{[[}(i_1) : \varphi & \ldots & \tau_{[[}(i_n) : \varphi \end{array}$

where $\{i_1, \ldots, i_n\} = \{j \in \delta(i, \tau) \mid \tau_{[[}(j) \neq \emptyset\}$ and this set is not empty.

$(\Box_\forall)$ $\quad i : [\tau]\varphi \to_{\Box_\forall} \tau_{[[}(i) : \varphi$

[26] This rule initiates the processing of disjunctions. The left branch introduces a new reference time symbol t which is initially constrained to i. The constraint for t may be narrowed down by successive rule applications. If a contradiction is detected then φ_1 is false for all elements in i and the right branch must be processed. $(i : \varphi_2 \vee \cdots \vee \varphi_n \mid t)$ is control information passed on to the $(\vee^k_\forall)$-rule and the $(\vee_\forall)$-closing condition.

[27] This rule continues the processing of disjunctions initiated by the $(\vee^1_\forall)$-rule. It is applied only when no other rule is applicable any more and the branch would normally be declared open. It triggers the investigation to figure out the subinterval of i where it is consistent to assume that φ_{k+1} holds. This is checked with a fresh reference time symbol t with initial constraint i. The constraint for i may subsequently be narrowed down. If a contradiction occurs then φ_{k+1} must be false in the entire interval i, and the right branch must be processed. The way we presented this rule suggests that the $(\vee^1_\forall)$-rule is applied again to $i : \varphi_1 \vee \cdots \vee \varphi_k \vee \varphi_{k+2} \vee \varphi_n$, processing $\varphi_1, \ldots, \varphi_k$ anew. With appropriate bookkeeping this can, of course, be avoided.

$(\tau_\vee^+)$ $i : \tau \vee \varphi \quad \rightarrow_{\tau_\vee^+} \quad i \setminus \tau_{[[}(i) : \varphi$
where $\tau \in \mathcal{CL}_{\mathcal{T}_\mathcal{U}}$.

$(\tau_\vee^-)$ $i : \neg\tau \vee \varphi \quad \rightarrow_{\tau_\vee^+} \quad \tau_{[[}(i) : \varphi$
if $\tau \in \mathcal{CL}_{\mathcal{T}_\mathcal{U}}$.

$(\tau_\top)$ $i : \tau \quad \rightarrow_{\tau_\top} \quad \emptyset$ if $\tau \in \mathcal{T}_\mathcal{U}$ and $i \subseteq \tau_{[[}(i)$.

$(\tau_{\neg\top})$ $i : \neg\tau \rightarrow_{\tau_{\neg\top}} \quad \emptyset$ if $\tau \in \mathcal{T}_\mathcal{U}$ and $i \cap \tau_{[[}(i) = \emptyset$.

$(\subseteq_\forall)$ $i : \tau_1 \subseteq \tau_2 \rightarrow_{\subseteq_\forall} \quad \emptyset$
if $\tau_1 \in \mathcal{T}_\mathcal{U}$, $\tau_2 \in \mathcal{T}_\mathcal{U}$, and for all $j \in \delta(i, \{\tau_1, \tau_2\})$, $\tau_{1[[}(j) \subseteq \tau_{2[[}(j)$.

$(\not\subseteq_\forall)$ $i : \tau_1 \not\subseteq \tau_2 \rightarrow_{\not\subseteq_\forall} \quad \emptyset$
if $\tau_1 \in \mathcal{T}_\mathcal{U}$, $\tau_2 \in \mathcal{T}_\mathcal{U}$, and for all $j \in \delta(i, \{\tau_1, \tau_2\})$, $\tau_{1[[}(j) \neq \tau_{2[[}(j)$.

$(\subseteq_\exists)$ $\begin{array}{l} t : \tau_1 \subseteq \tau_2 \\ t \in i \end{array} \rightarrow_{\subseteq_\exists} \begin{array}{l} t : \tau_1 \subseteq \tau_2 \\ t \in (i \setminus j) \end{array}$
if $\tau_1 \in \mathcal{T}_\mathcal{U}$, $\tau_2 \in \mathcal{T}_\mathcal{U}$, and $j \in \delta(i, \{\tau_1, \tau_2\})$ with $\tau_{1[[}(j) \neq \tau_{2[[}(j)$.

$(\not\subseteq_\exists)$ $\begin{array}{l} t : \tau_1 \not\subseteq \tau_2 \\ t \in i \end{array} \rightarrow_{\not\subseteq_\exists} \begin{array}{l} t : \tau_1 \neq \tau_2 \\ t \in (i \setminus j) \end{array}$
if $\tau_1 \in \mathcal{T}_\mathcal{U}$, $\tau_2 \in \mathcal{T}_\mathcal{U}$, and $j \in \delta(i, \{\tau_1, \tau_2\})$ with $\tau_{1[[}(j) = \tau_{2[[}(j)$.

- **Closing conditions**

$(\emptyset)$ $\dfrac{t \in \emptyset}{closed}$

$(\neg)$ $\dfrac{\begin{array}{c} t : p \\ t : \neg p \end{array}}{closed}$

$(\cap_\neg)$ $\dfrac{\begin{array}{c} i : p \\ j : \neg p \end{array}}{closed}$ if $i \cap j \neq \emptyset$.

$(\Diamond)$ $\dfrac{\begin{array}{c} t \in i \\ t : \langle\tau\rangle\varphi \end{array}}{closed}$ if $\tau_{[[}(i) = \emptyset$.

$(\tau_\forall)$ $\dfrac{i : \tau}{closed}$ where $\tau \in \mathcal{T}_\mathcal{U}$ and $i \not\subseteq \tau_{[[}(i)$.

$(\tau_{\neg\forall})$ $\dfrac{i : \neg\tau}{closed}$ where $\tau \in \mathcal{T}_\mathcal{U}$ and $i \cap \tau_{[[}(i) \neq \emptyset$.

$(\vee_\forall)$ $\dfrac{\begin{array}{l} i : \varphi_1 \vee \cdots \vee \varphi_n \\ \quad (i :| t_1, \ldots, t_l) \\ \qquad t_1 \in i_1 \\ \qquad \vdots \\ \qquad t_l \in i_n \end{array}}{closed}$ if $i_1 \cup \cdots \cup i_n \neq i$.[28]

[28]This rule finishes the processing of disjunctions started with the $(\vee_\forall^1)$-rule and continued with the $(\vee_\forall^k)$-rule.

$(\tau_{\neg\forall})$ $\dfrac{i : \neg\tau}{closed}$ where $\tau \in \mathcal{T}_{\mathcal{U}}$ and $i \cap \tau_{[[}(i) \neq \emptyset$.

$(\vee_{\forall})$ $i : \varphi_1 \vee \cdots \vee \varphi_n$
$(i :| t_1, \ldots, t_l)$
$t_1 \in i_1$ if $i_1 \cup \cdots \cup i_n \neq i$.[28]
$\vdots$
$\dfrac{t_l \in i_n}{closed}$

$(\subseteq_{\perp})$ $\dfrac{i : \tau_1 \subseteq \tau_2}{closed}$
if $\tau_1 \in \mathcal{T}_{\mathcal{U}}$, $\tau_2 \in \mathcal{T}_{\mathcal{U}}$, and for some $j \in \delta(i, \{\tau_1, \tau_2\})$, $\tau_{1[[}(j) \not\subseteq \tau_{2[[}(j)$.

$(\not\subseteq_{\perp})$ $\dfrac{i : \tau_1 \not\subseteq \tau_2}{closed}$
if $\tau_1 \in \mathcal{T}_{\mathcal{U}}$, $\tau_2 \in \mathcal{T}_{\mathcal{U}}$, and for some $j \in \delta(i, \{\tau_1, \tau_2\})$, $\tau_{1[[}(j) \subseteq \tau_{2[[}(j)$.

If A is a set of tableau entries, let $Tab(A)$ be the set of open branches resulting from exhaustive application of the tableau rules.

Example 18.33 (For a tableaux proof). We illustrate the whole procedure with a simple example. It nevertheless shows most of the aspects of our system. Suppose we have the following statements:

(i) last week it was rainy or foggy;
(ii) from Monday till Wednesday last week it was not rainy;
(iii) from Wednesday till Sunday last week it was not foggy.

This is clearly unsatisfiable because on Wednesday it was neither rainy nor foggy.

First of all, we have to fix a reference time line. To simplify the numbers, point 0 of the reference time line is assumed to be at the beginning of the last week. The time unit *day* is arranged at the reference time line such that day 0 (Monday last week) covers the interval $[0, 10[$, day 1 (Tuesday last week) covers the interval $[10, 20[$, etc. Time unit *week* is arranged that week 0 (last week) covers the interval $[0, 70[$ etc. The current point in time (now) is supposed to be at reference time point, say, 105, which is Thursday this week, day number 10.

The above statements are encoded in calendar logic as follows:

(1) $\langle 10, day\rangle : [x_{week} - 1](r \vee f)$
(2) $\langle 10, day\rangle : \langle x_{week} - 1\rangle[\{Mo(x_{week}), Tu(x_{week}), We(x_{week})\}]\neg r$
(3) $\langle 10, day\rangle : \langle x_{week} - 1\rangle[\{We(x_{week}), \ldots, Su(x_{week})\}]\neg f$.

$Mo(x_{week})$ is specified as the first day in the week, $Tu(x_{week})$ as the second day in the week, etc.

First of all the statements (1), (2) and (3) have to be turned into tableau entries using the replacement rules of Definition 18.31. The result is

[28]This rule finishes the processing of disjunctions started with the $(\vee_{\forall}^{1})$-rule and continued with the $(\vee_{\forall}^{k})$-rule.

(4) $t_1 : [x_{week} - 1](r \vee f)$
(5) $t_1 \in [100, 110[$
(6) $t_2 : \langle x_{week} - 1\rangle[\{Mo(x_{week}), Tu(x_{week}), We(x_{week})\}]\neg r$
(7) $t_2 \in [100, 110[$
(8) $t_3 : \langle x_{week} - 1\rangle[\{We(x_{week}), \ldots, So(x_{week})\}]\neg f$
(9) $t_3 \in [100, 110[$.

Processing the formulae top down we start by applying the $(\Box_\exists)$-rule to entries (4) and (5). $\delta([100, 110[, x_{week} - 1) = [100, 110[$ because the week-coordinate is 1 for all time points between 100 and 110 (this week is week 1). $(x_{week} - 1)([100, 110[) = [0, 70[$. (Week 0 ranges from time point 0 till time point 70.) The result of the $(\Box_\exists)$-rule application is therefore

(10) $[0, 70[: (r \vee f)$.

In a very similar way we apply the $(\Diamond_\exists)$-rule to (6) and (7) and to (8) and (9):

(11) $t_4 : [\{Mo(x_{week}), Tu(x_{week}), We(x_{week})\}]\neg r$
(12) $t_4 \in [0, 70[$
(13) $t_5 : [\{We(x_{week}), \ldots, So(x_{week})\}]\neg f$
(14) $t_5 \in [0, 70[$.

Using the definition of $Mo(_{week})$ etc., we apply the $(\Box_\exists)$-rule to (11), (12) and to (13), (14). The result is

(15) $[0, 30[: \neg r$
(16) $[20, 70[: \neg f$.

The next step is to initiate the investigation of the disjunction (10) with the $(\vee^1_\forall)$-rule.

We get the two branches

(10) $[0, 70[: (r \vee f)$	(20) $[0, 70[: f$
(17) $([0, 70[: f, t_6)$	
(18) $t_6 : r$	
(19) $t_6 \in [0, 70[$.	

The right branch (20) is immediately closed by the $(\cap_\neg)$-closing condition. Using (15) and (18) in the left branch, (19) is narrowed down by the $(\backslash)$-rule to

(21) $t_6 \in [30, 70[$.

No further rule except $(\vee^1_\forall)$ is applicable. This rule yields

(10) $[0, 70[: (r \vee f)$
(22) $([0, 70[:, t_6, t_7)$
(23) $t_7 : f$
(24) $t_7 \in [0, 70[$.

Using (16) and (23), (24) is narrowed down by the $(\backslash)$-rule to

(25) $t_7 \in [0, 20[$.

The closing condition $(\vee_\forall)$ now applies to (22), (21) and (25) because $([0,20[\cup [30,70[) \neq [0,70[$. The missing part where $r \vee f$ cannot hold is $[20,30[$, which was Wednesday last week, as expected.

18.3.2.1 *Termination, soundness and completeness*

Theorem 18.34 (Termination of the tableau system). *The application of the tableau rules (Definition 18.32) to a finite set of tableau entries eventually terminates.*

Proof Termination is proved by showing that the tableau rules can be applied only finitely many times. First of all we need to show that only finitely many reference time symbols can be generated by the tableau rules. The key observation is that all rules which generate new reference time symbols, and these are the $(\vee^1_\forall)$-rule, the $(\vee^k_\forall)$-rule, the $(\Diamond_\exists)$-rule and the $(\Box_\exists)$-rule, generate tableau entries $t : \varphi$ where the φ is smaller in size than the formulae in the statements to which the rule is applied. Furthermore there are no rules at all which generate formulae which are larger in size than the formulae to which the rule is applied. Therefore the generation of new reference time symbols is bound to terminate eventually.

Since the maximal number of reference time symbols ever generated in one branch is finite, the $(\forall_\vee)$-rule can be applied only finitely often.

The $(\vee^1_\forall)$-rule and the $(\vee^k_\forall)$-rule together process each disjunction only once and therefore terminate once a disjunction is 'worked off'.

All the other rules destructively replace complex entries with entries which are strictly smaller in size than the original formulae to which they are applied. Therefore their application terminates as well. □

Theorem 18.35 (Soundness). *If a finite set A of calendar logic statements is satisfiable then $Tab(A)$ has some open branches.*

Proof Soundness is proved by induction on the number of rule applications. To this end we have to show that if a tableau interpretation $\mathcal{E} = (\Im, \mathbb{T})$ satisfies a branch B of a tableau before the application of one of the tableau rules in Definition 18.32 then either $\mathcal{E}$ or some modification $\mathcal{E}'$ satisfies one of the successor branches of the rule application.

We check each rule separately.

$(\wedge_\exists)$-rule: **$(\wedge_\forall)$-rule**, **$(\vee_\exists)$-rule**: Obvious.

$(\forall_\vee)$-rule: If $\mathcal{E} \models i : \varphi_1 \vee \varphi_2$ and $\mathcal{E} \models t \in j$ and $i \cap j \neq \emptyset$ then either $\mathbb{T}(t) \in (i \setminus j)$, in which case $\mathcal{E} \models t : \varphi_1$ or $\mathcal{E} \models t : \varphi_2$, or $\mathbb{T}(t) \in (j \setminus i)$, which means $\mathcal{E} \models t \in (j \setminus i)$.

$(\vee^1_\forall)$-rule: If $\varphi_1 \vee \cdots \vee \varphi_n$ holds for all $t \in i$ then either φ_1 holds for some $t \in i$, in which case assigning $\mathbb{T}(t) \stackrel{\text{def}}{=} s$ satisfies the left branch, or φ_1 holds for no $t \in i$, in which case $\varphi_2 \vee \cdots \vee \varphi_n$ must hold for all $t \in i$. Therefore one of the successor branches is satisfied.

$(\vee^k_\forall)$-rule: The argument is analogous to the argument in the previous case.

($\Box_\forall$)-rule: If $\mathcal{E} \models i : [\tau]\varphi$ then for all $s \in i$, $\Im, s \models [\tau]\varphi$, i.e. for all $s' \in \tau_{[[}(s)$, $\Im, s' \models \varphi$. This means $\mathcal{E} \models \tau_{[[}(i) : \varphi$.

($\backslash$)-rule and all the remaining rules are straightforward to check. □

Lemma 18.36 (Closed branches). *A closed branch in a tableau generated with the rules in Definition 18.32 cannot have a model.*

Proof A branch B is closed if one of the closing conditions applies. For the closing conditions ($\emptyset$), ($\neg$), ($\cap_\neg$) and ($\Diamond$) it is obvious that the branch cannot have a model.

($\tau_\forall$)-condition: $\forall t \in i : t \in \tau_{[[}(t)$ is impossible if $i \not\subseteq \tau_{[[}(i)$.

($\tau_{\neg\forall}$)-condition: $\forall t \in i : t \notin \tau_{[[}(t)$ is equivalent to $\neg\exists t \in i : t \in \tau_{[[}(t)$ which is impossible if $i \cap \tau_{[[}(i) \neq \emptyset$.

($\vee_\forall$)-condition: This condition is triggered in a situation where for each $k \in \{1, \ldots, n\}$ $(i \backslash i_k) : \varphi_k$ has been refuted, and therefore the original constraint $t_k \in i$ introduced by the ($\vee^1_\forall$)-rule and the ($\vee^k_\forall$)-rule has been reduced to $t_k \in i_k$. Therefore for some $t \in i \setminus (i_1 \cup \cdots \cup i_n)$ $\neg\varphi_1 \wedge \cdots \wedge \neg\varphi_n$ must hold, which contradicts that for all $t \in i$, $\varphi_1 \vee \cdots \vee \varphi_n$ must hold.

($\subseteq_\perp$)-condition and **($\not\subseteq_\perp$)-condition**: Obvious. □

Proposition 18.37 (Constrained reference time symbols). *If a set A of tableau entries contains no reference time symbols, then all reference time symbols in $Tab(A)$ are constrained by a unique $t \in i$ entry in the corresponding branch.*

Proof The tableau rules which generate reference time symbols are the ($\vee^1_\forall$)-rule, the ($\vee^k_\forall$)-rule, the ($\Diamond_\exists$)-rule and the ($\Diamond_\forall$)-rule. They introduce constraints for the reference time symbols. The ($\cap$)-rule restricts these constraints to the smallest possible intervals. In $Tab(A)$ they are therefore unique. □

Lemma 18.38 (Canonical model for open branches). *If a set A of tableau entries contains no reference time symbols then all open branches in $Tab(A)$ have a model.*

Proof Let B_m be an open branch in $Tab(A)$. The canonical model $\mathcal{E} \stackrel{\text{def}}{=} ((\Im_{\mathcal{T}_\mathcal{U}}, \mathcal{P}), \mathbb{T})$ for B_m is defined as follows

If $t_1 \in i_1, \ldots, t_n \in i_n$ are all the membership relations contained in B_m then we choose one element $n_k \in i_k$ such that all the n_k are different, and assign $\mathbb{T}(t_k) = n_k$. This is possible because (i) none of the i_k is empty because otherwise the closing condition ($\emptyset$) would have been applied; (ii) the time line is dense such that there are enough points to choose from, and (iii) by the ($\cap$)-rule, the interval i_k is uniquely determined for each $t_k \in i_k$.

For the predicate symbols p we assign

$$\mathcal{P}(p) = \{\mathbb{T}(t) \mid `t : p\text{'} \in B_m\} \cup \bigcup_{`i:p\text{'} \in B_m} i.$$

If $t_1 \in i_1, \ldots, t_n \in i_n$ are all the membership relations contained in B_m then we choose one element $n_k \in i_k$ such that all the n_k are different, and assign $\mathbb{T}(t_k) = n_k$. This is possible because (i) none of the i_k is empty because otherwise the closing condition ($\emptyset$) would have been applied; (ii) the time line is dense such that there are enough points to choose from, and (iii) by the ($\cap$)-rule, the interval i_k is uniquely determined for each $t_k \in i_k$.

For the predicate symbols p we assign

$$\mathcal{P}(p) = \{\mathbb{T}(t) \mid `t : p' \in B_m\} \cup \bigcup_{`i:p' \in B_m} i.$$

We have to show that $\mathcal{E}$ really satisfies all entries. Let $B_1 \to \cdots \to B_m$ be the sequence of tableau branches derived from the initial set B_1 of tableau entries and ending with the open branch B_m. We show by induction on the structure of the $\mathcal{CL}_{\mathcal{T_U}}$-formulae that $\mathcal{E}$ satisfies all entries in *all* branches $B_1, \ldots, B_m$.

Base cases:

(i) Membership relations $t \in i$:
If '$t \in i$' $\in B_m$ then $\mathbb{T}(t) \in i$ holds by construction. If '$t \in i$' $\in B_k, k \neq m$, then '$t \in i'$' $\in B_m$ for some $i' \subseteq i$. This is the case because none of the rules removes a membership relation. But the rules ($\setminus$), ($\cap$), (τ_+) and (τ_-) may shrink the interval i. None of the rules increases the interval i. Since $\mathbb{T}(t) \in i'$ holds by construction, and $i \subseteq i'$, $\mathbb{T}(t) \in i$ must hold as well.

(ii) Positive predicate symbols p:
For '$t : p$' $\in B_m$, $\mathbb{T}(t) \in \mathcal{P}(p)$ is true by construction. None of the rules eliminates entries $t : p$. Therefore this holds for the entries '$t : p$' $\in B_k, k \neq m$, as well. The same arguments apply to entries '$i : p$'.

(iii) Negative predicate symbols p:
Consider '$t : \neg p$' $\in B_m$, '$t : p$' $\notin B_m$, because otherwise the closing condition ($\neg$) would have been applied. We also have to check that '$i : p$' $\notin B_m$, with $\mathbb{T}(t) \in i$. '$t : \neg p$' is accompanied by a constraint $t \in j$ for some j (Proposition 18.37). The ($\setminus$)-rule makes sure that j and i are disjoint. Therefore $\mathbb{T}(t) \in j$ and $\mathbb{T}(t) \notin i$. Thus, $\mathbb{T}(t) \notin \mathcal{P}(p)$, i.e. $\mathcal{E} \models t : \neg p$.
Now consider '$i : \neg p$' $\in B_m$. Owing to the closing condition ($\cap_\neg$) there is no '$j : p$' $\in B_m$, with $j \cap i \neq \emptyset$. Again, owing to the ($\setminus$)-rule there is no '$t : p$' $\in B_m$ with constraint $t \in j$ which has a non-empty intersection with i. Therefore, by construction, $\mathcal{P}(p) \cap i = \emptyset$, and this means $\mathcal{E} \models i : p$.
For entries '$t : \neg p$' $\in B_k$, and '$i : \neg p$' $\in B_k, k \neq m$, the same argument as in the positive case applies.

(iv) Time terms:
Consider '$t : \tau$' $\in B_k$ for some $\tau \in \mathcal{CL}_{\mathcal{T_U}}$ and $1 \leq k \leq m$. t is constrained by some '$t \in i$' $\in B_k$. The (τ)-rule then yields '$t \in (i \cap \tau_{[[}(i))$' $\in B_l$ for some $l \geq k$. As in the case above, we can conclude that $\mathbb{T}(t) \in (i \cap \tau_{[[}(i))$ and therefore $\mathbb{T}(t) \in \tau_{[[}(t)$, i.e. $\mathcal{E} \models t : \tau$.
Now consider '$i : \tau$' $\in B_k$. We must have $i \subseteq \tau_{[[}(i)$ because otherwise the closing condition ($\tau_\forall$) would apply. This means $\mathcal{E} \models i : \tau$.

The arguments for negated time terms are analogous.

(v) Subsumption relations:
Consider $t : \tau_1 \subseteq \tau_2$ with constraint $t \in i$. The $(\subseteq_\exists)$-rule guarantees that all the time points t in the interval i where $\tau_{1[[}(s) \not\subseteq \tau_{2[[}(s)$ are successively eliminated. The $(\not\subseteq_\exists)$-rule achieves the same effect for inequations $t : \tau_1 \subseteq \tau_2$.
Statements $i : \tau_1 \subseteq \tau_2$ and $i : \tau_1 \neq \tau_2$ are either valid, in which case the $(\subseteq_{\forall\top})$-rule and the $(\neq_{\forall\top})$-rule eliminate them, or not valid, in which case the closing conditions $(\subseteq_\perp)$ or $(\not\subseteq_\perp)$ apply.

Induction steps

$\wedge$ Consider '$t : \varphi_1 \wedge \varphi_2$' $\in B_k$. The $(\wedge_\exists)$-rule generates '$t : \varphi_1$' $\in B_l$ and '$t : \varphi_2$' $\in B_l$ for some $l \geq k$, to which the induction hypothesis applies. The same argument holds for entries '$i : \varphi_1 \wedge \varphi_2$' $\in B_k$.

$\vee$ Consider '$t : \varphi_1 \vee \varphi_2$' $\in B_k$. The $(\vee_\exists)$-rule generates '$t : \varphi_1$' $\in B_l$ or '$t : \varphi_2$' $\in B_l$ for some $l \geq k$, to which the induction hypothesis applies.
Now consider '$i : \varphi_1 \vee \cdots \vee \varphi_n$' $\in B_k$. The $(\vee_\forall^1)$-rule and the $(\vee_\forall^k)$-rule, together with the rules which shrink constraints for reference time symbols, generate the following entries: '$t_1 \in i_1$' $\in B_m$, '$t_1 : \varphi_1$' $\in B_{k_1}, \ldots,$ '$t_n \in i_n$' $\in B_m$, '$t_n : \varphi_n$' $\in B_{k_n}$, $k_l \geq k$. The induction hypothesis applies to all these entries. Owing to the $(\vee_\forall)$-closing condition, $i_1 \cup \cdots \cup i_n = i$.
Let $s \in i$. We have to show that $\mathcal{E}, t \models \varphi_1 \vee \cdots \vee \varphi_n$. If $s = \mathbb{T}(t)$ for some reference time symbol t, then '$t \in j$' $\in B_k$ and, by the $(\forall_\vee)$-rule, either '$t \in (j \setminus i)$' $\in B_{k'}$, which cannot be the case because $s = \mathbb{T}(t)$ and $s \in i$, or '$t : \varphi_l$' $\in B_{k''}$ for some $1 \leq l \leq n$ and $k'' \geq k$, to which the induction hypothesis applies. Therefore $\mathcal{E}, t \models \varphi_1 \vee \cdots \vee \varphi_n$.
If there is no t with $s = \mathbb{T}(t)$, we can argue as follows. Since $i_1 \cup \cdots \cup i_n = i$, $s \in i_k$ for some $1 \leq k \leq n$ must be the case. Furthermore there is '$t_k \in i_k$' $\in B_m$. In the construction of $\mathbb{T}$, we assigned $\mathbb{T}(t_k)$ to an arbitrary point in i_k. The only restriction was that the assignment must be different to the assignment of the other reference time symbols. Since there is no other t with $\mathcal{E}(t) = s$, we can assume that the assignment for t_k is just $\mathbb{T}(t_k) = s$. Together with the induction hypothesis, which guarantees $\mathcal{E} \models t_k : \varphi_k$, we can then conclude that $\mathcal{E}, t \models \varphi_1 \vee \cdots \vee \varphi_n$.

$t : \langle\tau\rangle$ Consider '$t : \langle\tau\rangle\varphi$' $\in B_k$ together with the constraint $t \in i$. $\tau_{[[}(i) \neq \emptyset$, otherwise the $(\Diamond)$-closing condition would apply. The $(\Diamond_\exists)$-rule generated '$t \in i_l$' $\in B_{k'}$, '$t_l : \varphi$' $\in B_{k'}$ and '$t_l \in \tau_{[[}(i_l)$' $\in B_{k'}$ for some $k' \geq k$ and $i_l \in \delta(i, \tau)$. This means $\mathbb{T}(t_k) \in \tau_{[[}(\mathbb{T}(t))$ and together with the induction hypothesis, $\mathcal{E} \models t : \langle\tau\rangle\varphi$.

$i : \langle\tau\rangle$ Consider '$i : \langle\tau\rangle\varphi$' $\in B_k$. $\tau_{[[}(i) \neq \emptyset$, otherwise the $(\Diamond)$-closing condition would apply. Let $s \in i$. We have to show $\mathcal{E}, t \models \langle\tau\rangle\varphi$, i.e. there is some $s' \in \tau_{[[}(s)$ with $\mathcal{E}, t' \models \varphi$.
The $(\Diamond_\forall)$-rule yields for all $i_l \in \delta(i, \tau)$: '$t_l : \varphi$' $\in B_{k'}$, '$t_l \in \tau_{[[}(i_l)$' $\in B_{k'}$ for $k' \geq k$, to which the induction hypothesis applies. Since δ partitions

$i : [\tau]$ Consider '$i : [\tau]\varphi$' $\in B_k$ The $(\Box_\forall)$-rule which generates $\tau_{[[}(i) : \varphi$ and the induction hypothesis guarantee $\mathcal{E} \models i : [\tau]\varphi$.

□

Theorem 18.39 (Completeness of the tableau system). *If a finite set A of calendar logic statements has no model, then $Tab(A)$ has only closed branches.*

Proof If there was an open branch B_m in $Tab(A)$ then the canonical model for B_m would also satisfy all the formulae in the initial tableau (Lemma 18.38), which is not possible because A has no model. □

18.4 Calendar logic plus modal logic

The non-temporal part of calendar logic is classical propositional logic, which is not very expressive. The next stage in expressivity can be obtained by combining the operators of calendar logic with (multi)modal logic. Modal logic is used for encoding many different notions, i.e. necessity, belief, knowledge, obligation, permission, or concepts in knowledge representation languages. Therefore the combination of calendar logic and modal logic can have many different interpretations. For example, [believe(I)]$\langle x_{day}+1\rangle raining$ may express 'I believe that it will rain tomorrow'. [obliged(you)]$[working\text{-}hours(x_{week})]work$ means 'you are obliged to work during the working hours of the week'. In a description logic interpretation, $\langle 1996\rangle(car \wedge \langle \text{sold}\rangle woman)$ denotes the set of cars sold to women in the year 1996. $[1996](woman \wedge employed \wedge \neg\langle has\text{-}child\rangle\top \wedge \langle x_{year}+1\rangle\langle has\text{-}child\rangle male)$ denotes the set of women employed all the time in 1996, who had a male child the year after.

The plan for this section is:

- To start defining the syntax and semantics of the combined calendar–multimodal logic.
- Then we show how the temporal operators can be eliminated from the combined logic. The resulting formulae can be treated with the mechanisms of the basic multi-modal logic.
- Since the elimination of the temporal operators is exponential, we develop an alternative inference system that avoids the exponential explosion. The framework for this alternative calculus is the *functional translation* into a predicate-logic-like language. This framework relies only on a few assumptions about the basic multimodal logic. Therefore the technical details can be formulated in a quite general way, covering many interesting multimodal logics.
- For the translated formulae we develop a unification algorithm as the basis for inference systems operating on the translated formulae.
- We present *resolution* as one particular inference system and show soundness, completeness and termination. Completeness and termination depend on assumptions about the basic multimodal logic. These assumptions are

predicate-logic-like language. This framework relies only on a few assumptions about the basic multimodal logic. Therefore the technical details can be formulated in a quite general way, covering many interesting multimodal logics.

- For the translated formulae we develop a unification algorithm as the basis for inference systems operating on the translated formulae.
- We present *resolution* as one particular inference system and show soundness, completeness and termination. Completeness and termination depend on assumptions about the basic multimodal logic. These assumptions are satisfied by many different systems, e.g. multimodal versions of K, KD, T, B, S5.
  The completeness proof follows the standard strategy for showing completeness for the resolution calculus in predicate logic. The technical details, however, have to be adapted to our case where in addition to terms and formulae, we have time regions, i.e. sets of finite intervals of real numbers, in the language. Therefore the well-known results from predicate logic, Herbrand's theorem for example, do not apply directly. They still hold, but the proofs are different to the original proofs.

18.4.1 *Syntax and semantics*

We define the syntax and semantics of the combined logic. The modal part is a standard propositional multimodal logic with a finite set of parameterized modal operators $[m]$ (box operator) and $\langle m\rangle$ (diamond operator).

Definition 18.40 (Syntax of the combined logic). If $\mathcal{T}_\mathcal{U}$ is a language of time terms (Definition 18.3), $\mathcal{CL}_{\mathcal{T}_\mathcal{U}}$ is calendar logic and $\mathbf{M}$ is the set of modal parameters, then the $\mathcal{CL}^{\mathcal{M}}_{\mathcal{T}_\mathcal{U}}$-formulae are built according to the following grammar:

$$\mathcal{CL}^{\mathcal{M}}_{\mathcal{T}_\mathcal{U}} := \mathcal{CL}_{\mathcal{T}_\mathcal{U}} \mid \neg\mathcal{CL}^{\mathcal{M}}_{\mathcal{T}_\mathcal{U}} \mid \mathcal{CL}^{\mathcal{M}}_{\mathcal{T}_\mathcal{U}} \wedge \mathcal{CL}^{\mathcal{M}}_{\mathcal{T}_\mathcal{U}} \mid \mathcal{CL}^{\mathcal{M}}_{\mathcal{T}_\mathcal{U}} \vee \mathcal{CL}^{\mathcal{M}}_{\mathcal{T}_\mathcal{U}} \mid \langle\mathbf{M}\rangle\mathcal{CL}^{\mathcal{M}}_{\mathcal{T}_\mathcal{U}} \mid [\mathbf{M}]\mathcal{CL}^{\mathcal{M}}_{\mathcal{T}_\mathcal{U}}.$$

A *statement* of $\mathcal{CL}^{\mathcal{M}}_{\mathcal{T}_\mathcal{U}}$ is either $\langle n, U\rangle : \varphi$ or $[n, U] : \varphi$ where U is a time unit, n is one of U's coordinates and φ is a $\mathcal{CL}^{\mathcal{M}}_{\mathcal{T}_\mathcal{U}}$-formula.

The standard Kripke semantics of normal modal logics is defined in terms of possible worlds and a binary accessibility relation R_m between worlds for each modal parameter m. The set of worlds together with the accessibility relations form a *frame*. Different classes of frames correspond to different modal logics. $[m]\varphi$ is true at a particular world w iff φ is true at *all* worlds R_m-accessible from w. Dually, $\langle m\rangle\varphi$ is true at w iff φ is true at *some* world R_m-accessible from w. The possible worlds w associate with each predicate symbol p a truth value. p can be either true or false at w. The truth value of p can change from world to world.

In the combined modal–calendar logic, we keep the basic structure of frames, but we stretch out each world to form a copy of the reference time line. The truth values of the propositional symbols therefore may vary when moving along a time line (world). A formula is interpreted at an actual time point s of an actual world

w. The temporal operators move along the actual time line, whereas the modal operators cause a jump to other worlds (time lines). Since all worlds are copies of the reference time line, we have two choices for jumps to other worlds, either the actual time point remains the same, or we jump to a different actual time point at a different world. The first option means that there is a global clock synchronizing the actual time points for all worlds. The second option means that the memory about the actual time may get lost when moving to another world. A formula $[\text{believe}]Friday(x_{week})$ may therefore be true at a Thursday if the 'believer' is not aware of the actual time. This is not the case if the actual time point is carried over to the new world. The calculus developed for the combined logic relies on the assumption that the time intervals generated by the temporal and modal operators are always finite. It would not work if we allowed for an operator which loses the temporal information completely when jumping to another world. Therefore the models for the combined logic will be restricted such that jumps to other worlds keep the actual time point. It is quite easy to define a new operator that jumps to another world and loses the information about the actual time to a certain degree. For example,$[believe]\langle x_{month}\rangle p$ expresses that p is believed to be true sometime this month. The combination of the believe operator with the temporal operator has the effect that a limited uncertainty about the actual time is introduced. The 'believer' is uncertain about the current time, but still knows which month it is.

Definition 18.41 (Semantics of the combined logic). A $\mathcal{CL}^{\mathcal{M}}_{\mathcal{T_U}}$-frame $\mathcal{F} \stackrel{\text{def}}{=} (\mathcal{W} \times \mathcal{T}, R)$ consists of a set $\mathcal{W}$ of *worlds*, which are stretched out to become copies of the reference time line $\mathcal{T}$, and for each modal parameter $m \in \mathbf{M}$ an accessibility relation $R_m : (\mathcal{W} \times \mathcal{T}) \to (\mathcal{W} \times \mathcal{T})$ with $R_m((w,t),(w',t')) \to t = t'$. R is the set of all accessibility relations.

A $\mathcal{CL}^{\mathcal{M}}_{\mathcal{T_U}}$-interpretation $\Im \stackrel{\text{def}}{=} (\mathcal{F}, \mathcal{P})$ consists of a time-dependent predicate assignment: for each $p \in \mathbb{P}$: $\mathcal{P}(p) \subseteq \mathcal{W} \times \mathcal{T}$.

For a world $w \in \mathcal{W}$ and a reference time point $t \in \mathcal{T}$ we define:

$$
\begin{array}{lll}
\Im, w, t \models p & \text{iff } (w,t) \in \mathcal{P}(p) & \text{where } p \in \mathbb{P} \\
\Im, w, t \models \tau & \text{iff } t \in \tau_{[[}(t) & \text{where } \tau \in \mathcal{T_U} \\
\Im, w, t \models \tau_1 \subseteq \tau_2 & \text{iff } \tau_{1[[}(t) \subseteq \tau_{2[[}(t) & \\
\Im, w, t \models \langle\tau\rangle\varphi & \text{iff } \Im, w, t' \models \varphi \text{ for some } t' \in \tau_{[[}(t) & \\
\Im, w, t \models [\tau]\varphi & \text{iff } \Im, w, t' \models \varphi \text{ for all } t' \in \tau_{[[}(t) & \\
\Im, w, t \models \langle m\rangle\varphi & \text{iff } \Im, w', t \models \varphi \text{ for some } w' \text{ with } R_m((w,t),(w',t)) & \\
\Im, w, t \models [m]\varphi & \text{iff } \Im, w', t \models \varphi \text{ for all } w' \text{ with } R_m((w,t),(w',t)) &
\end{array}
$$

$$
\begin{array}{ll}
\Im, w \models \langle n, U\rangle : \varphi & \text{iff } \Im, w, t \models \varphi \text{ for some } t \in U_{[[}(n) \\
\Im, w \models [n, U] : \varphi & \text{iff } \Im, w, t \models \varphi \text{ for all } t \in U_{[[}(n).
\end{array}
$$

The definitions for the classical connectives are the usual ones. Figure 18.2 illustrates such frames.

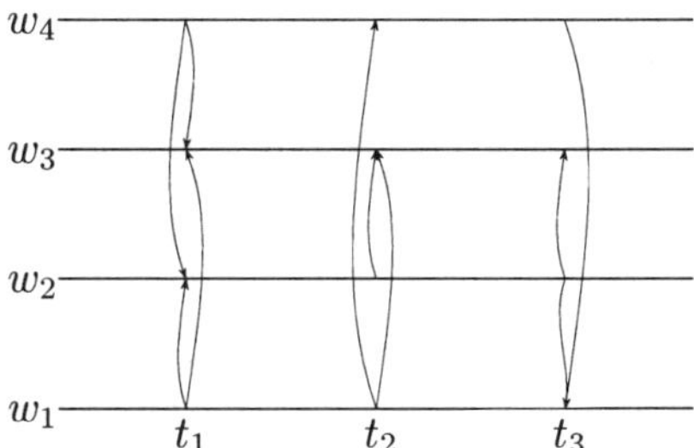

FIG. 18.2. A frame in the combined logic. The arrows indicate an accessibility relation R_m.

that the combination of calendar logic with a particular propositional multimodal logic inherits the main features from the multimodal logic. In particular, if the multimodal logic is decidable, then the combination is decidable as well.

The elimination of the temporal operators is done via an intermediate system $\mathcal{CL}_{\mathcal{T}u}^{\mathcal{M}I}$ containing besides the other operators the two time-region operators $[i]$ and $\langle i \rangle$ (Definition 18.17). $[i]\varphi$ is true in a world w (independent of the actual time point) iff φ is true at *all* time points in i and in the world w. $\langle i \rangle\varphi$ is true in a world w (also independent of the actual time point) iff φ is true at *some* time point in i and in the world w.

Definition 18.43 (Translation of $\mathcal{CL}_{\mathcal{T}u}^{\mathcal{M}}$ into $\mathcal{CL}_{\mathcal{T}u}^{\mathcal{M}I}$). We define rewrite rules for translating the temporal operators into time region operators.

The rewrite rules for $\mathcal{CL}_{\mathcal{T}u}^{\mathcal{M}}$-statements are

$$[n, U] : \varphi \rightarrow_I [U_{[[}(n)]\varphi$$
$$\langle n, U \rangle : \varphi \rightarrow_I \langle U_{[[}(n)\rangle\varphi.$$

The rewrite rules for $\mathcal{CL}_{\mathcal{T}u}^{\mathcal{M}}$-formulae are defined for the negation normal form (Definition 18.42)

$$[i]\Xi\lceil[\tau_k]\varphi_k, \langle\tau_l\rangle\psi_l\rfloor \rightarrow_I \textstyle\bigwedge_{r=1}^{h} [i_r]\Xi\lceil[\tau_{k[[}(i_r)]\varphi_k, \langle\tau_l(i_r)_{[[}\rangle\psi_l\rfloor$$
$$\langle i\rangle\Xi\lceil[\tau_k]\varphi_k, \langle\tau_l\rangle\psi_l\rfloor \rightarrow_I \textstyle\bigvee_{r=1}^{h} \langle i_r\rangle\Xi\lceil[\varphi_k(i_r)]\varphi_k, \langle\tau_l(i_r)_{[[}\rangle\ psi_l\rfloor$$

together with the simplification rules for empty time regions

$$[\emptyset]\varphi \rightarrow_I \top$$
$$\langle\emptyset\rangle\varphi \rightarrow_I \bot$$

where:

- $[i]$ and $\langle i \rangle$ are time region operators;
- $\Xi\lceil[\tau_k(i_r)]\varphi_k, \langle\tau_l(i_r)\rangle\psi_l\rfloor$ is a formula in negation normal form whose top-level temporal formulae are $\{[\tau_k]\varphi_k \mid 1 \leq k \leq n\}$ and $\{\langle\tau_l\rangle\psi_l \mid 1 \leq l \leq m\}$; these temporal formulae may occur in the scope of other modal operators, but not in the scope of other temporal operators;

- $\{i_1, \ldots, i_h\} \stackrel{\text{def}}{=} \delta(i, \{\tau_1, \ldots\})$ and $\{\tau_1, \ldots\}$ are all the time terms occurring at top-level temporal formulae in Ξ.

Proposition 18.44. *The translation $\rightarrow_I$ is an equivalence-preserving transformation.*

The proof is, in principle, analogous to the proof of Proposition 18.19. It uses the fact that the temporal operators are actually blind to the modal operators. The time region determined by a time term τ in a temporal formula $[\tau]\varphi$ is only determined by the actual time region i which itself is determined by the next enclosing temporal operator. It is not influenced by any embracing modal operator.

The $\mathcal{CL}^{\mathcal{M}I}_{\mathcal{T}_\mathcal{U}}$-formulae can be further simplified with the simplification rules in Lemmas 18.21 and 18.22. These rules can be applied in the same way as in pure calendar logic. The result of this simplification is formulae without time terms and subsumption relations at formula positions, and without nested time region operators. This means that between two modal operators, there is at most one time region operator. For example, $[m](\langle i\rangle([m']\langle i'\rangle p))$ is possible, but $[m](\langle i\rangle(\langle i'\rangle p))$ would be simplified to $[m](\langle i'\rangle p)$.

Furthermore, we can apply the atomic decomposition (Definition 18.23) in the same way as in pure calendar logic. The result is a *decomposed normal form* where all time operators $[a]\varphi$ and $\langle a\rangle\psi$ for different atomic time regions a are completely independent of each other.

In the last step we eliminate the time region operators completely. The technique is very similar to the technique for pure calendar logic (Definition 18.25) where time region formulae $[a]\varphi$ and $\langle a\rangle\psi$ are replaced with predicate symbols $b_{a\varphi}$ and $d_{a\psi}$. The only difference is that the time region operators in the combined logic may occur in a modal context. The same time region formula in a different modal context must be replaced with a different variable because the same formula in a different modal context may have different truth values. For example, $[a]\varphi \wedge \langle m\rangle([a]\varphi)$ must be translated into $b_{a\varphi} \wedge \langle m\rangle(b_{am\varphi})$. The axiomatization of the new variables must also take into account the modal context. For example, the axiom for $b_{am\varphi}$ has to be embedded into an $[m]$-operator to get the right modal context: $[m](b_{am\varphi} \rightarrow R_a(\varphi))$. The technique is in fact very similar to the well-known renaming technique, where complex subformulae of modal formulae are replaced with new predicate symbols, and these symbols are axiomatized separately.

Definition 18.45 (Elimination of time region operators). Let $R_{a\psi}(\varphi)$ and $R_{a\psi}(\varphi)$ be the same functions as in Definition 18.25.

The translation function $PL0$ introduces for each formula $[a]\varphi$ in a modal context M a new predicate symbol $p_{aM\varphi}$ and for the formula $\langle a\rangle\psi$ in C a new predicate symbol $d_{aM\psi}$:

$$M[a]\varphi \rightarrow_{PL0} b_{aM\varphi} \tag{18.19}$$

$$M\langle a\rangle\psi \rightarrow_{PL0} d_{aM\psi}. \tag{18.20}$$

M denotes the embracing non-temporal modalities of the time region formula $[a]\varphi$ and $\langle a\rangle\psi$. For example, for the formula $[m_1]\langle m_2\rangle(p \vee [a]\varphi)$ $M \stackrel{\text{def}}{=} m_1 m_2$ is the embracing modal context, and $[a]\varphi$ is replaced with $b_{am_1m_2\varphi}$

Let $PL0'(\eta)$ be the result of applying the rewrite relation $\rightarrow_{PL0}$ exhaustively outside in.

The translation function $PL0$ is defined as

$$
\begin{aligned}
PL0(\eta) \stackrel{\text{def}}{=} PL0'(\eta) & \\
& \wedge \bigwedge_{b_{aM\varphi}} M(b_{aM\varphi} \Rightarrow R_a(\varphi)) && (18.21)\\
& \wedge \bigwedge_{d_{aM\psi}} M(d_{aM\psi} \Rightarrow R_{a\psi}(\psi)) && (18.22)\\
& \wedge \bigwedge_{b_{aM\varphi}, d_{a\psi}} M(b_{aM\varphi} \Rightarrow R_{a\psi}(\varphi)). && (18.23)
\end{aligned}
$$

The modal contexts M in (18.21), (18.22) and (18.23) are sequences of box operators. For example, the modal context for $b_{am_1m_2\varphi}$ is $M = [m_1][m_2]$.

Lemma 18.46 (Soundness and completeness of $PL0$). *A decomposed formula η is satisfiable in the combination $\mathcal{CL}^{\mathcal{M}I}_{\mathcal{T}\mathcal{U}}$ of calendar logic and a particular multimodal logic if and only if the translated formula $PL0(\eta)$ is satisfiable in the multimodal logic alone.*

The proof is analogous to the proof of Lemma 18.26.

Theorem 18.47. *Satisfiability of a formula in a combination of calendar logic and some multimodal logic is decidable if and only if satisfiability of the formulae in the multimodal logic alone is decidable.*

Proof As we have seen, the temporal operators can be eliminated from $\mathcal{CL}^{\mathcal{M}I}_{\mathcal{T}\mathcal{U}}$-formulae in a satisfiability-preserving way. Therefore satisfiability of a formula in the combined system can be decided by deciding satisfiability of the translated formula in the basic multimodal system. □

18.4.3 *Functional translation*

The elimination of the temporal operators from $\mathcal{CL}^{\mathcal{M}}_{\mathcal{T}\mathcal{U}}$-formulae may again increase the size of the formulae exponentially. This is, therefore, not an optimal approach for an inference system. Unfortunately there is no straightforward extension of the tableau approach for calendar logic to a tableau system for the combined logic. The reason is that certain combinations of temporal and modal operators cause loops. Consider the statement 'today I believe that it rains sometime today'. If 'today' is day 10 then the encoding of this statement is $[10, day] : \langle b\rangle\langle x_{day}\rangle raining$. Suppose the current moment t in the reference time line of the actual world w lies within the day 10, and we can use this statement to derive $\langle b\rangle\langle x_{day}\rangle raining$. The first diamond operator causes the introduction of a new world w' where $\langle x_{day}\rangle raining$ holds. The temporal operator $\langle x_{day}\rangle$ now

extension of the tableau approach for calendar logic to a tableau system for the combined logic. The reason is that certain combinations of temporal and modal operators cause loops. Consider the statement 'today I believe that it rains sometime today'. If 'today' is day 10 then the encoding of this statement is $[10, day] : \langle b \rangle \langle x_{day} \rangle raining$. Suppose the current moment t in the reference time line of the actual world w lies within the day 10, and we can use this statement to derive $\langle b \rangle \langle x_{day} \rangle raining$. The first diamond operator causes the introduction of a new world w' where $\langle x_{day} \rangle raining$ holds. The temporal operator $\langle x_{day} \rangle$ now generates a new time point t' with the constraint that t' lies within the current day, which is day 10. Therefore the original statement is again applicable, and we generate another world w'' and another time point t'' etc. The effect is similar to the effect in modal logic S4 where $\Box\Diamond$ sequences causes the tableau system to loop. In the S4 case, these loops can be detected and stopped. It is an open problem whether this is also possible for our combined logic.

To get around these difficulties, we use the *functional translation* into predicate logic and provide a resolution system for the combined logic. The *functional translation* for modal and multimodal logics into predicate logic is a refinement of the standard (relational) translation. The relational translation uses the semantics definitions for the modal operators as translation rules. In the mono-modal case the semantics definitions for the two modal operators $\Box$ and $\Diamond$ are

$$\begin{array}{l} \Im, w \models \Box\varphi \text{ iff } \forall u\ R(w,u) \rightarrow \Im, u \models \varphi \\ \Im, w \models \Diamond\varphi \text{ iff } \exists u\ R(w,u) \wedge \Im, u \models \varphi. \end{array}$$

From these definitions one can derive translation rules T_r which translate the modal formulae into first-order predicate logic:

$$\begin{array}{l} T_r(\Box\varphi, w) = \forall u\ R(w,u) \rightarrow T_r(\varphi, u) \\ T_r(\Diamond\varphi, w) = \exists u\ R(w,u) \wedge T_r(\varphi, u). \end{array}$$

In addition there are the standard translation rules for the classical connectives, and the rule for predicate symbols p: $T_r(p, w) = p(w)$.

Different modal systems are characterized by different properties of the accessibility relation R. Examples are:

system	characteristic axiom	
D	$\forall x\ \exists y\ R(x,y)$	(seriality)
T	$\forall x\ R(x,x)$	(reflexivity)
B	$\forall x,y\ R(x,y) \rightarrow R(y,x)$	(symmetry)
4	$\forall x,y,z\ R(x,y) \wedge R(y,z) \rightarrow R(x,z)$	(transitivity)
5	$\forall x,y,z\ R(x,y) \wedge R(x,z) \rightarrow R(y,z)$	(Euclideaness)

The completeness theorems for modal logics with respect to first-order properties of the accessibility relation can be turned into completeness theorems for the relational translation:

Theorem 18.48 (Completeness of relational translation). *A modal formula φ is a theorem in a modal system M with first-order frame properties iff the characteristic axioms for the modal system M entail $\forall w\ T_r(\varphi, w)$ in first-order predicate logic.*

The relational translation for multimodal logics is essentially like the relational translation for monomodal logic. The only difference is that for each modality one has a different predicate for the accessibility relation.

The relational translation can be employed to make use of predicate logic inference systems and theorem provers for proving modal theorems. The problem is, however, that the standard versions of these inference systems are not necessarily decision procedures, although the modal logic itself is usually decidable. Special strategies for guiding the application of the inference rules are needed to terminate the proof search [Zamov, 1989].

The functional translation [Ohlbach, 1991] arranges the information about which world is accessible from which other in a more compact way such that terminating proof procedures can be designed more easily. It uses the fact that binary relations can be decomposed into sets $AF^{\Im}$ of *accessibility functions*

$$R(x, y) \Leftrightarrow \exists \gamma : AF^{\Im}\ \gamma(x) = y. \qquad (18.24)$$

Accessibility functions γ map worlds to accessible worlds. As an example, consider the relation R_1:

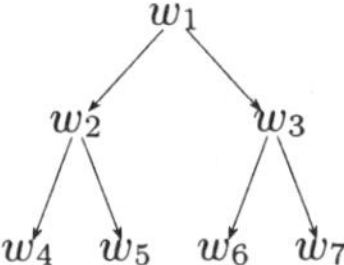

There are at least four different ways to decompose R_1 into sets of two functions $\{\gamma_1, \gamma_2\}$:

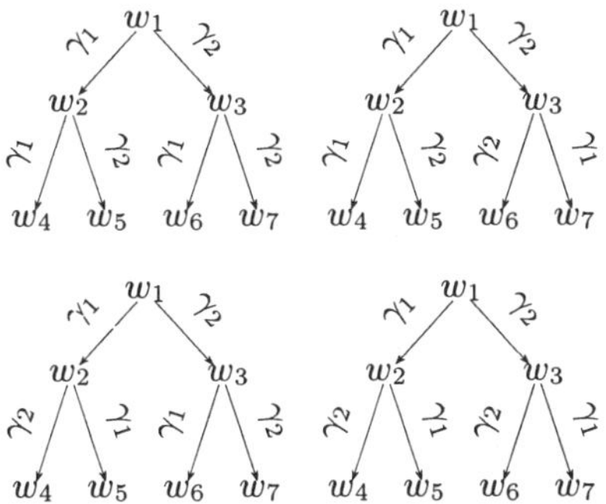

Since the relation is not serial, and there are dead-ends, all these γ_1s and γ_2s are *partial functions*. On the other hand, if the accessibility relation is serial, i.e. for each world there is an accessible world, it is sufficient to consider *total* accessibility functions. This is the case where the functional translation works best. In standard predicate logic, functions are always total. Therefore an extra mechanism is needed for covering the non-serial case. The trick is to introduce a special predicate *de* which is true for 'dead-end' worlds—that means worlds which have no further accessible worlds. The correspondence between accessibility relations and accessibility functions is then

$$R(x,y) \Leftrightarrow (de(x) \vee \exists \gamma : AF^{\Im}\ \gamma(x) = y). \tag{18.25}$$

The value of $\gamma(x)$ does not matter in worlds x where the *de*-predicate is true.

The functional translation can now be obtained as a composition of the relational translation and an application of (18.25) as a rewrite rule. In the multi-modal case, *sorted* variables are used to distinguish accessibility functions corresponding to the different accessibility relations. Therefore for each modality m, a sort s_m is introduced. A variable x^{s_m} denotes an accessibility function which maps worlds to m-accessible worlds. The 'dead-end' predicate *de* is also marked with the sort s_m. de_{s_m} is true in a world x if there are no m-accessible worlds from x.

The functional translation can now be formulated as

$$T_f([m]\varphi, \pi) = de_{s_m}(\pi) \vee \forall x^{s_m}\ T_f(\varphi, \pi x) \tag{18.26}$$
$$T_f(\langle m \rangle\varphi, \pi) = \neg de_{s_m}(\pi) \wedge \exists x\ T_f(\varphi, \pi x) \tag{18.27}$$

together with the standard rules for the classical connectives and the rule $T_f(p,\pi) = p(\pi)$. The de_{s_m}-part can be omitted if the m-accessibility relation is serial. For example, a formula $[m]\langle r \rangle p$ is functionally translated into

$$de_{s_m}([]) \vee \forall x^{s_m}\ (\neg de_{s_r}(x) \wedge \exists y^{s_r}\ p(xy)).$$

This is already an optimized translation. The most natural one would be

$$\forall w\ de_{s_m}(w) \vee \forall x^{s_m}\ \neg de_{s_r}(x(w)) \wedge \exists y^{s_r}\ p(y(x(w)))$$

in a second-order notation. It turns out that the variable w can be omitted if the negation of the modal formula is to be checked for satisfiability (e.g. to prove theoremhood of a formula by refuting its negation). In this case, w becomes existentially quantified. The existential quantifier gets skolemized and becomes a Skolem constant. This Skolem constant is always the same innermost argument for the *accessibility terms*. It can therefore be ignored. Moreover, nested function applications $y(x(w))$ can be represented as a string xy. These strings (usually called *world-paths*) represent sequences of transitions to accessible worlds. If w is the actual world, then xy denotes an x-transition followed by a y-transition. The sorts of x and y indicate the accessibility relations, along which the transitions take place. An empty world-path [] also has a meaning. It denotes the 'empty' transition, i.e. the identity function.

The functional translation can be further optimized for propositional modal logics by moving existential quantifiers over accessibility functions out of the

behind this is that the actual world used to interpret p is determined by xa, and this depends on x. In the original version where the existential quantifier is within the scope of the universal quantifier, skolemization yields $xf(x)$ with an unnecessary double dependence of x. (Notice that this optimization is not allowed for functionally translated *quantified* modal logic formulae. In this case, the truth value of a formula at a particular world may depend on the path used to get to this world. The arguments of the Skolem functions are then needed to distinguish different paths.)

18.4.4 *Functional translation for the combined logic*

We extend the functional translation method to cope with the temporal operators of calendar logic. From an abstract point of view, the combined logic is a multimodal logic itself. Therefore in principle the same mechanisms as for arbitrary multimodal logics can be used. 'Accessibility variables' and constants labelled with sorts are used to express transitions from worlds to worlds and jumps to particular time points. For the temporal operators $[\tau]$ and $\langle\tau\rangle$ we need sorts s_τ, and for the statements $[n, U] : \varphi$, where $[n, U]$ corresponds to a concrete interval i in the reference time line, we need sorts s_i.

For example, we may have a sort $s_{February(x_{year})}$. It denotes the set of (temporal) accessibility functions which map time points in each year to some time point in the February of this year. A sort $s_{[100,200[}$, on the other hand, denotes the set of functions mapping *all* time points into the interval $[100, 200[$ in the reference time line.

The functional translation for the combined logic could be specified by defining a translation function similar to T_f in (18.26) and (18.27). It turns out, however, that the technical details and the proofs become simpler and clearer by using an intermediate language which combines the elements of the original $\mathcal{CL}^{\mathcal{M}}_{\mathcal{T}\mathcal{U}}$-syntax with the world-paths of the functional translation. To this end we introduce a language with *labelled formulae* $\pi : \varphi$ where π is a world-path and φ is essentially a $\mathcal{CL}^{\mathcal{M}}_{\mathcal{T}\mathcal{U}}$- formula. $\pi : \varphi$ represents a partially translated formula. π stems from the already translated modal and temporal operators and φ still has to be translated. The translation ends up with labelled formulae of the kind $\pi : p$ or $\pi : \neg p$ where p is a predicate symbol. In this system one can prove soundness and completeness by showing that the translation rules are equivalences in the labelled language.

The formula part of the labelled formulae is essentially $\mathcal{CL}^{\mathcal{M}}_{\mathcal{T}\mathcal{U}}$, but we need a small extra. Consider the statement

$$\langle 1998, \mathit{year}\rangle(\mathit{holiday} \wedge (\mathit{off\text{-}season}(x_{year}) \rightarrow \mathit{save\text{-}money}))$$

where *off-season*(x_{year}) is a temporal notion defined accordingly. The formula expresses that in the year 1998, I have a holiday, and if the holiday happens to be not in the main season then I save money. The $\langle 1998, \mathit{year}\rangle$ operator causes the generation of a constant a^i where i is the interval corresponding to the year 1998. The temporal formula *off-season*(x_{year}) now constraints the interval i to

$$\langle 1998, year\rangle(holiday \wedge (\textit{off-season}(x_{year}) \rightarrow \textit{save-money}))$$

where $\textit{off-season}(x_{year})$ is a temporal notion defined accordingly. The formula expresses that in the year 1998, I have a holiday, and if the holiday happens to be not in the main season then I save money. The $\langle 1998, year\rangle$ operator causes the generation of a constant a^i where i is the interval corresponding to the year 1998. The temporal formula $\textit{off-season}(x_{year})$ now constraints the interval i to the part of the year which does not lie in the main season. Therefore in the translation we need an expression $a \in j$ at formula position. It expresses that the time point a actually lies in the interval j. The translated formula is

$$a^{s_i} : holiday \wedge (a^{s_i} \in j \vee a^{s_i} : \textit{save-money})$$

where j is the non-off-season time interval, i.e. the main season in the year. The meaning is: sometime in the interval i (year 1998) I have a holiday, and either t is in the main season (interval j), or otherwise I save some money.

Definition 18.49 (Labelled formulae). We define labelled formulae $\pi : \varphi$ where π is a world-path and φ a $\mathcal{CL}^{\mathcal{M}}_{\mathcal{T_U}}$-formula *in negation normal form*. The additional syntax elements consist of the following:

- A set $\mathbb{S}_M$ of *modal sort symbols*, a set $\mathbb{S}_R$ of *time region sort symbols*, a set $\mathbb{S}_T$ of *time term sort symbols*. $\mathbb{S}_M$ contains for each modality m a sort symbol s_m. $\mathbb{S}_R$ contains for each non-empty time region i a sort symbol s_i, $\mathbb{S}_T$ contains for each time term $\tau \in \mathcal{T_U}$ a sort symbol s_τ.
  Let $\mathbb{S} \stackrel{\text{def}}{=} \mathbb{S}_M \cup \mathbb{S}_R \cup \mathbb{S}_T$.
- A binary membership relation $\in$.
- Variable symbols $\mathbb{V}$ for each sort. To indicate that a variable x is of sort s_α, we write s_α. If $s_\alpha \in \mathbb{S}_M$ is a modal sort symbol, then x^{s_α} is a *modal* variable, and if $s_\alpha \in \mathbb{S}_R \cup \mathbb{S}_T$ is a temporal sort symbol, then x^{s_α} is a *temporal* variable.
- For each sort s there is at least one constant symbol, a^s.[29]

The language $\mathcal{LCL}^{\mathcal{M}}_{\mathcal{T_U}}$ of labelled $\mathcal{CL}^{\mathcal{M}}_{\mathcal{T_U}}$-formulae in *negation normal form* can now be defined as follows:

- a *world-path* is a string of modal and temporal variable symbols, $WP ::= [] \mid WP\mathbb{V}$;
- $\mathcal{LCL}^{\mathcal{M}}_{\mathcal{T_U}} ::= \mathcal{CL}^{\mathcal{M}}_{\mathcal{T_U}} \mid WP \in Int(\mathcal{T}) \mid \mathcal{LCL}^{\mathcal{M}}_{\mathcal{T_U}} \wedge \mathcal{LCL}^{\mathcal{M}}_{\mathcal{T_U}} \mid \mathcal{LCL}^{\mathcal{M}}_{\mathcal{T_U}} \vee \mathcal{LCL}^{\mathcal{M}}_{\mathcal{T_U}} \mid WP : \mathcal{CL}^{\mathcal{M}}_{\mathcal{T_U}} \mid \exists\mathbb{V}\ \mathcal{LCL}^{\mathcal{M}}_{\mathcal{T_U}} \mid \forall\mathbb{V}\ \mathcal{LCL}^{\mathcal{M}}_{\mathcal{T_U}}$.

A *T-block* is a world-path or a segment of a world-path consisting of temporal accessibility variables only.

An *M-block* is a world-path or a segment of a world-path consisting of modal accessibility variables only.

[29]This is a standard assumption. It ensures that the sort is not empty in a Herbrand interpretation.

A world-path is therefore a sequence of T-blocks and M-blocks.

It is important that the labelled formulae are in negation normal form. Otherwise there would be tautologies which make no sense. As an example, consider the formula

$$x : \textit{nice-weather} \vee \neg x : \textit{nice-weather}.$$

According to the standard semantics of $\vee$, this is definitely a tautology. Suppose, however, x is of sort $s_{\textit{leap-day}(y_{year})}$, and $\textit{leap-day}(y_{year})$ is a time term mapping a year-coordinate y_{year} to the day-coordinate of the leap day in the year y_{year}, or to the empty set for non-leap years. In this case the formula expresses 'at this year's leap day either it is nice weather, or it is not nice weather', which makes no sense if there is no leap day this year.

$$x : \textit{nice-weather} \vee x : \neg\textit{nice-weather},$$

on the other hand, is only a true statement if there is a leap day. To avoid problems like this, negated labelled formulae are forbidden.

Notation: In the remaining parts of this chapter, we use the convention that sorts s_m or s_n are always modal sorts in $\mathbb{S}_M$, sorts s_τ are temporal sorts in $\mathbb{S}_T$, and sorts s_i and s_j are temporal sorts in $\mathbb{S}_R$.

Definition 18.50 (Semantics of the labelled language). The semantics of $\mathcal{LCL}^{\mathcal{M}}_{\mathcal{T}u}$-formulae is an extension of the semantics of the $\mathcal{CL}^{\mathcal{M}}_{\mathcal{T}u}$-formulae.

There are the same kinds of frames $\mathcal{F} = (\mathcal{W} \times \mathcal{T}, R)$ as in the $\mathcal{CL}^{\mathcal{M}}_{\mathcal{T}u}$-case (Definition 18.41).

An *interpretation* $\Im \stackrel{\text{def}}{=} (\mathcal{F}, \mathcal{P}, \mathcal{V})$ consists of a $\mathcal{CL}^{\mathcal{M}}_{\mathcal{T}u}$-interpretation $(\mathcal{F}, \mathcal{P})$, and in addition a variable assignment $\mathcal{V}$ which maps constants and variables x^s of sort s to *accessibility functions* $\mathcal{V}(x^{s_m}) \in AF^{\Im}_s$ where for modal sorts $s_m \in \mathbb{S}_M$:

$$AF^{\Im}_{s_m} \stackrel{\text{def}}{=} \{\gamma : \mathcal{W} \times \mathcal{T} \mapsto \mathcal{W} \times \mathcal{T} \mid \text{if } \exists w' \text{ with } R_m((w,t),(w',t)) \\ \text{then } \gamma(w,t) \text{ is defined } \textit{and} \text{ if } \gamma(w,t) = (w',t) \text{ then } R_m((w,t),(w',t))\},$$

for temporal sorts $s_i \in \mathbb{S}_R$:

$$AF^{\Im}_{s_i} \stackrel{\text{def}}{=} \{\gamma : \mathcal{W} \times \mathcal{T} \mapsto \mathcal{W} \times \mathcal{T} \mid \gamma(w,t) = (w,t') \text{ with } t' \in i\}$$

and for temporal sorts $s_\tau \in \mathbb{S}_T$:

$$AF^{\Im}_{s_\tau} \stackrel{\text{def}}{=} \{\gamma : \mathcal{W} \times \mathcal{T} \mapsto \mathcal{W} \times \mathcal{T} \mid \forall t \text{ with } \tau_{[[}(t) \neq \emptyset : \\ \gamma(w,t) = (w,t') \text{ with } t' \in \tau_{[[}(t)\}.$$

Note that the accessibility functions can be *partial functions.* $AF^{\Im}_{s_m}$, however, contains total functions only if the accessibility relation for the m-modality is serial.

Let $\Im[x/\gamma]$ be like $\Im$, except the variable x is mapped to γ.

The interpretation of world-paths is

$$\Im([])(w,t) \stackrel{\text{def}}{=} (w,t) \qquad (18.28)$$

$$\Im(\pi a)(w,t) \stackrel{\text{def}}{=} \begin{cases} \mathcal{V}(a)(\Im(\pi)(w,t)) & \text{if } \mathcal{V}(a)(\Im(\pi)(w,t)) \text{ is defined} \\ \text{undefined} & \text{otherwise.} \end{cases} \quad (18.29)$$

The interpretation of $\mathcal{LCL}_{\mathcal{T}u}^{\mathcal{M}}$-formulae which are not $\mathcal{CL}_{\mathcal{T}u}^{\mathcal{M}}$-formulae is

$$\Im, w, t \models \pi \in \emptyset \text{ iff } \Im(\pi)(w,t) \text{ is undefined} \quad (18.30)$$

$$\Im, w, t \models \pi \in i \text{ iff } \Im(\pi)(w,t)_{|T} \in i \quad (18.31)$$

$$\Im, w, t \models \pi : \varphi \text{ iff } \Im(\pi)(w,t) \text{ is defined and } \Im, \Im(\pi)(w,t) \models \varphi \quad (18.32)$$

$$\Im, w, t \models \exists x^s\ \varphi \text{ iff } \Im[x/\gamma], w, t \models \varphi \text{ for some } \gamma \in AF_s^{\Im} \quad (18.33)$$

$$\Im, w, t \models \forall x^s\ \varphi \text{ iff } \Im[x/\gamma], w, t \models \varphi \text{ for all } \gamma \in AF_s^{\Im} \quad (18.34)$$

$$\Im, w, t \models \pi : (\varphi_1 \wedge \varphi_2) \text{ iff } \Im, w, t \models \pi : \varphi_1 \text{ and } \Im, w, t \models \pi : \varphi_2 \quad (18.35)$$

$$\Im, w, t \models \pi : (\varphi_1 \vee \varphi_2) \text{ iff } \Im, w, t \models \pi : \varphi_1 \text{ or } \Im, w, t \models \pi : \varphi_2. \quad (18.36)$$

The world-paths in labelled $\mathcal{LCL}_{\mathcal{T}u}^{\mathcal{M}}$-formulae consist of sorted variables. The sorts of the temporal variables contain very concrete information about the time regions under consideration. For example, in a labelled formula $\forall x^{s_{[100,200[}}\ x : p$ the sort $s_{[100,200[}$ indicates that p holds in all time points between 100 and 200. In a labelled formula $\forall x^{s_{[100,200[}}\ \exists y^{s_{February(z_{year})}}\ xy : p$ the sorts $s_{[100,200[}$ and $s_{February(z_{year})}$ indicate that p holds at some point in February of the year containing the interval [100, 200[, or, if [100, 200[intersects with several years, then p holds in every February of these years. The information about these time regions is very useful, and it can easily be computed from the sorts of the variables.

Definition 18.51 (Time regions of world-paths). For a time region i we define

$$\begin{array}{lll} [\,]_{[[}(i) & \stackrel{\text{def}}{=} i & \\ (\pi a^{s_m})_{[[}(i) & \stackrel{\text{def}}{=} \pi_{[[}(i) & \text{where } s_m \in \mathbb{S}_M \\ (\pi a^{s_j})_{[[}(i) & \stackrel{\text{def}}{=} \begin{cases} j \text{ if } \pi = [\,] \text{ or } \pi_{[[}(i) \neq \emptyset \\ \emptyset \text{ otherwise} \end{cases} & \text{where } s_j \in \mathbb{S}_R \\ (\pi a^{s_\tau})_{[[}(i) & \stackrel{\text{def}}{=} \tau_{[[}(\pi_{[[}(i)) & \text{where } s_\tau \in \mathbb{S}_T. \end{array}$$

Notice that $(a^{s_j}\pi)_{[[}(i) = \pi_{[[}(j)$ for every time region i. Therefore the argument (i) does not matter if the world-path $\pi = a^{s_j}\pi$ starts with a variable of sort s_j, and we can simply write $\pi_{[[}$ instead of $\pi_{[[}(i)$.

Functionally translated $\mathcal{CL}_{\mathcal{T}u}^{\mathcal{M}}$-statements always start with a variable of sort s_j. Let us call them *determined* world-paths.

Obviously $\pi_{[[}(i)$ is only determined by the temporal variables in π. Therefore the trailing non-temporal variables can be ignored for computing $\pi_{[[}(i)$.

Proposition 18.52. *If for world-path π, π_T is defined to be the initial sequence of π, up to the rightmost temporal variable, then*

$$\pi_{[[}(i) = (\pi_T)_{[[}(i) \quad (18.37)$$

holds for all time regions i.

The proof of this proposition is trivial. Not quite so trivial is the proof that the above definition for $\pi_{[[}$ is optimal in the sense that it computes the narrowest limits for the time regions a world-path accesses.

Proposition 18.53 ($\pi_{[[}$ is an exact approximation). *For all world-paths $\pi = x_1^{s_1} \ldots x_n^{s_n}$, for all time regions i, for all interpretations $\Im$ and for all worlds w:*

(a) for all $t \in i$: $(\Im(\pi)(w,t))_{|T} \in \pi_{[[}(i)$ and
(b) for all time points $t' \in \pi_{[[}(i)$ there is some $t \in i$ and accessibility functions $\gamma_l \in AF^{\Im}_{s_l}, 1 \le l \le n$, with $(\Im[x_1/\gamma_1, \ldots, x_n/\gamma_n](\pi)(w,t))_{|T} = t'$.

Proof By induction on the length of world-paths.
(a) Base case: $(\Im([])(w,t))_{|T} = t \in i = []_{[[}(i)$. (Defs 18.51 and 18.28)
Induction step: The induction hypothesis is: if $\Im(\pi)(w,t) = (w',t')$ then $t' \in \pi_{[[}(i)$, otherwise $\pi_{[[}(i) = \emptyset$.

Suppose $\Im(\pi)(w,t) = (w',t')$ is defined.

Case (i) $\Im(\pi x^{s_j})(w,t) = \mathcal{V}(x^{s_j})(w',t') \stackrel{\text{def}}{=} (w',t'')$ and, because $\mathcal{V}(x^{s_j}) \in AF^{\Im}_{s_j}$, $t'' \in j$. Since $(\pi x^{s_j})_{[[}(i) = j$ (Definition 18.51), we have $(\Im(\pi x^{s_j})(w,t))_{|T} \in (\pi x^{s_j})_{[[}(i)$.

Case (ii) $\Im(\pi x^{s_\tau})(w,t) = \mathcal{V}(x^{s_\tau})(w',t') \stackrel{\text{def}}{=} (w',t'')$ and, because $\mathcal{V}(x^{s_\tau}) \in AF^{\Im}_{s_\tau}$, $t'' \in \tau_{[[}(t') \in \tau_{[[}(\pi_{[[}(i)) = (\pi x^{s_\tau})_{[[}(i)$. Thus, $(\Im(\pi x^{s_\tau})(w,t))_{|T} \in (\pi x^{s_\tau})_{[[}(i)$.

Case (iii) $\Im(\pi x^{s_m})(w,t) = \mathcal{V}(x^{s_m})(w',t') \stackrel{\text{def}}{=} (w'',t')$. Since $t' \in \pi_{[[}(i) = (\pi x^{s_m})_{[[}(i)$, we have $(\Im(\pi x^{s_m})(w,t))_{|T} \in (\pi x^{s_m})_{[[}(i)$.

Now suppose $\Im(\pi)(w,t) = (w',t')$ is not defined and therefore $\pi_{[[}(i) = \emptyset$. All three cases in the definition of $\pi a_{[[}$ (Definition 18.51) yield $\pi a_{[[} = \emptyset$.

(b) Base case: $t' \in []_{[[}(i) = i$. If we choose $t = t'$, then $(\Im([])(w,t))_{|T} = t'$.
Induction step: The induction hypothesis is $\forall t' \in \pi_{[[}(i)\ \exists t \in i$ with $\Im(\pi)(w,t)_{|T} = t'$ where $\Im$ contains the appropriate assignments for the variables in π.

Case (i) Let $t' \in (\pi x^{s_j})_{[[}(i) = j$. Since $\pi_{[[}(i) \neq \emptyset$ (otherwise $(\pi x^{s_j})_{[[}(i) = \emptyset$), we can choose any $t'' \in \pi_{[[}(i)$ and find some $\gamma \in AF^{\Im}_{s_j}$ with $\gamma(w',t'') = (w',t')$. According to the induction hypothesis, one can find some $t \in i$ with $\Im(\pi)(w,t)_{|T} = t''$. Thus, $\Im(\pi x^{s_j})(w,t)_{|T} = t'$.

Case (ii) Let $t' \in (\pi x^{s_\tau})_{[[}(i) = \tau_{[[}(\pi_{[[}(i))$. Let $t'' \in \pi_{[[}(i)$ with $t' \in \tau_{[[}(t'')$. We can find some $\gamma \in AF^{\Im}_{s_\tau}$ with $\gamma(w',t'') = (w',t')$. Now the induction hypothesis can be used in the same way as in case (i) to prove the induction step.

Case (iii) Let $t' \in (\pi x^{s_m})_{[[}(i) = \pi_{[[}(i)$. This time the induction hypothesis can be applied immediately to prove the induction step. □

We now prove a number of equivalences which form the basis for the functional translation.

Lemma 18.54 **($\mathcal{LCL}^{\mathcal{M}}_{\mathcal{T}\mathcal{U}}$-equivalences).** *The following equivalences hold in $\mathcal{LCL}^{\mathcal{M}}_{\mathcal{T}\mathcal{U}}$:*

$$[n, U] : \varphi \Leftrightarrow \forall x^{s_{U_{[[}(n)}}\ x : \varphi \tag{18.38}$$
$$\langle n, U\rangle : \varphi \Leftrightarrow \exists x^{s_{U_{[[}(n)}}\ x : \varphi \tag{18.39}$$
$$\pi : (\varphi_1 \wedge \varphi_2) \Leftrightarrow \pi : \varphi_1 \wedge \pi : \varphi_2 \tag{18.40}$$
$$\pi : (\varphi_1 \vee \varphi_2) \Leftrightarrow \pi : \varphi_1 \vee \pi : \varphi_2 \tag{18.41}$$
$$\pi : \langle m\rangle\varphi \Leftrightarrow \exists x^{s_m}\ \pi x : \varphi \tag{18.42}$$
$$\pi : [m]\varphi \Leftrightarrow \forall x^{s_m}\ \pi x : \varphi \tag{18.43}$$
$$\pi : \langle\tau\rangle\varphi \Leftrightarrow \exists x^{s_\tau}\ \pi x : \varphi \tag{18.44}$$
$$\pi : [\tau]\varphi \Leftrightarrow \forall x^{s_\tau}\ \pi x : \varphi \tag{18.45}$$
$$\pi : \tau \Leftrightarrow \pi \in \bigcup_{i\in\delta(\pi_{[[},\tau)} (i \cap \tau_{[[}(i)) \tag{18.46}$$
$$\pi : \neg\tau \Leftrightarrow \pi \in \bigcup_{i\in\delta(\pi_{[[},\tau)} (i \setminus \tau_{[[}(i)) \tag{18.47}$$
$$\pi : \tau_1 \subseteq \tau_2 \Leftrightarrow \pi \in \bigcup_{i\in\delta(\pi_{[[},\{\tau_1,\tau_2\}),\tau_{1[[}(i)\subseteq\tau_{2[[}(i)} i \tag{18.48}$$
$$\pi : \tau_1 \not\subseteq \tau_2 \Leftrightarrow \pi \in \bigcup_{i\in\delta(\pi_{[[},\{\tau_1,\tau_2\}),\tau_{1[[}(i)\not\subseteq\tau_{2[[}(i)} i. \tag{18.49}$$

Proof The proofs are not difficult, but very technical. The sequence of arguments is indicated by listing the line numbers at the left margin and reasons for the argument at the right margin.

We use the symbol $\succ$ as a metaimplication symbol.

(18.38) ('$\Rightarrow$')

(1) Suppose $\Im, w, t \models [n, U] : \varphi$.
(2) $\succ \Im, w, t' \models \varphi$ for all $t' \in U_{[[}(n)$. (Def. 18.41)
(3) Let $\gamma \in AF^{\Im}_{s_{U_{[[}(n)}}$ and $\Im' = \Im[x/\gamma]$.
(4) $\succ \gamma(w, t)_{|T} \stackrel{\text{def}}{=} t' \in U_{[[}(n)$. (Def. of $AF^{\Im}_{s_{U_{[[}(n)}}$)
(5) $\succ \Im', \Im'(x)(w, t) \models \varphi$. (because of (2) and (4))
$\succ \Im, w, t \models \forall x^{s_{U_{[[}(n)}}\ x : \varphi$. ($\gamma \in AF^{\Im}_{s_{U_{[[}(n)}}$ arbitrary)

('$\Leftarrow$')

(6) Suppose $\Im, w, t \models \forall x^{s_{U_{[[}(n)}}\ x : \varphi$.
(7) Let $t' \in U_{[[}(n)$.
(8) $\succ$ There is some $\gamma \in AF^{\Im}_{s_{U_{[[}(n)}}$ with $\gamma(w, t) = (w, t')$.
(9) $\succ \Im[x/\gamma], \Im[x/\gamma](x)(w, t) \models \varphi$. (using (6))
(10) $\succ \Im[x/\gamma], w, t' \models \varphi$ (using (8), $\gamma(w, t) = (w, t')$)
(11) $\succ \Im, w, t' \models \varphi$. ($x \notin \varphi$)
$\succ \Im \models [n, U] : \varphi$. ((7) and (11), Def 18.41)

The proof for (18.39) is dual to the proof for (18.38).

The proofs for (18.40) and (18.41) are trivial.

(18.42) ($\Rightarrow$)

(12) Suppose $\Im, w, t \models \pi : \langle m \rangle \varphi$.
(13) $\succ \Im(\pi)(w,t)$ is defined. (Def. 18.50)
(14) Let $\Im(\pi)(w,t) \stackrel{\text{def}}{=} (w', t')$.
(15) $\succ \Im, w', t' \models \langle m \rangle \varphi$. (semantics of $\pi : \varphi$)
(16) $\succ \exists w''\ R_m((w', t'), (w'', t'))$ and $\Im, w'', t' \models \varphi$. (semantics of $\langle m \rangle$)
(17) $\succ \exists \gamma \in AF^{\Im}_{s_m}\ \Im[x/\gamma](\pi x)(w,t) = (w'', t')$. ($w''$ is R_m-accessible from w')
(18) $\succ \Im[x/\gamma], w, t \models \pi x : \varphi$. ($\Im, w'', t' \models \varphi$)
(19) $\succ \Im, w, t \models \exists x^{s_m} \pi x : \varphi$. ($\gamma \in AF^{\Im}_{s_m}$)

($\Leftarrow$)

(20) Suppose $\Im, w, t \models \exists x^{s_m} \pi x : \varphi$.
(21) $\succ \exists \gamma \in AF^{\Im}_{s_m}\ \Im[x/\gamma], w, t \models \pi x : \varphi$. (semantics of $\exists x^{s_m}$)
(22) $\succ \Im[x/\gamma](\pi x)(w,t)$ is defined.
(23) Let $\Im(\pi)(w,t) \stackrel{\text{def}}{=} (w', t')$.
(24) $\succ \Im[x/\gamma], \Im[x/\gamma](\pi x)(w,t) \models \varphi$ and (semantics of $\pi x : \varphi$)
(25) $R_m((w', t'), \gamma(w', t'))$. ($\gamma \in AF^{\Im}_{s_m}$, Def. 18.49)
(26) $\succ \Im, w', t' \models \langle m \rangle \varphi$.
(27) $\succ \Im, w, t \models \pi : \langle m \rangle \varphi$. (20)

The proof for (18.43) is dual to the proof for (18.42).

(18.44) ('$\rightarrow$')

(28) Suppose $\Im, w, t \models \pi : \langle \tau \rangle \varphi$.
(29) $\succ \Im(\pi)(w,t)$ is defined. (semantics of $\pi : \varphi$)
(30) Let $(w', t') \stackrel{\text{def}}{=} \Im(\pi)(w,t)$.
(31) $\succ \Im, w', t' \models \langle \tau \rangle \varphi$. (using (28))
(32) $\succ \exists t''\ t'' \in \tau_{[[}(t')$ and
(33) $\Im, w', t'' \models \varphi$. (semantics of $\langle \tau \rangle$)
(34) There is some $\gamma \in AF^{\Im}_{s_\tau}$ with $\gamma(w', t') = (w', t'')$. (using (32))
(35) $\succ \Im(\pi x)(w,t)$ is defined.
(36) $\succ \Im, \gamma(w', t') \models \varphi$. (using (33))
(37) $\succ \Im[x/\gamma], w', t' \models x : \varphi$.
$\succ \Im[x/\gamma], w, t \models \pi x : \varphi$.
$\succ \Im, w, t \models \exists x^{s_\tau}\ \pi x : \varphi$. (using (30) and (37))

($\Leftarrow$)

(38) Suppose $\Im, w, t \models \exists x^{s_\tau}\ \pi x : \varphi$.
(39) $\succ$ There is some $\gamma \in AF^{\Im}_{s_\tau}$ with $\Im[x/\gamma] \models \pi x : \varphi$.
(40) $\succ \Im(\pi x)(w,t)$ is defined.
(41) Let $\Im(\pi)(w,t) \stackrel{\text{def}}{=} (w', t')$.
(42) $\succ \gamma(w', t') \stackrel{\text{def}}{=} (w', t'')$ is defined and
(43) $\Im[x/\gamma], w', t' \models x : \varphi$. ($\Im, w, t \models \exists x^{s_\tau}\ \pi x : \varphi$ and (41))
(44) $\succ \Im[x/\gamma], w', t'' \models \varphi$. (using (39))
$\succ \Im, w', t' \models \langle \tau \rangle \varphi$. ($\gamma \in AF^{\Im}_{s_\tau}$, i.e. $(\gamma(w', t'))_{|T} \in \tau_{[[}(t')$)
$\succ \Im, w, t \models \pi : \langle \tau \rangle \varphi$. (using (41))

The proof for (18.45) is dual to the above proof.

(18.46) ($\Rightarrow$)

(45) Suppose $\Im, w, t \models \pi : \tau$.
(46) $\Im(\pi)(w,t)$ is defined.
(47) Let $\Im(\pi)(w,t) \stackrel{\text{def}}{=} (w', t')$.
(48) $\quad \Im, w', t' \models \tau$ and (using (45))
(49) $\quad t' \in \pi_{[[}$. (Prop. 18.53(a))
(50) $\succ t' \in \tau_{[[}(t')$. (Def. 18.41)
(51) $\quad t' \in i \in \delta(\pi_{[[}, \tau)$. (using (49))
(52) $\succ t' \in \tau_{[[}(i)$ ((50), τ is constant over i)
(53) $\succ t' \in (i \cap \tau_{[[}(i))$.
(54) $\succ t' \in \bigcup_{i \in \delta(\pi_{[[}, \tau)}(i \cap \tau_{[[}(i))$.
$\quad \succ \Im, w, t \models \bigcup_{i \in \delta(\pi_{[[}, \tau)}(i \cap \tau_{[[}(i))$. (from (47) and (54))

($\Leftarrow$)

(55) Suppose $\Im, w, t \models \pi \in \bigcup_{j \in \delta(\pi_{[[}, \tau)}(j \cap \tau_{[[}(j))$.
(56) $\succ \Im(\pi)(w,t) \in \bigcup_{i \in \delta(\pi_{[[}, \tau)}(i \cap \tau_{[[}(i))$.
(57) $\succ \Im(\pi)(w,t)$ is defined and not empty.
(58) Let $\Im(\pi)(w,t) = (w', t')$.
(59) $\succ t' \in i$ and $t' \in \tau_{[[}(i)$ for some $i \in \delta(\pi_{[[}, \tau)$. (from (55))
(60) $\succ t' \in \tau_{[[}(t')$. ($\tau$ constant over i)
$\quad \succ \Im, w, t \models \pi : \tau$. (using (60), (58))

The proofs for (18.47), (18.48) and (18.49) are analogous to the above proof. □

Theorem 18.55 (Functional translation). *If $TF_1(\varphi)$ is defined to be the result of applying the equivalences (18.38)–(18.49) exhaustively as rewrite rules from left to right to φ then $TF_1(\varphi)$ terminates and is an equivalence transformation.*

Proof Rewriting φ reduces the size of the formulae to which the rewrite rules can be applied. Therefore ithe rewriting process terminates. Each rewrite is equivalence preserving (Lemma 18.54), and therefore the transformation is equivalence preserving. □

Proposition 18.56. *All world-paths occurring in the TF_1-translated $\mathcal{CL}^{\mathcal{M}}_{\mathcal{T}u}$-statements are determined.*

Proof $\mathcal{CL}^{\mathcal{M}}_{\mathcal{T}u}$-statements are of the form $[n, U] : \varphi$ or $\langle n, U \rangle : \varphi$. They are translated using (18.38) and (18.39). Both rewritings introduce a variable $x^{s_{U_{[[}(n)}}$ with $s_{U_{[[}(n)} \in \mathbb{S}_R$ as the initial world-path. All other rewritings append variables to this world-path. □

The TF_1-translated formulae are almost like predicate logic formulae. One difference is that the predicate $\in$ in the literals $\pi \in i$ has a special meaning. This special meaning can either be axiomatized in *PL1*, or be represented in the form of special inference rules. We shall follow the second and more efficient option.

The string syntax for world-paths has advantages. It allows for considerable further simplifications of TF_1-translated formulae. Most of them are well known from the functional translation of standard modal formulae. First of all, we notice that the set of world-paths occurring in TF_1-translated formulae has a characteristic syntactic property. Each variable x in a world-path π stems from one particular occurrence of a modal or temporal operator. This operator may occur nested within a particular list of other operators which also get translated into variables. These other variables occur in exactly the same order as π in front of x as the nesting of the corresponding operators. For example, the functional translation of the modal formula $\Diamond\Box(p \vee \Diamond q)$ is $ax : p \vee axb : q$. The variable x stems from the $\Box$-operator. Both occurrences of x, one in $ax : p$ and one in $axb : q$, have the same prefix a, because the $\Box$-operator occurs within the $\Diamond$-operator which got translated into a. a comes before the x in ax and axb because the $\Box$ is preceded by the $\Diamond$-operator.

This *prefix stability* or *unique prefix property* is the basis for non-trivial further simplifications.

Definition 18.57. In a world-path $\pi x \pi'$ we call π the *prefix* of x in this world-path.

A set WP of world-paths is *prefix stable* iff all variables occurring in WP have the same prefix. (This property is also called the *unique prefix property.*)

Proposition 18.58. *The set of world-paths occurring in* $TF_1(\varphi)$ *is prefix stable.*

Proof Formulae are essentially trees. A partially TF_1-rewritten formula corresponds to a tree where the top part consists of classical connectives and quantifiers, then come nodes of the form $\pi : \psi$, and the lower parts, i.e. the ψ-part, are $\mathcal{CL}^{\mathcal{M}}_{\mathcal{T}_\mathcal{U}}$-formula.

The proof is therefore done by induction on the depth of the occurrence of the $\pi : \psi$-nodes in the tree.

The base case, depth 0, represents the original formula. There is no world-path at all; therefore prefix stability holds vacuously. In the induction step we can assume that the prefix stability holds for all trees where the $\pi : \psi$-nodes occur at depth n or smaller (induction hypothesis). This means that all variables occurring in the world-paths $\pi : \psi$ in these trees have the same prefix. Except for the rules for the modal and temporal operators none of the rewrite rules change the world-paths. Therefore they preserve prefix stability. The rules for the modal operators do not change the position of previously introduced variables. The newly introduced variables occur at different positions, but they are appended to the same world-path. Therefore their prefix is the same. □

A consequence of the prefix stability is that prefix stable world-paths π are *linear.* All variables occur at most once in π.

Another, even more important consequence is the possibility to move existential quantifiers stemming from diamond operators in front of the universal quan-

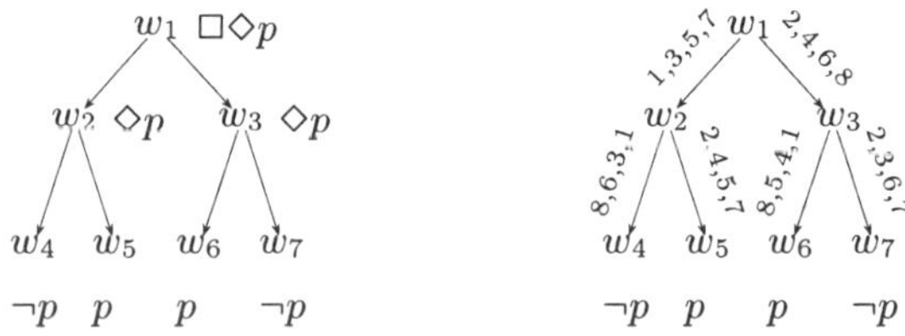

FIG. 18.3. Quantifier exchange in functional frames.

tifiers stemming from box operators. Skolemizing existential quantifiers which are not in the scope of universal quantifiers yields Skolem constants. Moving all existential quantifiers to the front and skolemizing them therefore leaves us with function-free translated $\mathcal{CL}^{\mathcal{M}}_{\mathcal{T}u}$-formulae. Their satisfiability is decidable because their Herbrand universe is finite. Moving the existential quantifiers to the front is therefore the first step in getting a decision procedure for the translated formulae.

Proposition 18.59 (Quantifier exchange). *Moving all existential quantifiers in $TF_1(\varphi)$ to the front is an equivalence-preserving transformation.*

The idea of the proof is very intuitive. It exploits the fact that all accessibility variables have the same prefix (Proposition 18.58).

As an example, consider the formula $\Box\Diamond p$ (in ordinary modal logic). Suppose $\Box\Diamond p$ is true in the world w_1 in the little frame depicted in the left-hand diagram of Fig. 18.3. In this model the relationally translated formula $\exists v \forall w\ R(w_1, w) \rightarrow R(w, v) \wedge v \models p$, in which we have swapped the existential quantifier $\exists v$ with the universal quantifier $\forall w$, is false.

But now consider the functional frame in the right-hand diagram. The numeric labels denote the possible accessibility functions. In the functional language we can express the fact that $\Box\Diamond p$ is true at w_1 by $\forall\gamma\exists\delta\ \delta(\gamma(w_1)) \models p$. For this model we can swap the $\exists\delta$ quantifier and the $\forall\gamma$ quantifier. $\exists\delta\forall\gamma\delta(\gamma(w_1)) \models p$ is also true at w_1, because the function 4 (as well as the function 5) maps the worlds w_2 and w_3 to a world where p holds. Moreover, regardless of which one of the worlds w_4, w_5, w_6 or w_7 makes p true, in this model there is always a function which maps w_2 and w_3 to the right worlds.

The details of the proof that moving existential quantifiers in front of universal quantifiers is, in this case, an equivalence-preserving transformation can be found in [Ohlbach and Schmidt, 1997]. It applies to the TF_1-translated $\mathcal{CL}^{\mathcal{M}}_{\mathcal{T}u}$-formulae with multimodal logic because the temporal operators get translated to accessibility functions in the same way as the modal operators.

Definition 18.60 (Translation and skolemization). Let $TF_2(\varphi)$ be the result of moving all existential quantifiers in $TF_1(\varphi)$ to the front, skolemizing them and deleting all the universal quantifiers.

Skolemizing existential quantifiers which are not in the scope of universal quantifiers yields Skolem constants. Therefore $TF_2(\varphi)$ has only Skolem constants and universally quantified variables, not explicit function symbols. Since skolemizing is only a satisfiability preserving transformation, TF_2 is also only satisfiability-preserving. This is sufficient for checking the satisfiability of $\mathcal{CL}^{\mathcal{M}}_{\mathcal{T}u}$-formulae and for proving theorems by refuting their negation.

18.4.4.1 *Substitution and unification* Unification is one of the key concepts in modern predicate logic inference systems. *Most general* unification of two terms t_1 and t_2 of literals involved in an inference operation is used to guide the instantiation of variables in such a way that too specific instances are avoided. For example, from $\forall x\ p(x,a) \to q(x)$ and $\forall y\ p(y,a)$ one could derive $q(a)$ by instantiating the two formulae to $p(a,a) \to q(a)$ and $p(a,a)$ and then using *modus ponens*. The *substitution* $\{x \mapsto a, y \mapsto a\}$ is a *unifier* for the two literals $p(x,a)$ and $p(y,a)$. A more general result can be obtained by instantiating y with x and then deriving $\forall x\ q(x)$. The unification algorithm for predicate logic terms generates in this case $\{y \mapsto x\}$ (or alternatively $\{x \mapsto y\}$), as the *most general* unifier for $p(y,a)$ and $p(x,a)$. Up to variable renaming there is at most one most general unifier, and the most general unifier can be computed in linear time [Paterson and Wegman, 1978].

The objects manipulated by most unification algorithms are *terms* and *substitutions*. In first-order predicate logic, terms are the usual term trees, and substitutions are represented as sets of pairs $\{x_1 \mapsto t_1, \ldots, x_n \mapsto t_n\}$. (As a mathematical object, a substitution is an endomorphism in the term algebra.) A substitution $\sigma = \{x_1 \mapsto t_1, \ldots, x_n \mapsto t_n\}$ is *applied to* a term t by replacing all occurrences of the variables x_i in t with t_i. For example, if $\sigma = \{x \mapsto f(y), z \mapsto g(a,y)\}$ and $t = g(x, g(z,x))$ then $t\sigma = g(f(y), g(g(a,y), f(y)))$.

The notions of substitution and unification have to be modified if terms carry more information than in standard first-order predicate logic. For example, in *sorted* first-order predicate logic, variables and terms carry *sort* information. If a variable x is of sort, say, $Integer$, and a term $sqrt(y)$ is of sort, say, $Real$, then instantiating x with $sqrt(y)$ may lead to false inferences. Therefore substitutions must be *well sorted.* A substitution pair $x \mapsto t$ is well sorted if the sort of t is equal to or smaller (in a subsort hierarchy) than the sort of x. Consequently, the unification algorithm for sorted logic has to avoid generating non-well- sorted substitutions by checking this condition, [Walther, 1987; Schmidt-Schauß, 1989].

Function symbols can also carry useful information for guiding inference systems. For example, if a given axiomatization contains the commutativity axiom $\forall x, y\ f(x,y) = f(y,x)$ then the information that the arguments of f can be exchanged can be attached to f directly. This allows the unification algorithm to unify $f(a,x)$ with $f(b,y)$, for example, where a and b are constants. The commutative unifier is $\{x \mapsto b, y \mapsto a\}$. For many equational theories corresponding (theory) unification algorithms have been developed [Kirchner, 1990]. Unification for some equational theories has similar properties as unification for free predicate logic. In the theory of Boolean rings, for example, there is also at most

enough, if f is commutative and associative, then there are again only finitely many most general unifiers [Stickel, 1981].

With respect to unification there are even worse theories than associativity. It is possible that there are infinite sequences of more and more general unifiers without ever reaching a most general unifier. Theories showing this kind of behaviour are said to be of *type 0*. An example is the theory of lists with the *append* function. If $first(x)$ denotes the first element of a list, and *nil* the empty list, then the unifiers for $first(x) = nil$ are

$$\begin{aligned}
\sigma_0 &= \{x \mapsto nil\} \\
\sigma_1 &= \{x \mapsto append(nil, x_1)\} \\
\sigma_2 &= \{x \mapsto append(append(nil, x_1), x_2)\} \\
&\vdots \\
\sigma_n &= \{x \mapsto append(\ldots(append(nil, x_1), x_2), \ldots, x_n)\}.
\end{aligned}$$

Every unifier σ_n can be obtained from σ_{n+1} by replacing x_n with *nil* and using $append(x, nil) = x$.

Unification algorithms for special theories are usually much more efficient than an axiomatic treatment of these theories. Therefore theory unification continues to be an active field of research.

18.4.4.2 *Unification of world-path for modal logics* The terms in functionally translated modal formulae are world-paths, which are strings of constant and variable symbols. A variable x denotes an accessibility function, i.e. a function which maps a given world to a world accessible in a single step. Substituting x with a constant a, which also denotes a function that maps a world to a world accessible in a single step, is therefore always an admissible operation. Substituting x with a longer world-path abc, which denotes a function mapping a world to a world accessible in three steps, is, therefore, only admissible if each world accessible in three steps can also be reached in one step, i.e. if the accessibility relation is transitive.

If the accessibility relation is reflexive, one can do a transition to an accessible world by just staying where one is. This means that a variable can be replaced with an empty world-path, provided empty world-paths are interpreted as identity functions.

The situation is a bit more tricky for symmetric accessibility relations. Symmetry means that each transition to an accessible world can be undone by just going back to where one came from. A world-path abx, for example, denotes a transition via a from a world w_1 to a world w_2, followed by a transition via b to a world w_3 and then a transition via x to a world w_4. Symmetry allows us to choose $w_4 = w_2$. Therefore there is an accessibility function available for x such that $abx = a$. It is convenient to use $\{x \mapsto b^{-1}\}$ and to assume $abb^{-1} = a$, although the accessibility functions need not be injective, and there need not be a unique inverse function. Therefore it is clearer to use a (quite unconventional) notation $\{abx \mapsto a\}$. The notations are in fact equivalent because, since world-paths are prefix stable (Proposition 18.58), all occurrences of x have the same prefix ab. Therefore whenever x is replaced with b^{-1}, the result is abb^{-1}, which can be simplified to a. A world-path acx, for example, which becomes acb^{-1} after replacing x, simply does not occur.

The kinds of substitutions admissible for an accessibility function variable depend on the properties of the modal logic. A general definition for propositional multimodal logics where the modality is encoded as sort information attached to variables and constants is:

Definition 18.61 (Admissible substitution). A substitution $x^{s_m} \mapsto \pi$ for a variable x^{s_m} with a world-path π is *admissible* for the modality m iff π denotes a function that maps worlds to m-accessible worlds.

Unification of world-paths for functionally translated propositional mono-modal logic KD (and K if the translation uses the *de*-predicate) is very simple. The algorithm goes through the two world-paths from left to right and compares the terms one by one. If one of them is a variable then the pair can be unified. If both are constant symbols then they must be equal, otherwise they cannot be unified. If the two world-paths have different lengths, there is no unifier either. For example, the two world-paths $xaby$ and $cazd$ have the unifier $\{x \mapsto c, z \mapsto b, y \mapsto d\}$. Since world-paths are linear (no variable occurs twice) and prefix stable there is no occurs-check problem as in standard unification.

The unification algorithm for other logics than K and KD exploit the fact that admissible substitutions can replace variables with world-paths which are shorter or longer than in the KD case. For example, in the logic T (reflexivity of the accessibility relation) $\{x \mapsto []\}$ is an admissible substitution. Therefore the T-unifier for axb and ab is $\{x \mapsto []\}$. If the accessibility relation is transitive then $axbc$ and $adey$ are unifiable and the unifier is $\{x \mapsto de, y \mapsto bc\}$. In the symmetric case, $abxc$ and ac are unifiable with $\{x \mapsto b^{-1}\}$.

For the purposes of this chapter it is sufficient to assume the existence of a suitable modal unification algorithm, which can be used for unifying two M-blocks. Details about unification of world-paths can be found in [Ohlbach, 1988a; 1988b; 1991; Schmitdt, 1997].

Definition 18.62 (Modal unification). In a multimodal system M a substitution σ is a *most general M-unifier* for two world-paths π_1 and π_2 iff

(i) $\pi_1\sigma = \pi_2\sigma$ (σ is a unifier),

(ii) $x^{s_m}\sigma$ is admissible for all variables x^{s_m},

(iii) all other unifiers λ for π_1 and π_2 are more special than σ, i.e. $\pi_1\lambda\sigma = \pi_1\lambda$ and $\pi_2\lambda\sigma = \pi_2\lambda$.

18.4.4.3 *Unification of world-path for calendar logic* The temporal accessibility variables and constants carry information of a different kind than the modal accessibility variables. For example, a variable x^{s_i} of sort s_i where i is a concrete time region (e.g. the year 2000) denotes a function mapping any time point to a

(iii) all other unifiers λ for π_1 and π_2 are more special than σ, i.e. $\pi_1\lambda\sigma = \pi_1\lambda$ and $\pi_2\lambda\sigma = \pi_2\lambda$.

18.4.4.3 *Unification of world-path for calendar logic* The temporal accessibility variables and constants carry information of a different kind than the modal accessibility variables. For example, a variable x^{s_i} of sort s_i where i is a concrete time region (e.g. the year 2000) denotes a function mapping any time point to a point within i. Therefore one can instantiate x^{s_i} with any world-path denoting a transition to a time point in i. This can be a constant a^{s_j} where $j \subseteq i$. It can also be a constant a^{s_j} where $j \subseteq i$ need not be true, but $a^{s_j} \in i$ can be derived from some other information. In this case $(j \cap i) \neq \emptyset$ must at least be true because otherwise $a^{s_j} \in i$, if it could be derived at all, would be false and therefore simplified to $\perp$. (Since i and j are concrete time regions, $j \subseteq i$ can, in fact, be computed and need not be deduced.)

Substituting x^{s_i} with a^{s_j} is admissible only if $a^{s_j} \in i$ can be derived, which may require a long proof. Therefore it is not possible to include the admissibility criterion directly in the unification algorithm. Instead of this we change the instantiation procedure. A substitution $\sigma = \{x^{s_i} \mapsto a^{s_j}\}$ causes the insertion of the condition (constraint) $a^{s_j} \in i$ into the instantiated formula. For example, $(x^{s_i} : p)\sigma$ becomes $a^{s_j} \in i \to a^{s_j} : p$. This way, there is no special algorithm needed for checking the admissibility of the substitution. It can be done by the general inference mechanism.

In a world-path cdx^{s_i} where the variable x occurs at the end or in the middle of a T-block, and c and d are other temporal accessibility terms, the whole T-block abx^{s_i} denotes a function mapping any time point into the time region i. The c and d are not relevant from this point of view. Thus, cdx^{s_i} can be replaced with a^{s_j}, provided $a^{s_j} \in i$ can be proved. cdx^{s_i} can also be replaced with a longer world-path π, provided $\pi \in i$ can be proved. A more general format for a substitution is therefore $\sigma = \{cdx^{s_i} \mapsto \pi\}$. The instantiation $(cdx^{s_i} : p)\sigma$ is $\pi \in i \to \pi : p$.

Proposition 18.63 (Instantiation of $\mathbb{S}_R$-variables). *If* $\pi_{[[} \neq \emptyset$ *and* $s_i \in \mathbb{S}_R$ *then*

$$(\vec{\forall}\forall x^{s_i}\ \varphi\lceil \pi x \rfloor) \to \vec{\forall}(\pi' \in i \to \varphi\lceil \pi' \rfloor)$$

is an $\mathcal{LCL}^{\mathcal{M}}_{\mathcal{T}\mathcal{U}}$*-theorem.*[30]

Proof Suppose $\Im, w, t \models \forall x^{s_i}\ \varphi\lceil \pi x \rfloor$.

Suppose $\Im', w, t \models \pi' \in i$ where $\Im' = \Im[x_1/\gamma_1, \ldots, x_n/\gamma_n]$, x_i are the free variables in π' and the γ_i some assignments.

$\succ \Im'(\pi')(w,t) \in i$

$\succ \exists \gamma \in AF^{\Im}_{s_i}$ with $\Im[x/\gamma](\pi x)(w,t) = \Im'(\pi')(w,t)$. Since $\Im[x/\gamma], w, t \models \varphi\lceil \pi x \rfloor$ we conclude by structural induction $\Im', w, t \models \varphi\lceil \pi' \rfloor$. □

[30]The notation $\varphi\lceil \pi \rfloor$ means that π is an initial segment of world-paths occurring in φ. The notation $\vec{\forall}\ \varphi$ denotes the universal closure of all free variables occurring in φ.

The situation is more difficult for variables x^{s_τ} where τ is a time term. In this case x denotes a temporal accessibility function which maps time points t to time points $\tau_{[[}(t)$. It is possible to replace $\mathbb{S}_T$-variables by $\mathbb{S}_R$-variables right from the beginning, and to shift all the problems with $\mathbb{S}_T$-variables back to the $\mathbb{S}_R$-variables:

Proposition 18.64 (Decomposition of $\mathbb{S}_T$-variables). *For every $\mathbb{S}_T$-variable s_τ and world-path π*

$$\vec{\forall}\varphi\lceil\pi x^{s_\tau}\rfloor \Leftrightarrow \vec{\forall} \bigwedge_{i\in\delta(\pi_{[[},\tau),\tau_{[[}(i)\neq\emptyset} (\pi \in i \rightarrow \forall y^{s_{\tau_{[[}(i)}}\ \varphi\lceil\pi y\rfloor)$$

is an $\mathcal{LCL}^{\mathcal{M}}_{\mathcal{T}_\mathcal{U}}$-theorem. Notice that if $\pi_{[[} = \emptyset$ or $\tau_{[[}(\pi_{[[}) = \emptyset$ then the right-hand-side formula becomes $\top$.

Proof '(⇒)' Suppose $\Im, w, t \models \varphi\lceil\pi x^{s_\tau}\rfloor$ ($\Im$ contains assignments for the variables occurring in π).

Suppose further, $\Im, w, t \models \pi \in i$ for some $i \in \delta(\pi_{[[}, \tau)$.

Let $\gamma \in AF^{\Im}_{s_{\tau_{[[}(i)}}$.

$\succ \gamma(\Im(\pi)(w,t))_{|T} \in \tau_{[[}(i)$.

$\succ \exists\nu \in AF^{\Im}_{s_\tau}\ \nu(\Im(\pi)(w,t)) = \gamma(\Im(\pi)(w,t))$.

$\succ \Im[x/\nu](\pi x)(w,t) = \Im[y/\gamma](\pi y)(w,t)$.

$\succ \Im[x/\nu], w, t \models \varphi\lceil\pi x\rfloor$ iff $\Im[y/\gamma], w, t \models \varphi\lceil\pi y\rfloor$. (by structural induction)

$\succ \Im[y/\gamma], w, t \models \varphi\lceil\pi y\rfloor$. (since $\Im, w, t \models \varphi\lceil\pi x^{s_\tau}\rfloor$)

$\succ \Im, w, t \models \forall y^{s_{\tau_{[[}(i)}}\ \varphi\lceil\pi y\rfloor$.

$\succ \Im, w, t \models \bigwedge_{i\in\delta(\pi_{[[},\tau)}(\pi \in i \rightarrow \forall y^{s_{\tau_{[[}(i)}}\ \varphi\lceil\pi y\rfloor)$.

('⇐') Suppose $\Im, w, t \models \forall \bigwedge_{i\in\delta(\pi_{[[},\tau)}(\pi \in i \rightarrow \forall y^{s_{\tau_{[[}(i)}}\ \varphi\lceil\pi y\rfloor)$.

$\Im(\pi)(w,t) \in i$ for some $i \in \delta(\pi_{[[}, \tau)$. (Proposition 18.53(a))

If $\tau_{[[}(i) = \emptyset$ then $\Im, w, t \models \varphi\lceil\pi y\rfloor$ iff $\Im, w, t \models \varphi\lceil\pi y : \bot\rfloor$ and therefore $\Im, w, t \models \varphi\lceil\pi x^{s_\tau}\rfloor$.

The proof for the case $\tau_{[[}(i) \neq \emptyset$ is just by reversing the arguments of the '→' case. □

This decomposition might be useful if the number of generated conjunctions is small, in particular it this number is just one. Then $\pi \in i$ becomes automatically $\top$ and the transformation can be simplified further:

Corollary 18.65 (Decomposition of $\mathbb{S}_T$-variables). *For every $\mathbb{S}_T$-variable s_τ and world-path π with $\delta(\pi_{[[}, \tau) = \pi_{[[}$*

$$\vec{\forall}\varphi\lceil\pi x^{s_\tau}\rfloor \Leftrightarrow \vec{\forall}\forall y^{s_{\tau_{[[}(i)}}\ \varphi\lceil\pi y\rfloor$$

is an $\mathcal{LCL}^{\mathcal{M}}_{\mathcal{T}_\mathcal{U}}$-theorem.

The decomposition can, however, also become quite large, in which case this transformation is not advisable. In this case $\mathbb{S}_T$-variables have to be treated in the inference procedure.

Variables x^{s_τ} or longer world-paths $\pi_1 x^{s_\tau}$ can also be instantiated with some world-path π_2, but the admissibility condition to be proved is more complex. The condition has to guarantee that if π_1 maps the initial time point to some time point t then π_2 must map the initial time point to $\tau_{[[}(t)$. The condition which turns out to be necessary and sufficient is that for all $i \in \delta(\pi_{1[[}, \tau)$, $\pi_2 \in \tau_{[[}(i)$ must hold. For $\sigma = \{\pi_1 x^{s_\tau} \mapsto \pi_2\}$ the instantiation $(\pi_1 x^{s_\tau} : p)\sigma$ is then

$$\left(\bigwedge_{i \in \delta(\pi_{1[[}, \tau)} \pi_1 \in i \to \pi_2 \in \tau_{[[}(i) \right) \to \pi_2 : p.$$

As an example, suppose somebody made the statement in January this year: 'there was a weekend in this month when it was raining all the time'. The $\mathcal{LCL}^{\mathcal{M}}_{\mathcal{TU}}$-encoding is $\langle 1, Month \rangle : [weekend(y_{week})]raining$. If we count the reference time in days such that the first of January is time point 1, and January is month 1, the translation is

$$a^{s[1,31[} x^{s_{weekend(y_{week})}} : raining.$$

The world-path $a^{s[1,31[} x^{s_{weekendy_{week}}}$ can be replaced with another world-path $\pi_2 = c^{s[1,365[} d^{s_{Saturday(z_{week})}}$ (which might come from a statement like 'Saturday I bought an umbrella', which was made sometime this year). The instantiation of the formula yields

$$\begin{aligned} &(a \in [1,7[\ \to \pi_2 \in \{[6,6[) \wedge \\ &(a \in [8,14[\ \to \pi_2 \in \{[13,13[) \wedge \\ &(a \in [15,21[\ \to \pi_2 \in \{[20,20[) \wedge \\ &(a \in [22,28[\ \to \pi_2 \in \{[27,27[)) \\ \to \ &\pi_2 : raining. \end{aligned}$$

In order to conclude $\pi_2 : raining$, i.e. it was raining this particular Saturday when I bought the umbrella, one has to check for each week in January where the first statement could have been made—that the Saturday in *this* week was meant.

The functional translation of more general $\mathcal{CL}^{\mathcal{M}}_{\mathcal{TU}}$-statements yields world-paths consisting of sequences of T-blocks and M-blocks. Temporal variables occurring in some T-block preceded by some M- and T-blocks can be instantiated in a similar way. The only difference is that the conditions generated by the instantiation must always contain the whole world-path up to the T-block containing the variable. For example, a world-path $\pi_T \pi_M abx^{s_i}$, where π_T is a T-block, π_M is an M-block and a and b are temporal terms, can be instantiated with a T-block π, i.e. $\sigma = \{abx^{s_i} \mapsto \pi\}$. The generated condition is $\pi_T \pi_M \pi \in i$, and not just $\pi \in i$ as in the simple examples above. Only the whole world-path determines the correct time for $\pi_T \pi_M \pi$ which must lie in i, and not the π fragment.

Since temporal variables can be instantiated with arbitrary T-blocks, unification of two T-blocks π_1 and π_2 becomes almost trivial. If the rightmost term in π_1 (or π_2) is a variable x then $x \mapsto \pi_2$ (or $x \mapsto \pi_1$ respectively) is a unifier. If the rightmost terms in both T-blocks consist of constants, then they can only be unified if the trailing sequence after the rightmost variable on one of the terms is identical to the corresponding trailing sequence in the other world-path. The variable is then instantiated with the corresponding initial sequence of the other world-path.

The basic idea for the world-path unification algorithm is now to unify T-block pairs and M-block pairs one by one, from left to right. For the M-blocks we use the corresponding unification algorithm for the basic multimodal logic, and for the T-blocks the one sketched above. There are, however, two special cases to be considered. If the modal unification is such that world-paths can be unified with the empty world-paths (e.g. in the modal logics T and B), then two T-blocks separated by an M-block can merge into one M-block.

Conversely, it may also be possible that a T-block can be unified with the empty world-path. If, for example, at some time point t *today* the statement 'the weather is nice all the day' is true, then 'the weather is nice' is true particularly at the time point t. Therefore the variable ranging over the interval corresponding to 'all the day' can be instantiated with the empty world-path. Instantiating a whole T-block with the empty world-path merges two M-blocks together. We provide special rules for each of these cases. The rule for unifying a T-block with an empty world-path is justified by the proposition below.

Proposition 18.66 (T-block elimination). *If* $\pi_{[[} \subseteq i$ *and* π' *is a T-block then*

$$(\vec{\forall}\forall x^{s_i}\ \varphi\lceil\pi\pi' x\rfloor) \to \vec{\forall}\varphi\lceil\pi\rfloor.$$

Proof Suppose $\pi_{[[} \subseteq i$ (which means $AF^{\Im}_{s_{\pi_{[[}}} \subseteq AF^{\Im}_{s_i}$) and $\Im, w, t \models \forall x^{s_i}$ $\varphi\lceil\pi\pi' x\rfloor)$.
$\succ \forall\gamma \in AF^{\Im}_{s_i}\ \Im[x/\gamma], w, t \models \varphi\lceil\pi\pi' x^{s_i}\rfloor$
$\succ \forall\gamma \in AF^{\Im}_{s_{\pi_{[[}}}\ \Im[x/\gamma], w, t \models \varphi\lceil\pi\pi' x^{s_i}\rfloor$.
Let $\gamma_0(\Im(\pi\pi')(w,t)) = \Im(\pi)(w,t)$.
$\succ \gamma_0 \in AF^{\Im}_{s_{\pi_{[[}}}$. (since $\Im(\pi)(w,t)_{|T} \in \pi_{[[}$ (Proposition 18.53))
$\succ \Im[x/\gamma_0], w, t \models \varphi\lceil\pi\pi' x\rfloor$.
$\succ \Im, w, t \models \varphi\lceil\pi\rfloor$. (since $\Im[x/\gamma_0](\pi x)(w,t) = \Im(\pi)(w,t)$).
□

18.4.4.4 *Substitution and unification: the technical details* The notion of substitution introduced below is slightly refined compared with the informal introduction above. In order to cover as many cases of modal unification as possible, modal unifiers are also allowed to replace longer world-paths ending in a variable with some other world-path. For example, in the modal logic B-case (symmetry of the accessibility relation) a substitution $ax \mapsto []$ is allowed and preferred to $x \mapsto a^{-1}$.

Instantiation with substitutions where several longer world-paths are replaced with other longer world-paths depend on the order of the replacements. Consider

with h, or just $efgy$ if the second replacement is tried first, which changes nothing. Since world-paths are unified from left to right, unification yields a natural ordering of the replacements. Therefore substitutions are represented as *ordered lists*, and not as sets as usual.

Definition 18.67 (Substitution). A *substitution pair* is a tuple $\pi_1 \mapsto \pi_2$ where either π_1 and π_2 are both either T-blocks or M-blocks or, π_2 is empty and π_1 ends with a variable. If π_1 and π_2 are M-blocks then $\pi_1 \mapsto \pi_2$ must be an *admissible modal substitution* (Definition 18.61). A *ground substitution pair* is a substitution pair $\pi_1 \mapsto \pi_2$ where π_2 does not contain variables. A *substitution* is a finite ordered list of substitution pairs. A *ground substitution* is a finite ordered list of ground substitution pairs.

Definition 18.68 (Instantiation). The instantiation of world-paths π with a *substitution pair* $\pi_1 \mapsto \pi_2$ is defined as

$$\pi(\pi_1 \mapsto \pi_2) \stackrel{\text{def}}{=} \begin{cases} \pi_0\pi_2\pi_3 & \text{if } \pi = \pi_0\pi_1\pi_3 \\ \pi & \text{otherwise.} \end{cases}$$

The instantiation of world-paths π with a *substitution* $\sigma = (\sigma_1, \ldots, \sigma_n)$ is

$$\pi\sigma \stackrel{\text{def}}{=} (\ldots((\pi\sigma_1)\sigma_2)\ldots\sigma_n).$$

The instantiation of a formula φ with a substitution pair $\pi_1 \mapsto \pi_2$ is defined as application of the following rules:

$$\begin{aligned}
\varphi\lceil\pi_0\pi x^{s_m}\rfloor(\pi x^{s_m} \mapsto \pi_2) &= \begin{cases} \pi_0\pi x \notin \emptyset \to \varphi\lceil\pi_0\pi_2\rfloor & \text{if } m \text{ is non-serial} \\ \varphi\lceil\pi_0\pi_2\rfloor & \text{otherwise} \end{cases} \\
\varphi\lceil\pi_0\pi x^{s_i}\rfloor(\pi x^{s_i} \mapsto \pi_2) &= \pi_0\pi_2 \in i \to \varphi\lceil\pi_0\pi_2\rfloor \\
\varphi\lceil\pi_0\pi x^{s_\tau}\rfloor(\pi x^{s_\tau} \mapsto \pi_2) &= \left(\bigwedge_{i\in\delta((\pi_0\pi)_{[[},\tau)} (\pi_0\pi \in i \to \pi_0\pi_2 \in \tau_{[[}(i))\right) \to \varphi\lceil\pi_0\pi_2\rfloor.
\end{aligned}$$

$\varphi\lceil\pi_0\pi_2\rfloor$ is a *direct instance*, whereas the literals in the premises of the implication are *constraint literals* or *residue literals*.

The instantiation of a formula φ with a *substitution* $\sigma = (\sigma_1, \ldots, \sigma_n)$ is

$$\varphi\sigma \stackrel{\text{def}}{=} (\ldots((\varphi\sigma_1)\sigma_2)\ldots\sigma_n).$$

Notice that constraint literals generated by the instantiation with σ_1 can be instantiated with σ_l, $l \geq 2$, which in turn may produce more constraint literals.

Lemma 18.69 (Soundness of instantiation). *For every closed $\mathcal{LCL}_{\mathcal{Tu}}^{\mathcal{M}}$-formula $\vec{\forall}\,\varphi$, substitution σ and interpretation $\Im$. if $\Im, w, t \models \vec{\forall}\,\varphi$ then $\Im, w, t \models \vec{\forall}\,(\varphi\sigma)$.*

Proof By induction on the number of substitution pairs in σ.

Suppose $\Im, w, t \models \vec{\forall}\,\varphi$.

Proof By induction on the number of substitution pairs in σ.

Suppose $\Im, w, t \models \vec{\forall}\, \varphi$.

The base case, σ is empty, is trivial because $\varphi\sigma = \varphi$. For the induction step we consider the three different cases in the definition of $\varphi(\pi_1 \mapsto \pi_2)$.

Case $\sigma = (\pi x^{s_m} \mapsto \pi_2)$: σ is an admissible modal substitution. Therefore the soundness follows from the soundness of modal instantiation.

Case $\sigma = (\pi x^{s_i} \mapsto \pi_2)$: $\varphi\lceil\pi_0\pi x\rfloor\sigma = (\pi_0\pi_2 \in i \to \varphi\lceil\pi_0\pi_2\rfloor)$.

Let $\Im' = \Im[x_1/\gamma_1, \ldots, x_k/\gamma_k]$ where $x_1, \ldots x_k$ are the variables occurring in $\varphi\sigma$. Suppose $\Im', w, t \models \pi_0\pi_2 \in i$. Then $\Im'(\pi_0\pi_2)(w,t) \stackrel{\text{def}}{=} (w', t')$ and $t' \in i$.

Let $\Im'' = \Im[x_1/\gamma_1, \ldots, x_k/\gamma_k, x^{s_i}/\gamma]$ with $\gamma \in AF^{\Im}_{s_i}$ and $\gamma(\Im''(\pi_0\pi)(w,t)) = (w', t')$. Such a γ exists because $t' \in i$ and πx^{s_i} and π_2 are T-blocks, which do not cause a transition to a different world. Therefore the world w' is determined by π_0. Since $\Im, w, t \models \vec{\forall}\varphi$, and the interpretation of $\pi_0\pi_2$ is the same as the interpretation of $\pi_0\pi x^{s_i}$, we find (by structural induction) that $\Im'', w, t \models \varphi\lceil\pi_0\pi_2\rfloor$. Since x^{s_i} does not occur in $\varphi\lceil\pi_0\pi_2\rfloor$ we get $\Im', w, t \models \varphi\lceil\pi_0\pi_2\rfloor$. Thus, $\Im, w, t \models \vec{\forall}\, \varphi\sigma$.

Case $\sigma = (\pi x^{s_\tau} \mapsto \pi_2)$:

$$\varphi\lceil\pi_0\pi x\rfloor\sigma = \left(\bigwedge_i \in \delta((\pi_0\pi)_{[[}, \tau)(\pi_0\pi \in i \to \pi_0\pi_2 \in \tau_{[[}(i)\right) \to \varphi\lceil\pi_0\pi_2\rfloor.$$

Let $\Im' = \Im[x_1/\gamma_1, \ldots, x_k/\gamma_k]$ where $x_1, \ldots, x_k$ are the variables occurring in $\varphi\sigma$. Suppose $\Im', w, t \models \pi_0\pi \in i \to \pi_0\pi_2 \in \tau_{[[}(i_{[[})$ for all $i \in \delta((\pi_0\pi)_{[[}, \tau)$. Whenever $\Im'(\pi_0\pi)(w,t) \stackrel{\text{def}}{=} (w', t')$ with $t' \in i$, then $\Im'(\pi_0\pi_2)(w,t) \stackrel{\text{def}}{=} (w', t'')$ with $t'' \in \tau_{[[}(i)$. (π and π_2 are both M-blocks and therefore do not indicate a jump to a different world.)

Let $\Im'' = \Im[x_1/\gamma_1, \ldots, x_k/\gamma_k, x^{s_i}/\gamma]$ with $\gamma \in AF^{\Im}_{s_\tau}$ and $\gamma(\Im''(\pi_0\pi)(w,t)) = (w', t')$. Such a γ exists because $t' \in i$ implies $t'' \in \tau_{[[}(i)$ and $\tau_{[[}$ is constant over i. The interpretations of $\pi_0\pi x^{s_\tau}$ and $\pi_0\pi_2$ are now the same. Therefore the same arguments as in the above case prove this case as well. □

The instantiation may generate conditions which can never be fulfilled. Consider, for example, the statement 'yesterday it was raining all the time'. This statement generates a temporal variable x^{s_i} where i represents the time interval of yesterday. According to the above definition, x can be instantiated with an accessibility constant a^{s_j} where j may be a time interval representing some time next year. Therefore the condition $a^{s_j} \in i$ can never be fulfilled. The instantiation is not false, because a false condition in an implication makes a formula true. But the instantiated formula is a useless tautology. In fact, it can easily be checked whether $a^{s_j} \in i$ can ever be satisfied. Since i and j are concrete time regions, one can just check whether $(i \cap j) \neq \emptyset$. If indeed $(i \cap j) = \emptyset$ then the instantiated formula can be dropped as useless.

In order to simplify the definitions and proofs, these optimizations are not included in the first version of the unification algorithm described below. For an implementation, however, it is important to test these conditions as early as possible. Their test is cheap, and it may reduce the number of useless derivations considerably.

Definition 18.70 (Unifier). A *unifier* for two world-paths π_1 and π_2 is a substitution λ with $\pi_1\lambda = \pi_2\lambda$ and the substitution pairs in λ are ordered from left to right with respect to π_1 and π_2. That means the substitution pairs $\pi_n x_n \mapsto \pi'_n$ in λ come in the same partial order as the variables x_n in π_1 and π_2.

A unifier σ for π_1 and π_2 is *most general* iff for all other unifiers λ for π_1 and π_2, $\pi_1\lambda = \pi_1\sigma\lambda' = \pi_2\lambda = \pi_2\sigma\lambda'$ for an appropriate substitution λ'.

Definition 18.71 (Unification). The algorithm for computing a most general unifier for two world-paths π_1 and π_2 is defined as a non-deterministic rule system. Starting with the tuple $\langle \pi_1 = \pi_2 \mid ()\rangle$ it unifies the world-paths passing them from left to right. Since the rules are non-deterministic, it generates a tree whose nodes are labelled with $\langle \pi'_1 = \pi'_2 \mid \sigma\rangle$ where $\pi'_1 = \pi'_2$ is the remaining unification problem, and σ is the already computed part of the unifier. The *solution leaf nodes* are nodes labelled with $\langle [] = [] \mid \sigma\rangle$. The *failure leaf nodes* are non-solution leaf nodes where no further rule is applicable. The rules are:

(eq) $\langle \pi_1\pi_2 = \pi_1\pi_3 \mid \sigma\rangle \mapsto \langle \pi_2 = \pi_3 \mid \sigma\rangle$;

$(\tau\tau)$ $\langle \pi_1 x^s \pi_2 = \pi_3\pi_4 \mid \sigma\rangle \mapsto \langle \pi_2 = \pi_4 \mid (\sigma, \pi_1 x^s \mapsto \pi_3)\rangle$
if
- (i) $s \in \mathbb{S}_R \cup \mathbb{S}_T$ is a temporal sort,
- (ii) $\pi_1 x^s$ and π_3 are initial segments of T-blocks,
- (iii) x^s is the rightmost variable in its T-block,
- (iv) the final segments of the T-blocks after x^s and π_3 are identical;

$(\tau[])$ $\langle \pi_1 x^s \pi_2 = \pi_3 \mid \sigma\rangle \mapsto \langle \pi_2 = \pi_3 \mid (\sigma, \pi_1 x^s \mapsto [])\rangle$
if
- (i) $s \in \mathbb{S}_R \cup \mathbb{S}_T$ is a temporal sort,
- (ii) $\pi_1 x^s$ is a T-block and π_2 and π_3 start with an M-block;

(mm) $\langle \pi_1\pi_2 = \pi_3\pi_4 \mid \sigma\rangle \mapsto \langle \pi_2 = \pi_4 \mid (\sigma, \rho)\rangle$
if
- (i) π_1 and π_3 are M-blocks, π_2 and π_4 are empty or T-blocks,
- (ii) $\pi_1\rho = \pi_3\rho$, i.e. ρ is a most general modal unifier for π_1 and π_2;

$(m[])$ $\langle \pi_1\pi_2 = \pi_3 \mid \sigma\rangle \mapsto \langle \pi_2 = \pi_3 \mid (\sigma, \rho)\rangle$
if
- (i) π_1 is an M-block, π_2 is empty or a T-block,
- (ii) $\pi_1\rho = []$, i.e. ρ is a most general modal unifier for π_1 and $[]$;

plus the corresponding symmetric rules.

Let $mgu(\pi_1, \pi_2)$ be the set of substitutions of the solution leaf nodes.

The above unification algorithm reduces the size of the remaining world-paths yet to be unified in each step. It therefore terminates, provided the number of most general modal unifiers for M-blocks is finite, which is the case for all known modal unification algorithms. In the next theorem we show that if the world-paths are unifiable, then it computes unifiers, and they are most general.

Theorem 18.72 (Soundness of the unification algorithm). *The set $mgu(\pi_1, \pi_2)$ as in Definition 18.71 is the set of most general unifiers for π_1 and π_2.*

Proof (i) Every $\sigma \in mgu(\pi_1, \pi_2)$ is a unifier for π_1 and π_2: this can be shown in a straightforward way by induction on the number of rule applications needed to compute σ.

(ii) For every unifier λ for π_1 and π_2 there is a unifier $\sigma \in mgu(\pi_1, \pi_2)$ which is more general than λ:

The substitution pairs in λ are ordered according to the variables occurring in π_1 and π_2. By induction on the position of variables replaced by λ we can find a σ and construct a λ' with $\pi_1\lambda = \pi_1\sigma\lambda' = \pi_2\lambda = \pi_2\sigma\lambda'$.

If x is the leftmost variable replaced by λ then the situation must look like:
$\pi_1 = \pi'_1\pi'_2x\pi'_3\pi'_4$
$\pi_2 = \pi'_1\pi'_5$ (or π_1 and π_2 exchanged),
where $\pi'_2x\pi'_3$ is an M-block or a T-block, π'_1 and π'_4 are empty or blocks of different type. Let $\pi'_2x\lambda \stackrel{\text{def}}{=} \pi$.

Case (i): $\pi'_2x\pi'_3$ is an M-block.
Case (ia): π'_5 starts with a T-block. Then $(\pi'_2x\pi'_3)\lambda = []$. That means $(\pi'_2x\pi'_3)$ unifies with the empty world-path. Since modal unification is assumed to compute most general unifiers, the $(m[])$-rule generates a more general unifier for $(\pi'_2x\pi'_3) = []$ than λ.

Case (ib): π'_5 starts with an M-block. This time the (mm)-rule generates a more general modal unifier than λ (again assuming that modal unification computes most general unifiers).

Case (ii): $\pi'_2x\pi'_3$ is a T-block.
Case (iia): π'_5 starts with an M-block. Then $(\pi'_2x\pi'_3)\lambda = []$. Thus, the last element in π'_3 must be a variable. In this case the $(\tau[])$-rule generates the component $\sigma = (\pi'_2x\pi'_3 \mapsto [])$. Thus, $(\pi'_2x\pi'_3)\lambda = (\pi'_2x\pi'_3)\sigma\lambda$, i.e. σ is a more general component than the relevant part of λ.

Case (iib): π'_5 starts with a T-block, i.e. $\pi'_5 = \pi'_6\pi'_7$ and π'_6 is a T-block. Since λ unifies π_1 and π_2, we have $(\pi'_2x\pi'_3)\lambda = \pi'_6\lambda$. Let y be the rightmost variable in either $\pi'_2x\pi'_3$ or π'_6. W.l.o.g assume $\pi'_6 = \pi'_8y$. Then the $(\tau\tau)$-rule generates a substitution $\pi'_6 \mapsto \pi'_2x\pi'_3$. $\pi'_6\lambda = (\pi'_2x\pi'_3)\lambda = \pi'_6\sigma\lambda$. Thus, σ is a more general component than the relevant part of λ.

Notice that the construction of σ is such that for T-blocks π,

$$\pi\lambda = \pi\sigma\lambda. \tag{18.50}$$

□

The fact that a unification algorithm computes most general unifiers is important for the completeness of all inference systems which use unification to control the instantiation of variables. Unfortunately, in our case this property is not sufficient. Since the instantiation with unifiers generates conditions, one must make sure that the instantiation with a most general unifier also generates the most general conditions. In concrete terms, this means that whenever a formula φ is instantiated with a most general unifier σ, the conditions c obtained from σ can be further instantiated to obtain the conditions c' for a less general unifier λ. The conditions c are *more general* than the conditions c' they *subsume* c'.

Lemma 18.73. *If λ is a unifier for two world-paths π_1 and π_2 and σ is a corresponding most general unifier (which exists according to Theorem 18.72) then $\varphi\sigma$ subsumes $\varphi\lambda$ for every formula φ containing π_1 and/or π_2.*

Proof Owing to the prefix stability of the world-paths, different blocks are independent of each other. For the M-blocks we assume that the property is guaranteed by the corresponding modal unification algorithm.

We therefore show the property for the case when π_1 and π_2 consist of single T-blocks. The property follows then by repeating the argument for each pair of T-blocks, and by using the fact that modal unification yields most general unifiers and no constraint literals.

Case (i): $\pi_1 = \pi_1' x^{s_i}$ and $s_i \in \mathbb{S}_R$.

$\succ \sigma = \pi_1' x \mapsto \pi_2$.

$\succ \varphi[\pi_1' x]\sigma = (\pi_2 \in i \to \varphi[\pi_2])$.

$\lambda = (\pi_1' x \mapsto \pi_2\lambda, \ldots)$ (because $\pi_1\lambda = \pi_2\lambda$)

$\succ \varphi[\pi_1' x]\lambda = (\alpha \wedge (\pi_2\lambda \in i)) \to \varphi[\pi_2\lambda]$

where α are some other constraints introduced by the instantiation with λ.

$\succ \varphi[\pi_1' x]\lambda = (\alpha \wedge \pi_2\sigma\lambda \in i) \to \varphi[\pi_2\sigma\lambda]$. (Theorem 18.72)

$\succ \varphi[\pi_1' x]\lambda = \varphi\sigma\lambda$.

Case (ii) $\pi_1 = \pi_1' x^{s_\tau}$ and $s_\tau \in \mathbb{S}_T$.

$\succ \sigma = \pi_1' x \mapsto \pi_2$.

$\succ \varphi[\pi_1' x]\sigma = (\bigwedge_{\ldots}(\pi_1' \in i \wedge \pi_2 \in \tau_{[[}(i))) \to \varphi[\pi_2]$.

$\lambda = (\pi_1' x \mapsto \pi_2\lambda, \ldots)$

$\succ \varphi[\pi_1' x]\lambda = (\alpha \wedge \bigwedge_{\ldots}(\pi_1'\lambda \in i \wedge \pi_2\lambda \in \tau_{[[}(i))) \to \varphi[\pi_2\lambda]$

where α are some other constraints introduced by the instantiation with λ.

$\succ \varphi[\pi_1' x]\lambda = ((\alpha \wedge \bigvee_{\ldots}(\pi_1'\sigma\lambda \in i \wedge \pi_2\sigma\lambda \in \tau_{[[}(i))) \to \varphi[\pi_2\sigma\lambda])$. (Theorem 18.72)

$\succ \varphi[\pi_1' x]\lambda = C\sigma\lambda$.

The T-blocks π_1 and π_2 may contain some common ground world-path π'' following $\pi_1' x$ and π_2. But this does not play any role for the unifiers involved.

□

As already mentioned, instantiation may generate conditions which are impossible to fulfil. One can use the temporal sorts as a rich source of information about the time regions involved to recognize and simplify many obvious cases.

Proposition 18.74 (Elementary tautologies). *If literals $\pi \notin i$ (with $i \neq \emptyset$) are normalized to $\pi \in \pi_{[[} \setminus i$ then a normalized literal L is a tautology iff L has the form:*

1. $[] \notin \emptyset$ *or*
2. $\pi \in \emptyset$ *with* $\pi \neq []$ *and* $\pi_{[[} = \emptyset$ *or*
3. $\pi \notin \emptyset$ *with* $\pi \neq []$ *and* $\pi_{[[} \neq \emptyset$ *and* π *do not contain modal accessibility terms corresponding to a non-serial accessibility relation, or*
4. $\pi \in i$ *with* $\pi \neq []$, $i \neq \emptyset$, *and* $\pi_{[[} \subseteq i$ *and* π *do not contain modal accessibility terms corresponding to a non-serial accessibility relation.*

These literals are called elementary tautologies. Their negations are elementary contradictions.

Proof Using Proposition 18.53, it is straightforward to check that all these literals are satisfied in all $\mathcal{LCL}_{\mathcal{T}u}^{\mathcal{M}}$-models.

By checking what syntactic forms literals can have, one can also prove that these are the only literals which are true in all models. □

Proposition 18.75 (Simplification rules). *The following equivalences hold for the TF_2-translated formulae:*

(a)	$L \Leftrightarrow \top$	*if L is an elementary tautology*
(b)	$L \Leftrightarrow \bot$	*if L is an elementary contradiction*
(c)	$\pi \in i \Leftrightarrow \pi \in (\pi_{[[} \cap i)$	*if* $i \neq \emptyset$
(d)	$\pi \notin i \Leftrightarrow \pi \in (\pi_{[[} \setminus i)$	*if* $i \neq \emptyset$
(e)	$\pi x^s \in \emptyset \Leftrightarrow \pi \in \emptyset$	*if* $s \in \mathbb{S}_R \cup \mathbb{S}_T$ *and* $\pi x_{[[} \neq \emptyset$ *or* $s = s^m \in \mathbb{S}_M$ *and m is a serial modality.*

If $CNF(\varphi)$ is defined as the result of applying the above equivalences as simplification rules, doing the standard propositional simplifications and converting the result into conjunctive normal form (clause form) then this is an equivalence-preserving simplification.

Proof These equivalences are all consequences of Propositions 18.74 and 18.53 ($\pi_{[[}$ is an exact approximation) and the fact that modal accessibility functions do not change the time and temporal accessibility functions do not change the world. □

The above simplification rules can be applied in particular to the conditions generated by the instantiation with most general unifiers. If they rewrite the conditions to $\bot$ then the instantiated formula becomes a useless tautology. It is therefore much more efficient to prevent the generation of these useless conditions in the unification algorithm. The unification rules in the definition below are refinements of the original rules in Definition 18.71. They produce the same kind of unifiers, but they contain extra checks to avoid the generation of unsatisfiable conditions.

Definition 18.76 (Optimized unification rules). The (*eq*)-rule and the modal rules (*mm*) and (*m*[]) are essentially the same as before. The temporal rules are refined by checking whether the condition associated with a substitution pair $\pi x \mapsto \pi'$ can be fulfilled. In order to compute the correct time information, it is also necessary to keep the previously unified initial part of the two world-paths. Therefore the intermediate states of the optimized algorithm are triples $\langle \pi_1 = \pi_2 \mid \sigma \mid \pi \rangle$ where, as before, $\pi_1 = \pi_2$ is the remaining unification problem, σ is the partial unifier and π is the already unified initial segment.

The temporal rules have different refinements for the two different kinds of temporal variables. Therefore they are split into two separate rules.

$(\tau\tau_i)$ $\langle \pi_1 x^{s_i} \pi_2 = \pi_3 \pi_4 \mid \sigma \mid \pi \rangle \mapsto \langle \pi_2 = \pi_4 \mid (\sigma, \pi_1 x^{s_i} \mapsto \pi_3) \mid \pi\pi_3 \rangle$
if (i) $s_i \in \mathbb{S}_R$ is a temporal sort;
(ii) $\pi_1 x^s$ and π_3 are initial segments of T-blocks;
(iii) x^{s_i} is the rightmost variable in its T-block;
(iv) the final segments of the T-blocks after x^{s_i} and π_3 are identical;
(v) $((\pi\pi_3)_{[[} \cap i) \neq \emptyset$ and $\pi\pi_3$ does not contain a modal variable corresponding to a non-serial accessibility relation.
This is the extra condition. If it is not fulfilled then the simplification rules 18.75(c), (e) and (b) rewrite the condition $\pi\pi_3 \in i$ to $\perp$.

$(\tau\tau_\tau)$ $\langle \pi_1 x^{s_\tau} \pi_2 = \pi_3 \pi_4 \mid \sigma \mid \pi \rangle \mapsto \langle \pi_2 = \pi_4 \mid (\sigma, \pi_1 x^{s_\tau} \mapsto \pi_3) \mid \pi\pi_3 \rangle$
if (i) $s_\tau \in \mathbb{S}_T$ is a temporal sort;
(ii) $\pi_1 x^s$ and π_3 are initial segments of T-blocks;
(iii) x^{s_τ} is the rightmost variable in its T-block;
(iv) the final segments of the T-blocks after x^{s_τ} and π_3 are identical;
(v) (a) if $(\pi\pi_3)_{[[} = \emptyset$ then $\tau_{[[}((\pi\pi_1)_{[[}) = \emptyset$ and
(b) if $(\pi\pi_3)_{[[} \neq \emptyset$ and $\pi\pi_3$ does not contain modal variables corresponding to a non-serial accessibility relation then $((\pi\pi_3)_{[[} \cap \tau_{[[}(i)) \neq \emptyset$ for all $i \in \delta((\pi\pi_1)_{[[}, \tau)$.

$(\tau[]_i)$ $\langle \pi_1 x^{s_i} \pi_2 = \pi_3 \mid \sigma \rangle \pi \mapsto \langle \pi_2 = \pi_3 \mid (\sigma, \pi_1 x^{s_i} \mapsto []) \mid \pi \rangle$
if
(i) $s^i \in \mathbb{S}_R$ is a temporal sort;
(ii) $\pi_1 x^s$ is a T-block and π_2 and π_3 start with an M-block;
(iii) $(\pi_{[[} \cap i) \neq \emptyset$.

$(\tau[]_\tau)$ $\langle \pi_1 x^{s_\tau} \pi_2 = \pi_3 \mid \sigma \mid \pi \rangle \mapsto \langle \pi_2 = \pi_3 \mid (\sigma, \pi_1 x^{s_\tau} \mapsto []) \mid \pi \rangle$
if
(i) $s \in \mathbb{S}_R \cup \mathbb{S}_T$ is a temporal sort;
(ii) $\pi_1 x^s$ is a T-block and π_2 and π_3 start with an M-block;
(iii) $(\pi_{[[} \cap (\pi\pi_1 x^{s_\tau})_{[[}) \neq \emptyset$.

The world-paths contain implicit temporal information which must sometimes be made explicit. To illustrate this phenomenon, consider the statement 'there was a year in the last decade when it was raining the whole leap day'. The $\mathcal{LCL}^{\mathcal{M}}_{\mathcal{TU}}$-encoded form is $a^{s_i} x^{s_{leap\text{-}day(z_{year})}} : raining$ where the time region i corresponds to the last decade. a^{s_i} denotes a particular time point t in this decade. $x^{s_{leap\text{-}day(z_{year})}}$ ranges over all time points in the leap day of the year containing t. Since not all years contain leap days, one can restrict the time region i to those years containing leap days. Therefore $a^{s_i} \in i'$ can be derived, where i' is the subset of i where $leap\text{-}day(z_{year})_{[[}(i')$ is defined.

This kind of reasoning works for the three kinds of literals which explicitly or implicitly enforce that the world-path denotes a defined time point: $\pi : p$, and $\pi \notin \emptyset$ and $\pi \in j$ with $j \neq \emptyset$.

In the last case one can derive an even stonger constraint. Consider the example 'one leap day in the last decade was a Friday', and its $\mathcal{CL}^{\mathcal{M}}_{\mathcal{TU}}$-formulation

$$\langle a^{s_i}\rangle[\mathit{leap\text{-}day}]\mathit{Friday}(z_{year})$$

where i again corresponds to the last decade. The functional translation is of the form $a^{s_i}x^{s_{leap\text{-}day(z_{year})}} \in k$ where k is the time region corresponding to all leap days in the last decade which were, in fact, Fridays. Since this happened only in certain years, one can derive a constraint $a^{s_i} \in i'$ where i' corresponds to those years in the last decade where the leap day was a Friday.

The next definition provides a function that extracts implicit temporal information from world-paths.

Definition 18.77 (Extracting information from world-paths). We define a rewrite relation $\rightarrow_e$ which reduces world-paths from right to left and computes corresponding temporal constraints.

$$\begin{array}{lll}
\pi x^{s_i} \notin \emptyset & \rightarrow_e \pi \neq \emptyset & \text{if } s_i \in \mathbb{S}_R \\
\pi x^{s_i} : \pm p & \rightarrow_e \pi \neq \emptyset & \text{if } s_i \in \mathbb{S}_R \\
\pi x^{s_i} \in j & \rightarrow_e \pi \neq \emptyset & \text{if } s_i \in \mathbb{S}_R \text{ and } j \neq \emptyset \\
\pi x^{s_\tau} \notin \emptyset & \rightarrow_e \pi \in \bigcup_{i \in \delta(\pi_{[[},\tau), \tau_{[[}(i) \neq \emptyset} i & \text{if } s_\tau \in \mathbb{S}_T \\
\pi x^{s_\tau} \in j & \rightarrow_e \pi \in \bigcup_{i \in \delta(\pi_{[[},\tau), \tau_{[[}(i) \subseteq j} i & \text{if } s_\tau \in \mathbb{S}_T \\
\pi x^{s_\tau} : \pm p & \rightarrow_e \pi \in \bigcup_{i \in \delta(\pi_{[[},\tau), \tau_{[[}(i) \neq \emptyset} i & \text{if } s_\tau \in \mathbb{S}_T.
\end{array}$$

For literals $\varphi\lceil\pi\pi'\rfloor$ of the form $\pi\pi' \in i$ with $i \neq \emptyset$ or $\pi\pi' \notin \emptyset$ or $\pi\pi' : \pm p$, let $\mathit{restrict}(\varphi\lceil\pi\pi'\rfloor\pi)$ be the result of applying the $\rightarrow_e$ rewrite relation until the result is of the form $\pi \in j$ or $\pi \notin \emptyset$.

Proposition 18.78 (*restriction*: soundness and completeness).

1. *The restrict-function performs a sound inference, i.e.* $\varphi\lceil\pi\pi'\rfloor$ *entails* $\mathit{restrict}(\varphi\lceil\pi\pi'\rfloor, \pi)$.
2. $\mathit{restrict}(\varphi\lceil\pi\pi'\rfloor, \pi)$ *computes the tightest approximation for the temporal constraint for* π. *That means, for every interpretation* $\Im, w, t$*: if* $\Im, w, t \models \mathit{restrict}(\varphi\lceil\pi\pi'\rfloor, \pi)$ *then there are accessibility functions* γ_i *for the constants and variables* x_i *occurring in* π' *such that* $\Im[x_1/\gamma_1, \ldots, x_n/\gamma_n](\pi\pi')$ (w, t) *is defined, and if* $\varphi\lceil\pi\pi'\rfloor = (\pi\pi' \in j)$ *then* $\Im[x_1/\gamma_1, \ldots, x_n/\gamma_n], w, t \models \pi\pi' \in j$.

Proof By induction on the length of π'. Both statements can easily be proved by checking the rewrite rules of the $\rightarrow_e$-relation. □

18.4.4.5 *Resolution* The discovery of the resolution principle by John Alan Robinson [1965] has been one of the major breakthroughs in automated theorem proving. The main impact, however, did not come from the structure of the resolution rule, which is a slight generalization of *modus ponens*, but from the unification principle, which was in fact discovered much earlier by Jacques Herbrand [1930] or even earlier by Emil Post. The resolution calculus for first-order predicate logic operates on clauses and consists of the two rules

$$\text{resolution: } \frac{\begin{array}{l} L \vee C \\ \neg L' \vee D \quad \sigma \text{ is a most general unifier for } L \text{ and } L' \end{array}}{(C \vee D)\sigma}$$

and

$$\text{factoring: } \frac{L \vee L' \vee C \;\; \sigma \text{ is a most general unifier for } L \text{ and } L'}{(L \vee C)\sigma}$$

where L and $\neg L'$ in the resolution rule are the *resolution literals*, $(C \vee D)\sigma$ is the *resolvent*. $(L \vee C)\sigma$ in the factoring rule is the *factor*.

Resolution is a *sound* inference principle, and it is *refutation complete*. Refutation completeness means that the empty clause can be derived if the initial clause set is unsatisfiable. If the clause set is satisfiable, resolution need not terminate.

The soundness of the resolution rule depends on the fact that the σ-instances of L and $\neg L'$ are contradictory. This is the case if $L\sigma$ and $L'\sigma$ are identical, as in the original formulation of resolution. But in a more general context, it is sufficient that $L\sigma \wedge L'\sigma$ are unsatisfiable in some background theory. If for example L is $x < 0$, L' is $a > 0$ and $\sigma = \{x \mapsto a\}$ then $L\sigma$ is $a < 0$ which contradicts $a > 0$ in the theory of arithmetic.

A more general resolution principle is therefore (total) theory resolution:

$$\frac{\begin{array}{l} L \vee C \\ L' \vee D \qquad L\sigma \wedge L'\sigma \models_T \bot \end{array}}{(C \vee D)\sigma}$$

where $L\sigma \wedge L'\sigma \models_T$ means that $L\sigma \wedge L'\sigma$ are unsatisfiable in some theory T.

Mark Stickel has generalized this idea further [Stickel, 1985]. The most general version is *partial theory resolution*

$$\frac{\begin{array}{l} L_1 \vee C_1 \\ \quad\vdots \\ L_n \vee C_n \qquad\qquad L_1\sigma \wedge \cdots \wedge L_n\sigma \models_T L \end{array}}{L \vee (C_1 \vee \cdots \vee C_n)\sigma.}$$

L is called the *residue*. It can be a single literal or an arbitrary complex formula. An example for a partial theory resolution in the theory of arithmetic is

$$\frac{\begin{array}{l} x \leq b \vee C \\ b \leq c \vee D \\ c \leq a \vee E \qquad\qquad \sigma = \{x \mapsto a\} \end{array}}{a = c \vee (C \vee D \vee E)\sigma.}$$

The resolution calculus for functionally translated $\mathcal{CL}^{\mathcal{M}}_{\mathcal{T}\mathcal{U}}$-formulae has to take into account the special meaning or the time regions and time terms. It is therefore a theory resolution calculus. The main rule (res_p) below is almost like the original resolution rule. There are only a few extras. The unification is, of course, world-path unification (Definition 18.71), and the instantiation generates the necessary conditions as residue literals. In addition we have to take into account that two labelled literals $\pi : p$ and $\pi : \neg p$ are not necessarily contradictory, even if the world-paths are identical. The conjunction of these literals can be satisfied if π is not defined, i.e. $\pi \in \emptyset$ holds. Therefore $\pi \in \emptyset$ must be generated as an additional residue literal.

As an example, consider the functionally translated $\mathcal{CL}^{\mathcal{M}}_{\mathcal{T}\mathcal{U}}$-encoded statements 'it was raining all day long on this year's leap day' and 'it was not raining all day long on this year's leap day'. These statements are not contradictory. Instead, one can conclude that there was no leap day this year. The quantification 'all day long on this year's leap day' is then vacuously true. The (res_p)-rule would derive the information 'no leap day this year' as a residue literal $x^{s_{leap\text{-}day(y_{year})}} \in \emptyset$.

The literals $\pi \in \emptyset$, on the other hand, are contradictory to three types of other literals: (i) $\pi\pi' \notin \emptyset$, (ii) $\pi\pi' \in i$ with $i \neq \emptyset$ and (iii) $\pi\pi' : p$ where π' is an M-block which does not change the time. Therefore a further resolution rule combines clauses containing literals $\pi \in \emptyset$ with these three kinds of other literals. Continuing the little example above, one could derive from 'no leap day this year' and 'there was sunshine sometime during this year's leap day' a contradiction. The statement 'there was sunshine sometime during this year's leap day' contains two different pieces of information: (a) there was a time point t in this year's leap day, and (b) the sun was shining at time point t; (a) contradicts 'no leap day this year', and this contradiction would be derived with the ($res_\emptyset$)-rule below.

The (res_i)-rule below combines the set-theoretic information contained in literals $\pi \in i$ and $\pi \in j$. From these two literals one can derive $\pi \in (i \cap j)$. Literals $\pi\pi' \in i$ and $\pi\pi'' \in j$, where π' and π'' are M-blocks (which do not change the time), also imply $\pi \in (i \cap j)$.

The first factoring rule (fac_p) is more or less the standard factoring rule as in predicate logic. The other rule (fac_i) is the set-theoretic counterpart of the (res_i)-rule: from $\pi \in i \vee \pi \in j$ one can derive $\pi \in (i \cup j)$.

The ($restr$)-rule extracts implicit information contained in world-paths. It should be applied only if the resulting literal is not an elementary tautology.

The ($inst_i$)-rule below eliminates constraint literals of the kind $\pi x^{s_i} \in j$ where $j \subset i$ by instantiating the variable x with a constant $a^{s_{j \setminus i}}$. The ($inst_\tau$)-rule performs the same kind of elimination for the $\mathbb{S}_T$-variables. These rules, together with the ($restr$)-rule, can eliminate constraint literals $\pi \in i$ by instantiating variables in π to become an elementary contradiction.

Definition 18.79 (Resolution). The resolution and factoring rules for TF_2-translated clauses are ($\pm$ below denotes either p or $\neg p$):

$$(res_p)\quad \frac{\begin{array}{ll}\pi_1 : p \vee C & \sigma \in mgu(\pi_1, \pi_2)\\ \pi_2 : \neg p \vee D & \end{array}}{CNF((\pi_1 \in \emptyset \vee C \vee D)\sigma)}$$

$$(res_\emptyset)\quad \frac{\begin{array}{ll}\pi_1 \in \emptyset \vee C & \sigma \in mgu(\pi_1, \pi_2)\\ \varphi\lceil \pi_2\pi_3 \rfloor \vee D & \varphi = (\pi_2\pi_3 \notin \emptyset) \text{ or } \varphi = (\pi_2\pi_3 \in i) \text{ (with } i \neq \emptyset)\\ & \text{or } \varphi = (\pi a^s_\tau \pi' : \pm p)\end{array}}{CNF((C \vee D)\sigma)}$$

$$(res_i)\quad \frac{\begin{array}{l}\pi_1\pi_1' \in i \vee C \;\; \sigma \in mgu(\pi_1, \pi_2), \pi_1' \text{ and } \pi_2' \text{ are M-blocks or empty}\\ \pi_2\pi_2' \in j \vee D \text{ where } i = \emptyset \; (j = \emptyset)\end{array}}{CNF((\pi_1 \in (i \cap j) \vee C \vee D)\sigma)}$$

$$(fac_p)\quad \frac{\pi_1 : \pm p \vee \pi_2 : \pm p \vee C \qquad \sigma \in mgu(\pi_1, \pi_2)}{CNF((\pi_1 : \pm p \vee C \vee D)\sigma)}$$

$$(fac_i)\quad \frac{\pi_1 \in i \vee \pi_2 \in j \vee C \qquad \sigma \in mgu(\pi_1, \pi_2)}{CNF((\pi_1 \in (i \cup j) \vee C)\sigma)}$$

$$(restr)\quad \frac{\begin{array}{ll}\varphi\lceil \pi\pi' \rfloor \vee C & \varphi = (\pi\pi' \notin \emptyset) \text{ or } \varphi = (\pi\pi' : \pm p)\\ & \text{or } \varphi = (\pi\pi' \in i), i \neq \emptyset\end{array}}{CNF(restrict(\varphi\lceil \pi\pi' \rfloor, \pi) \vee C)}$$

$$(inst_i)\quad \frac{\pi x^{s_i} \in j \vee C \;\; s_i \in \mathbb{S}_R, j \subset i, x \text{ is a variable}}{C(x \to a^{s_{i\backslash j}})}$$

The constant $a^{s_{i\backslash j}}$ can be one particular constant for the sort $s_{i\backslash j}$. If x occurs in C, it is just replaced by $a^{s_{i\backslash j}}$.

$$(inst_\tau)\quad \frac{\pi x^{s_\tau} \in j \vee C \qquad s_\tau \in \mathbb{S}_T, j \subset \tau_{[[}(\pi_{[[}), x \text{ is a variable}}{C(x \to a^{s_{\tau_{[[}(\pi_{[[})\backslash j}})}$$

The constant $a^{s_{\tau_{[[}(\pi_{[[})\backslash j}}$ can again be one particular constant for this sort.

Example 18.80 (For a resolution refutation). Example 18.33 was used to illustrate the tableau proof procedure. We now use a slight modification of the same example to illustrate the resolution procedure. The informal specification is:

(i) last week it was rainy or foggy;
(ii) it is known that from Monday till Wednesday last week it was not rainy;
(iii) it is known that from Wednesday till Sunday last week it was not foggy.

We use $\Box$ for the 'is known' operator. $\Box$ corresponds to a T-modality (reflexivity of the accessibility relation).
The $\mathcal{CL}^{\mathcal{M}}_{\mathcal{T}u}$-encodings of the statements are:

(1) $\langle 10, day\rangle : [x_{week} - 1](r \vee f)$
(2) $\langle 10, day\rangle : \Box\langle x_{week} - 1\rangle[\{Mo(x_{week}), Tu(x_{week}), We(x_{week})\}]\neg r$
(3) $\langle 10, day\rangle : \Box\langle x_{week} - 1\rangle[\{We(x_{week}), \ldots, So(x_{week})\}]\neg f$.

If the same reference time line is used as in Example 18.33 then the functional translation yields

$C_1\ a_1^{[100-110[}y^{x_{week}-1} : (r \vee f)$
$C_2\ a_2^{[100-110[}v_1 c^{x_{week}-1} z^{\{Mo(x_{week}),\ldots,We(x_{week})\}} : \neg r$
$C_3\ a_3^{[100-110[}v_2 d^{x_{week}-1} u^{\{We(x_{week}),\ldots,Sunday(x_{week})\}} : \neg f.$
v_1 and v_2 are modal variables derived from the $\Box$-operator.
We have $\delta([100,110[, x_{week}-1) = [100,110[$. Therefore the resolvent between C_1 and C_2 (with $v_1 \mapsto []$) is
R_1: $(a_1^{[100-110[} \in$
$[100,110[\ \rightarrow a_2^{[100-110[}c^{x_{week}-1} z^{\{Mo(x_{week}),\ldots,We(x_{week})\}} \in [0,70[) \rightarrow$
$a_2^{[100-110[}c^{x_{week}-1} z^{\{Mo(x_{week}),\ldots,We(x_{week})\}} : f.$
This gets simplified to $R_1' = a_2^{[100-110[}c^{x_{week}-1} z^{\{Mo(x_{week}),\ldots,We(x_{week})\}} : f.$
Resolution with C_3 ($v_2 \mapsto []$) yields
$(a_3^{[100-110[}d^{x_{week}-1} \in [0,70[\ \rightarrow a_2^{[100-110[}c^{x_{week}-1} z^{\{Mo(x_{week}),\ldots,We(x_{week})\}} \in$
$[20,70[) \rightarrow \perp.$
This in turn can be simplified to $a_2^{[100-110[}c^{x_{week}-1} z^{\{Mo(x_{week}),\ldots,We(x_{week})\}} \in$ $[20,30[$ and with the $(inst_i)$-rule resolved to the empty clause.

Theorem 18.81 (Soundness of resolution and factoring). *The resolution and factoring rules (Definition 18.79) are sound.*

Proof Unification and instantiation is sound (Theorem 18.69). Therefore we can always assume that the conditions inserted by the instantiation operation are satisfied, otherwise the resolvents and factors are tautologies anyway. Since $CNF()$-simplification is also sound (Proposition 18.75), we need only check the main part of the rules. For each rule we check that whenever an interpretation $\Im, w, t$ satisfies the σ-instances of the parent clauses then $\Im, w, t$ satisfies the resolvents and factors.

(res_p) If $\Im(\pi_1\sigma)(w,t)$ is defined then $\pi_1\sigma : p$ and $\pi_2\sigma : \neg p$ cannot both be true. In this case either $C\sigma$ or $D\sigma$ must be satisfied, which means the resolvent is satisfied.
If $\Im(\pi_1\sigma)(w,t)$ is not defined then $\Im, w, t \models \pi_1\sigma \in \emptyset$ and the resolvent is also satisfied.

$(res_\emptyset)$ $\Im, w, t \not\models \pi_1\sigma \in \emptyset$ then $\Im, w, t$ must satisfy $C\sigma$, which means the resolvent is satisfied. If $\Im, w, t \models \pi_1\sigma \in \emptyset$ then $\Im(\pi_1\sigma)(w,t)$ is not defined. Therefore neither $(\pi_2\pi_3)\sigma \notin \emptyset$ nor $(\pi_2\pi_3)\sigma \in i$ with $i \neq \emptyset$, nor $(\pi a_\tau^s \pi')\sigma : \pm p$ can be satisfied. In this case $D\sigma$ must be satisfied, which means the resolvent is satisfied.

(res_i) If either $C\sigma$ or $D\sigma$ is satisfied then the resolvent is satisfied as well. If both are not satisfied, then the resolution literals must both be satisfied. If both i and j are not empty then the residue literal $\pi_1\sigma \in i \cap j$ must be satisfied because the trailing M-blocks π_1' and π_2' do not change the time points.
If for example i is empty then $\pi_1\pi_1' \in \emptyset$ can only be satisfied if $\Im((\pi_1\pi_1')\sigma)(w,t)$ is undefined. In this case π_1' does not contain variables (otherwise the rule is not applicable). Therefore $\Im((\pi_1)\sigma)(w,t)$

must already be undefined, and the residue literal $\pi_1\sigma \in \emptyset$ is satisfied. The same argument holds if j is empty.

(fac_p) The proof is trivial.

(fac_i) The proof is purely set theoretic.

($restr$) Soundness of this rule has been proved in Proposition 18.78.

($inst_i$) If $\Im, w, t$ satisfies the universal closure of the original clause, then $\Im[x/\mathcal{V}(a^{s_i\setminus j})], w, t$ must satisfy the clause itself. But it cannot satisfy the literal $\pi x^{s_i} \in j$. Therefore it must satisfy $C(x \rightarrow a^{s_i\setminus\tau[\![(\pi[\![)})$.

($inst_\tau$) The argument is the same as in the previous case. □

18.4.4.6 *Completeness of the temporal resolution calculus* The resolution calculus for first-order predicate logic is *refutation complete*. This means that the empty clause can be derived in a finite time if the original clause set is unsatisfiable. We show that our version of the resolution calculus is refutation complete as well. Owing to the built-in datatype of time regions, it is not possible to use the well-known results from predicate logic. Fortunately one can use the methodology developed to prove refutation completeness for predicate logic resolution. Each step in this proof, however, has to be adapted.

The standard methodology for proving refutation completeness works as follows:

1. Prove Herbrand's theorem: a clause set is unsatisfiable if and only if it has a *finite* unsatisfiable set of ground instances.
   In order to do this,
   (a) define the notion of *Herbrand interpretations* (*canonical model*) and
   (b) show that a clause set is satisfiable if and only if it has a Herbrand model;
   (c) define *semantic trees* as a representation of *all* Herbrand interpretations;
   (d) show that a finite part of the semantic tree is sufficient to falsify a finite subset of the ground instances.
2. Prove the completeness of resolution for ground clause sets.
   In this step the semantic tree is used to guide the resolution sequence.
3. Prove the lifting lemma: each inference step with ground instances of some clauses C corresponds to an inference step with C such that the derived clause at the ground level is an instance of the derived clause with C.
   The lifting lemma need not hold for all inference steps, but only for those needed to derive the empty clause. The structure of the semantic tree which was used to guide the resolution towards the empty clause can be used to prove the lifting lemma only for the really necessary inference steps.

Some of the construction and proof steps, which are quite simple for predicate logic, are non-trivial in our case. The main problem is the notion of a Herbrand interpretation. In predicate logic, a Herbrand interpretation has as its domain the set of terms. The function symbols are interpreted as term constructor functions, and the predicate symbols are interpreted by listing all the literals which are supposed to hold in the Herbrand interpretation. Each set of positive ground literals therefore determines some Herbrand interpretations in a straightforward way.

Herbrand interpretations for $\mathcal{LCL}_{\mathcal{T}u}^{\mathcal{M}}$-clauses can not only be specified by the syntax elements. The (infinite) reference time line must always be part of the interpretation, otherwise it is not a proper $\mathcal{LCL}_{\mathcal{T}u}^{\mathcal{M}}$-interpretation. Nevertheless we still use sets of ground literals as the basis for a Herbrand interpretation. The relation between these sets of ground literals and the Herbrand interpretation itself, however, is a bit more complicated than in the predicate logic case.

Once Herbrand interpretations are properly defined, the remaining construction and proof steps follow quite naturally, although the technical details are still different from the predicate logic case.

The sets HB of ground literals which form the basis of an Herbrand interpretation must (i) be consistent and (ii) provide a maximum of information the syntax supports. In the predicate logic case consistency is guaranteed by the simple condition that no literal and its negation are together in HB. Condition (ii) means that for *each* ground literal L either L or its negation is in HB.

Consistency in the $\mathcal{LCL}_{\mathcal{T}u}^{\mathcal{M}}$-case is more difficult to ensure. For example, if a ground literal $\pi \in \emptyset$ is in HB then $\pi\pi' \notin \emptyset$ cannot be in HB for any π'. Literals $\pi\pi' : p$ can also not be in HB if '$\pi \in \emptyset$' $\in HB$, neither can $\pi\pi' : \neg p$. Nevertheless, there are sufficient syntactic criteria which determine consistent sets of ground literals as the basis for a Herbrand interpretation (Definition 18.83).

Condition (ii), which requires HB to provide the maximum information that the syntax supports, would be much easier to ensure if the $\mathcal{LCL}_{\mathcal{T}u}^{\mathcal{M}}$-syntax allowed for literals $\pi \in t$ where t is a concrete point on the reference time line. Then we could locate each world-path precisely on $\mathcal{T}$. Since the $\mathcal{LCL}_{\mathcal{T}u}^{\mathcal{M}}$-syntax only allows for literals $\pi \in i$ where i is a time region, there is always some uncertainty about the concrete location of π on $\mathcal{T}$. The best we can do is to locate π in the smallest time region which is available in the syntax of the clause set for which the Herbrand interpretation is to be defined.

But how can we obtain a 'smallest time region'? If $\pi \in i$ and $\pi \in j$ are allowed expressions in our language, then $\pi \in (i \cap j)$ is also an allowed expression. Therefore smallest time regions are obtained by intersecting all explicitly or implicitly mentioned time regions.

To get a proper definition, we take the Boolean algebra of all explicitly or implicitly mentioned time regions. The atoms in this Boolean algebra then give us the smallest time regions.

Definition 18.82 (Boolean algebra). For a finite clause set C let $BA(C)$ be the Boolean algebra generated from the time regions occurring explicitly or implicitly in C. These are:

- all time regions occurring in literals of the form $p \in i$ in C,
- all time regions occurring in the sorts of the accessibility terms in C,

- all time regions $\pi_{[[}$ for all initial segments of world-paths π occurring in C,
- all elements $i \in \delta(\pi_{[[}, \tau)$ for all initial segments πx^{s_τ} of world-paths occurring in C.

Since the number of time regions occurring in a finite clause set is finite, $BA(C)$ is a finite atomic Boolean algebra.

For a time region i let $Batoms(i, C)$ be the set of atoms below i in the Boolean algebra $BA(C)$, and let $Batoms(C)$ be the set of all atoms in the Boolean algebra $BA(C)$.

Note that $i = \bigcup_{k \in Batoms(i,C)} i$.

As a simple example of what the Boolean algebra construction gives us, consider the clause corresponding to the statement 'last month I spent a weekend with friends'. The time regions, contained explicitly and implicitly in this statement, are: i_1 = last month, i_2 = the first weekend last month, i_3 = the second weekend last month, i_4 = the third weekend last month, and i_5 = the fourth weekend last month. The Boolean algebra generated by these time regions contains more elements: for example, the complement of last month, the first week last month without the first weekend, etc.

The Boolean atoms consist of the sets: the complement of last month as a whole, the first week last month without the first weekend, the first weekend last month, and so on up to the fourth week last month without the fourth weekend, and the fourth weekend last month. These are the most specific choices which make sense in this example. For every world-path it is sufficient to locate it in one of these atomic time regions.

We are now ready to define a *Herbrand base* as a maximal consistent set of ground literals which provides as much information as the syntax allows. Herbrand bases are constructed iteratively, starting with '$[] \notin \emptyset$' and adding more and more ground literals with longer and longer world-paths. The rules for adding more literals ensure that the set remains consistent. If there are different choices, then each particular choice leads to a different Herbrand base.

Definition 18.83 (Herbrand base). A *Herbrand base HB* for a clause set C is a minimal set of ground atoms $\pi \in i$, $\pi \notin \emptyset$, $\pi : p$ or $\pi : \neg p$ closed under the following rules:

(1) All ground elementary tautologies (Proposition 18.74) are in HB.

(2) If '$\pi \in \emptyset$' $\in HB$ then
'$\pi a \in \emptyset$' $\in HB$ for all accessibility terms a.
If '$\pi \notin \emptyset$' $\in HB$ then:

Case[31] a^{s_m}

either '$\pi a \in \emptyset$' $\in HB$ (if m is a non-serial modality[32])
or '$\pi a \notin \emptyset$' $\in HB$ and
'$\pi a \in j$' $\in HB$ if '$\pi \in j$' $\in HB$.[33]

Case	a^{s_j}	
	then	'$\pi a \not\in \emptyset$' $\in HB$[34]and
		'$\pi a \in k$' $\in HB$ for one $k \in Batoms(j, C)$.[35]
Case	a^{s_τ}	and $\pi \neq []$ (we want only determined world-paths)
Subcase	$\tau_{[[}(i) = \emptyset$ where '$\pi \in i$' $\in HB$	
	then	'$\pi a \in \emptyset$' $\in HB$.
Subcase	$\tau_{[[}(i) \neq \emptyset$ where '$\pi \in i$' $\in HB$	
	then	'$\pi a \neq \emptyset$' $\in HB$ and
		'$\pi a \in k$' $\in HB$ for one $k \in Batoms(\tau_{[[}(i), C)$.

- If '$\pi \not\in \emptyset$' $\in HB$ then either '$\pi : p$' $\in HB$ or '$\pi : \neg p$' $\in HB$, but not both.[36]

For this construction it is important that for each Boolean atom $k \in atoms(C)$ there is a constant a^{s_k} with $s_k \in \mathbb{S}_R$.

We define a Herbrand interpretation for a Herbrand base HB and show that in fact it satisfies HB. The Herbrand interpretation must contain a frame $((\mathcal{W} \times \mathcal{T}), R)$ where the reference time line $\mathcal{T}$ is fixed. The worlds $\mathcal{W}$ and the accessibility relations in R are constructed from the Herbrand base. The worlds are world-path π ending in a modal accessibility term, and with '$\pi \not\in \emptyset$' $\in HB$. The initial world is just $w_{[]} \stackrel{\text{def}}{=} []$.

The information for constructing the accessibility relations is taken indirectly from the world-paths, via the interpretation of the accessibility terms. Therefore the interpretation of the accessibility terms has to be explained first.

Accessibility functions map worlds to worlds and time points to time points. Whereas literals like '$a \in k$' $\in HB$ leave some uncertainty about which time point $t \in k$ is actually meant, the interpretation $\mathcal{V}(a)$ for an accessibility term a must be precise about this. That means, when constructing the accessibility function $\mathcal{V}(a)$, it is necessary to choose a time point $t \in k$ and define $\mathcal{V}(a)(w_{[]}, t_{[]})$ such that $\mathcal{V}(a)(w_{[]}, t_{[]})_{|T} = t$. To this end, we select a grid of time points t_π for each π with '$\pi \in k$' $\in HB$ such that $t_\pi \in k$, and all t_π are different. (Since $\mathcal{T}$ is dense, there are enough time points to choose from.) The interpretation of the accessibility terms can now be defined such that they map each time point to one of the preselected time points t_π only.

Figure 18.4 illustrates the situation. The reference time line $\mathcal{T}$ is partitioned into the atomic components of the Boolean algebra generated by all explicitly

[31] Each case represents a possibility to append an accessibility constant a to the world-path π. Since there are three types of sorts, $\mathbb{S}_M$, $\mathbb{S}_R$ and $\mathbb{S}_T$, there are three cases.

[32] If m is a modality with serial accessibility relation then πa must denote a transition to an accessible world. Therefore πa cannot be empty.

[33] Since j is not empty, and '$\pi \not\in \emptyset$' $\in HB$, the new world-path must denote a defined transition.

[34] This is because '$\pi \not\in \emptyset$' $\in HB$ and j is not empty.

[35] a^{s_j} denotes a transition into the time region j. Locating πa in one of j's Boolean atoms is the finest choice which provides the maximum of information that makes sense for the given clause set.

[36] If '$\pi \in \emptyset$' $\in HB$ then π does not denote a defined transition. Therefore neither $\pi : p$ nor $\pi : \neg p$ can be satisfied.

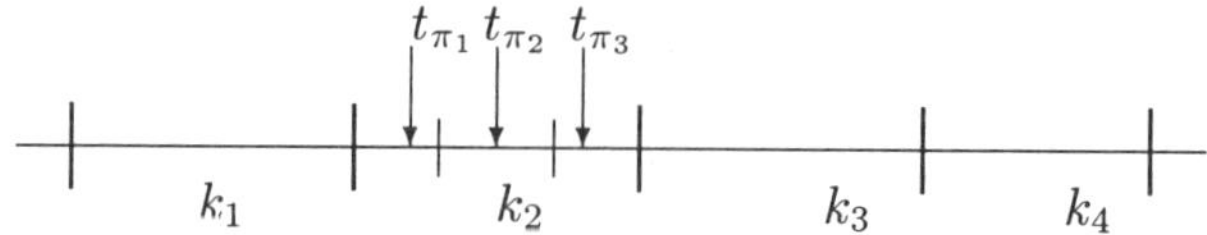

FIG. 18.4. Partitioning of the reference time line.

and implicitly mentioned time regions $k \in BA(C)$. For each atomic time region k there are (possibly infinite) world-paths with $\pi \in k$. For each π a time point t_π is chosen for a (yet to be defined) interpretation $\Im(\pi)(w_{[]}.t_{[]})$. The set $\{t_\pi \mid `\pi \in k' \in HB\}$ itself generates a partitioning of k into subregions each containing only a single t_π.

The partitioning of k yields an equivalence relation $t \sim t'$ iff t and t' are in the same partition. Since there is exactly one t_π in each segment of the partitioning, there is for all $t \sim t'$ a t_π with $t \sim t_\pi$ and $t' \sim t_\pi$.

The $\sim$-equivalence classes form the basic blocks of the Herbrand interpretation. The construction will be such that $\sim$-equivalent parts cannot be distinguished.

Interpretation of the accessibility constants a^{s_i}, $s_i \in \mathbb{S}_R$. These terms are interpreted such that they map $\sim$-equivalent parts to one reference time point. In the initial world $w_{[]}$, $\mathcal{V}(a^{s_i})$ maps the time points t_π where π is a T-block to time points $t_{\pi a}$. The construction of HB guarantees that '$t_{\pi a} \in k$' $\in HB$, for some Boolean atom k below i, and therefore $t_{\pi a} \in i$. All other time points $t \sim t_\pi$ are also mapped to $t_{\pi a}$.

In other worlds π_w, the function $\mathcal{V}(a^{s_i})$ maps time points $t_{\pi_w \pi}$ (π a T-block) to $t_{\pi_w \pi}$ and the same for all $t \sim t_{\pi_w \pi}$ (cf. Fig. 18.5).

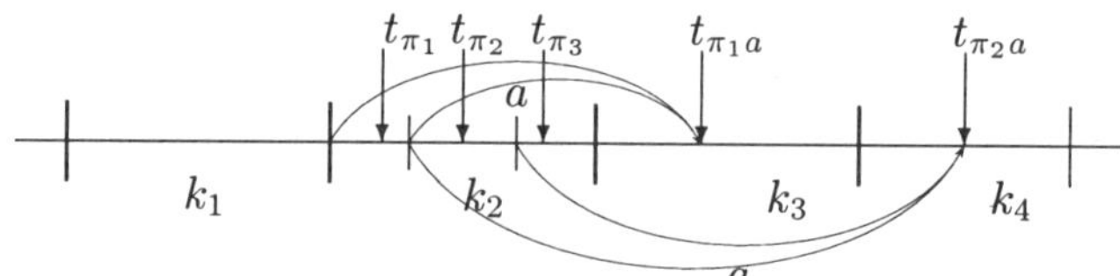

FIG. 18.5. Interpretation of the $\mathbb{S}_R$-constants.

Interpretation of the accessibility constants a^{s_τ}, $s_\tau \in \mathbb{S}_T$. These constants are interpreted almost in the same way as in the previous case. This is possible because the literals '$\pi a \in k$' $\in HB$ are used in the same way as above to locate $\mathcal{V}(a)$ in $\mathcal{T}$. The only difference is that for the time points t, $\tau_{[[}(t)$ must be defined. In this case $\mathcal{V}(a)(t)$ must also be defined. The construction of the Herbrand base guarantees that these cases can be recognized and checked with literals '$\pi a \notin \emptyset$' $\in HB$.

Interpretation of the accessibility constants a^{s_m}, $s_m \in \mathbb{S}_M$. A constant a^{s_m} is interpreted such that $\mathcal{V}(a)(\pi_w, t_{\pi_w \pi})$ (π a T-block) is mapped to $(\pi_w a, t_{\pi_w \pi})$

and time points $t \sim t_{\pi_w\pi}$ are mapped to the same world $\pi_w a$ while staying at the time point t.

Once the world-paths have an interpretation, the accessibility relation is defined in a natural way, by following $\mathcal{V}(a^{s_m})$-transitions. In order to get an accessibility relation which falls into the class of accessibility relations corresponding to the basic multimodal logic, a corresponding closure operation needs to be performed (reflexive closure to get a T-frame, transitive closure to get a K4-frame, etc.)

Finally the predicate symbols are interpreted according to the $\pi : p$-literals in HB. Again, $\sim$-equivalent time points are not distinguished by the interpretations $\mathcal{P}(p)$.

Definition 18.84 (Herbrand interpretation). For each Herbrand base HB for a clause set C we define a corresponding Herbrand interpretation $\Im = ((\mathcal{W} \times \mathcal{T}, R), \mathcal{P}, \mathcal{V})$ as follows.

Let $\mathcal{W} \stackrel{\text{def}}{=} \{[]\} \cup \{\pi \mid \pi$ ends in an M-block, '$\pi \notin \emptyset$' $\in HB\}$.

For each ground world-path π with '$\pi \in k$' $\in HB$, choose some *selected* $t_\pi \in k$ such that (i) t_π and $t_{\pi'}$ are different for different π and π' and (ii) $t_\pi = t_{\pi a^s_m}$ for all π and for all modal accessibility constants a^{s_m}; (i) is possible because $\mathcal{T}$ is dense. There are always enough time points to choose from.

Now partition each $k \in Batoms(C)$ such that each part in the partitioning contains exactly one t_π. Let $t \sim t'$ iff t and t' are in the same part of the partitioning. $\sim$ is an equivalence relation.

Since $Batoms(C)$ partitions $\mathcal{T}$ and each atom $k \in Batoms(C)$ contains at least one selected t_π (because there is some '$a^{s_k} \in k$' $\in HB$), there is for each $t \in \mathcal{T}$ exactly one t_π with $t \sim t_\pi$.

Let $\mathcal{V}(a^{s_i})(\pi_w, t) \stackrel{\text{def}}{=} (\pi_w, t_{\pi_w\pi a})$ where $t \sim t_{\pi_w\pi}$.

$$\text{Let } \mathcal{V}(a^{s_\tau})(\pi_w, t) \stackrel{\text{def}}{=} \begin{cases} (\pi_w, t_{\pi_w\pi a}) & \text{where } t \sim t_{\pi_w\pi} \\ & \text{and '}\pi_w\pi a \notin \emptyset\text{'} \in HB \\ \text{undefined} & \text{otherwise.} \end{cases}$$

$$\text{Let } \mathcal{V}(a^{s_m})(\pi_w, t) \stackrel{\text{def}}{=} \begin{cases} (\pi_w\pi a, t) & \text{where } t \sim t_{\pi_w\pi} \\ & \text{and '}\pi_w\pi a \notin \emptyset\text{'} \in HB \\ \text{undefined} & \text{otherwise.} \end{cases}$$

For the modality m let

$$R'_m \stackrel{\text{def}}{=} \{((w,t), \mathcal{V}(a^{s_m})(w,t)) \mid \mathcal{V}(a^{s_m})(w,t) \text{ is defined}\}.$$

If M is the multimodal logic under consideration, let R_m be the M-closure of R'_m.[37]

Let $(\pi_w, t) \in \mathcal{P}(p)$ iff $t \sim t_{\pi_w\pi}$ and '$\pi_w\pi : p$' $\in HB$.

This definition is possible because the $\sim$-equivalence classes contain only one single time point t_π with '$\pi \in k$' $\in HB$ where the interpretation of the predicate

[37]The M-closure of R'_m for modal logic T, for example, is the reflexive closure of R'_m; for modal logic S4 it is the reflexive transitive closure etc.

symbols p is determined by the Herbrand bases. For all other time points one can choose the interpretation freely.

The next lemma and its corollary show that the construction of the Herbrand interpretation is indeed such that $\sim$-equivalent time points are not distinguished.

Lemma 18.85. *For the Herbrand interpretation $\Im$ for the Herbrand base HB, if $t \sim t_{\pi_w\pi_1}$ and $\Im(\pi)(\pi_w, t) = (w_1, t_1)$ and $\Im(\pi)(\pi_w, t_{\pi_w\pi_1}) = (w_1, t_2)$ then:*

(i) $w_1 = (\pi_w\pi_1\pi)_{|M}$ where $(\pi_w\pi_1\pi)_{|M}$ denotes the initial segment of $\pi_w\pi_1\pi$ up to the last modal term, and

(ii) $t_2 = t_{\pi_w\pi_1\pi}$ and $t_1 \sim t_2$.

Proof By induction on the length of π.

We extend the definition of $\sim$ to tuples: $(w_1, t_1) \sim (w_2, t_2)$ iff $w_1 = w_2$ and $t_1 \sim t_2$.

If $\pi = []$ then $\Im([])(\pi_w, t) = (\pi_w, t)$ and $\Im([])(\pi_w, t_{\pi_w\pi_1}) = (\pi_w, t_{\pi_w\pi_1})$. Thus, $t_1 \sim t_2$ and (i) holds trivially.

In the induction step we need to consider the different sorts of accessibility terms.

Case a^{s_i}, $s_i \in \mathbb{S}_R$: (i) follows directly from the induction hypothesis, because the world is not changed by a temporal a^{s_i}-transition.

$\Im(\pi a)(\pi_w, t)$
$= \mathcal{V}(a)(\Im(\pi)(\pi_w, t))$
$= \mathcal{V}(a)((\pi_w\pi_1\pi)_{|M}, t_1)$
$= (\pi_w\pi_1\pi)_{|M}, t_{\pi_w\pi_1\pi a})$ (induction hypothesis, $t_1 \sim t_{\pi_w\pi_1\pi}$ and definition of $\mathcal{V}(a)$)

$\Im(\pi a)(\pi_w, t_{\pi_w\pi_1})$
$= \mathcal{V}(a)(\Im(\pi)(\pi_w, t_{\pi_w\pi_1}))$
$= (\pi_w\pi_1\pi)_{|M}, t_{\pi_w\pi_1\pi a})$.

Case a^{s_τ}, $s_\tau \in \mathbb{S}_T$: We can assume $\Im(\pi)(\pi_w, t)$ is defined and '$\pi_w\pi a \not\subseteq \emptyset$' $\in HB$, otherwise both expressions are undefined and the statement holds vacuously. The proof is then the same as in the previous case.

Case a^{s_m}, $s_m \in \mathbb{S}_M$:

$\Im(\pi a)(\pi_w, t)$
$= \mathcal{V}(a)(\Im(\pi)(\pi_w, t))$
$= \mathcal{V}(a)((\pi_w\pi_1\pi)_{|M}, t_1)$
$= (\pi_w\pi_1\pi a, t_1)$ (definition of $\mathcal{V}(a)$)

$\Im(\pi a)(\pi_w, t_w)$
$= \mathcal{V}(a)(\Im(\pi)(\pi_w, t_w))$
$= (\pi_w\pi_1\pi a, t_{\pi_w\pi_1\pi})$.

Thus, (i) holds because $w_1 = \pi_w\pi_1\pi a$ and (ii) is in fact the induction hypothesis: $t_1 \sim t_{\pi_w\pi_1}$. □

Corollary 18.86. *For the Herbrand interpretation $\Im$ for the Herbrand base HB, if $t \sim t_{\pi_w\pi_1}$ then $\Im, w, t \models \pi : p$ iff $\Im, w,_{\pi_w\pi_1} \models \pi : p$.*

This is a consequence of Lemma 18.85 and the definition of $\mathcal{P}(p)$ for a Herbrand interpretation, which does not distinguish $\sim$-equivalent time points.

Lemma 18.87. *The Herbrand interpretation (Definition 18.84) for the Herbrand base HB satisfies HB.*

Proof By induction on the length of the world-paths π occurring in the elements of HB we show that $\Im, w_{[]}, t$ satisfies each literal in HB for every $t \in \mathcal{T}$.

Base case: $\pi = []$:

$\Im, w_{[]}, t \models [] \notin \emptyset$ because $(\Im([])(w_{[]}, t))_{|T} = t \qquad (\notin \emptyset)$.

Induction step:

The induction hypothesis is, for each world-path π of length n:

If '$\pi \in i$' $\in HB$ then $\Im, w_{[]}, t \models \pi \in i$.

If '$\pi \notin \emptyset$' $\in HB$ then $\Im, w_{[]}, t \models \pi \notin i$.

Case '$\pi \in \emptyset$' $\in HB$:

$\succ$ '$\pi a \in \emptyset$' $\in HB$ for all accessibility terms a (first part of case 2).

Since $\Im, w_{[]}, t \models \pi \in \emptyset$, $\Im(\pi)(w_{[]}, t)$ is not defined. Therefore $\Im(\pi a)(w_{[]}, t)$ is not defined either. Thus, $\Im, w_{[]}, t \models \pi a \in \emptyset$.

Case '$\pi \in i$' $\in HB$ for some non-empty i:

Subcase πa^{s_m}, $s_m \in \mathbb{S}_M$:

If '$\pi a^{s_m} \in \emptyset$' $\in HB$ then $\pi a^{s_m} \notin \mathcal{W}$.

Therefore $\Im(\pi a)(w_{[]}, t)$ is not defined, which means $\Im, w_{[]}, t \models \pi a \in \emptyset$.

If '$\pi a^{s_m} \notin \emptyset$' $\in HB$ then $\pi a^{s_m} \in \mathcal{W}$.

Therefore $\Im, w_{[]}, t \models \pi a \notin \emptyset$.

Since '$\pi \in i$' $\in HB$ we have '$\pi a^{s_m} \in i$' $\in HB$ as well (construction of HB). Because $\Im, w_{[]}, t \models \pi \in i$ (induction hypothesis) and $\mathcal{V}(a)$ does not change the time, $\Im, w_{[]}, t \models \pi a \in i$.

Subcase πa^{s_j}, $s_j \in \mathbb{S}_R$:

$\succ$ '$\pi a \in k$' $\in HB$ for some $k \in atoms(j, C)$.

$\succ \mathcal{V}(a)(\Im(w_{[]}, t)) = \mathcal{V}(a)(w', t_\pi)$. (Lemma 18.85)

$\succ \mathcal{V}(a)(\Im(w_{[]}, t)) = (w', t_{\pi a})$ with $t_{\pi a} \in k$. (definition of $\mathcal{V}(a)$)

Therefore $\Im, w_{[]}, t \models \pi a \in k$.

Subcase πa^{s_τ}, $s_\tau \in \mathbb{S}_T$:

If '$\pi a \in \emptyset$' $\in HB$ then $\tau_{[[}(i) = \emptyset$ (construction of HB, '$\pi \notin \emptyset$' $\in HB$).

If '$\pi a \in k$' $\in HB$ for some $k \in atoms(\tau_{[[}(i), C)$ and '$\pi \in i$' $\in HB$. Then $\mathcal{V}(a)(\Im(w_{[]}, t)) = \mathcal{V}(a)(w', t_\pi)$ (Lemma 18.85) and therefore $\mathcal{V}(a)(\Im(w_{[]}, t)) = (w', t_{\pi a})$ with $t_{\pi a} \in k$ (definition of $\mathcal{V}(a)$).

Thus, $\Im, w_{[]}, t \models \pi a \in k$.

$\Im, w_{[]}, t \models \pi : p$ iff '$\pi : p$' $\in HB$, by definition. □

The next lemma states that every interpretation determines a Herbrand base, and therefore a Herbrand interpretation. There is a corresponding lemma in predicate logic, stating that a satisfiable clause set has a Herbrand interpretation. In the predicate logic case, one chooses just those ground literals satisfied by the interpretation as a representation for the Herbrand interpretation. The construction for the $\mathcal{LCL}^{\mathcal{M}}_{\mathcal{T}\mathcal{U}}$-case follows the same ideas. Ground literals satisfied by the interpretation are chosen for the Herbrand base. By following the iterative definition of a Herbrand base we make sure that the result is really a Herbrand base.

Lemma 18.88. *Every interpretation $\Im, w, t$ determines a Herbrand base, i.e. for every interpretation $\Im, w, t$ there is a Herbrand base satisfied by $\Im, w, t$.*

Proof Following the iterative construction of Herbrand bases, we use $\Im$ as a guideline to solve the non-determinisms in Definition 18.83. By induction on the length of world-paths, we show that $\Im$ satisfies the constructed HB.

1. By definition, $\Im, w, t$ satisfies all elementary tautologies.
2. Case '$\pi \in \emptyset$' $\in HB$ and $\Im, w, t \models \pi \in \emptyset$:
   Then '$\pi a \in \emptyset$' $\in HB$ for all accessibility terms a and $\Im, w, t \models \pi a \in \emptyset$.
   Case '$\pi \notin \emptyset$' $\in HB$:
   $\succ \Im(\pi)(w,t) = (w', t')$ is defined (induction hypothesis).

   Subcase a^{s_m}:
   If $\Im(\pi a)(w,t)$ is undefined then $\Im, w, t \models \pi a \in \emptyset$ and we put $\pi a \in \emptyset$ into HB; if $\Im(\pi a)(w,t)$ is defined then $\Im, w, t \models \pi a \notin \emptyset$ and we put $\pi a \notin \emptyset$ into HB.
   If in addition '$\pi \in j$' $\in HB$, then '$\pi a \in j$' $\in HB$ as well. Since $\Im, w, t \models \pi \in j$ (induction hypothesis), and because modal accessibility functions do not change the time, $\Im, w, t \models \pi a \in j$.

   Subcase a^{s_j}:
   Then $\Im(a)(w', t') = (w', t'')$ and $t'' \in j$, and therefore $t'' \in k \in Batoms(j, C)$. This means $\Im, w, t \models \pi a \in k$ (put $\pi a \in k$ into HB) and $\Im, w, t \models \pi a \notin \emptyset$ (put $\pi a \notin \emptyset$ into HB).

   Subcase a^{s_τ} and $\pi \neq []$:
   Sub-subcase $\tau_{[[}(i) = \emptyset$ where '$\pi \in i$' $\in HB$:
   Then $\Im, w, t \models \pi a \in \emptyset$. Put $\pi a \in \emptyset$ into HB.
   Sub-subcase $\tau_{[[}(i) \neq \emptyset$ where '$\pi \in i$' $\in HB$:
   Then $\Im(a)(w', t') = (w', t'')$ and $t'' \in \tau_{[[}(i)$, and therefore $t'' \in k \in Batoms(\tau_{[[}(i), C)$. This means $\Im, w, t \models \pi a \in k$ (put $\pi a \in k$ into HB) and $\Im, w, t \models \pi a \notin \emptyset$ (put $\pi a \notin \emptyset$ into HB).

□

The first really important theorem in this section states that a clause set are satisfiable if the set of all ground literals is satisfiable. The proof makes use of the results about our version of Herbrand interpretations.

Theorem 18.89. *If all ground instances C_{gr} of a clause set C is satisfiable then C itself is satisfiable.*

Proof If C_{gr} is satisfiable then it is satisfiable by a Herbrand interpretation $\Im$ (Lemma 18.88). Consider for a clause c with variables $x_1, \ldots, x_n$ the interpretation $\Im' \stackrel{\text{def}}{=} \Im[x_1/\gamma_1, \ldots, x_n/\gamma_n]$. We show that the interpretation of the world-paths with the given bindings can be $\sim$-approximated by suitable ground world-paths. This gives us the corresponding ground instance, satisfied by $\Im$. Since Herbrand interpretations do not distinguish $\sim$-equivalent time points, we then can conclude that $\Im$ satisfies c.

For each world-path πx_l occurring in c with $\Im'(\pi x_l)(w_{[]}, t_0) \stackrel{\text{def}}{=} (w, t)$ there is some selected t_{π_l} with $t_{\pi_l} \sim t$. Therefore, by Lemma 18.85, for every ground world-path π', $\Im'(\pi x_l \pi')(w_{[]}, t_0) \sim \Im'(\pi_l \pi')(w_{[]}, t_0)$.

This way we can find for all variables x_l with prefix $\pi_0 \pi$, some ground world-path π_l such that for $\sigma \stackrel{\text{def}}{=} (\pi x_1 \mapsto \pi_l, \ldots)$, for all world-paths π occurring in c, $\Im'(\pi\sigma)(w_{[]}, t_0) \sim \Im'(\pi)(w_{[]}, t_0)$.

We have to show that the constraints generated in $c\sigma$ are satisfied. That means for a substitution pair $\pi x^{s_i} \mapsto \pi_l$, $\pi_0 \pi_l \in i$ must be satisfied ,and for a substitution pair $\pi x^{s_\tau} \mapsto \pi_l$, $\bigwedge_{i \in \delta((\pi_0 \pi)_{[[},\tau)} (\pi_0 \pi \in i \rightarrow \pi_0 \pi_l \in \tau_{[[}(i))$ must be satisfied.

Since for the binding x^{s_l}/γ_l, $\gamma_l \in AF_{s_l}^{\Im}$ is required, we have $\Im'(\pi_0 \pi x_l^{s_i})(w_{[]}, t_0)_{|T}$ $\in i$, and therefore $\Im'(\pi_0 \pi_l)(w_{[]}, t_0)_{|T} \in i$ as well.

The argument for $x_l^{s_\tau}$ is similar.

Thus, one of the direct instances $L\sigma$ of $c\sigma$ must be satisfied by $\Im$ (which satisfied all ground instances). With Corollary 18.86 we can conclude‘ that $\Im' \models L$ and therefore $\Im \models C$. □

The next step for the proof of ground completeness of resolution is the introduction of semantic trees. Semantic trees are compact representations of the set of all Herbrand interpretations. They are a convenient means to get a handle on a usually infinite quantification ‘for all interpretations’.

Definition 18.90 (Semantic tree). A *semantic tree* for a finite set of $\mathcal{LCL}_{\mathcal{T}\mathcal{U}}^{\mathcal{M}}$-clauses is a tree where each node is labelled with ground literals such that the branches of the tree represent all Herbrand bases.

We follow the construction rules for Herbrand bases (18.83), such that each non-deterministic rule defines corresponding branches in the tree.

The root node is labelled with all elementary tautologies.

Let L be the current leaf node and B_L the set of labels of the branch from the root node leading to L. L is extended as follows:

Case ‘$\pi \in \emptyset$’ $\in B_L$: For an accessibility term ‘a’, extend L:

L (‘$\pi \in \emptyset$’ $\in B_L$)

|

$\pi a \in \emptyset.$

Case ‘$\pi \notin \emptyset$’ $\in B_L$:

Subcase a^{s_m}: Extend L to

$$L \; (\text{`}\pi \notin \emptyset\text{'} \in B_L)$$

$$\pi a \in \emptyset \qquad \begin{array}{l} \pi a \notin \emptyset \\ \pi a \in j \text{ if `}\pi \in j\text{'} \in B_L. \end{array}$$

Subcase a^{s_j}: $\{k_1, \ldots, k_n\} = Batoms(j, C)$: Extend L to

$$L \; (\text{`}\pi \notin \emptyset\text{'} \in B_L)$$

$$\begin{array}{ccc} \pi a \in k_1 & \ldots & \pi a \in k_n \\ \pi a \notin \emptyset & \ldots & \pi a \notin \emptyset. \end{array}$$

Subcase a^{s_τ}, sub-subcase $\tau_{[\![}(i) = \emptyset$ where '$\pi \in i$' $\in B_L$: Extend L to

$$L \; (\text{`}\pi \notin \emptyset\text{'} \in B_L)$$

$$\pi a \in \emptyset.$$

Sub-subcase $\tau_{[\![}(i) \neq \emptyset$ where '$\pi \in i$' $\in B_L$: $\{k_1, \ldots, k_n\} = Batoms(j, C)$: Extend L to

$$L \; (\text{`}\pi \in i\text{'} \in B_L)$$

$$\begin{array}{ccc} \pi a \in k_1 & \ldots & \pi a \in k_n \\ \pi a \notin \emptyset & \ldots & \pi a \notin \emptyset. \end{array}$$

If '$\pi \notin \emptyset$' $\in B_L$ then extend L to

$$L \; (\text{`}\pi \notin \emptyset\text{'} \in B_L)$$

$$\pi : p \qquad \pi : \neg p.$$

The construction of the semantic tree follows exactly the definitions of Herbrand bases (Definition 18.83). Therefore semantic trees represent all Herbrand bases.

Lemma 18.91 (Semantic trees and Herbrand models). *The set of branches of a semantic tree (Definition 18.90) for a clause set C represents all Herbrand models (Definition 18.83) for C.*

The first main result in this section is the $\mathcal{LCL}^{\mathcal{M}}_{\mathcal{T}_\mathcal{U}}$-counterpart of Herbrand's theorem:

Theorem 18.92 (Herbrand's theorem). *An unsatisfiable clause set C has a finite unsatisfiable set of ground instances.*

Proof A clause set C is unsatisfiable iff the set C_{gr} of its ground instances is unsatisfiable (Theorem 18.89). This means that for every Herbrand interpretation $\Im$ there is a ground clause c all of whose literals are false in $\Im$. By Lemma 18.91 this means that each branch B in a semantic tree for C falsifies all literals in c. c has only finitely many literals. Therefore a finite part of the branch B is sufficient to falsify c. Let B_c be the topmost node in B needed to falsify c. B_c is called a *failure node*. Each branch has a failure node at finite depth. Since the branching rate of the semantic tree is finite, there are only finitely many failure nodes $B_{c_1}, \ldots, B_{c_m}$ (König's lemma). The set $\{c_1, \ldots, c_m\}$ of ground instances is false in all Herbrand interpretations, and therefore in all interpretations. □

Now we can start putting the results together to obtain the completeness of the resolution calculus, first for the ground case, and then by lifting it to the general case.

Theorem 18.93 (Completeness of ground resolution). *An unsatisfiable ground clause set C can be refuted using the resolution rules in Definition 18.79 and the simplification rules in Proposition 18.75.*

Proof By the Herbrand theorem 18.92 there is a finite unsatisfiable subset of C, and a finite part of the semantic tree for C, all of whose leaf nodes are failure nodes. A node, all of whose subnodes are failure nodes, is called an *inference node*.

Inference nodes are used to guide the resolution steps such that the clause set plus the new resolvents have a smaller semantic tree, until eventually the semantic tree is empty and the clause set contains the empty clause.

Following the definition of semantic trees, we analyse every potential inference node and show that either some higher node must already be an inference node, or the inference node triggers a resolution step.

The root node is labelled with all elementary tautologies. These labels falsify all elementary contradictions. The simplication rules in Proposition 18.75 together with the ($restr$)-rule, the ($inst_i$)-rule and the ($inst_\tau$)-rule reduce them to $\perp$.

Case '$\pi \in \emptyset$' $\in B_L$: The potential inference node is

$$\begin{array}{l} L \ (\text{`}\pi \in \emptyset\text{'} \in B_L) \\ \quad | \\ \pi a \in \emptyset. \end{array}$$

$\pi a \in \emptyset$ refutes $\pi a \pi' \notin \emptyset$, $\pi a \pi' \in i$ for all non-empty i, and $\pi a \pi' : \pm p$. All these cases are already refuted by the node labelled with $\pi \in \emptyset$. Therefore L cannot be an inference node.

Case '$\pi \notin \emptyset$' $\in B_L$:

Subcase a^{s_m}: L is

$$L\ (\text{`}\pi \notin \emptyset\text{'} \in B_L)$$

$$\pi a \in \emptyset \qquad \pi a \notin \emptyset$$
$$\pi a \in j \text{ if `}\pi \in j\text{' } \in B_L.$$

The left node, $\pi a \in \emptyset$, refutes
(a) $\pi a \pi' \notin \emptyset$ and
(b) $\pi a \pi' \in i$ for all non-empty i and
(c) $\pi a \pi' : \pm p$.
The right node refutes
(d) $\pi a \in \emptyset$ and
(e) $\pi a \in j'$ where '$\pi \in j$' $\in B_L$ and $j \cap j' = \emptyset$.
$\pi a \in j'$, however, is already refuted by the node labelled with $\pi \in j$. Therefore (e) can be excluded.
(d) with all other cases triggers the $res_\emptyset$-rule.
In all cases L becomes a failure node for the resolvents.

Subcase a^{s_j}: $\{k_1, \ldots, k_n\} = Batoms(j, C)$: L is

$$L\ (\text{`}\pi \notin \emptyset\text{'} \in B_L)$$

$$\pi a \in k_1 \ \ldots\ \pi a \in k_n$$
$$\pi a \notin \emptyset \ \ldots\ \pi a \notin \emptyset.$$

Each subnode l refutes
(a) $\pi a \pi' \in j_l$ with $restrict(\pi a \pi' \in j_l, \pi a) =$ '$\pi a \in j_l'$' and $j_l' \cap k_l = \emptyset$
(b) $\pi a \in \emptyset$.
Case (b) triggers the simplification rule 18.75(e) (since '$\pi \in \emptyset$' $\in HB$, it must be $\pi_{[[} \neq \emptyset$.)
Literals $\pi a \pi' \in j$ are refuted by all nodes $\pi a \in k$ with $k \cap j = \emptyset$. If all failure nodes refute the same clause, this clause has been simplified by rule 18.75(c). The literal became $\pi a \in \emptyset$, and we actually have case (a).

If there are at least two clauses with literals $\pi a \pi_1' \in j_1$ and $\pi a \pi_2' \in j_2$, this triggers the the *restr*-rule which generates the residue $\pi a \in j_1 \cap j_2$. This literal is refuted by the union of the nodes which refuted $\pi a \in j_1$ and $\pi a \in j_2$. Therefore one clause less is needed to be refuted by all failure nodes. Repeating this step eventually makes L a failure node.

Subcase a^{s_τ}, sub-subcase $\tau_{[[}(i) = \emptyset$ where '$\pi \in i$' $\in B_L$, $i \neq \emptyset$: L is

$$L\ (\text{`}\pi \notin \emptyset\text{'} \in B_L)$$

$$\pi a \in \emptyset.$$

For all literals refuted by this node one can apply the res_τ-rule. The resulting literal $\pi \in \emptyset$ is refuted by the node labelled $\pi \in i$.

Sub-subcase $\tau_{[[}(i) \neq \emptyset$ where '$\pi \in i$' $\in B_L$: $\{k_1, \ldots, k_n\} = Batoms(j, C)$: L is

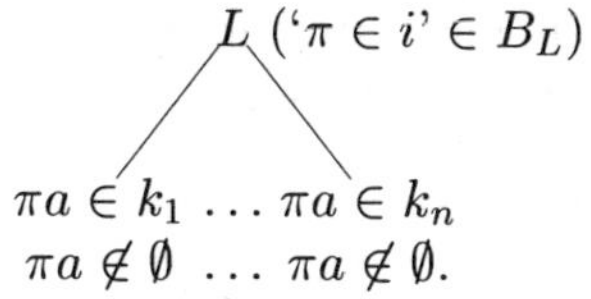

$$
\begin{array}{c}
L \ (\text{`}\pi \in i\text{'} \in B_L) \\
\swarrow \qquad \searrow \\
\pi a \in k_1 \ \ldots \ \pi a \in k_n \\
\pi a \notin \emptyset \ \ldots \ \pi a \notin \emptyset.
\end{array}
$$

The argument is the same as in the above case where the res_i-rule was used.

If L is

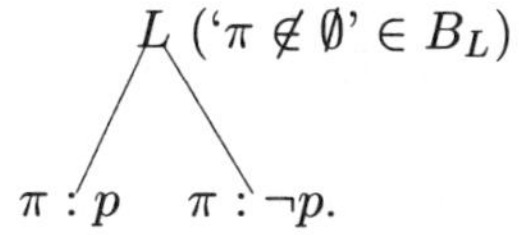

$$
\begin{array}{c}
L \ (\text{`}\pi \notin \emptyset\text{'} \in B_L) \\
\swarrow \qquad \searrow \\
\pi : p \qquad \pi : \neg p.
\end{array}
$$

then this situation triggers the res_p-rule. The residue literal $\pi \in \emptyset$ is refuted by the node labelled with $\pi \notin \emptyset$. □

Lemma 18.94 (Lifting lemma). *If C_{gr} is an unsatisfiable set of ground instances of an unsatisfiable clause set C then every resolution step determined by the corresponding semantic tree (Theorem 18.93) can be lifted to the general level.*

Proof Using Lemma 18.73 one can easily conclude that for each resolution step involving ground instances $c\lambda$ of some general clauses, one can find a most general unifier σ such that the corresponding resolvent with σ subsumes the resolvent with the ground instance, provided the resolution literals are *direct instances* of literals in c, and not residue literals introduced by λ.

The nodes in the semantic tree are ordered such that longer world-paths occur at deeper levels in the tree than shorter world-paths. Since the ground resolution steps are triggered by the deepest inference nodes, literals with longer world-paths are chosen first for a resolution step. Therefore the proof reduces to showing that the residue literals for a ground resolvent with largest world-path are, in fact, direct instances of the residue literals of the corresponding general resolvent.

The residue literals are generated from T-blocks, $\pi x^s i$ or $\pi^s \tau$ which were instantiated with some π'. In the first case the corresponding residue literal is $\pi' \in i$. For a most general unifier σ, $\pi' \in i$ is in fact the only residue literal for $\pi x^s i$. For the ground unifier λ, $\pi'\lambda \in i$ is one residue literal. Other residue literals may be $\pi'' \in j$ where π'' is an initial segment of $\pi'\lambda$. The important observation is that π'' is shorter than $\pi'\lambda$. Therefore the direct instance $\pi'\lambda$ is chosen for a further resolution step before the literal with π''.

Thus, each resolution step can be lifted to the general level. □

Corollary 18.95 (Completeness of resolution). *A finite unsatisfiable clause set C can be refuted using the resolution rules in Definition 18.79.*

Proof C has a finite unsatisfiable set of ground instances (Theorem 18.92). This can be refuted on the ground level (Theorem 18.93). To each step at the ground level there corresponds a step at the general level (Lemma 18.94), which eventually generates the empty clause. □

18.4.4.7 *Termination of resolution* In general, resolution is not a complete inference procedure. It may not terminate if the original clause set is satisfiable. In this case resolution may produce deeper and deeper nested terms, or in the $\mathcal{LCL}^{\mathcal{M}}_{\mathcal{T}u}$-case longer and longer world-path, or it may produce longer and longer clauses. If the satisfiability of the logic or the fragment of the logic under consideration is decidable, there is usually a resolution strategy which is complete and terminates the resolution. But this strategy may be arbitrarily complex and not of any practical value.

The situation is better if we can show that the number of different clauses is bounded, which is, for example, the case for propositional logic. In this case resolution eventually can no longer produce new clauses. This is, in fact, the optimal case to get a terminating resolution procedure. It means that all *refutation complete* resolution strategies automatically lead to a decision procedure, and there are quite a number of them.

Renate Schmidt has investigated the termination behaviour of resolution for *path logic* [Schmidt, 1997], which is essentially the target clause logic of the functional translation. As a consequence of the prefix stability or world-path, she showed that if the length of the world-paths can be bounded then the length of the *condensed* clauses is bounded as well. Condensed clauses are clauses which are not *subsumed* by one their own factors. (A clause C is subsumed by a clause D if $D\sigma \subseteq C$ for a substitution σ. This notion of subsumption is applied to $\mathcal{LCL}^{\mathcal{M}}_{\mathcal{T}u}$-clauses as well, with the appropriate instantiation procedure.) This means that if all initial clauses and resolvents with a subsuming factor are always replaced by this factor then the size of clauses stays finite, provided the length of clauses is bounded.

Theorem 18.96 (Termination of resolution for path logic ([Schmidt, 1997])). *In path logic (i.e. the fragment of predicate logic which corresponds to the functional translation of modal logics) there are only finitely many different condensed clauses with world-path of bounded size.*

As a consequence, resolution and condensing on path formulae terminates if the length of the world-paths can be bounded. This is automatically the case for systems like K, KD, T, B, S5 where world-path unification does not increase the size of the world-paths, or more general modal systems with a finite model property and a known upper bound for the number of worlds.

Whether the length of the world-paths is bounded depends on the structure of the unifier for the world-paths, and this depends on the properties of the modal logic. In the K4 and S4 cases (transitivity of the accessibility relation), variables can be replaced with longer world-paths. Therefore their length can only be bounded via the finite model property. Unification of world-paths or

K, KD, T and B-type modalities always replaces variables with world-paths of length at most 1, which means that the world-paths never become larger during resolution.

Instantiated T-blocks of world-paths of $\mathcal{LCL}_{\mathcal{T}u}^{\mathcal{M}}$-clauses can become larger than the original world-paths. The key observation is that a unifier σ for the T-blocks π_1 and π_2 instantiates them in such a way that the length of $\pi_1\sigma$ and $\pi_2\sigma$ is the maximum of the length of π_1 and π_2. Therefore T-blocks can never become larger than the length of all concatenated T-blocks in the initial clause set (since some M-blocks can be replaced with empty world-paths, T-blocks can be concatenated).

Lemma 18.97 (Bounded length of temporal world-paths). *If modal unification keeps the length of the world-paths bounded, then resolution and factoring (Definition 18.79) keeps the length of world-paths bounded.*

Proof For a clause set C, let C' be the superset of C' containing for each clause $c \in C$ all the clauses obtainable by deleting M-blocks. Since there are finitely many T-blocks in each clause, C' is finite. Let $max(C)$ be the maximum length of T-blocks in C'. We show that $max(C)$ does not increase by adding resolvents or factors. This has to be proved by checking the unification rules. The only non-trivial case is the $(m[])$-rule where two T-blocks are merged. The length of the instance is the sum of the length of the two T-blocks. The effect of this unification step can be achieved by using instead of the given clause $c \in C$ the corresponding clause in C' where the corresponding M-blocks are already deleted. Therefore the $(m[])$-rule is, in principle, not necessary. All other rules do not increase the length of the world-paths. □

Since T-blocks have bounded lengths, Renate Schmidt's result can be applied to $\mathcal{LCL}_{\mathcal{T}u}^{\mathcal{M}}$-clauses as well. As a corollary of Theorem 18.96 we therefore obtain a termination result for resolution.

Corollary 18.98 (Termination of resolution). *If resolution and condensing for the functionally translated basic multimodal logic formulae terminates then resolution and condensing terminates for every TF_2-translated $\mathcal{CL}_{\mathcal{T}u}^{\mathcal{M}}$-formula.*

Theorem 18.99 (Decision procedure). *If resolution terminates for functionally translated multimodal logic formulae then resolution is a decision procedure for checking the satisfiability of a finite set of clauses.*

Proof This is a consequence of soundness (Theorem 18.81), completeness of resolution (Corollary 18.95), and the termination proof (Corollary 18.98). □

The number of different $\mathcal{LCL}_{\mathcal{T}u}^{\mathcal{M}}$-clauses, although finite, can be very large. Resolution as a decision procedure therefore need not be the optimal choice. Extensive tests and comparisons with other procedures, mainly for modal logics K and KD, however, have shown that it performs very well, even if a general predicate logic theorem prover is used, without any special tuning for this class of formulae [Hustadt and Schmidt, 1997].

18.5 Summary

A rigorous non-idealized formalization of temporal notions is one piece in the big puzzle to make computers 'understand' our world and to communicate with us human beings in our language. In contrast to most notions used in natural language, which are fuzzy or at least context dependent and therefore difficult to formalize, most temporal notions are precise in their meaning and regular enough to allow for a relatively simple formalization.

With calendar logic we provide a specification language which contains only a few primitive concepts, but which allows one to define an infinite variety of temporal notions, and to link them to standard time systems and calendars. Using the time term language $\mathcal{T}_{\mathcal{U}}$ one can check whether a given point in time lies within a time region specified by a time term, e.g. whether it is lunch time right now. It is also possible to check whether two time terms subsume each other, e.g. whether lunch time is in the afternoon or in the morning.

Calendar logic itself has more expressivity. With the operators $\langle\tau\rangle$ and $[\tau]$ one can express for example 'sometime next week': $\langle x_{week} + 1\rangle \ldots$, or 'always last Wednesday': $[(Wednesday(x_{week}) < y_{day})?Wednesday(x_{week}) : Wednesday(x_{week} - 1)] \ldots$, or 'every Monday this year': $[1998, year]$ $(Monday(x_{week}) \rightarrow \ldots)$.

Since calendar logic is a logic, one can draw inferences, prove theorems, find counter examples in a very similar way to many other logics. We presented three different ways to do this: by translating the formulae into classical propositional logic and using a propositional logic inference procedure, or by a special tableau calculus, or by translating them into predicate logic and using resolution. The 'translation and resolution' method also works for a combination of calendar logic with other propositional modal logics. It might also work for a combination with quantified modal logic or ordinary predicate logic, but this has not been investigated.

The question is whether the methodology is restricted to time and temporal notions. In the technical details of the algorithm we make use of the underlying time structure, real numbers, and the algorithms only need to deal with finite sets of intervals of real numbers. In Euclidean space, for example, the underlying structure must be $\mathbb{R}^2$ or $\mathbb{R}^3$. The adequate formalization of spatial notions in higher dimensional space is much more complicated than in $\mathbb{R}^1$. But this is not a principal problem.

There is another aspect which makes the definition of temporal notions quite unique, compared with other areas of natural language. There is a multitude or temporal notions, which can be specified with a few primitive concepts in such a way that it is possible to reduce all computations to computations with floating point numbers. Most other areas of natural language, in particular spatial notions, are far more complex and (the greatest obstacle) have a much less precise meaning.

The essential features which characterize the methodology presented in this chapter, however, might be found in other areas as well. These are that some syntactic notions correspond to subsets of a global set, and it is possible to compute a finite representation of these subsets (in the temporal logic case these are the finite sets of temporal intervals). Other syntactic notions correspond to functions mapping elements of these sets to, again finitely representable, subsets, and, most importantly, these functions are constant over a finite partitioning of the finitely representable sets. For such a notion and for each finitely representable subset it must be possible to compute the partitioning again, as a finitely representable set of subsets.

The particular language for specifying such notions will most likely be very different to $\mathcal{T}_{\mathcal{U}}$, but the methods developed for calendar logic and its combination with multimodal logic should still work.

REFERENCES

[Abadi and Lamport, 1992] M. Abadi and L. Lamport. An old-fashioned recipe for real time. In J.W. de Bakker *et al.*, editors, *Real-time': theory in practice. REX workshop, Mook, The Netherlands, June 3–7, 1991, proceedings*, Lecture notes in computer science, no. 600. Springer-Verlag, 1992.

[Abadi and Manna, 1989] M. Abadi and Z. Manna. Temporal Logic Programming. *Journal of Symbolic Computation*, **8**, 277–295, 1989.

[Abiteboul *et al.*, 1995] S. Abiteboul, R. Hull and V. Vianu. *Foundations of databases.* Addison-Wesley, 1995.

[Abiteboul *et al.*, 1996] S. Abiteboul, L. Herr and J. Van den Bussche. Temporal versus first-order logic to query temporal databases. In *Proc. ACM Conference on Principles of Database Systems, PODS*, pages 49–57, 1996.

[Abrahamson, 1979] K. Abrahamson. Modal logic of concurrent nondeterministic programs. In *Lecture notes in computer science, no. 70*, Springer-Verlag, 1979.

[Abrahamson, 1980] K. Abrahamson. *Decidability and expressiveness of logics of programs.* PhD thesis, University of Washington, 1980.

[Agha, 1986] G. Agha. *Actors – a model of concurrent computation in distributed systems.* MIT Press, 1986.

[Aiken *et al.*, 1992] A. Aiken, J. Widom and J. M. Hellerstein. Behavior of database production rules: termination, confluence and observable determinism. In *Proceedings of the ACM SIGMOD International Conference on the Management of Data*, pages 58–67, 1992.

[Alechina, 1995] N. Alechina. *Modal quantifiers.* ILLC Dissertation series 1995–20. Institute for Logic, Language and Information, University of Amsterdam.

[Alechina and van Lambalgen, 1996] N. Alechina and M. van Lambalgen. Generalised quantification as substructural logic. *Journal of Symbolic Logic*, **61**, 1006–1044, 1996.

[Allen, 1981] J. F. Allen. An interval-based representation of temporal knowledge. In *Proc. 7th IJCAI*, 1981.

[Allen and Hayes, 1985] J. F. Allen and P. Hayes. A commonsense theory of time: the longer paper. Technical report, University of Rochester, 1985.

[Allen and Koomen, 1983] J. Allen and J. Koomen. Planning using a temporal world model. In *Proc. 8th IJCAI*, 1983.

[Allen *et al.*, 1991] J. Allen, H. Kautz, R. Pelavin' and J. Tenenberg. *Reasoning about Plans.* Morgan Kaufman, 1991.

[Andreka *et al.*, 1995] H. Andréka, J van Benthem and I. Németi. Back and forth between modal logic and classical logic. *Logic Journal of the IGPL*, **3**, 685–720, 1995.

[Aqvist, 1979] L. Aqvist. A conjectured axiomatization of two-dimensional Reichenbachian tense logic. *Journal of Philosophical Logic*, **8**, 1–45, 1979.

[Banâtre and Le Métayer, 1990] J. Banâtre and D. Le Métayer. The Gamma model and its discipline of programming. *Science of Computer Programming*, **15**, 55–77, 1990.

[Banâtre and Le Métayer, 1993] J. Banâtre and D. Le Métayer. Programming by multiset transformation. *Communications of the ACM*, **36**, 98–122, 1993.

[Banâtre *et al.*, 1988] J. Banâtre, A. Coutant and D. Le Métayer. A parallel machine for multiset transformation and its programming style. *Future Generation Computer Systems*, **4**, 133–144, 1988.

[Banieqbal *et al.*, 1987] B. Banieqbal, H. Barringer and A. Pnueli, editors. *Temporal logic in specification.* Lecture notes in computer science, no 398. Springer-Verlag, 1987.

[Barringer, 1985] H. Barringer. Up and down the temporal way. Technical report UMCS-85-9-3, Department of Computer Science, University of Manchester, 1985.

[Barringer, 1987] H. Barringer. The use of temporal logic in the compositional specification of concurrent systems. In A. Galton, editor, *Temporal logics and their applications.* Academic Press, 1987.

[Barringer *et al.*, 1984] H. Barringer, R. Kuiper and A. Pnueli. Now you may compose temporal logic specifications. In *Proc. 16th ACM Symp. on Theory of Computation*, pages 51–63, 1984.

[Barringer *et al.*, 1989] H. Barringer, M. Fisher, D. M. Gabbay, G. Gough and R. P. Owens. METATEM: a framework for programming in temporal logic. In *REX Workshop on Stepwise Refinement of Distributed Systems: Models, Formalism, Correctness*, Mook, Netherlands, pages 94–129. Lecture notes in computer science, no. 430, Springer-Verlag, 1989.

[Barringer *et al.*, 1991] H. Barringer, M. Fisher, D. Gabbay and A. Hunter. Meta-reasoning in executable temporal logic. In *Second Conference on the Principles of Knowledge and Reasoning*, pages 40–49, Morgan Kaufmann, San Mateo, California, 1991.

[Barringer *et al.*, 1996] H. Barringer, M. Fisher, D. Gabbay, R. Owens and M. Reynolds, editors. *The imperative future — principles of executable temporal logic.* Research Studies Press, 1996.

[Batini *et al.*, 1992] C. Batini, S. Ceri and S. Navathe. *Conceptual database design — an entity-relationship approach.* Benjamin Cummings, 1992.

[Baudinet *et al.*, 1991] M. Baudinet, M. Niezette and P. Wolper. On the representation of infinite temporal data and queries. In *10th ACM Symp. on Principles of Database Systems*, pages 280–290, 1991.

[Baudinet *et al.*, 1993] M. Baudinet, J. Chomicki and P. Wolper. Temporal deductive databases. In [Tansel *et al.*, 1993], chapter 13, pages 294–320. Benjamin Cummings, 1993.

[Belnap, 1996] N. Belnap. Agents in branching time. In J. Copeland, editor. *Logic and reality: essays in pure and applied logic, in memory of Arthur Prior*, pp. 239–272. Oxford University Press, 1996.

[Belnap and Green, 1994] N. Belnap and M. Green. Indeterminism and the red thin line. In *Philosophical Perspectives, 8, Logic and Language*, pages 365–388. 1994.

[Ben-Ari *et al.*, 1981] M. Ben-Ari, Z. Manna and A. Pnueli. The temporal logic of branching time. In *8th ACM Symp. on Principles of Programming Languages*, pages 164–176. Williamburg, VA, 1981.

[Bencivenga, 1986] E. Bencivenga. Free logic. In D. M. Gabbay and F. Guenthner, editors, *Handbook of philosophical logic*, Vol. 3. Kluwer, 1986.

[Bennett, 1996] B. Bennett. Modal logics for qualitative spatial reasoning. *Journal of the Interest Group in Pure and Applied Logic (IGPL)*, **4**, 1996.

[van Benthem, 1983] J. van Benthem. *The logic of time.* Reidel, 1983.

[van Benthem, 1995] J. van Benthem. Temporal logic. In D. Gabbay, editor, *Handbook of logic in artificial intelligence and logic programming*, Vol. 4. Oxford University Press, 1995.

[van Benthem, 1996] J. van Benthem. *Exploring logical dynamics.* Cambridge University Press, 1996.

[van Benthem and Alechina, 1996] J. van Benthem and N. Alechina. Modal quantification over structured domain. In M. de Rijke, editor, *Advances in Intensional Logic*, pp. 1–27. Kluwer, 1996.

[Bölen *et al.*, 1996] M. H. Bölen, J. Chomicki, R. T. Snodgrass and D. Toman. Querying TSQL2 databases with temporal logic. In *Advances in Databases (EDBT'96)*, pages 325–341, 1996.

[Boole, 1948] G. Boole. *The mathematical analysis of logic, being an essay towards a calculus of deductive reasoning.* Blackwell, 1948.

[Boolos and Jeffrey, 1989] G. S. Boolos and R. C. Jeffrey. *Computability and Logic.* Cambridge University Press, third edition, 1989.

[Brough *et al.*, 1996] D. Brough, M. Fisher, A. Hunter, R. Owens, H. Barringer, D. Gabbay, G. Gough, I. Hodkinson, P. McBrien and M. Reynolds. Languages, meta-languages and METATEM: a discussion paper. *Journal of the IGPL*, **4**, 255–273, March 1996.

[Brownston *et al.*, 1985] L. Brownston, R. Farrell, E. Kant and N. Martin. *Programming expert systems in OPS5: an introduction to rule-based programming.* Addison-Wesley, 1985.

[Büchi, 1962a] J. R. Büchi. On a decision method in restricted second order arithmetic. In *Logic, Methodology, and Philosophy of Science: Proc. 1960 Int. Congress*, pages 1–11. Stanford University Press, 1962.

[Buchi, 1962b] J. R. Buchi. On a decision method in restricted second order arithmetic. In *Logic, Methodology, and Philosophy of Science. Proc. 1960 Int. Congress*, pages 1–11. Stanford University Press, 1962.

[Büchi, 1983] J. R. Büchi. State strategies for games in $F_{\sigma\delta} \cap G_{\sigma\delta}$. *Journal of Symbolic Logic*, **48**, 1171–1198, 1983.

[Büchi and Landweber, 1969] J. R. Büchi and L. H. Landweber. Solving sequential conditions by finite state operators. *Transactions of the AMS*, **138**,

295–311, 1969.

[Burgess, 1979] J. P. Burgess. Logic and time. *Journal of Symbolic Logic*, **44**, 566–582, 1979.

[Burgess, 1982] J. P. Burgess. Axioms for tense logic I: "Since" and "Until". *Notre Dame Journal of Formal Logic*, **23**, 367–374, October 1982.

[Burgess, 1984] J. P. Burgess. Basic tense logic. In D. Gabbay and F. Guenthner, editors, *Handbook of philosophical logic*, volume II, pages 89–133. D. Reidel, 1984.

[Burgess and Gurevich, 1985] J. P. Burgess and Y. Gurevich. The decision problem for linear logic. *Notre Dame Journal of Formal Logic*, **26**, 566–582, April 1985.

[Büttner and Simonis, 1987] W. Büttner and H. Simonis. Embedding boolean expressions into logic programming. *Journal of Symbolic Computation*, 1987.

[Castro, 1990] J. Castro. Temporal-causal logic: A marriage of convenience. Manuscript 1990.

[Castro, 1991] J. Castro. A pluralistic approach to distributed system specification. In *Proceedings of Second IEEE Symposium on Parallel and Distributed Processing*, 1991.

[Castro and Kramer, 1990a] J. Castro and J. Kramer. Constructing distributed system specification: a temporal-causal approach. In *Proceedings of XVII SEMISH-integrated Software and Hardware Seminar, Congress of the Brazilian Computing Society*, July 1990.

[Castro and Kramer, 1990b] J. Castro and J. Kramer. Temporal-causal system specifications. In *Proceedings of IEEE International Conference on Computer Systems and Software Engineering (CompEuro90)*, May 1990.

[Chargov and Zakharayaschev, 1997] A. Chargov and M. Zakharayaschev. *Modal logic*, Vol. 35 of *Oxford logic guides.* Oxford University Press, 1997.

[Chaudron and de Jong, 1996] M. Chaudron and E. de Jong. Schedules for multiset transformer programs. In J.-M. Andreoli and C. Hankin, editors, *Coordination programming: mechanisms, models and semantics.* Imperial College Press, 1996.

[Chellas, 1980] B. F. Chellas. *Modal logic: an introduction.* Cambridge University Press, 1980.

[Chellas, 1992] B. Chellas. Time and modality in the logic of agency. *Studia Logica*, **51**, 485–517, 1992.

[Chen, 1976] P. Chen. The entity-relationship model: toward a unified view of data. *ACM Transactions on Database Systems*, **1**, 9–36, 1976.

[Chomicki and Imieliński, 1988] J. Chomicki and T. Imieliński. Temporal deductive databases and infinite objects. In *7th ACM Symposium on Principles of Database Systems*, pages 61–73, Austin, 1988.

[Clark, 1978] K. Clark. Negation as failure. In H. Gallaire and J. Minker, editors, *Logic and data bases*, pages 293–322. Plenum Press, 1978.

[Clarke and Emerson, 1981] E. Clarke and E. Emerson. Synthesis of synchronization skeletons for branching time temporal logic. In *Proc. IBM Workshop on Logic of Programs, Yorktown Heights, NY*, pages 52–71. Springer-Verlag, 1981.

[Clarke *et al.*, 1986] E. Clarke, E. Emerson and A. Sistla. Automatic verification of finite-state concurrent systems using temporal logic specifications. *ACM Toplas*, **8**, 244–263, 1986.

[Cleaveland *et al.*, 1989] R. Cleaveland, J. Parrow and B. Steffen. The concurrency workbench: a semantics based tool for the verification of concurrent systems. Technical report ECS-LFCS-89-83, Department of Computer Science, University of Edinburgh, 1989.

[Codd, 1970] E. F. Codd. A relational model for large shared data banks. *Communications of the ACM*, **13**, 377–387, 1970.

[Codd, 1972] E. F. Codd. Relational completeness of data base sublanguages. In R. Rustin, editor, *Data base systems*, pages 65–98. Prentice-Hall, 1972.

[Courcobetis and Yannakakis, 1990] C. Courcobetis and M. Yannakakis. Markov decision processes and regular events. In *17th Colloquium on Automata, Languages and Programming*, M. S. Patterson, editor. Lecture notes in computer science, no. 85, Springer-Verlag, 1990.

[Creveuil, 1991] C. Creveuil. *Techniques d'analyse et de mise en œuvre de programmes Gamma.* PhD thesis, INRIA, 1991.

[Creveuil, 1992] C. Creveuil. Implementation of Gamma on the connection machine. In *Proc. of the Workshop on Research Directions in High-level Parallel Programming Languages*, Lecture notes in computer science, no. 574, pp. 219–230, Springer-Verlag, 1992.

[Darlington and While, 1987] J. Darlington and L. While. Controlling the behaviour of functional programs. In *Third Conference on Functional Programming Languages and Computer Architecture*, 1987.

[De Morgan, 1864] A. De Morgan. On the syllogism, no. iv, and on the logic of relations. *Transactions of the Cambridge Philosophical Society*, **10**, 331–358, 1864.

[Dean and McDermott, 1987] Dean and McDermott. Temporal database management. *Artificial Intelligence*, **49**, 1–55, 1987.

[Dershowitz and Reingold, 1997] Nachum Dershowitz and Edward M. Reingold. *Calendrical calculations.* Cambridge University Press, 1997.

[Di Maio and Zanardo, 1994] M. Di Maio and A. Zanardo. Synchronized histories in Prior–Thomason representation of branching time. In D. Gabbay and H. Ohlbach, editors, *Temporal Logic, Proceedings of ICTL '94*, Lecture notes in artificial intelligence, no. 827, pages 265–282. Springer-Verlag, 1994.

[Kirchner, 1990] Claude Kirchner, editor. *Unification theory.* Academic Press, 1990.

[Elmasri *et al.*, 1993] R. Elmasri, G. T. J. Wuu and V. Kouramajian. A temporal model and query language for *eer* database. In [Tansel *et al.*, 1993]. '

[Emerson, 1996] E. Emerson. Automated temporal reasoning for reactive systems. In F. Moller and G. Birtwistle, editors, *Logics for concurrency*, pages 41–101. Springer-Verlag, 1996.

[Emerson and Clarke, 1980] E. Emerson and E. Clarke. Characterizing correctness properties of parallel programs using fixpoints. In *Proceedings 7th Int. Colloquium on Automata, Languages and Programming*, Lecture notes in computer science, no. 85, pages 169–181. Springer-Verlag, 1980.

[Emerson and Clarke, 1982] E. Emerson and E. Clarke. Using branching time temporal logic to synthesise synchronisation skeletons. *Sci. of Computer Programming*, **2**, 241–266, 1982.

[Emerson and Halpern, 1986] E. Emerson and J. Halpern. 'Sometimes' and 'not never' revisited: on branching versus linear time. *Journal of the ACM*, **33**, 1986.

[Emerson and Sistla, 1984] E. Emerson and A. Sistla. Deciding full branching time logic. *Information and Control*, **61**, 175–201, 1984.

[Enderton, 1972] H. B. Enderton. *A mathematical introduction to logics.* Academic Press, 1972.

[Fagin, 1982] R. Fagin. Horn clauses and database dependencies. *Journal of the ACM*, **29**, 952–985, 1982.

[Fiadeiro and Maibaum, 1994] J. Fiadeiro and T. Maibaum. Sometimes "tomorrow" is "sometime" – action refinement in a temporal logic of objects. In D. Gabbay and H. Ohlbach, editors, *Temporal logic, Proceedings of ICTL '94*, Lecture notes in artificial intelligence, no. 827, pages 48–66. Springer-Verlag, 1994.

[Filipič, 1988] B. Filipič. *PROLOG user's handbook.* Ellis Horwood, 1988.

[Fine and Schurz, 1991] K. Fine and G. Schurz. Transfer theorems for stratified multimodal logics. Unpublished manuscript, 1991.

[Finger, 1992] M. Finger. Handling database updates in two-dimensional temporal logic. *Journal of of Applied Non-Classical Logic*, **2**, 201–224, 1992.

[Finger, 1994] M. Finger. *Changing the past: database applications of two-dimensional temporal logics.* PhD thesis, Imperial College, Department of Computing, 1994.

[Finger and Gabbay, 1992] M. Finger and D. M. Gabbay. Adding a temporal dimension to a logic system. *Journal of Logic Language and Information*, **1**, 203–233, 1992.

[Finger and Gabbay, 1996] M. Finger and D. Gabbay. Combining temporal logic systems. *Notre Dame Journal of Formal Logic*, **37**, 204–232, 1996. Special Issue on Combining Logics.

[Finger and McBrien, 1996] M. Finger and P. McBrien. On the semantics of 'current-time' in temporal databases. In *XI Brazilian Database Symposium (SBBD96)*, pages 324–337, 1996. Also available at ftp://ftp.ime.usp.br/pub/mfinger/SBBD96.ps.gz.

[Finger *et al.*, 1991] M. Finger, P. McBrien and R. Owens. Databases and executable temporal logic. In *Annual Esprit Conference*, Comission of the European Communities, Brussels, 25–29 November, pages 288–302, 1991.

[Finger *et al.*, 1993] M. Finger, M. Fisher and R. Owens. METATEM at work: modelling reactive systems using executable temporal logic. In *Sixth International Conference on Industrial and Engineering Applications of Artificial*

Intelligence and Expert Systems, Edinburgh, 1–4 June, 1993.

[Fisher, 1991] M. Fisher. A resolution method for temporal logic. In *Proceedings of the International Joint Conference on Artificial Intelligence*, August 1991.

[Fisher, 1992] M. Fisher. A normal form for first-order temporal formulae. In *Proceedings of the Eleventh International Conference on Automated Deduction.* Springer-Verlag, June 1992.

[Fisher, 1993] M. Fisher. Concurrent METATEM – a language for modelling reactive systems. In *Parallel architectures and languages, Europe (PARLE)*, 1993.

[Fisher and Barringer, 1991] M. Fisher and H. Barringer. Concurrent METATEM processes—a language for distributed AI. In *Proceedings of the European Simulation Multiconference*, June 1991.

[Fisher and Owens, 1992] M. Fisher and R. Owens. From the past to the future: executing temporal logic programs. In *Proceedings of Logic Programming and Automated Reasoning.* Springer-Verlag, July 1992.

[Gabbay, 1981a] D. M. Gabbay. An irreflexivity lemma. In U. Monnich, editor, *Aspects of philosophical logic*, pages 67–89. Reidel, 1981.

[Gabbay, 1981b] D. M. Gabbay. Expressive functional completeness in tense logic. In U. Monnich, editor, *Aspects of philosophical logic*, pages 91–117. Reidel, 1981.

[Gabbay, 1985] D. Gabbay. N-Prolog, part 2. *Journal of Logic Programming*, **5**, 251–283, 1985.

[Gabbay, 1989] D. M. Gabbay. Declarative past and imperative future: executable temporal logic for interactive systems. In B. Banieqbal, H. Barringer and A. Pnueli, editors, *Proceedings of Colloquium on Temporal Logic in Specification, Altrincham, 1987*, Lecture notes in computer science, no. 398, Springer-Verlag, 1989.

[Gabbay, 1992] D. M. Gabbay. Fibred semantics and combinations of logics. Manuscript, Imperial College, 1992, published in *Journal of Symbolic Logic*, **61**, 1057–1120, 1996.

[Gabbay, 1994] D. M. Gabbay. Labelled Deductive Systems; principles and applications. Vol 1: Basic Principles. 1st draft Manuscript 1989, 2nd intermediate draft, University of Munich , CIS Bericht 90-22, 1990. Third intermediate draft, Max Planck Institute, Saarbrucken, Technical Report, MPI-I-94-223, 1994. Finally published as [Gabbay, 1996].

[Gabbay, 1996] D. M. Gabbay. *Labelled Deductive Systems.* Oxford University Press, 1996.

[Gabbay, 1998] D. M. Gabbay. *Fibring logics.* Oxford University Press, 1998.

[Gabbay, 1998b] D. M. Gabbay. *Elementary Logics: A procedural perspective*, Prentice Hall, 1998, 380 pp.

[Gabbay and McBrien, 1991] D. M. Gabbay and P. McBrien. Temporal logic and historical databases. In *17th Conference on Very Large Databases*, pages 423–430, Barcelona, September 1991.

[Gabbay and Olivetti, 1999] D. M. Gabbay and N. Olivetti. *Goal directed algorithmic proof*, APL Series, Kluwer, 1999.

[Gabbay and Guenthner, 1982] D. Gabbay and F. Guenthener. A note on systems of n-dimensional tense logics. In T. Pauli, editor, *Essays dedicated to L. Aqvist*, pages 63–71, Uppsala, 1982.

[Gabbay *et al.*, 1980] D. M. Gabbay, A. Pnueli, S. Shelah and J. Stavi. On the temporal analysis of fairness. In *7th ACM Symposium on Principles of Programming Languages, Las Vegas*, pages 163–173, 1980.

[Gabbay *et al.*, 1994] D. Gabbay, I. Hodkinson and M. Reynolds. *Temporal logic: mathematical foundations and computational aspects, Volume 1.* Oxford University Press, 1994.

[Gadia and Vaishnav, 1985] S. K. Gadia and J. H. Vaishnav. A query language for a homogeneous temporal databases. In *Proc. ACM Principles of Database Systems, PODS'85*, pages 51–56, 1985.

[Gelerter *et al.*, 1985] D. Gelerter, N. Carriero, S. Chandran and S. Chang. Parallel programming in Linda. In *International Conference on Parallel Processing*, August 1985.

[Gottlob *et al.*, 1996] G. Gottlob, G. Moerkotte and V. S. Subramanian. The park semantics for active rules. In *Proceedings of EDBT'96 (Advances in Database Technology)*, pages 35–54, 1996.

[Gurevich, 1985] Y. Gurevich. Monadic second-order theories. In J. Barwise and S. Feferman, editors, *Model-theoretic logics*, pages 479–507. Springer-Verlag, 1985.

[Gurevich and Harrington, 1982] Y. Gurevich and L. A. Harrington. Trees, automata, and games. In *Theory of Computing* (*Proc. 14th Annual ACM Symposium, San Francisco*), *Association for Computing Machinery*, pages 60–65, 1982.

[Gurevich and Shelah, 1983] Y. Gurevich and S. Shelah. Rabin's uniformisation problem. *Journal of Symbolic Logic*, **48**, 1105–1119, 1983.

[Gurevich and Shelah, 1985] Y. Gurevich and S. Shelah. To the decision problem for branching time logic. In G. Dorn and P. Weingartner, editors, *Foundations of logics and linguistics: problems and their solutions.* Plenum Press, 1985.

[Gurevich and Shelah, 1986] Y. Gurevich and S. Shelah. Fixed-point extensions of first-order logic. *Annals of Pure and Applied Logic*, **32**, 265 – 280, 1986.

[Hafer and Thomas, 1988] T. Hafer and W. Thomas. Computation tree logic ctl* and path quantifiers in the monadic theory of the binary tree. In T. Ottman, editor, *Proc. of ICALP 87*, Lecture notes in computer science, no. 267, pages 267–279. Springer-Verlag, 1988.

[Hale, 1987] R. Hale. Using temporal logic for prototyping: the design of a lift controller. In [Banieqbal *et al.*, 1987], pages 375–408, 1987.

[Halpern and Shoham, 1986] J. Halpern and Y. Shoham. A propositional modal logic of time intervals. In *Proceedings, Symposium on Logic in Computer Science*, IEEE, Boston, 1986.

[Halpern and Vardi, 1989] J. Halpern and M. Vardi. The complexity of reasoning about knowledge and time. I. Lower bounds. *Journal of Computer and System Science*, **38**, 195–237, 1989.

[Hankin *et al.*, 1992] C. Hankin, D. Le Métayer and D. Sands. A calculus of Gamma programs. Technical report, Imperial College, 1992.

[Harel, 1983] D. Harel. Recurring dominoes: making the highly undecidable highly understandable. In *Conference on Foundations of Computing Theory*, pages 177–194, Lecture notes in computer science, no. 158, Springer-Verlag, 1983.

[Harel, 1986] D. Harel. Effective transformations on infinite trees, with applications to high undecidability, dominoes, and fairness. *Journal of the ACM*, **33**, 224–248, 1986.

[Harel and Pnueli, 1985] D. Harel and A. Pnueli. On the development of reactive systems. Technical report CS85-02, Department of Applied Mathematics, The Weizmann Institute of Science, Revohot, Israel, January 1985.

[Hennessy, 1988] M. Hennessy. *Algebraic theory of processes.* MIT Press, 1988.

[Herbrand, 1930] J. Herbrand. *Récherches sur la théorie de la démonstration.* Travaux de la Société des Sciences et des Lettres de Varsovie, Classe III sciences mathematiques et Physiques, no 33, 128 pp. PhD thesis, 1930.

[Hirsch, 1993] R. Hirsch. *Relation algebras of intervals.* PhD thesis, Imperial College, 1993.

[Hirsch, 1996] R. Hirsch. Relation algebras of intervals. *Artificial Intelligence Journal*, **83**, 1–29, 1996.

[Hirsch and Hodkinson, 1997] R. Hirsch and I. Hodkinson. Step by step—building representations in algebraic logic. *Journal of Symbolic Logic*, **62**, 225–279, 1997.

[Hoare, 1985] C. Hoare. *Communicating sequential processes.* Prentice Hall, 1985.

[Hodges, 1985] W. Hodges. Logical features of Horn clauses. In *Handbook of Logic in Artificial Intelligence and Logic Programming,*' Vol. 1, D. M. Gabbay, C. J. Hogger and J. A. Robinson, eds. pp. 449–504. Oxford University Press, 1985.

[Hodkinson, 1989] I. Hodkinson. Decidability and elimination of fixed point operators in the temporal logic USF. Technical report, Imperial College, 1989.

[Hodkinson, 1995] I. Hodkinson. On Gabbay's temporal fixed point operator. *Journal of Theoretical Computer Science*, **139**, 1–25, 1995.

[Hollenberg, 1994] M. Hollenberg. Negative definability in modal logic. Technical report, Department of Philosophy, Utrecht University, 1994.

[Humberstone, 1979] I. L. Humberstone. Interval semantics for tense logics. *J. Philosophical Logic*, **8**, 171–196, 1979.

[Hustadt and Schmidt, 1997] Ullrich Hustadt and Renate Schmidt. On evaluating decision procedures for modal logics. In *Proceedings of IJCAI'97*, 1997.

[ISO, 1982] ISO. Concepts and terminology for the conceptual schema and the information base. Technical report ISO/TC9/SC5/WG3, 1982.

[Jensen *et al.*, 1992] C. S. Jensen, J. Clifford, S. K. Gadia, A. Segev and R. Snodgrass. A glossary of temporal database concepts. *SIGMOD RECORD*, **21**, 35–43, September 1992.

[Jónsson, 1991] B. Jónsson. The theory of binary relations. In H. Andréka, J. Monk, and I. Németi, editors, *Algebraic logic (Proc. Conf. Budapest 1988)*, pages 245–292. North-Holland, 1991.

[Kabanza *et al.*, 1990] F. Kabanza, J. M. Stevenne and P. Wolper. Handling infinite temporal data. In *Proc. ACM Symposium on Principles of Database Systems*, pages 392–403, 1990.

[Kahn *et al.*, 1987] K. Kahn, E. Tribble, M. Miller and D. Bobrow. Vulcan: logical concurrent objects. In B. Shriver and P. Wegner, editors, *Research directions in object-oriented programming*, pages 75–112. MIT Press, 1987.

[Kamp, 1968] H. Kamp. Seminar notes on tense logics. *Journal of Symbolic Logic*, 1968.

[Kamp, 1971] H. Kamp. Formal properties of now. *Theoria*, **35**, 227–273, 1971.

[Kamp and Reyle, 1993] H. Kamp and U. Reyle. *From discourse to logic*, Kluwer, 1993.

[Kanelakis *et al.*, 1990] P. C. Kanelakis, G. M. Kuper and P. Z. Revesz. Constraint query languages. In *Proc. 9th ACM Conference on Principles of Database Systems, PODS'90*, pages 299–313, 1990.

[Kolaitis and Papadimitriou, 1991] P. Kolaitis and C. Papadimitriou. Why not negation by fixpoint? *Journal of Computer and System Sciences*, **43**, 125–144, 1991.

[Kowalski and Sergot, 1986] R. Kowalski and M. Sergot. A logic-based calculus of events. *New Generation Computing*, **4**, 67 – 95, 1986.

[Koymans, 1987] R. Koymans. Specifying message buffers requires extending temporal logic. In *Proceedings 6th ACM Symposium on Principles of Distributed Computing*, pages 191–204, 1987.

[Kozens, 1982] D. Kozens. Results on the propositional μ-calculus. In *9th ICALP*, pages 340–359, Lecture notes in computer science, no. 140, Springer-Verlag, 1982.

[Kozens, 1983] D. Kozens. Results on the propositional μ-calculus. *Theoretical Computer Science*, **27**, 333–354, 1983. see also [Kozens, 1982].

[Kracht and Wolter, 1991] M. Kracht and F. Wolter. Properties of independently axiomatizable bimodal logics. *Journal of Symbolic Logic*, **56**, 1469–1485, 1991.

[Ladkin and Maddux, 1994] P. Ladkin and R. Maddux. On binary constraint problems. *Journal of the ACM*, **41**, 435–469, 1994.

[Ladner and Reif, 1986] R. Ladner and J. Reif. The logic of distributed protocols. In J. Halpern, editor, *Theoretical Aspects of Reasoning about Knowledge (Proc. 1986 Conference)*. Morgan Kaufman, 1986.

[Lambert and van Fraassen, 1972] K. Lambert and Bas C. van Fraassen. *Derivation and counterexample.* Dickinson, 1972.

[van Lambalgen, 1996] M. van Lambalgen. Natural deduction for generalised quantifiers. In J. van der Does and J. van Eijck, eds, *Quantifiers, Logic and*

Language, pp. 225–236. CSLI Lecture Notes, Cambridge University Press, 1996.

[Lamport, 1977] L. Lamport. Proving the correctness of multiprocess programs. *IEEE Transactions on Software Engineering*, **SE-3**, 125–143, 1977.

[Lamport, 1980] L. Lamport. "Sometimes" is sometimes "not never". In *7th ACM Symposium on Principles of Programming Languages*, pages 174–185, 1980.

[Lamport, 1983] L. Lamport. What good is temporal logic? In R. Mason, editor, *Information processing 83 IFIP*. North-Holland, 1983.

[Lamport, 1994] L. Lamport. The temporal logic of actions. *ACM Transactions on Programming Languages and Systems*, **16**, 872–923, May 1994.

[Laroussinie and Schnoebelen, 1994] F. Laroussinie and Ph. Schnoebelen. A hierarchy of temporal logics with past. In *Proc. STACS'94, Caen, France*, pages 47–58. Lecture notes in computer science, no. 775, Springer–Verlag, 1994.

[Läuchli and Leonard, 1966] H. Läuchli and J. Leonard. On the elementary theory of linear order. *Fundamenta Mathematicae*, **59**, 109–116, 1966.

[Lieberman, 1986] H. Lieberman. Concurrent object-oriented programming in act 1. In A. Yonezawa and M. Tokoro, editors, *Object-oriented concurrent programming*, pages 9–36. MIT Press, 1986.

[Lloyd, 1987] J. W. Lloyd. *Foundations of logic programming.* Springer-Verlag, second edition, 1987.

[Loucopoulos *et al.*, 1990] P. Loucopoulos, P. McBrien, U. Persson, F. Schumacker and P. Vasey. TEMPORA — integrating database technology, rule based systems and temporal reasoning for effective software. In Commission for European Communities, editor, *The Annual ESPRIT Conference*, pages 388–411, Brussels, November 1990. Kluwer.

[INMOS Ltd., 1984] INMOS. Ltd. *Occam programming manual.* Prentice Hall, 1984.

[Lyndon, 1950] R. Lyndon. The representation of relational algebras. *Annals of Mathematics*, **51**, 707–729, 1950.

[Maddux, 1991] R Maddux. The origin of relation algebras in the development and axiomatization of the calculus of relations. *Studia Logica*, **3/4**, 421–455, 1991.

[Maier, 1983] D. Maier. *The theory of relational databases.* Pitman, 1983.

[Makanin, 1977] G. S. Makanin. The problem of solvability of equations in a free semigroup. *Soviet Acad. Nauk SSR*, **233**, 1977.

[Manna and Pnueli, 1981] Z. Manna and A. Pnueli. Verification of concurrent programs: The temporal framework. In R. S. Boyer and J. S. Moore, editors, *The Correctness Problem in Computer Science*, pages 215–273. Academic Press, 1981. International Lecture Series in Computer Science.

[Manna and Pnueli, 1992] Z. Manna and A. Pnueli. *The temporal logic of reactive and concurrent systems: specification.* Springer-Verlag, 1992.

[Manning and Torsun, 1989] K. J. Manning and I. Torsun. The application of temporal logic to PAYE tax regulation. Technical report, Department of Computing, Bradford University, 1989.

[Marx and Reynolds, 1997] M. Marx and M. Reynolds. Undecidability of compass logic, *Journal of Logic and Computation*, **9**, 897–914, 1999.

[Marx and Venema, 1997] M. Marx and Y. Venema. *Multi-dimensional modal logic*. Kluwer, 1997.

[McBrien, 1993] P. J. McBrien. Principles of implementing historical databases in RDBMS. In *Proceedings of the 11th British National Conference on Databases*, pages 220–237. Lecture notes in computer science, Springer-Verlag, 1993.

[McBrien *et al.*, 1992] P. McBrien, A.-H. Seltveit and B. Wangler. An entity-relationship model extended to describe historical information. In *International Conference on Information Systems and Management of Data – CISMOD'92*, pages 244–260, Bangalore, India, 1992.

[McDermott, 1982] D. McDermott. A temporal logic for reasoning about process and plans. *Cognitive Science*, **6**, 101–155, 1982.

[McKenzie and Snodgrass, 1991] L. E. McKenzie, Jr. and R. T. Snodgrass. Evaluation of relational algebra incorporating the time dimension in databases. *ACM Computing Surveys*, **23**, 501–544, December 1991.

[McNaughton, 1966] R. McNaughton. Testing and generating infinite sequences by finite automata. *Information and Control*, **9**, 521–530, 1966.

[Milner, 1989] R. Milner. *Communication and concurrency*. Prentice Hall, 1989.

[Mokkedem and Méry, 1994] A. Mokkedem and D. Méry. A stuttering closed temporal logic for modular reasoning about concurrent programs. In D. Gabbay and H. Ohlbach, editors, *Temporal logic, Proceedings of ICTL '94*, Lecture notes in artificial intelligence, no. 827, , pages 382–397. Springer-Verlag, 1994.

[Monk, 1964] D. Monk. On representable relation algebras. *Michigan Mathematical Journal*, **11**, 207–210, 1964.

[Moszkowski, 1986] B. Moszkowski. *Executing temporal logic programs*. Cambridge University Press, 1986.

[Muchnik, 1984] A. A. Muchnik. Games on infinite trees and automata with dead-ends: a new proof of the decidability of the monadic theory of two successors. *Semiotics and Information*, **24**, 17–40, 1984. (In Russian.)

[Mussat, 1992] L. Mussat. Parallel programming with bags. In *Proc. of the Workshop on Research Directions in High-level Parallel Programming Languages*, Lecture notes in computer science, no. 574, Springer-Verlag, 1992.

[Németi, 1995] I. Németi. Decidable versions of first-order logic and cylindric-relativised set algebras. In L. Csirmaz, D. M. Gabbay and M. de Rijke, eds. *Logic Colloquium 92*, CSLI–FOLLI publication, pp. 177–241, 1995.

[Nishimura, 1979] H. Nishimura. Is the semantics of branching structures adequate for chronological modal logics? *Journal of Philosophical Logic*, **8**, 469–475, 1979.

[Noël, 1992] P. Noël. Syntactic approach to the synthesis of temporal specifications. November 1992.

[Nute, 1986] D. Nute. Defeasible reasoning and decision support systems. *Decision Support Systems*, **4**, 97–110, 1988.

[Ohlbach, 1988a] H. J. Ohlbach. *A resolution calculus for modal logics.* PhD thesis, FB Informatik, Universität Kaiserslautern, 1988.

[Ohlbach, 1988b] H. J. Ohlbach. A resolution calculus for modal logics. In E. Lusk and R. Overbeek, editors, *Proc. of 9th International Conference on Automated Deduction, CADE-88 Argonne, IL*, Lecture notes in computer science, no. 310, pages 500–516, Springer-Verlag, 1988.

[Ohlbach, 1991] H. J. Ohlbach. Semantics based translation methods for modal logics. *Journal of Logic and Computation*, 1(5):691–746, 1991.

[Ohlbach and Schmidt, 1997] H. J. Ohlbach and R. Schmidt. Functional translation and second-order frame properties of modal logics. *Journal of Logic and Computation*, **7**, 581–604, 1997.

[Øhrstrøm and Hasle, 1996] P. Øhrstrøm and P. F. V. Hasle. *Temporal logic: from ancient ideas to artificial intelligence*, volume 57 of *Studies in Linguistic and Philosophy.* Kluwer, 1996.

[Paterson and Wegman, 1978] M. S. Paterson and M. N. Wegman. Linear unification. *Journal of Computer and System Sciences*, **16**, 158–167, 1978.

[Persson and Wohed, 1991] U. Persson and R. Wohed. The Sweden post case study. Technical report, SISU–Swedish Institute for Software Research, 1991.

[Pnueli, 1977] A. Pnueli. The temporal logic of programs. In *Proceedings of the 18th Symposium on Foundations of Computer Science*, pages 46–57, 1977. Providence, RI.

[Pnueli, 1985] A. Pnueli. Linear and branching structures in the semantics and logics of reactive systems. In *12th ALP*, pages 15–32, 1985.

[Pnueli, 1986] A. Pnueli. *Applications of temporal logic to the specification and verification of reactive systems: a survey of current trends.* Lecture notes in computer science, no. 224, Springer-Verlag, 1986.

[Pnueli and Rosner, 1989a] A. Pnueli and R. Rosner. On the synthesis of a reactive module. In *Proceedings of the Sixteenth Symposium of Principles of Programming Languages*, pages 179–190. ACM, 1989.

[Pnueli and Rosner, 1989b] A. Pnueli and R. Rosner. On the synthesis of an asynchronous reactive module. In *ICALP89*, Lecture notes in computer science, no. 371, pages 652–671. Springer-Verlag, 1989.

[Prasad, 1991] K. Prasad. A calculus of broadcasting systems. In S. Abramsky and T. Maibaum, editors, *Proceedings of the International Joint Conference on Theory and Practice of Software Development*, Lecture notes in computer science, no 493, Springer-Verlag, 1991.

[Pratt and Francez, 1997] I. Pratt and N. Francez. A decidable temporal logic for temporal propositions, 1997. *Proc. ICTL'97*, 1997.

[Pratt, 1976] V. Pratt. Semantical considerations on Floyd–Hoare logic. In *Proc. 17th IEEE Symp. on Foundations of Computer Science*, pages 109–121, 1976.

[Prior, 1967] A. Prior. *Past, present and future.* Oxford University Press, 1967.

[Prior, 1968] A. Prior. Now. *Nous*, **2**, 101–119, 1968.

[Queille and Sifakis, 1982] J. Queille and J. Sifakis. Specification and verification of concurrent programs in CESAR. In *Proc. 5th Int. Symp. on Programming*,

Lecture notes in computer science, no. 137, Springer-Verlag, 1982.

[Rabin, 1969] M. O. Rabin. Decidability of second order theories and automata on infinite trees. *American Mathematical Society Transactions*, **141**, 1–35, 1969.

[Ramakrishnan *et al.*, 1987] R. Ramakrishnan, F. Bancilhon and A. Silberschatz. Safety of recursive Horn clauses with infinite relations. In *Proc. Sixth Symposium on Principles of Database Systems*, pages 328–339, 1987.

[Ramsey, 1930] F. P. Ramsey. On a problem of formal logic. *Proc. LMS*, **30**, 264–286, 1930.

[Rao and Georgeff, 1993] A. Rao and M. Georgeff. A model-theoretic approach to the verification of situated reasoning systems. In *Proc. IJCAI*, 1993.

[Reichenbach, 1947] H. Reichenbach. *Elements of symbolic logic*, Macmillan, 1947.

[Reisig, 1985] W. Reisig. *Petri nets: an introduction.* EATCS. Monographs on theoretical computer science. Springer-Verlag, 1985.

[Reynolds, 1992] M. A. Reynolds. An axiomatisation for Until and Since over the reals without the IRR rule. *Studia Logica*, **51**, 165–194, 1992.

[Reynolds, 1995] M. A. Reynolds. Towards first-order concurrent metatem. In M. Fisher and R. Owens, editors, *Proceedings of IJCAI Workshop on Executable Modal and Temporal Logics, Chambery, France 1993*, Lecture notes in artificial intelligence, no. 897, pages 118–143. Springer-Verlag, 1995.

[Reynolds, 1996] M. A. Reynolds. Temporal semantics for Gamma. In J.-M. Andreoli, C. Hankin and D. Le Métayer, editors, *Coordination programming: mechanisms, models and semantics*, pages 141–170. Imperial College Press, 1996.

[Reynolds, 1997] M. A. Reynolds. Metatem in intensive care. Technical report tr-97-01, King's College, 1997.

[Robinson, 1965] J. A. Robinson. A machine-oriented logic based on the resolution principle. *Journal of the ACM*, 12(1):23–41, 1965.

[Robinson, 1971] R. Robinson. Undecidability and nonperiodicity for tilings of the plane. *Inventiones Math*, **12**, 177–209, 1971.

[Safra, 1988] S. Safra. On the complexity of ω-automata. In *Proceedings of 29th IEEE Symposium on the Foundations of Computer Science*, 1988.

[Schmidt, 1997] R. A. Schmidt. *Optimised modal translation and resolution.* PhD thesis, Max-Planck-Institut für Informati, Saarbrücken, 1997.

[Schmidt-Schauß, 1989] M. Schmidt-Schauß. *Computational aspects of an order-sorted logic with term declarations*, Lecture notes in artificial intelligence, no. 395, Springer-Verlag, 1989.

[Schmidt-Schauß and Smolka, 1991] M. Schmidt-Schauß and Gert Smolka. Attributive concept descriptions with complements. *Artificial Intelligence*, **48**, 1–26, 1991.

[Segerberg, 1973] K. Segerberg. Two-dimensional modal logic. *Journal of Philosophical Logic*, **2**, 77–96, 1973.

[Shelah, 1975] S. Shelah. The monadic theory of order. *Annals of Mathematics*, **102**, 379–419, 1975.

[Snodgrass, 1995] R. T. Snodgrass, editor. *The TSQL2 temporal query language.*

[Snodgrass, 1995] R. T. Snodgrass, editor. *The TSQL2 temporal query language.* Kluwer, 1995.

[Snodgrass and Ahn, 1985] R. Snodgrass and I. Ahn. A taxonomy of time in databases. In *ACM SIGMOD International Conference on Management of Data*, pages 236–246, Austin, May 1985.

[Spaan, 1993] E. Spaan. *Complexity of modal logics.* PhD thesis, Free University of Amsterdam, 1993.

[Sripada, 1990] S. M. Sripada. A basis for historical deductive databases. Internal report, Imperial College, Department of Computing, March 1990.

[Stevens *et al.*, 1974] W. Stevens, G. Myers and L. Constantine. Structured design. *IBM Systems Journal*, **13**, 115–139, 1974.

[Stickel, 1981] M. E. Stickel. A unification algorithm for associative-commutative functions. *Journal of the ACM*, **28**, 423–434, 1981.

[Stickel, 1985] M. E. Stickel. Automated deduction by theory resolution. *Journal of Automated Reasoning*, **1**, 333–356, 1985.

[Stirling, 1992] C. Stirling. Modal and temporal logics. In S. Abramsky, D. Gabbay, and T. Maibaum, editors, *Handbook of logic in computer science, Volume 2*, pages 477–563. Oxford University press, 1992.

[Stirling and Walker, 1989] C. Stirling and D. Walker. Local model checking in local mu-calculus. In *Proceedings TAPSOFT*, 1989.

[Strulo *et al.*, 1993] B. Strulo, D. Gabbay and P. Harrison. Temporal logic in a stochastic environment. Manuscript, March 1993.

[Tansel *et al.*, 1993] A. Tansel, J. Clifford, S. Gadia, S. Jajodia, A. Segev and R. Snodgrass, editors. *Temporal databases: theory, design and implementation.* Database Systems and Application Series. Benjamin/Cummings , 1993.

[Tarski, 1941] A Tarski. On the calculus of relations. *Journal of Symbolic Logic*, **6**, 73–89, 1941.

[Thomason, 1984] R. Thomason. Combinations of tense and modality. In D. Gabbay and F. Guenthner, editors, *Handbook of philosophical logic* Vol II, pages 135–165. Reidel, 1984.

[Thomason, 1980] S. K. Thomason. Independent propositional modal logics. *Studia Logica*, **39**, 143–144, 1980.

[Turski and Maibaum, 1987] W. Turski and T. Maibaum. *The specification of computer programs.* Addison-Wesley, 1987.

[Tuzhilin and Clifford, 1990] A. Tuzhilin and J. Clifford. A temporal relational algebra as a basis for relational completeness. In *Proc. 16th Conference on Very Large Databases*, Brisbane, 1990.

[Ullman, 1988] J. D. Ullman. *Principles of database and knowledge-base systems*, Vol. I. Computer Science Press, 1988.

[Vardi and Wolper, 1986] M. Vardi and P. Wolper. Automata theoretic techniques for modal logics of programs. *Journal of Computer Systems Science*, **32**, 183–221, 1986.

[Venema, 1990] Y. Venema. Expressiveness and completeness of an interval tense logic. *Notre Dame Journal of Formal Logic*, **31**, 529–547, 1990.

and Computation, **1**, 453–476, 1991.

[Venema, 1992] Y. Venema. *Many-dimensional modal logic.* PhD thesis, University of Amsterdam, 1992.

[Venema, 1993] Y. Venema. Derivation rules as anti-axioms in modal logic. *Journal of Symbolic Logic*, **58**, 1003–1034, 1993.

[Venema, 1994] Y. Venema. Completeness through flatness in two-dimensional temporal logic. In D. Gabbay and H.-J. Ohlbach, editors, *Temporal logic, Proceedings of ICTL '94*, Lecture notes in artificial intelligence, no. 827, pages 149–164, Springer-Verlag, 1994.

[Vilain and Kautz, 1986] M. Vilain and H. Kautz. Constraint propagation algorithms for temporal reasoning. In *Proc. AAAI*, 1986.

[Walther, 1987] C. Walther. *A many-sorted calculus based on resolution and paramodulation.* Research Notes in Artificial Intelligence. Pitman, 1987.

[Wangler, 1992] B. Wangler. Syntax and semantics of the entity-relationship-time model. Technical report, SISU – Swedish Institute for Software Research, 1992.

[Wangler, 1993] B. Wangler. *Contribution to functional requirements modelling.* PhD thesis, Department of Computer Systems and Sciences, Royal Institute of Technology and Stockholm University, 1993.

[Widom and Ceri, 1996] J. Widom and S. Ceri, editors. *Active database systems—triggers and rules for advanced database processing.* Morgan Kaufmann, 1996.

[Xu, 1988] M. Xu. On some U, S-Tense Logics. *Journal of Philosophical Logic*, **17**, 181–202, 1988.

[Xu, 1994] M. Xu. Decidability of deliberative *stit* theories with multiple agents. In D. Gabbay and H. Ohlbach, editors, *Temporal logic, Proceedings of ICTL '94*, Lecture notes in artificial intelligence, no. 827, pages 332–348. Springer-Verlag, 1994.

[Yakhnis and Yakhnis, 1990] A. Yakhnis and V. Yakhnis. Extension of Gurevich–Harrington's restricted memory determinacy theorem: a criterion for the winning player and an explicit class of winning strategies. *Annals of Pure and Applied Logic*, **48**, 277–297, 1990.

[Yonezawa, 1990] A. Yonezawa, editor. *ABCL: an object-oriented concurrent system.* MIT Press, 1990.

[Zamov, 1989] N. K. Zamov. Modal resolutions. *Izvestiya VUZ. Mathematika*, **33**, 22–29, 1989. Also published in *Soviet Mathematics*, Allerton Press.

[Zanardo, 1985] A. Zanardo. A finite axiomatization of the set of strongly valid Ockamist formulas. *Journal of Philosophical Logic*, **14**, 447–468, 1985.

[Zanardo, 1990] A. Zanardo. Axiomatization of 'Peircean' branching-time logic. *Studia Logica*, **49**, 183–195, 1990.

[Zanardo, 1996] A. Zanardo. Branching-time logic with quantification over branches: the point of view of modal logic. *Journal of Symbolic Logic*, **61**, 1–39, 1996.

[Zaniolo, 1986] C. Zaniolo. Safety and compilation of non-recursive Horn clauses.

[Zaniolo, 1986] C. Zaniolo. Safety and compilation of non-recursive Horn clauses. In *Proc. First Int. Conf. on Expert Database Systems*, pages 237–252, 1986.

[Zeitman, 1993] S. Zeitman. Unforgettable forgetful determinacy. *Journal of Logic and Computation*, **4**, 273–284, 1994.

[Zhou, 1987] C. Zhou. Specifying communicating systems with temporal logic. In [Banieqbal *et al.*, 1987], pages 304–323, 1987.

Index

LIST OF SYMBOLS